ECOLOGY AND BIOGEOGRAPHY IN INDIA

MONOGRAPHIAE BIOLOGICAE

Editor

J. ILLIES

Schlitz

VOLUME 23

DR. W. JUNK b.v. PUBLISHERS THE HAGUE 1974

ECOLOGY AND BIOGEOGRAPHY IN INDIA

Edited by

M. S. MANI, M. A., D.Sc.

Emeritus Professor of Entomology
St. John's College
Agra

DR. W. JUNK b.v. PUBLISHERS THE HAGUE 1974

ISBN 90 6193 075 8
© 1974 by Dr. W. Junk b.v., Publishers, The Hague
Cover Design M. Velthuijs, The Hague
Printed in the Netherlands
Zuid-Nederlandsche Drukkerij N.V., 's-Hertogenbosch

CONTENTS

CHAPTERS' CONTENTS

* Recently renamed Shri Lanka.

XIV

PREFACE

This book describes the outstanding features of the ecology and bio-geography of the Indian region, comprising former British India, Nepal, Bhutan, Ceylon and Burma. It summarizes the results of nearly four decades' studies and field explorations and discussions with students on the distribution of plants and animals, practically throughout this vast area and on the underlying factors. A number of specialists in geology, meteorology, botany, zoology, ecology and anthropology have also actively collaborated with me and have contributed valuable chapters in their respective fields.

India has an exceptionally rich and highly diversified flora and fauna, exhibiting complex composition, character and affinities. Although the fauna of the Indian region as a whole is less completely known than its flora, we are nevertheless fairly well acquainted with at least the salient features of its faunal characters to enable us to present a meaningful discussion on some of the outstanding peculiarities of the biogeography of India. A general synthesis of the available, though much scattered, information should prove useful to future students of biogeography throughout the world. Such a review, to be really useful, must include not only summaries of the broad trends in the general ecology of plants and animals, but also cover an analysis of the present-day physical features of the region, the stratigraphy and tectonics, the orogeny of the Himalaya, climate, limiting factors in distribution, the routes and barriers to dispersal, composition, ecological characters, affinities and distributional patterns of important groups of plants and animals, the ecology, character flora and fauna of various natural regions and a comprehensive synthesis of the evidence from geology, meteorology, botany, zoology and anthropology.

The central concept throughout this book is that biogeographical and geomorphological evolution of India constitutes an integral whole and the flora and fauna and distributional peculiarities that we observe today represent a dynamic phase of this complex evolution. From the stand-point of biogeography, the flora and fauna of a large region behave as if they were a single organism. Just as we study the evolution of any given organism, we may also study the evolution of the fauna of a region and often be able to correlate it to the underlying factors. Biogeographical interpretations must, therefore as far as possible, deal with the whole complex of flora and fauna and must not be restricted to isolated and specialized groups, however peculiar their distribution may appear to us, when studied separately.

XVII

This is the first attempt at a comprehensive monograph on ecology and biogeography in India and to this fact must be attributed most of its shortcomings. In writing this book, my aim has largely been to bring together the basic facts and indicate the broad trends, in the hope that it will stimulate further research in a most interesting field of study.

I take this opportunity of expressing my cordial thanks to my numerous pupils for fruitful discussions and the various specialists who have collaborated with me and placed at my disposal their valuable advice and criticism, and contributed the chapters on geology, meteorology, botany and other topics. I am particularly indebted to my research collaborators Messrs. O. P. DUBEY, B. K. KAUL and G. G. SARASWAT for invaluable assistance in the preparation of the manuscript for press. My thanks are also due to MR. P. SAHADEVAN, Artist, for his willing help in redrawing and preparing figures from rough sketches furnished by different contributors. I must also thank my wife for her constant encouragement and suggestions.

M. S. MANI

AUTHORS' ADDRESSES

HOLLOWAY, J. D., Abbotsbury Givons Grove, Leatherhead, Surry, England.

JAYARAM, K. C., Zoological Survey of India, Southern Regional Station, Kandasami, Gromani 51, Madras-600004, India.

KURUP, G. U., Zoological Survey of India, Southern Regional Station, Madras 600004, India.

† KRISHNAN, M. S.

LAL, P., Human Ecologist, Anthropological Survey of India, Indian Museum, 27 Chowringhee Road, Calcutta 700013, India.

MUKHERJEE, A. K., Zoological Survey of India, Indian Museum, 27 Chowringhee Road, Calcutta-700013, India.

PRAKASH, I., Central Arid Zone Research Institute, Jodhpur, India.

RAMDAS, L. A., Former Director, Meteorological Department of India, Emeritus Scientist, National Physical Laboratory, New Delhi 110012, India.

RAO, A. S., Botanical Survey of India, Eastern Regional Station, Shillong, India.

RAU, M. S., Botanical Survey of India, Gangetic Plains Station, Dehra Dun, India.

SEN-SARMA, P. K., Division of Entomology, Indian Forest Research Institute, New Forest, Dehra Dun, India.

SINGH, S., School of Entomology, St. John's College, Agra 282002, India.

SUBRAMANYAM, K. & M. P. NAYAR, Botanical Survey of India, Madan Street, Calcutta 700013, India.

I. INTRODUCTION

by

M. S. MANI

The biogeographical area of India, as defined by BLANFORD (1901), includes the whole of former British India, with the addition of Ceylon, Maladive Islands, Nepal, Sikkim and Bhutan. In addition to India proper, British India embraced also Baluchistan and Burma. In 1937, Burma was separated and ten years later India was partitioned into the Dominions of India and Pakistan. In 1950, the Dominion of India became a Republic and since then the boundaries of the old 'native states', which were formerly ruled by princes and the Provinces of British India have been greatly altered and the whole country has been divided into a number of administrative units called states. The Andaman, Nicobar and Laccadive Islands belong to the Republic of India. The former French and Portugese enclaves in India are now integral parts of the Republic.

For a meaningful discussion of the biogeographical evolution of India, the beginnings of which must indeed be sought in the far off Madagascar, Indo-China and Malaya, we must follow more or less the limits defined by BLANFORD. Throughout this book when we speak of India, we include, therefore, not only the areas contained within the limits of the Republic of India, but also Pakistan with Baluchistan and parts of Afghanistan and Tibet, Nepal, Sikkim, Bhutan, Bangladesh, Burma, Ceylon and the Maladive and Seychelles Islands. Defined in this manner, the mainland of India stretches east-west nearly 3800 km, from 61° to nearly 100° EL and about 3000 km south-north and lies entirely north of the Equator. Cape Kumarin (= Comorin), the southernmost point of India is 8° north of the Equator and the Tropic of Cancer roughly cuts the country into two halves. The northern frontiers reach nearly to 37° NL, so that they are approximately about the same distance from the Tropic of Cancer as Cape Kumarin is from it. Although thus nearly half of India lies outside the tropics, in the middle latitudes and within the temperate zone, it is customary to speak of India as a tropical country, mainly because the region is shielded off by the Himalaya in the north from the rest of Asia and has nearly the same general type of tropical monsoon climate almost throughout the land. Nevertheless, the variety in elevation and local climate is extremely remarkable and includes transitions from the rainless deserts of Sind to the rainiest place on earth, Mausingam, in the Garo Hills of Assam (see Chapter V), from Jacobabad the hottest place on earth to alpine and arctic conditions on the Himalaya, from the geologically stable and ancient areas of the Peninsula with senile

1

topography to the geologically unstable and recent areas of youthful topography on the Himalaya. Within the limits of India, the Palaearctic, Ethiopian, Indo-Chinese, Malayan and endemic (Indian) floras and faunas meet and intermingle, giving rise to characteristic distributional patterns and contributing to the outstanding peculiarities of its bio-geographical evolution. We have here the derivatives of the ancient Gondwana Floras and Faunas and those of the younger Euro-Asiatic intrusive ones. India has thus an exceptionally rich and highly diversified flora and fauna, exhibiting complex composition, character and affinities.

The ecology of nearly the whole of India, with perhaps the exception of the higher Himalaya above the timberline, particularly to the west of Nepal, is dominated by the rhythm of the monsoon-rainfall climate. The ecology of only the high altitudes of the Himalaya is characteristically temperature dominated. Vast areas of monsoon-dominated ecology support tropical flora and fauna, but contain also numerous remarkable pockets of temperate floras and faunas. The dominance of the monsoon-rainfall as a factor in the ecology of India rests primarily on the channelling effects of the Himalaya on the monsoon currents (see chapter V) but also partly on the topographic peculiarity of the Peninsula, which may also in turn be traced back to events leading to the uplift of the Himalaya. In large areas the general ecology is also a relict of the influence, which the Pleistocene glaciations on the Himalaya exerted nearly to the extreme south of the Peninsula. Finally the ecology of plants and animals in the Indo-Gangetic Plains of north India and in large parts of the Peninsula has been very profoundly modified and altered, within historical times, by the effects of extensive destruction of natural habitats in the course of the advance of civilized man and recent rapid urbanization. The changes in character of the flora and fauna and the distributional patterns and ranges of plants and animals, brought about by the extensive destruction of natural habitats by civilized man are mirrored by the changes in the ecology and status of the primitive communities of man himself in India. With the influx of civilized races of man from the northwest, and under increasing pressure of their continued advance, aboriginal man in India steadily receded to the small isolated refugial pockets in dense and inaccessible forests, where he is found today as the tribal man. A study of the ecology of the primitive communities (see chapter XI) in India throws considerable light on the magnitude of human influence on ecology and biogeography in India, and provides valuable clues to some of the complex problems of distributional patterns of plants and animals. Like the primitive communities of human beings, the present-day flora and fauna of India represent merely the impoverished relicts of a formerly much larger and more widely distributed complex.

Field ecology and biogeographical researches unfortunately seem to have been looked upon in India with considerable misgivings, as belonging to the realm of speculation. Even among the few workers, who apparently ventured to 'speculate', it seems to have been fashionable merely to divide the country into biogeographical regions, subregions, divisions and subdivisions or at least discuss such divisions. With perhaps the notable exceptions of BLANFORD (1901) and HORA (see references in chapter XXIV), from the earliest contributions of JERDON in 1862 on the distribution of birds and of GÜNTHER (1858, 1864) on distribution of reptiles, zoologists seem to have been busy only partitioning India somewhat quite unnecessarily into a bewildering series of often conflicting 'zoogeographical' divisions. On the pattern of the distribution of land Mollusca, BLANFORD in 1870 divided India into four major provinces, viz. the Punjab Province, the Indian Province, the eastern Bengal Province and the Malabar Province with southern Ceylon. The Indian Province was further subdivided into a number of sub-provinces. Six years later, BLANFORD (1876) made a valuable contribution to the biogeography of India by drawing attention to the presence of African elements in our fauna.

In 1873, ELWES recognized that, India with the Malayan Peninsula forms a single biogeographical unit, which he named Indo-Malayan Region, and divided this region into three more or less well defined subregions, viz. the Himalayan or the Himalo-Chinese subregion, the Indian subregion and the Malayan subregion. WALLACE (1876) divided the Oriental Region into Hindustan or the Indian subregion, consisting of the whole of the Peninsula of India, from the foot of the Himalaya to the north of Seringapatam and Goa; Ceylon and South India; Himalayan or Indo-Chinese subregion, comprising the Himalaya as far west as Kashmir from the base of the mountains to an elevation of about 2800–3050 m and the area to the east of the Bay of Bengal, Assam, Burma, southern China, Thailand and Cochin-China; and the Indo-Malayan or the Malayan subregion, consisting of the Malay Peninsula and the Malay Archipelago.

In 1888, BLANFORD proposed the divisions 1. Tibetan, 2. Himalayan, 3. Indian, 4. Malabar or Ceylonese, 4. Burmese and 5. South Tenasserim. In 1901, BLANFORD published his admirable monograph on the distribution of Vertebrata in India, Ceylon and Burma, tabulating the distribution of genera and describing the principal zoögeographical divisions (Fig. 1). Somewhat earlier, SHARP proposed the divisions of 1. Indian Peninsular subregion, 2. Indo-Malayan subregion, 3. Indo-Chinese subregion, 4. Himalo-Malayan subregion and 5. the Himalo-Chinese subregion. About the same time, NEWTON (1893) and GADOW (1893), on the basis of the distribution of birds, and SCLATER (1899) on the basis of distribution of mammals, followed WALLACE more or less closely, but united the Indian and Ceylonese areas into a single unit.

On the basis of the distribution of Lepidoptera, mainly Rhopalocera,

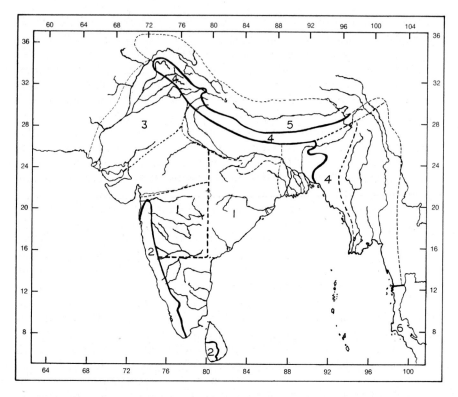

Fig. 1. The zoogeographical subdivisions of India proposed by BLANFORD. 1. The Cis-Gangetic subregion, 2. The Malabar coast tract and Ceylon Hill tract, 3. The Punjab tract, 4. Trans-Gangetic subregion, 5. Tibetan Subregion.

TALBOT (1939) recognized eight major divisions of India, viz. 1. Ceylon, 2. Peninsular India, 3. Northwest India, 4. West Himalaya, 5. Northeast India, 6. Burma, 7. Andaman Islands and 8. Nicobar Island. Each of these were further subdivided. He rightly considered that Ceylon is faunistically part of South India, but with a number of peculiar forms, some of which are related to those of the Malayan subregion. The butterflies of his Peninsular division show affinities partly to the Malayan and partly to Palaearctic forms. TALBOT's Northwest India shows affinities partly to the Palaearctic and partly to the Oriental, with infiltrations from the south.

ALCOCK (1910) recognized six territories in the distributional pattern of fresh-water Crustacea, 1. the Western Territory, 2. the Western Himalayan Territory, 3. the Northeastern Frontier or the Eastern Himalayan or the Eastern sub-Himalayan Territory, 4. Burma-Malay Territory, 5. the Peninsular Territory and 6. the Indo-Gangetic Plain Territory. Discussing the distribution of fresh-water sponges, Polyzoa,

4

etc., ANNANDALE in 1911 largely confirmed ALCOCK's divisions, but divided the Peninsula into 1. the area east of the Western Ghats and 2. the Malabar zone to include the Western Ghats, from the R. Tapti to Cape Comorin and eastwards to the sea and considered Ceylon, separate from South India, as a distinct Territory. According to CHRISTOPHERS (1933), the distributional pattern of Culicidae suggests the following divisions: 1. the Trans-Indus area, 2. the Indo-Gangetic area, 3. the Peninsular area, 4. Malabar-Ceylon area, 5. Assam-Burma area and 6. Himalayan area. PRASHAD (1942), who reviewed the earlier schemes, proposed the divisions 1. the Western Frontier Territory including Baluchistan and Northwest-Frontier Province and parts of the Punjab; 2. the Himalaya consisting of the upper Indus Valley with Ladak, Gilgit, western Himalaya from Hazara to the western limits of Nepal and the eastern Himalaya to the Mishmi Hills above the Assam Valley; 3. Assam and Burma, comprising the greater part of Lower Brahmaputra drainage system and the Burmese Territory including Tenasserim; 4. the Gangetic Plain to the east of Delhi and including the whole of the Uttar Pradesh, Bengal and parts of Assam up to the base of the Assam hills together with the plain of the R. Brahmaputra as far as Goalpara and Cachar, Sylhet and the plains of Tipperah; and 5. the Peninsular India with the Malabar zone as a distinct division and Ceylon.

STEPHENSON (1921, 1923) found that the Oligochaeta of India belong to the major divisions recognized by ANNANDALE, with however a narrow southern end of the Peninsula, south of Goa and south of the 15th north parallel from the eastern shore, as distinct region.

MALCOLM A. SMITH (1931, 1935, 1943) recognized that the reptiles of Thailand, Indo-China, south China, Burma, Assam and the Eastern Himalaya are very closely related, so that these areas must be considered as constituting a single region. The fauna of this vast area is very closely related to that of the Malabar tract of the Western Ghats within the Peninsula. He recognized the following subregions (Fig. 2): 1. the Indian subregion 2. Indo-Chinese subregion and 3. the Malaysian subregion. The Indian subregion was subdivided to seven tracts, viz. 1. the Desert Tract of the northwest, including Baluchistan, Northwest-Frontier Province, the Punjab, western Rajasthan as far as the Aravalli Range and Sind; 2. the Gangetic Plain Tract extending from the R. Indus to the right bank of the R. Brahmaputra; 3. Central India Tract including the area between the Gangetic Plain, Deccan, Aravalli and Chota-Nagpur; 4. Kashmir and western Himalaya to Nepal; 5. the Deccan Tract including the central tableland of the Deccan between the 12th and 21st north parallels and a part of the Western Ghats; 6. the Mountains of the Malabar Tract and Ceylon; 7. the Chota-Nagpur Tract to include Bihar south of the Gangetic Plain north Orissa and east of Madhya Pradesh. The differences between BLANFORD and SMITH may be summarized as follows:

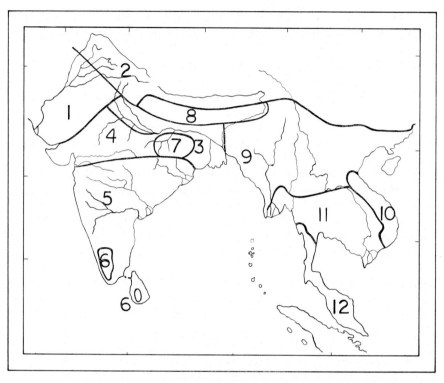

Fig. 2. The zoogeographical subdivisions of India proposed by SMITH. 1. Desert area, 2. Kashmir, West Himalaya, 3. Gangetic Plain, 4. Central India, 5. The Deccan, 6. Malabar mountains and Ceylon 7. Chota-Nagpur, 8. The Eastern Himalaya, 9. Trans-Himalayan, 10. Annam, 11. Plain of Indo-China, 12. Malayan subregion.

SMITH	BLANFORD
Desert area	Punjab Tract, excluding Indus Plain
Kashmir, West Himalaya	West Himalayan Tract
Gangetic Plain	UP, Bihar, Indus Plain and Bengal
Central India	Rajasthan and Central India Tract
The Deccan	Carnatic or Madras Tract
Malabar mountains and Ceylon	Malabar Coast Tract, Ceylon Tract
Chota-Nagpur area	Bihar-Orissa Tract

The Indo-Chinese subregion of SMITH comprises five areas, viz. 1. the Eastern Himalaya of the Assam Tract of BLANFORD, extending from the western border of Nepal to the bend of the R. Brahmaputra; 2. the Trans-Himalayan mountainous area corresponding to BLANFORD's Upper Burma Tract, the hills of Assam east of the R. Brahmaputra, Burma except the lowlands in the south, south Yunnan, north Indo-China and north Thailand; 3. Annam; 4. the Great Plain of Indo-China to include the Burmese lowlands south of Prome and Toungoo and at the mouth of the R. Salween, the plains of Thailand, Cambodia, Cochin-

China (corresponding to the Pegu Tract of BLANFORD) and 5. Tenasserim and Peninsular Thailand corresponding to Tenasserim and South Tenasserim of BLANFORD.

The herpetological studies of MAHENDRA (1939) largely confirmed SMITH's scheme. While perhaps he nearly achieved the rare merit of recognizing the natural amphitheatres of faunal differentiations, MAHENDRA succumbs to the temptation of 'partitioning' and proposes ten unnatural subdivisions, viz. 1. the arid or the semi-arid province of the northwest, 2. the Western Himalaya, 3. Trans-Gangetic province, 4. South Burmese Province, 5. Gangetic Plain up to 20° NL, 6. South India, 7. Travancore, 8. Ceylon, 9. Andaman Islands and 10. Nicobar Islands.

Not only zoologists have thus been busy subdividing India, botanists too have not lagged far behind (see chapters VII & IX). Commencing with HOOKER, hardly any botanist, who devoted any serious attention to the phytogeography of India, has wholly been able to resist this urge of partitioning the phytogeographic area and eventually ending by confusing the various subdivisions thus created with natural centres of origins, differentiations and radiations of floras. It must of course be obvious that nearly all the phytogeographical divisions, proposed so far, bear no relation whatever to the faunistic subdivisions, because botanists and zoologists have been working with complete independence in India. Not only the zoologists and botanists have had separate and mutually contradictory ideas in this direction, but as we have seen above, even among the zoologists, students of different groups have proposed wholly independent schemes for partitioning the region. Malacologists, herpetologists, dipterists, lepidopterologists, ichthyologists, ornithologists and others have, for example, each proposed their own 'zoogeographical' subdivisions of India! It is needless to point out that all the schemes of phytogeographical and zoogeographical divisions of India, which have been proposed so far, present only partial pictures of the present-day distributional patterns of isolated and often taxonomically and ecologically highly specialized and frequently geographically greatly localized groups of plants and animals, with over-emphasis on the present-day climatic conditions, but with no or perhaps only a casual reference to the geological history or to the earlier patterns. While the so called phytogeographical divisions, recognized at present, are mostly vegetation divisions, or strictly speaking, they mark out climatic-vegetational areas and have at least proved useful as such, the divisions proposed by zoologists were not even ecologically defined homogeneous units, but often represent merely the areas from which the material was studied by the worker. Occasional attempts at discussing the biogeographical affinities of a region or of a group of animals and plants suffer from the fact that nearly all of them are based on very limited studies on specialized forms, like for example, the torrential-stream fishes. It is,

7

therefore, not to be wondered at that practically none of the so called biogeographical subdivisions has any relation whatsoever to the centres of differentiation, radiation, evolution and affinity. They are nearly all more or less artificial meteorological – actually monsoon-rainfall divisions rather than biogeographical.

With so much emphasis on the biogeographical partitioning of India, there has been a most deplorable confusion regarding the component elements and their sources and the biogeographical nature of the flora and fauna, the changes in their compositions, the continually changing distributional patterns and the underlying factors, which have influenced these changes. It also seems astonishing that most workers should have considered biogeography as a static phenomenon and should have so completely overlooked the possibilities of gradual changes in the distributional patterns of plants and animals with passage of time – evolution of distributional patterns. The close relation between biogeography and the geomorphological evolution of a region has also been totally ignored or largely misunderstood. We find thus the anomaly that instead of interpreting distributional patterns and compositions of floras and faunas against the background of geomorphological evolution, zoologists have in all seriousness conceived of rivers where none ever flowed and of mountains where no orogenic activity has been detected by geologists. The supporters of the Indo-Brahm (see chapter II) and of the Satpura hypothesis (see chapters XVIII & XXIV) have, for example, gone to the grotesque extent of refitting the geomorphology of India to their pet ideas on the origins of the present-day distribution of certain animals.

The composition, size, ecological characters, affinities and the range of not only the flora but also the fauna are continually changing, *pari passu* with each other, so that the flora and fauna constitute an indivisible whole. The present-day flora and fauna of India are indubitably the product of such a continual change in the past; they represent the modified descendents of past floras and faunas. The evolution of the flora and fauna must be interpreted in terms of the continual changes in the size, location, configuration, topography, stratigraphy, drainage patterns and other tectonic changes which have taken place in the region in the past. The conditions which prevail today are in no sense the cause of such changes in the past and cannot, therefore, explain the present-day biogeographical characters. On the other hand, the present-day climate, the flora, fauna and the complex patterns of distribution of plants and animals may all be traced back to factors which operated in the past.

The ecology and biogeography of India are thus essentially part of a complex mosaic and the events in India are an integral phase of the larger biogeographical evolution of Asia. The biogeographical and geomorphological evolution of India involves, therefore, a number of problems of great fundamental importance, based on ideas of continental drift and cannot be studied in isolation. The present-day ecology of

India may be traced back to the complex series of events of continental drift, which climaxed in the uplift of the Himalaya. The Himalayan uplift holds likewise the key to the problems of the characteristic composition, affinities and distributional patterns of plants and animals throughout the region. The Himalaya might thus most appropriately be described as presiding over ecology and biogeography in India.

REFERENCES

ALCOCK, A. W. 1910. Catalogue of Indian Decapoda Crustacea collections in the Indian Museum. *Rec. Indian Mus.*, 1(2): 9–14.

ANNANDALE, T. N. 1911. Freshwater sponges, Hydroids and Polyzoa. Fauna British India, pp. 1–251.

BLANFORD, W. T. 1870. Notes on some Reptilia and Amphibia from Central India. *J. Asiatic Soc. Bengal*, 39(2): 335–376.

BLANFORD, W. T. 1876. Notes on the 'Africa-Indien' of A. VON PELZEN, and on the Mammalian fauna of Tibet. *Proc. zool. Soc. London*, 631–634.

BLANFORD, W. T. 1876. The African element in the fauna of India: A criticism of MR. WALLACE's views as expressed in the 'Geographical Distribution of Animals'. *Ann. Mag. nat. Hist.*, (4) 18: 277–294.

BLANFORD, W. T. 1888–1891. Primates, Carnivora, Insectivora. Fauna British India, Mammalia. 1: 1–250 (1888); Chiroptera to Edentata. 2: 251–617 (1891).

BLANFORD, W. T. 1895–1898. Birds. Fauna British India. 3: 1–450 (1895); 4: 1–500 (1898).

BLANFORD, W. T. 1901. Distribution of Vertebrate animals in India, Ceylon and Burma. *Philos. Trans. R. Soc. London*, (B) 194: 335–436.

CHRISTOPHERS, S. R. 1933. Family Culicidae: Anophelini. Fauna of British India, Diptera. 4: 1–371.

ELWES, H. J. 1873. On the geographical distribution of Asiatic birds. *Proc. zool. Soc. London*, pp. 645–682.

GADOW, H. 1893. Bronns Klassen und Ordnungen der Tiere. Vogel. 296.

GÜNTHER, A. C. L. 1858. On the geographical distribution of reptiles. *Proc. zool. Soc. London*, pp. 373–398.

GÜNTHER, A. C. L. 1864. Reptiles of British India. p. vi.

JERDON, T. C. 1862–1864. The birds of India, being a natural history of all the birds known to inhabit continental India. Calcutta. 3 vols.

MAHENDRA, B. C. 1939. The zoogeography of India in the light of herpetological studies. *Sci. Cult.*, 4(7): 1–11.

NEWTON, A. 1893. Dictionary of Birds. p. 356.

PRASHAD, B., 1941. The Indo-Brahm or the Siwalik river. *Rec. geol. Surv. India*, 74(4): 555–561 (1939).

PRASHAD, B. 1942. Zoogeography of India. *Sci. Cult.*, 7(9): 421–427.

SCLATER, P. L. 1858. On the general geographical distribution of the members of the class Aves. *J. Proc. Linn. Soc. London*, (Zool.) 2: 130–145.

SCLATER, P. L. 1891. On the recent advances in our knowledge of the geographical distribution of birds. *Ibis*, (6) 3: 514.

SCLATER, W. L. & P. L. 1899. The Geography of Mammals.

SMITH, M. A. 1931/1935/1943. Loricata, Testudines. Fauna British India. 1: 1–105. (1935); Sauria, 2: 1–440 (1935); Serpentes, 3: 1–583 (1943).

STEPHENSON, J. 1921. Contributions to the morphology, classification zoogeography of Indian Oligochaeta. I. Affinities and systematic position of the genus Eudichogaster Michs. and some related questions. II. On polyphyly in the Oligochaeta. III. Some general considerations on the geographical distribution of Indian Oligochaeta. *Proc. zool. Soc. London.*

STEPHENSON, J. 1923. Oligochaeta. Fauna British India, pp. 1–518.

TALBOT, G. 1939. Papilionidae, Pieridae. Fauna British India. 2nd edition. 1: 1–600.

WALLACE, A. R. 1876. Geographical Distribution of Animals. London. 2 vols.

II. PHYSICAL FEATURES

by

M. S. MANI

On geomorphological grounds, we may recognize in India a Peninsular and an Extra-Peninsular Division. As we shall see later, these primary divisions are also correlated with climatic, historical and biogeographical considerations.

1. The Peninsula

The Peninsula of India is a stable mass of Archaean and Pre-Cambrian formations, exposed over more than half the area at present. The rest of the Peninsula is covered by Gondwana and later formations and by the Deccan Lava flows. The major mountain-building disturbances in the Peninsula ceased in the Pre-Vindhyan (Pre-Cambrian) times, but some minor folding, block-faulting and epeirogenic movements affected the region in Post-Cambrian times. As the geology of India is discussed in sufficient detail by KRISHNAN in Chapter III, we shall confine ourselves here to only the broad features of relief.

The Peninsular Plateau is highest in the south and west and slopes eastwards. Large areas in the south exceed 600 m in elevation and sometimes even 900 m. The western edge of this plateau forms the escarpment of the socalled Western Ghats or the Sahyadri Mountains. The eastern edge is much broken and is known as the Eastern Ghats, but SPATE (1957) prefers the term Eastern Hills for them. The Eastern and Western Ghats meet south of Mysore, from which the lofty plateau of the Nilgiri is separated by a deep valley. Further to the south and separated from the Nilgiri by the Palghat Gap, the Cardamom Hills constitute the divide between the east and west coasts. The Peninsular Plateau is flanked by a narrow coastal strip on the west and by a much broader coastal area on the east. In the north some lines of mountains rise above the general surface of the plateau, from the west to the east. The Satpura is the most important of these and forms an important biogeographical barrier. Two other lines of mountains, viz. the Vindhya and the Ajanta Ranges reinforce the Satpura barrier. The plateau slopes northward from the Vindhya Range gradually to the Indo-Gangetic Plains of North India, but in the northwest and interrupting the slope extends the Aravalli Range. In the northwest of the Peninsular Plateau, the Aravalli Hills extend from the southwest to the northeast, through Rajasthan. The general surface of the Peninsular Plateau is deeply dissected by river erosion.

The Peninsula of India is a compact natural unit of geomorphological and biogeographical evolution. Yet at the present time, it is merely a relic of a once much larger landmass, the major part of which now lies concealed under the alluvium of the northern plains and thrust under the high Himalaya and Tibet. The Peninsula consists at present of a block of plateau, with a general slope to the east and characterized by its pronounced senile topography. In addition to the portions that lie concealed under the alluvium and thrust under Asia, parts of the Peninsula appear also in detached blocks, both in the extreme northeast (the Shillong Plateau) and in the northwest; Ceylon is likewise a detached portion of the Peninsular Block.

In the north, the border outlines of the Peninsula are not simple, owing to the very ancient Aravalli folding and the strain, on the Peninsular continental (Gondwana) block, of the tangential forces which gave rise to the Himalayan uplift. The northern boundary of the Peninsular Block is generally set along an imaginary line extending from Cutch over the western flank of the Aravalli Ranges to within the environs of Delhi, and thence eastwards nearly parallel to the R. Yamuna and the R. Ganga, as far as the Rajmahal Hills and curving south to the west of the Delta of Ganga in Bengal. The alluvium of the Indo-Gangetic Plain often penetrates far south of this imaginary line in many places. The ancient peninsular rocks are also comparatively close to the surface in the alluvium at some places, as for example, in a northerly wedge indicated by the Kirana Hills in the Punjab. From the biogeographical point of view, the northern boundary of the Peninsula is a transitional zone rather than a sharply defined topographical line. This transitional zone extends in reality to the foot of the Himalaya, but for most practical purposes may be considered as disappearing gradually in the present-day course of the R. Ganga, the lowest area in the Peninsular-Himalayan foredeep. Defined in this manner, the Peninsula would correspond to the Cis-Gangetic Subregion of GADOW (1893) and BLANFORD (1901). BLANFORD included in the subregion the whole of the Peninsula and Ceylon taken together, after excluding the Punjab Tract (see fig. 1).

The principal elements of the geomorphology of the Peninsula comprise 1. the great plateau of granite-gneiss, occupying nearly all the south and the east; 2. the mesa-like area of the Deccan Lavas in the west centre; 3. the old shallow troughs of the Krishna, Godavari and Mahanadi Valleys; 4. the much worn Aravalli Ranges and 5. the Vindhyan scarplands of the north, with the R. Narmada-Son and R. Tapti rifts. The geomorphology of the Peninsula is, on the whole, marked by its advanced maturity or even senility, except perhaps along the escarpment of the Western Ghats and a few hillier localities. The erosion surfaces present more than one cycle and reveal important and relative recent changes of level, although mostly of a negative nature. The deposits of alluvium, about 150 m thick, in the R. Narmada trough and somewhat less in the

R. Tapti trough, occupy definite rocky basins, and indicate faulting. The straightness and the relative steepness, with two waterfalls, of the lower 480 km length of the R. Narmada, from Handia to the sea, indicate a relatively recent origin.

The Peninsular divisions are: 1. the Peninsular Plateau, 2. the Peninsular Foreland, 3. the coastal regions and 4. Ceylon.

1.1. THE PENINSULAR PLATEAU

The Peninsular Plateau covers 1. the Western Ghats, including Coorg; 2. the Southern Block of the Nilgiri, Anamalai and the Cardamom Hills; 3. the Deccan Lavas; 4. Karnataka or southern Bombay, Deccan and Mysore; 5. the upper Mahanadi and the adjacent basins (the Wainganga Valley, the Chhatisgar, the Upper Brahmani and the Jamshedpur Gap); 6. Telangana or the southeast Hyderabad and the Madras Deccan; 7. Anantapur-Chittor Basins and 8. the Eastern Hills of Orissa and Bastar, the Cuddapah ranges and valleys.

The Southern Block lies near the Palghat Gap. On either side of this gap are the highest levels of the Peninsula – the Nilgiri to the north of the Gap and the Anamalai-Palni-Cardamom Hills to the south. These elevated areas are great horsts that correspond to similar structures in Ceylon (see Plate 1). The Nilgiri or the Blue Mountain forms a compact plateau of about 2600 sq. km area, with a summit-level of 1800–2500 m above mean sea-level, rising with extreme abruptness on all sides. The fall on the eastern slope is 1800 m in 3 km. On the north it is cut off from the Mysore Plateau (900–1200 m above mean sea-level) by the deep straight gorge of the R. Moyar, the narrow bed of which lies at 300–600 m above mean sea-level.

The Anamalai-Palni-Cardamom Hills are more complex than the Nilgiri Hills. The highest peak in the Peninsula rises to an elevation of 2695 m above mean sea-level in the Anamalai Hills. The front to the Palghat Gap is extremely steep and straight in the east. The southeast flanks of the Palni Hills, overlooking the Upper Vaigai Re-entrant, are abrupt. This feature is also true of the Cardamom Hills. Between the 10° north parallel and the Shencottah Gap, the streams of the exposed Arabian Sea Front have pushed the watershed back to nearly 6 km of the eastern edge of the hills. The transition from the forest-clad mountains to the plains of the Tamilnad is abrupt here.

The Deccan Lavas country is also known as the Maharashtra. Karnataka is really Carnatic, a name that has often been misapplied, especially by zoologists, to the Madras Littoral area. This region extends from the Deccan Lavas in the north to the R. Moyar in the south. The crest of the Western Ghats is its western limit and in the southeast are the border hills and the scars of the Mysore Plateau. In the northeast the high plateaus break down into isolated basins south of the R. Pennar and the

Cuddapah Ranges and in the north the Raichur Doab between the rivers Krishna and Tungabhadra. The area is greater Mysore. The plateaus are 450–600 m above mean sea-level in the north and 900–1200 m in the south.

The area to the east of the Deccan Lavas and between the Maikal Range and Chota-Nagpur to the north and the Orissa Hills to the south constitutes the Upper Mahanadi Basin (Chhatisgarh). In the west, the Wainganga Valley is separated from the Upper Mahanadi Basin by a low continuous system of hills, extending south from the Maikal Range. The Deccan Lavas give place, east of Nagpur, to undulating Archaean terrain (300–930 m above mean sea-level), broken by small disconnected hills. The valley is densely forested. The Upper Mahanadi Basin proper, between the Maikal and the Orissa Hills, is about 130–160 km wide. In its centre is a great basin of Cuddapah rocks. The rivers and streams here are extraordinarily mature, though there are indications of local rejuvenations. The plain lies mostly to the west and north of the R. Mahanadi and is drained by the R. Sheonath. On the northwest is the Maikal Scarp and to the south are the jungle-clad Bastar Hills.

Telangana, representing the rest of the Peninsular interior, within the bordering hills, is mostly peneplain, developed on Archaean gneisses. The area between the scarp of the Palkonda Range and the higher Mysore plateau forms the Anantapur-Chittoor Basins. The Eastern Hills (Eastern Ghats of the older authors) are by no means comparable with the Western Ghats. In the north elevated and highly dissected peneplains are cut across by the Gondwana Mahanadi-Brahmani trough. To the south of the lower Godavari trough there are different elements of the Cuddapah Ranges and further south we have the Javadi, Shevroy Hills, cut off from the Mysore Plateau by the middle Palar-Ponnaiyar trough.

1.2. THE PENINSULAR FORELAND

The Peninsular foreland covers the Aravalli Ranges, the series of scarped plateaus and troughs representing the buckling of the northern edge of the Peninsular Plateau under the stress of the Himalayan uplift, the peneplain of Chota-Nagpur, the Maikal Block, the much dissected gneissic terrain of Bundelkhand and the narrow salient of the Deccan Lavas in the Malwa. The Thar is almost a semi-desert area between the R. Indus and the R. Sutlej and the eastern edge of Aravalli Range and includes Bikaner, Jodhpur, Jaisalmer, the eastern half of Bahawalpur, most of Kairpur and the Thar Parkar in Sind. This is essentially a peneplain, covered by sand, from which project Vindhyan, Jurassic and Tertiary mostly sandstone inliers and Lower Gondwana Talchir Boulders. Bare hills of granites and rhyolites in Jodhpur are extrusions on to the old Aravalli surface. Recent deposits of calcareous conglomerates in Luni are evidence of more humid conditions in former times. The limestone ridge on the Jaisalmer-Jodhpur border, north of Luni, is somewhat sheltered from the sand-drift. Impervious clays beneath the sand near Kaipur hold up ground-water. There are a number of saline lakes (*dhands*) in the area. The desert seems to be gaining eastward by about 130 sq.km. annually.

Disregarding the Thar Desert, proceeding northwards from the Deccan, we have 1. the faulted trough of the R. Tapti, 2. the Satpura Block linked in the east by the Mahadeo Hills to the Maikal Range, 3. the Chota-Nagpur peneplains continuing the Satpura-Maikal trends further east and carrying the higher Hazaribagh Range and the faulted Damodar trough, 4. the Narmada-Son trough bound on the north by 5. the Vind-hya Hills (Deccan Lavas) and 6. the Bhanrer-Kaimur Hills (Vindhyan Rocks), 7. the slope of the Deccan lavas in the R. Chambal Basin north of the Vindhyan Hills masking in the south a continuation of 8. the triple outward-facing Vindhyan-rock scarps of north Malwa (the boundary fault marking them from the Aravalli), 9. the Aravallis, 10. the Gneissic Bundelkhand east of Malwa and 11. the scarped plateau of the Vindhyan sandstones in Rewa between the Bundelkhand Gneiss and the Narmada-Son furrow.

This area is divided into 1. the Aravallis, 2. Malwa and the Vindhyan Hills, 3. Bundelkhand, 4. Rewa Plateau, 5. Narmada-Son furrow, 6. Satpura-Mahadeo-Maikal Hills, 7. Chota-Nagpur and 8. the R. Rapti Valley.

The Aravalli: The Aravallis are one of the oldest mountain systems in the world, retaining some relief even at the present time. The closely packed synclinoria in quartzites, schists, etc. of the Delhi-Dharwar (Algonkian-Huronian) age were probably uplifted in Pre-Vindhyan (Pre-Torridonian) times. The main southwest-northeast strike is re-markably regular for a distance of 700 km, from Gujarat to Delhi. The steep front to the Thar is formed of discontinuous and echelonned ridges. The highest point at the present time is the granitic mass of Mt. Abu (1721 m above mean sea-level) on the main axis, in the extreme south-west. The Aravallis rise to elevations of 1070–1200 m in great nodes of spurs and curving ridges around Udaipur. From here a series of ridges strike off east-northeast, along the Great Boundary Fault, enclosing alluvial basins of the R. Tonk. These link with the western axis in the tangled area of small quartzite hills, half-smothered in the Gangetic alluvium, north of the saddle between Jaipur and Jodhpur. The area was peneplaned in the late Mesozoic and warped afterwards. Another peneplain on softer schists and gneisses, on the plains east of the western axis and in the strike valleys, is largely covered by thin layer of older alluvium, which has in turn been recently peneplaned. The hills are dissected by dry nullahs, some of which are occasionally filled by torrents, and surrounded by piedmont-fans.

Malwa and the Vindhyan Hills: Malwa forms a triangle between the

15

Vindhyan Hills, the Great Boundary Fault of the Aravallis and the scarp overlooking Bundelkhand. There are three major Vindhyan scarps (450–600 m), formed in massive sandstones and separated by shales, facing south-southeast between the rivers Banas and Chambal and east over Bundelkhand. Strong scarp in the northwest flanks the left bank of the R. Chambal. Beyond this a scarped block occupies Dholpur and Karauli. The nearly horizontal Vindhyan rocks are folded and faulted by the rigid Aravallis that are overthrust onto them along the Boundary Fault. This contact seems to have been responsible for the warping of the Mesozoic-peneplain in the Aravallis. The warping and displacement of the Boundary Fault are 1200–1500 m, but diminishing northeast and southwest. In the southwest is an area of very irregular and dissected Deccan Lavas. The Deccan Lavas of the south abut on the outer scarp. The eroded edge suggests that the main lineaments of the underlying Vindyan are similar to those exposed on the north. They form a tableland, rising gently to 760 m in great brow, overlooking the R. Narmada. This scarp is the so called Vindhyan Hills. The soil covering the area is only a few centimetres thick or the ground is of peneplaned rock surfaces of poor grassland, open acacia-scrub. The area is mainly drained by the R. Chambal and its right-bank tributaries, but in the southeast by the upper courses of the rivers Ken and Betwa. The rivers Chambal, Ken and Betwa, rising within about 6 km of the R. Narmada, appear as consequents on the Mesozoic surface, superimposed on the scarps. The R. Chambal cuts straight across them, with subsequent tributaries on the softer shales. The rivers Kunu and Kunwari appear to be subsequents. The former cuts through the innermost scarp at Nayagaon, but it may have developed as a consequent on the older surface and reached its present position by lateral shifting down the dip. The R. Chambal and its tributaries Kali, Sindh and Parbati have formed a triangular alluvial basin, about 200–270 m above the narrow trough of the lower Chambal in Kota.

Bundelkhand: Bundelkhand is a dissected upland mass of rounded hummocky hills, exhibiting the typical exfoliation-weathering in the reddish Bundelkhand Gneiss. It is cut across by white quartzite dykes, from veins of only a few centimetres to massive walls.

The Rewa Plateau: The area, about 500–650 km long and 80 km wide, between Bundelkhand and the R. Son is a series of wall-sided plateaus, ending abruptly in the Kaimur Scarp in the south and less abruptly in the north. The Vindhyans are massive sandstones, with some limestones and shales. The Kaimur Crest (450–600 m above mean sea-level) rises to 300 m above the bed of the R. Son. Except for a narrow strip along the R. Son, the drainage is towards the R. Ganga, either through the R. Ken or through the R. Tons. These rivers escape over the northern scarp in a series of waterfalls and cascades. The plateau is mature and much of it is high alluvial plain of black loamy soil. The margins and much of the higher ground within the plateau are covered by forest.

The R. Narmada-Son Furrow: The R. Narmada rises on the Amarakantak Plateau on the Maikal Hills. Its course is complex as far as the Marble Rocks of Jabalpur, below which it enters the alluvial Fault trough. The Vindhya-Kaimur Scarp flanks this trough to the north for about a thousand kilometres. The steepness and the straightness of its lower course show that the river originally flowed out on the Tapti-line through the Burhanpur Gap. The Vindhyan Hills rise steeply to 300 m above the floor of the valley. The Satpura and the Mahadeo Scarps to the south are less well marked and also less continuous. The valley floor is 30–65 km wide. The river is counter-sunk by about 6–34 m within it. The R. Son flows about 480 km, close under the Kaimur Scarps that rise to 150–300 m above the narrow floor of the valley. There is little alluvium along the river, which is sunk in a low terrace disappearing westward in Rewa. There are apparently aeolian alluvial and unstratified loam patches further upwards.

The Satpura-Mahadeo-Maikal Scarps: The R. Narmada is flanked on the south by a series of scarped plateaus, 600–900 m above mean sea-level. In the west, the Satpuras are simply the steep-sided Deccan Lavas Block, sinking to elevations of about 360 m between Burhanpur and Khandwa, in which the R. Narmada must once have flowed. The Deccan Trap Horst or the Gawilgarh Hill is found in the angle between the upper course of the R. Tapti and its tributary Purna. The Mahadeo Hills, further to the east, are a great window of Archaeans and Middle Gondwanas, thick masses of red sandstones forming small plateaus, with precipitous scarps. Beyond the Jabalpur Gap in the east, the Maikal Range is dwarfed by the Amarakantak (1065 m above mean sea-level), on which the R. Narmada rises. The Amarakantak is mostly a dissected plateau, drained by deep-cut valleys into the R. Narmada. In the extreme east, the Johilla tributary of the R. Son follows a course, separated from the broader parallel valley of the upper course of the R. Son by a narrow ridge of Deccan Lavas. The plateau is tilted to the northwest, falling from 900–1000 m in the abrupt scarp overlooking Chhattisgarh to about 600 m about Mandla, where the head-streams of the R. Narmada converge.

The Chota-Nagpur Plateau and the Damodar Basin: East of the ridge separating the rivers Johilla and Son is the area of Deogarh Hills (1026 m), on the Son-Mahanadi watershed. This area is formed largely of Gondwana rocks, with patches of Archaean rocks and Deccan Lavas and marine Permian beds in Umaria. There are also about 10400 sq km of mainly Archaean gneisses that form rolling peneplains, bisected longitudinally by the fault-trough of the R. Damodar. The Hazaribagh peneplain (about 990 m above mean sea-level) north of the R. Damodar is crossed by a higher plateau, with some monadnocks, the socalled Hazaribagh Range. The plateau is generally open, with numerous irregular spurs and outliers, and falls abruptly into the Gangetic Plain. The so-called Rajmahal Hills in the northeast are highly dissected plateaus of Gondwana basalts that rise steeply from the alluvium, in the great bend of the R. Ganga. The Damodar Basin occupies a relatively small area of about 19500 sq. km, with the principal streams flowing west-east. The Gondwana rocks of this basin present generally low undulating terrain.

The Tapti Valley: This valley comprises the main Tapti Trough, continued in that of the Purna and the Upper Tapti Valley in its northeast-southwest course through the Burhanpur Gap. The floor of the valley lies at elevations of 200–300 m above mean sea-level, but the river is entrenched as much as 15–20 m below this floor. In the north is the steep face of the Satpura and the Gawilgarh Hill. To the south we have the Ajanta Hills of the Sahyadri Mountains. Except for the alluvial filling of the trough, the whole area is Deccan Lava Terrain.

1.3. The coastal regions

The coastal regions of the Peninsula are the Western Littoral Region or the Malabar Coast and the Eastern Littoral or the Coromandel Coastal region.

A. The Western Littoral Region

The Western Littoral Region embraces 1. Cutch and Kathiawar, 2. Gujarat, 3. Konkan, 4. Goa and Kanara and 5. Kerala.

The Western Littoral Region is remarkable for the almost complete absence of river-captures, in spite of the conditions favourable for such a phenomena. This is attributed to large-scale regional subsidence at a recent date. Except perhaps south of Goa, the great fault-line scarp of the Western Ghats is continued as a remarkably sharp feature on the Archaean. Although the watershed recedes from the coast as we proceed southwards, the Ghats also recede and the watershed and the crest of the Ghats are generally never very far from each other, except in the breach to the south of Goa. Although the land would appear to be on the ascendent at present, the coast of Konkan suggests a plane of marine erosion. The submerged forests of Bombay and the appearance of the Deccan Lavas coast are evidence of recent depression. We may also observe some submergence in the north, a seaward advance of the land in the mudflats of Rann that link Kathiawar with the mainland, the prograding shores of the Gulf of Cambay, subsidence of the Arabian Sea in Konkan, emergence plane of marine erosion with some sinking followed by a still-stand and uplift assisted by prograding on a low shoreline of emergence in the south. The sagging of the region is attributed to be due to the loading by the Deccan Lavas. The macro-faulting is perhaps the result of Miocene Himalayan uplift. In the south the uplift is perhaps connected with the punching-up of the Nilgiri, Anamalai-Palni and Ceylon Horsts.

The area of Cutch and Kathiawar lies between the Rann of Cutch and the Gulf of Cambay, and merges in the north with the Thar Desert. The Rann is a vast expanse of tidal flats, with saline efflorescences, representing a broken anticline. Cutch presents a discontinuous series of Jurassic-Miocene sandstones, with intrusive and interbedded basalts (270–3300 m thick), flanked by alluvial and aeolian deposits. We find here flat-topped and steep-edged plateaus, greatly dissected around margins, and small alluvial basins. Kathiawar is formed of Deccan Lavas, intersected by Trap Dykes. Except the Jurassic sandstones of Drangadhara-Wadhwam Plateau in the north, the basalt platform is flanked by Tertiary clays, blown sand of largely foraminiferal casts, in a calcareous matrix, attaining a thickness of about 60 m in Junagarh (forming the well-known building stone of Porbandar). A discontinuous strip of laterite marks the edge of the Deccan Lavas here.

Gujarat comprises a great tract of alluvium, formed by the R. Sabarmati, R. Mahi and other streams and actively propagating into the Gulf of Cambay. We may recognize the alluvial piedmont between the highlands and the plain, the coastal marshes and the shelf of firm alluvium between them.

Konkan is the coastal lowland, much broken by hills, as far south as Goa, about 530 km long and 80 km wide. In the north, there is a belt of alluvium, hardly 13 km wide, along the coast. To the east of this alluvial belt is a series of parallel ridges (600 m above mean sea-level). Rivers like Amba, Ulhas and Vaitarni have their nearly parallel courses in this area, before reaching the coast. We may observe laterite-capped residual plateaus in south Ratnagiri. The scarps of the Western Ghats rise to elevations of 1000 m and are fretted into canyons at the heads of the valleys. Somewhat north of Goa, the Deccan Lavas are replaced by Archaeans in the Western Ghats. This is marked by a series of breaches in the mountain-wall, so that the rivers Kalinadi, Gangavali-Bedti, Tadri and Sharavati have encroached on the Krishna-Tungabhadra drainage. The watershed is here about 240 km from the sea-coast, instead of the usual 40–55 km that we find in the north.

North Kanara is essentially highland and the real lowland littoral is restricted to pockets along the lower course of the rivers that break the Ghats. The Ghats are here greatly dissected as a result of differential erosion, faulting, lithological characters, etc.

South Kanara is an embayment of lowland, about 75 km in the widest part in the Netravati Valley near Mangalore. The alluvium is better developed here than in the North Kanara and is backed by low lateritic plateau. The laterites and the alluvium are broken by ridges and isolated hills of Archaean gneisses and granites.

19

Kerala exhibits, in the erosion surfaces in the laterite, evidence of perhaps two phases of upward movement. There is a threefold longitudinal division of 1. the alluvial coastland, 2. the low lateritic plateaus and the foothills and 3. the gneissic highlands. The alluvium is abundantly developed in the areas of lagoons and backwaters, the largest of which widens into the Vembanad Lake, to the south of Cochin. Spurs of the Anamalai-Cardamom Hills project into the laterites.

B. The Eastern Littoral Region

The Eastern Littoral Regions are strikingly different from the Western. The lowland is much wider here and much of it is also true coastal plain in its structure, with infacing coestas in Cretaceous and Tertiary epeirogenetic deposits. In other places, it is formed of the deltas of the rivers Mahanadi, Godavari, Krishna and Cauvery. The coastal lowland is 100–130 km wide in the south and is backed by the broken Tamilnad Hills (Javadi, Shevroy Hills) and by low plateaus (300–450 m above mean sea-level) around the middle course of the R. Cauvery.

The subregion of the Deltaic Orissa and the Northern Circars in the north covers the rivers Vaitarni, Brahmani and Mahanadi. Orissa Delta is a great alluvial salient, about 195 km across the Chilka-Lake-Balassore Base and about 80 km wide. The R. Mahanadi is one of the most active depositing streams in India at present. Its delta is formed of swampy jungles on the prograding sea-face, a zone of firm older alluvium about 65 km wide and finally the lateritic shelves on irregular upland margin, and outliers of gneissic hills often reach nearly to the sea. The Chilka Lake varies in area from 900 to 1200 sq km and is alternately salty and fresh-water. Only a few metres deep, it is cut off from the sea by a long spit. The Northern Circars area is climatically a transitional belt. The Tamilnad subregion covers an area of nearly 130000 sq km between the Bay of Bengal and the Deccan Plateau, from the R. Krishna to Cape Comorin. The Nellore area is transitional and the Cauvery Delta makes a great breach in the continuity of the emergent lowland. The area to the south of the R. Cauvery is strikingly different from that north of it. Six subregions are generally recognized: 1. the Coastal Coromandel Plain, 2. the Tamilnad Hills, 3. Ponnaiyar-Palar Trough, 4. Kongunad, 5. the Cauvery Delta and 6. area southeast of the Vaigai Basin (Madura, Ramnad, Tinnevelly black-soil plain and the Tambarabarni Basin).

The Coromandel* Coast Plain is structurally a true Coastal plain. The lowland below an elevation of 150 m is about 80–100 km wide and comprises 1. the peneplaned gneisses below the hills, 2. the remnants of marine mainly Cretaceous-Eocene deposits, 3. the Cuddalore Sandstone

---

* Coromandel is a corruption of Cholo Mandal, named after the Chola Kings of South.

20

and laterite shelf, 4. the young alluvial plain and 5. the very recent alluvial strand-plain that is still prograding locally. The older rocks are close to the sea south of Madras, where the plain is much broken by many abrupt Charnockite Inselbergen (200–240 m high) and with northeast-southwest trend. The coast follows the main strike of the Deccan Charnockites, with a change in the trend at 16° NL. There is considerable evidence of both emergence and submergence. Between the rivers Palar and Cauvery, the coastal plain is backed by bold hills of Javadi, Shevroy, Kalrayan and Pachamalai. The more important ridges of these hills are Charnockites, intruded as sills along the foliation planes of the older gneisses. Though the general limits of the trough between the Mysore ghat and the Tamilnad Hills are well defined, the area in the south around Salem presents certain uncertainties.

1.4. CEYLON*

Shaped roughly like a pear, the island of Ceylon lies between 5°55′ and 9°50′ NL. Its longest diameter 535 km is north-south. Its area is about 65630 sq. km. The island consists of a central mass of mountains, rising to a mean elevation of about 1800 m above mean sea-level, and broad coastal plains. The peak Pidurutalagala is 2527 m above mean sea-level, Kirigalpotta is 2393 m, Adams Peak is 2243 m. The rivers are short and radiate from the central mass of mountains.

Though Ceylon is now separated by shallow and narrow Palk Strait, it was part of the mainland of the Peninsula even during the late Pleistocene times.

Ceylon is indeed a detached portion of the Peninsula, with land-forms, soils, climatic regions and natural vegetation reminiscent of South India. With the exception of the Jaffna Peninsula, the northwest coast and some other local coastal steps, the whole of the island consists of crystalline Pre-Cambrian rocks. The Khondalite Series, occupying the interior of the island, are similar to the same series in Madras and Orissa. It is made up of metamorphosed sediments, largely quartzites, schists and crystalline limestones. The Vijayan Series consists mainly of biolite-gneisses and schists, with discontinuous outcrop fringing that of the Khondalite Series. There are certain other ancient intrusive rocks, including Charnockites, similar to those found on the mainland of the Peninsula. Important outcrops of sedimentaries are confined to the Jaffna Peninsula and the northwest coast, where are almost horizontal Miocene limestones. Gondwana deposits are limited to one or two small outcrops of sandstones.

The island consists, as a whole, of ridges separated by deep valleys, occasional marshy or grassy plains surrounded by mountains. The

* Recently renamed Shri Lanka.

principal areas are Adam's Peak Ridges, the Hatton Plateau, the High Plains, the Uva Basin, the Lunugala Region, the southern Platform, the Piduru Ridge, the Kandy Plateau, the Dolosbage group of hills, the Northwestern Upland, the Matale Valley, the Knuckles group of hills and the Sabaragamuwa Hill. The lowland belt is in most parts rolling country, not above 300 m above mean sea-level and with laterite red soil. The Southwest Lowland is a wet area under the influence of the southwest monsoon. The Hambantola or the Southeast is the Dry Zone of deficient rainfall.

2. The Extra-Peninsular Area

The Extra-Peninsular area is geologically young and has been subjected to intense mountain-building activity during the Cretaceous, Tertiary and Pleistocene times. The mountain belts here are characterized by complex folding, overthrusts and nappes of great dimension, involving horizontal compressions of the crust for hundreds of kilometres. There are three principal mountain arcs viz. the Baluchistan Arc, the Himalayan Arc and the Burmese Arc. In each of the arcs, the convex side faces the stable mass of the Peninsula, so that the thrusts are directed towards the south in the Himalaya, west in the Burmese Arc and east in the Baluchistan Arc. These mountain arcs comprise in each case a series of mountain ranges, one behind the other and sometimes gathered into a series of the socalled 'festoons' as in Baluchistan.

The Himalaya extends from the Pamirs Knot in the extreme northwest, as an unbroken chain of mountain-wall, in a smooth curve of about 2500 km length, to the east. In the northwest the mountain chain is rather complex and consists of many ranges. For some distance from the Pamirs, the Hindu Kush forms the boundary between India and northeastern Afghanistan. A great tangle of hills and high mountains merges, between the Punjab and north Baluchistan, into the Sulaiman Range. The Himalaya is relatively simple in the east. The mountains between India and Burma form a continuous curve from the northeast Assam to the Cape Negrais. In the north the curve comprises comparatively simple and narrow divide of the Patkoi (or also Patkai) Hills, to broaden out into the Naga Hills and the Plateau of Manipur, from where a branch extends into Assam. Starting as the Barail Range, this branch is separated by a col from the Jaintia, Khasi and Garo Hills. The Lushai and Chin Hills extend southward from Manipur as the narrow Arakan Yoma.

The Himalaya (from the Sanskrit *hima* = snow *alaya* = abode) is the name applied in the ancient India to the Great Snowy Range of mountains, visible in the north from the Indo-Gangetic Plain. As now understood, the Himalaya embraces the complex system of nearly parallel ranges of Tertiary mountains, extending over 3200 km from north of

Burma in the east to nearly Afghanistan in the west (approximately between east longitudes 72° and 91° and north latitudes 27° and 37°). The width of the Himalayan System is extremely variable and is in places only 80 km and in other places exceeds 300 km. Though the Himalaya proper appears to terminate in the east at the southward bend of the R. Brahmaputra and in the west similarly at the bend of the R. Indus, the Himalayan system continues much further westward into the mountains of Afghanistan. Structurally and geologically, the Himalayan system is likewise continued into the meridional mountains of Burma. While the Himalayan system rises nearly abruptly from the plains of India, it is continued in the north as a great series of folds, of which the Kuen Lun represents perhaps the northernmost. Between Kuen Lun and the Himalaya proper lies the bleak high plateau of Tibet, at elevations between 4750 and 4880 metres. The physical unity of Tibet with the Himalaya is evident in the general geological structure, in the parallelism of the Kailas Range, the Ladak Range, the great and Lesser Himalayan Ranges, all of which change the directions of their trendlines together.

2.1. THE RANGES OF THE HIMALAYA

The ranges of the Himalaya fall under two major groups, viz. the Cis-Himalayan and the Trans-Himalayan. The former group of ranges lies south of the Great Himalaya (the main range) and comprise the Siwalik Ranges and Lesser Himalayan Ranges. The Trans-Himalayan ranges lie north of the main range and include Zaskar, Ladak and Karakoram ranges.

The Siwalik Range separates the Himalaya proper from the Indo-Gangetic plain and is in reality the southern border-range of the Himalayan System. Though its upheaval was accompanied by movements of the Himalaya also and perhaps also by increases in the elevation of the main Himalayan range, the Siwalik is of more recent origin than the great mountains in the north. With the exception of a short distance of about 80 km, opposite the basins of the R. Teesta and the R. Raidak, the Siwalik Range lies with remarkable uniformity, in front of the Himalaya throughout its whole length, from the bend of the R. Brahmaputra to that of the R. Indus. There is a break at the passage of the R. Sutlej in the alignment and the two lengths of the range appear to overlap, so that the range to the north of the Sutlej is not in direct prolongation of the one to the south. In some places the Siwalik Range is pressed against the outer Himalayan ranges and in other places it is separated from these ranges by distances of 30 to 80 km, to enclose the characteristically shaped longitudinal valleys called 'duns', filled with deposits of rounded stones, gravel and sand, brought down from the Himalaya. The Siwalik Range is strongly developed opposite Dehra Dun, with steep southern slope and gentle northern slope.

23

The mountainous region, about 150 km wide, between the Great Himalaya and the Siwalik, constitutes an intricate system of the Lesser Himalayan Ranges. The Lesser Himalayan Ranges have been compressed horizontally, but are the result of a series of crustal movements, with more complex history than that of the Siwalik Ranges. After their uplift, the Lesser Himalayan Ranges appear to have been forced to change their direction, so that the whole region has been subjected to successive compressions and the general wrinkling process seems to be still active. In parts of Nepal and in Kashmir, the outer ranges and the flat alluvial valleys behind the Lesser Himalayan Ranges are distinct. In the Kumaon such high level and flat valleys are absent. The Lesser Himalayan Ranges comprise two distinct groups: 1. the ranges that branch off from the Great Himalaya proper and 2. the ranges that are separate folds. The branch ranges stretch obliquely across the mountain area. The separate folds follow the curvilinear alignments parallel to the Great Range. The Great Himalaya bifurcates at points where there is a change in the alignment and each successive branch adopts the alignment forsaken by the trunk range. There are seven Lesser Himalayan Ranges, viz. the Nag Tibba, the Dhauladhar, the Pir Panjal, the North Kashmir, the Mahabharat, the Mussurie and the Ratan Pir.

The Zaskar Range branches off from the Great Himalaya near Nampa and its well known peak is Mt. Kamet (7770 m above mean sea-level). The Ladakh Range extends from Assam to Baltistan, but its continuity north of the Great Himalaya is not distinct throughout. North of Assam, the Ladakh Range is strongly developed and forms the water-parting between the Tibetan and the Indian sections of the R. Brahmaputra. Westwards from Nyang basin, for a distance of about 320 km, the Ladakh Range is parallel to the Great Himalaya and the intervening trough is occupied by the R. Arun. North of its bifurcation from the Great Himalaya at Dhaulagiri, the mean elevation of the Ladakh Range increases. North of the R. Karnali Basin, the Ladakh Range is strongly developed. South of the Lake Mansarovar is the peak Gurla Mandhata, west of which the continuation of the Ladakh Range becomes somewhat vague. It is generally believed that the Ladakh Range has risen subsequent to the birth of the R. Indus. West of the lake Mansarovar, the relations of the Ladakh Range to the R. Indus are extremely peculiar. For the first 290 km from its source, the R. Indus flows parallel to it along the trough north of the Ladak Range. It then bends at a right angle to cut across the Ladakh Range and to flow for about 480 km along southern flank of the range. Shortly before its confluence with the R. Shyok, the Indus crosses back to the north of the range and flows for about 160 km before again it cuts across the range for the third time.

The Karakoram and the Hindu Kush Ranges are, strictly speaking, different sections of the same crustal fold that stretches from the southeast to the northwest, curves round Hunza and Gilgit, passing north of

24

Chitral and entering Afghanistan from the northeast to southwest. The eastern part of the fold is known as the Karakoram and the western part is the Hindu Kush. The Karakoram forms a mass of rock and ice, extending some 400 km long from the R. Shyok to the R. Hunza. The Aling Kangri Peak is generally considered to mark its extreme eastern limit. The mean elevation of the Karakoram Range in Tibet is between 5000 and 5500 m, in other words hardly 300–600 metres above the general level of the plateau. We have here perhaps the greatest assemblage of high peaks in the world; there are over 30 peaks rising to elevations above 7300 m and the Gasherbrum Summits rise to elevations over 7925 m and K_2 or Mt. Godwin Austen is an irregular cone of ice and limestone on a granite-gneiss base, rising to an elevation of 8610 m. None of the passes in the region are lower than the elevation of Mt. Blanc. The Karakoram is believed to be much older than the Himalaya. The absence of Tertiary sediments between Ladakh and the northern flanks of the Kun Lun Range is considered to favour this view. The rivers Shyok, Hanza, Gilgit and Kunar drain the trough north of the Karakoram. The R. Nubra rises on the Karakoram and the glacier at its source has cut a notch in the crestline of the mountain range.

The Kailas Range is a Trans-Himalayan mountain range that is parallel to the Ladakh Range, but lies about 80 km north of it. Mt. Kailas is the highest of a cluster of peaks, exceeding 6000 m in elevation in the area of the Lake Mansarovar. Opposite Mt. Kailas is the peak Gurla Mandhata on the Ladakh Range. East of the 85° meridian, the Kailas Range bifurcates and for a distance of nearly 240 km the R. Raga flows in the trough between the two branches. Immediately after the bifurcation, the branch range rises to an elevation of 6000 m. It appears to conjoin with the Ladakh Range near the lake Yamdrok (Karo-La Range). The main Kailas Range extends east, with peaks 6000 m above mean sea-level. The Kailas Range appears to end in the Sajum Peak (6099 m) near the Pongong Lake, but it really continues further west and forms the water-parting between the R. Shyok on the south and the R. Nubra on the north. West of the rivers Nubra and Shyok, the Kailas Range is parallel to the Karakoram and the long troughs between the Kailas and the Karakoram Ranges are occupied by the Biafo, Hispar and Chogo Lugma Glaciers. Opposite the bend of the R. Indus, the Kailas Range has the Haramosh Peak (7390 m); its highest peak is Rakaposhi, about 15 km from the intersection by the R. Hunza.

2.2. THE GEOGRAPHICAL AND BIOGEOGRAPHICAL DIVISIONS OF THE HIMALAYA

Geographically, the Himalaya is divided into 1. the Eastern or the Assam Himalaya, 2. the Central or the Nepal Himalaya, 3. The Kumaon or the western Himalaya and 4. the Northwest or the Pubjab Himalaya.

The Assam Himalaya, approximately 720 km long, comprises the portion between the Namcha Barva Peak (7750 m), east of which the R. Brahmaputra curves southward, and the R. Teesta in the west. In this division, there is very little of the sub-Himalayan tract so that the Himalaya rises rather abruptly from the plain. In Sikkim, where the Assam Himalaya passes into the Nepal Himalaya there is also a change of alignment, with the ranges on the west extending from east a little north of west. The lower and outer ranges also disappear here and paired spurs, the Singalila Ridge from the Kanchenjunga and the Chola Ridge from Pauhurni extend southward. The snowline in the region is at an elevation of 4875 m, but glaciers come down by about 1000 m below these limits. The former glaciers descended much lower, up to nearly 2680 m above mean sea-level. There are a number of hot springs like the Phut Spring (with the water at a temperature of 38 °C), Ralong Spring (55 °C), Yeumtung Spring with the water saline, Momay Spring at an elevation of 4880 m (about 1600 m below the Kanchenjunga glacier) (47 °C).

The Nepal Himalaya, between the R. Teesta in the east and the R. Kali in the west, is about 900 km long. A number of well known peaks like Mt. Everest (8848 m), Kanchenjunga (8579 m), Makalu (8470 m), Dhaulagiri (8425 m), Annapurna, Gosainthan (8010 m), etc., are situated in this division. The Great Himalaya bends and bifurcates near the Dhaulagiri Peak, west of which the mean elevation diminishes, so that none of the peaks rise to over 6700 m. Near the western end of the basin of the R. Karnali are the Api-Nampa group of peaks, where there is another bifurcation. The southern branch is the Great Himalaya, with Nanda Devi (7820 m) and the Badrinath Peak (7060 m), and the northerly branch, with Mt. Kamet (7770 m), is the Zaskar Range. The southern boundary lies in the terai. Beyond the Siwalik ridges, of which the most important is the Churia Ghati that bars the way to Kathmandu, there is a broad zone at 900–2750 m elevation, with west-northwest-east-southeast trends. The north is occupied by spurs from the Main Himalaya, like the Dhaulagiri Massif, the Singalila Ridge, etc. Orographically the region comprises 1. the lowland at the foot of the hills, a narrow belt of the Nepal Terai (16–50 km wide); 2. the sandstone range, with duns, 180–240 m higher than the Terai zone and a continuation of the Siwaliks; 3. from the northern extremity of the duns, the Sub-Himalayan zone of mean elevation of 3050 m, with the valleys at an elevation of 1200 m and 4. the mountain zone (the Main Himalaya). Three natural regions, divided by lofty ridges, are named after the rivers that drain them. The Western Division is Kauriala (Karnali) or the Gogra Basin. The Central Division is the Gandak Basin and Western Division is divided into two unequal parts by the R. Kali (Sarda) that forms the boundary between Nepal and Kumaon. The Eastern Division is the Kosi Basin.

The Kumaon Himalaya extends for about 320 km between the R. Kali and the great defile of the R. Sutlej. Naini Tal, Almora and Garhwal of the Uttar Pradesh State are within this division. Nanda Devi, Trisul, Mana, Badrinath, Kedarnath, Gurla Mandhata, Gangotri and Bandar Punch are some of the better known high peaks of the Kumaon division. The Kumaon Himalaya is much corrugated. The Northwest Himalaya is the division west of the defile of the R. Sutlej, approximately 560 kilometres long to Mt. Nanga Parbat, west of which the R. Indus curves round south westward.

Biogeographically, the defile of the R. Sutlej is a most important landmark. To the east of it the Himalaya forms a zoögeographical unit that is distinct and fundamentally different from the Northwest Himalaya to the west of the defile. At the defile of the R. Sutlej there is an abrupt break in the general trendline of the mountain chains. While east of the defile the ranges are mostly east-west, the trend of the ranges in the Northwest Himalaya is more southeast to the northwest. While the rest of the Himalaya lies between 27° and 29° NL., the Northwest Himalaya stretches from about 30° to 36° NL. and has thus a much greater width, embracing an extensive Sub-Himalayan tract. West of the defile of the R. Sutlej there is an abrupt fall in the mean elevation of all the ranges. In the rest of the Himalaya the high peaks rise to mean elevations of 7600 m and many even above 8000 m; very few peaks are above 6700 m in the Northwest Himalaya. There are, however, numerous peaks with a mean elevation of 6000 m in the Northwest Himalaya. Two notable high peaks are the Nun Kun twin Peaks (7130 and 7086 m) and the great Nanga Parbat (8126 m). While many river gorges cut through the main range east of the defile, no rivers pierce the main range in the Northwest Himalaya. The main water-parting between India and Tibet lies just north of the Great Himalaya in the divisions east of the defile, but in the Northwest Himalaya the crestline of the main range actually coincides with the water-parting. North of the Northwest Himalaya are the Karakoram-Pamir mass and Tibet lies to the northeast. The succession of ranges from the south are the Siwalik, the Nag Tibba, the Dhauladhar, the Pir Panjal, the Great Himalaya, the Zaskar, the Ladak and the Karakoram. These ranges are essentially secondary undulations on a great broad elevated arch, the span of which reaches from the plains of the Pubjab in the south to beyond the R. Indus in Tibet.

In Kashmir there are two mountain masses, viz. the Korakoram and the Himalaya, with the R. Indus in between. On the southern flanks of the main Himalaya lies the valley of Kashmir proper and is walled in by the Pir Panjal Range. Biogeographically the Kashmir region includes also a narrow strip, about 8–20 km, of the Pubjab Plains. The Siwalik Hills rise to elevations of 600–1200 m and are largely anticlinal and overlook a series of duns, succeeded in turn by the Miocene sediments of sandstones and Eocene Nummulitic limestones, at elevations of 1800–

2400 m. The Siwalik zone, about 6000 m thick, has undergone very recent folding and faulting and early Pleistocene thrusting. The outer ridges are clothed with sparse dry scrub and the inner ones have *Pinus longifolia* forests. The Pir Panjal bifurcates from the Main Himalaya, but structurally and lithologically, it is most complex. Most of the contemporary glaciers are on the northern slope, in contrast to those of the Himalaya, which is rather arid on the Tibetan side. The snow on the Pir Panjal Range is mostly derived in winter from the west or the northwest. The general river pattern, particularly the great bends of the R. Chenab and the Jhelum and their tributaries at these bends carrying on the line of the lower main streams, suggest that the drainage was formerly to the southeast.

The Vale of Kashmir is a basin, about 135 km by 40 km, at an elevation of about 1550 m in the Jhelum flood plain. The southern flanks fall relatively gently from the crest of the Pir Panjal. The northern wall is dissected by the Sind and other rivers antecedent to the bordering hills. One of the most striking features of the Vale of Kashmir are the flat-topped terraces of the Karewas or the Pleistocene sediments of clays, sands and silts of lacustrine origin, in which bands of marl and loessic silt, together with lenticles of conglomerates form old deltaic fans. The existing lakes, of which the Wular is the largest, are not strictly relics of the former lakes, but are enlarged old oxbows and abandoned courses of the R. Jhelum. The ponding by detrital fans from the hills seems also to partly account for these lakes. The Karewas are in places eaten into great bluffs by the R. Jhelum and the terrace lies at about 125 m above the river at present. In the south, the Karewa beds extend from about 1575 to 2700 m and are quite steeply tilted. With the Pir Panjal uplift, the deposits were folded and faulted in places. The term Karewa is applied really to the level surface between the incised streams dissecting the terraces, the flanks of which are generally steep.

The Indus-Kohistan is a land of mountains in a mountain world. The Mt. Nanga Parbat is almost separated from the main mass of the Himalaya by the valleys of the R. Krishan Ganga and the R. Astor, between which is the Burzil Pass. To the north and east is the deep gorge of the Indus. On the Nanga Parbat Massif are about 270 sq. km of snowfields that drain into glaciers, descending nearly 2400 m below the snowline. There is intense erosion round the massif. The R. Indus flows round this massif, in gorges 4575–5180 m deep and hardly 18–24 km wide. The floor is relatively wide and flat, hot and arid and has been most appropriately described as a 'desert embedded between icy gravels'.

3. *The Indo-Gangetic Plains*

The Indo-Gangetic Plain of north India, about 250–450 km wide, extends from end to end more than 3000 km, from the Arabian Sea to the Bay of Bengal. This great plain is remarkable for the dead flatness, the gentle slope seaward and the immense thickness of the alluvium, which the present-day rivers Indus, Ganga and Brahmaputra could not certainly have laid down, at the speed of the present day deposition.

The alluvial morphology of the Indo-Gangetic Plain is described by GEDDES (1960). Its relation to the Himalaya is discussed by HAYDEN (1913). The alluvium from the delta of the R. Indus to that of the Ganga-Brahmaputra, represents the filling of the foredeep, warped down between the stable northward drifting Gondwana Block of the Peninsula and the advancing Laurasia. It was formerly believed that the alluvial filling was about 4525 m thick and was deposited in a trough sinking beneath its own weight. Recent data on gravity anomalies indicate, however, a maximum depth of about 1980 m and the alluvial filling is of unequal thickness. The Indo-Gangetic trough does not correspond to the full extent of the Indo-Gangetic Plain. Outcrops of older rocks indicate that the alluvial cover is thinner in the Indus than in the Ganga area. There is also evidence of concealed ridges, prolonging the Aravalli-axis between Delhi and Hardwar and also northwest from Delhi towards the Salt Range in the Punjab. The floor of the Ganga Delta is still sinking.

The plain is topographically homogenous for hundreds of kilometres, the only noticeable relief being that of the floodplain bluffs and belts of ravines and badlands, formed by gully erosion along some of the larger streams like the lower course of the R. Chambal. Two important surface differences must, however, be observed. Along the outer slopes of the Siwaliks, there is commonly steep gravel talus slope, called the *bhabar*, in which all but the larger streams lose themselves, but seep out lower down in marshy and jungly *terai*. The older or the Pleistocene alluvium, called *bhangar*, occupies generally higher ground than the recent *khadar* that grades into the most recent delta silts. The alluvium is on the whole fairly stiff clay, with some sand. The bhangar includes irregular lime concretions of *kankar*, often with 30% calcareous matter in some places, but in the drier areas of Uttar Pradesh, Pubjab and Rajasthan there are stretches of barren saline efflorescences called *reh* or *kallar*.

4. The Soil

The older authorities recognized the following types of soils:

1. *The alluvial soils:* The alluvial soils cover an area of about 780000 km² in the Indo-Gangetic Plains. The deltas in Madras, Gujarat and Kerala are also alluvial in nature. The alluvium of the Indo-Gangetic Plains grades from coarse material of the piedmont *bhabar* to the fine silts of the Ganga Delta. As already pointed out, there is considerable difference between the older *bhangar* and the long fingers of the more recent *khadar* in the main floodplains, often refreshed by newer silt.

2. *The regur soils:* The regur soils are perhaps best developed on the

Deccan Lavas, but some of it is also redeposited in the valleys of the streams flowing from the Lavas. Regur soil is also found on the Archaean gneisses and other rocks, particularly in Madras. The alluvium of Gujarat and parts of the Coromandal Plains includes also considerable areas of regur soils. The moisture-retentive qualities of the regur soils and their aeration by deep hot-weather cracking are characteristic features. When wetted, the regur soils swell up, thus ensuring thorough mixing of the soil particles. The black colour of regur soils, which was formerly supposed to be due to the presence of humus, is now generally attributed to the presence of finely divided iron particles. There is also a high proportion of calcium and magnesium carbonates. The Deccan-Lavas regur soil is, at its best form of development, very deep but on higher ground it is thinner and grades into reddish-brown or red soils. On the Archaean rocks the regur soil is often underlain by *kankar* horizon.

3. *Red soils:* Red soils are best developed on Archaean crystalline rocks and are sometimes either brown, grey or even black. The colour differs within wide limits, depending on the nature of the parent rock, climatic conditions and local factors of the terrain. In the uplands, the red soils are on the whole poorly developed and may be no more than gravels. In the depressions or the soil-wash traps, they are good loams.

4. *Laterite soils:* These are essentially only lateritic.

The scheme of SCHOKALSKAYA (1932) considers the soil in large parts of the Indo-Gangetic Plains to be of the *steppe-like serozem* type. The deep and the medium-black soils of Maharashtra are compared to the *simkins*. The other types of soils, according to this scheme, include the *steppe-serozem-desert* soil; meadow and bog soils, coastal sands, swamp-soils and saline marshes; soils of the vertical zones of the Himalaya, comprising most diversified soils varying from the mountain-meadow soils to the *zheltzem* (yellow soil) and the *krasnozem* (red soil). Ignoring the montane vertical zones, limestone soils, the red soils and regur, the non-lateritic soils lie mostly at the extremes of the humidity-scale. We have thus the swamp or the forest soils, steppe soils, sub-steppe soils, etc. Humid soils are generally restricted in occurrence. The rain-forest areas are mostly montane and the largest areas are on the very permeable Cretaceous-Tertiary sandstones and limestones in Assam-Burma mountains. Swamp, peat-bog and muck soils are generally confined to the deltas of great rivers and are particularly developed in the Sundarbans of Bengal, in the Sibi and Nara tracts of Sind and in small areas on the fans of streams debouching from the Suleiman Ranges. Peat-bogs are confined to the Nilgiri Hills and other southern high hills. The terai soils are sometimes considered as approaching this type, but are really meadow types that are slightly podsolised. *Solonchaks* fringe the sea-face and major tidal channels of the deltas. The swamp soils, with the exception of the solonchaks, are fertile on being cleared first, but the very clearing results in the cessation of supplies of new humus material and thus leaching sets in rapidly. The arid soils occupy larger areas than the humid soils and cover most of the Indo-Gangetic Plains alluvium, to the west of Patna (85° EL) and grade from the prairie soils, with considerable calcareous content (*kankar*) into the sandy *serozem* of Central Thar desert and the Sind-Sagar Doab area. The soils of the Central Pubjab, Sind and the Aravalli piedmont are saline *serozems* of the Schokalskaya scheme, but with patches of very saline *solonets* and *solonchaks*.

True laterites and lateritic formations are generally restricted. The expression laterite was first applied by Buchanan to clayey rock that hardens on exposure to air in Malabar. It is, however, now used for soils formed with 90–100 % of iron, aluminium, titanium and manganese oxides. The virtual leaching-out of silica and consequent concentration of the oxides of the metals mentioned above are characteristic of laterites. The fully developed lateritic soils have 1. a few centimetres of soil retaining some organic material on the surface crust of iron oxide, forming extensive pebbly-gravel layers; 2. red and yellow, more or less crumbly and sticky subsoil, with iron concretions around old roots; 3. deeper less weathered red and yellow mottled clays, with vermiform iron nodules enclosing clay; 4. more iron pans along fracture-planes or quartz-veins and 5. a white decomposition zone passing into the parent rock. Laterite occurs at high and low elevations, up to 900–1500 m above mean sea-level on the Western Ghats and 15–60 m on the low dissected plains of Konkan, cuestas of Madras, etc. It is generally found in between on flat surfaces, but does not cover slopes between the levels. It is really a capping, with a thin-bevelled-off veneer, reaching slightly down the top of the slope. Eocene or perhaps later laterites are found in the Nummulitic Series in Baluchistan.

Tropical weathering of soils differs from that in the temperate regions in several respects. In the hot climates, the temperatures are higher by 10–20 °C and all the chemical reactions, therefore, proceed at least two to four times faster than in the temperate regions. There is also no actual interruption of the soil weathering due to winter. In the monsoonal humid regions, the seasonal reversal of the ground-water movements is also an important feature in the weathering of the soil. Higher temperatures involve rapid decomposition of the humus-content in tropical weathering. The high intensity of tropical rainfall and the resulting high proportion of the runoff in sheet-flow or gully-erosion is of great significance in soil weathering in tropical regions. Slopes of any steepness in the humid areas rarely retain much of the precipitation. Soil-creep is also too rapid to be strictly termed creep, so that the slopes are generally regions of relative aridity. The foothills are on the other hand zones of deep soil, with relatively high water-table and soil humidity.

5. The Drainage

Like the other aspects of the geomorphology, the natural drainage of India is of considerable importance in its ecology and biogeography. Considerable literature exists on the drainage patterns, hydrology, source and evolution of the rivers of India. CHIBBER (1946) has recently given a useful review of our knowledge of the age, origin and classification of Indian rivers. LAW (1968) has summarized the information about them from ancient Sanskrit literature. The general hydrology of the rivers is discussed by VIJ et al. (1968). The riddle of the sources of the R. Indus, Sutlej, Ganga and Brahmaputra has recently been satisfactorily solved by the researches of SWAMI PRANAVANADA (1939, 1968). The Peninsular rivers are described by a number of workers like VERMA (1968) and RADHAKRISHNA (1968). The other important contributions on our rivers include BHATTASALI (1941), BOSE (1968), BOSE (1968), DESAI (1968), GULHATI (1968), SEN (1968) and SINGH (1968).

The rivers of India fall into two natural major groups, *viz.* the Peninsular rivers and the Extra-Peninsular rivers, differing fundamentally from each other in their history and other characters.

5.1. THE PENINSULAR RIVERS

The Peninsula is characterized by its typical radial drainage pattern. To this pattern belong the north-flowing rivers Chambal, Banas, Sindh, Betwa, Ken and Son, the east-flowing Damodar, the southwesterly-flowing Swarnarekha, the east-flowing Mahanadi and its tributaries, the south-flowing Wainganga, Wardha (tributaries of the R. Godavari), the westerly-flowing Narmada and Tapti. The Chambal, Sindh, Betwa and Son are typical anterior-drainage pattern rivers; they are much older than the R. Yamuna and Ganga, into which they flow. In the region of the steep scarp on one side and gradual slope as plateau on the other side, we have the typical drainage pattern associated with such uniclinal structure. In the Vindhya south of Mirzapur, we have examples of this pattern. These patterns include trellis drainage, forming cascades and waterfalls, and the escarpment drainage from the scarp of a plateau, short streams with hanging valleys and waterfalls. The best examples of these are observed in the westerly drainage, south of Bombay in the case of the rivers Amba, Kundalika, Savitri, Vasishta, Shastri, Kajvi and Vaghotan, all of which have their courses in the Deccan Lavas area.

The Peninsular rivers are entirely fed by the monsoon rains and are, therefore, often dry more or less completely in summer. We may recognize two groups of Peninsular rivers: 1. the coastal and 2. the inland rivers. The coastal rivers are relatively small streams, over six hundred, on the west Coast, from Saurashtra in the north to Cape Kumarin (= Cape Comorin) in the south. They drain the western side of the Western Ghats, cut across the narrow plains before emptying into the Arabian Sea. The inland rivers comprise the west-flowing Narmada and Tapti and the east-flowing Mahanadi, Godavari, Krishna and Cauvery. The latter rivers are characterized by the absence of braiding, aggrading or degrading, so typical of the Ganga-Brahmaputra river system. The west-flowing inland rivers flow between mountain ridges, so that their catchments are elongate and narrow and lack delta formation at their mouths. The east-flowing inland rivers have wide and fan-shaped catchment areas and have extensive deltaic deposits. The main water-parting is formed by the Western Ghats. The patterns of the Peninsular river drainage are of great age and show no evidence of reversal or diversion of an original west-flowing drainage. The Palghat Gap is regarded as the ancient valley of a river that flowed either from the east or from the west, before the subsidence of the Arabian Sea. The wide and senile valleys (Pl. 70) of the east-flowing rivers are almost graded on their heads, practically within sight of the Arabian Sea, in sharp contrast to the youthful gorges-like forms of the west-flowing streams. The latter have hardly 80 km length to fall about 600 m to the base-level, as compared to the 480–960 km in straight-line distances for the rivers flowing in the Bay of Bengal. There does not seem to have been time enough on the Deccan Lavas for any large-scale river captures. The deep canyons suggest that the streams are still eroding vertically faster than they are cutting back the valley sides.

The rivers Narmada and Tapti flow westward in comparatively deep and narrow valleys to the Arabian Sea and divide the sloping area of the Vindhyan System from the Deccan Lavas and the Plateau of the central region. The two valleys are themselves separated by the densely forested Satpura Range that culminates in the Amarakantak Plateau (1057 m). The upper course of the R. Narmada is confined within remarkably narrow and steeply enclosed valley, set deep between the scarps of the Vindhya on the north and the spurs of the Satpura on the south. The Narmada is a clear stream here that breaks into cascades near the source and into leaping waterfalls, where the Marble Rocks of Jabalpur enclose the river. It widens out further below to nearly 1.5 km and enters its estuary about 25 km wide below the city of Broach. Its length is about 1300 km. The R. Tapti, the second largest of the west-flowing inland rivers, is 725 km long and drains an area of 64750 km². It arises at an elevation of 760 m above mean sea level near Multai in the Betul District and flows between the Satpura Hills on the north and the Gawilgarh Hills on the south. After confluence of its principal tributary, it flows between the Satpura and Ajanta Hills. The east-west direction of the R. Narmada and R. Tapti may be explained by assuming that they occupy two rifts, formed by the sag-faulting in the north of the Peninsula at the time of the stress caused by the Himalayan orogeny. The R. Tapti occupies a trough between the Ajanta and Satpura Ranges.

The R. Narmada occupies a trough between the Satpura and the Vindhya Ranges. Both these rivers flow into the Arabian Sea. To the north of the Vindhya Mountains, the drainage is mostly towards the R. Ganga. The R. Godavari, Krishna and Cauvery and their numerous tributaries rise on the slopes of the Western Ghats and the R. Mahanadi is characterized by its more restricted course in the north-east of the peninsular plateau. All these rivers find their way to the east coast and their passage through the Eastern Ghats is generally marked by rapids.

The R. Mahanadi drains an area of about 132100 km² and has a length of 840 km. Its upper basin is a saucer-shaped depression (Chhatisgarh in the Bastar Hills), where it arises near Sihawa (442 m above mean sea-level) and first flows northwards for nearly 260 km. After receiving its major tributary, Sheonath on the left, it then flows more or less eastwards. From Sambalpur it flows south to the confluence with its right-bank tributary Tel and then it empties into the Bay of Bengal. It has a large delta and it brings down considerable silt that is spread as a long bar in the mouth of the river.

The Godavari the largest of the east-flowing Peninsular rivers, rises in the Nasik Hills hardly 80 km from the Arabian Sea and follows a generally southeasterly course to a total length of over 1500 km. The upper reaches of the river are comparatively shallow, wide and gentle stream. Its principal tributaries are the R. Wardha, Penganga and Wainganga. When joined by the tributaries Indravati and Sabari, it develops into a wide channel with many islets. The channel, contracts however, later between hills and breaks through a gorge, hardly 60 m wide, before spreading out again on the eastern coast plains, about 95 km from the Bay of Bengal.

The Krishna, the second largest of the east-flowing inland river, rises at an elevation of 1340 m near Mahabaleshwar, about 60 km from the western sea coast, passes southward and eastward. It is fed by Bhima from the north and Tungabhadra from the south. It flows over rocky bottom and breaks out through the Eastern Ghats. Its length is 1400 km.

The R. Cauvery, the fourth largest, rises at an elevation of 1340 m in Coorg and its upper course in Mysore is very tortuous, with rocky bed and steep banks. It encloses two islands, Srirangapatam and Sivasamudram. Around the latter there is a waterfall that cascades some 100 metres. Its delta is a vast alluvial plain, 3200 km², with a northern main channel called Coleroon and the southern channel retaining the name Cauvery and serving as source of extensive irrigation. The Cauvery proper finally becomes reduced to an insignificant stream before entering the Bay of Bengal at Cauveripattinam, north of Tranquebar. The length of the river is 805 km, its drainage area is 80290 km² and its delta is over 160 km long.

The minor rivers of the Peninsula are Pennar, Palar and Vaigai in the south.

5.2. The extra-peninsular rivers

The Extra-Peninsular rivers belong to the drainage system of the Himalaya. Unlike the Peninsular rivers, they do not depend on the monsoon rains, but are fed by the melting of snow on the Himalaya. Though the volume of water in these rivers may fluctuate greatly, they never dry up completely. Most of them traverse as slow streams that wander lazily across the alluvial plains of North India. A fact of considerable biogeographical importance is that the Himalayan drainage, in a large measure, is not consequent drainage. Its formation was not consequent upon the relief of the Himalayan mountains, but there is undisputed evidence to show that the principal rivers are older than the mountains. During the long process of mountain building, the old rivers kept very much to their own channels, though working at an accelerated rate. The increased momentum acquired by the upheaval of the mountains was expended in eroding their channels at faster rate. This explains the peculiarity, that the antecedent rivers drain not only the southern but also the northern slopes of the Himalayan ranges, so that the waterparting lies far north of the crestline of the main range. The drainage of the northern slopes flows in longitudinal valleys, parallel to the mountain ranges in Tibet.

The major drainage patterns of the Extra-Peninsular rivers may be summarized as follows: 1. Insequent drainage includes the ancient, and thrust-superimposed drainages. The antecedent rivers are the Brahmaputra, Sutlej, Indus and the Jahnavai rising north of the main crestline of the Great Himalaya and older than the Himalaya. They have kept their channels open *pari passu* with the mountain uplift. 2. The consequent drainage pattern of the Great Himalayan rivers like the Ganga and its tributaries Yamuna, Sarda, Gogra, Gandak and Kosi. These rivers have their origin on the southern slopes of the Great Himalaya. These rivers have assumed their present form in the Pliocene or even during the Pleistocene times. The Lesser Himalayan rivers like Bagmati, Rapti, Ramganga, Koh and some of the tributaries of the Indus such as Beas, Ravi, Chenab and Jhelum and the Siwalik rivers viz. the Hindon, Solan, the rivers of the Indo-Gangetic Plain like Gumti and Varuna are also of this drainage pattern. The R. Bagmati and Manohra in the Kathmandu Valley and adjoining Nepal Himalaya are of the centripetal drainage pattern. The migratory drainage (CHIBBER 1949) is associated with westerly drift of the rivers, particularly the R. Bhagirathi in Bengal the R. Kosi, Gandak, Gogra, Son, Sutlej, Ravi and Indus. The R. Kosi has shifted its course 120 km in the past 200 years. The Burhi Gandak marks the old channel of the R. Gandak and represents about 145 km westward migration of the river (see Fig. 3). Of the Extra Peninsular rivers, the Indus, Ganga and Brahmaputra are the most important. The principal tributaries of the R. Indus are Shyok, Jhelum, Chenab, Ravi,

Beas, Sutlej, Gilgit, Swat, Kabul and the Kurram. The R. Sutlej is a relatively young river, which developed by the collapse along a line of weakness – the Gondwana trough continued by the line of the R. Ghaggar. The cleft, about 1525–2100 m deep and about 160 km long in a straight line from north of Simla to Shipki, shows strong evidence of its youth. The upper course of the Sutlej in Tibet is a broad arid basin, at elevations of 4670–5270 m above mean sea-level, filled with detritus, in which the glacier-fed river has cut a canyon, about 1000 m deep in some places. In one place the Sutlej-Karnali watershed is level alluvium. The fall of the river is steep, about 1525 m in a distance of about 320 km from the Rakas Tal to Shipki. It has cut gorges 180–210 m deeper than the neighbouring rivers Beas and Giri (Yamuna).

The Indus (Sanskrit *Sindhu*), 2880 km long, arises at an elevation of 5182 m above mean sea-level in the springs of the Singge Khambab northeast of Mt. Kailas and 85 km from Parkha in Tibet (PRANAVANDA 1968). It is this river that has given the name India to the country; the land of the Hindus of the people who settled in the valley of the R. Hindu (a corruption in Persian for sindhu). From its source, it flows first northwest around the Lake Mansarovar, takes a turn southwards in the Haramosh Mountain (7407 m). Opposite Attock (in Pakistan) it is joined by the R. Kabul. It then flows parallel to the Sulaiman Range and receives the accumulated waters of the five rivers of the Punjab, the eastern tributaries, at 805 km up from its mouth. It empties into the Arabian Sea by many mouths near Karachi; its delta extends for nearly two hundred kilometres along the sea coast and has an area of nearly 7770 km².

Of its left-bank tributaries, the R. Jhelum (Sanskrit *Vitasta*) arises in the Verinag Spring, at the bottom of a scarp of the Pir Panjal Range in Kashmir. It is joined by the R. Chenab (Sanskrit *Askini*), formed by the confluence of the Bhaga and Chandra. After receiving the water of the R. Ravi, it joins the R. Sutlej. The R. Beas (Sanskrit *Vipasa*) also joins the Sutlej. The R. Sutlej (Sanskrit *Satadru*) rises at an elevation of 4630 m above mean sea-level in the springs near Dulchu Gompa, 35 km west of Parkha in Tibet. First it flows in a northwesterly course along the southern slopes of the Kailas Range, then turns southwest and enters the plains of north India near Rupar.

The R. Ganga has for several generations been confused with the R. Sutlej and thus arose the myth that the Ganga arose from the Lake Mansarovar and the Mt. Kailas. The Hindu Puranas and a host of modern geographers, as well as Hindu pilgrims to the Holy Kailas, have made this mistake. This is mainly due to the fact that the R. Sutlej has the Indian name Ganga in the Tibetan Purana. The outlet of the Mansarovar Lake is called in Tibet as Ganga-Chhu. The pilgrims and the geographers, who have gone to Mt. Kailas and to the Lake Mansarovar, have to cross the Ganga-Chhu and have readily confounded it with the famous Ganga they have known back in India. These facts have recently been adequately explained by PRANAVANDA (1968).

The source streams of the R. Ganga are five, viz. the Bhagirathi, Mandakini, Alaknanda, Dhauli-Ganga and the Pindar. The Alaknanda is the main tributary of the Bhagirathi, which meet together below Devprayag and acquire the name Ganga. The Mandakini arises from the Chorbari Glacier near Kedarnath and joins the Alaknanda at Rudra-prayag. The Pindar is the eastern-most feeder and arises in the Pindari Glacier in the shadow of Mt. Nanda Devi (7817 m) and joins the Alak-nanda at Karnaprayag. The Dhauli-Ganga joins Alaknanda at Vishnu-prayag. The other lesser source streams include Jahnavi rising in the Nilang Glacier and the Saraswathi rising near the Mana Pass. The R. Bhagirathi is considered as the main source stream of R. Ganga. The Alaknanda arises from the glaciers Bhagirath Kharak and Satopanth on the eastern slopes of the Chaukhamba Massif. It flows past the Badrinath shrine.

The combined water of the rivers Alaknanda, Dhauli-Ganga, Pindar and the Mandakini join the R. Bhagirathi at Devprayag, to form the R. Ganga, which then emerges finally from the Himalayan mountains into the north Indian plains at Hardwar. The Ganga then flows through the Indo-Gangetic Plains of North India to empty into the Bay of Bengal, near the Sagar Island (south of Calcutta), where the puranas tell us the mighty Bhagirath brought the Ganga from Gaumukh to pour upon the ashes of the six thousand sons of the king Sagar, who had been scorched to ashes by the wrath of the sage Kapil Muni.

Of the tributaries of the R. Ganga, the Yamuna, Gogra, Sarda, Rapti and the Gandak arise on the Himalaya and except the Yamuna, are all on the left bank of the Ganga. Though rising on the Himalaya, Yamuna is a right-bank tributary of the Ganga. The other right-bank tributaries like the R. Son, arise on the Peninsular tableland.

The R. Yamuna has its source in the Yamunotri Glacier on the western slopes of the Mt. Bandarpunch (6387 m). It originated in the Post Middle Miocene times, the consequent of the second (the main) phase of the uplift of the Main Himalayan Range. Its largest Himalayan tributary is the Tons that arises on the northeast slopes of the Mt. Bandarpunch and brings down nearly twice as much water as the Yamuna.

The R. Brahmaputra is known as the Tsan-po in Tibet and Dihang in the gorge through the Himalaya. Though much longer than the R. Ganga, it is of little importance to India, because much of its course within India is shorter and because its narrow valley is an area well watered by heavy monsoon rainfall. The R. Brahmaputra (Sanskrit meaning son of Brahma, the creator) has the unique privilege of being called in India in the masculine gender, while all the other rivers are in the feminine. It is about 2900 km long, but the greater part of it is outside India. Its source is at an elevation of 5150 m in the Tamchok Khambab, at the head of the Chema-Yungdung at the Tamchok Khambab Chhorten 148 km from Parkha in Tibet (PRANAVANDA 1968). It belongs to the east-flowing river system of Hwang-ho, Yangtze-kiang and Mekong. In Tibet it flows eastwards for nearly 1100 km, parallel to the Himalaya, to meet the first tributary, Ranga Tsanpo, near Lhatse Dzong. Thence it is a navigable river for nearly 640 km, at an elevation of 3650 m the only navigable channel at that altitude in the world. It enters India across Sadiya in Assam, where it is joined by R. Dibang and R. Luhit to become the Brahmaputra; at the point of entry it is known Dihang. It flows then west for nearly 720 km, oscillating from side to side and forming many smaller or larger islands. Traversing round the spurs of the Garo Hill, the river flows 270 km before joining the R. Ganga; the combined Ganga-Brahmaputra river is known at Padma.

5.3. Changes in the course and evolution of the drainage patterns of the extra-peninsular rivers

The changes in the courses of the Extra-Peninsular rivers, which have occurred within historical and recent times and the earlier evolution of their drainage patterns from the Pleistocene to Recent are of considerable biogeographical interest. A number of workers like BHATTASALI (1941), CHIBBER (1946, 1949, 1950), GREGORY (1925), OLDHAM (1893), OLDHAM (1896), PASCOE (1920), PILGRIM (1919) and SEN (1968) have contributed materially to our knowledge of this problem.

There is abundant evidence to believe the rivers Chenab, Ravi, Beas and Sutlej have greatly changed their courses within the past few centuries. For example, the R. Chenab flowed, as late as 1245 AD, to the east of city of Multan (Pakistan) and the R. Beas occupied its old bed near Diplpur. The R. Jhelum, Chenab and Ravi met southwest of Multan and joined the Beas 45 km south of the city. The R. Sutlej also had an independent outlet to the sea in the Rann of Kutch. About 1000 AD it was a tributary of the R. Hakra and flowed into the Eastern Nara, but about 1245, it took a more northerly course and the Hakra dried up. About 1593 the Sutlej again changed its course and it assumed its present channel in 1796.

A number of major and numerous other minor changes in the course of the river Ganga, particularly in the Middle Gangetic Plain, are well established facts. The reader will find a detailed discussion of these changes in SINGH (1968) and SEN (1968).

A fact of considerable ecological and biogeographical significance is the eastward shifting of the R. Ganga from the original main outlet, along the western margin of Bengal, to the present-day Padma-Meghna course. The diverse streams like Ichamati, Jalangi, Matabhanga, Gorai, etc. represent perhaps the various intermediate positions of the most important channel. It is not, however, fully known whether the shifting

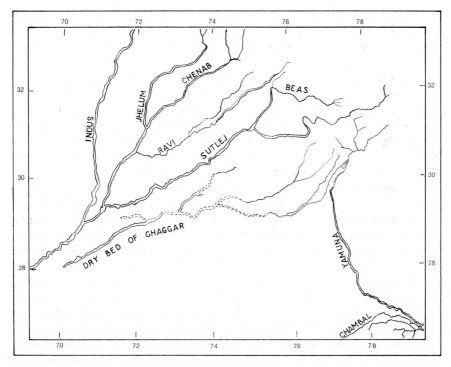

Fig. 3. Map of the area between the R. Indus and the present course of the R. Yamuna, showing the various dry beds of the former courses of the Yamuna, which formerly flowed either direct to the Arabian Sea or was part of the Indus system.

of the river was due to largely or mainly the alluviation at the heads of the successive main spillways, to tectonic changes or to the shifts in the balance of the delta as a result of changes somewhere else. The precise sequence of events is not also completely understood at present. It seems, however, obvious that the R. Bhagirathi, or at least one of its several branches like the Hoogly, Saraswathi or Adi Ganga (Tolly's Nullah in Calcutta), was the most important distributory channel during the seventeenth century (BHATTASALI 1941). The R. Hoogly is at present fed also by the rivers like Damodar and Rupnarayan from the Peninsula. The R. Tista has recently diverted into the R. Brahmaputra, resulting in a relative decline in the old Brahmaputra course to the east of Dacca. The main Tista-Brahmaputra outlet by the Jamuna and the water and silt brought down to the R. Padma backed up the waters of the Ganga and opened up the R. Gorai.

The Yamuna drainage Basin has two major divisions, viz. the Himalayan and the Peninsular. The Himalayan Basin has undergone only minor changes in direction and course, but the Peninsular division has undergone great changes. Originally the R. Yamuna was member of the

39

R. Indus System and flowed into the Arabian Sea, from the middle Miocene till Recent times. In Sub-Recent times it became a tributary of the R. Ganga, due to the subsidence of the Ganga Delta and the tilting uplift of the Sutlej-Yamuna Divide. To begin with, it was a twin stream, with the R. Saraswathi that arose between the Yamuna and the Sutlej. The twin stream flowed southwestwards, to combine near Suratgarh, north of Bikaner. The combined waters of the two rivers then flowed as the R. Ghaggar (R. Hakra) through Bahawalpur to join the R. Indus. The dry bed of the R. Ghaggar still exists (Fig. 3). Some authorities have considered that the R. Saraswathi flowed independently to the Rann of Kutch (KRISHNAN 1952, OLDHAM 1893, 1896). OLDHAM (1893) and KRISHNAN (1952) consider, for example, that the R. Yamuna either 'flowed into what was the Saraswathi or at least shared much of its waters with that river'. Due to the subsidence of the Ganga-Delta, a tributary of the Ganga (Fig. 4) began working headward to capture the Yamuna and subsequently the later uplift of the Aravalli-Delhi Axis and the gradual rise of the Eocene sea floor of Rajasthan Desert completed the drama of the Yamuna becoming tributary of the Ganga, at the confluence at Allahabad. As the combined streams of Saraswathi and Yamuna had been formerly flowing into the single river Saraswathi or the R. Ghaggar, the belief arose that the waters of the Saraswathi have also gone into the confluence at Allahabad and hence the popular name Triveni Sangam or confluence of three rivers to Allahabad. Fox (1942) and WADIA (1953) strongly support the view of the shift of the Yamuna from the Indus to the Ganga System.

The evolution of the longitudinal courses of the rivers Indus, Sutlej and Tsanpo on the Tibetan Plateau and the enormous gorges these and other Himalayan rivers have cut right across the Great Himalaya, in the vicinity of great peaks, of considerable relevance to the biogeography of India, is at present only very imperfectly understood. The sands, gravels and boulder beds that make up the Siwalik Hills are attributed to fan-deposition on a vast scale. The alluvial fans of the streams debouch from the Himalaya and coalesce to form a great piedmont apron. The variations in thickness and lithology and the conformable grading into Recent fluviatile deposits indicate that throughout the Siwalik times, the main drainage lines were substantially as they are today.

PASCOE (1920) and PILGRIM (1919) supposed however that these deposits were laid down in the valley of a hypothetical Indo-Brahm River (of PASCOE) or the Siwalik River (of PILGRIM) that is supposed to have flowed to the northwest, between the rising Himalaya and the Gond-wanaland. This river was to the north of the present Ganga, so that the whole system gradually shifted south by the advancing outer Himalayan foothills and the active deposition of the tributaries themselves from the north, forcing the R. Yamuna and the Ganga against the northern flanks of the Peninsular Mass. The Indo-Brahm river is believed to have

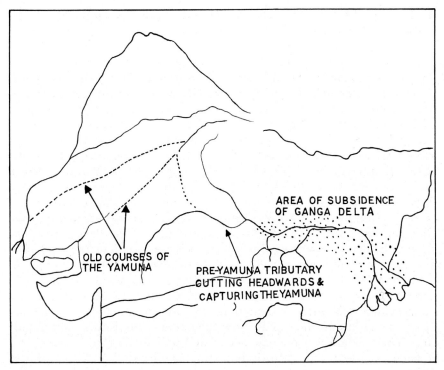

Fig. 4. Sketch map (not to scale) of a part of the Indo-Gangetic Plains, showing the probable changes in the course of the R. Yamuna and the evolution of its present confluence with the R. Ganga. The Yamuna originally belonged to the Indus system from Post-Mid Miocene to the Recent, but was captured by the Ganga system within historical times.

flowed into the Arabian Sea, more or less along the line of the lower course of the R. Indus, but perhaps somewhat to the west. The fact that the Siwalik deposits border the Sulaiman and Kirthar Ranges seems, however, to have been completely overlooked by the authors of the Indo-Brahm hypothesis. This hypothesis does not at the same time explain the relation of Siwaliks in the Sibi-Re-entrant to the old estuary. It also ignores the fact that the area of what is now lower course of the Indus was occupied, during the Eocene times, by a Sind Gulf that has subsequently been filled up by river deposits. The present-day river layout was ascribed to earth movements, damming of the Siwalik river northwest of Kangra, assisted by the cutting-back of the powerful rivers flowing south from the Rajmahal-Shillong watershed and in the alliance with the uplift in the Ganga-Brahmaputra Delta. The northern limb of most of V-shaped forks of rivers in the Indo-Gangetic Plain are regarded by PASCOE as the remnants of old righthand tributaries of the Indo-Brahm river becoming more deeply impressed and permanent, with the rise of

41

the Siwalik Hills. The middle part of Indo-Brahm is supposed to have in the meanwhile been attacked from two directions viz. from the south-west by the left-bank tributaries of the lower Indo-Brahm itself and from the southeast by the headstreams of the Ganga, now diverted into the Bay of Bengal. About the time the Garo-Rajmahal Hills Gap is supposed to have been breached and the Ganga-Brahmaputra drainage was supposed to become diverted into the Bay of Bengal. These events would of course imply rejuvination and that the activated western headstreams gradually annexed the right-bank Indo-Brahm tributaries in the Gogra-Yamuna area. The annexation of the Yamuna would mark the last phase of the disappearance of the Indo-Brahm river. The R. Yamuna flowed in the course now marked by the dried up Ghaggar depression in Rajasthan until perhaps well into historic times (this was the R. Saraswathi of ancient times). The Upper Sutlej may have also debouched into the Ghaggar bed until a late capture by a tributary of the Beas-Sutlej. The main difficulty in accepting this idea is that the subsidence of the Bay of Bengal area would lead to rejuvination and consequent increase in the cutting-back capacity of the streams. Though this would at first sight appear to support the Indo-Brahm hypothesis, these recent events should show evidence of such rejuvination, at least in the upper Mahanadi Basin, even if not in the rapidly adgraded area of the Rajmahal-Shillong Gap. There is however no evidence of such a rejuvination.

These ideas of the Indo-Brahm river overlook also the conclusive evidence for the general persistence of east-coast and its character, at least further south, of a raised plain of marine erosion. It has been assumed that the alluvium in the Garo-Rajmahal Hills Gap is thin, but it is actually hundreds of metres thick. Headward erosion in really hard rocks is of doubtful efficiency, without structural assistance; earth movements are more likely to have been agents of the change.

PASCOE also envisaged a west-flowing Tibetan river, from Pemakoi to Gilgit. The furrow of the Tsanpo-Mansarovar Lake-Sutlej-Gartang-Indus line, partly filled with Ladakh Nummulites, seems to have some structural continuity, either geosynclinal zone or a belt of soft rocks or faulted. In Tibet many of the larger feeders of the Tsan-po like Kyi, Rong, Nyang and Shabki have a westerly course that strongly suggests reversal of the main stream. The same argument would also favour the idea of reversal of the Indus in view of the southeasterly trend of the Shigar, Nubra and the Upper Shyok. PASCOE suggested that the Tibetan river may have flowed by the Photu Pass (only 76 m higher than the Tsan-po Valley), by the Karnali, the upper Sutlej or the upper Indus. The plateau section of the Indus is twice as steep as that of Tsan-po. At Bunji the R. Indus is 1036 m lower than the R. Tsan-po at the point where the latter leaves the furrow, so that it is the more active of the two rivers. Its transverse gorge on the other hand has a relatively gentle gradient and is cut in hard rocks, so that it cannot be very young.

42

Discussing the evidence of the similarity of the character species of animals of the R. Ganga and Indus, adduced by the supporters of the Indo-Brahm hypothesis, ANNANDALE (1914) explained the similarity on the hypothesis of a more or less broad marine strait, between the Peninsula and the rest of Asia that gradually became narrower and narrower to be eventually obliterated by the advance of the Himalaya and by the filling up by the products of erosion, the penultimate stage being one of constantly shifting lagoons. He does not, therefore, subscribe to the Indo-Brahm hypothesis. Recently DE TERRA (1934) re-orientated the problem by focussing the attention on the longitudinal valleys of the Karakoram-Pongong area. This pattern is ancient and the longitudinal valleys antedate the transverse sections. He holds the view that in Preglacial times the drainage of the Karakoram-Ladakh region flowed in the south-east and east along the Tsan-po furrow into eastern Tibet and Szechwan. The Shigar-Nubra-Upper-Shyok trend supports this view. He does not believe in the Indo-Brahm river hypothesis but put forward the view that the Siwalik deposits are local precipitates of an antecedent slope drainage, successive fan and basin sediments and their origin differing in no way from that of other foredeep filling. Some workers hold that the Brahmaputra is an old river and if it flowed in an east-west direction, its mouth must have perhaps been somewhere in the China Sea and may have continued eastward in the R. Yangtse-kiang. If this view is correct, the interruption would have brought about by elevation of the mountains that now fill the gap between the two rivers and then the Brahmaputra would have to be thought of as an old river that dates back to the Palaeozoic times. It is, however, far more likely that this river did not at all exit before the uplift of the Himalaya and it took its origin from the drainage of the northern slopes of the newly uplifted mountains and flowed eastward in the longitudinal valleys. The southward bend in Tibet is conditioned by the general structure of the area and the same condition would also explain the westward bend in Assam.

6. The Natural Regions

It does not seem to have been satisfactorily decided as to what constitutes a natural region. Geographers, biologists and meteorologists may mean different things when speaking of a natural region. The confusion is partly the result of lack of uniformity in the criteria on which these divisions are based. Geomorphology, topography, climate, vegetation, fauna and even semi-political boundaries have each been considered as decisive factors in determining the limits of a natural region, but the result has not always been satisfactory. The value of different criteria in defining the limits of a natural region varies within wide limits, depending on a complex set of conditions. Moreover, with rare exceptions, most natural regions do not have sharply defined boundaries, but only transitional margins, where other adjoining natural regions overlap. The problem is also somewhat complicated in India by the colossal deforestation and the vast changes in the general topography brought about by human agency. There is hardly an area in India, except perhaps in the more inaccessible and inhospitable higher elevations of the Himalaya, that has not been altered in some way by man. Strictly speaking, a natural region should present a synthesis of the geomorphological, physiographical, climatic, floristic and faunistic characters and should be capable of being recognized by its distinctive structure, climate, flora and fauna. The character fauna of a natural region is generally bound up with the development of natural vegetation – its character flora. The flora of a region is in turn reflective of the characteristic climate and in the case of India, the flora and the rainfall together constitute a more or less well defined and constant interdependent character of a natural region.

There is considerable diversity of opinion regarding the major divisions and the boundaries of the subdivisions of India. The early contributions of BAKER and STAMP have in recent years been modified by a number of workers like AHMED (1941) and PITHAWALA (1939). The last mentioned author has, in particular, emphasized the great importance of physiographical characters, but his map of the divisions and subdivisions is not wholly physiographically uniform and not always consistent with his text. There are besides several other anomalies in the treatment of the entire Irrawaddy Basin as a single natural unit of the Extra Peninsular mountains. The boundaries in the plains between the R. Indus and R. Ganga are also somewhat confusing. In 1942 PITHAWALA published a somewhat modified scheme of physiographic divisions. He is of the view that the socalled natural regions need to be replaced by the expression physiographic regions. He takes into consideration 1. the structural features of the land, its geological condition including the kind and nature of the rock, their initial stage before erosion acted on them; 2. the process of erosion, depending to a great extent on the structure of the rock and the forces at work; and 3. the stage of erosion at the present time and the cycle of the erosive changes that have taken place on the original structure of the rocks. The basis is thus geomorphological.

We have already mentioned the geomorphological divisions of India as 1. the Peninsular Block, 2. the Extra-Peninsular Mountains and 3. the Indo-Gangetic Plain. These major divisions are subdivided by Pithawala into 16 Provinces, depending on the nature of the rock, the land-forms and their erosional history. The Provinces are further subdivided into 54 sections, according to the stage to which a particular land-form has reached and other physical characters like the soil, climate, drainage, mineral wealth, etc. The sections are again divided into subsections, 57 in number.

The following is a brief synopsis of PITHAWALA's revised scheme
PITHAWALA (1942):

I. THE EXTRA-PENINSULAR MOUNTAINS

Laterally compressed, complex series of greatly folded rocks, an unstable mountain-wall of frequent earthquakes, with a plutonic core.

Province 1. The Western Highland: the Western extension of the
Himalaya (lower elevations) and folded Tertiary rocks; the arid part of
India.
Section 1. Kirthar-Sulaiman Mountains: Very dry area of barren
mountains, nearly 1800 m above m.s.l., of folded Tertiary rocks and with
scanty soil on the top; subaerial denudation pronounced and gives rise to
small plateaus and gradually passing into Iran.
Section 2. Kohistan, an area of lower ranges of mountains, 1000 m,
with somewhat more rainfall than Section 1 (including the winter
precipitation); subaerial denudation more pronounced, resulting in
broad anticlinal valleys; number of hotsprings due to fracture in the
folds of the Tertiary rocks; topography typically dryland, limestone
country; river valleys mostly dry; but underground water present.

Province 2. Greater Himalaya that forms the axis of the mountains,
covered by permanent snow.
Section 1. Northern Himalaya: the region of the crestline of the Great
Himalaya, with the giant peaks like Mt. Everest and Mt. K2.
Section 2. Southern Himalaya, with the windward southern side
wetter and receiving the full force of both the monsoons, covered alpine
vegetation and an area of newer and contorted rocks outcrops in many
places.

Province 3. Middle Himalaya: (below the snowline, at elevations of
about 4500–6020 m).
Section 1. Northwest Drylands, hot, dry, very cold in winter, with
marine rocks exposed and includes the Kabula and Kurram river basins.
Section 2. The Dun section: longitudinal tectonic valleys between the
high ranges (4500–6100 m) and including also Kashmir (a structural
valley with numerous lakes and glacial material and cut by the R.
Jhelum). Temperate forests on the monsoon windward slopes with rainfall about 62–75 cm.
Section 3. The Lesser Himalaya.

Province 4. The Sub-Himalayan Region of outer belt of the foothills
and longitudinal valleys.
Section 1. the Potwar plateau of erosion, about 300 m above mean

sea-level, bounded in the south by the Salt Range; a crumpled geosyncline.

Section 2. Siwalik Section. Low hills between the Himalaya and the Indo-Gangetic Plain, hills about 1000–1500 m above mean sea-level; Tertiary fluviatile deposits; Sal and subtropical forests; the foothills of the Himalaya formed by gravel, boulders, sands, etc. brought down by the rivers; the rocks often locally overthrust and folded.

Province 5. Eastern Highlands: Eastern flanks of the Himalaya, detached portion of the Peninsular Block, ranges of Tertiary rocks, denuded plateau of Archaean rocks; wetter parts of India and Burma.

Section 1. Shillong Plateau. Rainfall heaviest in India, over 1250 cm; tropical forests.

Section 2. the Yomas: related to the Western Highlands but wetter and consisting of folded Tertiary rocks. The Andaman-Nicobars belong here.

Section 3. the Irrawaddy Valley and broad belt of lowland valley, a geosyncline formed from Tertiary folded rocks containing oil shales and draining the Chindwin Valley; rainfall over 200 cm.

Section 4. the Shan Plateau, a denuded plateau. The mountain ranges of Archaean rocks, separated by plains and dried-up valleys, continued into Tenasserim, Thailand and Malay Peninsula.

Section 5. Irrawaddy Delta.

Section 6. Kuladam Valley lowland occupied by the R. Kuladam.

Section 7. Salween Basin, the monsoon side forested with evergreen; covers portion of the Salween Delta.

II. The indo-gangetic plain

Filled with detritus alluvium from the mountains, a sagged area between the Himalaya and the Peninsular Massif, with a hidden range along the southern parallel of 23 NL (suspected).

Province 1. Lower Indus Valley, an arrogated valley below the confluence of the rivers and the Indus delta.

Section 1. Western valley of old alluvium, with seasonal hill torrents and springs, rainfall scanty.

Section 2. Eastern Valley, new alluvium and crossed by the Dhoroes (old river channels), shifting river banks and scanty rainfall, with salt lakes and salt deposits.

Section 3. Indus Deltaic section, uncultivated swamp and sandy in part, with changing mouths of the river.

Province 2. Upper Indus Valley (area between the rivers of the Pubjab).
Section 1. Doab section of higher ground, at elevation of about 300 m.

Section 2. Punjab proper: lowlands of khadar formed by the detritus from the mountains in the north, rich soil drift more clayey than in Sind.

Province 3. Desert Province.
Section 1. Pat section, covered with clay or silt and with longitudinal *bhits* (sandhills), old dry valleys; dhands or salt lakes formed in valleys.
Section 2. Thar Section of sandhills (SW–NE trend).

Province 4. Upper Gangetic Valley (synclinorium filled with alluvium).
Section 1. Indo-Gangetic watershed, the region between the old Yamuna and the present course of the R. Yamuna.
Section 2. the Doab Section, with patches of *reh* or *usar* (salt deposits).
Section 3. Piedmont zone of damper and higher wooded country, with rainfall upto 100 cm; the hill slopes of gravel and coarse sand, from 200–150 m, often covered with tall and coarse grass near the flat plain.
Section 4. Trans-Yamuna Tract of the Terai slopes of the hills, belonging to Rajasthan uplands on the southern side, with good discharge from the Chambal and other streams.

Province 5. The Middle Gangetic Valley, a plain sloping from about 150 m.
Section 1. the Bhangar section, clayey and *kankar* soil, cut up by numerous streams, damper than the Upper Gangetic Province, but drier than the Eastern Section, rainfall about 130 cm, densely populated region.
Section 2. *Khadar* section of newer alluvium, more sandy and damper.

Province 6. The Lower Gangetic Valley, an aggraded valley and the Brahmaputra-Ganges land.
Section 1. Lower Brahmaputra valley, leeward side of the Shillong Plateau, less rain, but soil with alluvium overlying ancient plateau rocks, a ramp-valley to a great extent aggraded and liable to floods.
Section 2. Old Ganges Delta: the Ganges delta has shifted gradually from the west to the east; the western part of the old river course, most thickly populated area, fine silt and sand with more than 130 cm rainfall, includes swampy parts, mangrove swamps near the coast.
Section 3. New Ganges Delta, the eastern part of the delta, marshes enclosing islands, the R. Brahmaputra probably flowed once here.

III. THE INDIAN PENINSULA

Comparatively stable block of oldest Archaean rocks and lavas, relict mountains.

Province 1. Rajasthan Uplands, highly degraded mountain system.

Section 1. Northwest section: the edges of the Peninsular Massif passing under the Indus Plain; sandy waste, the R. Luni of salt water the only watershed; topography arid desert, piedmonts, with outcrops of the Vindhya and Dharwar rocks extending up to Agra-Delhi.

Section 2. Marwar Peneplain, irregular, gneissic, with outcrops of Dharwar rocks, synclinorium of tectonic origin.

Section 3. Southeastern Section of Pathar and Uparmal and triple plateau of concentric scarps of the Vindhyan sandstones in the northeast corner.

Province 2. Deccan Trap Region, plateau of Deccan, basalt lavas, denuded into mountains and valleys; soil black; topography relict.

Section 1. Central India Tableland, basaltic lava rocks forming part of the Narmada and Tapti Valleys in the south and the Vindhyan mountains; flat-topped hill topography.

Section 2. Western Ghats, the highest parts of the denuded tableland, residual mountains, rising to elevation of 1525 m, trap of terrace topography, fissure eruption, differential denudation, rainfall good, with monsoon forests on the windward side, flat-topped hills and intervening tableland, cut by river valleys.

Section 3. Bombay-Deccan the leeward side of the Ghats, a true plateau, the crateriform Lonar Lake with soda deposits, rainfall less, regur or black soil is lava product, laterite caps here and there.

Section 4. Western Peneplain showing surfaces of ancient and recent peneplains with complex structure. Rings of Archaean, Deccan Trap, folded Jurassic, Cretaceous and Tertiary rocks ending in shore facies of post-Tertiary in Cutch and Kathiawar; marine denudation prominent. Monsoon scanty.

Section 5. Konkan Coast, a faulted edge of the Deccan Trap, plain of marine denudation.

Province 3. Northeastern Tableland of mixed denuded rocks forming the Eastern Ghats; red soil region, topography inselberg type; Mahanadi-Godavari areas.

Section 1. Mahanadi Basin of Gondwana trough faults; Archaean gneisses and schists, Dharwar-Cuddapah-Vindhya and Gondwana rocks; double monsoon effects; rainfall 100–130 cm; ends in alluvial fan of the delta and the backwater Chilka Lake.

Section 2. Godavari Basin, the Gondwana rocks, full of trough faults; rainfall between 75 and 100 cm; rift-valley with sand-banks.

Section 3. the Eastern Ghats a residual mountain with Archaean rocks highly and unevenly denuded; rainfall 130–200 cm due to double monsoon; topography inselberg, with isolated hill ranges, ending in deltaic plain.

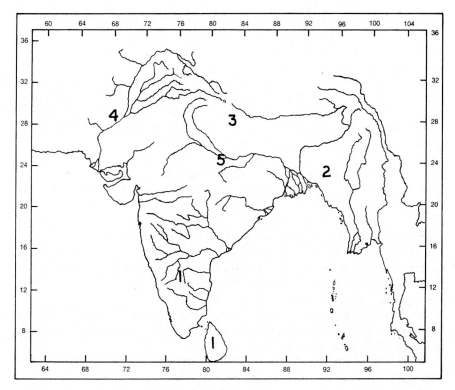

Fig. 5. Map of India, showing the principal natural divisions followed in this book: 1. The Peninsula, 2. The Eastern Borderlands, 3. The Himalaya, 4. The Western Borderlands and 5. The Indo-Gangetic Plains. The divisions 2, 3 and 4 from the Extra-Peninsular area; division 5 is a transitional area between the Peninsular and Extra-Peninsular divisions and in which the northern boundary of the Peninsula grades off to the foot of the Himalaya.

Section 4. Golkonda Coast shore facies of the Eastern Ghats and including the deltas of the R. Mahanadi, R. Godavari and Krishna; topography inselberg type.

Province 4. South India Region: The most ancient block of Archaean rocks, with inselberg topography and raised beaches; folded and contorted strata; has suffered block-faulting and horst uplifts, with rejuvination of rivers.

Section 1. Cuddapah section of Cuddapah rocks and Vindhyan age rocks; retreating monsoon effective; horizontal flat hills.

Section 2. Bellary Section: Leeward part of the Ghats, erosion plateau, scrubland, rain-shadow area.

Section 3. Nilgiri Hills: Dharwar rocks, forested. Topography youthful, with gorges, canyons and waterfalls after block-faulting uplift.

Section 4. Tamil Section of Archaean rocks and considerable de-

49

nudation; gneisses and schists; double monsoon. Includes the Palghat Gap, Tambaravarni basin, Palar Basin, Cauvery Basin alluvial fans on the eastern boundary of the hills.

Section 5. Eastern slopes: Cretaceous and Older Tertiary rocks; formerly submerged, but uplifted later; shore-line emergence, marine denudation, with platforms in parallel belts falling from 300 m to sea-level.

Section 6. Malabar coastland of shore facies of the Southern Block, crystalline belts with sandhills, lake lagoons, foothills and hill slopes; alluvial area cut by short and rapid rivers.

Section 7. Coromandel Coastland or the Carnatic Region, broad, coastal plain of marine denudation, falling from 300 m and reaching the east coast, with parallel belts from the Ghats to the Cauvery Delta and other river deltas; lagoons and back waters; double monsoon.

Province 5. Ceylon an isolated portion of the Peninsula.

Section 1. the Central Massif or the fundamental gneisses in strike conforming with those of South India; sharp fault escarpments with waterfalls, rejuvinated by block uplift, mountains about 2500 m; climate hot, damp, forest evergreen.

Section 2. Southwest Wet lowlands under the influence of double monsoon, subaerial denudation pronounced, laterite, sandy soil, sand dunes with lagoons and backwaters.

Section 3. North-northeast Dry lowlands, part of the northern plain covered with limestone, about 60–90 m thick, some dry peninsulas.

While structure is without doubt of considerable importance in determining the natural regions and their boundaries, climatic and positional factors cannot also at the same time be totally ignored. SPATE (1957) has recently summarized the divergent views on this subject. The following is a synopsis of his revised scheme:

A. THE MOUNTAIN RIM

I. BALUCHISTAN
1. Northern Ranges
   a. Sulaiman Range
   b. Loralai-Zhob arcs
   c. Toba-Khakar Ranges
   d. Quetta node
2. Southern Ranges
   a. Kirthar-Kalat Plateau and Valley
   b. Makran
      1. Eastern limestone/lava folds
      2. Western flysch
3. Interior Plateaus
   a. Desert basins
      1. Mashkel
      2. Lora
   b. Chagai Hills-Koh-i-Sultan volcanoes

50

II. NORTHWESTERN HILLS
  1. Southern Transverse Zone
     a. Waziristan
     b. Kurram Valley
     c. Safed Koh
        1. Safed Koh Range
        2. Kabul Valley
  2. Northern Longitudinal Zone
        1-3. Chitral, Panjkora, Swat valleys

III. SUBMONTANE INDUS
  1. Trans-Indus Basins
     a. Vale of Peshawar
     b. 1. Kohat Valley
        2. Bannu
  2. Potwar Plateau – Kala Chitta Dhar, Chach
  3. Salt Range
        1. Cis-Indus
        2. Trans-Indus – Kurram watergap, Pezu windgap
Regions I-III. Western Borderlands

IV. KASHMIR
  1. Punch and Jammu
     a. Siwalik zone
     b. Sub-Himalayan zone
        1. Foothills
        2. Mid-Chenab Valley
  2. Pir Panjal Range
  3. Vale of Kashmir
  4. Main Himalayan Mass
     a. Nanga Parbat Massif
     b. Great Himalaya
     c. Upper Chenab Valley
     d. Zaskar Range
        1. The Zaskar range proper
        2. Deosai Plains
        3. Rupshu
  5. Gilgit-Hunza
     a. 1. Astor valley
        2. Indus Kohistan (Indus gorge)
     b. 1. Gilgit-Hunza Valleys
        2. Hindu Kush

V. KARAKORAM
  1. Ladakh
  2. Karakoram
     a. Baltistan
     b. Shyok-Nubra Valleys
     c. Karakoram Massif
     d. Tibetan Plateau
        1. Depsang and Oingzi-tang Plains
        2. Pangong Rift

VI. CENTRAL HIMALAYA
  1. Himachal Pradesh

    a. Siwalik zone
    b. Sub-Himalayan zone
    c. Upper Sutlej
       1. Spiti
       2. Hundes
  2. Kumaon
    a. Siwalik zone
       1. Siwaliks
       2. Dehra Dun
    b. Sub-Himalayan zone Yamuna, Ganges, Kali Valleys
    c. High Bhotiya Valleys
  3. Nepal
    a. Siwalik zone Dundwa, Sumesar, Churia Ghati ranges
    b. Pahar Katmandu valley minor duns
    c. high Himalaya

VII. EASTERN HIMALAYA
  1. Kosi basin
      1. Siwaliks and longitudinal valleys
      2. Arun gorge
      3. Everest Massif
  2. Darjeeling-Sikkim
      1. Tista Valley
      2. Chumbi Valley
  3. Bhutan and Assam Himalaya

VIII. ASSAM-BURMA RANGES
  1. Border Hills: Patkoi, Naga, Chin, Lushai, Chittagong Hills
  2. Barail Range
(Regions VIII grouped with XIV and XX En Borderlands)

B. THE INDO-GANGETIC PLAINS

IX. SIND
  1. Sind:
    a. Sind Kohistan
    b. Lower Indus Valley
      1. Sewistan (Sibi or Kacchi)
      2. Indus/Nara Doab
      3. Indus Delta

X. PUNJAB
  1. Punjab Plains
    a. Derajat Indus floodplain
    b. Thal (Sind Sagar Doab)
    c. Sub-Siwalik (winter rain) zone
    d. central Doabs
    e. Bahawalpur

XI. INDO-GANGETIC DIVIDE
  1. a. sub-Siwalik zone
    b. Sirhind (Hariana)

XII. GANGES PLAINS
  1. Upper Ganges Plains

a. Sub-Siwalik zone bhabar, terai
b. Yamuna/Ganges Doab
c. Rohilkhand-Oudh doabs
d. Trans-Yumuna alluvial veneer
2. Middle Ganges Plains
a. Sub-Siwalik zone
b. Tirhut Kosi floodplain
c. Trans-Ganges alluvial veneer
1. Son delta
2. South Bihar

XIII. BENGAL
1. a. Duars (= terai)
b. Northern Paradelta (Ganges/Brahmaputra Doab) Barind, Tista floodplain
c. Western Margins
1. Rarh lateritic doabs, paddy floodplains
2. Damodar deltaic area
3. Contai coastal plain
d. Eastern Margins
1. Surma-Meghna Valley
2. Chittagong coastal fans
e. Delta proper
1. moribund-jhils
2. mature
3. active Sundarbans

XIV. ASSAM VALLEY
1. Brahmaputra Valley Kapili/Dhansiri re-entrants; detrital terraces, floodplains

C. THE PENINSULA

XV. THAR DESERT
1. a. Pat
b. Thar proper Bikaner irrigated area, Luni wadi, Aravalli daman; duns, monadnocks
(The numerous hills of old rock protruding through the aeolian veneer indicate that the Thar is part of the Peninsular mass; but most of it is covered with superficial deposits, and the boundaries are hence ill-defined, except where the desert is banked against the Aravalli).

XVI. ARAVALLIS
1. Aravalli Range – Delhi Ridges, Jodhpur-Jaipur saddle; Godwar (daman); Mt. Abu, Lake Sambhar
2. Udaipur Hills
1. Mewar
2. Bagar

XVII. CENTRAL VINDHYAN COUNTRY
1. Malwa
a. Vindhyan rock zone-scarplands, Dholpur-Karauli plateau
b. Deccan Lava zone
1. Malwa plateau
2. Vindhyan Hills scarp (overlap with XVII. 3b(1))
2. Gneissic Bundelkhand

3. Vindhyan 'Ranges' and Plateaus
   a. Rewa plateau
   b. scarps of the
      1. Vindhyan Hills (= XVII. 1b(2))
      2. Bhanrer-Kaimur hills
4. Narmada-Son furrow
   a. Narmada Valley
      1. Lower gorges
      2. Rift floor
   b. Son Valley

XVIII. SATPURA-MAIKAL
1. Ranges
   a. Satpura range
      1. Satpuras proper
      2. Gawilgarh hills
   b. Mahadeo ranges intermont basins Jubbulpur gap; Marble Rocks
   c. Maikal dissected plateau upper Narmada Valley Maikal scarp; Amarkantak
2. Khandesh
   a. Tapti-Purna valley
      1. Lower Tapti gorges
      2. Rift floor-Purna/Wardha watershed

XIX. CHOTA NAGPUR
1. Upper Son-Deogarh uplands
2. Chota Nagpur
   a. Hazaribagh Range
   b. Peneplains
      1. Hazaribagh
      2. Ranchi
   c. Gondwana trough (Keol-Damodar basins); Parasnath
   d. Rajmahal hills-daman, upper valleys

XX. SHILLONG PLATEAU
1. Shillong Plateau (Garo, Khasi and Jaintia Hills)

XXI. CUTCH AND KATHIAWAR
1. Cutch
      1. Rann mudflats
      2. Lava/sandstone plateaus
      3. Alluvial/aeolian margins
2. Kathiawar
   a. central platform
      1. Drangadhra-Wadhwan sandstone plateaus
      2. Northern and southern lava plateaus Gir Hs, Girnar
   b. lowland margins
      1. Halar coast creeklands
      2. Dwarka foreland-Okha Rann
      3. Sorath coast-Bhadar-Ojat and Shetrunji valley miliolite zone; Cambay coast
      4. Gohilwad – Nal depression

XXII. GUJARAT
1. Gujarat Plains
      1. Cambay coastal marshes

54

2. Central alluvial shelf-Charotar
3. Eastern alluvial veneer

XXIII. KONKAN
1. Konkan coastal lowland
   a. North Konkan
      1. Northern lowland: longitudinal ridges and valleys (Vaitarni, Amba); coastal alluvium and mangrove
      2. Ulhas basin-foothills to Ghats Matheran mesa, Kalyan lowland, Salsette Is. (dune and rock coast, alluvial lateritic shelf, central hills, creeklands) Bombay
   b. Kolaba-Ratnagiri
      1. Indented coast-mangrove flats
      2. Hilly lowland; paddy valley-floors, lateritic interfluves, Chiplun amphi-theatres

XXIV. GOA AND KANARA
1. Konkan-Kerala transition
   a. Goa
      1. Ilhas deltaic zone-Ilha de Goa
      2. Lowland-Bardez, Marmagoa peninsula
      3. Foothills-Braganza Ghat
   b. N Kanara
      1. Discordant coast
      2. Ghats breaches zone
   c. S Kanara
      1. Netravati (Mangalore) lowland: alluvial coast, lateritic shelves-Ghats outliers

XXV. KERALA (MALABAR)
1. Kerala coastal plain
      1. Littoral – dunes- and -lows, backwaters
      2. Alluvium/laterite shelf
      3. Gneissic lowlands Palghat approaches Nagercoil valleys

XXVI. WESTERN GHATS
1. Deccan Lava Ghats, scarp and crest-Dangs and Peint forests, Koyna and upper Krishna valleys
2. Archaean Ghats, scarp and crest
   a. Ghats breaches zone
   b. Higher southern zone
      1. Contact zone along crest
      2. Coorg coulisses
      3. Wynaad plateau

XXVII. MAHARASHTRA
1. Deccan Lava country
   a. Maval
   b. plateau
      1. Wardha valley
      2. Ajanta Hills
      3. Godavari valley: Nasik basin
      4. Balaghat 'Range'
      5. Bhima valley
      6. Bijapur dry zone

XXVIII. KARNATAKA (SOUTH DECCAN PLATEAUS)
1. Bombay Karnatak
   a. Belgaum marginal zone
   b. Dharwar peneplains
2. Mysore Karnatak
   a. Malnad-sub-Ghats strip (evergreen forest – Babu Bhudan Hills)
   b. Maidan-peneplains, Mysore Ghat

XXIX. SOUTHERN BLOCKS
1. Nilgiri – Moyar trench, Nilgiri plateau
2. Anaimalais:
   1. Anaimalai/Palni Hills
   2. Cardamom Hills: upper Periyar Valley
   3. Varushanad/Andipatti Hills
   4. Comorin Hills: Shencottah gap

XXX. NORTHEAST DECCAN (MAHANADI BASIN AND ANNEXES)
1. Wainganga Valley (Eastern flank)
   Wainganga/Mahanadi watershed; haematite monadnocks
2. Chhattisgarh
   1. Northern (sub-Maikal) margins
   2. Seonath/Mahanadi doab
   3. Raigarh basin
3. 3. Sankh/S Koel/Brahmani basins
   4. Jamshedpur basin Subarnarekha valley

XXXI. TELANGANA
1. Lower Godavari trough
2. Telangana proper
   1. Hyderabad
   2. Bellary peneplains
   3. Raichur (Kistna/Tungabhadra) Doab
3. Anantapur/Chittor basins
   a. interior basins
      1. Chatravati
      2. Papagni
      3. Cheyyur-Bahuda
   b. Transitional zone
      1. Suvarnamukhi valley
      2. Nagari Hills – intermont basins

XXXII. EASTERN HILLS
1. Orissa/Bastar mass
   a. Orissa Hills
      1. Hill massifs
      2. Brahmani/Mahanadi trough
   b. Khondalite zone
      1. Dissected peneplains
      2. Tel/Sileru trough
2. Cuddapah ranges and basins
   a. Western arcs (Erramalai-Sesachalam-Palkonda Hills)
      Palkonda scarp, gorges (Cheyyur, Papagni)
   b. Central (Kunderu) basin
      1. Kurnool-Cuddapah plain
      2. Razampeta corridor – lateritic piedmont, Cheyyur Shinglespread

    c. Eastern ridges
      1. Nallamalais
      2. Central (Sagileru) valley
      3. Velikondas

XXXIII. ORISSA DELTAS
1. Mahanadi/Brahmani Deltas – Baitarni valley-Balasore gap, delta seaface, Lake Chilka, Mahendragiri gap

XXXIV. NORTHERN CIRCARS AND NELLORE
1. Vizag-Ganjam lowland – Rushikulya, Languliya, Vamsadhara Valleys – lateritic foothill zone – Waltair Highlands
2. Godavari/Kistna Deltas
    1. Godavari/Krishna breach
    2. Godavari delta – sub-deltaic margins, delta proper-seaface, Colair Lake
    3. Krishna delta
3. Nellore
    a. Nellore lowlands
      1. Archaean low peneplain
      2. Coastal alluvium – cuestiform marine deposits, Pulicat lagoon

XXXV. TAMILNAD
1. Coromandel coastal plain
    1. Archaean low peneplains – monadnocks, Cretaceous-Eocene inliers
    2. Cuddalore/laterite shelf: Red Hills, Capper Hills
    3. Young alluvial zone – embayments, strandplain; Korteliyar, Cooum, Adyar, Palar Valleys Madras
2. Tamilnad Hills
    a. Javadis – Agaram – Cheyyur trough Yelagiri, Ponnaiyar gap
    b. Shevaroys, Kalroyans, Pachamalais, Salem monadnocks
3. Palar/Ponnaiyar trough
    1. Lower shelf of Mysore Ghat
    2. Baramahal
    3. Southern margins (Salem area, Chalk Hills magnetite monadnocks)
4. Coimbatore plateau (Kongunad) Bhavani, Noyil, Amaravati Valleys, Palghat sill, Coimbatore Hills
5. Cauvery delta
    a. Delta head
    b. Delta Proper
      1. Valar/Coleroon doab
      2. Coleroon/Cauvery doab – Srirangam Island, floodplains
      3. Main delta plains – higher western margins (Vallam Tableland), older irrigated area
      4. Seaface – marshy low, dune belt, Pt. Calimere, Vedaranyam salt swamp
6. Dry Southeast
    a. Upper Vaigai
      1. Varushanad Valley
      2. Kambam valley
      3. Dindigul col
    b. Madura/Ramnad shelf
      1. Alluvial piedmont zone – Monadnocks (Sirumalais)
      2. Laterite/old alluvium panfan (from Varshalei to Vaippar), tank country, coastal strip, Pamban Island (old reefs, Adam's Bridge)
    c. Black Soil area
    d. Tinnevelly

1. Alluvial zone
2. Red soil zone – teris, coastal dunes
3. Tamprabarni basin – foothills, Chittar valley, Tambrabarni valley

D. THE ISLANDS

XXXVI. MALADIVES AND LACCADIVES

XXXVII. ANDAMANS AND NICOBARS
1. Andamans
2. Nicobars

XXXVIII. CEYLON

In our discussions on ecology and biogeography, we recognize in this book the following major divisions (Fig. 5): 1. The Peninsula (embracing the regions XV–XIX, XXI–XXXV and XXXVIII in the above synopsis); 2. the Eastern Borderlands (regions VIII, XIV, XX and XXXVII); 3. the Himalaya (regions IV–VII); 4. the Western Borderlands (regions I–III) and 5. the Indo-Gangetic Plains (regions IX–XIII).

REFERENCES

AHMED, K. 1941. Physiographic division of India. *Indian geogr. J.*, 16(3): 257.

ANNANDALE, T. N. 1914. The African element in the fresh-water fauna of British India. Proc. IX. internat. Congr. Zool. Monaco, pp. 579–588.

BHATTASALI, N. K. 1941. Antiquity of the Lower Ganges and its course. *Sci. & Cult.*, 7(5): 233–239.

BLANFORD, W. T. 1901. The distribution of Vertebrate animals in India, Ceylon and Burma. *Philos. Trans. R. Soc. London*, (B) 194: 335–436.

BOSE, N. K. 1968. The Ganga. In: Mountains and Rivers of India; 21 internat. Congr. Geogr. New Delhi, pp. 356–360.

BOSE, S. C. 1968. Source rivers of the Ganga. In: Mountains and Rivers of India; 21 internat. Congr. Geogr. New Delhi, pp. 361–375.

CHIBBER, H. L. 1946. The age, origin and classification of the rivers of India. *Bull. nat. geogr. Soc. India*, 9: 1–19.

CHIBBER, H. L. 1949. Westerly drift of rivers of northern India and Pakistan. *Bull. nat. geogr. Soc. India*, 12: 1–16.

CHIBBER, H. L. 1950. Easterly drift of the Son between Rohtas and Dehri, Sahabad District, Bihar. *Bull. nat. geogr. Soc. India*, 14: 64–65.

DESAI, H. J. 1968. The Brahmaputra. In: Mountains and Rivers of India; 21 internat. Congr. Geogr. New Delhi, pp. 431–437.

DE TERRA, A. 1934. Physiographic results of the recent survey in Little Tibet. *Geogr. Rev.*, 42(1): 12.

FOX, C. S. 1942. Physical Geography for Indian Students. London: MacMillan & Co. (on p.319).

GADOW, A. 1893. Vogel. In: Bronns Klassen und Ordnungen der Tiere. 296.

GEDDES, A. 1960. The alluvial morphology of the Indo-Gangetic Plain. *Trans. Papers Inst. British Geogr.*, 21: 262–263.

58

GREGORY, J. W. 1925. The evolution of the river system of southeastern Asia. *Scottish geogr. Mag.*, 41: 129–141.

GULHATI, N. D. 1968. The Indus and its tributaries. In: Mountains and Rivers of India; 21 internat. Congr. Geogr. New Delhi, pp. 348–355.

HAYDEN, H. H. 1913. Relation of the Himalaya to the Gangetic Plains. *Rec. geol. Surv. India.* 43: 1.

KAUSHIC, S. D. 1968. The Yamuna. In: Mountains and Rivers of India; 21 internat. Congr., New Delhi, pp. 369–412.

KRISHNAN, M. S. 1952. Geological history of Rajasthan. Proc. Symp. Rajputana Desert; Nat. Inst. Sci. India, New Delhi, pp. 27–29.

LAW, B. C. 1968. Rivers of India in Ancient Literature. In: Mountains and Rivers of India; 21 internat. Congr. Geogr. New Delhi, pp. 187–210.

OLDHAM, C. F. 1893. The Saraswathi and the 'lost' river of the Indian Desert. *J. Asiatic Soc. Bengal*, (NS) 25: 49–76.

OLDHAM, R. D. 1886. On probable changes in the geography of the Pubjab and its rivers. An historico-geographical study. *J. Asiatic Soc. Bengal*, 55.

OLDHAM, R. D. 1917. The structure of the Himalaya and the Gangetic Plains. *Mem. geol. Surv. India*, 42: 2.

PASCOE, E. H. 1920. Early history of the Indus, Brahmaputra and Ganges. *Quart. J. geol. Soc.* 75: 136–155.

PILGRIM, G. E. 1919. Suggestions concerning the history of the drainage of Northern India, arising out of a study of the Siwalik Boulder Conglomerates. *J. Asiatic Soc. Bengal*, (NS) 15: 81–99.

PITHAWALA, M. B. 1939. The need of uniformity in the physiographic division of India. *J. Madras geogr. Assoc.*, 14(4).

PITHAWALA, M. B. 1942. Physiographic division of India, Burma and Ceylon. *Sci. Cult.*, 7(1): 533–543, map 1.

PRANAVANANDA, SWAMI, 1939. The sources of the Brahmaputra, Indus, Sutlej and Karnali. *Geogr. J.* 39: 126–135.

PRANAVANANDA, SWAMI, 1968. New light on the sources of the four rivers of Holy Kailas and Mansarovar. In: Mountains and Rivers of India; 21 internat. Congr. Geogr., New Delhi, pp. 221–230.

PRASHAD, B. 1941. The Indo-Brahm or the Siwalik River. *Rec. geol. Surv. India*, 74(4): 555–561 (1939).

RADHAKRISHNA, B. P. 1968. The Cauvery. In: Mountains and Rivers of India; 21 internat. Congr. Geogr., New Delhi, pp. 427–430.

SCHOKALSKAYA, Z. J. 1932. Contributions to the Knowledge of soils of Asia. Leningrad.

SEN, S. 1968. Major changes in river courses in recent history. In: Mountains and Rivers of India; 21 internat. Congr. Geogr., New Delhi, pp. 211–220.

SEN, S. 1968. Bhagirathi-Hoogly Basin. In: Mountains and Rivers of India: 21 internat. Congr. Geogr., New Delhi, pp. 384–395.

SINGH, U. 1968. Middle Ganga. In: Mountains and Rivers of India; 21 internat. Congr. Geogr., New Delhi. pp. 376–383.

SPATE, O. H. K. 1957. India and Pakistan: A General and Regional Geography. London: Methuen & Co. pp. xxv–829, fig. 160.

VERMA, P. (Miss) 1968. The Mahanadi. In: Mountains and Rivers of India; 21 internat. Congr. Geogr., New Delhi, pp. 420–426.

VERMA, P. (Miss) 1968. The Narmada. In: Mountains and Rivers of India; 21 internat. Congr. Geogr., New Delhi, pp. 413–419.

VIJ, G. K. & R. C. SHENOY, 1968. Hydrology of Indian rivers. In: Mountains and Rivers of India; 21 internat. Congr. New Delhi, pp. 258–283.

WADIA, D. N. 1938. The post-Tertiary hydrography of northern India and the changes in the courses of its rivers during the last glacial epoch. *Proc. Nat. Inst. Sci. India*, 4: 387–394.

WADIA, D. N. 1953. Geology of India. London.

III. GEOLOGY

by

M. S. KRISHNAN†

The MS of this Chapter was received on April 2, 1970. Dr. M. S. KRISHNAN passed away on 24 April 1970, after surgical operation, at the age of 72, at Thanjavur in South India. This chapter is his last scientific contribution.

Dr. KRISHNAN was born in 1898. He took the M.A. degree from the University of Madras in 1919 and four years later he was awarded the Ph.D degree of the London University. Joining the Geological Survey of India in 1924, he carried out extensive field investigations in the geology of the Indian Peninsula. He worked for some time as a lecturer in geology in the Indian Forest College at Dehra Dun and in the Presidency College, Calcutta. He then returned to the Geological Survey of India, where he served in various capacities and finally became its Director for four years before retiring. He was also Director of the Indian Bureau of Mines; Joint Secretary to the Government of India, Ministry of Natural Resources; Director, Indian School of Mines and Head of the Department of Geophysics, Andhra University. He was elected President of the Geology Section, Indian Science Congress, in 1935. He was a Fellow of the National Institute of Sciences of India, Indian Academy of Sciences, Geological and Mining Society, etc. He represented India in 1949 at the UN Conference on the Conservation and Utilization of Natural Resources and in 1954 presided over the ECAFE Meeting at Bangkok. His publications include, in addition to numerous technical papers, the books *Geology of India and Burma, Structural and Tectonic History of India*, etc. He was widower at the time of his death.   M. S. MANI

A satisfactory interpretation of the biogeographical peculiarities of India is best attempted against a background of the outstanding features of its geomorphological evolution. An attempt is made in this chapter to present a brief outline of the salient features of the general structural and tectonic history of India from the earliest to the Recent. The reader will find excellent accounts of the geology of India in COATES (1935), DE TERRA(1936), DUNN (1939), FERMOR (1930), FOX (1931), GEE 1926), HERON (1934, 1935), KRISHNAN (1952, 1953, 1968), PASCOE (1950, 1961, 1965), WADIA (1931, 1938, 1942) and others. The following outline is necessarily brief and leaves out much of details.

1. General Structure

The salient structural features (Fig. 6) may be considered under 1. the ancient stable massif, of which the central portion forms the Deccan, together with later rocks that mask much of the surface and edges; and 2. the belt of fold mountains, which wrap around the ancient block and which owe their existence to compression against its resistant edge.

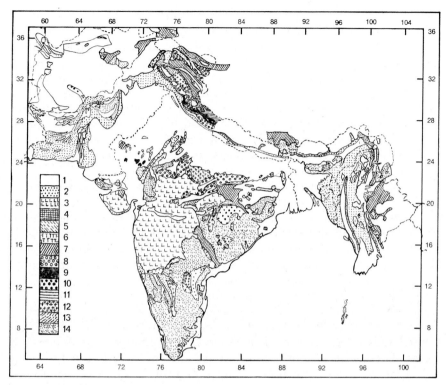

Fig. 6. Map of India illustrating the salient features of its geology. 1. Pleistocene, Recent, 2. Tertiary sediments (coastal Tertiary, Siwaliks etc.), 3. Deccan Trap (Mesozoic-Lower Tertiary), 4. Granites and Ortho-Gneisses (Post Mesozoic), 5. Cretaceous sediments, 6. Trap (RAJMAHAL & SYLHET) (Upper Mesozoic), 7. Gondwanas of the Peninsula/Marine deposits of Mesozoic and U. Palaeozoic, 8. Vindhyan sediments, 9. Malani volcanics & Granite, 10. Cuddapah, Delhi etc. sediments, 11. Lower Palaeozoic of Extra-Peninsular region, 12. Granites, 13. Dharwars, 14. Unclassified crystalline rocks of Archaean age.

1.1. THE PENINSULAR AREA

The basal complex of the ancient block of the Peninsula consists of highly metamorphosed rocks, like gneisses and schists of the Archaean System. By far the larger part of the Peninsula, but particularly the central and southern portions, is occupied by this ancient crystalline complex. We find these rocks also in large areas in the northeast of the Peninsula in Chota-Nagpur, Orissa and in Madhya Pradesh and in the northwest in the Aravallis and other parts of the Rajasthan. Granites, of which the Charnokite Series are of particular interest, have intruded into this complex. The rocks of Dharwar Series occupy long troughs and hollows in the basal complex. These are highly folded and metamorphosed sediments that now appear as phyllites, slates, schists and marbles.

A group of highly folded and altered sediments of slates and schists is known as the Cuddapah Series, which were folded into the already complex Archaean and Dharwar Series during the Pre-Cambrian times. The Vindhyan System consists of vast thickness of sandstones, shales and limestones of perhaps Pre-Cambrian age, resting in an almost undisturbed condition on the surface of the older rocks. The earlier view that the gneisses formed a floor, on which the sedimentaries were deposited, is now no longer considered valid, without considerable modifications. It is known, for example, that much of the gneiss is intrusive into the Dharwar rocks. Nevertheless, the Peninsula has been a stable landmass since very ancient times, at least the Pre-Cambrian*.

At a later period, the Peninsula was part of the Gondwana Continent, in the hollows of which series of fresh-water deposits of sandstones and shales were laid down. From that remote period down to the present times, the Deccan area has remained a continental mass. During Jurassic times, marine conditions prevailed over large areas of what are now Rajasthan and over parts of the coastal areas of Madras, as witnessed by the fossiliferous Trichninopoly beds. During Cretaceous times, basaltic lavas (the socalled Deccan-Trap of the older writers) covered extensive areas of what we know as Deccan. These Lava flows cover at present an area of over 520,000 km² and are recognized by the characteristic flat-topped hills of the region. The Deccan rocks reappear in Assam and Delhi. The Archaean Formations are gneisses and schists and Pre-Cambrian sediments and igneous rocks, metamorphosed in a variety of ways. Of the three distinct granitic intrusions generally recognized, the Peninsular gneiss is the earliest, the augen-gneiss being next and the upper Pre-Cambrian Closepet gneiss of Mysore being the third. The metamorphosed and schistose Pre-Cambrians are known by diverse names. The earliest gneisses are intruded by the Charnockites, which are hypersthene granulites, well developed particularly in the Eastern Ghats and in the Western Ghats, from south Mysore and Coorg to the southern tip of the Peninsula. These are of the Pre-Cuddapah or the Pre-Algonkian age.

Regional trends in the Archaean rocks are shown by the Aravalli strike, the Dharwarian strike, the Eastern Ghats strike and the Satpura strike. The Aravalli strike is northeast-southwest, observed in the Aravalli mountain belt in Rajasthan, and may be clearly followed from Delhi to Champaner in Gujarat. In Gujarat, a part of the strike is directed towards Mysore, where the Dharwarian rocks have a dominant north-northwest to south-southeast strike. Although the vast intervening area is covered by the Deccan Lavas, there seems to be a distinct connection

---

* The age data used in this Chapter are taken from the papers published by A. R. CRAWFORD (1969), excepting those relating to Bihar and Orissa for which references are given in the text.

between the Archaeans of the two areas. The Aravalli tends to turn southeast and east further east in Gujarat. The Aravalli strike is continued into Garhwal in some of the older rocks. There seems to have occurred some rejuvination of the northern portion of the Aravalli during the Tertiary times. The Aravalli-Champaner trend is continued to the south into the Laccadive Islands, through the Banks in the Gulf of Cambay. The Laccadive, Maladive and the Chagos Islands lie along this alignment and rise from platforms, 1800–2000 fathoms deep.

The Dharwarian strike is a south-southeast trend of a part of the Aravalli in Gujarat and this trend continues perhaps under the Deccan Lavas into Mysore and adjacent parts of the Andhra State, where the trend is north-northwest-south-southeast to northwest-southeast. High metamorphism of these rocks is observed near Mysore, where they also assume a north-south strike, turn southwest and west-southwest to adjust their trends to that of the Eastern Ghats and the Nilgiris.

The Eastern Ghats strike is an east-northeast-west-southwest strike in parts of Nilgiri-Coimbatore-Salem. In Malabar the strike is north-north-west-south-southeast. The trend of the Eastern Ghats is typically northeast-southwest, from the northeast Orissa to the R. Krishna. The northern part of the Eastern Ghats occupies a broad zone, in the western half of which the trend is north-northeast-south-southwest. Near Nellore a sigmoidal curve borders the Cuddapah Basin, with the slight convexity to the east. There is a further turn to the southwest near Madras and west-southwest in Coimbatore-Nilgiri. This belt cuts across and is superposed by the Dharwarian trend in the south. It is generally believed that it continues into Assam Plateau in the extreme northeast.

The Satpura strike is east-northeast-west-southwest, as observed in the Narmada-Son drainage basin. A southerly branch continues in Nagpur-Chindwara-Balaghat area and Gangpur and Singhbhum east. The area in between is occupied by granites and gneiss of the same general trend. In the western end it appears to merge into the Arvalli trend.

The thick deposits of the Cuddapah and the Vindhyan (Algonkian age) were laid down on the Archaean basement. The remnants of these deposits are seen in the Cuddapah Basin, Chhatisgarh-Mahanadi area and Orissa and the Great Vindhya Basin. The Aravalli and the Delhis have been uplifted along a fault that marks the western margin of the Vindhya and thrust against the Vindhyan during the Mesozoic times.

The Dharwar and the Aravalli formations are remarkable for the greatly diversified lithological character, with a high degree of metamorphism. The Aravalli Range apparently arose during the close of the Dharwar times, was then denuded extensively, uplifted during the Cambrian and again perhaps before the Permo-Carboniferous times. The Aravalli Range must indeed be described as the oldest mountain system on the earth that still retains some relief. The orogenic movements that gave rise to the Aravalli and Dharwarians were followed by a period of

erosion and subsidence during the Eparchaean Interval, which is believed by competent authorities to have been as long as the sumtotal of
all the succeeding geological periods. A very pronounced unconformity
separates the Dharwarian from about 6000 m thickness of slates, quartzites and limestones that form the Cuddapah System, deposited perhaps
in geosynclinal basins. The Cuddapah rocks are preserved mainly east
of Deccan, between the R. Krishna and the R. Pennar and also in the
valley of the upper Mahanadi river. Except perhaps in the long border
ridges of the Nallamalai and Velikonda Hills, these are very little disturbed. The Delhi quartzites occur in narrow tightly-packed belts in the
centre of the Aravalli synclinorium and constitute rocky echelonned
ridges, terminating in the Delhi Ridge.

The Vindhyan rocks* overlie the Cuddapah rocks in the lowest part
of the R. Krishna-Pennar trough. They occur, however, mainly in a belt
along the northern flank of the Peninsula, from the R. Chambal to the
R. Son, broken by the expanse of the ancient Bundelkhand gneiss near
Jhansi. To the west of the Aravalli Hills, there are patches of lava of
Lower Vindhyan times near Jodhpur. The Lower Vindhyan System
comprises marine shales, limestones and sandstones. Above these are
great thickness of nearly horizontal fluviatile and estuarine sandstones,
including red sandstones. Except in isolated patches west of the Aravalli,
the Vindhyan rocks are very little disturbed or metamorphosed. The
scarp that marks the northern flank of the R. Narmada and Son Valleys,
is perhaps the most striking feature formed by the Vindhyan rocks. In
the west this is formed largely of Deccan lavas, but the Vindhyan rocks
occur between Bhopal and Itarsi and dominate farther east in the continuous scarp of the Kaimur Hills, overlooking the R. Son.

The Gondwanas consist of immense thickness of fluviatile and lacustrine
deposits of sandstones, with shale and clay, of continental origin, laid
down in great troughs, formed by tensional faults and subsidence on the
old plateau-like surface. There is a remarkably striking parallelism
between them and those of similar age in South Africa, Australia and
South America, in the presence of glacial basal conglomerates, formed
perhaps by the glaciers radiating from the Aravalli, and in the *Glossopteris*-flora. There are also more or less continuous belts of the Gondwana rocks along the lower Penganga and Godavari rivers and between
the R. Mahanadi and R. Brahmani, from Talchir to the head-streams of
the Narmada and Son, with a string of outcrops in the Damodar Valley.
The Gondwanas synchronized with the Permian glaciation. Fluviatile and

---

* There is a difference between the geographical and geological expressions Vindhya;
the Vindhya mountains extend from about 75 to 78 EL and are mostly Deccan lavas,
but eastward the general line is continued by the Bhanrer and Kaimur Hills, formed
of the Vindhyan rocks. The Vindhyan rocks also occur in the Bhima Valley, between
Sholapur and Raichur and also perhaps underlie much of the Deccan lavas. M. S. MANI.

lacustrine strata, containing *Glossopteris*-flora, were laid down over the first tillites during the Permo-Carboniferous and Permian times. Then followed semi-arid conditions, marked by red sandstones with fossil amphibia, reptiles, etc. Moist conditions returned during the Upper Gondwana (Jurassic times), with fluviatile and lacustrine sediments, containing the *Thinnfeldtia-Ptilophyllum* flora. Marine incursions appeared on the eastern coast. Then followed uplift and erosion. The principal phase of block-faulting in the Gondwana strata occurred about the upper Triassic to the Lower Jurassic. A series of fractures, associated with the block-faulting, radiate to the west, northwest and north from the area of Ranigunj. These fractures are perhaps contemporaneous with the fissures, through which the Rajmahal Trap (lavas) erupted. Other possible fractures, trending northeast from the same focus towards Bhutan, may perhaps be the fore-runners of the present Garo-Rajmahal Hills Gap. The Gondwanas also underwent some folding during the Cretaceous times, affecting the R. Son Valley more than perhaps the R. Damodar Valley. The base of the Gondwanas lies at 60 m in Bengal and 1800 m above msl in Sohagpur and Chhindwara.

The earliest changes, after the deposition of the earlier Peninsular sedimentaries, seem to be the folding of the Aravallis during the earlier Vindhyan Period. The Upper Vindhyan sandstones appear to have formed from the debris from these mountains. The more disturbed parts of the Nallamalai and Velikonda Hills also appear to have been elevated about the same time. These changes seem to have been followed by the peneplanation of the Aravallis and a rejuvination at the beginning of the Gondwana times. Although little or no compressional orogenic activity took place in the Peninsula since then, other movements of a less tangential nature seem to have had important effects and peneplanation seems to have been followed by more than one rejuvination. The highlands of Ceylon, the Palni and the Nilgiri Hills are not merely stumps of an eroded plateau, but are also great horsts, uplifted during Post-Jurassic times or perhaps even during the Tertiary times. These periods are close to the uplift of the Himalaya, the Deccan Lava flows and the subsidence of the Arabian Sea area from the Western Ghats. The long straight edge of the Western Ghats, developed on practically horizontal Deccan Lavas and on ancient gneisses, is strong evidence of faulting and subsidence on a large scale. The Deccan Lavas extend down to at least 600 m below the sea-level at present.

The Aravallis are now no more than mere stumps of a once lofty range of mountains. They reach their maximum elevation in Mt. Abu (1721 m) in the southwest, but diminish to low hills in Jodhpur-Jaipur Saddle, to rise again to the northeast, before disappearing in a series of echelonned ridges, half buried in the Indo-Gangetic alluvium and reaching as far as the Ridge in Delhi. East of the Aravalli, the lower course of the R. Chambal is believed to occupy a strike-valley in the Vindhyan scarplands. North of Kotah it is superimposed, cutting across the scarps and its uppermost course is more nearly consequent on the Deccan lavas. The Malwa Scarps of the Vindhyan rocks face the south and east (elevation 450–510 m). The Vindhya (Deccan lavas) and the Kaimur (Vindhyan sandstones) hills form a great scarp, overlooking the Narmada and Son Valleys. Their drainage is nearly wholly north-westwards to the R. Yamuna and the Ganga and the R. Narmada and the R. Son do not have any important north-bank tributaries. In the Son Valley there is evidence of a drainage superimposed from a higher plateau level, the main outlines of which did not differ from what we see today. The Vindhya-Kaimur Scarp exceeds 600 m in elevation in certain places.

The Gneissic plateau of Chota-Nagpur rises to an elevation of 1680 m in the Hazaribagh Range, but most of Ranchi Plateau is peneplain, about 600 m above mean sea-level, with a few monadnocks. The Peninsula may be said to terminate at the Rajmahal Hills (largely Gondwana basalt), but it seems probable that a sill of the old rock, relatively near the surface of the Gangetic alluvium, connects it with the outlying Shillong Plateau. South of the Rajmahal Hills lie the coal-bearing Gondwana basins of the Damodar Valley, with sandstone ridges striking east-west in a synclinal trough. South of Ranchi Plateau there is a corridor at just over 300 m from the Ganges Delta to the Brahmani and Mahanadi Basins, between the plateau and the broken forested hills of Orissa.

The Satpura-Mahadeo Hills lie between the R. Narmada and Tapti and represent perhaps an ancient tectonic range, but are at present merely scarped blocks, steeper on the R. Tapti than on the Narmada side, and covered largely with the Deccan Lavas, but with some gneissic inliers. From their eastern continuation in the Amarakantak Plateau (the Maikal Hills), a mixed Deccan Lavas and gneissic upland, there radiate the headwaters of the rivers Narmada-Son-Mahanadi and those of the R. Wainganga (a tributary of the R. Godavari). Except for the northeast-southwest trend of the Aravalli-lower-Chambal area, the northern section of the Peninsula is dominated by a strong east-west trend, with however a slight northeast-southwest strike in the Maikal and Hazaribagh Hills, which are perhaps themselves influenced by the buckling and sagging of the northern flanks of the old block, under the stress of the Himalayan orogeny (see Fig. 6). In the west, Kathiawar is largely Deccan lavas, with some marine Jurassic and Tertiary fringe, particularly in Kutch. These two areas, form a small dissected plateau and scarpland, linked to the Peninsula by alluvial plain of Gujarat, the subsidence of which has formed the salt-marshes and large mudflats of the Rann.

To the south of the eastuary of the R. Tapti begin the Western Ghats (locally known as the Sahyadri Range) and rise to an elevation of 900–1200 m. There is a rather steep and much dissected fall to the undulating and narrow coastal lowland of Konkan, but once over the crest, the broad mature and even senile valleys of plateau appear almost at once. The Deccan Lavas form the Ghats to a little north of Goa and the steep seaward face is like a great wall, dissected by deep-canyon-like valleys into mesas, buttes and pinnacles. The old gneisses and granites appear south of Goa and for about 320 km the crest sinks to elevations below 900 m, but to rise again to the great gneissic boss of the Nilgiri Hills (2760 m). This boss is a much-worn massif, elevated and redissected, with very steep drops on all sides. South-westwards across the Palghat Gap, the wider and more forested Cardamom, Anamalai and Palni Hills are also similar in origin. Anaimudi (2965 m) is the highest summit in the Anamalai Hills. The Palghat Gap is perhaps of tectonic origin. The summit of the Gap is a broad tableland, not much above 300 m in elevation. Ignoring the rather narrow Shenkottah Gap in the extreme south, this is the only really easy passage across the Western Ghats from the R. Tapti to Cape Comorin, a distance of nearly 1410 km.

The socalled Eastern Ghats are much less strongly marked than the Western Ghats and seem to disappear for a distance of about 160 km, between the rivers Godavari and Krishna. In the north there are some dissected massifs of the older Peninsular rocks. Relics of ancient mountains like the Nallamalai, Velikonda, Palkonda in the middle and south of the R. Krishna, the gneiss bosess of the Shevroys and Pachamalai Hills in the south belong to the Eastern Ghats. The expression Eastern Hills is to be preferred to the Eastern Hills for the north; Cuddapah Ranges for the middle and Tamilnad Hills for the southern portion.

The Deccan Lavas seem to have erupted from numerous fissures in the crust, during a period of tension. The estimated thickness of the Lavas is about 1830 m near West Goa, but much less elsewhere. They seem to have extended westwards beyond what is now India, but these parts faulted down during the Miocene times. The intercalations of the basic lavas, found in the sediments in the mountains of Sind-Baluchistan border, are also considered to belong to the Deccan Lavas. The lavas appear to be of the uppermost Cretaceous to the Lower Eocene in Kutch, Kathiawar and Gujarat. They are about 750 m thick in Kutch and have been exposed by subaerial weathering and denudation, with alteration of the topmost layers to laterites, before the Nummulitic Laki Series were deposited over them. The lavas in most areas are, however, of Eocene age. There are also later eruptive centres in Kathiawar and Narmada Valley, from which acid volcanics have erupted and plutonic intrusives have been contributed. The Deccan Lavas erupted and spread over an uneven pre-existing land surface that was already in an advanced stage of maturity and their base is now found at various levels, both above and below the sea-level. It lies, for example, at 600 m above mean sea-level near Belgaum, 300 m near Nagpur, 500 m on the flanks of the Maikal Range, 750 m south of Sholapur and in the Ranchi plateau and over 1830 m in Jashpur, 150–550 m below sea-level in some places near Bombay and in the Narmada Valley due to faulting and folding. Though they have also suffered gentle warping in certain places, they are mostly practically horizontal.

The coastal-plains of the Coromandel Coast and of the Circars are typical upland plains of marine erosion, with inland-facing cuestas, isolated granite or gneissic hills, representing old offshore islands and coastal lagoons. Deltaic formations mark the mouths of the larger rivers. Within the frame of the Eastern Hills, the Cuddapah, the Western Ghats and Satpura-Maikal Ranges and the Hazaribagh Hills lie the true Deccan. The Peninsular rivers find their way from the broad well-graded upper basins to the sea by relatively narrow corridors through the Eastern Hills. The close correspondence of the lower Godavari and the gap shared by the R. Mahanadi and the R. Brahmani with the belts of the Gondwana rocks would appear to suggest a tectonic-trough origin. The general lithology and the stratigraphy of the marine deposits on the east coast seem to indicate that since the latter part of the Palaeozoic Era, the general run of the coastline has never been very far from its present position.

1.2. The extra-peninsular area

The disposition of the extra-peninsular mountains to the north, west and east is the result of the intense squeezing out of the Tethyan geosyncline between Laurasia, advancing from the north, and the Indian Peninsular (Gondwana) Block from the south. The opposing fronts squeezed the soft contents of the Tethyan geosyncline into the east-west Himalaya. Since however the Indian Block is much narrower than the Laurasian mass, its advance threw the sediments on either side into the north-south folds of Baluchistan and Burma. Laurasia overrode the Peninsular Block and the latter also dived under Laurasia and led to the elevation of the Tibetan Plateau. The same movements also brought about a northward tilting of the Peninsular Block and carried down the Vindhyan-Satpura Mountains. The intense mountain building movements produced the foredeep in front of the convex side of the Himalayan Arc, by the bending down and underthrust of the northern edge of the Peninsular Block, which thus came into opposition with the Asiatic continental mass. This foredeep is not, however, a continuous depression throughout the length of the Himalaya, but consists of three strips, with the same alignment as the Himalaya and separated from each other by a transverse ridge-like structure west of Delhi and another to the east of Cooch-Behar. This foredeep is underlain, at least in the northern part, by Tertiary and older rocks, dipping down into it from the Himalayan side. The rocks on the Peninsular side continue into it from the south.

During the middle of the Tertiary times, the areas that are now Tibet and the Himalaya were covered by an extension of the Tethys Sea, in which deposition of immense sediments had continued for a vast period of time.

The Tethys Sea separated Eurasia from the southern Gondwana land-mass. Except the Altai of eastern Turkestan and the Aravalli Ranges of India, no other high mountains existed. The sedimentation, accumulated from the Palaeozoic Era, attained a thickness of perhaps 15,200 m and was accompanied by slow sinking of the sea bed. During a period of crust movements, the floor of the sea began to rise gradually and was thrown into series of long, parallel, wave-like folds. The crests of the waves were eroded by rain and weather and the rising land became much broken and irregular. Drainage basins came to be carved out of the flanks of the folds and river systems, composed of transverse valleys, were gradually formed. As the uplift progressed, the troughs of the folds emerged to form a series of longitudinal valleys, at right angles to the transverse valleys and parallel to the longitudinal axis of the crustal folds. A combination of concurrent uplift and erosion thus gave rise to the mountain systems of the Himalaya and Middle Asia. As denudation proceeded, deeper and deeper parts of the crust were exposed, but the forms of many folds may even now be traced and the trends of the longitudinal axis may clearly be followed over long distances. Folds were superimposed on folds, arches were overturned until almost horizontal and the whole region became greatly distorted and crumpled. The uplift of the mountains in this region has thus been brought about by horizontal pressure of the crust, acting in a meridional direction through long periods, right down to the present times. The wrinkling of the crust has taken diverse forms. The plateaus were wrinkled into ranges and the folded surfaces were wrinkled and these wrinkled mountains have in their turn been corrugated to form smaller folds. The intervening troughs were filled detritus from the mountains. In places where the stress exceeded the breaking strain of the crust, the rocks have fractured and have greatly complicated the structural features. Great parts of the crust have subsided and have moved horizontally. Considerable molten material has been forced up from below in places of weakness and fracture and has partly absorbed the original sediments also.

The Himalaya is, therefore, the result of a series of great orogenic movements, separated by periods of relative quiescence. The deformation seems to have been initiated during the Upper Cretaceous times and continued through the Middle Miocene, end of the Pliocene, Pleistocene and Sub-Recent times. The Middle Miocene times represent perhaps the period of maximum uplift, when the great masses of granites were intruded into the axial region of the Main Himalaya. The succession of mountain ranges from the orogenic activity that transformed the Tethyan geosyncline is thus marked by three major phases, viz. 1. the elevation of the central axis of the ancient crystalline and sedimentary rocks during the Oligocene times. The Nummulitic limestones were deposited in a series of basins, especially in Ladak. 2. The Miocene movement that folded the Murree sediments of the Potwar Basin and 3. The Post-Pliocene phase that affected Mio-Pleistocene Siwalik sediments and which apparently has not yet ceased. Initial disturbances probably long preceded the first phase. The region of the Karakoram has for example no marine Tertiary and may perhaps have been uplifted during the Cretaceous. There is a definite southward shift in the orogenic activity, welding the successive belts of geosynclinal sediments onto the Middle Asiatic Core.

Current ideas on the uplift of the Himalaya are based on interpretations, analogous to the nappe theory of the Alps (DE TERRA 1936). The socalled boundary faults, regarded by earlier geologists as steeply-dipping reversed faults, marking successive southern limits of the mountain-building and the northern limits of the Tertiary sedimentation, are now interpreted as great thrust-planes. This border zone is now considered as the old surface of erosion, over which the older Himalayan formations were thrust and through the gaps of which they advanced in huge arch-shaped waves. The deposition of 4575–6000 m thickness of the Siwalik beds seems to have been made possible by the tectonic downwarp. The conditions were perhaps similar to those prevailing in the present-day Gangetic alluviation, but the foredeep was farther north than today and was pushed southwards by the tectonic advances, involving successive detrital accumulations.

The following is a brief outline of the general scheme of thrusts and uplifts: 1. imbricated marginal thrusts (Simla-Kumaon); 2. interior secondary thrust-sheets; 3. the Main Central Thrust Mass, with deep-rooted injected crystallines, 3050–6100 m thick and covered with 3050–4575 m thickness of Algonkian-Mesozoic sediments; 4. Palaeozoic and Mesozoic sediments thrust and recumbently folded onto the back of the main root; 5. the 'exotic' Tibetan thrust (the Kiogar Klippen); and 6. the Flysch-zone south of the Trans-Himalaya, with a possible weak counter-thrust northwards. The Mt. Everest and Mt. Kanchenjunga are carved out of the back of an enormous nappe, which may be a continuation of the Main Central Thrust Mass.

Thrusting was not the only type of orogenic activity, but isostatic uplift seems to have played an important part, at least in the last phases. The problem of the Himalayan compensation is closely bound up with the origin of the Indo-Gangetic Trough. Recent evidence shows the uplift of some 1800 m of the Pir Panjal since the middle of the Pleistocene and such uplifts have affected the entire Himalaya and the Karakoram. The unloading due to the shrinkage of the Himalayan glaciers resulted in isostatic uplift. The effects of unloading brought about by the removal of vast quantities of erosion products by the south-flowing rivers must also be stressed. We cannot, however, minimize the rôle of horizontal compressional forces. The high peaks like Everest, Kanchenjunga and Dhaulagiri are opposite the great foredeep of the Gangetic Plain and are evidence of the expression of a balance-movement in regions of the greatest exchange of load. Thrusting seems to be still in progress in the border regions and the vertical uplift both in the Pir Panjal and in the inner Himalaya.

From the Pamirs-Complex fan out east and west the Tien-Shan-Kun-Lun-Karakoram and the Alai-Hindu-Kush. To the south of the Alai and the Hindu Kusj are the lower ranges of Afghanistan and Baluchistan, which are in turn looped around the Sibi-re-entrant. North of this, the Sulaiman Range rises abruptly from the Indus Plains. There is a mass of echelonned ridges in Kalat, between the north-south Kirthar Ranges (on the Sind-Baluchistan border) and the east-west Chagai Range, sinking to the Seistan Depression and swinging around east-west, parallel to the sea-coast in Mekran. Each arc is in reality a series of concentric arcs, connected at the extreme ends, with arid depressions between them. These mountains are of simple anticlinal structure are and developed for the most part in relatively soft Cretaceous and Tertiary sandstones and limestones, with flysch facies in the north. There is close parallelism between the Sibi-re-entrant and the greater re-entrant north of the Punjab. As the northwest syntaxial area culminates in the great peaks of the Karakoram Range and in the Nanga Parbat, the highest summits between Safed Koh (34° NL) and the sea are to be found in the angle around Quetta. The presence of a concealed projection of the Gondwana (Peninsular) Block seems to explain these facts.

The Tertiary folding has wrapped itself around a projection of the Gondwana Block, as indicated by the outcrop of old rocks in the Kirana Hills. Fronting the Punjab Plains is the great monoclinal scarp of the Salt Range, largely due to thrusting. Behind this and between the rivers Indus and Jhelum, is the Potwar Plateau, a peneplain formed on the folded Murree and Siwalik Beds, largely masked by the loess-like silt.

The northwest syntaxis of the Himalaya forms the great knee-bend, about 490 km deep, and affects the strike of the mountain ranges as far as the foot of the Pamirs. The extension of the old Gondwana Peninsular Block is indicated by the Kirana outcrop (32° NL), far north of the Aravalli and hardly 95–115 km from the Salt Range. The steep front of the Salt Range from the plains and the long dip behind it in the north and its thrusts showing a horizontal movement of some 30 km and its curiously twisted alignment show that it is very largely controlled by the resistance of the Peninsular Archaean mass hidden beneath the Punjab. The stability and the competence of the basement rocks of the Peninsular Foreland, underlying the Tertiary sediments of the Potwar Trough, are evident from the fact that the mantle of the Murrees and Siwaliks is merely wrinkled up on the basement and not metamorphosed or indurated. The Murree sediments are strikingly, different from those of the Siwalik Hills and are derived from iron-bearing Peninsular rocks rather than from the rising Himalaya.

The influence of the Gondwana (Peninsular) Block on the alignment of the Himalaya is, therefore, very profound. Round this Block, the ranges of the Himalaya are wrapped in loops. The strike of the rock systems parallels that of the planes of thrusting onto the Peninsular Foreland. The Great Himalaya represents the original axis of the uplift of the Tethyan geosyncline, bending sharply southwards at each end, into the Baluchistan and into the Assam-Arakan Ranges, where the pressure of the Gondwana Block suddenly ceases. The Himalayan compression is not simply an expression of outward creep from Middle Asia, but largely due to the extensive under-thrusting from the ocean floors and a definite northward drive of the old blocks.

The great intensity of the compression of the Tethyan geosyncline is evident in the great recumbent folds and thrust sheets of the Himalaya. The lateral Baluchistan and Burmese Arcs were formed about the same time by the sediments and the sides being comparatively mildly thrust over the northeastern and northwestern regions. Though the folding in the lateral arcs is considerable, it is not so violent as in the Himalayan Arc, where the sediments have been piled up to form the highest mountain ranges in the world. Baluchistan region has suffered more than the Burmese region, due to the presence of two wedges of the Peninsula, distorting the smoothness of the arc. The thrusts around the tips of Kashmir and Assam wedges are of considerable interest. In Kashmir the formations almost run a complete circle over to the west and southwest, the thrusts being directed everywhere at right angles to the strike of the rocks. The rocks have literally flowed around the tips of the Peninsular wedges, anti-clockwise in Kashmir and clockwise in Assam.

A. The Himalayan Arc

The Himalayan Arc extends from the Mt. Nanga Parbat in the west to the Namcha Barwa Peak in the east. The following four longitudinal stratigraphic zones are generally recognized. The outermost zone, bordering the Indo-Gangetic Plain, consists of the Siwalik Hills, rising to elevations of 1200 m, often 45–50 km wide and composed of Tertiary sandstones and shales. This zone is characterized by conglomerates, sandstones and clays of Tertiary age. The immensely thick Upper Siwalik beds are composed of loosely aggregated conglomerates and soft earthy deposits. Below these lie very considerable thicknesses of soft sandstones, resting on harder sandstones of the Nahan stage. Pleistocene high level terraces of the Karewa beds in Kashmir and the ossiferous rocks of Nagri Korsam also belong to this zone. Two series, the Sirmur Series and the Siwalik Series, are generally recognized. Between the Siwalik Series and the older deposits, there are the socalled reversed faults or the main boundary fault, in which the older rocks have been thrust up over the younger. The Siwalik Series do not overlap the boundary fault line and are also never found among the mountain ranges further north. The total thickness of the Siwalik Series is 4875 m. The Sirmur Series are not generally observed east of the R. Yamuna.

The Sub-Himalayan (often also called the Lesser Himalayan) zone, about 75–80 km wide and rising to elevations of 2135–3050 m above mean sea-level, but in some parts to much higher altitudes, is composed of sediments ranging from the Pre-Cambrian upward to the Mesozoic and mostly unfossiliferous. There are here a number of overthrusts and nappes, in which the recumbent folds and inverted sequences are common. This is taken as representing the border of the former landmass of Indian Peninsula and the marine basin that lay beyond.

To the north of this comes the Central Himalayan zone, about 70 km wide and in which are all the high snow-covered peaks of the Himalaya. This zone consists of some sedimentary and metamorphic rocks largely Pre-Cambrian and Palaeozoic and large masses of igneous intrusive rocks. The intrusive granite of this zone is of a different age, but largely Cenozoic and partly Mesozoic.

North of this lies The Tethyan (or Tibetan) zone, composed of sediments of all ages, from the Cambrian to the late Tertiary, formed in the Tethyan geosyncline and it is here that we have the valleys of the R. Indus and R. Brahmaputra. The deposits are of immense thickness, often exceeding 6000 m of almost entirely marine sediments. These sedimentaries are in contact with the granite axis of the Great Himalaya in the south.

B. The Burmese Arc

The Himalayan formations curve southward to continue as the Burma-Arakan Ranges or the Burmese Arc. The northeasterly strike of the eastern end of the Himalaya turns southeast beyond Sadiya in Assam, then southwest and finally south. The Shan States of Burma are geologically a part of Yunan-Indo-China. Its proximity to India is of Post-Cretaceous origin.

The Burmese Arc sweeps in a broad curve through Arakan and the Andaman Islands to Sumatra and beyond. It is convex towards India, but somewhat concave in Arakan, perhaps as a result of part of the Indian Peninsular shield hidden under the Ganga-Brahmaputra delta. The southern part of the Burmese Arc is largely submerged in the Bay of Bengal. The Andaman and the Nicobar Islands represent the unsubmerged peaks of the ridges of this southern part of the Burmese Arc. The mountains of the Andamans are composed of folded Mesozoic and Tertiary rocks, intruded by granitic and ultrabasic rocks. In the core of this are Triassic and Cretaceous rocks, folded in the Upper Cretaceous and Tertiary times.

The Naga and Haflong-Disang overthrusts, in addition to other minor thrusts and reversed faults, are found on the Assam side of the Burmese

Arc. All these thrusts are directed from Burma towards India in a north-westerly direction.

Inside the Burmese Arc and parallel to it is a zone of Upper Tertiary to Recent volcanoes, continuing with the volcanic zone of Sumatra and Java. This volcanic zone lies in the faulted junction of the eastern border of the main Burma-Arakan Range and the western border of the Median Tertiary belt of Burma. Most of the volcanoes in this zone were active during the Upper Miocene to the Pleistocene times and many in the Indonesian part of the Arc are active even today.

The Tertiary rocks are faulted against the more ancient rocks of the belt of Shan States Plateau to the east. Another line of volcanoes is found in this fault zone. The Shan belt shows Pre-Cambrian, Palaeozoic and some Mesozoic beds, intruded by Pre-Cambrian and Mesozoic (Jurassic) granites. The granite belt passes through Bhamo and Mogok in the north and through Tenasserim into Malay States in the south and turns east into Central Borneo.

C. The Baluchistan Arc

The general northwesterly trend of the formations in Kashmir bends sharply round southwest of the Nanga Parbat and spreads out further south, partly through Hazara into the Safed Koh Mountains and Afghanistan and also into the Suleiman and Kirthar Ranges in the Sind-Baluchistan border and into Mekran and Eastern and Southern Iran. In sharp contrast to the smooth broad curve of the Burmese Arc, the Baluchistan Arc is characterized by three socalled 'festoons', sharp kinks or re-entrants, where the strata are gathered up in sheaf-like fashion.

The overthrust in the Baluchistan Arc is from the northwest and west towards India. The sharp re-entrances are attributed to the presence of concealed wedge-like promontories of the Pre-Cambrian shield of the Peninsula, underlying the alluvium of the R. Indus. These wedges have apparently been able to push back the sediments into a series of Arcs when the Peninsula drifted northwards during the late Mesozoic and Tertiary times. The apices of these wedges in the Baluchistan Arc are located at the western end of Salt Range, the Gomal Pass and near Quetta. On the convex side of the Arc lie Mesozoic sediments and Tertiaries of flysch type. Two sedimentary facies lie side by side north of Hazara and have a general southwest trend. The northwestern facies is of the Himalayan (Spiti) type and the southeastern facies is the calcareous zone, continuing into Sind-Baluchistan. The calcareous zone exposes the Permo-Carboniferous, Upper Triassic, Jurassic, Cretaceous and Eocene rocks, striking into the sea near Karachi and turning west-southwestwards towards the Oman Coast. As in the case of the Burmese Arc, the ultrabasic intrusives are of Upper Cretaceous age. The formations spread out into south Iran, south of Quetta and west of Kirthar Range. The Tertiary rocks contain intermediate and basic volcanics. Some volcanic cones were active during the Upper Tertiary and one is still active. This region is more highly folded and faulted than the corresponding Tertiary belt of Burma.

2. Pre-Cambrian Eras

The Indian Peninsula exposes rocks of Pre-Cambrian ages over more than half of its area, the rest being covered by rocks of Mesozoic and Tertiary ages. The latter comprise the volcanic flows of the Deccan Trap formation, covering a large part of Central and Western India, the Gondwana formations of Central and East Central India, and the Tertiaries forming a narrow fringe along the coast.

The Pre-Cambrians include crystalline gneisses and schists, as well as sedimentary rocks restricted to basinlike structures, which are underlain by gneisses. The oldest Pre-Cambrian rocks, forming the Archaean divisions, of more than 2400 millions years, are restricted to limited exposures in Kerala, Mysore, South Bihar and the Aravalli belt of Rajasthan. Radiometric dating has reavealed the existence of some gneisses and schists in Southern Kerala and in Southern Mysore, showing ages of 3000–3200 million years. The Older Metamorphic Series of the southernmost part of the Singhbhum District in Bihar are also of this age. They comprise garnetiferous gneisses, mica and chlorite schists, hornblende gneisses and calc gneisses. Zircons, isolated from some of the gneisses near Jaipur in Rajasthan, have indicated ages round 3600 million years and it is likely that patches of similar rocks will be found amongst the other gneisses of Eastern Rajasthan. It is also likely that the Eastern Ghats of Orissa and Visakhapatnam contain rocks 2500–3000 million years old. These Archaean rocks have limited areal extent, as they have been subjected to disruption and assimilation by igneous rocks of later ages.

The Post-Archaean formations of Pre-Cambrian age are widely distributed. They are described in the following paragraphs region-wise.

2.1. KERALA

The state of Kerala, forming the western part of southernmost India, is dominated by the hill ranges of the Western Ghats, constituted by Charnockites (HERON 1934), gneisses and schists. Most of the gneisses exposed here appear to be of great antiquity, around 2500 million years or older. The Charnockites have been given ages between 2500 and 2800 million years (CRAWFORD 1969). There are, however, younger intrusives with ages of about 2000–2100, 1000 and 450–600 million years. Though the actual rocks have not been dated, monazite, allanite, zircon, etc. found in the river beds and beach sands of Kerala show the presence of rocks of these ages.

2.2. MADRAS

This state is composed entirely of ancient crystalline rocks, except for the Cauvery basin and the coastal fringe containing Cretaceous and Tertiary formations. The Charnockite, masses forming the Nilgiri mountains and the hills of Pallavaram and St. Thomas Mount, have radiometric ages of 2550 to 2700 million years. It is presumed that the Charnockites of the Shevaroy, Palni and Varshanad Hills are also of about the same age. The rest of this state is composed of gneisses and schists. The limited age data available indicate that the Peninsular gneisses may be as old as the Charnockites, while they contain also later intrusives like the gneisses of Madurai and Ramnad (about 700 million years), the nepheline syenite of Sivamalai (1100 to 1200 million years), the biotite-pyroxenite associated with carbonatite at Sevathur in Arcot (720 million years) and the tourmaline granite of Ramnad (700 million years).

2.3. MYSORE

A large part of the State of Mysore has been mapped and the geology is fairly well known (RAO 1940, 1964). The basement gneisses here are of granodioritic composition and are 2500 to 2600 million years old. They contain several bands of highly folded schistose rocks, which appear to be the axial parts of a once continuous anticlinorium trending NNW–SSE in the north, N–S in the south turning to the southwest in the extreme southwest, where they are associated with bodies of Charnockites and ultramafic rocks. The oldest members of the schistose series appear to be metamorphosed basic lavas whose ages range from 2350 to perhaps 2500 million years. It is likely that they may be underlain by still older rocks, which may be the oldest members. These schistose rocks constitute the Dharwar System, showing the succession given in Table I.

Although no rocks of the Dharwar System, except the basic meta lavas have been dated radiometrically, they are believed to have been deposited between 2300 and 3000 million years. They are usually divided into three major groups, the oldest being dominantly igneous and the other two sedimentary. The middle division contains banded ferruginous formations (both hematite-quartzites and magnetite-quartzites), which have given rise to bodies of rich iron ores. These sedimentary formations are mainly argillaceous, with subordinate development of dolomite and limestone. The folds plunge to N.N.W. The northern part contains rocks of low grade metamorphism (green schist facies), while the southern part shows higher metamorphic grades (amphibolite and granulite facies). In the eastern part of Mysore there is a broad band of granitic to granodioritic gneisses, usually referred to as the Closepet Granite. The Dharwars and the Peninsular Gneisses are intruded by younger granites, pegmatites, porphyrites, etc.

76

Table I. Classification of the Dharwar System

Post-Dharwar	Basic Dykes Porphyry and Felsite dykes and some Granites
Upper Dharwar	Meta-sediments including mica-schists Calcareous rocks and quartzites
Middle Dharwar	Intrusive granite and pegmatite Porphyry dykes and ultramafic dykes Metasediments Conglomerates and quartzites
Lower Dharwar	Champion Gneiss (?) Porphyry and rhyolite Meta-basalts, etc. (amphibolites and hornblende-schists) *Base not seen* Peninsular Gneiss and Charnockite

Though the Peninsular gneisses and charnockites were formerly thought to be younger than the Dharwars, they are now known to be much older. The Peninsular gneisses give ages round 2400–2600 million years. They enclose patches of older schistose rocks, samples of which gave ages of 2950 and 3100 million years. Peninsular gneiss from Lal Bagh at Bangalore gave an age of 2565 millions years. The Closepet granite, which is mainly metasomatic, is rather complex as it contains some inclusions of older gneisses and also later granitic intrusions. Their age is around 2300 million years, but the later intrusives are about 2100 million years old. Dykes of porphyry traverse the Peninsular gneisses. Those at Seringapatam, near Mysore, gave whole rock ages of 1200–1250 million years. The granite massif of Chamundi Hill (near Mysore) is younger, being only about 800 million years. These later intrusives are obviously Post-Dharwar in age. The Peninsular gneisses also give evidence of a metamorphic episode at about 2100 million years, as such an age is indicated by biotite in them in the Lal Bagh exposures. There are also granites of this age in the Godavari Valley. There are numerous basic dykes in the gneisses which, though not dated, must be younger. They may be of Cuddapah age or even much later.

2.4. RAJASTHAN

The dominant physiographic feature of Rajasthan is the Aravalli Mountain chain, which traverses it from near Delhi in the northeast to northern Gujarat in the southwest. As described by HERON (1935), the oldest rocks here are the banded gneissic complex and the Bundelkhand granite. The latter is well exposed in the Berach River Valley. Several samples of this granite from different places have given a good isochron age of 2550 million years. The banded gneissic complex probably contains granitic intrusives of different ages and also encloses patches and streaks of older schists. These are overlain by the Aravalli System, which is exposed mainly in the area east of the mountain axis. It contains different types of schists, some of which are garnetiferous, and there are also some basic igneous rocks, especially in the lower part. The metalavas and amphibolites give an age of 2300–2400 million years. Post-Aravalli granites have ages of 2100 and 1900 million years. The Aravallis have suffered folding at about 1900 million years.

There are clear indications that there are schistose rocks older than the Aravallis. Such rocks, which occur as patches and inclusions in the gneisses near Udaipur, have been called the Bhilwara Series by RAJA RAO (1967). Detrital zircon in some Pre-Aravalli schists gave Pb-isochron age of 3600 million years (VINOGRADOV & TUGARINOV 1964).

The Delhi System succeeds the Aravalli System and forms well marked synclinoria along the mountain axis. It comprises a lower Alwar Series and an Upper Ajabgarh Series, composed of meta lavas, conglomerates and sandstones, pelitic schist and some limestones. The Delhi System is now known to range in age from 1900–1300 million years, but there are indications of intrusive activity at about 1400 million years, which is the age given for the alkali syenite of Kishengarh. The Delhis were subjected to folding at about 1200 million years and again at 900 millions years. These periods are also marked by granite intrusions, the latter being of the same age as the Erinpura granites (937 million years). The Delhis are overlain by the Raialo Series, which are mainly limestones, exemplified by the Makrana marbles near Jaipur.

The youngest group amongst the Pre-Cambrians of Rajasthan is the Vindhyan System, which occupies a vast basin in Central India, as well as in Eastern Uttar Pradesh and Bihar. Some exposures also occur in Western Rajasthan, associated with the Malani rhyolites and ignimbrites, which have an age of about 750 million years. These are the effusive equivalents of the Jalor and Siwana Granites of the same age.

The Vindhyan System occupies a large area around the Bundelkhand granite exposure of Central India. This granite has now been proved to be of the same age (2550 million years) as the Berach granite of Rajasthan, both of which were correlated by the earlier geologists on purely lithological grounds.

The Vindhyans are divided into two major groups. The Lower Vindhyans, well exposed in the Son Valley, are composed of lavas, pyroclastics, sandstones, limestones and shales, dominantly marine in character. Glauconites in the sediments indicate an (K-Ar) age between 1400 and 1100 million years for the Lower Vindhyans. The Upper Vindhyans occupy large areas of Central India and Rajasthan, being dominantly sandstone formations with subordinate shale beds. They form a plateau-like region, with well marked sandstone scarps, these being in fact called the Vindhya Mountains. They are considered as deltaic and floodplain deposits laid down in very shallow waters and are characterized by current bedding. They are divided into three formations – the Kaimur, Rewa and Bhander, the upper two being mainly red sandstones. The Bhander Series in Rajasthan is associated with beds of gypsum, and bears much resemblance to the Purple Sandstone Stage of the Punjab Salt Range in Pakistan. It is considered to be of Lower Cambrian age, because the Purple Sandstones are overlain by Middle Cambrian fossiliferous beds. The Upper Vindhyans are believed to cover the time interval between 1000 and 600 million years. It is of interest to note that the diamond-bearing kimberlite pipe of Panna in Central India, which intrudes the Kaimur Sandstones, has given an age of 1140 million years (K-Ar age for Phlogopite) although the age of glauconite from the Kaimurs is around 950 million years.

There are two other formations about whose ages there was uncertainty till recently. One of these is the Bijawar Series, comprising basic lavas, sandstones etc., which have now been more or less satisfactorily correlated with the Aravallis, as the lavas give ages between 2400 and 2500 million years. The Gwalior Series in and around Gwalior city contains basic lavas, whose age has been satisfactorily determined as between 1850 and 1950 millions years and can therefore be correlated with the lower part of the Delhi System.

2.5. SOUTHERN BIHAR AND ORISSA

A large part of the Ranchi Plateau is occupied by the Chota-Nagpur granite-gneiss. To its south is the Singhbhum shear-zone, which contains the well known copper and uranium deposits, now under active exploitation. South of the shear-zone, which runs east-west in Northern Singhbhum and gradually turns to the southeast in Eastern Singhbhum, is a large area occupied by the Iron Ore Series and by a huge batholithic mass of granite. The Iron Ore Series has been tightly folded along northeast-southwest axis, but it has been bent round parallel to the shear-zone and its vicinity. The Iron Ore Series consists of metalavas at the base, followed by Lower Shales and Sandstones, banded hematite-quartzites and Upper shales. Their age was originally given as between 2100 and 2400 million years, based on the fact that the Singhbhum Granite gave K-Ar ages of 2000–2100 million years (SARKAR et al. 1964, 1969). The age of the Singhbhum granite has recently been revised to 2700 million years, based on new determinations and the Iron-Ore Series has consequently been given an older age of 2700–3000 million years. There are some small exposures of the older Metamorphic Series in South Singhbhum, the exposures being isolated and partly assimilated by the Singhbhum granite. Amphiboles and micas from the Older Metamorphics, which consist of amphibolites, metadolerites, calc-gneisses, chlorite, talc schists, etc. have given ages of 3200 to 3400 million years and these rocks are, therefore, amongst the oldest known in India.

To the west of the Iron Ore Series of Singhbhum is the unit called the Gangpur Series, which forms an anticlinorium plunging eastward under the Iron Ore Series. The Gangpur Series consists of dolomites, limestones and carbonaceous quartzites and phyillites, with basic lavas at the top. These lavas are considered by SARKAR et al. (1969) as the equivalents of the Dalma lavas of Eastern Singhbhum. These authors regard the rocks north of the Singhbhum shear-zone as younger than the Iron Ore Series and 2000 to 1700 million years old, while the Dalma lavas are 1700 to 1600 million years old.

The Singhbhum shear-zone is intruded by soda-granites and grano-phyres, which are of the same age as the Romapahari granite of Mayur-bhanj and the mica-pegmatites of Hazaribagh and Gaya Districts, the age being 950 to 850 million years. The geologic history of the region north and northeast of the shear-zone is yet to be worked out in detail. The mica-pegmatites are the source of the rich muscovite deposits of Bihar and they also contain uraninite, allanite, triplite, beryl, tantalite and lithium minerals. The schistose rocks which are the hosts of the pegmatites should represent sediments laid down in a basin during the period 1500 to 1000 million years ago.

2.6. THE EASTERN GHATS

The Eastern Ghats form a range of mountains extending parallel to the coast in a northeast-southwest direction from the Mahanadi Valley to a little beyond the Krishna valley. They are composed of various types of sediment, which have generally been highly metamorphosed. They consist of quartz and micaschists, manganiferous sediments and crystalline limestones. A conspicuous member of the series is the rock type known as Khondalite, which is a quartz-garnet-sillimanite-graphite schist. The whole series has been intruded by phosphorus-rich granite and by charnockite and locally by alkali-syenites and ultramafics. The khondalites, which are generally rich in alumina, often give rise to ferruginous laterite and bauxite on weathered outcrops. The manganiferous sediments have been converted into a rock called kodurite as a result of assimilation by phosphorus-rich granite. The hybrid rock has often been decomposed near the surface to form secondary manganese ore deposits such as those of the Srikakulam and Vizag Districts.

Age data reveal that the charnockites are probably of different ages 1950, 1840 and 1350 million years. Some calc-granulites have given an age of 1600 million years. There are also rocks in the Eastern Ghats, which give an age of 2300–2400 million years. Several minerals from pegmatites show ages around 1600–1750 million years. It may, therefore, be taken that the period of sedimentation in this region extended from 2400–1800 million years and was interrupted by periods of intrusion by igneous rocks.

Further south in the Nellore and Guntur Districts, there are granites of 2300, 2100 and 1500 million years, the last are closely allied to the mica bearing pegmatites of the Nellore District, which have given ages around 1460 million years.

2.7. Assam

The Shillong Plateau and the Mikir Hills form an isolated Pre-Cambrian province. The rocks resemble those of Bihar and comprise amphibolites, micaschists, sillimanite schists, crystalline limestones and ferruginous rocks. They are probably mainly of Middle and Upper Pre-Cambrian age. The only age data available relate to Mylliam granite, which is a late intrusive, only about 765 million years old. It is believed, on structural and stratigraphic ground, that the Shillong Plateau was displaced eastward through a distance of about 200 km from its original position north of the Bihar-Mica belt, during the Miocene and Pliocene phases of the Himalayan orogeny.

2.8. Cuddapah basin

Proterozoic rocks are exposed in a crescent-shaped basin, extending parallel to the eastern coast, from south of the R. Godavari to near Madras City. It contains two distinct groups (See Table II).

Table II. Cuddapah & Kurnool Systems

Kurnool System (500–3000 m)	Kundair Series Paniam Series Jammalamadugu Series Banganapalli Series
Cuddapah System (6000–6300 m)	Kistna Series Nallamalai Series Cheyair Series Papaghni Series

The older, called the Cuddapah system, comprises four series namely the Papaghni, Cheyair, Nallamalai and Kistna. The Cuddapah rocks overlie the gneisses with a marked unconformity, the lowest beds being conglomerates and quartzites. There are contemporaneous basic lavas and pyroclastics in the lowest division. These have given ages around 1350 million years, which may be taken as more or less the beginning of the Cuddapah period. The sedimentary rocks are mainly quartzites and shales with subordinate dolomitic limestone. Dolorite dykes of probable Kistna age are also found traversing the Cuddapah rocks. As they do not penetrate the younger rocks of the Kurnool System, they are believed to have been intruded towards the close of the Cuddapah period. The Cuddapah rocks of Nallamalai age are penetrated by a diamond-bearing kimberlite pipe at Chelima in Kurnool district. The age of this pipe is indicated as about 1300–1350 million years, but the dolorites which are stratigraphically of the same age give only 980 to 1000 million years. It may be inferred from this that the kimberlite was actually formed in early Cuddapah times, but was intruded into the Nallamalai shales only towards the end of the deposition of the Cuddapahs. The Cuddapah System may, therefore, be taken as spanning the interval between 1400 and 980 million years.

The rocks of the Kurnool System were deposited in the western and northwestern portions of the Cuddapah Basin, after the Cuddapahs were raised up and were subjected to some folding. They are mainly shales and limestones. They are also divided into the four series, the Banganapalli, Jammalamadugu, Paniam and Kundair, from below upwards. Their age seems to be between 900 and 550 million years or thereabouts.

The Eastern margin of the Cuddapah Basin is highly disturbed, showing the gneisses on the east thrust over the sedimentary rocks on the west. The Nallamalai strata along the disturbed zone are highly metamorphosed and sheared. The age of this disturbance is probably Post-Cuddapah and Pre-Kurnool, as the Kurnool rocks have not been affected by it. It is possible that this disturbance coincided with a period of uplift of the Eastern Ghats. The Cuddapah Basin must originally have extended for a considerable distance to the east of its present eastern margin, but the rocks in the eastern portion have been eroded away. There are some indications that the Cuddapah sediments were subjected to mild metamorphism at about 450 to 500 million years ago; this age is indicated by mica in the phyllitic rocks of the Cuddapahs and in the charnockites of the Kondapalli Hills near Vijayawada.

It will be noticed that deposition of sediments in the Vindhyan Basin of north India and Cuddapah Basin of South India were roughly contemporaneous. The Cuddapahs cover the same time span as the Semri Series, while the Kurnools are of the same age as the Upper Vindhyans. During these periods another large sedimentary basin was present in Madhya Pradesh and Orissa. This has been isolated by erosion into comparatively small exposures now found in Raipur, Bastar and Sambhalpur areas. The age of the rocks of the Barapahar Hills of Chhattisgarh, west of Sambhalpur is believed to be Cuddapah, but needs confirmation. Recent work indicates that the rocks of the Raipur and Bastar areas are mainly of Kurnool age, as they bear good lithological resemblance to the Kurnools. The presence of diamonds, derived from the Barapahar area, may possibly be due to rocks of Kurnool age occurring there, for the Kurnool succession shows two intra-formational conglomerate beds containing diamonds.

Rocks of Cuddapah and Kurnool ages also occur along the Godavari Valley, Andhra and Madhya Pradesh. They have been assigned to the Pakhal and Sullavai Formations.

2.9. The himalayan area

Pre-Cambrian rocks are exposed at various places along the Sub-Himalayan zone in the Himalaya. This zone contains several thrust-sheets overlapping each other, all having been thrust from the north to the south in a direction at right angles to the main curvature of the mountain system (see above). They have been studied in some detail only in five or six regions. In the Kashmir-Himalaya they are represented by the highly metamorphosed Salkhala Series and by the younger slate series like the Hazara Dogra and Attock Slates. In the Simla region the Jutogh and Chail Series and the Simla Slates are of Pre-Cambrian age while further east are the Mandhali, Chandpur and the Deoban Series. As in Kashmir these are also associated with gneissic rocks in places. In the Darjeeling region the Daling schists are of this age, while the Darjeeling gneiss is regarded as the granitised representative of the same formations. Similar rocks are also found in Nepal to the west and Bhutan to the east. There are also Pre-Cambrian rocks in the Central Himalaya, but they have not been described in any detail. Heim and Gansser (1939) have described, for instance, certain schistose rocks under the names Martoli Series and Budhi schists in the U.P. Himalaya, to which they assign a Pre-Cambrian age. Some of the peaks like Nanda Devi, Nandakot, etc. are composed of Pre-Cambrian gneisses and schists. Because of the difficulties of terrain and structural complexities, the Himalayan region is still poorly known and it will be many years before reasonably accurate geological maps of the different accessible areas become available.

There are no Pre-Cambrian rocks in the Baluchistan and the Burmese Arcs, but they are known in the Shan Plateau beyond and east of the Tertiary basin of Burma, where they have been described under the names Mong Long schists and Chaung Magyi Series composed of quartzites, limestones and schists of various types. Similar rocks are apparently present on the borders of Afghanistan and Iran, west of the Tertiary basin of Northwest-Frontier Province and Baluchistan.

3. The Palaeozoic Era

Sedimentary rocks of Palaeozoic Era are absent from Peninsular India, except certain continental formations representing the late Carboniferous and Permian periods. These form the lowest part of the Gondwana Formations, which occur in a series of exposures in the east, central and northeastern part of the Peninsula. They are, however, fairly well represented in the Tethys-Himalayan zone, in parts of the Central Himalaya and in the Shan-Tenasserim Belt of Burma.

It is only in Kashmir that fossiliferous Palaeozoic rocks are seen south of the Central Himalayan axis. This is due to the fact that the more or less continuous Tethyan-Himalayan zone has been cut across by the axis of the central Himalaya, which developed during the Tertiary Era as a result of compression of the Tethys Basin between the Indian shield and the southern Asia (see above). The presence of the Kashmir Wedge at the northwestern end of the Indian Shield has been responsible for the bend in the general trend in the north western part of the Himalaya.

3.1. LOWER PALAEOZOIC

Cambrian rocks are well developed in the Punjab Salt Range, which forms the southern limit of the Potwar Plateau in West Pakistan. The Salt Range forms a scarp being composed of Mesozoic and Tertiary rocks, which dip gently to the north to form the Potwar Plateau. The Cambrian rocks generally overlie the Saline Series, which comprises beds of dolomite, gypsum, salt marl and rocks salt with occasional bituminous material. The contact between the Saline Series and the Cambrian beds is generally a plane of overthrust. Although there has been much controversy about the age of the Saline Series, it appears to be late Pre-Cambrian or early Cambrian, as there are some undisturbed contacts between the Cambrian and the Saline Series*.

The Cambrian Beds are about 300 m thick and consist of purple sandstone at the base, overlain successively by Neobolus Beds, magnesian sandstones and salt-Pseudomorph shales. The Purple Sandstones, which are unfossiliferous and appear to be deposits of an arid environment, are strikingly similar to the Bhander sandstone (Uppermost Vindhyan) of Rajasthan. They may be Lower Cambrian in age. The Neobolus Beds contain numerous fossils, particularly Trilobites and Brachiopods, including *Redlichia, Ptychoparia, Neobolus, Lingulella, Mobergia, Orthis*, etc., which indicate a Middle Cambrian age. The magnesian sandstone is poor in fossils, while the topmost beds are shales, which show excellent impressions of salt crystals along bedding planes, because of which they are called Salt-Pseudomorph shales. Similar shales have also been found in the Upper Vindhyans of Rajasthan. The Cambrian rocks are succeeded by tillites of late Carboniferous age.

---

* Microfossils of plants and insects described by SAHANI and MANI indicate, however, Eocene ages for the Saline Series, M. S. MANI.

84

Cambrian, Ordovician and Silurian rocks are exposed in the northern parts of the Kashmir Valley and the Lidar valley, forming a series of anticlines. The Cambrian strata are found along the cores, while the others are exposed on the flanks. Fossils characteristic of the different ages are found in these rocks. They indicate a deep sea habitat in Kashmir, in contrast to the shelf facies found in the Salt Range and an intermediate facies in the Spiti Valley of the Punjab Himalaya.

In the Spiti Valley, the Cambrians form part of a thick sedimentary succession of quartzites, slates, shales and dolomites of a total thickness of 1200 to 1400 m, called the Haimanta System. The lower part of this system, which is 600 to 900 m thick, is unfossiliferous and somewhat metamorphosed. It is considered to be late Pre-Cambrian, but may be partly Lower Cambrian. There may be slight sedimentary break between the Lower and the Upper Haimantas, as the Upper division covers only Middle and Upper Cambrian Beds and contains several zones rich in fossils. Lower Palaeozoic beds are also well developed in Northern Kumaon (Kali River Valley), where they have been described under the names Garbyang, Shiala and Variegated Series. They comprise slates, red and black shales and dolomitic limestone, the last being sometimes a crinoidal limestone. The thickness in the Kali Valley is as much as 6000 m or more. The same beds are also found in the Zanskar Range to the northwest and in Nepal to the southeast. The thickness is variable in different areas because of folding and faulting. Both in Kashmir and Spiti, the Lower Palaeozoics are succeeded by the Muth Quartzites, generally with a distinct unconformity.

Lower Palaeozoic rocks are well developed in the Shan States of Burma, where they have been given local names such as Naung-Kangyi and Hwe-Maung Beds (Ordovician); Mawson, Orthoceras and Graptolite Beds (Silurian). The strata are fairly rich in fossils and are easy to correlate with the formations of Europe.

3.2. Devonian system

The Devonian System is represented in the Himalayan area by the Muth Quartzites, which are generally unfossiliferous and often lie unconformably over the Silurian beds. They are 100–200 m thick, but in parts of Central Himalaya may be as much as 800 m thick. Their base may be upper Silurian or lower Devonian, but they cover most of the Devonian age. The upper part may show dolomite, where the strata attain a large thickness. In parts of Spiti, Kanaur and elsewhere in the Central Himalaya, the Devonian is represented by a richly fossiliferous limestone facies, containing characteristic Brachiopods and corals. This facies extends to Chitral State at the northwestern extremity of the Himalaya and beyond.

In Burma there is a large expanse of a calcareous facies called the

Plateau Limestone in the Shan States. Though fossiliferous, most of the fossils have been spoilt during the dolomitisation of the limestone. This formation covers also the Carboniferous and Lower Permian. There are two exposures near Wetwin, where a limestone facies and a shale facies are separately developed within a few kilometres of each other.

3.3. Carboniferous system

Strata of this age are well developed in Kanaur in the Himalaya, where they are named the Lipak and Po series. They attain a total thickness of 1200 m. The Lipak Series contains the Trilobite *Phillipsia*, the Pteropod *Conularia* and several Brachiopods. The succeeding Po Series contains *Rhacopteris* and other plants in the lower part and several species of Brachiopods (*Productus*, *Spirifer*, etc.) in the upper part.

In Kashmir the System is represented by the Syringothyris Limestone and Fenestella Shales, which contain also other fossils. A full marine development is found in Chitral.

In Burma, the upper part of the Plateau Limestone is of Carboniferous age. In Lower Burma (Tenasserim) the local beds are called the Moulmein Limestone, which contains the characteristic Brachiopods of this age.

Peninsular India contains strata of Late Carboniferous age, these being generally glacial tillites overlain by greenish shales and sandstones. They are called the Talchir Series, which is regarded as the oldest strata of the Gondwana group.

There are also unfossiliferous Palaeozoic strata in Kashmir, and other parts of the Himalaya. A quartzite formation, called the Tanawal Series which is found in Hazara, may be of Upper Palaeozoic age. The Nagthat Series of Garhwal may represent some part of Palaeozoic.

There is a well marked stratigraphic gap in the Peninsula as well as in the Sub-Himalayan region below the Upper Carboniferous. This marks the period of the great Hercynian mountain building movements in Europe and of the commencement of Gondwana Glaciation in the continents of the Southern Hemisphere, including India. Because of this important geological gap we shall deal with the Permian strata in a subsequent section after describing the Gondwana group.

3.4. The gondwana group

This has the status of a Super-Group, as it covers a long period of time, commencing with the Upper Carboniferous and ending with the Lower Cretaceous. The formations are widely distributed in all the Southern continents and have a great deal of resemblance in lithology and fossil content, so that it is thought that the lands in which they occur, viz. Africa, South America, Australia, India and Antarctica formed one large super-continent. Practically all the formations are of fluviatile and lacustrine character and contain both fossilized plants and land animals. There are good evidences to indicate that this super-continent was centered around the South Pole and that the pole migrated within it during most of the time. The Gondwana continent was split into its component parts during the Jurassic and Cretaceous and these moved away from each other until they attained their present positions during late Tertiary times. The Indian Peninsula, which was a part of the super-continent, was originally located in the higher southern latitudes, but drifted to its present position from about the Upper Cretaceous times.

Whatever may have been the distribution of the continental masses in Pre-Permian times, two great supercontinents existed during the Permian. These were Laurasia in the northern hemisphere and Gondwanaland in the southern hemisphere. A large tropical Mediterranean ocean separated the two super-continents. This has been called the Tethys (see above). When it was compressed as a result of the movements of the continents on either side towards each other, mountain ranges were formed along their borders. They now constitute the Pyrenees, Northern Alps, Caucasus, Elburz, Hindukush and Karakoram on the one hand and the Atlas, Apennines, Southern Alps, Hellenic Mountains, Taurus, Zagros, Oman Mountains and the Himalayan system together with Baluchistan and Burmese Arcs on the other hand. The two systems come close together in the Alps and in the Himalayan region where the original Tethys has been obliterated. The present Mediterrenean Sea is a small remnant of that great Mesozoic ocean.

The rocks of the Gondwana group in India are distributed along trough faults which form a triangular pattern. On the northern side is the Narmada and Son Valleys, continuing into the Damodar Valley. On the southwestern side is the Godavari Valley. The third side of the of triangle is the eastern coast along which only Upper Gondwana rocks are found. There is a subsidiary trough inside the triangle along the Mahanadi Valley. It is inferred that the Gondwana rocks originally covered a very large part of this triangular area and also an area to its northeast in Eastern Bihar and Western Bengal, for we find coal-bearing Lower Gondwana rocks along the Siwalik zone, where they occur in thrust-sheets generally overlain by younger rocks. They have been found in southern Kashmir, as well as in a series of exposures in the Himalayan foothill zone from Nepal to Assam.

The Gondwana formations of India are subdivided as shown in Table III. The main divisions are shown on the left hand side and some of the equivalents on the right hand side.

Talchir Series: The Talchir formations begin with a glacial tillite, which is found at the base of the Gondwanas in several areas. The tillite consists of a mixture of pebbles and boulders of different sizes, held together by a matrix of fine sand, silt and clay. Some of the boulders as

Table III. Classification of the Gondwana Group

Standard Division	Major Indian Divisions	Equivalents
Cretaceous	Umial Series	Gollapilli-Raghavapuram-Tirupati; Athgarh; Sriperumbudur, etc.
Jurassic	Rajmahal-Jabalpur Chaugan Kota	
Triassic	Mahadeva Panchet	Pachmarhi; Maleri Bhimaram Yerrapalli
Permian	Raniganj Series Barren Measures Barakar Series Karharbari Series	Kamthi; Himgir; Chintalpudi Ironstone Shales
Upper Carboniferous	Talchir Series Talchir Tillite	

well as the basement gneisses show characteristic striae due to movements of rock fragments over the basement during the flow of glaciers. The tillite is generally overlain by fine dark green to greenish-brown sandstones and shales, which constitute the Talchir formations. The upper part of this shows a few plant impressions of *Gangamopteris, Noeggerathiopsis, Vertebraria*, etc. They are the earliest plants of the *Glossopteris*-flora.

Tillites have also been found at the base of the Permian strata in the Pubjab Salt Range and at a few places along the eastern part of the Himalayas. The Blaini Bounder beds in the Simla Hills and the Tanakki Bounder Bed in Hazara are also taken to be of the same age.

Marine Permo-Carboniferous beds of comparatively small thickness have been found near Umaria and Manendragarh in Central India, directly overlying the Pre-Cambrian rocks and underlying Lower Gondwana formations. They contain *Productus, Spiriferina, Reticularia, Eurydesma, Aviculopecten*, etc. Similar beds occur also in Sikkim and the Assam Himalaya, either alone or in association with Lower Gondwanas. In Rajasthan (Jaisalmer), the Birmania Formation, which is of Permo-Carboniferous age, overlies some unfossiliferous sedimentary strata, which are believed to be Lower Palaeozoic. These can be connected up with the occurrences of the Salt Range, which were formed along the same coast to the west of the Pre-Cambrian shield of Rajasthan.

Karharbari Stage: This stage is well developed in the Giridih coalfields in Southern Bihar, where it contains workable coal seams. This formation is found also in several other coalfields. It is only slightly younger than the Talchir stage and contains some species of *Glossopteris* and *Schizoneura* in addition to those found in the Talchirs. Their equivalents in Southern Kashmir are the *Gangamopteris*-Beds, which are associated with pyroclastics and agglomerates. The *Gangamopteris*-beds also contain some amphibian and fish remains.

Barakar Series: The succeeding Barakar Series is extensively developed in all the above mentioned Gondwana troughs as well as along the Himalayan foothills. It is about 1000–1200 m, thick and always contains coal seams as well as streaks and pockets of carbonaceous material. It was formed in fresh-water lakes and river basins under conditions of slow sinking and uplift. The Jharia coalfield in Bihar contains at least twenty-five coal seams, deposited during repetitive cycles of sedimentation. The proportion of coal to the full thickness of the strata is of the order of 8 to 12%. Careful comparison of the succession in adjoining basins shows that each basin was an independent area of sedimentation. The coal seams are of good bituminous quality but generally high in ash, the minimum ash content being about 8 or 9%.

Barren Measures: The Barren Measures succeed the Barakars and attain a thickness of over 300 m in the Jharia coalfields. They also occur in the Raniganj coalfields, but contain beds of siderite, which have been converted into limonite when altered near the surface. These ironstone shales were formerly used as iron ores in the smelting furnaces at Kulti.

Raniganj Stage: This is best developed in the Raniganj coalfield, which is the eastern-most in the Damodar Valley. The strata are similar to the Barakars and contain several coal seems in the Raniganj and Jharia fields, but become progressively barren further westwards, where they are represented by reddish sandstones. This series is over 1000 m thick and the sandstones are somewhat finer grained than those of the Barakars.

Beds of the same age are known as Kamthi near Nagpur, Himgir in northwest Orissa and Chintalpudi in the Godavari Valley. The Raniganj beds contain numerous plant fossils and also some remains of amphibia.

Panchet Series: These are mainly sandstone formations of Lower Triassic age. They are best developed in the Panchet Hill near the Raniganj coalfield. Some shales occur in the Upper part. They have yielded remains of amphibia and crustacea.

Mahadeva Series: This succeeds the Panchet stage, but is well developed in the Pachmarhi Hills of Madhya Pradesh. It is about 760 m thick and consists of massive red and buff sandstone with clay beds near the top and bottom, and thin layers of platy hematite at various horizons. There is no carbonaceous matter although some leaf impressions are present.

Elsewhere this is represented by the Maleri, Bagra and Denwa Beds.

The Maleri Beds have yielded remains of reptiles and fishes which show that they are of Carnic to Noric age. These are succeeded by the Kotah Beds whose dinosaurian and fish fossils indicate Lower Jurassic age.

Rajmahal Series: Rajmahal strata are well developed in the Rajmahal Hills at the head of the Ganges Delta, where they are intercalated with the basaltic flows of the Rajmahal Trap. The traps have a thickness of about 500–600 m and are probably of Upper Jurassic age. The sedimentary strata are rich in plant fossils, which belong to the *Ptilophyllum*-flora and may be of late Jurassic to early Cretaceous age.

Umia Series: This is developed in Kathiawar (Saurashtra) and Kutch, where an area of 2500 km² are occupied by these rocks as well as the Dhrangadhra sandstones, which are slightly older and equivalent to the Jabalpur Beds. They are mainly sandstones and shales containing some plant fossils like *Ptilophyllum*. They are overlain by the Wadhwan sandstones, which may be of Middle Cretaceous age. The Umia Series is composed largely of Barren Sandstones, but contains some intercalations showing *Trigonia* and other fossils. In the Upper portion there are plant bearing beds, followed by some marine beds containing Aptian Ammonites. The continental sedimentation of the Upper Gondwana period is terminated by the marine incursion of Aptian age.

Along the eastern coast there are four or five distinct patches of Upper Gondwana age, corresponding mainly to the Rajmahal stage and somewhat younger. These areas are near Athgarh (near Bhuvaneshwar) in Orissa, near Raghavapuram in the Godavari Valley, near Vemaveram in the Guntur District, near Sriperumbudur in Chingleput District and near Sivaganga in Ramnad District in Madras. The strata are usually described under three divisions. They are all believed to be of Lower Cretaceous age and in some cases show marine intercalations. The presence of these strata along the eastern coast of India leads to the inference that this coast began to take shape in the Upper Jurassic, when the Rajmahal traps were errupted following the initial disruption of Gondwana land. Ceylon must also have been temperarily separated from the Indian Peninsula at this time for Upper Gondwana rocks of similar character are found near Tabbowa and Andigama near the northwestern coast of the Island.

The Gondwanas are the chief storehouse of coal deposits in India. The total resources of coal of India are of the order of 100,000 million tonnes, more than three quarters of which is of Gondwana age. As already stated, these deposits are confined to the northeastern and central parts of the Peninsula.

3.5. The permian system

Permian strata are of continental character in the Peninsula, except in two or three places in Madhya Pradesh, where there are marine intercalations. They are represented by the Lower Gondwanas – the Barakar Barren Measures and Ranigunj Series – which have already been described. Marine Permian strata are well developed in the Tethys-Himalayan zone all along the Himalaya. In Kashmir they are called the Zewan Beds, while in the Salt Range they form the *Productus*-limestone, which are exposed as scarps at several places. They are rich in characteristic fossils, such as Brachiopods, Gastropods, bivalves and Cephalopods. Similar strata are found in the Spiti Valley of the Punjab Himalaya, where they are called the Kanawar System. They have been described also from the Mr. Everest region, Sikkim and parts of Assam.

4. The Mesozoic Era

4.1. TRIASSIC SYSTEM

The Triassic, the lowest of the three systems in the Mesozoic, has been studied in several places along the Himalaya. It is certainly present in the Burmese Arc, but has not been studied because of the inaccessibility of the terrain. It is well developed in some places in the Baluchistan Arc, although to a less extent than the younger systems.

Excellent sections of the Triassic have been studied near Lilang in the Spiti Valley and at Painkhanda and Byans. Near Lilang the system attains a thickness of over 1200 m, nearly 500 m of this being of Rhaetic age grading into the Lias. All the divisions namely the Scythic, Muschel-kalk, Ladinic, Carnic, Noric and Rhaetic are well developed and highly fossiliferous, because of which they are easily correlated with similar formations in other parts of the world. The thickness of the different division varies considerably in different areas. In the Salt Range the facies indicates shallow seas so that the fossil assemblage is different and the thickness is much less. In the Baluchistan Arc Triassic rocks attain a large thickness, locally being represented by greenish shales and dark limestones. On the Burmese side, the Himalayan facies is known to be present in the Indo-Burma frontier Ranges, but details are lacking. On the Shan Plateau, however, they belong to a different basin of deposition, the fossils being rich in Lamellibranchs. These are called the Napeng Beds that continue into Thailand.

4.2. JURASSIC SYSTEM

The Jurassic System is quite well developed in the Extra-Peninsular region. In most of the Peninsula it is developed as part of the Upper Gondwanas. In Kutch and Rajasthan, they are well developed as a marine sequence, especially from the Callovian upwards. The Lower part of the Jurassic beds in Kutch are shallow coastal deposits containing *Trigonia*, *Corbula* and some corals. They are succeeded by excellently developed marine rocks, mainly limestones and shales, with rich ammonite fauna, which can be correlated almost zone by zone with those occurring in Madagascar and East Africa. The Jurassic Strata attain a thickness of nearly 2000 m and continue into the Cretaceous.

Rajasthan and the western part of the Salt Range also contain a similar facies, but less uniformly marine. Some exposures are found in Baluchistan also. There is an excellent development of the Jurassic rocks in the Tibetan zone, these apparently extending over large areas of Tibet. The well developed Kioto limestone of Spiti, which is mainly Rhaetic, extends into the Lower Jurassic. There is a break in sedimentation below and above the Callovian. The strata marking part of the break being a black ferrugenous colite, characterized by *Belemnopsis sulcacutus*. The beds above these are a very conspicuous shale facies exposed over great length in the Himalaya and their eastern extensions. They are the well known Spiti Shales, ranging from Upper Oxfordian to the Lower Cretaceous. They are rich in fossils, which are enclosed in nodules and are carried down the different rivers, reaching the plains through Uttar Pradesh and Nepal*. The Spiti Shales are also exposed over the Triassic rocks in the Mount Everest region and further east. They may be present also in the Burmese Arc. On the Shan Plateau, they are represented by the Namyau Beds and Loian Series the former of which is a shallow marine facies, while the latter is continental and plant-bearing.

4.3. Cretaceous system

Strata of this system are found both in the Peninsula and in the Extra-Peninsula. In the latter, they are partly of the flysch facies, associated with volcanic rocks, as the Tethys Basin began to experience compression at this period. Part of the Baluchistan Arc probably formed an island Arc, for a large part of the Cretaceous strata is intercalated with pyroclastics and tuffs, while the upper part shows typical association with radiolation cherts, ultramafics and basalts.

Marine Cretaceous rocks are found at several places in the Tibetan zone. They are represented by the Giumal and Chikkim Series in Spiti; by limestones, sandy shales, radiolarites and volcanic rocks in Northern Kumaon and adjacent parts of Tibet. In the Hundes region of Tibet pink and white limestones, representing the whole succession from the Permian to the Cretaceous, are found in thrust-sheets, but not *in situ*. This facies has an extraordinary resemblance to the rocks in the Eastern Alps of Australia, both in lithology and in fossil content. It is obvious that the original exposures of this facies has been covered over as a result

---

* They are the famous saligram used in worship by devout Hindus. M. S. Mani.

of the intense folding and compression of the Himalayan region, only a small part of it being brought to the surface in a thrust-sheet.

A fully Cretaceous succession is found in central and Eastern Tibet, overlying the Jurassic rocks near Kampa Dzong. They have been called the Kampa System, which include the Giri Limestone, Kampa Shale, the Hemiaster Beds, Scarp Limestone and Tuna Limestone which attain a total thickness of 500 m and pass upwards into the Eocene.

In Baluchistan, the Cretaceous rocks occur in a calcareous facies along the main arc, whereas a shaly and sandy facies is found to its west and northwest. The Samana Range shows sandstones and limestones with Brachiopods, Mollusca and a few Ammonites. In the calcareous zone the strata are better developed and comprise shales, sandstones and limestones, the last being dominant. Most of the strata are fossiliferous and some are intercalated with ash-beds and pyroclastics. At the top are found serpentines associated with chromite deposits.

Cretaceous rocks occur also in the Arakan region, as well as in the Shan States. In northern Burma there are Cretaceous limestones and shales in association with serpentines and Jadeite bearing rocks.

Marine Cretaceous basins, in which a fairly full succession is found from the Albian upwards, occur in southern Madras and in the Narmada valley. These two basins show different fossil assemblages, but in the Upper Cretaceous they show much greater resemblance apparently because a marine connection was established around Cape Kumarin when India drifted northwards. The fauna of the Trichinopoly (South Madras) Cretaceous is more related to that of Madagascar, Natal and West Australia than to the Tethyian region. The mingling of the two faunas and homogenization took place only during the Tertiary Era.

4.4. DECCAN TRAP

Towards the end of the Cretaceous, Western and Central India were covered by large sheets of basic lava. These formations are called the Deccan Trap. The volcanic activity continued into the Lower Eocene and there are indications of renewed activity about the end of the Eocene, particularly in Western India. The lava flows have a total thickness of about 1800 m along the Bombay coast and much thinner along the margins in Central, Eastern and Southern India. They are believed to have been erupted through a series of fissures in the crust and their large extent may be attributed to the high temperature and a large degree of super-heat possessed by lavas during eruption. The lower parts of the Trap contain a few sedimentary beds laid down in lakes during the cool intervals between eruptions. These sedimentary beds contain plant remains, algae and Mollusca. Ash Beds are more common in the Upper Traps which also contain differentiated types such as rhyolite, pitchstone, porphyry, and ultramafic types such as limburgites, oceanites and ankaramites.

The individual Trap flows may vary in thickness from a few metres to 30 m or so. They now form an elevated country with characteristic features namely a series of horizontal platforms with step-like edges. The flows are generally somewhat vesicular at the bottom and at the top. The top often contains a lateritised clayey material. The vesicles contain secondary minerals like chalcedony, agate and other forms of silica, and zeolites. Some of these, especially when found in cavities in the trap rocks, are of gem quality like amethyst, carnelian, rock crystal, etc. and beautifully crystalised zeolites.

The Deccan Traps are to a large extent extra-ordinarily uniform in chemical composition, being doloritic to basaltic. Other petrologic types are found only locally. On weathering, they give rise to beds of bauxite and ferruginous laterite at the top. The majority of the bauxite deposit in India are derived from the trap rocks.

5. The Tertiary Era

The Tertiary Era was marked by great crustal disturbances in the Himalayan region, as the Tethyian Basin was compressed and raised up into mountain chains, which became the Himalayan system. The initial disturbance was experienced at the end of the Cretaceous. Others followed in the Upper Eocene, in the Middle Miocene and in the Plio-Pleistocene (see above). These are reflected in the breaks in the sedimentation at these periods. The sea retreated from the Himalayan area at the end of the Eocene, for marine Eocene rocks are found only along the southern border of the Himalaya from Kashmir to Uttar Pradesh. Some Eocene rocks are found in the Tibetan zone, but they are mainly fluviatile sandstones and conglomerates. Along the two lateral arcs, however, there existed gulfs, which were gradually pushed back to the south by sediments deposited in them during the whole of the Tertiary period. From the Oligocene onwards a deep furrow was developed between the Peninsula and the Himalayan region, and this was gradually filled up by brackish and fresh-water sediments of great thickness. These formations enclose the remains of a very rich mammalian fauna, most of which suffered extinction at the onset of glaciation in the Pleistocene. The Tertiary formations are developed in the Himalayan, Baluchistan and Burmese arcs, in the Andaman and Nicobar Islands, in the Ganges-Brahmaputra Delta south of the Assam Plateau, in the Indus Delta, in Kutch, Kathiawar and along the coasts of the Peninsula.

5.1. EOCENE SYSTEM

Marine Eocene rocks show excellent development along the southern

94

border of the western Himalaya and along the two lateral arcs. They are generally richly fossiliferous and contain a variety of fossils, the most important groups being Foraminifera, Echinoids and Mollusca. They are divided into three major formations – the Ranikot, Laki and Kirthar Series. Where they are represented by estuarine or fluviatile facies, they contain carbonaceous shales with lignite and some coal of low rank, as also pyritous shales and in some places evaporites including gypsum and rock salt, as in the Tras-Indus Salt-Range. Eocene lignite and coal deposits are found in Upper Assam along the southern border of the Assam Plateau, in the Salt-Range and in parts of the Baluchistan Arc. The marine deposits are mainly limestones with subordinate shales and sandstones. The Eocene is the most important petroleum-bearing formations of the Indian Region. Although some of the Eocene rocks contain petroleum deposits, the oil has in some cases migrated to younger formations during later earth movements. In the region of Cambay, Ahmedabad, Surat and Broach in Western India, a full Tertiary succession has been encountered in bore holes above the Deccan Trap. Eocene is about 500 m thick and is oil-bearing. Oil fields have been developed near the above mentioned places during the last decade and are now producing over three million tonnes per annum. The Eocene is overlain by comparatively thin Oligocene strata. These are overlain by thick Miocene, Pliocene and Pleistocene, with an aggregate thickness of 1500 m. In Assam, the Eocene is represented by the Jaintia Series and part of the Barail Series. The latter contains both coal and petroleum deposits in different areas, the new oil fields of Nahor-Katiya, Moran, etc. being in the Barails. A well marked unconformity throughout Assam occurs above the Barails in the Oligocene. These rocks continue into the Bengal basin, lying under several thousand metres of younger rocks. Burmese Eocene in the Chindwin and Irrawady valleys is more or less similar to that of Assam. It consists of fresh-water sediments in the Lower portion becoming completely marine in the Upper portion. Eocene is succeeded by the Pegu Series of Oligocene and Lower Miocene ages, which is also petroliferous. Eocene Strata have been found also in the Cauvery Basin in Southern Madras, following the Cretaceous beds.

5.2. Oligocene and lower miocene

At the end of the Eocene important crustal movements took place in the Extra-Peninsular region. These were followed by the deposition of Oligocene and Lower Miocene strata. In the Himalaya, these are represented by the Murree Series, which are red sandstones derived from a well oxidised terrain. In the Baluchistan Arc, the Nari and Gaj Series represent these divisions. They are composed mainly of limestones and sandstones containing rich fauna. Strata of the same age show a flysch facies in Baluchistan beyond the mountain ranges of the Sind-Baluchistan frontier. These are generally unfossiliferous, greenish sand-stones and shales, called the Khojak formation. On the Burmese side also, the Oligocene is well developed forming part of the Pegu Series.

In most of the other regions in India, the Oligocene shows a regressive phase and the marine strata are comparatively thin. Such is the case in Western India and along the coasts. In Assam, the Upper part of the Barail Series and the Surma Series are of Oligocene age.

5.3. MIDDLE MIOCENE TO LOWER PLEISTOCENE

The third phase of the Himalayan orogeny took place about the Middle Miocene. The changes in Himalayan region were spectacular as the major part of the region was raised into high mountains, with the accompaniment of intrusions of great batholiths of granite and grano-diorite. It was during this period that the Assam Plateau, which lay in the region of Eastern Bihar was moved to its present position along the Dauki Fault, which is continued to the northeast by the Haflong-Disang Thrust. It was after the uplift of the Himalaya that the Indo-Gangetic basin received sediments from the north, these constituting the Siwalik system. This basin became deeper but was gradually filled up from the Upper Miocene onwards.

The Siwalik system is predominantly formed of conglomerates, sand-stones and silts and attains an aggregate thickness of almost 5000 m. They are fresh-water deposits, contributed by numerous rivers which flowed in directions transverse to the newly risen mountain chains, so that the individual beds are of uneven thickness and show prominent current bedding. The basin of deposition extended from the north-eastern corner of Assam to Jammu and the Potwar Plateau and ap-parently consisted of a series of large lakes and flood-plains, which were connected with each other. The whole region must have been covered by tropical forests and marshes with lush vegetation, which was well suited to support a rich Mammalian fauna of great diversity. This fauna devel-oped from the Eocene onwards and attained its greatest development during the Miocene and Pliocene. Much of this fauna became extinct with the onset of glaciation in the Pleistocene.

The Siwalik System is divided into the Kamlial, Chinji, Nagri, Dhok Pathan, Tatrot and Pinjor stages, followed by the Boulder-Conglom-erates of Lower Pleistocene age. Some of these stages contain red sandstone and shale indicating that the sediments were well oxidised. In addition to mammalian fossils, they contain also silicified wood in some horizons. The Chinji stage contains, besides some mammals, several genera of reptiles. The richest horizons of mammalian remains are in the Nagri and Dhok Pathan Stages. Primates appear in the Nagri stage and also several genera of the pig family. Proboscideans, giraffes and bovids are found in great abundance in the Dhok Pathan Stage. There was a period of uplift and erosion just after the Dhok Pathan period. The succeeding Tatrot period was one of good rainfall. At the end of the Pinjor period there were further earth movements. The Pir Panjal was apparently raised up during this time. The strata deposited thereafter are coarse conglomerates mixed with boulders, which must have been laid down by glaciers, in part. This was the beginning of Pleistocene glaciation.

The Siwaliks are represented in Sind by the Manchhar Series, which is mainly composed of conglomerates, sandstones and brown clay. In Southern Baluchistan they are called the Makran Series, which is mainly of marine character. Siwalik strata are found all along the foothill zone of the Himalaya, while beds of the same character constitute the Irrawaddy System of Burma and the Disang Series of Assam. These are also mainly sandstone formations with clay beds.

Along the coast of the Peninsula the Upper Tertiary Beds are called Warkalli (Varkala) Beds, the Cuddalore Sandstone and the Rajamahendri Sandstone. They contain beds of lignite in some places e.g. near Calicut, Varkala, Cuddalore and Pondicherry. Recent borings for petroleum both in the Ganges delta and in the Cauvery delta have revealed the existance of a full Tertiary succession, as indicated by the fossil remains. Mio-Pliocene Beds occur also along the northwest coast of Ceylon, so that it is clear that the Gulf of Mannar between Madras and Ceylon is underlain by large thickness of Tertiary sediments, amounting to 5000 m or more.

6. Pleistocene and Recent

Strata of these ages are widespread along all the great river valleys and in the deltas of the larger rivers. They form the older and newer alluvium of Indo-Gangetic basin and the alluvial deposits in the deltas and elsewhere. They are composed of sands and clays and are the chief agricultural areas of the Indian region. The alluvial deposits have different thicknesses in different areas, depending upon the amount of sediments, which have accumulated during the last two million years or so. In some places they contain peat beds. The sand layers are excellent aquifers. At a few localities Pleistocene fossils have been found. There are evidences of the existence of early man in the Potwar Plateau of Pakistan, in the Kashmir Valley and in a few places in Burma and in Indian Peninsula.

The final uplift of the Himalaya took place sometime during the Pleistocene. The Karewa formations of Kashmir have been lifted and tilted, as seen on the flanks of the Pir Panjal Range. Along the Central Himalaya this uplift was accompanied by the intrusion of great masses of white tourmaline-bearing granite.

Many changes are taking place in the configuration of the region at the present day. The Indus and the Ganges-Brahmaputra deltas are reclaiming land from the sea. Even the smaller deltas like those of the Narmada, Mahanadi, Godavari, Krishna and the Cauveri are also growing and many former settlements on the sea-coast have become buried under alluvium, on the other hand some parts of the coast have experienced erosion and the sea has encroached over them.

REFERENCES

COATES, J. S. 1935. Geology of Ceylon. *Ceylon J. Sci.*, (B) 19(2): 101–187.

CRAWFORD, A. R. 1969. India, Ceylon and Pakistan: new age data and comparisons with Australia. *Nature*, 223(5204): 380–384.

DE TERRA, H. 1936. Himalayan and Alpine orogenies. XVI internat. geol. Congr., 2: 859–871.

DUNN, J. A. 1939. Post-Mesozoic movements in the northern parts of the Peninsula. *Mem. geol. Surv. India*, 73: 137–142.

FERMOR, L. L. 1930. On the age of the Aravalli Range. *Rec. geol. Surv. India*, 62: 391–402.

FOX, C. S. 1931. Gondwana System of India. *Mem. geol. Surv. India*, 58.

GEE, E. R. 1926. The geology of the Andaman and Nicobar Islands. *Rec. geol. Surv. India*, 59: 208–232.

HEIM, A. 1956. The geological structure of the Himalaya compared with the Alps. *Proc. nat. Inst. Sci. India*, 22(A): 228–235.

HEIM, A. & A. GANSSER, 1939. Central Himalaya, geological observations of the Swiss Expedition 1936. *Denkschr. schweiz. naturf. Ges.*, 73(1).

HERON, A. M. 1934. Sketch of the geography and geology of the Himalaya Mountains and Tibet. Part 4 2nd Ed. DEHRA DUN.

HERON, A. M. 1935. Synopsis of the Pre-Vindhyan geology of Rajputana. *Trans. nat. Inst. Sci. India*, 1(2).

HOLLAND, T. H. 1900. The Charnockites series. *Mem. geol. Surv. India*, 28(2).

KRISHNAN, M. S. 1952. Geological history of Rajasthan. Proc. symp. Rajasthan Desert, Nat. Inst. Sci. India, New Delhi, pp. 27–29.

KRISHNAN, M. S. 1953. The structural and tectonic history of India. *Mem. geol. Surv. India*, 81: 137, fig. 21.

KRISHNAN, M. S. 1968. Geology of India and Burma Madras: Higginbothams.

PASCOE, E. H. 1950–1961–1965. Manual of Geology of India and Burma, 3 vols. Calcutta: Government of India Press.

RAO, RAJA, C. S. 1967. On the age of the Pre-Cambrian group of Rajasthan. *J. Min. Met. & Fuels*, Calcutta, 15(9): 306–309.

RAO, RAMA, B. 1940. The Archaean Complex of Mysore. *Bull. Mysore geol. Depart.*, 17.

RAO, RAMA, B. 1964. Handbook of the geology of Mysore. Bangalore: Bangalore Press.

SARKAR, S. N., A. POLKANOV, E. K. GERLING and F. V. CHUKROV, 1964. Precambrian geochronology of Peninsular India a synopsis. *Sci. & Cult.*, 30: 527–537.

SARKAR, S. N., N. K. SAHA and J. A. MILLER, 1969. Geochronology of the Precambrian rocks of Singhbhum and adjacent regions, Eastern India. *Geol. Mag.*, 106(1): 15–45.

VINOGRADOV, A. P. and A. I. TUGARINOV 1964. Geochronology of the Indian Pre-cambrian. XXII internat. geol. Congr. Rep.

WADIA, D. N. 1931. The syntaxis of the Northwest Himalaya, its rocks, tectonics and orogeny. *Quart. J. geol. min. Soc. India*, 65: 189–220.

WADIA, D. N. 1938. Progress of geology and geography in India during the last twenty-five years. In: Progress of Science in India during the Past twenty-five years. Edited by B. PRASHAD. Indian Sci. Congr. Assoc. Silver Jubilee 1938. Calcutta. pp. 86–132.

WADIA, D. N. 1942. The making of India. *Sci. Cult.*, 7 (Suppl.): 1–10.

IV. WEATHER AND CLIMATIC PATTERNS

by

L. A. RAMDAS

1. Introduction

This chapter contains a brief outline of the salient features of the weather and climatic patterns in India, with a special reference to the behaviour of the monsoon and the incidence of other major phenomena like cyclonic storms, depressions, thunderstorms, etc., which also influence the seasonal and regional distribution of the annual rainfall in India. Some reference is also made to phenomena like cold and heat waves, frost, high wind, hailstorm, etc. (GOVINDASWAMY 1953, RAMDAS 1946, RAMDAS et al. 1954, RAMDAS 1960, 1968).

The systematic recording of daily observations of meteorological factors with standard instruments was commenced in India only by 1875 and the network of meteorological stations, installed in the seventies of the last century has been steadily growing. These stations supply the data for the preparation of weather charts, on the basis of which the daily forecasts of weather are issued. The work of these surface observatories has been supplemented since the twenties of the present century by a growing network of upper-air observatories, which send up hydrogen-filled baloons for recording the wind velocities, temperature and humidity of the atmosphere up to sixteen kilometres or more above the surface. The estimates of climate, based on these observations, represent the general or macroclimate or the weather phenomena in several kilometres of the troposphere. The first two metres immediately above the ground are usually avoided in recording the macroclimatic data, on account of great disturbances or turbulent fluctuations of air density, temperature, humidity and wind velocity gradients occurring in the surface layer of air. In recent years considerable attention is also being paid to a careful study of this disturbance-zone, close to the ground surface, leading to the development of an important branch of microclimatology. Microclimate deals with the large vertical gradients of temperature, wind, humidity, etc. of the layer of air, which is close to the ground and in which man, animals and general vegetation thrive. In the following pages, we shall first briefly outline the general macroclimatic or the large-scale weather phenomena and then discuss some of the outstanding facts of microclimate.

The Indian region is remarkable in having for its northern boundary the highest and the most extensive mountain system in the world, viz. the Himalayan System. The Himalaya obstructs the moisture-laden winds

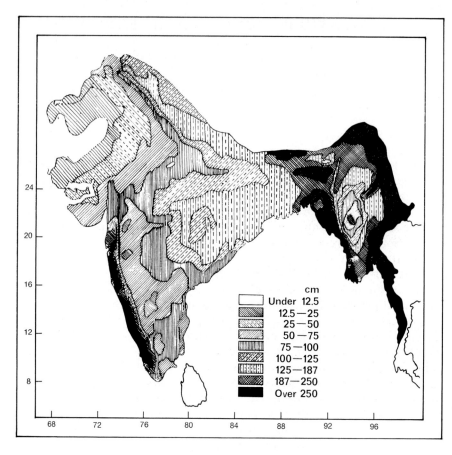

Fig. 7. Rainfall distribution map of India.

from the south, causing them to shed their moisture as copious rainfall along the submontane areas, north of the Indo-Gangetic Plains, and as snowfall on the mountains further north. This great mountain barrier is also equally effective in protecting India from the direct invasion of extremely cold winds from the north temperate and frigid regions of Tibet, Mongolia and Siberia. The Peninsular India, flanked by the Arabian Sea on the west and by the Bay of Bengal in the east, abuts into the vast seaboard of the Indian Ocean. The Peninsula is also flanked by the Western Ghats along its West Coast and though less effectively by the Eastern Ghats along its east Coast. These two minor barriers play a most significant rôle in ensuring plentiful orographic rains on their windward sides.

100

2. Normal Seasonal and Annual Rainfall

The general pattern of distribution of normal annual rainfall is shown in fig. 7. The areas of very heavy rainfall of 250 cm or more a year are found on the windward sides of the Western Ghats and in the hills of Assam. Along the submontane tracts of the Himalaya the precipitation may be as high as 287.5 cm. These areas are the watersheds from which the major rivers arise. In the Deccan Plateau, the plains of North India and of the South, the effects of orography are less pronounced or even nearly completely absent and the rainfall in these areas is only moderate. Most of Northwest India, the desert tracts of Rajasthan and the adjoining parts of Pakistan constitute the arid zone of our region (see Chapter XI).

Meteorologists recognize the following four seasons in India: 1. winter from December to February, 2. summer or pre-monsoon from March to May, 3. southwest monsoon from June to September and 4. post-monsoon from October to November. The normal rainfall, in different seasons during the whole year in thirty typical rainfall subdivisions are summarized in table I. In columns 2 to 5 the figures in parenthesis are the seasonal rainfall, expressed as percentages of the total annual rainfall. The data presented in the table show that the monsoon season represents the most important rainfall period over the largest part of the country. Kashmir, Baluchistan, the old Northwest-Frontier Province and Madras southeast are, however, exceptions to this conclusion; in the extreme north, a good part of the annual rain is received during winter, while in southeast Madras nearly half the annual rainfall occurs during the post-monsoon (the socalled northeast monsoon) period, after September.

The major phenomenon of the monsoon thus dominates the entire country from June to September. We shall discuss briefly below its characteristics, from year to year, over a long period of years. We shall also refer to other phenomena of relatively shorter duration, but often quite violent, such as, for example, cyclonic storms of the pre- and post-monsoon periods, the depressions characteristic of winter and those that occur even during the monsoon and contribute to bring about an equitable distribution of the non-orographic type of rainfall in the interior of the country.

3. The Indian Monsoon

What we generally call the Indian monsoon is the result of the influence of the characteristic layout of the mountain systems on the pattern of wind circulation (SIMPSON 1921). The mean air-flow over India and the surrounding areas near the surface during July, when the monsoon is at its height, is shown in fig. 8. SIMPSON (1921) has rightly emphasized the importance of the rôle played by the general pressure distribution over Asia in summer and the building up of the main monsoon current from

Table I. Normal seasonal rainfall in mm in thirty rainfall-subdivisions

Subdivision	Winter		Summer		Monsoon		Postmonsoon		Annual
1. Assam	60.45	(2.4)	636.52	(25.7)	1632.20	(65.8)	151.38	(6.1)	2480.55
2. Bengal	38.83	(2.0)	315.47	(16.5)	1422.65	(74.5)	131.32	(6.1)	1908.30
3. Orissa	46.23	(3.2)	142.75	(9.9)	1130.05	(78.2)	126.49	(8.8)	1445.52
4. Chota-Nagpur	65.28	(5.0)	92.46	(7.1)	1084.83	(83.4)	57.40	(4.4)	1299.98
5. Bihar	35.81	(2.9)	83.82	(6.8)	1040.38	(85.0)	64.52	(5.3)	1224.53
6. *Uttar Pradesh (E)	38.86	(3.9)	28.45	(2.9)	874.78	(88.0)	51.82	(5.2)	993.91
7. Uttar Pradesh (W)	57.66	(6.0)	34.54	(3.6)	837.69	(87.8)	24.64	(2.6)	954.53
8. Punjab (E & N)	70.10	(11.9)	48.01	(8.1)	463.04	(78.4)	9.40	(1.6)	590.55
9. Punjab (SW)	32.51	(13.7)	34.54	(14.5)	167.13	(70.4)	3.30	(1.4)	237.48
10. Kashmir	231.65	(22.1)	230.89	(22.0)	563.63	(53.7)	23.88	(2.3)	1050.05
11. Northwest-Frontier Province	85.34	(20.0)	106.17	(24.9)	219.71	(51.5)	15.75	(3.7)	426.96
12. Baluchistan	88.90	(45.6)	51.56	(26.4)	48.01	(24.6)	6.60	(3.4)	195.07
13. Sind	17.02	(10.4)	10.41	(6.4)	134.11	(82.4)	2.03	(1.2)	163.57
14. Rajasthan (W)	15.75	(4.8)	14.22	(4.3)	298.20	(90.0)	3.05	(0.9)	331.22
15. Rajasthan (E)	24.38	(3.8)	29.82	(3.1)	581.91	(90.9)	13.97	(2.2)	640.07
16. Gujarat	5.59	(0.7)	6.10	(0.7)	799.08	(96.2)	19.56	(2.4)	830.33
17. Central India (W)	21.59	(2.5)	11.94	(1.4)	801.62	(93.8)	19.05	(2.2)	854.20
18. Central India (E)	36.58	(3.7)	20.07	(2.0)	890.27	(90.9)	33.02	(3.4)	979.94
19. Berar	25.65	(3.1)	24.38	(3.0)	713.74	(87.4)	52.58	(6.4)	816.35
20. **Central Provinces (W)	37.34	(3.2)	28.96	(2.5)	1042.42	(90.4)	44.70	(3.9)	1153.42
21. Central Provinces (E)	40.13	(3.0)	53.34	(4.0)	1177.80	(89.1)	50.55	(3.8)	1321.82
22. Konkan	7.11	(0.3)	46.99	(1.7)	2602.23	(93.7)	120.65	(4.3)	2776.97
23. Bombay-Deccan	12.95	(1.7)	54.10	(6.9)	620.01	(79.1)	97.03	(12.4)	784.10
24. Hyderabad (N)	17.02	(1.9)	38.86	(4.4)	749.55	(84.5)	81.28	(9.2)	886.71
25. Hyderabad (S)	14.48	(1.9)	53.34	(7.0)	593.85	(78.1)	98.55	(13.0)	760.22
26. Mysore	18.54	(2.0)	138.94	(15.2)	565.66	(61.8)	37.72	(20.9)	914.66
27. Malabar	69.34	(2.6)	320.29	(12.2)	1815.34	(68.9)	430.02	(16.3)	2634.99
28. Madras (SE)	120.90	(13.6)	115.06	(12.9)	305.05	(34.2)	350.52	(39.3)	891.53
29. Madras-Deccan	18.80	(3.0)	61.47	(9.9)	387.86	(62.3)	2.29	(24.8)	622.82
30. Madras North Coast	42.93	(4.2)	87.38	(8.9)	635.76	(62.3)	254.00	(24.9)	1020.07

* Formerly United Provinces; ** Now renamed in part as Madhya Pradesh.

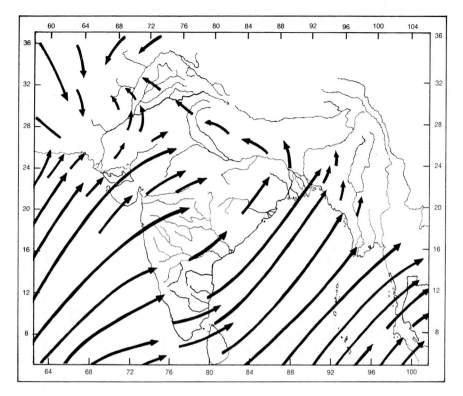

Fig. 8. Mean surface air-flow in the month of July in the Indian region.

the southeasterly trades crossing the equator and turning into the south-westerlies over southeast Arabian Sea and the Bay of Bengal. The intense precipitation, associated with the monsoon, occurs only when these moisture-laden air currents actually come to be uplifted, as they meet the Western Ghats in the Peninsula, the Arrakan Yoma in the Burmese west coast and later, when after deflection westwards these air currents impinge over the Himalaya.

3.1. THE DATES OF ESTABLISHMENT OF THE MONSOON

Establishing itself first along the West Coast of the Peninsula and of Burma about the beginning of June, the monsoon rapidly advances inland during June and is in full swing over the entire country by the Middle of July. It starts retreating from the Indus Valley by the beginning of September and is well out of North India by the middle of October. Its further retreat southwards through the south of the Peninsula, transforming itself into the retreating or the socalled northeast monsoon during November-December, is a peculiarity of the climate

103

Table II. Date of establishment of the southwest monsoon along the West Coast of India.

Year	Travancore-Cochin	S. Kanara	Ratnagiri	Kolaba
1891	May 27	June 3	June 19	June 21
1892	May 22	May 24	May 29	May 31
1893	May 22	June 4	June 10	June 10
1894	June 1	June 2	June 7	June 7
1895	June 8	June 12	June 14	June 15
1896	May 30	May 31	June 1	June 1
1897	May 30	June 5	June 7	June 7
1898	June 2	June 3	June 8	June 8
1899	May 23	June 7	June 9	June 10
1900	June 6	June 8	June 9	June 9
1901	June 1	June 4	June 7	June 7
1902	May 31	June 6	June 7	June 12
1903	June 8	June 11	June 12	June 12
1904	May 29	June 1	June 7	June 8
1905	June 6	June 8	June 9	June 10
1906	June 3	June 6	June 7	June 8
1907	May 31	June 5	June 11	June 11
1908	June 8	June 10	June 11	June 11
1909	June 1	June 2	June 3	June 3
1910	May 28	June 2	June 3	June 3
1911	June 1	June 2	June 4	June 4
1912	June 4	June 6	June 12	June 12
1913	May 24	June 1	June 6	June 7
1914	May 28	June 5	June 13	June 13
1915	June 3	June 12	June 17	June 18
1916	May 26	May 27	May 31	June 1
1917	May 26	May 29	June 4	June 5
1918	May 7	May 15	May 22	May 25
1919	May 16	May 26	June 4	June 6
1920	May 27	June 2	June 6	June 6
1921	June 1	June 3	June 10	June 12
1922	May 25	May 31	June 10	June 12
1923	June 4	June 11	June 12	June 13
1924	May 31	June 3	June 10	June 12
1925	May 27	May 28	May 29	May 29
1926	May 28	June 5	June 9	June 10
1927	May 23	May 27	June 10	June 10
1928	May 31	May 31	June 5	June 7
1929	May 29	May 30	June 1	June 6
1930	May 21	June 7	June 8	June 9
1931	May 23	May 29	June 14	June 14
1932	May 14	June 2	June 3	June 3
1933	May 22	May 28	June 1	June 1
1934	June 6	June 6	June 10	June 10
1935	June 10	June 10	June 12	June 14
1936	May 20	May 22	May 29	June 1
1937	June 3	June 10	June 11	June 12
1938	June 1	June 2	June 2	June 4
1939	June 6	June 6	June 7	June 9

Table II (continued)

Year	Travancore-Cochin	S. Kanara	Ratnagiri	Kolaba
1940	June 7	June 13	June 16	June 18
1941	May 23	June 3	June 14	June 16
1942	June 4	June 8	June 12	June 13
1943	May 12	May 14	May 21	May 21

of India. It is during the northeast monsoon that these southern areas receive their main rainfall of the year. It must not, however, be supposed that these features are by any means regular; they are subject to considerable variations, though within known limits, from year to year. In Table II we have given the actual dates of the establishment of southwest monsoon in four well defined areas, along the West Coast of the Peninsula, during the years 1891–1943. It may be seen from this table that there is considerable variation not only in the date of establishment but also in the speed with which the monsoon current moves from Kerala in the south to Kolaba in the north. Table III summarizes the main features of Table II. During the period of over fifty years considered here, the monsoon has set in as early as May 7 in 1918 and as late as June 10 in 1935 in Kerala. The general variability is shown by the standard deviation in the third column in Table III, which is of the order of seven days in the above area and five days in Kolaba.

Table III. Mean features of the dates establishment of the Southwest Monsoon along the West Coast of the Peninsula

Area	Mean date	Standard deviation in days	Earliest date	Latest date
Kerala	May 29	7.0	May 7	June 10
South Kanara	June 3	5.7	May 15	June 12
Ratnagiri	June 7	5.4	May 22	June 19
Kolaba	June 8	5.2	May 25	June 21

3.2. The behaviour of the southwest monsoon during 1875–1950

In discussing the behaviour of the monsoon, year by year from 1875 to 1950, it would be necessary first to take an overall view of the entire phenomenon, so as to avoid losing ourselves in details. To achieve this,

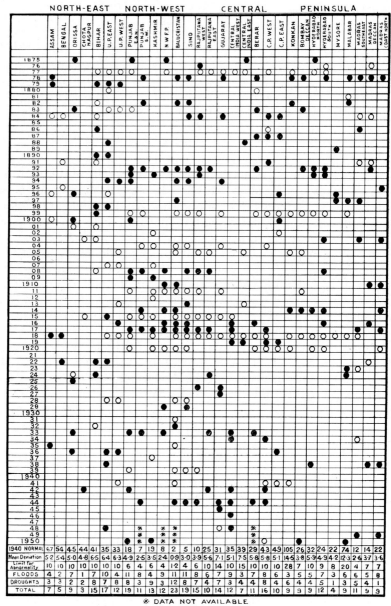

● FLOODS and ○ DROUGHTS in INDIA.

Fig. 9. Pattern of distribution of flood and drought in thirty major rainfall divisions of India during the period 1875–1950.

106

we distinguish here between *drought, flood* and *normal;* a year in which the actual rainfall of a particular subdivision is less than the difference between the mean (normal rainfall) and twice the mean deviation during the entire series of years is designated as *drought. Flood* is defined as the year in which the actual rainfall is more than the sum of the normal and twice the mean deviation; fig. 9 shows the incidence of *floods* (black circles) and *droughts* (open circles) in the thirty rainfall subdivisions, during each of the years 1875 to 1950. This figure shows the major abnormalities (floods and droughts) and the frequencies of their occurrence, both in time (yearwise) and space (subdivision-wise). The blanks in the diagram represent the years and subdivisions, in which the seasonal rainfall lies between the limits of abnormality, viz. M+2d and M—2d and may, therefore, be defined as more or less normal.

Taking fig. 9 as a whole, we find that the frequency of abnormalities is higher in areas of scanty rainfall, like for example the Northwest India, Rajasthan and Gujarat, than in areas of heavy rainfall like Malabar, Konkan, Bengal and Assam. The occurrence of floods or droughts in any subdivision in a series of years is more or less random, but in a few years there is a tendency for many and sometimes most of the subdivisions to experience the same abnormality, flood or drought as the case may be. Almost the whole of India was thus subjected to severe drought during 1877, 1899 and 1918. Such country-wide failure of the monsoon rainfall may occur, on an average, once in twenty years, while the failure of the rainfall over parts of the country may occur in three or four out of every twenty years. About fifteen out of every twenty years may, therefore, be expected to have a reasonably good monsoon rainfall distribution. Almost country-wide floods occurred in 1878, 1892 and 1917. Like the country-wide droughts, such large-scale and country-wide floods may also therefore be expected once in every twenty years. In about six or seven years out of every twenty, only parts of the country have experienced excessive rainfalls and the remaining twelve or thirteen years have tended to be more or less normal. There does not seem to be striking evidence of any periodicity in the incidence of these major abnormalities. We find that sometimes alternate years (1878 flood, 1877 drought; 1917 flood and 1918 drought) may be afflicted by floods and droughts respectively on a country-wide scale. We often also find that while some parts of the country may have suffered severe drought, other parts have had heavy floods. It is remarkable that there have been a few years when the entire region was normal, for example 1885, 1906, 1921, 1930, 1943 and 1947, or in other words roughly one year out of every ten.

3.3. RAINFALL WEEK BY WEEK DURING THE YEAR

We may observe that the commencement and termination of the wet season in each subdivision is defined by the normal rainfall being more

107

RAINFALL OF MALABAR WEEK BY WEEK.

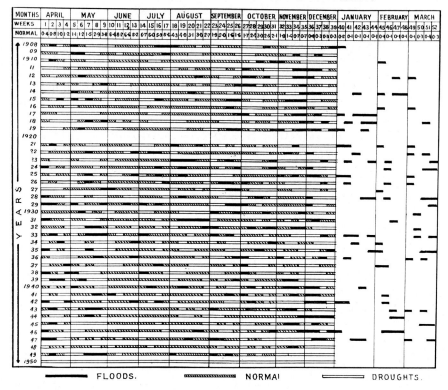

Fig. 10. Rainfall distribution, week by week, in Malabar, a typical wet subdivision.

than 5 mm. The duration of the wet season is naturally prolonged in the wetter areas of Kerala and the northeast India and shortest in the arid parts of northwest. In the latter areas there is also a secondary wet season, representing the winter rains that commence some time after the cessation of the monsoon rains. Outside the wet season, where either no rain or very little of it falls during the normal dry weather, the normal rainfall is less than the limiting value of 5 mm. In such dry seasons the lines indicating drought are entirely absent, but occasions of untimely or unseasonal rains, exceeding twice the limiting normal of 5 mm, are marked. In the year 1946–1947, there were numerous weeks with unseasonal rains, particularly in the central parts of the country and in the north Deccan. Such unseasonal precipitation during the dry parts of the year are often followed by incidence of outbreaks of various insect Pests on crops and human diseases in epidemic forms. Similar diagrams have been prepared for the years 1908–1950, from which GOVINDASWAMY (1953) constructed a series of subdivisions diagrams, showing the ab-

108

RAINFALL OF GUJARAT WEEK BY WEEK.

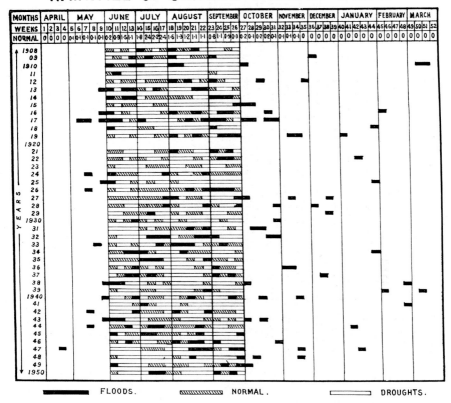

Fig. 11. Rainfall distribution, week by week in Gujarat, a typical dry subdivision.

normalities for all the years under consideration. These have proved extremely useful in judging the frequency of floods and droughts in a series of years, to which the subdivision is liable. Fig. 10 & 11 are typical examples for Malabar (a wet subdivision) and Gujarat (a dry subdivision).

3.4. Contemporary relationships of the monsoon rainfall in fifteen divisions

Of considerable importance is the problem of relation between deficiency of the monsoon rainfall in one part and the possible compensation by excessive rainfall in another. The data summarized in table IV show the contemporary correlation coefficients for fifteen major divisions. A positive coefficient demonstrates that the two areas are likely to be associated with increase in both, but a negative coefficient

109

Table IV. Intercorrelation of the monsoon rainfall (June–September) between pairs of divisions during 1875–1918

Division	2	3	4	5	6	7	8	9	10	11	12	13	14	15
1. Burma	−0.47	−0.14	−0.05	−0.01	−0.09	−0.21	+0.29	−0.07	−0.26	+0.01	−0.01	−0.32	−0.15	−0.33
2. Assam		+0.42	+0.27	+0.01	−0.16	−0.10	−0.13	−0.17	−0.24	−0.27	−0.40	−0.01	−0.11	+0.14
3. Bengal			+0.56	−0.07	−0.07	−0.05	+0.04	−0.24	−0.31	−0.22	−0.45	−0.18	−0.15	−0.12
4. Behar & Orissa				+0.31	+0.13	+0.06	+0.06	+0.05	−0.10	+0.04	−0.27	−0.01	−0.07	−0.16
5. Uttar Pradesh					+0.49	+0.31	+0.06	+0.55	+0.30	+0.68	+0.33	+0.37	+0.07	+0.22
6. Punjab						+0.82	+0.56	+0.87	+0.59	+0.79	+0.59	+0.61	+0.11	+0.35
7. Northwest-Frontier Province (Pakistan)							+0.65	+0.73	+0.59	+0.64	+0.46	+0.63	+0.22	+0.45
8. Sind								+0.50	+0.65	+0.34	+0.34	+0.40	+0.24	+0.40
9. Rajasthan									+0.57	+0.81	+0.54	+0.57	+0.21	+0.40
10. Bombay										+0.46	+0.58	+0.72	+0.53	+0.66
11. Central India											+0.61	+0.40	+0.50	+0.21
12. Central Provinces												+0.55	+0.07	+0.18
13. Hyderabad													+0.38	+0.61
14. Mysore														+0.80
15. Madras														

would indicate a decrease in one area associated with increase in the other. We find that a copious monsoon rainfall in Burma shows a distinct tendency to be associated with subnormal monsoon rainfall in India and vice versa. To a lesser extent the excess of rainfall in the northeast India is associated with a more or less pronounced deficiency elsewhere. The correlation coefficients are generally positive in the Northwest and Central India and in the Peninsula, so that fluctuations from the normal tend to be similar for the greater part of the country and the same abnormal patterns prevail in a number of adjoining rainfall subdivisions.

3.5. Regional peculiarities of distribution of rainfall

On the basis of the normal rainfall, the entire region may be divided into more or less climatically homogenous zones, which show, however, numerous local peculiarities, especially in regard to the variability in space and time, from station to station and over a number of consecutive days in different subdivisions. Table V summarizes the results of analysis of the variability of rainfall in July 1942 in twenty stations, selected at random from each of these areas. The ratios of variance are

$$\frac{\text{Variance between stations}}{\text{Residual variance}} \quad \& \quad \frac{\text{Variance between days}}{\text{Residual variance}}.$$

If the variability between stations, between days and the residual are all of the same order of magnitude, the ratio F is not significant. The variability between stations is significantly larger than that attributable to random variability in the Punjab, West Bengal and especially Rajasthan and Kerala. In Malabar this variability between stations is due to orographical peculiarities and in Rajasthan we have a real *climatic non-homogeneity* and it would be impossible to say where rain would fall during a spell.

4. *Other Important Weather Phenomena that bring rainfall*

4.1. The eastern depression

The intensity of the monsoon is punctuated by a series of depressions, which originate mostly at the head of the Bay of Bengal, but often also as far east as the Pacific Ocean or the China Sea, weakening over land, but reviving again when they enter the north Bay of Bengal. These monsoon depressions travel in a northwesterly direction towards the Northwest India and cause heavy precipitation along their track. The frequency of these depressions is three or four each month during June–September. But for them, the monsoon rains tend to be largely orographic and confined to the hills and mountains. They are, therefore, extremely useful in bringing rains into the plains of north and central India and the arid zone of Rajasthan and Gujarat.

Table V. Analysis of variance of rainfall in July 1942

Area	Due to	Degree of freedom	Sum of squares	Mean sq (Variance)	Stand. Dev.	Variance ratio: F	Rainfall in mm per day	Coefficient of variability
1. Punjab	Stations	19	21.1334	1.1123	1.55	3.73(S)	—	680
	Days	30	19.8421	0.6614	0.81	2.72(S)	5.79	356
	Residue	570	160.8981	0.2981	0.55	—	—	241
	Total	619	210.8736	0.3407				
2. Uttar Pradesh	Stations	19	16.1105	0.8479	0.92	1.55	—	197
	Days	30	114.5280	3.8143	1.95	6.97(S)	11.84	418
	Residue	570	312.1042	0.5475	0.74	/	/	159
	Total	619	442.7427	/				
3. Central Provinces	Stations	19	45.5754	2.3987	1.55	1.35	17.70	222
	Days	30	270.4394	9.1465	3.02	5.16(S)	—	433
	Residue	570	613.9953	1.7715	1.33	—	—	191
	Total	619	929.9901	1.5024				
4. Bengal	Stations	19	41.4545	2.1818	1.48	3.62(S)	10.72	350
	Days	30	43.2902	1.4430	1.20	2.39(S)	—	284
	Residue	570	343.4197	0.6025	0.78	—	—	185
	Total	619	428.1644	0.6917				
5. Rajasthan	Stations	19	108.8973	5.7314	2.39	6.92(S)	10.59	573
	Days	30	46.2551	1.5418	1.24	1.86	—	297
	Residue	570	471.9398	0.8280	0.91	—	—	218
	Total	619	627.0922	1.0131				
6. Malabar	Stations	19	85.4648	4.4981	2.12	5.53(S)	—	159
	Days	30	382.1749	12.7392	3.57	15.68(S)	—	268
	Residue	570	463.1872	0.8126	0.90	—	33.81	68
	Total	619	930.8269	1.5038				

S means significance at 1% level

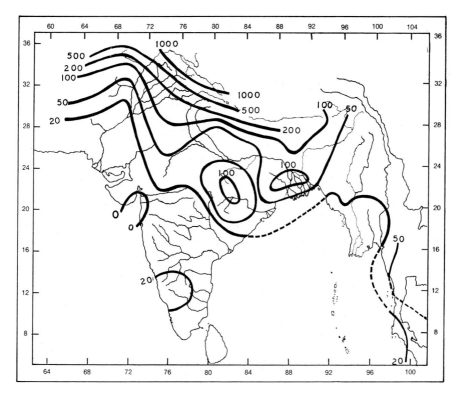

Fig. 12. Frequency of days of with hailstorms in hundred years in the Indian region.

4.2. The western depressions

During November–May a series of western depressions enter India through Baluchistan and the Northwest-Frontier Province and travel eastwards across North India towards the northeast. These depressions, some of which can be traced from as far west as the Mediterranean Sea or also the Atlantic Ocean, result in cloudy weather and light rainfall in the plains and snowfall in the Himalaya during the north Indian winter. Their passage across a region is followed by incidence of northerly to northwesterly cold winds. The frequency of the western depressions is on an average two in November, four or five from December to April and two in May. The winter crops of north and Central India benefit from the winter rains caused by the western depressions.

4.3. Cyclonic storms

The cyclonic storms, which are much more severe than the depressions, usually form in the Bay of Bengal and in the Arabian Sea during the

113

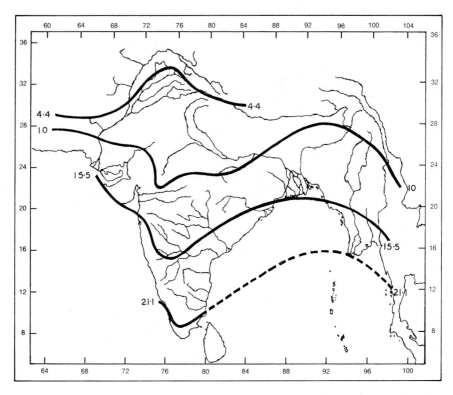

Fig. 13. Mean daily minimum temperature in °C in January in the Indian region.

transition months of April–June and October–December. They move inland and cause considerable havoc due to heavy rains, accompanied by high winds. Occasionally tidal waves may also occur in the coastal areas. On an average, one or two severe cyclonic storms may be expected in the pre-monsoon period and two or three in the post-monsoon period. The mode of occurrence of these storms, their favourite tracks and the incalculable damage due to heavy rains, followed by high winds and gales that lay waste crops, are discussed in numerous reports and research papers, published by the India Meteorological Department.

4.4. THUNDERSTORMS AND HAILSTORMS

As compared with cyclones and depressions, thunderstorms are local and sporadic phenomena, lasting hardly for a few hours, but often accompanied by severe squalls up to 128 km or even 160 km per hour of a few minutes duration. Severe thunderstorms often leave a trail of devastation and damage along their track. They usually occur during the summer or the pre-monsoon months, as well as after the withdrawal of

114

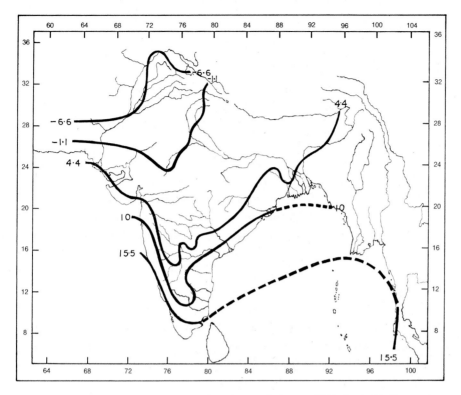

Fig. 14. Lowest minimum temperature in °C recorded up to 1920, in the Indian region.

the southwest monsoon. In their mode of occurrence there are points of similarity between thunderstorms, dust-storms and hailstorms. In the absence of sufficient moisture in the atmosphere, a dust-storm results, but if enough moisture is present we have a thunderstorm. A hailstorm is a particularly violent thunderstorm, with the formation of hailstones. In many parts of the country, like for example, the Deccan and the adjoining areas, the pre-monsoon and the post-monsoon thunderstorms are the main source of moisture for vegetation. Fig. 12 shows the frequency of hailstorms in India. They are most frequent along the Himalayan slopes, Bengal and the Chota-Nagpur Plateau; in Deccan Plateau, Mysore and other parts of the Peninsula their frequency is less.

5. Cold and Heat Waves

Clear weather sets in over the Northwest India by October and extends over practically the rest of the country, except South India, by the beginning of December. As already mentioned, cold winds set in in the wake of each of the western disturbances that move across north India

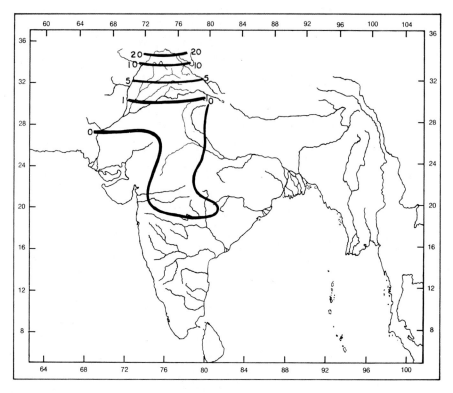

Fig. 15. Frequency of days with Frost in November in the Indian region.

during the winter. During period of severe cold waves, the minimum atmospheric temperature may fall below normal by as much as 11.1 °C in north India and 5.5 °C in the central parts of the country and the north Deccan Plateau. Fig. 13 shows the mean daily minimum temperature in January, which is typical of the winter. Fig. 14 shows the lowest minimum temperature recorded up to 1920. During periods of cold waves, frosts are likely to occur near the ground and the frequency of such freezing temperatures during November–February is given in figs. 15, 16, 17, 18. Table VI gives the mean daily minimum temperature in thirty sub-division in different months during the year. The maximum temperatures are summarized similarly in table VII. It may be observed from these tables that on many days the actual maximum temperatures may be much higher than the means. From charts showing the frequencies of days with abnormally high maximum temperatures exceeding 37.7°, 40.4°, 43.1 °C, etc., it is evident that the centre of high frequency of high maximum temperatures of 37.7 °C and above lies in the Deccan and central parts of India in March, April and May. The frequency increases from twenty days in March to thirty in May in Deccan, but

116

Table VI. Mean daily minimum temperature in °C in different subdivisions

Subdivision	April	May	June	July	Aug.	Sept.	Oct.	Nov.	Dec.	Jan.	Feb.	March
1. Assam	20.0	22.2	25.0	25.0	25.5	24.4	21.7	16.1	11.6	10.5	12.2	16.1
2. Bengal	23.3	25.0	26.1	26.1	25.6	25.8	23.3	16.7	12.7	12.7	15.0	19.4
3. Orissa	25.0	26.7	26.7	25.6	25.6	25.0	23.2	18.3	13.9	11.1	17.8	21.1
4. Chota-Nagpur	22.7	25.0	25.0	24.4	23.8	23.3	20.0	14.5	10.5	10.5	13.2	17.2
5. Bihar	22.2	25.0	26.1	26.1	25.6	25.6	22.2	15.0	10.5	10.0	12.2	17.2
6. Uttar Pradesh (E)	21.7	25.6	27.2	26.6	26.1	25.0	20.0	12.7	8.9	8.9	11.1	15.6
7. Uttar Pradesh (W)	20.5	25.0	26.6	26.1	25.6	23.3	17.8	11.1	7.8	7.8	10.0	14.5
8. Punjab (E)	19.4	24.4	28.8	26.6	26.1	23.9	17.2	10.5	6.6	6.1	8.9	13.9
9. Kashmir	3.3	8.3	11.1	14.5	14.5	10.0	3.3	−2.8	−7.2	−8.9	−8.3	−2.2
10. Rajasthan (W)	21.6	26.1	27.7	26.6	25.0	23.9	18.3	13.3	10.0	9.5	11.1	16.1
11. Rajasthan (E)	22.2	26.6	27.7	25.6	24.4	23.3	18.9	13.3	9.5	8.9	11.1	16.1
12. Saurastra & Cutch	22.7	25.6	26.6	26.1	25.0	23.9	22.2	17.7	14.5	13.3	15.0	18.8
13. Gujarat	23.3	26.1	26.6	25.6	25.0	24.5	22.2	18.8	15.0	13.9	15.6	18.8
14. Madhya Bharat & Bhopal	22.2	25.0	25.0	23.3	22.2	21.7	18.8	12.8	10.0	10.0	11.7	16.1
15. Vindhya Pradesh	22.2	26.6	28.3	26.1	25.0	23.9	18.8	12.2	8.3	8.9	11.1	16.1
16. Berar	24.5	27.2	25.6	23.9	22.7	22.7	20.0	16.1	12.8	13.9	15.6	19.3
17. Central Provinces (W)	22.7	26.1	25.6	23.9	23.3	22.7	18.8	13.9	10.5	11.7	13.2	17.7
18. Central Provinces (E)	23.3	26.1	25.0	23.3	22.7	22.2	19.3	20.0	10.5	11.7	13.9	18.4
19. Konkan	25.0	26.6	25.0	24.5	24.5	23.9	23.9	21.6	20.0	19.5	19.5	22.2
20. Bombay-Deccan	21.6	22.2	22.7	21.6	21.1	20.5	19.4	15.6	12.8	13.3	15.0	18.4
21. Hyderabad (N)	23.9	25.6	23.9	22.7	22.2	21.6	20.0	16.1	12.8	13.3	16.1	19.4
22. Hyderabad (S)	25.0	26.6	24.5	23.3	22.7	22.2	21.1	17.7	15.6	16.1	18.4	21.1
23. Mysore	21.1	21.1	20.0	19.4	18.8	18.8	18.8	17.2	15.0	14.4	16.7	18.8
24. Malabar	26.1	25.6	23.9	23.3	23.9	23.9	23.9	23.3	22.2	22.2	22.7	24.5
25. Madras SE	25.6	26.1	25.6	25.6	24.5	24.5	23.3	22.2	20.5	20.0	21.1	23.3
26. Madras-Deccan	25.6	26.6	25.6	24.5	24.5	23.3	22.7	19.4	17.2	17.2	19.4	22.2
27. Madras Coast (N)	25.6	27.2	27.2	26.1	26.1	25.6	24.5	21.6	18.8	18.3	20.5	22.7
28. Bangla Desh	22.7	23.9	25.0	25.6	25.6	25.6	23.3	18.3	13.3	12.2	14.5	18.8
29. Pakistan	16.1	21.1	25.0	26.1	24.5	21.1	14.5	7.8	3.3	4.4	7.2	12.2

117

Table VII. Mean daily Maximum temperature in °C in different subdivisions

Subdivisions	April	May	June	July	Aug.	Sept.	Oct.	Nov.	Dec.	Jan.	Feb.	March
1. Assam	29.4	30.5	31.1	31.6	31.6	31.1	30.5	27.2	24.4	23.9	25.5	28.3
2. Bengal	35.5	35.0	33.3	31.6	31.6	31.6	31.1	28.3	24.5	25.0	27.7	32.2
3. Orissa	36.1	36.6	33.8	31.6	31.1	31.6	31.6	28.8	26.6	27.2	30.0	33.3
4. Chota-Nagpur	37.7	38.8	35.0	31.6	30.5	31.1	30.5	27.2	24.4	25.5	27.2	32.2
5. Bihar	37.2	37.7	35.0	32.7	31.6	31.6	31.6	27.7	23.9	23.4	26.1	32.2
6. Uttar Pradesh (E)	37.7	40.0	37.7	33.2	32.2	32.7	32.2	28.3	23.4	23.4	26.1	31.6
7. Uttar Pradesh (W)	36.6	40.0	37.7	33.2	32.2	32.7	32.7	27.7	23.4	21.6	25.5	30.0
8. Punjab (E)	36.1	40.0	40.0	36.1	34.5	35.0	33.8	27.7	22.9	20.5	23.4	29.5
9. Kashmir	15.5	22.16	25.5	27.8	27.8	23.8	19.4	11.6	4.4	1.1	3.3	8.8
10. Rajasthan (W)	37.7	40.5	40.0	36.1	33.8	34.5	35.5	31.1	26.1	25.5	26.7	32.2
11. Rajasthan (E)	37.2	40.0	38.8	33.8	31.6	32.7	33.8	30.0	24.5	23.9	26.1	31.6
12. Saurastra & Cutch	35.0	36.1	35.0	31.6	30.5	31.6	33.8	31.7	28.8	27.2	28.8	32.2
13. Gujarat	38.8	38.8	36.1	35.0	31.6	32.7	35.5	33.3	30.5	30.0	31.1	35.5
14. Madhya Bharat & Bhopal	37.7	39.4	36.1	30.5	38.8	30.5	32.2	29.4	26.7	26.1	28.3	33.3
15. Vindhya Pradesh	38.3	41.1	38.2	32.2	30.5	31.6	31.7	28.3	23.9	23.9	26.7	32.7
16. Berar	41.1	42.2	36.6	31.6	30.5	31.6	33.3	30.5	29.4	30.0	32.7	36.6
17. Central Provinces (W)	38.8	41.1	36.6	30.5	29.3	30.5	32.2	28.8	26.7	26.6	28.8	34.4
18. Central Provinces (E)	38.3	40.0	35.5	30.0	29.3	30.0	30.0	27.7	26.1	27.2	30.0	34.4
19. Konkan	32.2	32.7	40.5	28.9	28.9	28.9	30.5	31.7	31.1	30.0	29.5	29.5
20. Bombay-Deccan	38.3	38.3	32.2	29.3	28.9	30.0	31.1	30.0	28.8	29.4	32.7	29.5
21. Hyderabad (N)	39.4	40.5	35.5	31.6	30.0	31.1	32.2	30.5	29.8	30.0	32.2	35.5
22. Hyderabad (S)	38.8	40.0	35.0	31.6	31.1	31.1	32.2	30.5	28.8	30.0	33.3	35.5
23. Mysore	34.5	33.3	28.9	27.2	27.2	28.3	28.3	27.2	26.7	27.7	31.1	33.3
24. Malabar	32.7	31.6	29.4	28.3	28.3	28.8	30.0	30.5	30.5	31.1	31.1	31.6
25. Madras Southeast	36.1	36.6	36.1	35.0	34.5	33.8	32.2	30.5	28.8	30.0	32.2	34.4
26. Madras-Deccan	39.4	40.0	36.6	34.5	33.3	33.3	32.7	31.1	30.0	30.5	34.4	37.7
27. Madras Coast North	33.8	36.1	35.0	33.3	32.7	32.2	31.7	28.8	27.7	27.2	30.0	32.7
28. Bangla Desh	33.3	32.7	31.7	31.6	30.5	31.1	30.5	28.3	25.5	25.0	27.2	31.6
29. Pakistan	31.6	37.7	40.0	38.2	36.6	35.5	32.2	25.5	20.0	18.8	21.6	32.2

118

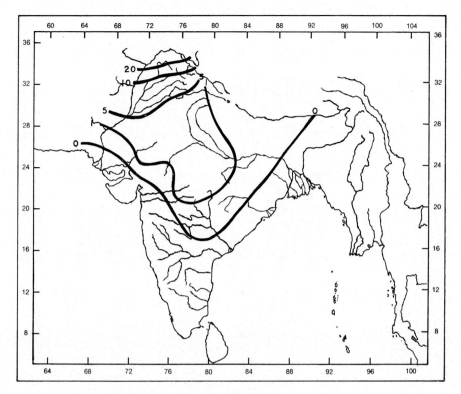

Fig. 16. Frequency of days with Frost in December in the Indian region.

once the monsoon sets in, the centre of high frequency shifts towards northwest India. In July and August, the southeast Madras too is another centre of high maximum temperatures. Fig. 19 shows the highest maximum temperature recorded up to 1930 in different parts of the region.

As is well known, the rate of evaporation is proportional to wind velocity and the saturation deficiency of the air. Fig. 20 shows the mean annual evaporation, as measured from the free surface of water, 30 cm above ground level by the standard (1.2 m diameter) USA type of evaporimeter. The area of highest evaporation occurs over Bombay-Deccan and as we move northwards, eastwards or southwards from here, the annual rate of evaporation decreases. Although we cannot alter the general circulation of the atmosphere over the country, it is possible to check the air movements near the ground by building a system of windbreaks and in this manner reduce the loss of water by evaporation. Reference may here be made to the simple inexpensive technique of suppressing excessive evaporation from exposed surfaces by spreading a monomolecular layer of cetyl alcohol on the water surface, leading to a

119

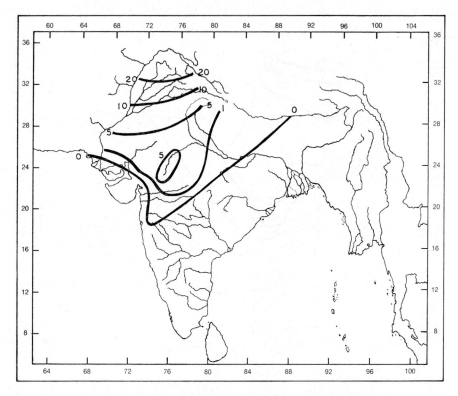

Fig. 17. Frequency of days with Frost in January in the Indian region.

reduction of evaporation by almost 30%. This method is also applicable to soil surface and for preventing vegetables and fruits from drying up (RAMDAS 1957, 1962).

6. Microclimatology

6.1. SOLAR AND TERRESTRIAL RADIATION

The visible radiation from the Sun

When the sun's radiation enters the uppermost layers of the earth's atmosphere, the atomic oxygen and ozone present there absorb the shortest wavelengths, from X-rays to the ultra-violet, thus effectively filtering off the radiations that are injurious to life. Solar radiation has most of its energy within the visible region of the spectrum, with the maximum energy in the bluish-green. During its downward passage through the atmosphere, the radiation undergoes depletion by diffraction by the molecules and aerosols (including dust) of the atmosphere,

120

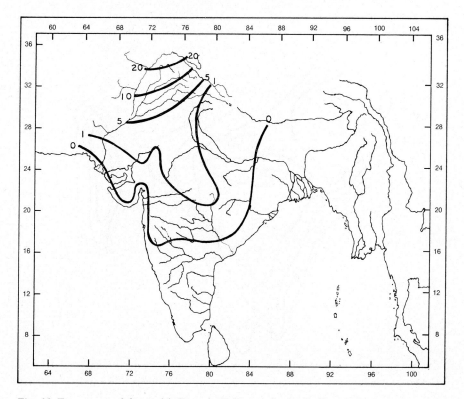

Fig. 18. Frequency of days with Frost in February in the Indian region.

so that only about 70–80 % of the visible radiation actually reaches the ground. The solar constant, or the radiation incident per square centimetre of a plane surface held normal to the rays outside the atmosphere, is estimated to be about 2 gr.-cal. per square centimetre per minute; of this about 1.6–1.7 gr.-cal. reach the ground surface. Depending upon the actual colour of the ground surface, the visible radiation from the sun and from the sunlit sky is partly reflected and partly absorbed. A perfectly white surface would reflect or scatter 100 %, but a perfectly black surface would absorb 100 % of the incident radiation. Table VIII gives the percentage absorption of the visible solar radiation by some typical surfaces.

Infra-red radiation

The small quantities of water vapour, carbondioxide and ozone present in the atmosphere absorb some of the outgoing thermal radiation emitted by the earth's surface at a mean temperature of 300 °A (27 °C). This infra-red or black-body radiation emitted by the earths surface has

121

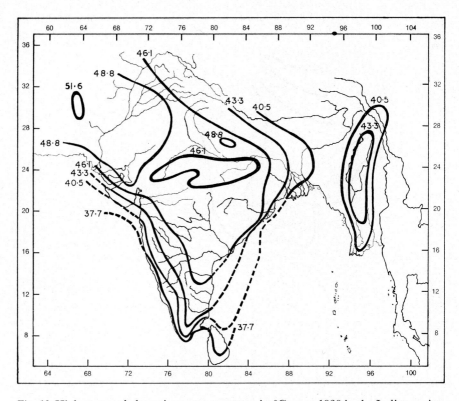

Fig. 19. Highest recorded maximum temperature in °C up to 1930 in the Indian region.

Table VIII. Percentage absorption of solar radiation by some common surfaces

Surface	Absorption %
French chalk	0.0
White paint	20.0
Aluminium foil	15.0
Aluminium paint	20.0
Quartz powder (white sand)	28.0
Grey alluvial soil	59.0
Brick	55.0
Concrete (cement)	60.0
Galvanized iron	65.0
Asbestos slate	81.0
Grass-covered lawn (green)	68.0
Black paint	96.0
Charcoal powder	96.0
Black-cotton soil	84.0

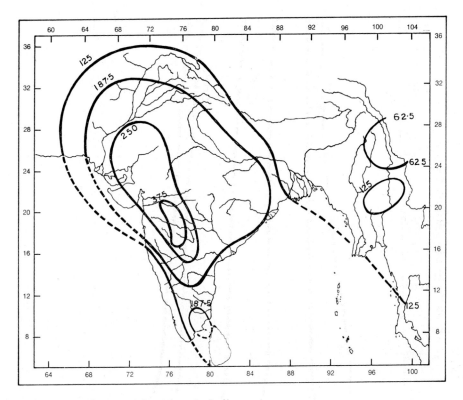

Fig. 20. Annual evaporation in cm in India.

its maximum energy at about 10 micra. According to the well known KIRCHHOFF's law, these gases also re-emit, in the same wavelengths as their characteristic absorption bands, a part of the emission being directed towards the ground. By returning a part of the outgoing radiation, the water vapour, carbon dioxide and ozone of the atmosphere exercise, as it were, a blanketing effect on the Earth's surface and the adjacent air layers, thus maintaining the temperature near the ground within limits that are optimal to organisms.

6.2. THE GROUND AS ACTIVE SURFACE

The surface of the ground plays a special rôle in the disposal of the energy, received as radiation from the sun and from the sunlit sky, by behaving as an active surface. The ground surface, exposed to insolation during the daytime and by absorbing part of the incident solar radiation, converts the absorbed part into heat and thereby becomes the source of heat to the air as well as the soil layers adjacent to it, supplying a part of this thermal energy through conduction and convection. If moisture

123

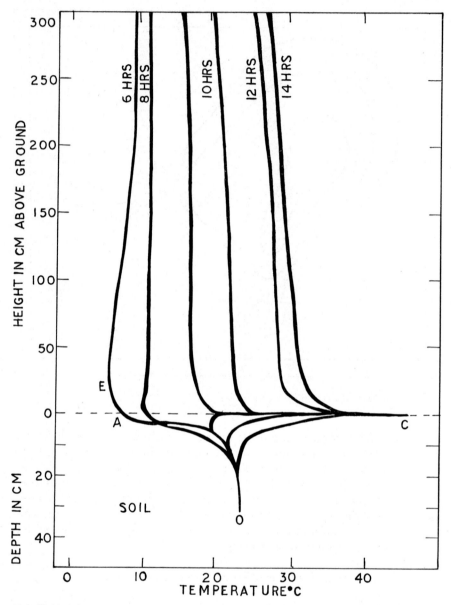

Fig. 21. Diurnal range of temperatures °C of the soil and the layer of air above ground on January 5, 1933.

is present in the soil surface, a part of the energy is utilized for evaporation. Further, the ground surface radiates fully like a blackbody in the infrared region of the spectrum throughout the day and night, at a rate equal to σ T^4, where σ is the well known STEFAN-BOLTZMANN constant

$(8.17 \times 10^{-11}$ cal. cm^{-2} min.$^{-1}$ degree$^{-4})$ and T is the surface tempera-ture in degrees absolute. A part of this outgoing thermal radiation is, as already mentioned, absorbed by the water vapour, carbon dioxide and ozone of the atmosphere and reradiated towards the ground surface, thus setting a limit to nocturnal cooling of the ground and the air layers above it by radiative exchange with the atmosphere.

6.3. THE DIURNAL TEMPERATURE RANGE

It would follow from the above discussion that the diurnal variations of temperature are naturally maximum at the active ground surface. The diurnal range of temperature decreases rapidly as the height in-creases above the surface or as we move down into the soil layer below the surface. Fig. 21 gives a typical example of the diurnal temperature range for clear season at Poona, where the diurnal swing of the tem-perature is largest at the ground surface, from A at the minimum tem-perature epoch to the point C at the maximum temperature epoch. It must be observed that with increase in the depth of the soil, the diurnal temperature range dies down rapidly to zero within a depth of about 30 cm. The decrease of the diurnal range of temperature in the air layers, with increase of height above the ground, is quite large, but not so very rapid as within the soil. A feature of great interest is the fact that the fall of temperature with height (the socalled lapse rate) is extremely high, as much as 200,000 times the adiabatic lapse rate of the free atmosphere above a few kilometres, at the maximum temperature epoch (see graph C Fig. 21). At the extreme left of the figure, we have the curve AEB, in which the portion AE represents the remnant of the lapse condition and above it the part EB shows the inversion layer, above which temperature increases with height. The inversion layer is one of great stability, in contrast to the tremendous instability during the intense lapse condition prevailing during the daytime, particularly at the maximum temperature epoch. These features have been discussed at great length in a number of contributions (RAMDAS, 1948, 1951).

6.4. SHIMMERING LAYER ABOVE GROUND

The well known phenomenon of shimmering is associated with the special type of up and down convections in air layers in contact with a hot surface, like for example, ground receiving insolation on clear days, when it is heated intensely by the incident sunrays. The air in contact with the heated surface rises up, breaking through the colder and denser layers of air above, in the form of rising filaments. At the same time, filaments of colder air from above come down as compensating currents in between the rising filaments of hot air. There is considerable variation of the air density and consequently also of the refractive indices of the air between the

125

uprising and the downcoming currents, so that when one looks at a distant object through such layer, a characteristic shimmering or optical distortion of the distant object is observed. Actual observations show that the thickness of the shimmering layer reaches its maximum value of about 60 cm in the afternoon, when the air temperatures are also maximum. Thereafter, as the sun goes down, the shimmering activity weakens rapidly and the layer collapses towards the ground. As mentioned earlier (see curve AEB in Fig. 21), even after a whole night of radiative cooling, at the minimum temperature epoch, the remnant of the shimmering layer may still be observed to persist from the ground level to a height of perhaps 30 to 60 cm (see part AE of the curve in Fig. 21). Under these circumstances, we have the coldest air layer at the level of E, the inversion layer, as it were, resting on the shimmering layer below. This cold-layer-above-ground phenomenon is reported and discussed in a series of papers (RAMDAS 1932, 1968).

6.5. INVISIBLE CONDENSATION AT SOIL SURFACE

If a quantity of air-dry soil is spread out evenly in a shallow dish and exposed flush with the ground level to the sun, sky and the atmosphere, it is found by weighing the dish at regular hourly intervals, that the soil sample in the dish gives up moisture by desorption or evaporation to the air layers above, from the minimum temperature epoch to the maximum, from about 7 AM to 2–3 PM, thus losing weight. Thereafter, during the late afternoon, evening and the night till 7 AM next morning, the same sample of soil reabsorbs water vapour from the air layers above it, so that it gains in weight. This exchange of water vapour between soil and the atmosphere goes on daily, during the entire clear season, without any significant change of the mean weight of the soil sample in the dish. The diurnal range of weight of soil is a function of the colloid or clay content of the soil (Fig. 22). The black cotton-soil of Deccan shows the maximum range, the red and partly silty soil of the central parts of the country come next and the grey alluvium of the Indo-Gangetic Plains shows the least diurnal variation. A sample of sand with no clay shows hardly any diurnal variation in weight. The daily exchange of water vapour between the soil surface and the air layer above it takes place throughout the country, as soon as the surface crust of the soil is so far desiccated after the cessation of the wet season and after the clear season has commenced, as to retain only hygroscopic moisture. The exchange phenomenon has also the effect of desiccating the air layers during the absorption regime, daily from about 2–3 PM to 7 AM of the next day. Conversely the air layers are enriched with the evaporated moisture from 7 AM to 2 PM daily (see Table IX). The absorption of water vapour by the soil surface from the air layers above it has been designated as invisible condensation of water molecules in the uncondensed vapour phase. This

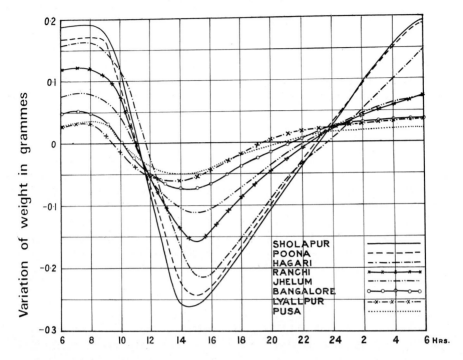

Fig. 22. Diurnal variation in the moisture content of typical soils of India (60 gramme samples spread over 12.6 cm²).

Table IX. Variations of vapour pressure with height above the ground at the epochs of maximum and minimum temperatures, averages for January during 1933–1937

Height above ground in cm	Vapour pressure in mm Hg	
	Mx. temp. epoch	Min. temp. epoch
0.75	8.6	5.9
2.50	8.2	6.0
7.50	8.0	6.0
15.00	7.8	6.1
30.00	7.7	6.2
60.00	7.6	6.4
90.00	7.5	6.6
122.00	7.5	6.8
183.00	7.4	7.1
244.00	7.4	7.1
304.80	7.3	7.4

is quite distinct from dew deposition, which can occur only when exposed objects in the open cool below the dewpoint by nocturnal cooling. Dewfall is, therefore, a less frequent phenomenon (RAMDAS 1948). It may be remarked that invisible condensation is a secondary source of moisture to all forms of dry vegetable matter and it is believed even to xerophytic plants growing in arid areas.

6.6. THERMAL BALANCE ON A CLEAR DAY

In the microclimatological studies one is fortunate in being able to utilize practically all the facilities of a laboratory for investigation of the various factors that control the thermal energy balance at the ground surface. Such studies, based on actual experimental measurements throughout the day, enable us to obtain an idea of the different special phenomena occurring near the ground. These problems are discussed in detail in my monograph on crops and weather in India (RAMDAS 1960). The following table (Table X) is reproduced from this monograph and gives the balance sheet of energy for a typical day at Poona (April 23, 1936).

There is thus a small carry-over of 11 grammes-calories. With this objective assessment of the thermal balance in view, we may next consider how far the thermal balance will alter if one of the factors is altered experimentally. The factors are interlinked with each other, so that if any one of them is altered, most other factors adjust themselves automatically, so as to maintain the overall thermal balance at the ground surface. If for example the ground surface is given a liberal irrigation, while the factor a' may alter slightly, the factor f' is eliminated, the factor c' decreases very much on account of the decreased temperature due to a large fraction of the energy being consumed by factor f' (evaporation) and factor d' (convection) heat loss is also decreased. As regards the factor e, owing to increase in heat conductivity of the wet soil, the diurnal wave of temperature penetrates somewhat deeper, but the nett transfer of heat does not change, since any loss during the day is returned by the night to the surface. If the ground surface is given a coating of white chalk so as to cover the black surface, the albedo tends towards 100%, the surface temperature becomes reduced considerably so that the factors c, d, e and f decrease correspondingly. On the other hand, if the soil is covered with a coat of black charcoal powder, a' tends to become zero, the surface temperature increases considerably so that the factors c, d, e and f also increase rapidly.

6.7. MICROCLIMATE OF THE OPEN AND SOME TYPICAL CROP-FIELDS

A series of investigations undertaken by the author (GOVINDASWAMY 1953, RAMDAS 1951, 1953) show considerable differences between the

Table X. Thermal balance at ground surface

Gain in grammes-calories per square centimetre per day		Loss in gramme-calories per square centimetre per day	
a. Visible radiation from the sun and sunlit sky during daytime	780	a'. Visible radiation from the sun and sunlit sky reflected by ground	125
b. Thermal or infra-red emitted by water vapour, carbon dioxide and ozone of the atmosphere during day and night	691	c. Thermal radiation emitted by the ground surface during day and night	950
f. Heat gain by condensation or absorption	20	d. Convective heat loss from the ground surface (mainly during daytime)	350
		e. Nett heat transfer between ground surface and soil layers below (during daytime and night) by conduction	35
		f'. Heat loss by evaporation	20
Total gain	1491	Total loss	1480

temperature, humidity, wind velocity, rate of evaporation, etc. in the open and inside fields of different crop plants. These differences depend on the density of the plant cover, the extent and vertical distribution of the foliage, by shading of the solar radiation by plants and the wind-break effect of the plants.

A. Some typical microclimates and their diurnal range

We may consider here the dry-bulb temperature in °C and the partial pressure of water vapour in mm of mercury, measured by sensitive portable instruments like the Assmann psychrometer (ventilated). These data have been recorded in the Central Agricultural Meteorological Observatory at Poona daily, both above a bare ground in the open and inside a number of typical plant communities, particularly well known crop plants, at the epochs of maximum and minimum temperatures. Figs. 23, 24 show the mean values of the dry-bulb temperatures and the vapour pressure at various standard levels above the ground, during the period December 22, 1945 to January 17, 1946 in the open ground and in fields of crops like cotton, betel-vine, wheat, double-bean and sugar-cane. The curves showing the values at the epoch of minimum temperatures are marked N, and those of maximum temperature X. The horizontal separation between N and X curves in each environment indicates the diurnal range of temperature and the vapour pressure respectively at each level. The heat content of a parcel of air depends on

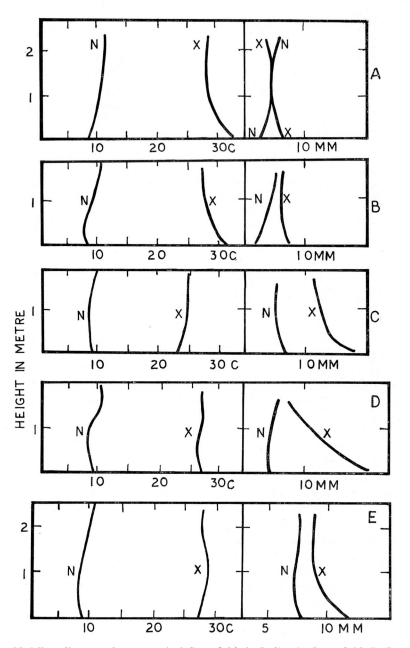

Fig. 23. Microclimates of some typical Crop-fields in India. A. Open field, B. Cotton field, C. Betel-vine, D. Wheat field, E. Double bean-field.

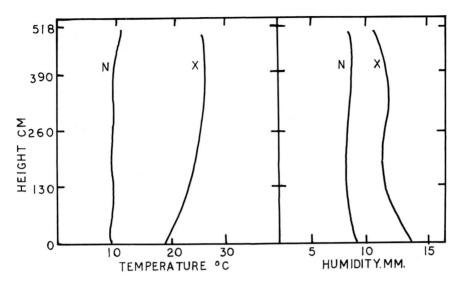

Fig. 24. Microclimate of the sugarcane field.

its temperature. The vapour pressure in mm of mercury also indicates its water content, for it can be shown that a vapour pressure of one mm of mercury corresponds to one gramme of precipitable water per cubic metre of the air sample. We can thus define the identity of air sample or its microclimate uniquely by its heat and water contents, that is to say by its temperature and vapour pressure, so that the curves in the figures directly indicate the exact difference between one environment and another and even the nature of the variations with the differences in the height of the plants within the same environment. In the open ground, the diurnal range of temperature is maximum at the surface and decreases with height. The vapour pressure, however, shows a diurnal variation only close to the ground, on account of the moisture exchange effects between the soil and air layers above it. There is hardly any diurnal variation of this factor at higher levels. In rather sparsely growing fields like cotton, the diurnal variation of temperature is only slightly different from that of the open, but the vapour pressure variation shows a significant increase, particularly at the lower levels. The betel-vine is a heavily irrigated crop, in which the creepers are supported by shade plants. From the figure it is evident that the diurnal temperature range shows a significant decrease at all levels, but the separation between the N and X curves of vapour pressure is quite large. Furthermore, we also note that both the curves have shifted to the right side, due to the high soil moisture and plant transpiration. Irrigated wheat fields show a reduced diurnal temperature range and a considerable rise of vapour pressure, particularly at the lower levels. Double-bean field shows a moderate diurnal temperature

131

Table XI. Mean wind velocity in the afternoon in different crop fields, expressed as percentages of the velocity at corresponding levels in the open

Height in cm above ground	Percentage wind velocity					
	Millet	Sugarcane	Cotton	Wheat	Tobacco	Double-bean
7.5	29	18	32	20	51	30
15.0	24	17	29	17	50	29
30.5	28	20	26	15	48	24
61.0	26	16	34	21	52	25
91.4	29	17	50	40	58	28
122.0	39	18	64	57	68	29
183.0	47	16	75	75	79	36
244.0	60	16	78	—	—	52

range and vapour pressure at all levels, up to the top of the plants. The sugarcane field shows the largest deviations from the open field. The crop is tall, often 4.6 m high, with dense growth and foliage to the canopy that now acts as the source of heat to the air layers above and completely shade the ground. The solar heating being thus confined to the canopy, it is much warmer than the sheltered regions below. We thus have actually an invertion layer from the cool ground to the level of the canopy, the lapse layer of the free atmosphere existing only above the canopy. This daytime inversion in a sugarcane field occurs at a time when in the open ground there are tremendous lapse rates of temperatures with accompanying severe shimmering and turbulence. This inversion with great stability, little wind and no turbulence is characteristic of all canopied vegetations, like forests. In conclusion, we may point out that the presence of plant communities tends to decrease the air temperatures; there is a compensating increase in the vapour pressure as well as its diurnal range, owing to the evaporation and transpiration from the plant.

B. Effect of plant communities on wind velocity

RAMAN (1943) has studied the wind-break effects of plant cover by different crops at Poona, by comparing the wind velocities in the open and inside various crops fields at different levels, using a sensitive hot wire anemometer. The observations are summarized in Table XI.

C. Effect of environment on the evaporating power of air layers near the ground

The problem of effect of different environments on the evaporating power of the air layers near the ground was investigated in Poona with

Table XII. Evaporation in different environments

Environment	Evaporation as percentage of that in the open at corresponding levels		
	122 cm	60 cm	30 cm
Winter millet field	62	43	37
Double-bean field	41	27	25
Sugarcane field	34	30	27
Betel-vine field	25	21	20
Actual evaporation in the open in cm	9.55	10.56	9.8

the help of Piche evaporimeters. It was found that the evaporation tends, in general, to increase with height. It is least in the betel-vine field, in which the wind velocity is also minimum and the humidity is very high. Next comes the sugarcane field, which is also a dense crop that is frequently irrigated. In the fields of double-beans and millets, the increase in evaporation with height is rather more pronounced. Table XII gives the mean values of evaporation, expressed as percentages of those observed in the open fields at corresponding levels. These data are based on the records for the period December 1940 to February 1941 and the actual values of evaporation in the open are given in cm at the bottom of the table.

REFERENCES

Govindaswamy, T. S. 1953. Rainfall abnormalities week by week. *Irrig. & Power J.*, 10(2–3).

Raman, P. K. 1943. The wind break effect of crops. *Indian J. agric. Res.*

Ramdas, L. A. & S. Atmanathan, 1932. The vertical distribution of air temperature near the ground during the night. *Gerl. Beitr. Geophys.*, 37: 116.

Ramdas, L. A. 1935. Frost hazard in India. *Curr. Sci.*, 3(3).

Ramdas, L. A. 1946. Rainfall of India – A brief review. *Emp. J. exptl. Agric.*, 14(54).

Ramdas, L. A. 1946. The microclimates of plant communities. *Indian Ecologist*, 1(1).

Ramdas, L. A. 1948. Some new instruments and experimental techniques developed by the Agricultural Meteorology Section at Poona. *J. Sci. Indust. Res.*, 7: 16–29.

Ramdas, L. A. 1948. The physics of the bottom layers of the atmosphere. Indian Sci. Cngr. Presidential Address: Physics Sec.

Ramdas, L. A. 1951. Microclimatic investigations in India. *Archiv. met. Geophysi. Bioclim.* (B) 3.

Ramdas, L. A. 1953. Convective phenomena near a heated surface. *Proc. Indian Acad. Sci.*, (A) 37.

RAMDAS, L. A. 1956. The movement of moisture through the soil. Proc. Symp. Groundwater Central Board of Geophysics, Publ. 4.

RAMDAS, L. A. 1956. Meteorology of the air layers near ground 1–3. *Tech. Notes Indian meteorol. Depart.*, 3, 9, 210.

RAMDAS, L. A. 1956. Phenomena controlling the thermal balance at ground surface. Proc. UNESCO Symp. Climat. & Microclim. Canberra.

RAMDAS, L. A. 1957. Evaporation control. *Indian J. Meteorol. & Geophysics*, 8 (Special number).

RAMDAS, L. A. 1960. Weather and crops in India. Special Monograph Indian Council of Agricultural Research, pp. 127, fig. 56.

RAMDAS, L. A. 1962. On the spreading of active organic compounds as monomolecular films on clean water surface and the use of some of them for evaporation control. Proc. UNESCO Symp. Water Evapor. Control, Poona, pp. 1–18.

RAMDAS, L. A. 1968. Monsoon and rainfall pattern in the Indian Subcontinent. Mountains and Rivers of India: 21 Internat. Congr. Geogr. pp. 231–257.

RAMDAS, L. A. 1968. The cold layer above ground during clear nights with little or no wind. Internat. Symp. Radiation including Satellite Techniques. Bergen. World Meteorol. Organiz.

RAMDAS, L. A., P. JAGANNATHAN & S. GOPAL RAO, 1954. Prediction of the date of the establishment of the southwest monsoon along the West Coast of India. *Indian J. Meteorol. Geophys.*, 5(4).

RAMDAS, L. A. & K. P. RAMAKRISHNAN, 1956. Wind energy in India. Proc. Symp. Wind & Solar Energy. UNESCO, New Delhi, pp. 42–55.

RAMDAS, L. A. & S. YAGNANARAYANAN, 1956. Solar energy in India. Proc. Symp. Wind & Solar Energy, UNESCO, New Delhi, pp. 188–197.

SIMPSON, G. C. 1921. *Quart. J. R. Meteorol. Soc. London*, 47 (199).

V. LIMITING FACTORS

by

M. S. MANI

Contrary to our expectation from the size and diversity of ecological conditions and habitats, the limiting factors that underlie the biogeographical characters of India are by no means complex. Two important factors, the geomorphological evolution of India and the spread of human settlements in India have played a dominant rôle in the biogeography of India. While geomorphological changes have largely contributed to the specific composition, unique characters and the area of our flora and fauna, the spread of human settlements and particularly recent human interference have largely influenced the characteristic distributional patterns. The principal event of the geomorphological evolution that has materially contributed to the general ecology and biogeography of the whole of India is the drift and breakup of the Gondwana landmass, in accordance with the theory of WEGENER, culminating in the uplift of the Himalayan System. From the biogeographical point of view, the immediate effects of the Himalayan uplift are 1. the establishment of direct land connection of the Gondwanaland relict of the present-day Peninsula of India with Asia and 2. the shaping of the present-day monsoon-climate pattern of the region.

The Himalayan uplift is indeed the key to all the peculiarities of the biogeography of India. Even at the present time, the Himalaya determines the climate, flora and fauna of the whole of India, including the extreme south of the Peninsula. Most students of biogeography of India seem to have, however, so far either greatly underestimated or even totally overlooked the tremendous importance of the part played by the Himalayan uplift, and have at the same time over emphasized the rôle of climate, especially the rainfall pattern, when describing and interpreting the faunal distribution of India. A great many peculiarities of distribution have consequently remained obscure and unexplained or have been explained on wild and fanciful lines.

The climatic conditions prevailing in India today, while naturally of considerable ecological importance, have in reality played only a minor rôle in the evolution of our flora and fauna and in determining the present-day distributional patterns and other biogeographical characters of India. The present-day climatic conditions and the biogeographical characters of India are to be traced, strictly speaking, to the profound influence of the geomorphological evolution of India during the past, particularly during the Tertiary times. The climate of a region, like everything else, is constantly changing. The climatic conditions that prevailed at the times when the principal component elements of our fauna spread to the areas, which they inhabit at the present time, were evidently very different from the present-day climate of India.

1. Present-day Climate and its relation to the Himalaya

The reader will find excellent accounts of the climate of India in BLANFORD (1889), HARWOOD (1923), RAMANATHAN et al. (1933), RIEHL (1954), THORNWHITE (1948) and WALKER et al. (1924). The major peculiarities of the monsoon rainfall patterns are described by RAMDAS in Chapter IV. In this Chapter, we shall draw attention to certain other characters of the climate of India, in relation to its general ecology, and deal in particular with the rôle of the Himalayan uplift in shaping these and other climatic peculiarities. The climatic divisions, on conventional lines, mainly on the basis of the prevailing temperature conditions, into the tropical, subtropical, temperate and alpine zones, result in the anomaly of an area falling in one zone because of the January isotherm and in another zone because of the mean annual temperature. Parts of Sind (Hyderabad) are, for example, subtropical on basis of the mean annual temperature, but tropical on basis of the January temperature. Except that the temperature conditions influence rainfall indirectly through changes in the atmospheric pressure, it is the differences in rainfall rather than temperature changes that are of primary importance in Indian ecology. The regional contrasts in rainfall are extremely striking. We have, for example, the mean annual rainfall of over 1000 cm, with a record of 2263 cm in two months, at Cherrapunji on the Khasi-Jaintia Hills while Jacobobad has a mean of only 100 cm and some parts in Sind may remain completely rainless in some years. Though most textbooks repeat that Cherrapunji, with an annual mean rainfall of 1250 cm, is the rainiest spot in the world, recent record shows, however, that this honour goes to Mawsynram, on the same hills, which has an annual mean rainfall of 1750 cm. Despite these extreme contrasts, there is an underlying rhythm of seasons throughout the region, dependent upon the monsoon. Although more than half the region lies north of the Tropic of Cancer, India is ecologically a land of tropical monsoon climates. Our divisions of the year into seasons are dominated by the monsoon, for example, 1. the season of the northeast monsoon, December to March; 2. the transitional hot weather during April–May; 3. the season of the southwest monsoon, June to September; and 4. the transitional period of the retreating southwest monsoon during October–November.

The extreme south of the Peninsula is sufficiently high to capture the necessary southwest monsoon and thus provide optimal conditions for the persistence of a cover of evergreen rain-forests. As is well known, while climate determines the vegetation and the nature of the forest cover, the latter also profoundly influence the climate. When a virgin forest is destroyed by man, as desastrously practised by civilized man or on a limited scale in shifting cultivation of primitive man (Chapter XXIII), even if he subsequently abandons agriculture in the deforested area, the original forest does not by any means regenerate. The climatic conditions may no doubt remain optimal for regeneration of forest for some time, but gradually deteriorate and the area becomes covered by savannah-type of vegetation. Most of Coorg was, for example, formerly covered by evergreen forest, but now almost completely deforested, the evergreen forest has not however, gained in parts vacated by man.

The climate of the extreme south of the Peninsula, as already mentioned, is more humid and the rainfall is more heavy than in the north. The west coast areas of the Peninsula receive at present a rainfall exceeding 300 cm annually and nearly all of this rainfalls during the period of the southwest monsoon. There are here parts with rainfall of over 500 cm. On the east coast, however, the rainfall is between 125 and 175 cm and much of this received during October–December. In the interior of the Peninsula, the rainfall is between 125 and 50 cm. The east coast area of the extreme south is sheltered from the Arabian Sea branch of the southwest monsoon by the Anamalai-Cardamom Block. The projection of the Cauvery Delta, low-lying as it is, shuts off some of the retreating monsoon rain, so that we have to the north of it remarkably low rainfall of only 60–75 cm for a coastal region in the subequatorial latitude. In the north of this coastal area, being in the full track of the Bay of Bengal branch of the monsoon, receives about 150 cm rainfall, but southern Orissa has a rainfall below 125 cm. Further southward lies the largest area of anomalous rainfall in India. In the Western Littoral Regions the rainfall diminishes from the south to the north. The area of Cutch and Kathiawar has a rainfall of 30–40 cm, sometimes no more than 3 cm. In Gujarat the rainfall is about 170 cm in the south and falls off to 100 cm in Surat and 70 cm in Ahmedabad.

Though tropical, the insular nature of Ceylon makes its climate equable, so that there is little difference (6.6 °C) between the day and night temperatures. The annual range is only 2.7 °C. The mean monthly temperatures in the lowlands are relatively high and vary but little from month to month. In the north and east, the mean temperatures are generally higher than in the southwest. The temperatures on the hills are naturally lower. At Nuwara Eliya (1880 m), the mean temperature is about 15.5 °C. The island is under the influence of the southwest monsoon from late May to August–September and of the northeast monsoon from November to January. The southwest of Ceylon, known as the Wet Zone, has a much higher rainfall than other parts. In the Dry Zone, the southwest monsoon brings very little rain, so that June–August are relatively dry, but there is some rain during September–November and a large portion of the rainfall of this zone is during the northeast monsoon months.

The Wet Zone is climatically analogous to the extreme south of the mainland of the Peninsula, particularly Travancore, but the seasonal contrast in the rainfall is less marked in Ceylon. The Dry Zone of Ceylon is climatically analogous with the Tamilnad, though in the latter area the little-monsoon effect is slight.

India is large enough to develop its individual monsoon-system and to display marked continentality of climates in the north. The most outstanding character of the meteorology of India is the alternation of seasons called the southwest monsoon and the northeast monsoon. Monsoon (from *mausim* = Season) refers to the great air-current of winter and summer, viz. the northeast and the southwest monsoons respectively. It must be remarked that these directions are true only for the winds over the Indian Ocean, but not strictly in case of the landmass of India. This difference is due, in part at least, to the excentric position of the area of the greatest pressure-reversal and general lie of the land and sea, mountain and lowland plains. The actual direction of the prevalent winds is southwest or northeast in relatively few localities.

1.1. THE MONSOON-DOMINATED SEASONS

The cold weather lasts from October to the end of February. During January there is a pronounced temperature gradient, from the north to the south. In Peshwar, for example, the mean temperature during January is 10 °C, in the Punjab Plains 12.7 °C, Benares 15 °C, at Madras 24 °C, Calicut 25.6 °C and Colombo 26 °C. During this month the days are generally warm but the nights are cold and there is slight frost in the Punjab Plains, but frost is unknown in Madras and further south. There is during this month a feeble high-pressure over the cold plains of the northwest India, from which the winds blow outward towards the Equatorial belt of low pressure, gathering force as they move. It must be pointed that there is no connection between the low-pressure area of the northwest India and that of Middle Asia, because of the effective Himalayan barrier projecting into the high atmosphere. India does not, therefore, suffer the intense cold winds that sweep across China during the winter months. The winds in India are on the other hand lighter and no more than three or four km per hour in north India. Relatively high pressures prevail over the northwest India by the end of December, so that the general movement of air in the north is down the R. Indus and the R. Ganga Plains, towards the low pressures prevailing just north of the Equator. The winds are easterly over the Peninsula, since the low pressures extend further north over the Arabian Sea than over the Bay of Bengal. The gradients are not steep enough to produce strong winds. This period is characterized by striking temperature contrasts. The temperature ranges from under 26 °C in the south to 10 °C in the Punjab and the humidity is generally low. The high nocturnal radiation results in low temperatures in the northwest. Insolation is correspondingly high during the day, so that in localities like Lahore the January maximum may be 20 °C and the minimum 4 °C. The absolute minimum may be even —4 °C. Ground fogs, specially for three or four hours from sunrise, due to temperature inversion, are frequent in the north and in the Deccan Plateau. The extreme south of the Peninsula is more humid than the rest of the country and in combination with the lower latitude has a much smaller temperature range. Trivandrum has, for example, the January maximum 34.4 °C and the minimum 22 °C (the mean annual temperature range is only 2.7 °C). The cold weather is generally rainless over most of India, but there may be occasional late-monsoon storms. The southeast littoral area receives some precipitation from the retreating southwest monsoon. In the northwest, the cold weather rainfall is not very heavy. The cold weather precipitation at higher elevations in the north and northwest is as snowfall and is associated with depressions, mostly of Mediterranean origin. Some secondary depressions developed over Iran may also perhaps contribute to it. In the extreme northwest, this cold weather rain is actually more than that of the usual rainy-season rain. Peshawar has, for example, during November–April nearly 20 cm of its annual 40 cm rains. These depressions bring some rain as far east as the western parts of the Uttar Pradesh. The rôle of the winter rains in the substenance of the Himalayan snow is considerable.

During the cold weather, the outwards winds are controlled directionally by the general topography of the land, so that they are westerly or northwesterly down the Ganga Valley, northerly in the Ganga delta and northeasterly over the Bay of Bengal. These offshore winds are mostly dry, so that January and February remain mostly cloudless over most parts of north India. Being within the influence of the Equatorial belt, Ceylon and the extreme south of the Peninsula are, however, marked by temporary northward migrants of the Equatorial low-pressure belt. This has the result of bringing considerable rainfall to Madras and the southeast parts of the Peninsula during November and December, i.e. in the period of the retreating southwest monsoon.

Cyclones, originating over the Mediterranean area during December–March, travel eastwards across Iran and Baluchistan or even Afghanistan and down over the Punjab Plains. They are only shallow depressions and are usually accompanied by light winds and provide for appreciable rains in north Punjab, but die out before reaching the lower parts of the Ganga Plains. Though the rainfall is small in comparison to southwest monsoon rains, these cyclones probably account for the bulk of the snowfall on the mountains of Kashmir and the northwest.

The mean temperatures rise and the pressures decrease from early March. The mean temperature is over 37.7 °C during March in the north. By May it rises to over 40 °C in Deccan, over 43 °C in the Punjab and 49 °C in Sind. Jacobadad has recorded the highest maximum temperature of 52.2 °C. In the humid Ganga Delta, the mean temperature during May may exceed 29.5 °C and over 32 °C in the middle course of the Ganga Plain*. The relative humidity may be as low as 1 % and the diurnal variations are large, especially in the interior. The diurnal range in Sind during May ranges from 24–28 °C to 40–43 °C. The extreme south of the Peninsula has, however, a more equable temperature regime. Along the sea coast, the general heat is mitigated by sea-breeze but at the same time higher relative humidities prevail in these areas. In the northwest India the relative humidity is low. The low-pressure is fully developed by May and light onshore breezes prevail along the coast. The northerlies are, however, still over the Arabian Sea.

In the south and along the east and in Assam-Bengal, the precursors of general rain or the socalled mango-showers, are associated with local depressions and conventional movements. There are violent thundery squalls of northwesters, often bringing considerable rain. In the Indo-Gangetic Plains, however, similar disturbances cause fierce dust-storms.

During April–May a feeble but distinct low-pressure area develops over India and the winds bring considerable rain to South India and southeast Ceylon. These rains fall during violent thunder-storms, developing in the late afternoons and continuing in the evening.

The monsoon 'bursts' about the middle of June and is marked by steady southwesterly winds, the goal of which is the low-pressure area over northwest India, which as explained earlier, has been in existence for a month or two even before the arrival of the monsoon. The normal equatorial low-pressure, towards which the southeast trade winds are drawn, is separated from the north Indian low-pressure area by a ridge of higher pressure. This barrier is suddenly overcome and the southeast trades are drawn to the north Indian area. The main mass of the maritime tropical air overwhelms the continental pressure that exists to the north. The rain-bearing wind is much stronger than the lighter cold-weather winds described earlier. The direction of the winds is controlled to a marked degree by the peculiarities of relief of the land. There are two principal streams, separated by the mass of the Peninsula. The influence of that stream over the Arabian Sea extends hardly north of the Gulf of Cambay. The Bay of Bengal stream blows up the Ganga Plain as an east-wind and reaches the Punjab from the southeast. The monsoon bursts first on the west coast and arrives later in other parts. The monsoon arrives at Bombay about June 5th and retreats about October 15, in Bengal arrives about June 15th and retreats about 15–30 October, and in the Punjab it arrives about the 1st of July and retreats about September 14–21. The peculiarities in the dates of arrival of the monsoon rains are discussed by RAMDAS in Chapter IV.

---

* In 1972, GAYA in the Gangetic Plain recorded a maximum summer temperature of 51 °C.

The bulk of the rainfall of India, with the exception of the Madras Coast, is derived from the southwest monsoon. However, even during the monsoon period and in the wettest area, it does not rain continually. Heavy downpours of rains are interspread with intervals of fine weather. Except in the driest areas of the Punjab and Sind, where the hot weather conditions continue even in June–July, the onset of the monsoon in the rest of the country (north India) is accompanied by a drop in the mean temperature.

The onset of the rains in the west coastal regions is marked by a characteristic abruptness. Strong winds from the sea, violent lashing rain and sudden drop in the temperature mark the onset of the monsoon rains. These changes are naturally much less sharp, but all the same well marked inland.

The monsoon winds blow from the west or south of the west over the Peninsula and are stronger on the Bombay-coast than elsewhere, so that the Peninsular rainfall is strongly orographical. The Western Ghats receive 200–250 cm or more of rainfall on their surface and show a well marked rain-shadow to the leeward. The conditions may be illustrated by the rainfall at Mahabaleshwar, which is situated partially on the crest of the Western Ghats. The rainfall here is about 652 cm, but at Panchgani hardly 16 km to the east the rainfall is only 170 cm and at Wai another 16 km further to the east, the rainfall is only 72 cm.

The conditions in the north India are, however, very different. The monsoon current over the Bay of Bengal is southerly. It is deflected by the Himalaya in the Assam-Bengal area, so that it flows up the R. Ganga from the southeast. Except in Assam and on the southern slopes of the Himalaya, there are no pronounced relief features athwart its course, so that there is a fairly steady diminution in the amount of rain from Bengal to the Punjab. A comparable decrease may also be observed on the southern slopes of the Himalayan ranges, though with higher total rainfall than on the plains. The low pressure trough extends southwest from the Indus low-pressure centre, along a line south of and roughly parallel to the R. Ganga, into Orissa. Here converge the Arabian Sea and the Bay of Bengal currents. This is, therefore, an area of unstable air and this area forms an avenue for depressions forming in the Bay of Bengal and thus heavy local rains result.

The pulsatory course of the monsoon rains is associated with the passage of depressions or the interlocking of new and old air-masses. In the northwest India most of the rainy season is actually dry, except perhaps when cyclones press into the Punjab and Rajasthan. The southern Punjab, western Rajasthan and Sind receive under 25 cm of rain and Jacobabad receives only 10 cm. Even this rain falls often in violent downpours, from either exceptionally strong depressions or local convections. As this area of most intense low pressure adjoins the Arabian Sea and due to the absence of relief barriers, there is extreme aridity.

On the north and west, the region is shut off by the Himalaya and the Afghan-Baluchistan ranges respectively. The air that may arrive here from the continental interior is warmed by the descent. Not much moisture is also left in the 'Bay of Bengal air' by the time it has traversed distance of some 1900 km of land up the Ganga Plains. As the area of maximum temperature and minimum pressure is some way inland, the Arabian Sea current is largely deflected around the low pressure centre and brings some rains to the Aravalli Hills, but not to the Thar Desert. The air that enters directly the Sind coast is derived mostly from originally dry currents, trending from the northwest across Makran and swinging northeast towards the pressure centre. This air picks up some slight moisture during its brief sea-passage and gives rise to the general humidity at Karachi. Condensation does not, however, take place because before it can occur the air mass meets a strong and dry anti-monsoonal upper air current from Baluchistan. This prevents cloud-formation, thus resulting in strong insolation at ground level, with consequent heating and loss of humidity. The southern slopes of the Shillong Plateau in Assam, standing arthwart the monsoon and immediately above the warm floodwaters of the Surma Valley, have rainfalls of over 1000 cm. This enormous rainfall is, however, restricted to a small area and diminishes rapidly in the rain-shadow area to the north, in Shillong, Gauhati and other places. To the northeast the Upper Assam Valley is a funnel into which the moist air is driven up the Brahmaputra itself and this area receives a total of about 250 cm of rainfall.

The rains become less during October, during which month there is also a slight rise in the mean temperature. The land is, however, still water-logged and the atmosphere is thus humid. During November–December there is a drop in the mean temperature in north India. Over the sea and land, the southwesterly air current ceases almost completely. Local variations of heat and moisture give rise to tropical cyclones, most of which originate near the Andaman Islands and move towards the west or northwest over the Bay of Bengal and bring rains to the Madras Coast. November–December are the wettest months in this region, but the rest of India remains almost dry.

The low pressure is breaking or broken up by October and thenceforward the low pressure centre shifts back to the equatorial latitudes. Depressions become weaker and less frequent over India as a whole. There is, however, no abrupt change as found at the time of the onset of the southwest monsoon rains. The drier conditions begin over most of north India, with westerly winds during September. Pressure remains low over the Bay of Bengal, forming cyclones that curve round to the north of the central low pressure and approach the east Coast of India from the west. This spasmodic activity brings rains to the northern coastal districts of Madras during October–November and to the southern districts during November–December. The latter area is rain-shadow with respect to the Arabian Sea branch and largely bypassed by the Bay of Bengal current earlier. The cold weather rainfall of the southeast is thus due to the retreating southwest monsoon and not as is often supposed to the northeast monsoon, which has not yet properly set in at this time.

The low air-current of the southwest monsoon withdraws from the Punjab about the third week of September, from the western parts of the Uttar Pradesh in the last week of September and from the eastern parts of the Uttar Pradesh and Bihar in the first or the second week of October and from Bengal, Upper and Central Burma about the third or the fourth week of October. The dryland-westerly wind extend eastward during this period, down the Ganga Plains and at the same time increase in intensity. These winds are, however, fully established over the northern parts of the Bay of Bengal, the whole of north and central India only by the end of October.

While these changes are taking place in the north and in the area of the Bay of Bengal, similar changes arise also in western India and in the Arabian Sea. The southwest and moisture-laden currents usually prevail at the beginning of September, but decrease in their strength, elevation and volume and retreat from the head of the Arabian Sea, northern Bombay and Rajasthan during the second-third week of September.

During the period of the southwest monsoon proper, pressure is lowest in a belt across the Persian Gulf, north Arabian Sea, Baluchistan, Sind, east Rajasthan and southern parts of the Uttar Pradesh. As the temperature falls first gradually, and more rapidly later to the middle of December, the air contracts over the cooling area so that there is a flux in the higher regions from the areas to the south (Indian Ocean). This double condition results in a continuous increase of pressure over the land-areas of India, maximum in places (northwest India), where the temperature decrease is maximum and the pressure was lowest. The monsoon current recurves at the head of the Bay of Bengal through north and northwest to the west, as at the beginning of the monsoon current, but now it is not due to the mountain barrier but to the special pressure conditions in the Bay of Bengal. The low-pressure usually passes out of the Bay of Bengal to the Equatorial belt during the third week of December.

1.2. The mechanism of the indian monsoon

Some meteorologists have attempted to explain the peculiarities of the Indian monsoon wholly on the basis of the theory of heat-flow over northwest India, thus completely disregarding the rôle of the Himalayan barrier. It is, however, obvious that with a mean elevation exceeding 6000 m and therefore rising almost to the lower half, at least to the meteorologically effective half of the atmosphere, the Himalayan System shuts off the Indian side from the air-masses generated over Middle Asia. Although these influences affect the upper air over India, it must nevertheless be remembered that the monsoon is confined strictly to the lower layer of the atmosphere, so that it is very profoundly influenced by the Himalayan barrier.

The explanation of the sudden burst of the monsoon in India, particularly in view of the gradual rise of temperature and pressure earlier, must therefore be sought for not only in the pressure conditions over the Indian Ocean as a whole, but also in the alignment of the Himalayan mountain barrier. If the Indian Peninsula did not exist, the movements of air between 35° SL and 35° NL should approximate to the conditions observed in the Pacific and Atlantic Oceans. The northeast winds would then be steady and permanent over the northern half and the southwest winds in the southern half. In narrow belt in between, near the Equator, but shifting somewhat northward or southward during the year, would prevail variable winds, with frequent squalls. As is well known, the mean annual temperature symmetry is not in respect to the Equator, but to the socalled 'heat-equator', situated about 5° NL, so that the southern hemisphere is slightly cooler than the northern hemisphere. A region of low pressure (the socalled equatorial trough) exists between these subtropical high-pressure zones. This trough is centred near 5° SL during January and at 12–15° NL during July; the trough thus migrates through about 20° latitude between seasons. This migration influences the seasonal march of cloudiness and rainfall and the formation of tropical storms. The conditions observed in the Atlantic and Pacific Oceans are due to the large and permanent differences of temperatures between the tropical and temperate zones. Atmospheric pressure is lowest near the Equator, but increases to a narrow belt of high pressure about 35–40° NL & SL.

The presence of the large landmass modifies, however, the movements of air and the pressure conditions very largely in the Indian Ocean. In the northern portion, the otherwise permanent air movement is converted into a periodic movement, viz. the monsoon. During one part of the year, Middle and southern Asia are much cooler but during the rest of the Year the mean temperatures are considerably higher than at corresponding latitudes on the Atlantic and Pacific Oceans.

When fully established, the southwest monsoon is fed by the southeast trade-winds, the goal of which is the equatorial low. During April and even May, a tract of relatively high pressure over the Arabian Sea extends from Oman to Cutch and thence southwards and south-eastwards over the Peninsula to the Cauvery Delta. The Arabian Sea lies indeed in between two great low-pressure zones, one of which is over the Punjab and the other over eastern Sudan. The winds round the Indian low-pressure zone trend mostly from the northwest. The air currents do not cross

144

the west coast during May, but their relative humidity is much less than that of similar currents during July. With the intense insolation during the months of May and early June, there is an intensification of the low-pressure zone over India and there is a steepening of the pressure-gradients. The change from the anticyclonic to cyclonic conditions takes place, however, without any large alteration in the circulation at elevations between 2000 and 8000 m above mean sea-level, west of 70° EL. There is in effect a transport of air from the north to the south. The rising air flows southwards to settle in the equatorial low zone. Once this fills up, the moisture-laden southeast trade-winds sweep unchecked towards the Indian low-pressure zone.

The northwest parts of India are an area of scanty rain, but intense insolation, so that the surface low-pressure zone is perpetuated and the gradients over India itself are much steeper during July than in May. Strong winds and depressions penetrate the interior of India. The pulsatory nature of the Indian monsoon is attributed to the interaction along local fronts of three types of air-masses, viz. the old-monsoon air, the fresh-monsoon air and old continental air from the northwest India. The relative humidity of the old-monsoon air is less at the surface than of the fresh monsoon air, but the former air is warmer because of its contact with land and has thus a higher absolute humidity. It may act as a warm air-mass, rising above the fresh-monsoon air or a wedge of continental air may persist between the old and the fresh monsoon air masses, at elevations of 2000–4000 m above mean sea-level as cold air.

In the western Uttar Pradesh and northern Rajasthan rainfall occurs by the obstruction of the easterly fresh-monsoon air by the continental air from the northwest that spreads either close to the ground or extends as a wedge between the southwesterly branch of the monsoon and its deflected eastern branch, coming up over the Gangetic Plain. Condensation also takes place purely as a result of the direct ascent of the new over the old air.

It is interesting to observe that the intensity of these complex conditions is greatly accentuated by the Himalaya, which influences the whole system of wind in India, up to an elevation of 6000 m above mean sea-level. The mean movement of the air is, therefore, parallel to the Himalaya during the winter and hot weather. During this period, there is maximum wind strength up to an elevation of 6000 m in the east of the Uttar Pradesh and Bihar. The localization of the monsoon-low within India is also due to the effect of the Himalaya. Very little of the immense air-mass poured into India, during the monsoon, trickles over the Himalaya, but most of it joins the complex circulation in the upper air or is perhaps also returned southwards as an anti-monsoonal current.

By the beginning of January, the northeast monsoon air movements over India and the sea are fully established. A belt of high pressure stretches from the west Mediterranean to Middle Asia and the northeast

China. Although this corresponds to the high-pressure belt in the $30°$ NL in the Pacific and Atlantic Oceans, it really lies further to the north. From this belt, the pressure diminishes southwards to a belt, somewhat south of the Equator and thence increases again to a broad belt, extending from southwest Australia across the Indian Ocean to the Cape of Good Hope. The northern high-pressure belt separates the region of storms of north Europe from the area of local storms of the Mediterranean, Iran and India and limits to the north the Indian monsoon region. The pressure conditions accompany and determine the northerly winds in the centre of the Indian Ocean. On the land area, the winds are modified by the trend and elevation of the Himalaya. The mean temperature begins to rise during February in Asia, although only slightly at first. The air movements and the pressure conditions remain, however, the same between $40°$ SL and $35°$ NL, with the air movements of two independent circulations. Because of the Himalayan barrier, the high-pressure zone of Middle Asia is ineffective in regard to the movements of the lower air over India. Clear skies, low humidities, large diurnal temperature ranges and light land winds are observed first in the Punjab during early October, but these conditions gradually extend eastward and southward and finally to about $8°$ NL by the end of December. Air movements are from the west to the east in the Gangetic Plains, curving northwest and north across Bengal and in the Bay of Bengal from the northeast to the east. The movements continue across the Peninsula, from the east and pass out into the Arabian Sea. The Western Ghats shield the west coast from these winds. It is practically opposite in direction in north India and in the Peninsula. In the intervening areas of parts of Kandesh and Berar, the winds are variable and these areas are the seats of maximum deviations from the normal weather conditions. An interesting aspect of the northern wind system is the return of the upper air current over India from the south, deflected to the southwest and west-southwest by the rotation of the earth. It is really a continuation of the ascending humid currents in the equatorial belt, previously passed over large area of sea, but because of the ascent deprived of much of moisture. This current is not, however, fully established till the end of December, when the southwest monsoon has already been replaced by the northeast monsoon.

Storms during the cold weather are due to upper-air currents. A succession of shallow storms passes eastwards across Iran and north India during the cold weather. These are not, however, continuations of the disturbances over Europe and are areally land-formed. In most areas there is a slight precipitation, but at higher elevations of the Hindu Kush and the Himalaya precipitation is large. A part of these conditions is the formation of a brief secondary depression in the Punjab and result in moderate to heavy rains in the plains and violent gales and heavy snow-storms on the Himalaya. In front of these storms there is dry clear weather and stronger and cooler winds prevail. In parts of Baluchistan and India, temperatures may fall as much as 11–16 °C within 48 hours as a result of the passage of these warm and cold waves.

Mention must be made of another interesting peculiarity of the south-west monsoon rains. While these rains set in generally over Burma in April–May, they are, however, retarded over India, where as already explained, the temperature rises very high during May. The first burst carries the monsoon rains nearly to the north of the Peninsula, but then the monsoon advances more gradually towards the Himalaya and the northwest parts of India. The rains last no more than two months in the Indus area, so that the mean rainfall variation is high from year to year.

The pattern of distribution of the pressure conditions near the sea-level over the north and south Indian Oceans is considered by some as pro-viding a clue to this peculiarity of the Indian monsoon. The channeling effect produced by the elevation and shape of the Himalaya should not also be overlooked. A considerable portion of the belt of westerlies circles the southern rim of the Himalaya in December. Even at elevations of 8000 m, the streamlines follow the contours of the Himalaya. A trough develops near 85–90 ° EL. The narrow and intense winter jet transports much air-mass and acts as the dominating influence during winter and on the precipitation over the south and central China. The westerlies retreat to the north of mountains during summer, so that a continuous north-south trough extends from the westerlies to the easterlies near the 75 meridion, a little to the west of the mountain mass and about 10 ° to the west of the winter position. Assuming that the line of subtropics lies at 35 ° NL, a column exists over northwest India. To the north of this, the westerlies move clockwise along the northern boundary of the Himalaya. Burma is situated east of the position of the mean upper-air trough, but India is situated to its west in winter and east in summer. The superposition of the high tropospheric air-flow pattern and its attendant field of pressure on the low-level circulations accelerate the monsoon over Burma and at the same time retard it over India, as long as the trough remains in the Bay of Bengal. The shifting of the trough to 75 ° EL is followed by southerly wind component at high elevations over India-Burma and reinforces the monsoon over the whole region.

The view relegates the theory of heat-flow over northwest India to a secondary rôle. The heat-flow results from the intense insolation under clear skies, which cannot be considered a primary factor. The sudden displacement is explained on the basis of analysis of the winds at elevations of 1525 m, 3050 m and 6000 m during May–June. The Equatorial shear-line or the forward edge of the monsoon is located at an elevation of 3000 metres.

In conclusion, it might be stated that the Himalaya exercises as dominating an influence on the meteorological conditions of India as of the vast areas to the north. It governs the circulation systems of air and water. The snow-covered mountains naturally have a very pronounced moderating influence on the atmospheric temperature and humidity over the whole of North India. By reason of the high altitude and the situation directly in the part of the monsoon winds, it is most favourably conditioned for the precipitation of moisture either as rain or as snow. The snowfields and glaciers feed the perennial rivers of India. The Himalaya has a major contributary factor in desiccation that is overspreading Middle and Central Asia. The most significant recent influence of this effect is the vast desert tract of Tibet and the Tarim Basin to the north of Tibet, the Tarim occupying an area almost as large as the Indo-Gangetic Plains. These are among the most desolate regions of the world at present. It is, however, well known that the aridity of all these areas is of very recent origin. In the case of Takla Makan Desert in the Tarim Basin, the desolate condition arose within historical times. These formerly well-watered and fertile areas have been fighting against adverse climatic conditions since end of the Pleistocene glaciations, and though forest and cultivation persisted as late as the early years of the Christian Era, they have eventually succumbed. The lakes in Tibet show low-level strands of former millenia and the water of most of the lakes is highly saline and many of them contain deposits of gypsum, borax, etc. Several ruins of former settlements that flourished 1000–4000 years ago may be seen along old river courses. The ever increasing desiccation of these areas, closely bound up with the uplift and interposition of the lofty Himalaya, has had a devasting effect on the river systems, which were formerly extensive and well developed. At present the river system is also decayed and withered to such an extent that the exceedingly few streams that still exist lose themselves completely in the sands and surface debris. The glaciers on the Kun Luen mountains are wasting away and their ice is retreating. The perennial north-flowing rivers that formerly arose from these glaciers disappear now near the foot of the mountains, in the piedmont gravels and shifting dunes of the Takla Makan Desert. The water brought by the monsoon winds is turned back to India by the mighty Himalayan System. The North India lies along the Tropic of Cancer. This belt, which is part of the Earth's desert zone, is not dry because of the configuration of the Himalayan mountain chain and its influence on rainfall. North India is also saved the gradual desiccation that has overspread Middle Asia.

It may, therefore, be concluded that the origin of the present-day climate of India is closely bound up with that of the Himalaya. The recorded meteorological data, the oldest of which hardly dates back to 1813 from the Madras Observatory, do not naturally reveal trends in periodicity or the long-term changes in the climate. Other indirect evidence enables us, however, to infer that the rise of the Himalaya has contributed to the gradual dominance of arid conditions not only to its north in Tibet and Middle Asia, but even in the south from the western parts of India. RAMDAS has shown in an earlier chapter, for example, that copious monsoon rainfall in Burma and the northeast India has a tendency to be associated with more or less pronounced failure of the monsoon in other parts, particularly the western parts of north India, although the Peninsula is not affected. With the continued rise in elevation of the Himalaya, particularly in its eastern portion, this effect is likely to be further accentuated in the western parts of the Indo-Gangetic Plains. RANDHAWA (1945) has shown, for example, that arid conditions have progressively increased and spread in northern India within historical times. This once again emphasizes the dominant rôle of the Himalayan uplift in shaping the general climate of at least the northern half of India, even if not of the southern parts of the Peninsula. We must also take into consideration the effect of the Himalayan glaciations during the Pleistocene epoch on the general climate of India. It also seems reasonable to expect that the present-day climatic pattern of the whole of India arose only since the end of the Pleistocene glaciations on the Himalaya. The Pleistocene lowering of the general atmospheric temperature was followed by aridity in India. Most Indian zoologists have, however, erroneously assumed that Pleistocene glaciations on the Himalaya resulted in higher atmospheric humidity and precipitation in India. On the other hand, the Himalayan Glaciers locked up so much of the atmospheric moisture that relatively dry conditions prevailed over much of the Indo-Gangetic plains and parts of the Peninsula. It might be remembered in this connection that the Permian Glaciation was followed by drier climate than the warm humid Carboniferous period.

2. The Relation between the Himalayan uplift and the biogeographical composition of India

Analogous to the origin of the present-day climatic patterns, discussed above, the biogeographical composition and distributional patterns are also largely the concomitant results of the Himalayan uplift. The composition and general character of the original Gondwana flora and fauna were very profoundly influenced and modified by the rise of the Himalaya, mainly through the formation of biogeographical routes of interchange and barriers with Eurasia, and to a limited extent with parts of Africa.

One of the earliest and perhaps the most important biogeographical effects of the Himalayan uplift was the establishment of a direct connection between the ancient Peninsular Block of India and Asia in the northeast, where, as explained in Chapter III, the Peninsular Block was rammed against the young Tethyan sediments of Asia, forcing these sediments to flow around in the form of the Himalayan and Burmese Arcs.

The Tethys Sea was obliterated first in the northeast and the older Peninsular Block came into contact with the young rocks, giving rise to the North Burma-Assam region. This region is the meeting-point where the faunas from the east and from the south Asia (now southeast Asia) converged before entering India and spreading on the Peninsula. This brought the Indian Mass into direct land-connection with the areas, which are to day South China, Indo-China, Malaya, etc. The immediate effect of this connection was the initiation of faunal movements from the newly formed areas, with their highly plastic and actively expanding Tertiary faunas, to the faunistically saturated Peninsula. The faunal influx took place through Assam, which may appropriately be described as the faunistic gateway of India – a threshold for the intermingling of an ancient and a young fauna. From Burma-Assam, the route bifurcated in India. A northern branch lay along the foothills and southern slopes of the Eastern Himalaya and a southeastern branch on the Peninsula. The Himalayan branch was connected southwards by a short secondary branch across the so-called Darjeeling-Monghyr Gap with the Peninsular northeast. Finally the Himalayan branch was again connected with the Peninsula through the Aravalli in the west. The faunal movements along the Himalaya have been largely from the east to the west. The southeastern branch on the Peninsula has, however, seen major movements, both from the east to the west and also from the west to the east.

The northwestern routes arose recently, only when the entire Tethys has been completely obliterated. In the course of the Himalayan uplift, the northern edge of the Peninsular Block buckled, down-warped and slid under the Asiatic mass, thus giving rise to the Peninsular foredeep that is now filled by alluvium. The filling-up of the Peninsular foredeep with the detritus from the Himalaya (and partly also with detritus from the northern escarpment of the Peninsular Block) extended gradually westwards from the Assam end. The connection with Asia in the west was eventually established and thus opened up possibilities of faunal interchanges between western Asiatic, North African and Peninsular areas. There is a Himalayan branch across Afghanistan and a more southern branch entering Sind. Through the northwestern routes have entered West Asiatic, southern European and African faunas into India. In Recent times, perhaps even during the Pleistocene, the faunal movements in the northwestern routes seem to have been both from the east to the west and from the west to the east.

The most important faunal routes are, therefore, the products of the vast changes consequent on the Himalayan uplift. There are thus two principal routes, through which faunal movements have occurred, viz. one in the northeast, the Assam Gateway and the other in the northwest. As mentioned above, the northeast route is the older and the first to be established; it is also perhaps the more important of the two. The faunal movements through the northeastern route have been mostly from the east to the west. The great influx of eastern faunas through this route has had far-reaching effects on the biogeography of India.

This influx introduced profound changes in the composition, character and distributional patterns of an otherwise stable faunal area. This influence continues to be observed even today at the extreme south of the Indian Peninsula and also in Ceylon. The autochthonous Peninsular fauna was partly displaced, retreating of the advancing new fauna and partly surrounded and isolated in pockets, until eventually it came to be restricted to the extreme south of the Peninsula.

The major barriers to faunal movements in India include 1. the Himalaya, 2. the Indo-Gangetic Plain, particularly the part that separates the hills of Assam-Burma from the northeast corner of the Peninsula, 3. the Deccan Lava-area and 4. the desert region of the Rajasthan-Sind region.

The Himalaya has nearly effectively prevented the movements of the Peninsular fauna to Middle Asia and has also at the same time confined the Middle Asiatic faunas to the higher elevations of the Himalaya. The Himalaya has thus served at the same time as route of faunal movement from the east and barrier from the north.

The Gangetic Plain between Assam and the northeast corner of the Peninsula separates the Garo Hills of Assam and the Rajmahal Hills of the Peninsula. It is the well-known Garo-Rajmahal Gap in Indian biogeography. This Gap lies in the major route of faunal movement from the east to the west or southwest on the Peninsula and has been considered by some zoologists as an important barrier (see chapter XXIV).

The formation of the Satpura Escarpment, which has served as a barrier since then, resulted from the down-warping of the northern parts of the Peninsular Block, which also created stresses and led to the consequent fissuring of the Block. Through these fissures flowed the Deccan Lavas. These lava flows have very profoundly altered the effective size of the biogeographical area, shifted and reduced the range of numerous species and generally contributed to the formation of a barrier, both physical and climatic, to the south.

As mentioned in earlier chapters, the Deccan lava-flows were intermittent, so that during periods of quiescence, plants and animals from the surrounding areas of the Peninsula colonized the lava-covered regions. Though these faunal movements were localized, they nevertheless created conditions of relative instability in the Peninsular autochthonous fauna. It has generally been assumed by most earlier workers that these lava-flows meant total annihilation of the flora and fauna. This assumption, however, completely overlooks the important fact that while numbers of individuals of species, most of which were then widely and continuously distributed in the Peninsular Block, were eliminated from the lava covered areas, there was by no means a large-scale destruction of the Peninsular fauna. The component species of this fauna were not necessarily restricted to the areas, which are now covered by lavas, but were widely distributed. Intermission in the lava-flow activity was followed by weathering of the lava rocks and soil formation. These changes were associated with rapid recolonization by plants and animals from outside the lava-covered areas. The evidence of fossil plants, insects, Molluscs and Vertebrates, described from the socalled Inter-Trappean sedimentary rocks of the Deccan Lavas, gives a clue to the rich and diversified flora and fauna of those times. The Deccan lava flows have influenced the biogeography of India in another way also. The lava area now acts climatically as a biogeographical barrier, at least for the moisture-loving tropical forest elements from the northeast and north. The lava-covered area thus contributes to isolate the relatively less disturbed southern part from the profoundly disturbed northern and northeastern parts of the Peninsula.

The block-fracturing and marine subsidence in the west coast of the formerly much larger Peninsular Plateau, leading to the formation of the escarpment now known as the Western Ghats, are also indirect consequences of the tectonic movements associated with the Himalayan uplift. The sum-total of these complex events is a drastic reduction of the biogeographical area.

3. The Relation between Distributional Patterns and Man

The influence which the activities of human beings have had on the distributional patterns in India has been largely neglected so far in discussions on zoögeography. We cannot, however, attempt here to deal with this problem in specific details, because it would lie largely outside the scope of the present volume; we shall, therefore, draw attention to some of the more salient effects of human activities on the present-day distributional patterns. Some illustrative cases of the effects of man on our fauna are described by MUKHERJEE in Chapter XII.

As already indicated, the human activities have influenced more the distributional patterns than on the composition and character of our fauna. The distributional patterns have been profoundly influenced, within historical times, by man in two ways, viz. 1. by large-scale and irreversible destruction of natural habitats nearly throughout the region and 2. by extermination of diverse species. In both the cases the general ecosystem has suffered irreversible changes, so that restoration of the patterns of distribution, which prevailed before the advent of man in India, is now wholly ruled out. The human influence on distributional patterns has thus been both direct and indirect.

3.1. DESTRUCTION OF HABITATS

The history of human colonization and spread of settlements in India is too well known to need any discussion here. It is, however, of interest to us to recollect that while the major faunal-floristic interchanges with Asia took place in the northeast, the gateway for the influx of *Homo sapiens* has in the main been in the northwest,though minor infiltrations occurred in the northeast also. It must also be pointed that before the advent of the settlements of the Aryan races in the Indus area, the Pre-Aryan human settlers, who constitute the aboriginal communities, have also come through the northwest.

Paradoxically, however, the early human settlements, including also the Aryan Indus area settlements, do not seem to have been of any significant influence in the biogeography of India. These settlements doubtless altered the qualitative and quantitative characters of the eco-systems, but these disturbances were largely localized in the Indo-Gangetic Plains and soon passed away, so that there was more or less complete recovery. Irreversible changes in the ecosystems were, however, induced by human agency much later and indeed within historical times. The available evidence seems to show that at least until the time of the Mahabharata War (5000 years ago), the ecosystems, flora and fauna, their compositions, characters and the distributional patterns were practically much the same as towards the end of the last Pleistocene glaciations on the Himalaya. These conditions prevailed over perhaps the greater part of the land and in any case the Peninsula and the major parts of the north India were still largely undisturbed. The general flora and fauna and the climax forests were not seriously interfered with by man nearly over the whole of India. Virgin forests, with climax vege-tations, were then far more extensive and nearly all species were also far more widely and nearly continuously distributed throughout the region than we find at present. Even as late as the time of the Moghul rulers, vast areas of Uttar Pradesh (Gangetic Plain) were densely forested and there are records of the Moghul emperors hunting the elephant, buffalo, bison, rhinoceros and the lion in the Ganga-Yamuna Doab (see Chapter XII). Tiger, bear and leopard are now enormously reduced in numbers and wild pig, rodent and monkey are now greatly restricted. Away from the Terai and narrow riverine strips, there is now very little real woodland. These are also relics and are often cultivated patches of *Butea, Bassia, Melia, Acacia arabica, Zizyphus*, etc. Even the secondary wild vegetation, after the cultivation has been abondoned, is markedly xerophytic. In these areas, the pressure of human population has caused the spread of savannah-like cover even in wetter areas in the east. The combined effects of human activities are gradual depression of natural vegetation from the original climax monsoon-deciduous forest types (see Chapter VI) to the open dry-grassland type, making up grazing tracts, with scattered relics of resistant woody plants. This effect is particularly marked over extensive tracts of laterites, where the destruc-tion of forest cover rapidly results in soil impoverishment, climatic deterioration and elimination of diverse habitats and ecosystems.

The destruction of natural habitats over ever-increasing areas pro-gressed rather slowly at first, but came to be abruptly greatly accelerated

153

within perhaps the last two or three hundred years. The principal human activities that led to irreversible disappearance of many natural habitats and deterioration of other habitats as chain reaction in the disturbed ecosystems, include progressive deforestation, primarily for use of human settlements. The primitive races and even to this day the primitive tribes in many parts of India practise shifting cultivation; an area of virgin forest is cleared, very often by starting devastating forest fires, cultivated for a year or more and then abandoned, in favour of another newly cleared forest tract (see Chapter XI). A forest thus cleared once and cultivated does not, however, recover, and regenerate if subsequently abandoned by man and instead of the original forest or other forest type, the tract becomes scrub, savannah or grassland. In course of time extended areas, stretching the whole of the Indo-Gangetic Plains were brought under permanent and intensive agriculture. This was followed by intense irrigation, reclamation of marshland and other steps that further accelerated the deterioration of powers of adjustments in the ecosystems. Lopping of forest for firewood, over-grazing, rapid urbanization, recent industralization and consequent industrial pollution of waters, railroad constructions, etc. contributed to the further destruction of habitats. The introduction and extensive cultivation of plantation crops like coffee, tea, rubber, eucalyptus, cacao and cashewnut have taken a heavy toll of even the relict forests of hills like the Nilgiri. The destruction of given habitats triggered off chain effects in other ecosystems, so that the effects of human interference in one area gradually diffused to far off parts of the country, not directly involved and gave rise to unsuspected remote effects. Deforestation and agriculture also resulted in general impoverishment of soil and altered the general vegetation over extensive tracts. Natural habitats have been totally destroyed nearly everywhere man has settled in India. The destruction has not, however, been so thorough or so large in areas where the socalled 'tribes' or primitive tribal man, a rapidly vanishing being himself, is precariously concentrated and isolated in refugial areas, not easily reached by cultivator races. Some of the inaccessible and less hospitable areas of the Himalaya, Assam and the extreme south of the Western Ghats have thus largely escaped direct destruction of habitats by man, but even in these areas there is abundant evidence of the indirect effects of chain-reaction, due to disturbances in ecosystem elsewhere. It is not possible to estimate the size of the area in which the natural habitats have more or less escaped the effects of human interference; in any case it is certainly less than one percent of its former size.

The destruction of natural habitats by man has been far more extensive and complete in India than perhaps anywhere else in the world. Of the various factors by which this has been brought about, deforestation must be considered as the most important. The forest has been systematically reduced and eradicated nearly over the whole of India (see Chapter VI).

154

The disappearance of the forest has so completely altered the entire ecosystem that the conditions are suboptimal for the greatest majority of species. The other factors, which we have enumerated above, greatly accelerated the effects of destruction of forests.

3.2. Extermination of fauna

Irreversible changes and deterioration of ecosystems were caused not only by the extensive and most thorough destruction of natural habitats, but also by direct extermination of many character species of animals. This extermination has been nearly complete in many cases and in others the final end can no more be prevented. Man has systematically and most ruthlessly hunted animals for food, fur and various other valuable products in India, with complete disregard for conservation of flora and fauna.

The reduction in the distributional range of an unexpectedly large number of species of animals in nearly all groups, including even the higher and actively migratory types like mammals and birds, has reached a high magnitude within historic times. Not only countless genera and species have either disappeared but their habitats have also been destroyed. During the past two hundred years, for example, at least two species of birds and about fifteen species of mammals have been practiclly exterminated. In another Chapter, MUKHERJEE has listed some of these vanishing forms. Among birds, he mentions the pink-headed duck *Rhodonessa caryophyllacea* (Fig. 26) as now almost exterminated, but in the beginning of the present century, it ranged from Manipur-Assam-Burma to the Uttar Pradesh, and all along the foothills of the Himalaya, to the Punjab and was even found in Madras. *Cairina scutulata*, the white-winged wood-duck, is now restricted to northeast Assam within India, but was formerly common throughout Assam (Fig. 27). *Ophrysia superciliosa* the mountain-quail is now restricted to elevations of 1500–2100 m in Mussoorie-Naini Tal area, but was formerly widely distributed even up to the borders of Kashmir. *Choriotis nigriceps* (Fig. 29) the great Indian bustard is now confined to semi-arid and arid parts of Rajasthan, but was formerly found in Deccan up to nearly south of Mysore. Among the mammals *Presbytis johni* the Nilgiri langur (Fig. 41) is confined to the southern parts of the western Ghats, the Anamalai and Palni Hills, but was formerly found throughout South India. *Rhinoceros unicornis*, the Indian one-horned rhino, is today restricted to the eastern (Fig. 31) Terai (Nepal) and small isolated pockets in Assam, but was formerly distributed throughout the whole of the Indo-Gangetic Plain, from beyond the R. Indus in the west to Assam in the east. These are only some of the more striking examples of vanishing fauna. MUKHERJEE has estimated that the reduction in the distributional range of several common Vertebrates of India amounts on an average to over 89 % of their former size. This must be considered as a conservative estimate. He has also shown that this reduction is the result of recent activities of man, in total disregard of game laws, most of which are flouted with impurity. The process of extermination of species became biogeographically significant about two hundred years ago in India. Since India became independent in 1947, partly the haste to break away from centuries-old taboos, restrictions and religious traditions that had most effectively insured the sanctity of flora and fauna, and partly the disastrous policy of civil disobedience preached by politicians during the independence struggle have widely encouraged a wholesome contempt for game laws and nature sanctuaries and government control regulations have only served to create a most flourishing illegal market in various animal and forest products. India has thus effectively destroyed in twenty years of freedom what Nature had endowed her as a result of millions of years of organic evolution. Even the relatively sheltered forms that are not directly influenced by human activities have been involved in these chain processes. The spread of man has no doubt directly affected relatively few species, but even a minor factor that influences relatively few species in a stable ecosystem, indirectly and inevitably involves the entire complex.

It is of considerable significance to note that not only the distribution of plants and animals has been very profoundly influenced by the advent and spread of agricultural and industrial communities of man, but also the distribution of the primitive nonagricultural communities of man (the aboriginals of India) has affected in an identical way. Advanced human communities are capable of largely modifying the physical environment to suit their special requirements or also capable of creating their particular environment. The primitive human communities, dependent largely or wholly on the forest for livelihood, are, however, directly influenced by changes in the physical environment, such as extensive disappearance of natural habitats. As shown by LAL in a subsequent chapter, the aboriginal inhabitants of India steadily retreated in front of the agricultural settlers of Aryan stock, who entered through the northwest faunal route, until at present they are reduced to the status of relict communities, isolated on relatively inaccessible hill forests refugial niches. The history of these primitive communities in India closely parallels that of the flora and fauna and the factors which have influenced them all are identical. The ecology of the primitive communities of man, discussed by LAL in a subsequent chapter, has therefore considerable relevance to the consideration of biogeographical problems.

4. Conclusion

The ecology of the Peninsula, in relation to its biogeography, is remarkable for the pronounced dominance, over the present-day climatic and other environmental factors, of the continuing influence of the effects of the Himalayan uplift and the increasing intensity of the pressure spread of man. These facts distinguish the ecology of the Peninsula at once from that of any other natural region of India. This dominance of historical over the present-day factors is largely due to and also accentuates the biogeographical effects of the senile topography of the Peninsula and the correlated relative stagnation of the evolution of its fauna in the past. The high ecological stability, the preponderance of climax communities of plants and animals, the high faunistic maturity and the stagnation that generally characterized the Peninsula at the end of Miocene times gave place to rapid deterioration of the optimal conditions that exercised a stabilizing influence. The influx of Asiatic faunas by way of the Assam Gateway combined with the disturbing effects of the Deccan Lava flows to radically modify the entire ecological character of the Peninsula. The present-day ecological conditions should be described as biogeographically suboptimal. Like its topography, the Peninsular ecology is remarkable for its senility and relict nature. The conditions became suboptimal, however, only within historical times. This ecological deterioration is the concomitant result of the long continued and steadily increasing pressure of human expansion, the disastrous methods of agriculture and settlements of human communities. As a result of the advance of human civilization in the Peninsula, there has been a progressive deterioration in the climate, elimination and destruction of habitats, impoverishment of flora and fauna and regression of distributional ranges.

That formerly more humid conditions especially humid tropical forests flourished over most of the Southern Block than now is evident from the fact that nearly 70% of the mammals now found in semi-arid areas of the Eastern Ghats are ecologically semi-humid elements. The only recognizable pockets of survival of the conditions that prevailed in former times, are met with in the high mountains of the Southern Block and to some extent in the wooded parts of the Chota-Nagpur Plateau. Everywhere else a combination of natural and human factors has completely altered the original ecology of the Peninsula.

The extreme south of the Peninsula, especially the high hills of the Malabar region and some parts of Ceylon, represent at the present time ecological islands, in which we may perhaps observe the conditions that prevailed before modifications set in. From these regions, there is a distinct ecological gradient of biogeographical optima to the north and the northeast. The peak of this gradient is a relic of the ecology that determined the character fauna of the Peninsula in the past. The ecological islands are by no means stable, but are at the present time gradually dwindling.

The ecological pockets introduce an important factor in biogeographical evolution, viz. isolation. The isolation is partly purely geographical, due to physical barriers like the Garo-Rajmahal Hills Gap, the gaps in the socalled Eastern Ghats, the Palghat Gap, etc. It is to a much greater extent the result of destruction of habitats due to deforestation and human settlements. The rôle of human civilization in accentuating the effects of isolation in the evolution of faunas and distributional patterns has so far been almost completely overlooked. Isolation has the effect of preserving the relict character of the Peninsular ecology and, as we shall see presently, biogeography. Interpretation of the present-day climatic and vegetational peculiarities of the Peninsula, without reference to the historical background, results in creating a number of biogeographical anomalies.

The predominant ecological conditions at the present time, taken in their sum-total, tend strongly to favour a pronounced impoverishment of the major components and character elements of our fauna and severely reduce the size or to break up their range into discontinuous patches. Most of the autochthonous and character elements of the Indian fauna are thus being rapidly reduced to the level of relicts. Biogeographically India may largely be described, therefore, as a land of vanishing relicts. In combination with the geological maturity, the senile topography, faunal stability and the relative evolutionary stagnation that marks the Peninsula, the effects of the present-day conditions tend greatly to accelerate and intensify these processes.

REFERENCES

BLANFORD, H. F. 1889. Climates and weather in India, Ceylon and Burma. New York: MacMillan Company.

HARWOOD, W. A. 1923. The atmosphere in India. *Mem. Indian meteorol. Deptt.*, 24(6): 167–216.

PALMER, C. E. 1952. *Quart J. R. meteorol. Soc.* 78: 126.

RAMNATHAN, K. R. & K. P. RAMAKRISHNAN, 1933. The Indian southwest monsoon and the structure of the depressions associated with it. *Mem. Indian Meteorol. Deptt.*, 26(2): 1–36.

RANDHAWA, M. S. 1945. Progressive desiccation of northern India in historic times. *J. Bombay nat. Soc.*, 45: 558–565.

RIEHL, H. 1954. Tropical meteorology. New York: McGraw-Hill Book Company, pp. x–392.

SUBRAHMANYAM, V. P., B. SUBBA RAO & A. R. SUBRAMANIAM, 1965. Koppen and Thornwhite system of climatic system of climatic classification as applied to India. *Ann. Arid Zone, Jodhpur*, 4(1): 46–55, fig. 3.

THORNWHITE, C. W. 1948. An approach towards a rational classification of climate. *Geogr. Rev.*, 38: 55–94.

WALKER, G. T. & J. C. K. RAO, 1924. Rainfall types in India in the cold weather period. *Mem. Indian meteorol. Deptt.*, 24(11): 347–354.

VI. THE FLORA

by

M. S. MANI

This chapter attempts to give a brief outline of salient characters, composition and ecological peculiarities of the general flora of India. The special features of the vegetation of the Peninsula, the Himalaya and the Eastern Borderlands are discussed by different specialists in the succeeding chapters.

1. General Characters of the Flora

Except perhaps in the higher elevations of the Himalaya and in the more arid parts of Thar Desert and in Baluchistan, the natural vegetation of India is essentially arboreal. At the present time, however, less 10% of the total area of India is covered by forest. There are, however, large areas of the socalled jungle and wild and uncultivated land or areas in which cultivation has been abandoned.

The original natural vegetational cover of India has been most profoundly modified by human activity. Over at least five thousand years of intensive cultivation, grazing by cattle and sheep, human expansion and settling and the recent attempts of rapid urbanization and industrialization, particularly since the second World War, have stripped the forest from nearly all the plains and from much of the lower hills and plateaus and have turned these areas into scrublands. In the Indo-Gangetic Plain, for example, woodland is at present severely restricted to narrow riverine patches. Vast areas of the Deccan are at present short-grass savannahs, with only scattered trees, especially *Acacia, Euphorbia,* etc. As in other tropical forests, there is a general absence of gregariousness in the few Indian forests that have survived. The floral landscape is on the whole lacking in absolute preponderance of one species or assemblage of species; the belts of *Rhododendron* on the Himalaya may above be said to approach this condition. The semi-desert vegetation of the northwest and the bamboos locally in the south, palms, acacias and *Shorea robusta* (the *sal* tree of Indian Forestry) often give a distinctive mark to the vegetational cover over large areas, but these must not be mistaken for assemblages.

The great diversity of the flora of India is largely attributable to the enormous geographical area, stretching over many degrees of latitudes, to the marked differences in elevation and to the climate, ranging from the tropical to the arctic conditions from high aridity to high humidity. The richness of the flora is traced to the immigration and colonization

159

of plant species from widely different bordering territories, particularly the Chinese and Malayan, European and African and the Tibetan-Siberian. Though India is richer in plant species than perhaps any other equally large area in the world, there is a striking poverty of endemic genera of plants. The Malayan floristic element is dominant, but there is also a large African element. Recently Sudan-affinities in the flora of Deccan and Rajasthan have also been reported (MEHER-HOMJI 1965). The Tibetan-Siberian elements are confined to the alpine and higher zones of the Himalaya. The Chinese-Japanese elements are evident in Burma and in the temperate zones of the southern parts of the Eastern Himalaya.

Mention must be made of the presence of certain interesting forms. There are a few species in India of genera like *Baeckea*, *Leptospermum*, *Melaleuca*, *Leucopogon*, *Casuarina*, *Stylidium* and *Helicia*, which have their original home in Australia. The extra-American species of the genus found in India is the solitary *Oxybaphus himalaicus*. *Pyrularia edulis* has its congeners in North America and another in Java. *Vogelia*, restricted to South Africa and Socotra, is also represented in India. There is no indigenous *Tilia*, *Fagus* and *Castanea* in the temperate zones of the Himalaya. The rhododendron-zone of the Eastern and parts of Central Himalaya is perhaps the only heath form in India. The Conifers of the Himalaya generally resemble those from the north. The Dipteracorpaceae are characteristic of Burma. The most conspicuous trees are *Shorea robusta*, *Dipterocarpus tuberculatus*, *Dalbergia sissoo*, *Acacia catechu* and *Acacia arabica*. The more common indigenous palms are *Corypha*, *Phoenix sylvestris*, *Borassus flabellifer*, *Cocos nucifera*, *Calamus* etc. Bamboos are common. Tree-ferns are abundant in the forests of the Eastern Himalaya, Central India, Vizagapatam, Burma, Malabar, Malay Peninsula and Ceylon. *Strobilanthes* is characteristic of the western hills in the Peninsula. *Impatiens* is abundant in humid tracts, but are not found in Malaya. The genus *Impatiens*, with over four hundred species, mainly from the mountains of tropical Africa, India, China and Malaya (with two species from America, one from Europe, two Siberia and four from Japan), is remarkable for the distributional patterns of species. Africa has fifty species and India has nearly two hundred species; in India the species fall into two groups. One group contains about one hundred and thirty species and the other contains about sixty. Only a single species of the first group occurs in the Northwest Himalaya, but not a single species of the second group is found either in the Peninsula or Ceylon. The Western Ghats, especially its extreme southern end, is particularly rich in endemic species, most of which have apparently differentiated in recent times. The genus *Strobilanthes* with over two hundred species, all Asiatic, particularly the Western Ghats of South India, is remarkable in that the plants flower once only and then wither away. Nearly all the plants of a species flower in the same season, so that there is a great outburst of flowers in a locality once in a few years, after which in the intervening years not a flower may be found. The genus has been treated by BEDDOME (1868–1879) and recently by SANTAPAU (1944, 1951). The Ericaceae include *Gaultheria* found in America, Australia, New Zealand, Malaya, Burma and occurring on the Himalaya and mountains of Burma; *Pieris* from Northeast America, Japan, Burma and Himalaya and *Rhododendron*. The common aquatic plants include *Nymphaea*, *Nelumbium speciosum*, *Utricularia*, *Podostemon*, etc. Marine distribution would account for the abundance of *Ipomoea biloba*, *Phoenix farinifera*, *Vigna lutea*, *Canavalia lineata*, *Launaea pinnatifida*, *Spinifex*, etc. Orchids are abundant and more than 2000 species of largely tropical epiphytes occur in the Eastern Himalaya, Burma and in Malaya. The dominant Natural Orders in India are Leguminosae, Gramineae and Euphorbiaceae, followed by Acanthaceae, Compositae, Cyperaceae, Labiatae and Urticaceae (including Moraceae).

Except the more arid margins, the whole of the Peninsula was formerly densely forested, but we find in most areas at the present time often only *Acacia*-scrub semi-desert. The thornforest in the west, the closed monsoon-forest of *Shorea* in Chota-Nagpur and the open deciduous forest in between are only relicts. The natural vegetation of the Western Ghats comprises evergreen forest from Ratnagiri southwards, but this forest has been destroyed in the north. This tropical evergreen forest of the Western Ghats differs from the eastern tropical evergreen forest of Burma in the sparseness or even absence of *Dipterocarpus*.

We have along the Western Ghats an equatorial type of forest, characterized by conditions of high humidity and temperature, favouring vigorous growth of trees that often attain a height of 60 m. The tropical wet evergreen and the semi-evergreen forests are typically rain-forests. The true evergreen is found on the windward side, in a narrow strip, at elevations between 460 and 1370 m along the Western Ghats. This is bordered by semi-evergreen on the drier leeward side. The littoral areas of Orissa were formerly covered by semi-evergreen forests. The deep shade in these forests has the result of bare forest-floor. The wetter western side of the Peninsular Plateau is covered by tropical deciduous monsoon forest, which is essentially a mischwald, with preponderance of *Tectona grandis* and *Santalum album*. In the drier interior, as for example, Deccan, we find the dry-forest or the thorn-forest or scrubland, with stunted spiny trees, having long roots; *Acacia* is common in these parts. These areas are essentially savannahs.

Phytogeographically, Deccan comprises two areas. In the area bounded on the south by the Ajanta Hills and the R. Godavari (excluding the Orissa Plains) and including the upper valleys of the rivers Narmada, Tapti, Mahanadi and Godavari, we have deciduous forests. *Euphorbia nivula*, *Sarcostemma*, *Sterculia urens*, *Boswellia serrata*, *Cochlospermum*, etc. are common species of the deciduous forests. The mischwald of the dry slopes have *Anogeissus latifolia*, *Ougeinia*, *Odina*, *Cleistanthus collinus*, *Zizy-*

161

phus xylopyra, Buchanania latifolia, Terminalia, Bauhinia, etc. We find also *Shorea robusta, Adina, Bassia latifolia, Diospyros, Symplocos racemosa, Ptero-carpus marsupium, Eugenia jambolanum, Wendlandia tinctoria,* etc. The woody plants belong to Leguminosae, Rubiaceae, Euphorbiaceae and Urticaceae. Though typically otherwise of the Deccan type, the flora of this area contains some eastern and some western Himalayan species also. Deccan in the restricted sense, or the area between the R. Godavari and R. Krishna and the lower course of the Cauvery in the south, has deciduous forest, but there are some evergreen forests on the coasts and slopes in the eastern aspect. The areas of black-cotton soil are covered by *Capparis divaricata, Acacia arabica, Prosopis spicigera, Parkinsonia aculeata, Balanites roxburghii, Cadaba indica, Zizyphus nummularia, Cassia auriculata, Calotropis procera, Jatropha glandulifera,* etc.

The Nilgiri Hills have been described as belonging to a botanic realm of their own that shows affinities, in some cases, with the Assam flora and with the flora of the southern slopes of the Himalaya. Greater part of the region is forested, where the dominant species are *Tectona grandis* and *Santalum album,* but there is considerable open rolling downland, interspersed with woods called sholas, containing *Rhododendron, Ilex,* ferns, bracken, tree-orchids, hill gooseberries, blackberries, wild strawberries, heliotropes, *Fuchsia, Geranium,* etc. The introduction of quick-growing *Eucalyptus* from Australia has displaced much of the natural vegetation of the Nilgiri.

The Malabar region is characterized by the presence of Guttiferae, Diptercarpeae, Myristicaceae, abundance of Malayan forms especially among Sterculiaceae, Tiliaceae, Anacardiaceae, Meliaceae, Myrtaceae, Melastomataceae, Gesneriaceae, Piperaceae, Orchidaceae, etc. The conifer *Podocarpus latifolia* is confined to the hills in Tinnevelly in the whole of the Peninsula and outside it is known only from Burma and Malaya. The palm *Bentinckia coddapana* is a native of Travancore and has a congener in the Nicobar Islands. There is an abundance of *Strobilanthes* and *Impatiens.* The principal shrubs are *Strobilanthes kunthianus, Berberis aristata, Hypericum mysorense, Sophora glauca, Crotalaria formosa, Rhododendron arboreum, Rubus, Osbeckia, Hedyotis, Helichrysum, Gaultheria,* etc. *Senecio, Anaphalis, Ceropegia, Pedicularis, Cyonotis* and *Lobellia excelsa* are the common herbs. The species found here and also known from the Eastern Himalaya include *Ternstroemia japonica, Hypericum hookerianum, Hypericum nepaulense, Eurya japonica, Rhamnus dahuricus, Photina notoniana, Rubus ellipticus, Rubus lasiocarpus, Carallia integerrima, Rhododendron arboreum, Gaultheria fragrantissima, Gardneria ovata, Meliosma, Rosa, Pygeum, Viburnum, Lonicera,* etc.

In the Coromandel Coast areas, there are mangrove swamps in the mouths of the rivers. The rest of the area is generally covered by thickets of thorny evergreens and deciduous trees and shrubs like *Flacourtia, Randia, Scutia, Diospyros, Mimusops, Garcinia, Sapindus, Pterospermum, Strychnos nux-vomica,* etc. The umbrella-shaped crown of *Acacia planifrons, Cocculus leaeba,*

Capparis aphylla, *Cassia ovata* and *Cassia angustifolia* are characteristic of the extreme south of this coast. The sandy beaches have *Spinifex squarossus*.

Although so near the mainland of the Peninsula and with such close floral affinities with Malabar and Deccan, Ceylon contains very little endemic genera and species. The principal endemic species belong to Dipterocarpaceae, Rubiaceae, Euphorbiaceae, Melastomataceae, Myrtaceae, *Strobilanthes*, *Eugenia*, *Memecylon*, etc. Over one-third of the Phanerogams found in Ceylon occurs also in the mainland, but the majority are either Burmese or Malayan. Conifers and Salicaceae are absent in Ceylon. The hills have temperate forms like *Agrimonia*, *Crawfurdia* and *Protium*, which are not found in the Peninsula. The Peninsular temperate elements on the hills of Ceylon are represented by *Anemone*, *Thalictrum*, *Berberis*, *Cardamine*, *Viola*, *Cerastium*, *Geranium*, *Rubus*, *Potentilla*, *Alchemilla*, *Sanicula*, *Pimpenella*, *Peucedanum*, *Galium*, *Valeriana*, *Dipsacus*, *Artemisia*, *Vaccinium*, *Gaultheria*, *Rhododendron*, *Gentiana*, *Swertia*, *Calamintha*, *Teucrium*, etc. The characteristic feature of the vegetation of the island is the grass and shrub-covered *patanas*, particularly in the southeast, from sea-level to elevations of about 1525 m. Part of this vegetation is natural, but most of it is succession after the destruction of forest by man and abandoned fields. *Ochlandra stridula* is a peculiar bamboo that covers much of these *patanas*. The natural vegetation of the Wet Zone has all but disappeared and the Dry Zone lowland is covered with tropophilous forest, largely secondary in origin. The eastern parts have grassland *patanas*.

The present-day flora of India, like its fauna, is remarkable for the ecological and geographical complexity. Unlike the fauna, the flora of India is, however, largely characterized by the dominance of relatively young intrusive elements, so that the ancient forms are comparatively of lesser importance. On the other hand like the fauna, the flora also shows unmistakable evidence of the profound influence of the Himalayan uplift and human interference, both on its composition and on its distributional patterns. Its characteristically complex composition is to be traced primarily to the influx of humid-tropical Asiatic elements through the Assam gateway, referred to in an earlier chapter. The flora received later considerable contribution of boreal and steppes elements from Asia and Europe and Mediterranean and tropical African elements through the northwest. A large number of exotic and particularly tropical American and Australian plants introduced by man within the past two centuries have spread widely throughout and have become naturalized in India; some of them have even differentiated into local races and varieties. We do not include in this category the numerous cultivated plants like the American corn, Egyptian cotton, potato, tomato, chilli, etc. The influx of the Asiatic, European and African floral elements greatly altered not only the ecosystems throughout, but also the distributional patterns in general. A number of typical Peninsular species of plants also spread and penetrated into the Extra-Peninsular areas and infiltrated deep into the Himalaya and extended their range outside to the areas east of India. The Pleistocene glaciations on the Himalaya saw the rapid spread of a number of European and boreal Asiatic species deep south into the Peninsula, where they were left behind isolated as relicts when the glaciers on the Himalaya retreated at the end of the Pleistocene. Both the ecological and floristic composition and the distributional patterns have since then been greatly altered by deforestation carried by man.

2. *The Major Floristic groups*

We shall give here a brief outline of the major floristic groups, with some examples of character genera and species. The major groups recognized here are 1. the exotic naturalized species, 2. the humid tropical Asiatic or the Indo-Chinese and Malayan species, 3. the temperate European species, 4. the steppes elements, 5. the Mediterranean elements, 6. tropical East African elements, 7. Peninsular and other Indian endemics, 8. Pleistocene relicts, and 9. some other miscellaneous types. Marked discontinuity characterizes the distribution of the humid tropical Asiatic species and of the Pleistocene relicts. The ranges of these floral components are broken up into a southern Peninsular and a north-eastern or into a Himalayan area.

2.1. Exotic naturalized plants

In addition to the economically useful and widely cultivated plants introduced into India from other regions, a large number of wild species were brought unwittingly by man from a variety of countries, but particularly tropical America and Australia, mostly during eighteenth and nineteenth centuries. Nearly all these species have spread widely throughout India, displaced many indigenous species from their eco-systems and have themselves become completely naturalized. A great many of them have even differentiated from the original types so far that in some cases they have attained subspecific levels. It is of course not possible to give a complete list of these species, but the following will serve to illustrate the phenomena.

The following are some of the species which have come from America: Cruciferae: *Senebiera didyma* a common winter weed in North Indian plains and naturalized on the Nilgiri Hills in South India. Sterculiaceae: *Guazuma tomentosa* and *Melochia nodiflora;* Papaveraceae: *Argemone mexicana;* Rutaceae: *Swietenia mahagoni;* Anacardiaceae: *Anacardium occidentale;* Leguminosae: *Caesalpinia coriaria, Parkinsonia aculeata, Cassia hirsuta, Cassia occidentalis, Cassia alata, Mimosa pudica, Desmanthes virgatus, Leucaena glauca, Acacia farnisiana* and *Enterolobium saman;* Myrtaceae: *Couroupita guianensis;* Turneraceae: *Turnera ulmifolia* and *Turnera trioniflora;* Passiflorae: *Passiflora foetida;* Cactaceae: *Cereus hexagonus, Opuntia coccinellifera*

164

(= *Nopalia cochinilifera*) introduced from South America into the Masuli-patam area on the Coromandel Coast in the eighteenth century for the cochneal dye coccid and several other species of Opuntia; Compositae: *Flaveria contrayerba, Tridax procumbens, Erigeron mucronatum, Eupatorium odoratum* and *Lagasca mollis;* Apocynaceae: *Thevetia neriifolia, Rauwolfia canescens* and *Lochnera rosea;* Convolvulaceae: *Quamoclit phoenicia, Q. pinnata* and *Ipomoea quinquefolia;* Solanaceae: *Datura metel, Nicotiana plumbaginifolia* and *Solanum seaforthianum;* Scrophulariaceae: *Scrophularia dulcis;* Bigoniaceae: *Bigonia megapotamica;* Pedaliaceae: *Martynia annua;* Verbenaceae: *Lantana aculeata* and *Clerodendron aculeatum;* Labiatae: *Hyptis suaveolens;* Amarantaceae: *Gomphrena globosa* and *Gomphrena decumbens;* Piperaceae: *Peperomia pellucida;* Euphorbiaceae: *Euphorbia prostrata* and *Croton sparsiflorus* introduced in 1898 from South America into Bengal has now spread to the southern parts of the Peninsula also; Urticaceae: *Pilea microphylla;* Pontederiaceae: *Eichhornia crassipes* and Anonaceae: *Anona squamosa* and *A. reticulata.*

The introductions from Australia include many species of *Acacia*, which have now become completely naturalized. Most of them have phyllodes, but as seedlings have bicompound leaves. These include *Acacia retinodes, A. longifolia, A. melanoxylon, A. decurrens* and *A. dealbata.* Among the other Australian species naturalized in different parts of India are *Flaveria australasia* (Compositae). Some of the African species naturalized include *Adansonia digitata** (Malvaceae), and *Bryophyllum pinnatum* (Crassulaceae). The Madagascan *Delonix regia* (Leguminosae) has also become naturalized.

2.2. TROPICAL ASIATIC ELEMENTS

The tropical Asiatic elements of our flora have largely Indo-Chinese and Malayan affinities and represent perhaps the most dominant component member of our present-day flora, not only on the lower slopes of the Himalaya but also deep south in the Peninsula. These elements have spread far west along the southern slopes of the Himalaya practically up to Kashmir and along the Eastern Ghats to the Peninsula. In many cases the species are identical in the Peninsula, Burma-Assam, Thailand and Malaya and in other cases local species have evolved in the Peninsula. The distribution of these elements presents nearly every degree of gradation from the completely continuous to the greatly disjunct isolates. The following are some of the common and striking examples of the humid tropical Asiatic elements in our flora: Anonaceae: *Uvaria* and *Goniothalamus;* Pittosporaceae: *Pittosporum glabratum* extending from Hong Kong to the Khasi Hills in Assam, *Pittosporum tetraspermum* occurring in the Nilgiri Hills and on the mountains of Ceylon, *P. nilghirense* on the

* Considered by my friend Mr. K. M. VAID of the Indian Forest Institute as the *Kalpak vriksha* of Sanskrit literature and worshipped in many parts of India.

165

Nilgiri, *P. ceylanicum* in Ceylon, *P. floribundum* on the Khasi Hills and subtropical parts of the Himalaya from the east westwards up to Garhwal and on the Western Ghats, *P. dasycaulon* on the Western Ghats and *P. eriocarpum* occurring on the Western Himalaya; Guttiferae: *Garcinia indica* on the Western Ghats, *G. echinocarpa* and *G. terpnophylla* in Ceylon and *G. ovalifolia* on the Eastern Ghats. The Leguminosae are represented by *Cassia siamea* occurring Thailand, Malay Peninsula, South India and Ceylon, *Saraca* occurring in the Himalaya from the east westwards up to Kumaon and in Ceylon and *Xylia xylocarpa* found in Malaya, Burma, Philippines and the Peninsula. *Terminalia catappa* and *Quisqualis indica* are two common Combretaceae extending from Java-Malaya into India. *Rhodomyrtus tomentosa* occurs in Malaya, Nilgiri and Palni Hills and in the mountains of Ceylon. *Decaspermum paniculatum*, another Combretaceae, occurs in Philippines, Australia, Java, Malaya, Burma and Assam. *Oxyspora* from Sumatra, E. Himalaya and Khasi Hills and *Medinilla* from Malaya, Khasi Hills, E. Himalaya and Ceylon are interesting genera of Melastomataceae. The Dilleniaceae contain *Tetracera laevis* found on the Western Ghats and in Ceylon and *T. assamensis T. euryandra* and *T. macrophylla* of the Eastern Ghats and *Acrotrema* of Malaya, Burma and the Eastern Ghats. The other examples include *Gymnopetalum* (Cucurbitaceae) from Malaya, Burma, Assam, Eastern Himalaya, Chota-Nagpur Plateau, Deccan and Ceylon; *Pentanura* (Asclepiadaceae) extending from Yunan to the Khasi Hills; *Myristica* (Myristicaceae) from Malaya, Andaman, Burma, Eastern Himalaya, Western Ghats and Ceylon; *Nepenthes* (Nepenthaceae) from Borneo, Sumatra, Malaya, Khasi Hills and Ceylon; the Lauraceae *Beilschmedia* from the Western Ghats, Central and Eastern Himalaya, Assam, Burma and Ceylon; *Cinnamomum, Machilus, Phoebe* and *Litsea* from Burma, Malaya, Eastern Himalaya, Western Ghats and Ceylon; the Rubiaceae *Sarcocephalus* from Philippines, Java, Malaya, Burma and Ceylon; *Anthocephalus* from Malaya, India and Ceylon; *Adina* from Burma, Assam, Himalaya up to Kumaon, hills of South India and Ceylon; *Uncaria* from Malaya, Burma, Assam, Eastern and Western Himalaya; *Hedyotis* from Burma, Assam, Himalaya, Deccan, Western Ghats and Ceylon; *Knoxia* from Malaya, Burma, Assam, Himalaya, Western and Eastern Ghats, Ceylon and also tropical Australia and *Lasianthus* from Malaya, Burma, Andaman, Assam, Eastern Himalaya, Eastern and Western Ghats and Ceylon. *Willoughbeia* (Apocynaceae) is also a Malayan genus that occurs in Burma, Assam and discontinuously in Ceylon.

2.3. THE TEMPERATE ZONE ELEMENTS

The greatest bulk of the species of plants of the temperate-zone origin are derived from Europe and their distribution is characteristically restricted to the Himalaya, the Khasi Hills in Assam, parts of the Eastern

Ghats like the Mahendragiri, Yercaud and Palni-Anamalai Hills (see Chapters VII, VIII) and the southern parts of the Western Ghats. The distribution is thus markedly discontinuous. There is also another extremely interesting feature regarding the phenology and occurrence of the temperate-zone species in India; they occur as mentioned on the mountains during summer and appear largely as weeds in the western parts of the Indo-Gangetic Plains of north India during the winter, after the receding of the southwest monsoon. The vegetation of these parts is remarkable for the fact that during the monsoon rains the areas are colonized by typically Peninsular and tropical species and during the winter by European and temperate-zone species (Fig. 149).

The temperate elements of our flora are exemplified by Ranunculaceae like *Clematis*, *Anemone*, *Thalictrum*, *Ranunculus* and *Caltha*, which occur on the Himalaya and in the higher reaches of the Western and Eastern Ghats. The Berberidaceae are represented by *Berberis* and *Epimedium* on the Himalaya. *Corydalis* and *Fumaria* (Fumariaceae) occur on the Himalaya and Nilgiri Hills. *Viola patrinii* occurs on the Himalaya, Western Tibet, Khasi Hills, Eastern and Western Ghats, Ceylon in addition to Afghanistan, Russia, north Asia and Japan. *Silene* (Caryophyllaceae) occurs both on the Himalaya and the Nilgiri Hills. The Geraniaceae include *Biebersteinia emodi*, *Geranium* and *Erodium*. The important temperate elements among the Leguminosae are *Thermopsis*, *Trifolium pratense* (Europe, Siberia, Afghanistan and the Himalaya), *T. repens* (Himalaya, Nilgiri, Ceylon, North Asia, Europe and North America), *Lotus*, etc. The Rosaceae comprise *Spiraea*, *Fragaria*, etc. *Sedum* (Crassulaceae) occurs on the Himalaya. The Caprifoliaceae *Sambucus* occurs on the Himalaya and Khasi Hills, and *Viburnum* and *Lonicera* occur on the Himalaya and the southern parts of the Western Ghats. *Rubia* occurs on the Himalaya, Western Ghats, Ceylon, Japan etc.; *Galium*, another Rubiaceae, occurs on the Himalaya, Burma, Nilgiri and Ceylon and *Asperula* is confined to the Himalaya. *Valeriana* (Valerianaceae) occurs on the Himalaya, Nilgiri, Palni and Anamalai Hills and Coorg as well as the higher mountains of Ceylon. The Compositae have a large temperate component and are represented by *Aster*, *Brachyaster*, *Erigeron*, *Leontopodium*, *Anaphalis*, *Achillea*, *Tanacetum*, *Senecio*, *Saussurea*, *Cichorium*, *Picris*, *Crepis* and *Taraxacum*. *Primula*, *Androsace* and *Lysimachia* (Primulaceae) are Himalayan and so are the Boragineae *Myosotis*, *Lithospermum* and *Arnebia*. *Pedicularis* (Scrophulariaceae), *Mentha*, *Thymus serphyllum*, *Calamintha*, *Dracocephalum* and *Teucrium* are also typically Himalayan plants.

Considerable part of the temperate elements are typically boreal forms, often with circumpolar distribution and present almost exclusively on the higher reaches of the Himalaya and occurring as Pleistocene relicts in the Nilgiri Hills in South India. These include the largely Himalayan Ranunculaceae: *Callianthemum*, *Trollius*, *Coptis*, *Isopyrum*, *Aquilegia*, *Del-*

phinium, Aconitum, Cimicifuga and *Paeonia*. The Berberidaceae *Podophyllum* that occurs in North America is also found on the Himalaya. The important boreal Cruciferae include *Parrya, Cheiranthus Arabis, Cardamine, Draba, Cochlearia, Eutrema, Braya* and *Thalaspe*. The boreal Caryophyllaceae *Dianthus, Acanthophyllum, Gypsophila, Lychnis,* and *Cerastium* are exclusively Himalayan and *Arenaria* and *Stellaria* occur both on the Himalaya and on the Nilgiri Hills. The boreal Leguminosae are rather sparse and belong to *Oxytropis* and *Lespedza* from the Himalaya. *Geum, Potentilla, Alchemilla, Agrimonia* and *Protium* are the Rosaceae, some of which like *Alchemilla* and *Proterium*, have extended to the Western Ghats and the mountains of Ceylon. *Saxifraga sibirica, S. cernua* and *S. flagellaris, Parnassia* sp. are the important boreal forms of Saxifragaceae, some of which like species of *Parnassia*, have extended to the southern parts of the Western Ghats. *Epilobium* (Onagraceae) occurs on the Himalaya and Khasi Hills and another genus *Circaea* occurs on the Himalaya and the Nilgiri and Palni Hills. The Umbelliferae *Seseli sibiricum* has an interesting distribution, extending from Siberia, Russia (European) to the Northwest Himalaya. *Pleurospermum*, also belonging to the same order, has a similar distribution and occurs in addition in the Eastern Himalaya also.

2.4. THE STEPPES ELEMENTS

The steppes species are mostly of Middle Asiatic origin and have their home in the lowlands of Turkestan, in the Pamirs, Afghanistan and occur also on the Northwest Himalaya. Some of them also come partly from west Asia and partly from the eastern Mediterranean and extend eastwards to the plains in Sind and the Punjab, rarely penetrating to the western parts of the upper Gangetic Plains of north India. One or two species have spread southwards to Deccan and the Nilgiri Hills. The steppes elements constitute minor constituents of the Indian flora. *Megascarpaea, Tauscheria* and *Euclidium* are the typical steppes Cruciferae found on Northwest Himalaya. We find here *Hippophaë rhamnoides* and *Myricaria* also, both of which are so characteristic of the Pamirs and Turkestan. The characteristic Leguminosae include *Guldenstaedtia* and *Astragalus. Triplostegia, Morina, Scabiosa* and *Dipsacus* occur on the Northwest Himalaya; the last mentioned genus is found in the Nilgiri and Palni also. The steppes Compositae are *Anthemis, Artemisia* and *Cnicus*, occurring on the Himalaya, with the last mentioned genus being found as well on the Nilgiri Hills. *Hyoscyamus* is confined to the Northwest and West Himalaya. Chenopodiaceae *Eurotia ceratoides* and *Axyris amaranthoides* are among the typical steppes species found on the Northwest Himalaya; *Haloxylon recurvum* another steppes form occurs in addition to the plains of Sind in Deccan also.

2.5. The mediterranean elements

The Mediterranean elements comprise partly the southern European and partly also the North African species of plants. Some of the latter come from Abyssinia, Sudan and Sahara and have been termed the Sudano-Deccan elements of our flora by some botanists (Meher-Homji 1965). They extend eastwards through Iraq, Arabia, Baluchistan and sometimes partly also through Afghanistan to the Northwest Himalaya, but more commonly to the plains of Sind, Punjab and occasionally further southwards to Deccan. Like the steppes elements, a number of the mediterranean forms has penetrated to the western margin of the Upper Ganga Plains of North India and southwards along the Western Ghats to Ceylon. Like the temperate and boreal plants, the Mediterranean species appear as winter weeds in the western parts of the north Indian plains, where they are, however, absent during the months of monsoon rains. They are nearly as strong as the temperate components in the flora. The following are some of the important species: Zygophyllaceae: *Fagonia arabica*, occurring in northwest India, Sind, Punjab and the Western Ghats and extending outside India to Egypt and *Fagonia brugieri* extending from Algeria to northwest India. The Geraniaceae include: *Mansonia senegalensis*, occurring in the western Ganga Plains and Deccan, and extending westward to Sind, Baluchistan, Arabia and North Africa; *Mansonia heliotropioides* extends from Egypt to Sind and eastern Punjab plains. The Leguminoseae are strongly represented and comprise the following: *Argyrolobium flaccidum* (Himalaya and northwest plains), *Ononis* (Northwest Himalaya), *Trifolium fragiferum* (Northwest Himalaya, North Africa and Europe), *Trigonella* (wild winter weed in the western parts of the Upper Ganga Plains and occurring commonly in the Himalaya), *Mellilotus* (Siberia, Europe, Northwest Himalaya, winter weed in the Upper Ganga Plains and occurring in the Western Ghats), *Medicago*, *Colutea*, *Traverniera*, *Ebenus*, *Alhagi*, *Prosopis*, etc. *Papaver rhoeas* and *P. dubium*, *Hypecoum* and the Cruciferae *Farsetia*, *Alyssum*, *Malcomia*, *Diplotaxis*, *Eruca*, *Moricandia*, *Capsella*, *Iberidella*, *Istais* and *Chorispora* are the common examples. The Capparidaceae are represented by a number of species of *Cleome*, some of which have penetrated to Ceylon. There are also two species of Umbelliferae *Eryngium caeruleum* and *E. billardieri* occurring in the Northwest Himalaya.

2.6. The tropical african and madagascan elements

The Tropical African, especially the East African and the Madagascan elements, constitute an important, though small, section of the Peninsular flora. Some of them have differentiated locally into diverse forms. Though the majority of them appear to be of older stock, some are apparently recent intrusions. It is not also often easy to distinguish the typical East

African elements from the Mediterranean (North African) species. It may, however, be noted that the typical tropical African plants have become naturalized and differentiated into endemic forms on the Western Ghats and Ceylon, but those of the Mediterranean origin are largely confined to the northwest and northern parts of the Deccan. Unlike the mediterranean elements, these grow and flower predominantly during the monsoon rains. Though small compared to the tropical Asiatic components, the tropical African elements are on the whole very characteristic in the Peninsular flora.

The following examples will serve to illustrate the African components: The Madagascan genus *Ochrocarpus* is represented by *O. longifolius* in the Western Ghats. *Erythroxylon monogynum* (Linaceae), occurring on the Western Ghats and Ceylon, is considered to be very close if not indeed identical with African species. *Mundulea*, a South African and Madagascan genus, occurs in the Western Ghats and Ceylon. Other Leguminosae include *Geissapsis, Leptodesmia congesta* (Madagascar and Nilgiri), *Delonix regia* (Madagascar), *Tamarindus indica, Acacia; Bryophyllum pinnatum, Plumbago capensis, Salvadora, Cryptostegia grandiflora, Sesamum, Pedalium murex, Ricinus communis; Pseudarthria, Hardwickia* and *Humboldtia* (Leguminosae); *Cephalandra, Ctenolepis, Gaertnera, Blepharis, Peristrophe bicalyculata* and the Compositae *Lasiopogon, Vicoa, Carthamus* and *Dicoma tomentosa.*

2.7. PLEISTOCENE RELICTS

The Pleistocene relicts are temperate and boreal species of the Himalaya that spread southwards, during the Pleistocene glaciations on the Himalaya, across the Aravalli Mountains to the Western Ghats and reached to the extreme southern end of the Peninsula, and in some cases even to Ceylon that was still connected with mainland of the Peninsula. With the retreat of the glaciers from the Himalaya in the Post-Pleistocene, they became isolated in the south from the main range in the Himalaya. The greatest bulk of these species are concentrated on the higher elevations of the Nilgiri, Anamalai, Palni and Cardamom Hills in the southern part of the Western Ghats, but some of them may be found in Mahableshwar in the northern end of the Western Ghats. Ranunculaceae: *Clematis* (Himalaya, Nilgiri, Shevroy, Anamalai and Ceylon); *Anemone rivularis* (Himalaya, Nilgiri and Anamalai) and *Ranunculus* (Himalaya and South Indian mountains). Fumariaceae: *Corydalis lutea* and *Fumaria;* Cruciferae: *Cardamine* and *Capsella*: Violaceae: *Viola patrinii* (Siberia, Russia, Japan, Himalaya, Eastern and Western Ghats especially the Mahendragiri and Nilgiri, Palni and Shevroy); Polygalaceae: *Polygala sibirica* (Himalaya, Khasi Hills, Siberia, China, Japan, Western Ghats from Nilgiri southwards to Ceylon); Caryophyllaceae: *Silene gallica, Stellaria paniculata, S. media* and *S. saxatilis* (the last mentioned species Siberia, Japan, Khasi Hills, Himalaya and Nilgiri), *Arenaria;* Geraniaceae:

170

Geranium nepalense (Himalaya, Nilgiri and Ceylon); Rhamnaceae: *Rhamnus virgatus* (China, Japan, Himalaya, Nilgiri and Palni); Leguminosae: *Ulex europeus, Cytisus scoparius, Trifolium, Parochaetus communis* and *Indigofera pulchella;* Rosaceae: *Prinsepia utilis* (Himalaya, Khasi Hills and Nilgiri), *Fragaria, Potentilla* and *Alchemilla;* Umbelliferae: *Bupleurum* and *Heracleum;* Caprifoliaceae: *Viburnum, Lonicera;* Rubiaceae: *Galium;* Compositae: *Erigeron alpinus, E. canadense;* Primulaceae: *Lysimachia;* Gentianaceae: *Exacum* and *Swertia;* Scrophulariaceae: *Veronica* and *Pedicularis;* Labiatae: *Teucrium;* Euphorbiaceae: *Euphorbia helioscopia;* Ulmaceae: *Celtis;* Juncaceae: *Juncus* and Araceae: *Arisaema.*

3. The Major Ecological groups

The principal criterion that underlies the divisions of the forest cover of India is the monsoon rainfall pattern, but on the Himalaya the temperature conditions also play an important part. The character and composition of the soil are on the whole of only secondary importance in determining the forest type; and topography may play some rôle, in a restricted sense. The forest types, recognized at present in India, are essentially the monsoon rainfall-vegetation zones. Ignoring the high mountainous regions, the following four rainfall-vegetational regions are generally recognized, viz. 1. the wet zone, with a rainfall of over 250 cm annually (evergreen forest); 2. the intermediate zone, with a rainfall between 200 and 100 cm (deciduous or the monsoon forest zone); 3. the dry zone with 100–50 cm rainfall (dry forest and scrubland) and 4. the arid zone (desert, semi desert) with less than 50 cm rainfall.

The types of forests recognized at present in India are based on the classification of forests proposed by CHAMPION (1936).

The following is a synopsis of the principal forest types:

I. Tropical forests
  1. Wet evergreen (rain) forest of dense, tall, entirely evergreen trees
  2. Semi-evergreen forest, with certain deciduous trees as dominants, but also with evergreen trees
  3. Moist deciduous forest, with mainly deciduous trees, lower storey of evergreens
  4. Dry deciduous forest, with entirely deciduous trees and top canopy light
  5. Thorn forest, with deciduous low thorny trees, xerophytes, and broken canopy
  6. Dry evergreen forest of sclerophyll leaved trees

II. Montane subtropical forests
  1. Subtropical wet hill-forest of broad-leaved, evergreen south-north forms
  2. Subtropical moist hill-forest, with pine-tree associations predominating
  3. Subtropical dry evergreen forest of xerophytes and scrubs

III. Montane temperate forests
  1. Wet temperate forest of evergreens and conifers
  2. Himalayan moist temperate forest of evergreen, sclerophyll *Quercus*-Conifers
  3. Himalayan dry temperate forest of open Conifers, sparse xerophyte undergrowth

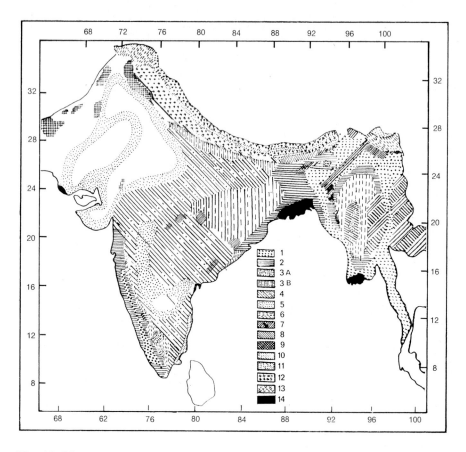

Fig. 25. Climatic Vegetation map of India. Key to shaded areas: 1. tropical wet-evergreen forest, 2. tropical semi-evergreen forest, 3A. tropical moist deciduous forest, 3B. tropical moist deciduous *sal*-forest, 4. tropical dry deciduous forest, 5. tropical thorn forest, 6. tropical dry-evergeen forest, 7. subtropical wet forest, 8. subtropical wet pine forest, 9. subtropical dry forest, 10. wet temperate forest, 11. moist temperate forest, 12. dry temperate forest, 13. alpine vegetation, 14. tidal forest.

IV. Alpine forests
   1. Alpine forest of stunted, deciduous and evergreens, with or also without conifers
   2. Moist alpine-scrub of low dense scrub
   3. Dry alpine scrub of xerophytes in open formation

V. Specialized and local forest types
   1. Tidal forests on esturine mud
   2. Beach forest on coastal sand
   3. Fresh-water swamp on wet alluvium
   4. Riverine forest on new alluvium that is periodically inundated

172

3.1. Tropical forests

The tropical wet evergreen and semi-evergreen forests are typical fain-forests, best developed in areas that receive an annual rainfall of over 300 cm and have a relatively brief dry season. The true evergreen type is found in a strip, at elevations of 450–1370 m, along the Western Ghats, to the south of Bombay and up to an elevation of about 1060 m on the Assam hills. This forest type is bordered on the drier side by the semi-evergreen forest, merging in its turn with the moist deciduous forest type. The littoral areas of Bengal and Orissa were formerly covered with semi-evergreen forests.

The rain-forest is generally dense and lofty, with the upper storey reaching to 35–45 m, but with individual trees often growing to heights of 60 m. The deep shade has the result of bare forest-floor. The edges of breaks in the canopy, especially along stream margins, have dense undergrowth of bamboos and palms also. A common feature in these forests in the butteresed trunks of the trees. There is also general abundance of epiphytes. The number of species is generally large, particularly on the Western Ghats. In Assam, the semi-evergreen forests often show a tendency to become gregarious, with erect bamboos and forming a transition to the moist-deciduous forest types.

The tropical deciduous or the monsoon forests constitute the natural vegetational cover over nearly all India, except between the Himalaya, the Thar and the Western Ghats. It must be remembered, however, that vast areas have already been denuded of the natural vegetation by man. This forest is developed in a long strip on the eastern side of the Western Ghats, in the northeast of the Peninsula (in Chota-Nagpur, Orissa, eastern Madhya Pradesh) and in a large strip along the Siwaliks Hills, in the *bhabar* and *terai* (see Chapter II) areas, from 77° to 88° EL. Though most trees shed their leaves for about six to eight weeks, during the hot weather, the leaf-fall period in different species differ, so that the forest is rarely ever absolutely leafless. The general aspect of the forest is, however, a characteristic burnt-out appearance during the hot weather. The undergrowth in moister localities is evergreen and is usually denser than in the rain-forests. Climbers and bamboos are common. In these forests, the more gregarious trees include *Shorea robusta*, *Tectona grandis*, *Santalum album*, *Dalbergia sissoo*, *Terminalia chebula*, *Bassia latifolia*, *Acacia catechu*, etc.

The thorn-forest is met with in areas receiving a rainfall of less than 75 cm. We find this type in the northwest and in the lee-side of the Western Ghats. The vegetation is open stunted trees, breaking down into xerophytic bush. This forest type grades into nearly complete desert in the northwest. Most trees are no taller than 6 m and are also mostly widely scattered. *Acacia* spp., with their wide spread of roots, are generally prominent and these trees are generally even widely spaced. *Acacia catechu* covers large areas, but *Acacia arabica* is more common than *A. catechu* in the lowlands of the northwest. *Euphorbia* is often also dominant and may often attain the size of small trees. *Phoenix sylvestris* is found in relatively damper localities. This originally poor forest has in most places deteriorated terribly in recent years as a result of overgrazing and fodder-cutting.

The tropical dry-evergreen forest is limited to the coastal areas, from Madras to the Point Calimere, where the rainfall is about 100 cm, received mostly during October–December and where the humidity is generally high. This type is characterized by closed, but low canopy of about 1–13 m height, with shrubby and spiny undergrowth.

3.2. MONTANE SUBTROPICAL AND TEMPERATE FORESTS

The subtropical and temperate forest types are met with in three widely separated areas, viz. the Nilgiri, Anamalai and the Palni Hills in South India and on the outer Himalaya in the north and Assam Hills in the northeast.

There are numerous unexplained affinities between the floras of the Nilgiri Hills and Assam-Manipur. The southern subtropical wet-hill forest is often not readily separated from the tropical rain-forest at lower elevations and from the temperate forest higher above. This forest type is found on the Nilgiri Hills, at elevations of 1070–1200 m and also on the Palni Hills.

It is essentially a stunted rain-forest and is not so luxurious as the true tropical evergreen forest. There are a number of subtypes on the higher reaches of the Western Ghats and on the summits of the Satpura-Maikal Ranges, and sometimes as far as Mr. Abu on the Aravalli Range. The southern wet-temperate forest subtype is met with at elevations above 1500 m on the Nilgiri-Anamalai-Palni Hills. This subtype is characteristic of rainfall of 150–650 cm or more. Wind acts as an important inhibitory agent, so that the forest is found in the lower or the sheltered aspects of open downland, presenting the general appearance of a savannah, with occasional peat-bogs. It is a dense, but a low forest of trees usually 15–18 m high, with much undergrowth abundant epiphytes, moss, magnolia and rhododendron, fern, laurel, etc.

The high altitudes of the Himalaya and the higher latitudes in the north introduce new climatic and topographic features, so that atmospheric temperature and aspect insolation become of great importance in determining the forest type. An eastern wetter and a western wetter subtype is generally recognized and their transition is about 86–88° EL. In the northern subtropical wet-hill forest, the trees are fairly high (20–30 m). Such dense forests cover elevations of 1000–1800 m on the Himalaya, east of 88° EL and at somewhat higher elevations in Assam. Evergreen *Quercus* and chestnut, with some ash and beech, generally predominate in these forests. *Shorea* occurs in favourable localities and

174

Pinus grows at higher elevations. Climbers and epiphytes are also common.

The subtropical moist-hill forest or the pine-forest occupies a long belt, from 73° to 88° EL, on the southern slopes of the Himalaya, at elevations of 900–1800 m. Patches of this type may also be found at higher elevations on the Khasi Hills and on other mountains of Assam-North Burma. *Pinus longifolia* is the dominant tree and often forms pure stands.

The subtropical dry-evergreen forest is restricted to elevations of 1050–1525 m on the foothills of the Himalaya, the Salt Range, Kashmir and Northwest Frontier Province and to patches in Baluchistan. The forest is low and scrubby and has a general resemblance to the Mediterranean maquis. *Olea cuspidata* and *Acacia modesta* are common, with large areas covered by dwarf creeping-palm *Nannorhops*.

The northern wet-temperate forest is a closed forest type of mainly *Quercus*, laurel and chestnut, with undergrowth of dwarf bamboo, at elevations of 1800–2900 m in the wetter areas having a rainfall of over 200 cm, to the east of the 88th East Meridian.

The Himalayan moist-temperate forest is the most widespread of the Himalayan forests. It occurs in rainfall-areas of 100–250 cm, on the outer ranges in the wetter east, where the broad-leaved evergreens are mixed with the dominant Conifers and become scarce to the west. The forest ranges from elevations of 1500 to 3050 m. The Conifers are generally absent on the southern slopes, where *Quercus* replaces them. Pines, cedars, silver-firs, spruce, etc. are the most important trees that form high but open forest, with scrubby undergrowth of *Quercus*, *Rhododendron*, laurel, bamboo, etc. To the east of 80° EL, *Cedrus deodara* forms large pure stands, in the belt of intermediate rainfall of 110–175 cm.

The Himalayan dry-temperate forest is a somewhat open xerophytic form, found on the inner Himalayan ranges of north Sikkim and Kashmir, with rainfall of about 100 cm (mostly as snowfall). There is a predominance of Conifers, including deodar and juniper, with some scattered *Quercus* and Ash.

3.3. Alpine forests

The alpine forest and scrub are found at elevations above 2895 m to 3500–3660 m on the Himalaya. These formations are characterized mostly by shrub silver-fir, juniper, pine, birch and rhododendron; the last mentioned plant may often grow to heights of 7 metres. Most of these plants are on the whole crooked. This type grades into low evergreen scrub, which in its turn passes into open xerophytic bush and *Salix* on the Tibetan side of the Himalaya.

3.4. Littoral forests

The tidal forests represent the best known of the specialized tropical forests. Bordering tidal channels, *Rhizophora* grows to high forest, but other mangroves are generally low and light. They are characterized by pneumatophores. The seaface of most deltas is fringed mangroves. In Bengal they are backed by the Sundarban. At higher levels are *Pandanus*, *Calamus*, palms especially *Nepa*, etc. Beaches and sand-dunes have fringes of *Casuarina*.

4. Phytogeographical Divisions

The common phytogeographical divisions of India, on the basis of Hooker, are, 1. the Himalayan Division, 2. the Eastern Division and 3. the Western Division. The Himalayan Division has rich tropical, temperate and alpine flora, with Conifers, *Quercus* and numerous Orchids. It also abounds in European and Siberian forms. The Eastern Area lacks the alpine-zone forms, but has sparse temperate forms, a few Conifers, more *Quercus*, palms and an abundance of Orchids. This area is also rich in Chinese and Malayan elements. The Western Area has extremely few Conifers, no *Quercus*, few palms and Orchids, but has many European, Oriental and African elements.

These three major areas are subdivided into nine regions, viz. 1. the Eastern Himalaya, from Sikkim to the Mishmi Hills in Assam; 2. the Western Himalaya from the Kumaon area to Chitral; 3. the Indus region, including the Punjab, Sind and Rajasthan west of the Aravalli Range and R. Yamuna, Cutch and north Gujarat; 4. the Gangetic Plain, from the Aravalli Range and the R. Jamuna to Bengal, Sundarbans, the Assam plains and Sylhet, the low parts of Orissa north of the R. Mahanadi, with an upper arid and a lower humid subregions; 5. Malabar (*sensu lat.*) for the humid belt of the hilly and mountainous area, extending along the western side of the Peninsula, from south Gujarat to the Cape Comorin; 6. the Deccan (*sensu lat.*) for the elevated dry tableland of the Peninsula, east of Malabar and south of the Indo-Gangetic Plain, with subregion of the Coromandel Coast for the low coastal strip from Orissa to Tinnevelly; 7. Ceylon and Maladive Islands; 8. Burma, Andaman Islands; and 9. Malay Peninsula from Kedah to Singapore and Nicobar Islands.

The boundaries of these phytogeographical regions are not sharp, but most of these regions correspond roughly to areas of comparative humidity-rainfall or dryness proposed by PRAIN (1903), viz. 1. *India deserta* for the Indus Plain, 2. *India diluvia* for the Gangetic plain, 3. *India aquosa* for Malabar, 4. *India vera* for Deccan, 5. *India subaquosa* for the Eastern Ghats and Coromandel subregion, and 6. *India littorea* for the Sundarbans of Bengal.

176

REFERENCES

BEDDOME, R. H. 1868–1874. Icones Plantarum Indiae Orientalis.

CHAMPION, H. G. 1936. A preliminary Survey of Forest Types of India and Burma. *Indian For. Rec.*, (NS) (Silv.) 1(1): 1–286.

CHATTERJEE, D. 1940. Studies on the endemic flora of India and Burma. *J. Asiatic Soc. Bengal*, 5: 19–67.

CHATTERJEE, D. 1962. Floristic patterns of Indian Vegetation. Proc. Summer School Botany, Darjeeling, 1960: 32–42.

CLARKE, C. B. 1898. Subareas of British Empire, illustrated by the detailed distribution of Cyperaceae in that Empire. *J. Linn. Soc. London*, 34: 1–146.

LEGRIS, P. 1963. Le Vegetation de l'Inde ecologie et fore. Pondichery.

MEHER-HOMJI, V. M. 1965. On the 'Sudan-Deccanian' floral elements. *J. Bombay nat. Hist. Soc.*, 62(1): 15–18, fig. 2.

PRAIN, D. 1903. Bengal Plants, 1, 2.

PURI, G. S. 1960. Indian Forest Ecology. New Delhi: Oxford vols. 2.

RICHARDS, P. W. 1952. Tropical Rain Forests. London.

SANTAPAU, H. 1944. The flowering of Strobilanthes. *J. Bombay nat. Hist. Soc.*, 44: 605–606.

SANTAPAU, H. 1951. The Acanthaceae of Bombay. *Univ. Bombay Bot. Mon.*, 2: 1–104.

TALBOT, W. A. 1909–1911. Forest Flora of the Bombay Presidency and Sind.

VII. VEGETATION AND PHYTOGEOGRAPHY OF THE WESTERN GHATS

by

K. SUBRAMANYAM AND M. P. NAYAR

1. Introduction

This chapter attempts to present on outline of the characteristic vegetation and phytogeographical peculiarities of the Western Ghats. The general features of the Western Ghats, the geological structure and tectonic history and the climatic characters of the region are discussed in sufficient detail by specialists in earlier chapters of this book. From the stand point of the present chapter, we may appropriately describe the Western Ghats as an important part of the monsoonland, where the vegetation is influenced more by the abundance and distribution of the seasonal rainfall than the atmospheric temperature. The western side of the Western Ghats is on the threshold of southwest monsoon and receives a rainfall of 203–254 cm, and the eastern side lies in the rain-shadow area of the Peninsula. The main types of soils met with in the Western Ghats are red soils, laterites, black soils and humid soils. The red soils are developed on the Archean crystallines and are brown, grey or black, is deficient in organic matter, phosphoric acid and nitrogen. Evergreen forest of *Calophyllum, Dipterocarpus, Hopea, Myristica* and *Xylia* are characteristic of red-soil areas. The laterites consist of 90–95 % of iron, aluminium, titanium and manganese oxides and are deficient in lime and organic material, an extend up to 1600 m in the Western Ghats. *Shorea* and *Xylia* are the dominant species in lateritic soils of Western Ghats. Black soils, formed out of the basaltic Deccan lava, are deficient in organic matter, nitrogen and phosphoric acid, but generally have enough lime and potash. The red and black soils also occur in various combinations and streak all along the Western Ghats. The humid soils occur in the peat bogs of Nilgiris and they are limited in extent. Shifting cultivation, grazing and indiscriminate lopping have resulted in total destruction of some of the virgin forests, which now survive only in some of the inaccessible mountain summit areas. With accelerated population growth, large areas have been recently brought under cultivation. The destruction of forests resulted in a pronounced imbalance in the effectiveness of precipitation, maintenance of the water-table and percentage of humidity and transpiration. The lack of maintenance of forest eco-system has resulted in floods and erosion. Introduction of plantation crops like tea, coffee, rubber and extension of teak in southern regions of the Western Ghats, and cultivation of *Eucalyptus* especially in Nilgiri, have resulted in unprecedented destruction of large virgin-forest areas. Recently, the establishment of large number of hydro-electric projects, resulting in the submersion of catchment areas rich in vegetation, have further accelerated in the regressive changes in the forest flora of the region.

2. History of Botanical Studies in the Western Ghats

Western Ghats, especially the Malabar Coast is well known in world history and commerce as an important and perhaps the sole centre of spice trade, especially pepper, ginger and cardamom. Greek, Arab and later European traders, lured by the spices, found their way to the Western Coast of India. The Portuguese settlement at Goa and the Dutch possessions of Malabar, interested in the exploration of flora of this region, contributed for the first time to the scientific study of the plants of the region. In 1565, GARCIA DE ORTA enumerated a list of medicinal plants of India. Between 1678–1703, HEINRICH VAN RHEEDE TOT DRAAKENSTEIN published *Hortus Malabaricus*, a monumental work on the plants of Malabar. ROBERT WIGHT, in the middle of the nineteenth century, published a series of books with illustrations on South Indian plants. Between 1872 and 1897, HOOKER, assisted by several botanists, published for the first time a comprehensive Flora of the British India. This flora accelerated the publication of provincial floras like the Flora of the Madras Presidency by J. S. GAMBLE and subsequently completed by C. E. C. FISCHER (1915–1936), and the Flora of the Presidency of Bombay by T. COOKE (1901–1908). Other workers like BLATTER, BOURDILLON, CLEGHORN, DALZELL, FISCHER, FYSON, GIBSON, LAWSON, SANTAPAU and TALBOT have made significant contributions to the knowledge of the flora of the Western Ghats. At present the Botanical Survey of India is engaged in a revision of the flora of this region.

3. Phytogeographical Regions and Vegetation

In 1855 HOOKER and THOMSON (1855), in their introductory essay to the *Flora Indica*, and later HOOKER (1907) in the Imperial Gazetteer of India, analysed the phytogeographical regions of India on the basis of the species-content of the families in each botanical province. HOOKER classified the botanical regions of the British India as follows: 1. Eastern Himalaya, 2. Western Himalaya, 3. Indus Plain, 4. Gangetic Plain, 5. Malabar, 6. Deccan, 7. Ceylon and Maladives, 8. Burma and 9. the Malay Peninsula. The mountain ranges of Western Ghats fall under HOOKER's botanical region Malabar, which includes the humid belt of hilly or mountainous country, extending along the western side of the Western

179

Peninsula, from the mouth of the R. Tapti to Cape Kumarin (= Comorin). CLARKE (1898) proposed the following phytogeographical provinces: 1. West Himalaya, 2. India deserta, 3. Malabarica, 4. Ceylon, 5. Coromandelia, 6. Gangetic Plain, 7. East Himalaya, 8. Assam, 9. Ava, 10. Pegu and 11. Malaya Peninsula. PRAIN (1903) classified the phytogeographical regions as follows on the basis of humidity or dryness: 1. India deserta, 2. India diluvia, 3. India aquosa, 4. India vera, 5. India subaquosa and 6. India littorea. According to him, India aquosa comprises the wet-forest tracts along the Western Ghats, from Gujarat to Travancore, which receive the full force of the southwest monsoon, and corresponds to the Malabar Province of HOOKER and CLARKE. CHATTERJEE (1940), after a study of the endemic species of the Dicotyledons, recognizes the botanical regions of 1. Western Himalaya, 2. Indus Plain, 3. Malabar, 4. Deccan, 5. Gangetic Plain, 6. Eastern Himalaya, 7. Assam, 8. Central Himalaya, 9. Upper Burma and 10. Lower Burma. He excludes, however, Malaya and Ceylon, as they have characteristic floras of their own. Further, since Assam has a uniquely interesting flora, CHATTERJEE follows CLARKE in considering Assam as an independent botanical province. He further divides the Himalaya into the three botanical regions: 1. Northern Himalaya, 2. Central Himalaya and 3. Eastern Himalaya. It is thus evident that all the above mentioned phytogeographers agree in placing the Western Ghats in the 'Malabar' of HOOKER, CLARKE and CHATTERJEE, and 'India aquosa' of PRAIN.

The botanical province 'Deccan' of HOOKER is adjacent to the Western Ghats, and the flora of the leeward side of the Western Ghats merges with the floristic elements of Deccan. The exact boundaries of the botanical provinces of Malabar and Deccan are not sharp, as large number of spurs of Western Ghats enter into Deccan and merge with the mountains of the Eastern Ghats. So also in the north the Vindhya and Satpura Ranges, Mahadeo Hills carry some of the deciduous floristic elements to Central India. HOOKER (1907), after a study of the flora of this botanical province Malabar, observed that the most distinctive characters of the Malabar flora, in contrast to those of Deccan, are primarily the presence of Guttiferae, Dipterocarpaceae, Myristicaceae, Palmae and Bambusae.

The most outstanding feature of the Malabar botanical province is the development of the tropical rainforest in the Western Ghats, prominently seen on the windward side of the southern Western Ghats, usually between 500 to 1500 m. The humid tropic belt of the Western Ghats possesses the following forest types: 1. tropical moist deciduous, 2. tropical semi-evergreen and 3. tropical evergreen. According to RICHARDS (1952), tropical rain forests have no marked summer and winter seasons, but only wet and dry seasons and the seasonal changes of temperature are quite insignificant in relation to the seasonal variations in rainfall. These forests are characterized by multistoried canopies of vegetation and the

180

various synusiae according to RICHARDS are 1. trees and shrubs, 2. herbs, 3. climbers, 4. stranglers and 5. epiphytes. Some of the features characteristic of this biological spectrum are the presence of tall trees, with prominent buttresses and trees-like *Myristica fatua* var. *magnifica* developing stilt roots, occurrence of lianas, canes and epiphytes and the development of cauliflory. The ground layer and the trees themselves are carpeted with mosses, ferns, orchids and lichens. The concept of dominance of one species over the other in tropical rain forest is fallacious, as there is no gregarious growth of one species strand; but associations of two or more species can be readily recognized.

In low rainfall areas and usually on black soil, moist deciduous type of vegetation is seen and these show transition to the semi-evergreen type where higher rainfall occurs. The climax type of rain forests results through a series of transitions from moist deciduous type to evergreen forms like *Myristica, Hopea, Calophyllum, Dipterocarpus* and *Palaquium*. AIYER (1932) has described the following associations: *Vateria-Cullenia, Vateria-Mesua, Poeciloneuron-Palaquium, Mesua-Calophyllum* and *Cullenia-Palaquium*. The moist deciduous associations are *Terminalia-Tectona-Adina* type, *Terminalia-Grewia* type and *Anogeissus-Terminalia-Phyllanthus* type, which gradually merges into *Xylia*-Bamboo mixed forests. Where there is high rainfall, with red soil of laterite-gneiss derivation, evergreen types like *Xylia-Olea-Cinnamomum* and *Diospyros-Cinnamomum-Xylia* associations develop. ARORA (1960) noted the following associations in the tropical wet evergreen North Kanara forests: 1. *Cinnamomum-Olea-Diospyros* type, 2. *Diospyros-Cinnamomum-Myristica* type, 3. *Hopea-Myristica* type, 4. *Myristica-Hopea-Diospyros* type and 5. *Myristica-Diospyros-Polyalthia* type.

Out of over 13,000 species of flowering plants so far described from India, about 3500 are reported from the Western Ghats. The ten dominant Natural orders are Gramineae, Leguminoseae, Acanthaceae, Orchidaceae, Compositae, Euphorbiaceae, Rubiaceae, Asclepiadaceae, Geraniaceae and Labiatae. Some of the genera, having more than fifteen species, are *Crotalaria, Impatiens, Diospyros, Ipomoea, Eugenia, Strobilanthes, Ficus, Desmodium, Habenaria, Grewia* and *Osbeckia*. Some of the endemic genera of the Western Ghats are *Poeciloneuron, Adenoon, Griffithella, Willisia, Meineckia, Pseudoglochidion, Baeolepis, Nanothamnus, Wagatea, Otonephelium*. The endemic genera common to the Western Ghats and Ceylon are *Chloroxylon, Haplanthodes, Kendrickia* and *Nothopegia*. The only indigenous Gymnosperm present in the Western Ghats is *Decussocarpus wallichianus*.

In the Western Ghats, the Guttiferae are represented by *Poeciloneuron, Mesua, Calophyllum, Mammea* and *Garcinia*. The endemic genus *Poeciloneuron*, with two species *P. indicum* and *P. pauciflorum*, occurs in the Western Ghats, from Mysore southwards. The well known timber tree *Mesua ferrea* occurs in the Western Ghats, also from Mysore southwards. The Dipterocarpaceae are represented by *Dipterocarpus, Hopea, Shorea, Vatica* and *Vateria*. *Dipterocarpus indicus*, occurring in the Western Ghats from Mysore southwards, is valued as one of the best timber trees. *Hopea utilis* is endemic to Tinnevelly (= Tirunelvelly) Hills. *Vateria indica*, the Indian copal tree, occurs in the evergreen forests of the Western Ghats, from Mysore southwards. The genus *Myristica* is essentially Malaysian, and is represented by four species in the Western Ghats; *Myristica malabarica* an endemic species, occurs from Mysore southwards; *M. fragrans*, the nutmeg tree, the kernal and aril of which are used as spices, is cultivated in the southern Western Ghats. The other species occurring in the Western Ghats are *Myristica dactyloides* and *M. fatua* var. *magnifica*. *Helicia*, a genus of the Australian Natural Order Proteaceae, is represented by the endemic species *H. travancorica* and *H. nilagirica* in the southern Western Ghats. Out of about twenty-one palms in the Western Ghats, *Bentinckia coddapanna*, *Pinanga dicksonii* and about nine species of *Calamus* are endemic to the southern Western Ghats. As stated by Hooker (1907), the Malayan Natural Orders especially Sterculiaceae, Tiliaceae, Anacardiaceae, Meliaceae, Myrtaceae, Melastomataceae, Vitaceae, Gesneriaceae, Piperaceae, Scitamineae, Orchidaceae and Araceae, are well represented in the Western Ghats. Bambusae are very conspicuous by their arborescent habit and there are six genera and about seventeen species in the Western Ghats. The genus *Ochlandra* is well represented in the Western Ghats and out of eight species in India, six occur in the southern Western Ghats, of which *O. travancorica* is the most abundant. Other common species are *Arundinaria wightiana*, *A. densifolia*, *Dendrocalamus strictus*, *Bambusa arundinacea* and *Oxytenanthera bourdilloni*. Among the herbaceous species the most conspicuous genus is *Impatiens*, which has its highest development in the humid southern Western Ghats and in the Eastern Himalaya. Of the 175 species from India, about 77 species occur in the southern Western Ghats and many are endemics. It is interesting to note the well marked discontinuity in the distribution of this genus; the Himalaya and Western Ghats having their own endemic species and not a single species is common to either of these regions. Other interesting Natural orders, which find full development are Podostemaceae, Umbelliferae, Loranthaceae and Acanthaceae.

According to Chatterjee (1940), who analysed the floristic divisions on the basis of endemic species, after the Himalaya, which has about 3169 endemic Dicotyledons, Peninsular India ranks second, with about 2045 endemic Dicotyledons. In the Western Ghats, there are about 1500 endemic species and a brief discussion on this will follow later.

The Western Ghats can be divided into four phytogeographical regions, viz. 1. the Western Ghats from the R. Tapti to Goa, 2. the Western Ghats from the R. Kalinadi to Coorg, 3. the Nilgiri, and 4. the Anamalai, Palni and Cardamom Hills.

As observed earlier, this botanical division is dominated by the mountain chains, rising to 1000 m abruptly within a short distance of 2–3 km. Along the western side there are deep ravines and canyons and on the eastern side there are flat-topped spurs intersected by valleys. The spurs lose height towards the east. This region receives the full blast of the monsoon rainfall from June to September. The following are the main types of vegetation: The scrub and dry semi-deciduous type occurs in the foothills, along the eastern side of the Western Ghats; the elevation ranges from 200 to 500 m and the rainfall from 37 to 61 cm. The character species include *Solanum surattense, Argemone mexicana, Barleria prionitis, B. cristata* var. *dichotoma, Eranthemum roseum, Hemigraphis latebrosa, Rungia repens, Dicliptera zeylanica, Justicia diffusa, Aerva sanguinolenta, Mimosa pudica, Acacia nilotica, Corchorus trilocularis, Jatropha curcas* and *Solanum nigrum.* Along the valleys and ravines we observe the following species of trees *Terminalia chebula, Albizia procera, A. lebbeck, Erinocarpus nimmonii, Turraea villosa, Strombosia zeylanica, Pouteria tomentosa, Pavetta indica, Lantana camara* var. *aculeata, Ixora brachiata, Carissa congesta, Hemidesmus indicus, Smilax zeylanica* and *Ventilago madraspatana.*

Dry deciduous hill forests are found on the eastern side of the Western Ghats, at elevations between 500 to 1166 m, with rainfall of 50 to 152 cm. The character species are *Diospyros montana, D. sylvatica, Eriolena quinquelocularis, Sterculia urens* and *Canthium dicoccum.*

Moist deciduous forest type is situated on the windward side of the Western Ghats, at elevations between 500 m to 833 m, with annual rainfall of 100 to 200 cm. These forests merge with the evergreen type, depending on the range of rainfall. Trees of great commercial importance like *Terminalia crenulata, Dalbergia latifolia, Anogeissus latifolia, Lagerstroemia lanceolata, Pterygota alata, Schleichera oleosa, Grewia tiliaefolia,* and *Pterocarpus marsupium* grow here. Other common species are *Bambusa arundinacea, Dillenia pentagyna,* several species of *Crotalaria* and *Desmodium, Canavalia gladiata, Garcinia indica, Careya arborea* and several species of Zingiberaceae like *Kampferia scaposa, Hitchenia caulina, Curcuma pseudomontana, Zingiber cernuum, Costus speciosus,* and several species of Araceae like *Cryptocoryne spiralis, Arisaema murrayi, Pothos scandens* and *Colocasia esculenta.* The common species seen along the swamps and marshy areas are *Hydrolea zeylanica, Hygroryza aristata, Sphenoclea zeylanica* and several species of *Eriocaulon.*

Qureshi (1965) remarks that the evergreen forests occurring in the Western Ghats of Maharashtra are not typical tropical evergreen forests. Though this region receives rainfall of 625 to 750 cm, the evergreen trees are characteristically dwarfish, with no tiers or canopies of tropical species. Hence they are classified as montane subtropical evergreen forests, recognized by the trees *Amoora lawii, Aphanamyxis rohituka, Walsura trijuga,*

Toona ciliata, Holoptelea integrifolia, Alstonia scholaris, Pongamia pinnata, Caryota urens, Tetrameles nudiflora, Terminalia chebula, and *Bridelia squamosa.* The second layer in these forests consists mainly of *Miliusa tomentosa, Murraya paniculata, Syzygium cumini, Meyna laxiflora, Mammea suriga, Gnetum ula* and *Calycopteris floribunda.*

The species of herbaceous plants, particularly conspicuous during the monsoon rains, include *Impatiens kleinii, I. balsamina, Chlorophytum tuberosum, C. glaucum, Begonia crenata, Commelina obliqua, Striga gesneroides, Elephantopus scaber, Cyanotis tuberosa, Geissapsis cristata, Balanophora indica. Burmannia pusilla* and several species of *Lindernia.* The Orchidaceae are represented by terrestrial genera like *Habenaria, Peristylus, Platanthera, Calanthe, Eulophia, Geodorum, Epipogum, Nervilia, Pachystoma, Liparis, Spiranthes, Zeuxine;* the genus *Habenaria* is well represented by about twenty-one species. The epiphytic or lithophytic genera include *Dendrobium, Pholidota, Cymbidium, Eria, Porpax, Oberonia, Sirhookera, Diplocentrum, Vanda, Rhynchostylis, Acampe, Aerides, Bulbophyllum,* etc. *Platanthera susannae, Vanda tessellata, Aerides crispum, Dendrobium lawianum* are some of the orchids of considerable beauty found in the Ghats.

3.2. THE WESTERN GHATS FROM THE R. KALINADI TO COORG

The Deccan lava gives way to the Archeans and the change is marked by the series of breaches in the mountain wall by the rivers Kalinadi, Gangavali-Bedti, Tadri and Sharavati. At these breaches the Maharashtra Lava Ghats of 800–1000 m elevation break down for a distance of about 320 km. The highest peak in this sector of Western Ghats is at Kudremukha (2071 m). The access to the interior is not easy, since the valleys are surrounded by deep gorges 3–5 km across and 300 m deep. The entire area is hot and humid. The heavy rainfall favours thick tropical forest growth. Faulting and differential erosion make this region an extremely dissected tract and in some areas the Ghat forests reach down to the sea. The upper evergreen zone supports the best teak plantations. The main types of vegetation observed here are scrub forests, moist deciduous forests and wet evergreen forests. According to CLEMENTS (1928), the tropical rain forest is a formation-type of a pan-climax and the rain forest of the Western Ghats belongs to Indo-Malayan rain forests type. High atmospheric humidity, warm temperature, rainfall ranging from 200–250 cm and non-seasonable climate favour the development of tropical rain forests. According to the classification of CHAMPION (1936), the southern Western Ghats comes under the group western tropical evergreen (W. Coast along the W. Ghats). The floristic components of rain forests along the southern Western Ghats differ basically from those of the rain forests of Assam and Andamans, though certain species of *Calophyllum,* and *Mesua ferrea* occur in all the three regions. The rain forests occur in different tiers and about three or four layers can be

184

demarcated on the basis of height. ARORA (1960), after a study of the vegetation of Coorg, recognizes the following layers: The first tier is composed of trees of 30–45 m height and some of the trees like *Tetrameles nudiflora, Elaeocarpus tuberculatus, Dipterocarpus indicus, Dysoxylum malabaricum, Diospyros microphylla* have buttresses. The second layer consists of trees having 15–23 m height and include *Alstonia scholaris, Hardwickia pinnata, Strychnos nuxvomica, Xylia xylocarpa, Xanthophyllum flavescens, Artocarpus lakoocha,* etc. The third layer consists of small trees about 10–15 m height, like *Callicarpa tomentosa, Flacourtia montana, Leea indica* and species of *Memecylon* and *Psychotria.* The undergrowths form dense thickets and are composed of shrubs and climbers, especially *Ancistrocladus heyneanus, Calycopteris floribunda, Entada pursaetha, Hemidesmus indicus, Naravelia zeylanica, Allophylus serratus, Calamus* sp., *Gnetum ula, Pothos scandens, Smilax zeylanica, Piper nigrum* and *Psychotria* sp.

The deciduous forests, between 666 m to 1000 m, which receive moderate annual rainfall ranging from 150–200 cm, are without any tiers of different species unlike the evergreen wet forests, and they reach a height of 12–24 m. The chief species of trees of these forests are *Adina cordifolia, Albizia* sp., *Bauhinia* sp., *Bridelia squamosa, Butea monosperma, Dalbergia latifolia, Dillenia pentagyna, Diospyros montana, Ehretia laevis, Emblica officinalis, Ficus* sp., *Garuga pinnata, Gmelina arborea, Grewia tiliaefolia, Kydia calycina, Lagerstroemia lanceolata, Lannea coromandelica, Mallotus philippinensis, Pterocarpus marsupium, Schleichera oleosa, Saccopetalum tomentosum, Shorea talura, Sterculia* sp., *Tectona grandis, Terminalia tomentosa, T. paniculata, T. Chebula, Xylia xylocarpa, Zanthoxylum rhetsa.* Among small trees and shrubs, the following species are more common: *Carissa congesta, Callicarpa tomentosa, Colebrookea oppositifolia, Holarrhena antidysenterica, Meyna laxiflora, Murraya koenigii, Lantana camara* var. *aculeata, Leea indica, Xeromorphis spinosa,* and species of *Solanum* and *Zizyphus.* The common climbers are *Asparagus racemosus, Calycopteris floribunda, Cissus* sp., *Cryptolepis buchanani, Diploclisia glaucescens* and *Entada pursaetha.* The undergrowth consists of *Carvia callosa, Crotalaria* sp., *Desmodium* sp., *Mimosa pudica, Moghania strobilifera, Sida rhombifolia* and *Urena lobata.* Grasses like *Themeda, Apluda, Eragrostis, Oplismenus* and bamboos like *Dendrocalamus* are commonly met with in these forests.

The scrub type too is distributed along the eastern belt, where a vegetation of dry deciduous species is typical. Rainfall in the areas, which support scrub forests, is, as may be expected, very low. The vegetation chiefly consists of thorny species with a few stunted, crooked and malformed trees. The chief components of the scrub forests are *Acacia catechu, Argyreia cuneata, Balanites aegyptiaca, Capparis* sp., *Carissa congesta, Cassia auriculata, Cipadessa baccifera, Dodonaea viscosa, Erythroxylum* sp., *Euphorbia* sp., *Flacourtia indica, Securinega* sp., *Gardenia* sp., *Maytenus* sp., *Ixora parviflora, Lantana camara* var. *aculeata, Pavetta indica, Xeromphis spinosa, Rhus mysorensis* and *Soymida febrifuga.* The species of trees, commonly observed in these forests, are *Anogeissus latifolia, Bauhinia racemosa, Bridelia squamosa, Buchanania lanzan, Careya arborea, Cassia fistula, Chloroxylon swietenia, Cochlospermum religiosum, Diospyros melanoxylon, Lagerstroemia* sp., *Pterocarpus marsupium, Semecarpus anacardium, Radermachera xylocarpa. Santalum album* seems to prefer these open jungles with scrubby associates. Amongst the undergrowth, grasses like *Apluda varia, Aristida* sp., *Eragrostis unioloides, Heteropogon contortus, Oplismenus* sp., and *Setaria glauca* and herbs like *Andrographis* sp., *Blepharis* sp., *Polycarpaea argentea, Clerodendrum serratum, Crotalaria* sp., *Alysicarpus* sp., *Goniogyna* sp., *Indigofera* sp., and *Barleria buxifolia* may be observed.

3.3. THE NILGIRI

The Nilgiri forms a compact plateau of about 2590 km², (Doddabeta 2920 m) an elevated and dissected much-worn massif, with swelling hills and rolling downs. On its southeastern side is the Palghat gap. The R. Moyar ditch cuts off Nilgiri from Mysore Plateau and this massif is islanded between the R. Moyar and R. Bhavani to the south. With equitable climate and rainfall ranging from 125–406 cm, the Nilgiri harbours interesting flora, which shows pronounced relationship with the Assam flora. The rolling downs are interspersed with wood and *shola* forests. The forest is evergreen, composed of tropical and sub-tropical vegetation. The grassy downs have a distinct vegetation consisting of *Strobilanthes* sp., *Berberis* sp., *Hypericum* sp., many Leguminosae, *Rubus* sp., *Hedyotis* sp., *Helichrysum* sp., and *Gaultheria* sp. Among the herbaceous plants are species of *Anaphalis, Senecio, Wahlenbergia* and *Campanula*.

The *sholas* are characteristically seen along the folds of rolling downs at a height of 1666 m and above; they are filled with evergreen forests with thick undergrowth. The most conspicuous shrubs and trees are of the scholas are *Hydnocarpus alpina, Michelia nilgirica, Berberis tinctoria, Mahonia leschenaultii, Garcinia cambogia, Gordonia obtusa, Ternstroemia japonica, Ilex denticulata, I. wightiana, Euonymus crenulatus, Microtropis ramiflora, Cinnamomum wightii, Meliosma wightii, M. microcarpa, Osyris wightiana, Pentapanax leschenaultii, Schefflera racemosa* and *Macaranga indica*. The undergrowth consists mainly of *Clematis wightiana, Viola serpens, Polygala arillata, Parthenocissus neilgherriensis* and *Osbeckia leschenaultiana*. In the *sholas* there is rich growth of orchids like *Calantha veratrifolia, Aerides ringens* and *Habenaria longicornu*. In the lower elevations of scholas *Hydnocarpus alpinia* reaches a height of 20 m and is accompanied by an extensive growth of species of *Memecylon, Psychotria, Maesa* and *Osyris*. The insectivorous plants like *Drosera* and *Utricularia* are common in the open downs above 2000 m. *Shola* forests, similar to those of the Nilgiri, also occur in Anamalai and Palni Hills, but since these are at lower elevations than on the Nilgiri they harbour a rich tropical vegetation. In the open downs, there are mainly herbaceous and shrubby species like *Anemone rivularis, Ranun-*

Plate 1. The *Shola* in upper Bhavani in the Nilgiri Hills

culus reniformis, Cardamine hirsuta, Viola patrinii, V. distans, Polygala sibirica, Hypericum mysorense, Impatiens nilgirica, I. pusilla, Crotalaria scabrella, C. ovalifolia, C. albida, C. barbata, C. madurensis, Indigofera pulchella, Smithia gracilis, Rubus moluccanus, Parnassia mysorensis, Rhodomyrtus tomentosa, Osbeckia cupularis, Bupleurum mucronatum, Heracleum rigens, H. hookerianum, Galium asperifolium, Rubia cordifolia, Vernonia sp., *Erigeron* sp., *Blumea* sp., *Anaphalis* sp., *Campanula fulgens, C. wightii, Lysimachia obovata, Swertia minor, S. corymbosa, Sopubia delphinifolia, Strobilanthes* sp., *Habenaria decipiens, H. heyneana* and *Satyrium neilgherriensis,* Peat bogs occur in the hills at about 2333 m and the chief constituents are grasses, sedges and mosses. The common species seen in this area are *Juncus prismatocarpus, J. bufonius, Eriocaulon brownianum, E. collinum, E. gamblei, Cyperus* sp., *Carex nubigena, Isachne kunthiana, Panicum repens, Xyris schoenoides, Cyanotis villosa, Utricularia graminifolia, U. striatula* and *Lindernia* sp. The flora of Nilgiri shows distinct floristic elements of Khasi and Naga Hills, and Eastern Himalaya. The trees and shrubs common to these localities include *Ternstroemia japonica, Hypericum hookerianum, Thalictrum javanicum, Turpinia nepalensis, Meliosma microcarpa, Cotoneaster buxifolia, Parnassia wightiana, Pentapanax leschenaultii, Lonicera ligustrina, Galium asperifolium, Lactuca hastata, Gaultheria fragrantissima, Rhododendron arboreum, Lysimachia obovata* and *Symplocos laurina.*

187

3.4. THE ANAMALAI, CARDAMOM AND PALNI HILLS

The topography of these hills is well illustrated by SPATE (1957) who states 'This remarkable group of hills is more complex than the Nilgiri, and in Anamudi itself they have the highest peak of the Peninsula, 8841 ft. The front to the Palghat Gap is remarkably steep and in the E remarkably straight; SE flanks of Palnis, overlooking the upper re-entrant, are also remarkably abrupt, as are the Cardamoms and their protrusions (Varushanad Hills) S of Vaigai. But to the NW the hills fray out into long SE–NW ridges; and indeed over much of the area this trend is most marked, the rivers (e.g. the Periyar) having longitudinal stretches of such straightness as to suggest control by faults, with transverse gorges producing a perfect trellis-pattern. Between 10°N and the Shencottah gap the active streams of the exposed Arabian Sea front have pushed the watershed back to within 4 or 5 miles of the eastern edge of the hills: here the change from jungle-clad mountains to the tank-pitted Tamilnad plains is very sudden'.

The type of vegetation of the Anamalai, Palni and Cardamom Hills may be briefly considered here. Starting from the plains, on the leeward side of Western Ghats, the dry semi-desert type occupies from the foot-hills, the elevation being about 400 m and the rainfall ranging from 45–53 cm. The following species are conspicuous: *Commiphora berryi*, *Dichrostachys cinerea, Acacia latronum* (generally very gregarious), a sprinkling of *Acacia planifrons*, *Opuntia dillenii*, *Dichoma tomentosa*, *Azima tetracantha*, *Solanum trilobatum*, *Euphorbia antiquorum* and *E. tortilis*. Other characteristic plants are *Euphorbia dracunculoides* and *Jatropha wightiana*.

The dry deciduous hill type is characteristic of the lower elevations, with annual rainfall ranging from 160–260 cm. Wherever there is heavy rainfall, evergreen trees like *Mallotus philippinensis, Ficus glomerata, Santalum album*, *Olea dioica, Limonia acidissima* and *Pavetta tomentosa* are seen. The character species are *Anogeissus latifolia*, *Ailanthus excelsa*, *Butea mono-sperma, Cassia fistula, Emblica officinalis, Grewia tiliaefolia, Dalbergia latifolia*, *D. paniculata, Pterocarpus marsupium, Melia composita, Tectona grandis*, *Terminalia bellerica, T. paniculata, T. crenulata, Albizia amara, Ficus glomerata*, *Bombax ceiba, Bridelia retusa, Mitragyna parvifolia, Cochlospermum religiosum*, *Shorea talura, Sterculia urens, Eriolaena quinquelocularis, Buettneria herbacea*, *Diospyros montana* and *Givotia rottleriformis*. Among shrubs, species of *Capparis, Grewia, Flacourtia, Securinega* and *Phyllanthus* are very common, followed by climbers like *Ventilago madraspatana, Dalbergia volubilis*, *Cayratia pedata, Canavalia gladiata, Clitoria ternatea, Argyreia pamacea* and *Glycine javanica*. There are also species of *Calamus* and *Phoenix;* the bam-boos are mainly represented by *Bambusa arundinacea* and *Dendrocalamus strictus*.

The moist deciduous forest types occur between 500–900 m, with a rainfall ranging from 240–350 cm. Many dry deciduous trees of lower

188

Plate 2. General view of the vegetation of the hills, adjoining the Anamudi Peak in the Anamalai Hills.

elevations and evergreen trees of higher elevations intrude into this zone. Timber-trees like *Tectona grandis, Dalbergia latifolia, Terminalia tomentosa, Anogeissus latifolia, Pterocarpus marsupium* and several others grow in this zone. Another conspicuous feature of this zone is the luxuriant growth of *Bambusa arundinacea.* Other important species are *Sterculia guttata, Bridelia retusa, Terminalia bellerica, Stereospermum personatum.* These form the top layer, reaching a height of about 20 m. The middle layer consists of *Kydia calycina, Clausena heptaphylla, Nothopodytes foetida, Litsea deccanensis, Cycas circinalis;* in addition many species of Scitamineae occur in this forest. The fern *Drynaria quercifolia* and a large number of orchids are seen on tree trunks and on rocks. The following climbers over trees form thick canopies: *Gouania microcarpa, Erythropalum populifolium, Diploclisia glaucescens, Dioscorea tomentosa* and *D. pentaphylla.*

The wet evergreen forest types are developed on elevations ranging from 500–2500 m along the windward side of the Western Ghats, where the rainfall ranges from 250–500 cm. At higher elevations shola forests or stunted evergreen forests are interspersed with rolling downs. The genus *Dipterocarpus* is represented in these forests by only a single species, though there are many species in wet evergreen forests of the Andamans and Nicobar Islands. They show a large number of species arranged in tiers. The top tier includes *Mesua ferrea, Vitex altissima, Aglaia roxburghiana, Elaeodendron glaucum, Polyalthia fragrans, Diospyros microphylla, Eugenia*

189

gardneri, Canarium strictum, Artocarpus lakoocha, A. heterophyllus, Bischofia javanica, Grewia nudiflora, Alstonia scholaris, Gordonia obtusa, Elaeocarpus glandulosus, Symplocos laurina, Vernonia monosis, Meliosma microcarpa, Ligustrum roxburghii, Elaeocarpus munroii, Apodytes beddomei and *Olea dioica.* The gigantic trees of timber value are *Mesua ferrea, Calophyllum tomentosum, Palaquium ellipticum,* and *Diospyros ebenum.* Lower storey consists of shrubs and trees, which also adapt themselves to shady conditions, like *Goniothalamus wightii, Garcinia* sp., *Holigarna beddomei, Eugenia munronii, Premna coriacea, Macaranga tomentosa, Pavetta hispidula, Psychotria anamalayana, Ardisia solanacea* and *Antidesma menasu.* Undergrowth is represented by *Strobilanthes* sp., *Orophaea zeylanica, Psychotria* sp., *Calamus* sp., and a large number of ferns and ground orchids. *Decussocarpus wallichianus,* which occurs in this zone, is the only South Indian conifer. In the evergreen forests of higher altitudes above 2500 m, there are species of *Rhododendron arboreum, Gaultheria fragrantissima, Rhodomyrtus tomentosa, Microtropis* sp., *Eurya japonica, Michelia nilgirica, Mahonia leschañaultii* and *Symplocos anamallayana.*

The grassland types are found in regions of elevations above 900 m. There are shrubby and herbaceous plants, showing an alpine affinity and generally resembling the flora of corresponding elevations in the Nilgiri. The main species observed here are *Hypericum japonicum, Fragaria indica, Osbeckia leschenaultiana, Dipsacus leschenaultii, Vernonia bourneana, Anaphalis aristata, Cnicus wallichii, Lysimachia leschenaultii, Pedicularis zeylanica, Strobilanthes kunthianus, Plantago major, Rumex nepalensis, Thesium wightianum, Lilium neilgherrense, Isachne gardneri, Arundinella wightiana, Exacum bicolor, Conyza stricta, Lobelia nicotianaefolia, Tephrosia tinctoria, Launaea acaulis, Plectranthus wightii, Leucas prostrata, Justicia procumbens, Laggera alata* and *Vernonia divergens.*

The flora of this division of the Western Ghats shows marked affinity with the Ceylon flora in having large number of species common to both; some of them are *Kendrickia walkeri, Filicium decipiens, Gyrinops walla, Gordonia obtusa, Kydia calycina, Myristica dactyloides, Polyalthia longifolia, Clematis gouriana, Naravelia zeylanica, Capparis grandis, Olax scandens, Gouania microcarpa, Zizyphus xylopyra, Tetrastigma lanceolarium, Meliosma microcarpa, Crotalaria nana, Pterocarpus marsupium, Acacia suma, A. ferruginea, A. caesia, Albizia amara, Osbeckia wightiana, Begonia malabarica, Neurocalyx calycinus, Hedyotis nitida, Tarenna asiatica, Maesa perrottetiana, Canscora wallichii, Premna tomentosa, Teucrium tomentosum, Apama siliquosa, Sarcandra irvingbaileyi, Litsea deccanensis, Helixanthera hookeriana, Schumannianthus virgatus, Molineria trichocarpa* and *Calamus pseudotenuis.*

From the point of distribution it is rather interesting to find that the monotypic genus *Kendrickia,* which includes the species *K. walkeri* (Melastomataceae) is restricted to Anamudi region of Anamalais in South India and Adam's Peak in Ceylon. It is also significant to find in this connection that while Anamudi is the highest peak in the Western Ghats of Penin-

Plate 3. Lobelia leschenaultiana from the Nilgiri Hills.

Plate 4. Cyathea crinita from the Anamalai Hills.

Plate 5. Griffithella hookeriana from Ambavane: Poona Dist.

sular India, correspondingly Adam's Peak is also the highest in Ceylon. It appears that *Kendrickia* supports the concept that in bygone times Ceylon and South India were united and this taxon represents one of the identical types occurring in the floras of the hills of both South India and Ceylon. It may also be mentioned that this taxon is not recorded from the lower elevations of the Western Ghats.

192

4. The Endemic Flora of the Western Ghats

It is seen that India has a high percentage of endemic species and it is comparable to oceanic islands. In oceanic islands the endemism may be due to either the fact that fortified by its insularity the flora of oceanic islands might have undergone an evolution giving rise to endemic species; or protected by the sea barrier the islands might have escaped the onslaught of unfavourable climatic changes. The first view supports the contention that islands are nurseries of speciation, while the second view considers the island floras as relict flora or estabilized floras and relatively free from at least some of the climatic changes and biological instabilities which affect larger land masses. The Himalaya has the largest number of endemic species, of which 3169 are Dicotyledons and about 1000 Monocotyledons. Peninsular India has about 2045 species of endemic Dicotyledons and about 500 endemic Monocotyledons (CHATTERJEE 1962). The Western Ghats, protected by the sea along the western side and Vindhya and Satpura Ranges on the northern side and the semi-arid Deccan plateau on the eastern side of India, behave like an oceanic island in the development of endemic species. VAN STEENIS (1962) proposed a theory of land bridges and indicated that the hill-top floras show similarities in the species content. After a study of the vegetation of Malaya and Indonesian islands, STAPF (1894) and later on VAN STEENIS (1962) proposed that since the hill top flora of Kinabalu in Borneo and other hill-top floras of Malaya and Philippines show similarities in the floristic elements, they contended that in the early Pre-Tertiary period these areas formed highlands of wide Indo-Malaysian-Australian continental areas. In the Western Ghats the hill-top floras of Nilgiri, Palni and Cardamom hills and Adam's Peak in Ceylon show similarities and this indicates they formed one land mass in ancient times.

The following are some of the endemic genera of the Western Ghats: *Adenoon* (monotypic), *Baeolepis* (monotypic), *Blepharistemma*, *Campbellia*, *Calacanthus* (monotypic), *Erinocarpus* (monotypic), *Frerea* (monotypic), *Griffithella* (monotypic), *Haplothismia* (monotypic), *Jerdonia* (monotypic), *Lamprochaenium* (monotypic), *Meteoromyrtus*, *Nanothamnus* (monotypic), *Octotropis*, *Otonephilum*, *Poeciloneuron*, *Polyzygus* (monotypic), *Pseudoglochidion*, *Wagatea* (monotypic) and *Willisia* (monotypic). Out of these, more than half represent monotypic genera.

Some of the characteristic endemic species of the Western Ghats are listed below:

Species	Natural order	Distribution in the Western Ghats
Acranthera grandiflora	Rubiaceae	Anamalai southwards
Actinodaphne lanata	Lauraceae	Nilgiri
Adenoon indicum	Compositae	General

Species	Natural order	Distribution in the Western Ghats
Aglaia anamallayana	Meliaceae	Shola forest of Nilgiri
Alphonsea zeylanica	Annonaceae	Kerala and Tirunelveli Hills
Amoora lawii	Meliaceae	Evergreen forests
Antiaris toxicaria	Moraceae	Evergreen forests
Antistrophe serratifolia	Myrsinaceae	Anamalai
Apama barberi	Aristolochiaceae	Tirunelveli Hills
A. siliquosa	–do–	Evergreen forest
Apodytes beddomei	Icacinaceae	Nilgiri southwards
Apollonias arnottii	Lauraceae	Evergreen forests
Aporosa lindleyana	Euphorbiaceae	Konkan southwards
Ardisia blatteri	Myrsinaceae	Palni southwards
Arenga wightii	Palmae	Mysore southwards
Baccaurea courtallensis	Euphorbiaceae	Mysore southwards
Baeolepis nervosa	Asclepiadaceae	Nilgiri
Beaumontia jerdoniana	Apocynaceae	Throughout
Begonia aliciae	Begoniaceae	Nilgiri southwards
Bentinkia coddapanna	Palmae	Tirunelveli and Kerala Hills
Blepharistemma membranifolia	Rhizophoraceae	Evergreen forests
Calacanthus grandiflorus	Acanthaceae	General
Calophyllum apetalum	Guttiferae	General
Campanula wightii	Campanulaceae	Nilgiri and Palni
Campbellia cytinoides	Orobanchaceae	Nilgiri and Palni Hills
Cassipourea ceylanica	Rhizophoraceae	General
Chilocarpus malabaricus	Apocynaceae	General
Chiloschista pusilla	Orchidaceae	Nilgiri and Malabar
Christisonia bicolor	Orobanchaceae	Nilgiri southwards
C. calcarata	–do–	–do–
C. lawii	–do–	–do–
Cymbopogon travancorensis	Gramineae	Kerala
Daphniphyllum neilgherrense	Daphniphyllaceae	Nilgiri
Didymocarpus ovalifolia	Gesneriaceae	Tirunelveli Hills
Diplocentrum congestum	Orchidaceae	Mysore southwards
D. recurvum	–do–	–do–
Dipterocarpus indicus	Dipterocarpaceae	General
Dysoxylum malabaricum	Meliaceae	Mysore southwards
Elaeocarpus tuberculatus	Elaeocarpaceae	Mysore southwards
Ellertonia rheedii	Apocynaceae	Mysore southwards
Emblica fischeri	Euphoribaceae	General
Erinocarpus nimmonii	Tiliaceae	General
Erythropalum populifolium	Erythropalaceae	Malabar southwards
Frerea indica	Asclepiadaceae	Southern regions
Garcinia cambogia	Guttiferae	Konkan southwards
Gluta travancorica	Anacardiaceae	Tirunelveli and Kerala Hills
Glyptopetalum lawsonii	Celastraceae	Nilgiri southwards
Griffithella hookeriana	Podostemaceae	Mountain streams
Gymnacranthera canarica	Myristicaceae	Mysore southwards
Haplothismia exannulata	Burmanniaceae	Anamalai
Hopea utilis	Dipterocarpaceae	Tirunelveli Hills
Hubbardia heptaneuron	Gramineae	Mysore southwards
Humboldtia bourdillonii	Caesalpiniaceae	Kerala
Indochloa oligocantha	Gramineae	Mysore southwards

194

Species	Natural order	Distribution in the Western Ghats
Indotristicha ramosissima	Podostemaceae	Mysore southwards
Inga cynometroides	Mimosaceae	Kerala
Isonandra lanceolata	Sapotaceae	Evergreen forests
Jerdonia indica	Gesneriaceae	Nilgiri southwards
Kingiodendron pinnatum	Caesalpiniaceae	Mysore southwards
Knema attenuata	Myristicaceae	Mysore southwards
Lamprochaenium microcephalum	Compositae	Konkan southwards
Ligustrum travancoricum	Oleaceae	Nilgiri southwards
L. walkeri	–do–	–do–
Linociera malabarica	–do–	General
Litsea nigrescens	Lauraceae	Travancore and Tirunelveli Hills
Meteoromyrtus wynaadensis	Myrtaceae	Malabar
Myxopyrum serratulum	Oleaceae	Kerala
Nanothamnus sericeus	Compositae	Konkan
Nothopodytes foetida	Icacinaceae	General
Nothopegia travancorica	Anacardiaceae	Mysore southwards
Octotropis travancorica	Rubiaceae	Kerala
Olax wightiana	Olacaceae	Mysore southwards
Ormosia travancorica	Papilionaceae	Mysore southwards
Otonephelium stipulaceum	Sapindaceae	Kerala
Oxytenanthera monostigma	Gramineae	Konkan southwards
Palaquium ellipticum	Sapotaceae	General
Pedicularis perrottetii	Scrophulariaceae	Nilgiri and Palni
Pittosporum dasycaulon	Pittosporaceae	Mysore southwards
Poeciloneuron indicum	Guttiferae	Mysore southwards
Polyzygus tuberosus	Umbelliferae	Mysore
Pseudoglochidion anamalayanum	Euphorbiaceae	Anamalai
Rapanea dephnoides	Myrsinaceae	Tirunelveli and Kerala Hills
R. wightiana	–do–	Nilgiri southwards
Schefflera capitata	Araliaceae	Nilgiri southwards
Tetracera akara	Dilleniaceae	Kerala
Vaccinium leschenaultii	Vacciniaceae	Nilgiri and Palni
Vateria indica	Dipterocarpaceae	Mysore southwards
Wagatea spicata	Caesalpiniaceae	General
Willisia selaginoides	Podostemaceae	Mountain streams

5. Acknowledgments

Our sincere thanks are due to Mr. A. N. HENRY and Mr. R. L. MITRA for helpful suggestions during the preparation of this manuscript.

195

REFERENCES

AIYER, T. V. V. 1932. The sholas of the Palghat Division. *Indian For.*, 48: 414–431; 473–486.

ARORA, R. K. 1960. The Botany of Coorg Forests. *Proc. Nat. Acad. Sci. India*, (B) 30: 289–305.

ARORA, R. K. 1960. Climatic climax along the Western Ghats. *Indian For.*, 86: 435–439.

CHAMPION, H. G. 1936. A preliminary survey of Forest Types of India and Burma. *Indian For. Rec.*, (N.S.) 1(1): 1–286.

CHATTERJEE, D. 1940. Studies on the endemic Flora of India and Burma. *J. Asiat. Soc. Bengal*, 5: 19–67.

CHATTERJEE, D. 1962. Floristic patterns of Indian vegetation. *Proc. Summer School of Botany, Darjeeling*, 1960: 32–42, New Delhi.

CLARKE, C. B. 1898. On the Soil Sub-Areas of British India. *J. Linn. Soc. London*, 34: 1–146.

CLEMENTS, F. E. 1928. Plant succession and indicators. New York.

COOKE, T. 1901–1908. The Flora of the Presidency of Bombay. London.

GAMBLE, J. S. 1915–1936. Flora of the Presidency of Madras. London. (Issued in 11 parts, of which 1–7 by J. S. GAMBLE, and 8–11 by C. E. C. FISCHER).

GARCIA D'ORTA, 1565. Os Coloquinos, Goa.

HOOKER, J. D. 1907. Sketch of the Flora of British India. *Imperial Gazetteer of India* (3) 1, (4): 157–212.

HOOKER, J. D. et al. 1872–1897. The Flora of British India. I–VII. London.

HOOKER, J. D. & T. THOMSON 1855. Flora Indica. London.

PRAIN, D. 1903. Bengal Plants, Vols. 1 & 2. Calcutta.

QURESHI, I. M. 1965. Tropical Rain Forests of India and their silvicultural and ecological aspects. Symposium on Ecological Research in Humid Tropics Vegetation, Kuching, Sarawak. 1963, 120–136.

RHEEDE TOT DRAAKENSTEIN, H. VAN 1678–1703. Hortus Indicus Malabaricus. Amsterdam.

RICHARDS, P. W. 1952. The Tropical Rain Forests. London.

SPATE, O. H. K. 1957. India and Pakistan. A General and Regional Geography. Ed. 2., London.

STAPF, O. 1894. On the Flora of Mt. Kinabalu in North Borneo. *Trans. Linn. Soc. Bot.*, 4: 69–263.

STEENIS, C. G. G. J. VAN 1962. The Land Bridge Theory in Botany with particular reference to tropical plants. *Blumea*, 11: 235–372.

VIII. THE VEGETATION AND PHYTOGEOGRAPHY OF THE EASTERN GHATS

by

M. S. MANI

1. Introduction

Unlike the Western Ghats, the Eastern Ghats are not by any means a range of mountains or escarpment, but represent the much broken and weathered relics of the Peninsular Plateau, marked by a series of isolated 'hills'. The Eastern Ghats mark the eastern borders of the Peninsular Plateau and thus extend from the extreme northeast to the south of the Chota-Nagpur Plateau, to the extreme southwest corner of the Peninsula. The eastern edge of the Nilgiri, Anamalai, and Palni Hills are also parts of the Eastern Ghats. The geology and general climatic conditions of this region have been outlined in foregoing chapters. An attempt is made here to present a broad picture of the characteristic vegetation and phytogeographical peculiarities of the Eastern Ghats.

Our knowledge of the vegetation of the Eastern Ghats is derived largely from the works of GAMBLE (1915–1936). The forest ecology was described by PURI (1960). We have also numerous contributions on the local flora of different sections of the Eastern Ghats by a number of workers like FYSON (1932), KAPOOR (1964), RAO (1958), SRINIVASAN et al. (1961) and others. SEBASTINE (1968) recently published brief notes on the vegetation of the Eastern Ghats.

The flora of the Eastern Ghats is essentially the attenuated complex of the Peninsula, the salient elements of which are today concentrated at higher levels of the Western Ghats. The vegetation of the Eastern Ghats is remarkable for the concentration of character species like *Shorea robusta, Shorea thumbuggaia, Pterocarpus santalinus, Terminalia pallida, Syzigium alternifolium* and *Santalum album* in certain well defined areas and for the presence of complex associations of tropical, subtropical and temperate species and of evergreens at elevations above 1070 m above mean sea-level. As a whole the vegetation is typically deciduous and scrub jungle in most places.

The two major phytogeographical divisions of the Eastern Ghats, generally recognized by botanists, are based on a restricted concept of the Ghats, viz. 1. the northern *sal* division and 2. the southern Deccan division. We consider here the extreme southern parts of the Peninsula, especially the eastern edge of the Palni and Nilgiri Hills, the trends of which coincide with those of the other sections, as the third phytogeographical division.

197

2. The Northern sal Division

The northern *sal* division, so named because of the dominance of *Shorea robusta* or the sal tree of Indian forestry, covers the extreme northern section and is bounded in the south by the wide Godavari-Krishna Gap. Forests of *Shorea robusta* are characteristic of the northeast of the Eastern Ghats. The maximum development of this forest is met with at the foot of the Ghat Range in the valleys of rivers Gullery and Mahanadi, chiefly in level ground, valleys and hill slopes. The upper areas are covered by deciduous forest and bamboo. The ravines abound in *Mangifera indica*. In the plains the principal associates of *Shorea robusta* include *Adina cordifolia, Anogeissus latifolia, Chloroxylon swietenia, Diospyros exsculpta* and sparsely also *Terminalia tomentosa*. The *sal* division contains the well known Mahendragiri (1500 m above mean sea-level), with a mean annual rainfall of 100–150 cm. The Ghumsur area is marked by extremes of cold and heat. Further south in the Visakapatnam District, the mean annual temperature fluctuations lie between 25° and 37°C and both the southwest and the northeast monsoons bring rain, though the bulk of the rain is during the southwest monsoon and the mean annual rainfall amounts to 114.3 cm.

In the socalled *kankar* tracts that frequently occur by side of *Shorea*-tract, we find tall growths of *Soymida febrifuga*. *Shorea* ascends to 600–700 m on the hill slopes, on the southwest, near Ghumsur in Ganjam District (Orissa) and particularly relatively wetter localities are remarkable for the tall trees of *Toona ciliata* and *Xylia xylocarpa*. The main hilly localities of the Eastern Ghats in this division are characterized by scrub jungles and *Shorea robusta* in the plains in between. In addition to *Shorea robusta, Adina cordifolia* and *Anogeissus latifolia*, the character species of this division belong to the Leguminosae, Compositae, Acanthaceae, Euphorbiaceae, Rubiaceae and Cyperaceae. The principal forest trees of this division include the following species: Dillaniaceae: *Dillenia pentagyna;* Malvaceae: *Bombax ceiba;* Rutaceae: *Chloroxylon swietenia;* Sapindaceae: *Schleichera oleosa;* Sterculiaceae: *Sterculia urens;* Burseraceae: *Garuga pinnata, Protium serratum;* Anacardiaceae: *Buchanania lanzan, Lannea coromandelica, Mangifera indica, Semecarpus anacardium;* Leguminosae: *Dalbergia latifolia, Pterocarpus marsupium;* Lythraceae: *Lagestroemia parviflora;* Combretaceae: *Anogeissus latifolia; Terminalia bellerica, Terminalia chebula, Terminalia tomentosa;* Myrtaceae: *Syzigium cumini;* Rubiaceae: *Adina cordifolia, Mitragyna parvifolia;* Verbenaceae: *Gmelina arborea;* Moraceae: *Ficus retusa;* Euphorbiaceae: *Bridelia retusa, Cleistanthus collinus.* We find in addition *Dendrocalamus strictus* and *Bambusa arundinacea* near water courses.

On the relatively steeper southern slopes, covered by shallow soil, there is a preponderance of the grey and white barked species, specialized for resisting the heat of summer, particularly *Anogeissus latifolia, Sterculia urens* and *Adina cordifolia. Dendrocalamus* is also characteristic of the scrub

198

jungles, with climbers like *Bauhinia vahlii, Millettia auriculata* and *Caly-copteris floribunda*. Like in the Nilgiri Hills, the upper regions of the Mahendragiri are characterized by sholas (see Chapters VI and VII) and the lower regions by forest. *Xylia xylocarpa* is the principal tree. The character vegetation of Mahendragiri is a complex of tropical, subtropical and temperate species.

The character species of Mahendragiri include the following: Ster-culiaceae: *Pterospermum xylocarpum;* Celastraceae: *Gymnosporia emarginata;* Sabiaceae: *Meliosma arvense;* Leguminosae: *Acacia catechu, Alysicarpus racemosa, Bauhinia vahlii, Crotalaria alata, Millettia auriculata, Pseudarthria viscida, Sophora glauca, Tephrosia roxburghiana;* Melastomataceae: *Osbeckia hispidissima, Memycylon umbellatum;* Samydaceae: *Homalium nepalense;* Rubiaceae: *Anotis calycina, Knoxia linearis, Pavetta breviflora, Wendlandia gamblei;* Compositae: *Anaphalis lawii, Blumea jacquemontii, Gymnura lyco-persifolia, Senecio candicans, Senecio corymbosus, Senecio nudicaulis, Vernonia divergens;* Ebenaceae: *Diospyros candolleana;* Oleaceae: *Linociera ramiflora;* Asclepiadaceae: *Gymnema sylvestre;* Gentianaceae: *Exacum perrottettii;* Convolvulaceae: *Ipomoea diversifolia;* Scrophulariaceae: *Lindenbergia grandiflora;* Acanthaceae: *Ecbolium viride, Eranthemum capense, Rungia parvi-flora monticola, Thunbergia fragrans hispida, Thunbergia fragrans vestita;* Labiatae: *Leucas montana, Ajuga macrosperma;* Lauraceae: *Litsea mono-petala, Litsea laeta;* Euphorbiaceae: *Macaranga peltata, Euphorbia rothiana, Gelonium lanceolatum.*

The principal temperate species include the following: Ranunculaceae: *Clematis roylei, Clematis wightiana;* Sabiaceae: *Meliosma arvense;* Umbelli-ferae: *Pimpenella heyneana, Bupleurum mucronatum;* Gentianaceae: *Exacum perrottetti;* Samydaceae: *Homalium nepalense;* Violaceae: *Viola patrinii;* Caprifoliaceae: *Viburnum acuminatum;* Urticaceae: *Pouzolzia bennettiana gardneri.*

In marked contrast to the Mahendragiri, the Parlakimedi Hills show a general similarity to the Deccan Division, but differ in the presence of *Shorea robusta*. In this area three vegetational types are generally rec-ognized, viz. 1. the low-hills jungle vegetation, 2. dry broken jungle vegetation and 3. the rocky hill vegetation. The low-hills jungle vege-tation is characterized by the presence of *Ailanthus excelsa, Crotalaria albida, Gonotheca ovatifolia, Hugonia mystax, Maba buxifolia, Murraya pani-culata*. The other character species include the following: *Cissampelos pareira* (Menispermaceae), *Flacourtia indica* (Bixaceae), *Pavonia odorata* (Malvaceae), *Aspidopterys indica* (Malpighiaceae), *Ceropegia tuberosa* (Asclepiadaceae), *Erycibe paniculata* (Convolvulaceae), *Morinda tinctoria* (Rubiaceae), *Oldenlandia nitida* (Rubiaceae), *Justicia glauca* (Acantha-ceae), *Leucas mollissima* (Labiatae), *Bridelia retusa* (Euphorbiaceae). The steeper slopes are covered by *Litsea glutinosa* and *Bidens biternata*. The character species of the dry broken jungles are *Adina cordifolia, Hybanthus enneaspermus* and *Antidesma diandrum*, with *Derris scandens, Mimosa rubicaulis,*

199

Pseudarthria viscida and *Knoxia sumatrensis* among the rocky outcrops. On the rocky slopes we come across *Helicteres isora* and *Blepharis mader-aspatensis* and in somewhat drier localities *Flacourtia indica, Blepharis molluginifolia* and *Indigofera glandulosa. Rauwolfia tetraphylla* and *Vitex pinnata* grow in valleys.

The Velikonda Hills are characterized by *Chloroxylon swietenia, Terminalia tomentosa* and *Xylia xylocarpa.*

The Godavari area of the Eastern Ghats is remarkable for the mixed deciduous vegetation of *Anogeissus latifolia, Bombax ceiba, Chloroxylon swietenia, Cochlospermum religiosum,* associated with *Dendrocalamus strictus.* The laterite localities have *Zizyphus xylopyra, Albizia amara, Bauhinia racemosa, Cassia fistula* and *Erythroxylon monogynum.* The open grasslands include some scattered *Dillenia pentagyna, Pterocarpus marsupium* and *Terminalia chebula.* Deciduous forest covers the valleys. Some of the evergreens, which are common here, include *Maba buxifolia, Manilkara hexandra* and *Memecylon umbellatum.*

3. The Deccan Division

South of the wide Godavari-Krishna Gap, where the Eastern Ghats are interrupted, lies the middle phytogeographical section of the Deccan division of the Eastern Ghats. The Godavari Gap is occupied in the northeast by the R. Godavari and in the southwest by the R. Krishna. The Ghats are continued south of the R. Krishna by the Nallamalai Hills, extending from Guntur through Karnool, Cuddapah and North Arcot to Salem Districts.

The vegetation of the Nallamalai Hills in Guntur District is characterized by the following dominant species: Sterculiaceae: *Helicteres isora;* Linaceae: *Erythroxylon monogynum;* Rutaceae: *Aegle marmelos, Atalantia monophylla;* Simarubaceae: *Balanites aegyptiaca;* Burseraceae: *Commiphora caudata;* Meliaceae: *Azadirachta indica, Soymida febrifuga;* Rhamnaceae: *Zizyphus mauritiana, Ventilago maderaspatana;* Sapindaceae: *Sapindus emarginatus;* Anacardiaceae: *Lannea coromandelica;* Leguminosae: *Acacia catechu, Acacia farnisiana, Acacia horrida, Acacia leucophloea, Cassia auriculata, Cassia tora, Hardwickia binata, Pongamia glabra, Prosopis spicigera, Pterolobium hexapetalum;* Alangiaceae: *Alangium salvifolium;* Rubiaceae: *Adina cordifolia, Canthium dicoccum, Gardenia latifolia, Ixora arborea, Mitragyna parvifolia, Morinda citrifolia;* Sapotaceae: *Manilkara hexandra;* Apocynaceae: *Wrightia tinctoria;* Ebenaceae: *Diospyros chloroxylon, Diospyros melanoxylon, Maba buxifolia;* Loganiaceae: *Strychnos nuxvomica;* Boraginaceae: *Cordia gharaf;* Bignoniaceae: *Dolichandrone falcata;* Verbenaceae: *Premna tomentosa, Vitex altissima;* Euphorbiaceae: *Bridelia montana, Cleistanthus collinus, Euphorbia antiquorum;* Moraceae: *Streblus asper.*

The dominant grasses include *Andropogon pumilus, Apluda mutica, Chloris barbata, Dactyloctenium aegyptium, Eragrostis tremula, Heteropogon*

200

contortus and *Setaria pallidefusoa*. Further southwest in the Karnool area, the Nallamalai Hills are characterized by dominance of *Anogeissus lati-folia, Hardwickia binata, Terminalia tomentosa* and *Tectona grandis*. The other deciduous trees are *Garuga pinnata, Givotia rottleriformis* and *Miliusa velutina*. We thus observe that the vegetation of the Nallamalai Hills varies from evergreen patches to dry deciduous and moist deciduous at higher levels. Scrub vegetation is also present and is characterized by *Atalanta monophylla, Dichrostachys cinerea, Zizyphus rugosa, Zizyphus oenoplia* and *Plectronia parviflora*. The typical moist deciduous species include *Clerodendron serratum, Costus speciosus, Glochidion leutinum, Tacca leonto-petaloides, Thunbergia laevis, Entada pursaetha* and *Bauhinia vahlii*.

The Cuddapah area of the Nallamalai Hills is remarkable for its mixed deciduous vegetation, with an abundance of *Hardwickia binata, Anogeissus latifolia* and *Pterocarpus santalinus* above the scrub jungle. The dominant climbers in Nallamalai area are *Acacia caesia, Acacia pennata, Jacque-montia paniculata, Merrimia hederacea, Pterolobium hexapetalum, Rivea hypo-crateriformis* and *Ventilago calycinus. Pterocarpus santalinus* occurs at elevations of 250–600 m, above which we come across *Shorea thumbaggaia* and *Syzigium alternifolium*.

On the Javadi Hills in North Arcot District there is an abundance of *Santalum album*. Botanists recognize three vegetational types on the Shevroy Hills (in Salem), viz. the southern Kutch thorn forest vegetation, the South Indian tropical dry deciduous vegetation and the subtropical evergreen vegetation. The southern Kutch thorn forest vegetation is extensive up to an elevation of 460 m. The dominant species here comprise: Leguminosae: *Acacia chandra, Acacia ferruginea, Acacia horrida, Acacia leucophloea, Albizia amara, Dichrostachys cinerea, Tamarindus indica;* Simarubaceae: *Ailanthes excelsa;* Rutaceae: *Chloroxylon swietenia;* Combretaceae: *Gyrocarpus americanus;* Rubiaceae: *Canthium dicoccum;* Loganiaceae: *Strychnos nuxvomica, Strychnos potatorum;* Rhamnaceae: *Zizyphus mauritiana*.

The South Indian tropical dry deciduous vegetation is common at elevations of 450–2070 m and though qualitatively not very different from the vegetation at lower levels in the thorn forest, this part of the forest is more dense. The common species here include *Shorea roxburghii, Syzygium cumini, Disopyros montana, Hardwickia binata, Dalbergia paniculata, Bauhinia racemosa, Albizia lebbek* and *Albizia odoratissima*. The subtropical evergreen types occur at elevations above 1070 m and resemble those on the slopes of the Nilgiri, Palni and Anamalai Hills and may be observed in the Sanyasimalai in Yercaud. *Aleodaphne semecarpifolia, Artocarpus lakoocha* and *Machilus macrantha* are common and are festooned with epiphytic orchids like *Dendrobium aquem, Diplocentrum recurvum, Luisia tenuifolia, Saccolobium pulchellum* and *Vanda testacea*. The ground orchid *Acanthophippium bicolor* is also common here. Over fifty species of ferns, including *Actinopteris dichotoma, Aeliantum cuneatum, Asplenium aethiopicum, Botrychium lanuginosum,*

Dichronopteris linearis, Dryopteris sparsa, Nephrolepis cordicolia, etc. are reported to occur here (SUBRAMANYAN 1961).

4. The Southern Division

The southern division comprises the eastern scarps of the Nilgiri, Palni and Anamalai Hills. This division is characterized by the presence of typically temperate plants, many of which are common to the higher and interior Western Ghats and the Mahendragiri division of the Eastern Ghats. The following are some of the more important species in this division: Ranunculaceae: *Clematis, Anemone rivularis, Thalictrum, Ranunculus muricatus;* Anonaceae: *Uvaria;* Berberidaceae: *Berberis tinctoria;* Fumariaceae: *Fumaria parviflora;* Cruciferae: *Cardamine;* Violaceae: *Viola patrinii;* Caryophyllaceae: *Cersatium indicum, Stellaria media, Arenaria* and *Sagina;* Hypericaceae: *Hypericum;* Geraniaceae: *Geranium nepalense* and *Impatiens;* Leguminosae: *Parochaetus;* Rosaceae: *Fragaria, Potentilla, Alchemilla* and *Cotoneaster;* Melastomataceae: *Osbeckia;* Umbelliferae: *Bupleurum, Pimpenella* and *Heracleum;* Caprifoliaceae: *Viburnum* and *Lonicera;* Compositae: *Erigeron, Cnicus* and *Picris;* Ericaceae: *Rhododendron;* and Primulariaceae: *Lysimachia.*

The temperate species representing Pleistocene relicts on high areas of the Eastern Ghats, which are also found in the Western Ghats and on the Himalaya, include the following: Ranunculaceae: *Clematis wightiana* occurring in Mahableshwar and Nilgiri on the Western Ghats and in the Shevroy, Palni and Kodaikanal Hills on the Eastern Ghats; *Clematis munroana* on the Nilgiri and Kodaikanal Hills; *Clematis gouriana* on the Shevroy Hills; *Anemone rivularis* occurs in the Himalaya, Nilgiri and Palni Hills; *Thalictrum* on the Himalaya and Anamalai Hills; *Ranunculus reniformis* is endemic in the Palni and Nilgiri Hills; *Ranunculus subpinnatus* on the Nilgiri and Palni Hills; *Ranunculus wallichianus* on the Nilgiri and Palni Hills; Berberidaceae: *Berberis tinctoria* on the Nilgiri and Palni Hills; Violaceae: *Viola patrinii* Mahendragiri, Shevroy, Nilgiri, Palni Hills and the Himalaya; Polygalaceae: *Polygala rosmarifolia* in the eastern parts of the Eastern Ghats and in the Shevroy Hills; *Polygala sibirica* on the Himalaya, Khasi Hills in Assam, China, Japan and Siberia, the Western Ghats south of Nilgiri and in the mountains of Ceylon; Caryophyllaceae: *Stellaria media* on the Himalaya, Nilgiri, Shevroy and Palni Hills and in the mountains of Ceylon; Geraniaceae: *Geranium nepalense* on the Himalaya, Khasi Hills in Assam, Indo-China, Nilgiri Hills and in the mountains of Ceylon; Rhamnaceae: *Rhamnus virgatus* on the Himalaya, China, Japan, Nilgiri, Palni and Tinnevelly Hills; Caprifoliaceae: *Viburnum acuminatum* in Mahendragiri, Shevroy, Nilgiri and Palni Hills and in the high hills of Travancore; Compositae: *Cnicus wallichii* on the Himalaya, Nilgiri and Palni Hills and *Taraxacum officinale* on the Himalaya, Kodaikanal and Palni Hills; Primulriaceae: *Lysimachia leschenaulti* and

Lysimachia deltoides on Nilgiri and Palni Hills and on the mountains of Ceylon; and Gentianaceae: *Exacum perrottetii* on Mahendragiri to the Nilgiri Hills; *Exacum atropurpureum anamalayanum* on the Anamalai Hills and *Exacum wightianum* on the Nilgiri and Palni Hills.

REFERENCES

FYSON, P. F. 1932. The Flora of the South Indian Hill Stations. Vols. 1–2, pp. 1–697, 1–611 (pls. in Vol. 2) Madras: Government Press.

GAMBLE, J. S. 1915–1936. Flora of the Presidency of Madras. London: Vols. 11.

KAPOOR, S. L. 1964. A contribution to our knowledge of the Flora of Mahendragiri Hills of Orissa. *J. Bombay nat. Hist. Soc.*, 61: 354–396.

PURI, G. S. 1960. Indian Forest Ecology. New Delhi. Vols. 2.

RAO, S. S. 1958. Observations on the vegetation of the Rampa and Gudem Agency Tract of the Eastern Ghats. *J. Bombay nat. Hist. Soc.*, 55: 429–449.

SEBASTINE, K. M. 1968. Natural vegetation of the Eastern Ghats. Mountains and Rivers of India, 21 Intern. geogr. Congr. India, pp. 153–166.

SRINIVASAN, K. S. & G. V. SUBBA RAO. 1961. The flora of Parlakimedi and its immediate neighbourhood. *J. Bombay nat. Hist. Soc.*, 58: 155–170.

SUBRAMANYAN, K. 1961. On a collection of ferns from Shevroy Hills Salem District, Madras State. *Bull. bot. Surv. India*, 2: 323–327.

IX. THE VEGETATION AND PHYTOGEOGRAPHY OF ASSAM-BURMA

by

A. S. RAO

1. Introduction

Assam and Burma are parts of the Eastern Borderlands (see Chapters II & XX), a region largely of Tertiary mountains, characterized by highly humid tropical climate and remarkable for the wealth and diversity of vegetation and flora. Indeed over half the total number of Phanerogams, described so far from India, occur in Assam. Biogeographically, Assam and north Burma represent a highly transitional region, where large-scale commingling of the Asiatic and Indian Peninsular Floras has occurred.

Assam comprises the Surma Valley, the Brahmaputra Valley or Assam proper and the intervening ranges of hills. The Surma Valley is a level plain, about 200 km long and 100 km wide, surrounded by hills on three sides. The R. Surma rises on the southern slopes of the mountains on the border of the Naga Hills and flows south through the Manipur Hills. The Brahmaputra Valley is an alluvial plain, about 750 km long and 80 km wide, enclosed by hills on all sides, except in the west. In the Assam Valley the R. Brahmaputra is a much broken sheet of water, with numerous islands. Except where the Mikir Hills project from the Assam Range, the valley is generally of uniform width, almost up to the southern sector of the R. Brahmaputra. There are outcrops of gneiss above the alluvium between Tezpur and Dhubri. Through most of its course in Assam, the R. Brahmaputra is bounded on either side by stretches of marsh and thick grass, but further inland the level rises. The plain is intensively cultivated.

The Assam Range that separates these two valleys projects almost east-west at right angles to the Burmese system of meridional mountains. At the western end, near Tura, it is about 1200 m above mean sea-level and the hills are much broken up into serrated ridges and deep valleys. Eastwards near Shillong, the range rises to an elevation of 1930 m. The Garo Hills constitute the western extremity of the Assam Range and rise sharply from the plains on the south. The hills attain a maximum elevation of 1468 m (Nokrek Peak) to the east of Tura. On the north, however, there is a succession of hills towards the R. Brahmaputra. The principal river is Someswari that rises north of Tura and falls into the R. Kangsa. There are numerous ridges, with deep gullies in between and the whole area is densely forested. The Khasi Hills rise in the north gradually from the Assam Plain by a succession of low hills. In the south the hills rise, rather abruptly from the level plain, to an elevation of about 1200 m above mean sea-level. The Jaintia Hills slope somewhat more gently to the plain than the Khasi Hills. The southern and central parts of the area constitute a wide plateau, the Shillong Plateau, about 1200–1930 m above mean sea-level. The highest point is Shillong Peak (1930 m). The Shillong Plateau is also known as Meghalaya (= the abode of clouds) and has the general appearance of undulating downs. The Shillong Plateau is a detached block of the Peninsula* and the Mikir and the Rengma Hills to the north are the more dissected outlier. The plateau is linked by a geologically complex saddle to the Assam-Burma ranges in the east. With the extensions in the Mikir Hills, the Shillong Plateau plays in the northeast the same rôle as the concealed continental block in the northwest. The Tertiary ranges are wrapped around it: the Arakan Arc consists of tightly packed parallel ridges and valleys, within a narrow belt of about 130–250 km width, formed of Cretaceous and Tertiary sandstones, limestones and shales. The steep southern edge of the plateau is straight and precipitous and rises to an elevation of 1500 m in a distance of 16–20 km. Scoured by the heaviest rainfall in the world, it is largely dissected and is covered by dense forest. The Lushai Hills rise to a mean elevation of 1200 m in the west and 1600 m in the east and in places even to 2400 m. The rocks are southwards continuation of those forming the Patkai Range and appear to have been laid down in the delta or estuary of a large river from the Himalaya during the Tertiary times. The principal rivers are Tlong or Dhaleswari, Sonai and Tuivol on the north and Koladyne on the east and Karnaphuli on the west. The Naga Hills is a region of narrow hills, mostly of Pre-Tertiary rocks overlain by Tertiary beds. The Barail Range extends into the area from the west. The Japvo Peak, situated a little to the south of Kohima, rises to an elevation of 3050 m. Here it is met by the meridional prolongation from the Arakan Ranges. From this point, the main range extends in a north-northeasterly direction. The principal river in the area is Doiang. The Patkai Hills are situated between 26° 30′ and 27° 15′ NL and 95° 15′ and 96° 15′ EL. The mean elevation of the hills is 1200 m, but some peaks rise to 2100 m above mean sea-level. The hills are composed of Upper Tertiary rocks.

Though the flora of Assam and Burma has attracted the attention of a number of workers, the region must be described as still largely unexplored botanically. Our knowledge of the vegetation and the floristic character and composition of Burma is also at present even more meagre and we can only deal with these problems in a general outline.

The earliest and perhaps the first characterization of the flora of Assam was by ROBINSON (1841) in his descriptive account of Assam.

---

* See Chapters II, V and XXIV.

He also gives an account of the discovery of the tea plant in Assam. This was followed by the pioneer explorations of HOOKER (1854) in the Khasi and Jaintia Hills. At the end of the prodigious labour, marking the completion of his monumental Flora of the British India, HOOKER (1872–1897) hoped, amongst other objectives, that the work will 'enable the phytogeographer to discuss the problems of distribution of plants from the points of view of what is perhaps the richest and is certainly the most varied botanical area on the surface of the globe and one which, in a greater degree than any other, contains, representatives of the floras of both the Eastern and Western Hemispheres'. A phytogeographical analysis of the flora of the vast Indian Empire has been attempted earlier by HOOKER (1855) himself and his later summary (HOOKER 1906) still remains a classic in the field. A galaxy of plant collectors have since then traversed the hills and plains of the region and BURKILL (1965) has done signal service to Indian botany by his methodical chronicling of the numerous plant collectors, who have contributed to our knowledge of the flora of India, including of course Assam. We are indebted in particular to GRIFFITH (1847), CLARKE (1889), BOR (1938, 1942), BISWAS (1941, 1943) and KINGDON WARD (1960) for our data on the Assam flora. DAS (1942) gave an account of the floristics of Assam, both the plains and the hilly tracts. BURKILL's (1925) Botany of the Abor Expedition is a major contribution to our knowledge of the distribution of plants. BOR (1938, 1942) and BISWAS (1941, 1943) made brief studies of the flora of the Aka Hills. CLARKE made comparative observations on the flora of Kohima in the Naga Hills and of Manipur. We have excellent accounts by FISCHER (1938) of the plants from the Mizo (Lushai) Hills and by CARTER (1921) of Lakhimpur. DEB (1960) studied the forests of Manipur and also carried out some preliminary studies on the flora of Tripura. BOR (1942) has also made a detailed study of the relict flora of the Shillong Plateau; he compared this flora with that of the Naga Hills in the light of CLARKE's observations on the affinities of these floras. NAIK and PANIGRAHI (1961) have described a collection of plants from Subansiri. PANIGRAHI and JOSEPH (1966) have enumerated the plants of Tirap District of the North East Frontier Agency (NEFA). RAO and JOSEPH (1968) have also made interesting observations on the flora of the Siang District and have completed a study of the flora of the Kameng District. RAO and RABHA (1971) have recently listed the species of vascular plants of the Kamrup District. The first detailed account of the flora of Assam is, however, that of KANJILAL (1934–1940), a work in five volumes (incomplete). The first four volumes cover the Dicotyledons, but chiefly the woody species and the fifth volume by BOR (1934–1940) deals with the Gramineae. The first volume contains also a discussion by KANJILAL on the ecology and vegetation of Assam. CHAMPION, (1936) who classified the vegetation of India into a number of major types and subtypes, included the Assam

206

plants also in his classification. A more recent survey by LEGRIS (1963) also similarly deals with vegetation of India as a whole, including Assam. ROWNTREE (1953) has outlined the vegetation, mainly the forest types, found in the Assam Valley. DAS and RAJKHOWA (1968) have described the woodlands of Assam. TURRILL (1953), in a review of the work of the pioneer plant geographer HOOKER, has also briefly analysed and recounted the efforts of later botanists in this field. CHATTERJEE (1940, 1962) has recently published an analysis of the Indian flora, including that of Assam, mainly from the point of view of the percentage of endemics as a measure of the distinctiveness of the Indian flora. In a recent review of the flora of the Republic of India, MAHESWARI and his collaborators (1965) have followed the analysis of CHATTERJEE, with further considerations on the distribution of Monocotyledons, Gymnosperms and Pteridophytes. No comprehensive survey of the vegetation of Assam proper has, however, been undertaken so far, but RAO (1970) has recently published a brief sketch of the vegetation of the northeast India.

With the revival of the Botanical Survey of India in 1956, the Eastern Circle was opened at Shillong. It began work with the old Assam Forest Herbarium as its nucleus and has in recent years greatly enlarged the original material. It has undertaken extensive explorations in different parts of Assam, including the little known NEFA.* As a result of these explorations and collections, RAO and PANIGRAHI (1961) have given a brief account of the salient features of the vegetation of Eastern India, but mainly of Assam and the Eastern Himalaya within NEFA. These explorations have brought to light over a score of species of plants new to science and has also recorded a number of plants not previously known from within these areas. A number of remarkable illustrations of discontinuous distribution have been discovered. Yet it must be observed here that over half the area of Assam still remains more or less completely unexplored and even the areas which have been investigated need further intensive study. No serious attempt has also been made so far to elucidate the intriguing problems of the phytogeography of Assam; all earlier references to Assam being incidental to discussions on the vegetation and phytogeography of India as a whole. Assam represents the floristic Gateway of India and a knowledge of its botany is of particular interest to discussions on the biogeography of India as a whole.

2. Vegetation of Assam

Assam abounds in forests, meadows, marshes and swamps, each with its characteristic plants, and special ecology. The vegetation of Assam may be broadly classified into three major types, viz. 1. the tropical, 2. the temperate and 3. the alpine, each comprising numerous subtypes.

* NEFA = North East Frontier Agency, since renamed ARUNACHAL.

207

2.1. The tropical vegetation

The tropical vegetation typically covers areas upto elevations of about 900 m. It embraces evergreen and semi-evergreen forests, deciduous forests (dry and moist) and grasslands, including the scattered riparian forests and swamps. Tropical evergreen forests are found in the Assam Valley, the foothills of the Eastern Himalaya and in the lower parts of Naga Hills and Manipur. The storied nature of these dense dark forests is rather difficult to discern, as the tall trees, with their close canopy cover, stifle the shorter trees. There is a bewildering wealth of species in these forests, not all of them being common to all the areas. The tallest trees are *Dipterocarpus turbinatus, Canarium resiniferum, Artocarpus chaplasha, Tetrameles nudiflora, Ailanthus grandis, Euphorbia longana, Kayea assamica, Terminalia chebula, Mesua ferrea, Phoebe goalparensis, Toona ciliata, Dysoxylum binectariferum, Dillenia indica* and *Duabanga grandiflora.* Other lower trees are *Amoora wallichii, Ficus rumphii, Lagerstroemia parviflora* and *Terminalia myriocarpa.* Some of the other interesting trees are *Pachylarnax, Alcimandra* and *Michelia* all of the Magnoliaceae. Of the numerous lianes intertwining the trees, species of *Bauhinia, Acacia, Derris, Vitis, Unona, Toddalia, Mezoneurum* and *Gnetum* are the more prominent. Several species of the prickly *Calamus* stretch for long distance, from tree to tree. A few other palms like *Caryota, Licuala, Arenga, Pinanga* and *Didymosperma,* are also conspicuous. Four species of *Musa* also occur often in gregarious patches. Another conspicuous element is the large bamboo *Dendrocalamus* sp. and *Bambusa* sp. often crowded in clumps. *Saurauja roxburghii, Antidesma* spp., *Pavetta indica, Maesa montana, Holarrhena antidysenterica* and a few others are the common short tree or large shrub species. Clumps of *Pandanus,* often associated with tall grass, are found near streams. The epiphytic climbers *Rhaphidophora* spp., *Pothos* and *Scindapsus officinalis* (Araceae); *Hoya* spp. often with beautiful bunches of star-like flowers and the peculiar *Dischidia rafflesiana* (Asclepiadaceae); *Aeschynanthes* (Gesneriaceae) are frequent. Stem parasites of the Loranthaceae and the holoparasite *Cuscuta reflexa* and *Cassytha capillaris* are not uncommon. The most conspicuous epiphytic elements are, however, the orchids, ferns and fern allies. Amongst the orchids *Dendrobium* spp. predominate, with *Cymbidium* coming next. The forest floor is dense, with a myriad herbs, a joyful sight when many of them are in bloom. *Impatiens* spp., *Pouzolzia* spp., *Elatostemma* spp. and others chiefly of Acanthaceae, Lauraceae and Papilionaceae are easily recognized. Of the rhizomatous Monocotyledons, which are also conspicuous, *Curcuma* spp., *Boesenbergia longiflora, Phrynium capitatum, Molineria recurvata, Costus speciosus* and *Zingiber zerumbet* are important. In some shaded areas associations of *Forrestia mollissima hispida* with *Colocasia antiquorum* and *Homalonema aromatica* can be seen.

Ferns are also profuse, the most conspicuous and elegant being the tree-fern *Cyathea* spp. and the equally large and handsome *Angiopteris*

Plate 6. Shorea robusta or the sal forest in Kamrup Dist., Assam.

evecta. Most of the others are Polypodiaceous. The tree trunks, wet boulders and moist banks are heavily plastered with matted liverworts and mosses and sometimes with fine growths of the filmy-ferns *Hymenophyllum* and *Trichomanes.* In the forest floor in some areas can be seen the rare *Helminthostachys* and in other places the elegant-leaved *Botrychium.*

Deciduous Forest: The deciduous forests include *Shorea* (the *sal*) forests, with the single species *Shorea robusta* dominating. Due to the yield of valuable timber, these forests have been extensively exploited and greatly disturbed. Such forests are found in the Districts of Goalpara, Kamrup, Nowgong and Darrang, in the northern lower slopes of the Khasi and Garo Hills of the Shillong Plateau and in some parts of North Cachar Hills. The associated species of trees are *Careya arborea, Kydia calycina, Sterculia villosa, Bombax ceiba, Grewia* spp., *Terminalia* spp., *Bauhinia* spp., *Acacia* spp., *Albizia* spp., *Adina cordifolia* and *Gmelina arborea.* The woody climbers are also few, the common ones being *Bauhinia vahlii* and *Combretum decandrum.* The herbaceous vegetation is less profuse and includes Oxalidaceae, Balsaminaceae, Acanthaceae, Asteraceae and Urticaceae, with sedges and grasses and in moist places *Eriocaulon.* There are not many ferns and fern allies.

Tropical grasslands or savanahs occur in riparian flats, inundated by flood water of the R. Brahmaputra. The grasses are tall and belong to species of *Saccharum, Anthistiria, Erianthus, Arundo donax* and *Phragmites communis.* The edaphic effects of floods are accentuated by the biotic factors of grazing and felling of the few small trees and shrubs. These grasslands are distinct from those at higher altitudes of the Shillong Plateau and the lower parts of North Cachar and Mikir Hills. The

Plate 7. Evergreen vegetation comprising *Tetrameles, Eugenia, Elaeocarpus, Euphorbia, Terminalia* and *Duabanga* among others.

Plate 8. Deciduous vegetation comprising *Terminalia, Bridelia, Sterculia, Bombax* and *Gmelina,* among others.

Plate 9. Swamp vegetation of Kaziranga (Note the rhinos in the water hole). The grasses are *Erianthus, Arundo, Saccharum, Phragmites*, etc. The trees in the distance include *Albizia, Bombax, Sterculia, Bridelia* and *Semecarpus*.

grassland area of Kaziranga, the home of the one-horned rhino (and now a National Park), represents a combination of grassland, swamp forest and marsh. There are tall grasses, reaching to 5 m, of which the prominent ones are *Erianthus longisetosus, Thysanolaena maxima, Imperata cylindrica, Arundo donax, Sclerostachya fusca, Saccharum spontaneum, Vetiveria zizanoides* and *Phragmites communis*.

There are dotted clumps or isolated trees of *Semecarpus anacardium, Albizia* spp., *Dalbergia* spp., *Lagerstroemia flos-reginae, Dillenia indica* and *Duabanga grandiflora*. Scattered here and there are small puddles or pools (in which the rhino loves to wallow), which contain either a gregarious growth of the water hyacinth, *Eichhornia crassipes* or other floating herbs like, *Ottelia, Jussiaea* and *Nymphaea*. At the edge may be found clumps of *Monochoria* and several sedges including *Cyperus* and *Kyllinga*. The spiny aroid *Lasia spinosa* is also often found with *Typhonium flagelliforme* and *Colocasia antiquorum*.

The grasslands of North Kamrup represent degraded and secondary vegetation, due to the combined effects of frequent heavy floods, fire, indiscriminate tree-felling and grazing. The original semi-evergreen or the moist deciduous forest has given place to grasslands, containing species of *Saccharum, Apluda, Themeda* and *Erianthus* mixed with Papilionaceous herbs and sparsely isolated *Leea, Clerodendrum* and *Melastoma*.

A singular feature of the tropical vegetation, in the warm humid, Assam

211

Valley is the swamp or marsh vegetation. There are innumerable stagnant ponds called *beels*, sometimes formed in obstructed, abandoned river channels, especially in Goalpara, Kamrup and North Lakhimpur. Members of the Nymphaeaceae, Lemnaceae, Araceae, Cyperaceae, Eriocaulaceae and Naidaceae are common in such marshes; *Typha elephantina*, *Arundo donax* and *Phragmites communis* also occur. Shrubs of *Crataeva lophosperma*, *Eugenia cuneata* and *Homonia riparia*, with stunted trees of *Salix tetrasperma*, form the other common elements of these swamps.

Subtropical mixed forest: Subtropical mixed forests occur in western Kameng, the inner valleys of Siang and Lohit Districts in the NEFA Himalaya and in parts of Tirap District, adjoining the Patkai Hills and the Burma border, in areas reaching upto 1500 m above mean sea-level. These forests include associations of *Castanopsis*, *Schima*, *Engelhardtia*, *Terminalia*, *Ficus*, *Michelia*, *Albizia*, *Bridelia*, *Cinnamomum*, *Lindera* and *Garcinia* along with a few palms, *Musa* spp. and in some places *Quercus* spp., *Acer* spp. with *Saurauja* and *Photinia* also occur.

2.2. THE TEMPERATE VEGETATION

The *temperate* vegetation occurs at elevations from about 1300 m to 2500 m in the Shillong Plateau, the Naga, Mizo (Lushai) and Mikir Hills and in the NEFA districts of the Assam Himalaya. In several suitable localities there is mixed temperate vegetation with tropical to subtropical vegetation. *Albizia*, *Acer*, *Juglans*, *Quercus*, *Magnolia* and *Michelia* with *Rhododendron*, *Rubus* spp. and a sprinkling of *Arundinaria* are present; still other species on somewhat higher slopes are *Alnus nepalensis*, *Cornus controversa* and *Ilex* spp. There is a gradual change in the composition and density of the species with the increase in altitude. *Rhododendron* spp. predominate with *Pyrus*, *Prunus*, *Spiraea*, *Eriobotrya* and some other Rosaceæ, gradually ending with coniferous vegetation with Tsuga-Picea-Abies association in places in the Assam Himalaya. The forest floor has often a gregarious growth of the ferns *Plagiogyria* and *Dryopteris*.

The temperate vegetation in the Khasi and Jaintia Hills of Shillong Plateau especially of the 'sacred forests' at Shillong Peak, Mawphlong, Mowsmai and some other places, was originally studied by HOOKER (1854) and more recently by BOR (1942). RAO (1969) has also described them in an account of the vegetation of the Khasi and Jaintia Hills. These represent relicts, amidst a much disturbed and altered vegetation, due to the devastating practice of *jhuming*, a kind of primitive agricultural practice, involving large-scale cutting down and burning of trees before planting (see Chapter XI for a discussion on *jhum* cultivation). The small pockets of 'sacred forests,' left untouched due to religious beliefs, afford us a glimpse of the original forest that must have once clothed these hills in prehistoric times. It is these pockets that contain a great 'profusion of species'. They are present in saucer-shaped depressions, amidst rolling

Plate 10. 'La Lyngdoh' or the Sacred forest at Mawphlong, Khasi and Jaintia Hills, Assam. The monoliths at the extreme left are old monuments to the dead.

grassland, and often have little mountain streams meandering through them. Fagaceae with *Quercus* spp. and *Castanopsis* spp., Rosaceae with *Rosa, Photinia, Eriobotrya, Pyrus, Prunus, Sorbus* and several other shrubby and herbaceous species, *Corylopsis* and *Exbucklandia, Albizia, Manglietia* with climbing *Schizandra* and *Kadsura; Acer* sp. with an occasional *Alnus, Engelhardtia, Mahonia,* and Vacciniaceae including *Agapetes* and *Vaccinium* mostly epiphytic, occur in them. The trees are heavily loaded with numerous other epiphytes, but principally the orchids, with ferns and fern allies and some aroids. The forest floor has a dense carpet of herbaceous vegetation, belonging to Ranunculaceæ, Rosaceæ, Begoniaceæ and Asteraceae. A comparable kind of temperate vegetation, agreeing even in the species content, occurs at comparable altitudes on the Naga Hills (Bor, 1938). An important result of human influence from prehistoric times in these hills, is the intrusion and spread of *Pinus insularis*. In the Khasi and Jaintia Hills of the Shillong Plateau this pine makes its appearance at about 900 m and forms extensive pure groves at higher elevations, giving the landscape a parkland appearance, with the interspersed rolling grasslands. These pinewoods may contain sometimes a sprinkling of *Symplocos, Schima wallichii* and *Schima khasiana*. The floor underneath is thick with a carpet of pine needles and as may be expected, is devoid of any plant growth, except in small clearings, where there may be shrubs of *Pieris ovalifolia* and some scattered *Anemone* sp. The grassland on these hills is partly a climatic climax and partly due to biotic factors. The numerous species of grasses found here are not tall, scarcely attaining height of a metre. Nearly eighty species have been collected in a small area around Mowsmai. Some of the common grasses are *Arundinella, Chrysopogon, Cymbopogon, Echinochloa, Eragrostis, Erianthus, Ischaemum, Panicum* and *Paspalum*. In this description of the vegetation of the Khasi

213

Plate 11. Low grass-covered hills near Cherrapunji, Khasi and Jaintia Hills, Assam.

Plate 12. A tree of *Castanopsis* studded with epiphytic orchids.

and Jaintia Hills, the situation at Cherrapunji, until recently famed as the wettest place on earth, needs special notice. The area looks disappointingly bleak and bare of wooded vegetation, due to the poor soil cover; all the soil being leached out by the heavy rains, leaving behind smooth, bare rocks. For vast distances all round, only dwarf grass-growth is visible. It is only in the comparatively sheltered depressions, as at Mamloo and Mowsmai, where there is a deposit of soil and humus, are there small islands of wooded vegetation, in an otherwise vast sea of grassland.

214

Returning to our discussion of the temperate vegetation, we should refer to the temperate pine forests on the higher slopes of Assam Himalaya. Here occur *Pinus wallichiana*, mixed with *Rhododendron*, *Quercus* spp. and *Castanopsis* spp., with *Pinus* predominating in some and *Quercus* and *Rhododendron* in other places. Alpine and subalpine vegetations have only a limited distribution. The subalpine vegetation occurs at altitudes of 3000 to 4500 m in the Aka Hills of Kameng District, NEFA; Naga Hills and in Manipur. The dominant tree is *Abies*, with *Tsuga* and *Picea* and protecting a dense bushy zone of *Rhododendron*, *Juniperus recurva*, *Berberis*, *Salix*, *Cotoneaster*, *Agapetes*, *Vaccinium*, *Sorbus* and *Rubus*. There is a profusion of herbaceous species, particularly of the Ranunculaceae and Rosaceae, with some Polygonaceae and Gentianaceae here and there.

2.3. The alpine vegetation

The alpine vegetation is limited to altitudes of 4500 m to 5500 m, at which point the vegetation becomes scarce. There is however no sharp divide between the subalpine and the alpine types of vegetation. The alpine vegetation occurs only at the higher elevations of the Assam Himalaya. There are no trees and the aspect is that of moorland or coarse meadow. The vegetation consists of stunted, gnarled, dwarf shrubs, with deep roots. Even the herbaceous species have characteristic habits, with deep roots and stunted foliage. *Rhododendron anthopogon* and *R. nivale* are frequent with *Rheum*, *Saussurea*, *Sedum* and *Saxifraga*. *Gentiana ornata* makes a splash of colour, with a few *Aconitum*, *Bromus*, *Stipa* and *Festuca*.

In the Kameng District of NEFA of the Assam Himalaya, on the Aka Hills there occur numerous species of *Rhododendron*, in clumps, which bloom more or less together, in a riot of colour. Another interesting aspect is the autumn colour of the *Quercus-Acer* woods along the R. Tenga, in the area of Sergaon and Jigaon, near Rupa. Amongst gregarious plants should be mentioned the bamboos: *Dendrocalamus* and *Bambusa* – which form extensive pure clumps in Cachar.

2.4. Some other interesting elements of the Assam flora

We may draw attention here to some interesting elements of the flora of Assam. *Cycas pectinata* occurs scattered in Kamrup, and *Gnetum gnemon* has been recorded from the Khasi, Jaintia and Naga Hills and in isolated areas in the Assam Valley. *Podocarpus nerifolius* also occurs on the Khasi Hills. *Nymphaea pygmaea* is an unusual aquatic plant, with very discontinuous distribution in Siberia and Khasi and Jaintia Hills. *Euryale ferox*, with giant floating leaves, is also seen in several *beels* in Kamrup and Sibsagar. Amongst interesting root-parasites may be mentioned *Sapria himalayana*, first discovered by Griffith in the Mishmi Hills and found later by Bor on the Aka Hills and by Deb in Manipur. This has also been found

215

Plate 13. Cycas pectinata near Gauhati, Assam.

Plate 14. Euryale ferox, with large round leaves covering the water surface of a pond or 'beel' near Gauhati, Assam.

Plate 15. Sapria himalayana from Mishmi Hills, Lohit Dist., NEFA, Assam.

217

Plate 16. Boschniaekia himalaica from Aka Hills, Kameng Dist., NEFA, Assam.

in Kamrup, and was again collected in the Mishmi Hills recently. *Balanophora dioica* is commonly found infesting the roots of several species of trees, but particularly *Ficus* in dark humid forests. *Boschniaekia himalaica* is a root parasite of *Rhododendron* in the alpine meadows of the Aka Hills. A common root parasite on grass is *Aeginitia indica*, easily spotted when in bloom. A very interesting recent discovery is the unusual Rafflesiaceae root parasite *Mitrastemon yamamotoi*, in the Mowsmai forests of Khasi and Jaintia Hills.

Amongst interesting saprophytic plants mention may be made of

218

Plate 17. Mitrastemon yamamotoi on the roots of *Castanopsis tribuloides* in the Mowsmai Forest, Khasi & Jaintia Hills, Assam.

Monotropa uniflora, Epipogium roseum and the giant orchid *Galeola falconeri.*

Amongst insectivorous plants we have *Drosera burmanni* and *Drosera peltata,* D EB has reported the occurrence of *Aldrovanda* from Tripura. *Nepenthes khasiana,* with its pretty large foliar pitchers is confined to the Shillong Plateau.

Two other unusual plants, observed in the course of recent exploration and study (J OSEPH, 1969) are *Salmonia aphylla* (Polygalaceae) and *Cotylanthera tenuis* (Gentianaceæ); the former was originally known from Burma, Borneo and Malacca and the latter from Java and Sikkim. Both these have been collected from near Nongpoh in the Khasi and Jaintia Hills. Some of the interesting plant species are listed in Table I, with their distribution.

Plate 18. Epipogeum roseum in the dark humid forest near Pynursla, Khasi & Jaintia Hills, Assam.

Plate 19. Nepenthes khasiana near Jorain, Khasi & Jaintia Hills, Assam.

Table I. Assam plants: rare, endemic or otherwise with interesting distribution. Plants marked by an asterisk indicate species recently described as new. (From the Kanjilal Herbarium of the Botanical Survey of India and various publications).

Family, genus, species and habitat	Original distribution	Maximum range	Assam locality
Nymphaeaceae	Siberia,	Siberia, N. China,	Nongkhrem
Nymphaea pygmaea	N. China	N.W. Himalaya;	1850 m K. & J. Hills
Aquatic *ca* 1500 m		Khasi & Jaintia Hills	
Annonaceae	Burma	Burma, NE.	Garampani in
Orophea polycarpa		Assam	Sibsagar District
Evergreen forests, up to 1000 m			
Popowia kurzii	Burma	Burma, NE.	North Cachar
Evergreen forests, up to 700 m		Assam & Andamans	
Uvaria lurida	Khasi Hills	Endemic to	Umran, K. J. Hills
Evergreen forests, up to 1000 m		Khasi Hills	
Magnoliaceae			
Alcimandra cathcartii	Sikkim	E. Himalaya,	Nongringkoh,
Evergreen forests, 1000–2000 m		K. & J. Hills & Naga Hills	Sohrarim K. & J. Hills
Magnolia gustavii			
Evergreen forests up to 1000 m	Lakhimpur	Endemic to Lakhimpur	Makum forests
M. pealiana			
Evergreen forests up to 1000 m	Assam	Assam	Assam
M. lanuginosa	Nepal	Central &	Kynshi,
Subtropical & temperate evergreen forest from 1500–2300 m		Eastern Himalaya K. & J. Hills	K. & J. Hills
Pachylarnax pleiocarpa	Lakhimpur	Lakhimpur	Digboi
Evergreen forests up to 500 m			
Illiciaceae			
Illicium cambodianum		Southern Indo-China, Southern Burma NEFA	Ziro to Begi Subansiri dist. NEFA
Fumariaceae			
Dicentra roylei	N.W. Himalaya	Simla to Bhutan	Khasi &
Temperate forests 1500 m		and Assam	Jaintia Hills
Flacourtiaceae			
Homalium schleichii	Burma	Burma & Assam	Cachar, Wah-Rang-
Evergreen forests 1000m			Ka, K. & J. Hills
Polygalaceae			
Salomonia aphylla	Malacca &	Malacca,	Nongpoh *ca* 1000 m
In shady places	Tenasserim	Tenasserim, Borneo, Assam	K. & J. Hills

222

Table I (continued)

Family, genus, species and habitat	Original distribution	Maximum range	Assam locality
Theaceae			
Anneslea fragrans	Burma	Burma & Nagaland	Naga Hills
Evergreen forests 1000			
Saurauja griffithii	Assam	Assam, Sikkim	Goalpara
Evergreen forest 1000			
Icacinaceae	Western Ghats of	Western Ghats of	Shangpung
Apodytes benthamiana	the Peninsular	the Peninsular,	
Evergreen forests 1500	India & Assam	India & Assam	
Iodes hookeriana	Assam	Assam & E.	Shangpung
Evergreen forests 1500		Pakistan	
Aquifoliaceae			
Ilex embelioides	Khasi Hills	Assam	Cherra, Nunklow,
Evergreen forests 1500			Dowki, K. & J. Hills
Celastraceae			
Euonymus echinatus	Eastern Himalaya	E. Himalaya	Sutynga,
In open places		K. & J. Hills	K. & J. Hills
1000–2000			
Hamamelidaceae			
Distylium indicum	Khasi Hills	Endemic to	Khasi Hills
Evergreen forests 1500		Khasi Hills	
Araliaceae			
Merrilliopanax cordifolia			
Evergreen forests 1500	Begi-Amjee 1540	Begi-Amjee *ca* 1540 Subansiri Dist., NEFA	Begi-Amjee Subansiri Dist., NEFA
Rubiaceae			
Mycetia mukerjiana	Makum Hill,	Makum Hill,	Makum Hill,
Evergreen forests 1500	Lakhimpur Dist. Assam	Lakhimpur Dist. Assam	Lakhimpur Dist.,
Nertera sinensis	Szechuan	Australia,	Subansiri Dist.
Wet rocks in stream	(China)	Newzealand, Formosa, China and Assam	NEFA
Ericaceae			
Rhododendron santapaui	Begi in Subansiri Dist. NEFA	Begi in Subansiri Dist., NEFA	Begi in Subansiri Dist., NEFA
Epiphytic			
Primulaceae			
Lysimachia santapaui	Amjee in Sabansiri Dist., NEFA (1050m)	Amjee in Subansiri Dist., NEFA	Amjee in Subansiri Dist., NEFA
Myrsinaceae			
Amblyanthus multiflorus	Assam	Endemic to Assam	Assam
Evergreen forests			
Ardisia quinquangularis	Khasi Hills	Endemic to Assam	Khasi Hills
Evergreen forests			
A. rhynchophylla	Khasi Hills	Endemic to Assam	Cherrapunje, K. & J. Hills

223

Table I (continued)

Family, genus, species and habitat	Original distribution	Maximum range	Assam locality
Evergreen forests			
Sapotaceae			
Palaquium polyanthum	Sylhet Chittagong	Sylhet, Chittagong & Cachar Burma	Cachar
Evergreen forest 800 m			
Styracaceae			
Alniphyllum fortunei	Yunnan, Amoy & West of Hupeh in China	Yunnan, Amoy & West of Hupeh in China, Subansiri Dist.	Begi-Amjee, Subansiri Dist., NEFA
Bruinsmia polysperma	Mahadeo, (1000 m) Khasia Hills.	Assam, Burma	Umsaw and Mahadeo, (1000 m) K. & J. Hills.
Huodendron biaristatum	Northeast of Burma & Tonkin	Yunnan, Kweichew, Kwangsi in China & Myitkyina in Northeast Burma, Subansiri Dist., NEFA	Hapoli, Subansiri Dist., NEFA
Asclepiadaceae			
Hoya manipurensis Epiphytic on *Ficus glomerata*	Manipur (1000 m)	Manipur (1000 m)	Manipur
Gentianaceae			
Cotylanthera tenuis In shady forest floor	Java	Java, Sikkim, Khasi Hills	Nongpoh (*ca* 1000) Khasi Hills
Solanaceae			
Pauia belladonna	Way to Wakka, 6 km away, (2,100 m) Tirap Dist., NEFA	Way to Wakka Tirap Dist., NEFA	Way to Wakka, Tirap Dist., NEFA
Orobanchaceae			
Gleadovia banerjiana Parasites on subterranean roots of *Strobilanthes discolor*	Koupru Hill (*ca* 2000 m) Manipur	Koupru Hill, Manipur	Koupru Hill, Manipur
Lentibulariaceae			
Utricularia pubescens On moist soil, amidst moss and grasses on hill slopes near margin of streamlet	Sierra Leone	South America, Tropical Africa & Assam	Barapani, K. & J. Hills
Gesneriaceae			
Beccarinda cordifolia	Upper Burma	Upper Burma to Yunnan and other parts of China Sirang, Siang Dist., NEFA	Sirang, Siang Dist., NEFA
Rhynchoglossum lazulinum	Krishna 36 km from	Krishna, Bhallukpong	Krishna, Bhallukpong

Table I (continued)

Family, genus, species and habitat	Original distribution	Maximum range	Assam locality
In shady moist places near the wayside on black humus soil	Bhallukpong (1250) on the way to Sessa Kameng Dist., NEFA	on the way to Sessa, Kameng Dist., NEFA	on the way to Sessa, Kameng Dist., NEFA
Polygonaceae			
Polygonum sarbhanganicum	Goalpara (1200), Assam	Goalpara, Assam	Goalpara, Assam
Sandy bank of R. Sarbhang			
Rafflesiaceae			
Sapria himalayana	Mishmi Hills	Mishmi Hills Aka Hills, NEFA Kamrup, Dist. Assam Manipur	Boko, Kamrup Dist. Assam
In dark humus covered forest floor			
Mitrastemon yamomotoi			
In Oak forests	Japan and Sumatra	Japan, Sumatra & Assam	Mawsmai, Khasi Hills
Euphorbiaceae			
Trigonostemon chatterjii	Dawki, Khasi Hills	Dawki, Khasi Hills	Dawki, Khasi Hills
Urticaceae			
Boehmeria tirapensis	Kothong, Tirap Dist., NEFA	Kothong, Tirap Dist., NEFA	Kothong, Tirap Dist., NEFA
Orchidaceae			
Acanthophippium sylhetense	Khasi Hills	Endemic to Assam	K. & J. Hills (610–1220 m)
Terrestrial			
Aphyllorchis montana	Ceylon	Khasi Hills Sikkim, Ceylon	K. & J. Hills (1070 m)
Terrestrial			
A. vaginata	Khasi Hills	Khasi Hills	K. & J. Hills (1525 m)
Terrestrial			
Apostasia wallichii	Sumatra, Java, N. Guinea	Assam, Nepal, Ceylon, Java Sumatra and N. Guinea	Khasi Hills Sibsagar, Tripura
Terrestrial			
Bulbophyllum listerii	Bhutan 330 m	Bhutan, K. & J. Hills	Nongpoh, K. & J. Hills
Epiphytes			
B. penicillium	Tenasserim	Assam, Tenasserim and Sikkim	Jorain, Jaintia Hills
Epiphytes			
B. piluliferum	Sikkim	Sikkim & Jaintia Hills	Near Jowai, Jaintia Hills
Epiphytes			
B. triste	Tenasserim	Burma, Sikkim, Western Himalaya & Assam	Nongpoh, K. & J. Hills
Epiphytes			
Coelogyne carnea	Perak	Perak, K. & J. Hills	Nongkhlaw, Mawsmai, Nongpoh, K. & J. Hills,
Epiphytes			
C. viscosa	Khasi Hills	Khasi Hills	Khasi Hills
Epiphytes			

Table I (continued)

Family, genus, species and habitat	Original distribution	Maximum range	Assam locality
Corybas purpureus Moss-cushion of rocks in shady thick forest floor	Shillong	Shillong, K. & J. Hills	Shillong, K. & J. Hills
Dendrobium bensoniae Epiphytes	Tongou, West of Prome in Pegu, Burma	Burma, Northern Thailand, Lower Siam	Mizo (Lushai) Hills
D. chrysotoxum Epiphytes	Arrakan, Burma	Arrakan, Burma; Mizo (Lushai) Hills, Assam, Manipur Tirap Dist., NEFA	Aijal Mizo (Lushai) Hills, Manipur, Niusa, Tirap Dist., Kimin, Subansiri Dist. NEFA
D. infundibulum Epiphytes	Burma, Tenasserim	Burma, Tenasserim	Mizo (Lushai) Hills
D. pendulum Epiphytes			Mizo (Lushai) Hills
D. podagraria Epiphytes	Burma, Tenasserim	Burma, Tenasserim, Assam, Cachar	Cachar, Assam
D. terminale Terrestrial	Sikkim and Tenasserim	Sikkim, Tenasserim & Assam	Nongpoh, K. & J. Hills
Epipogium roseum Epiphytes	West Africa, Java, Australia	Nepal, Sikkim W. Africa, Java, Australia	Khasi Mts. Nongpoh Pynursla K. & J. Hills
Eria barbata Epiphytes	Khasi Hills	Khasi Hills	Khasi Hills (*ca* 1300 m)
E. biflora Epiphytes	Burma & Sikkim	Burma, Sikkim & Khasi Hills	Nongpoh, Khasi Hills
E. crassicaulis Epiphytes	Khasi Hills	Khasi Hills	K. & J. Hills (1220–1525 m)
E. fragrans Humus-debris	Burma & Sikkim	Burma, Sikkim & Assam	Nongpoh, K. & J. Hills
Galeola lindleyana Terrestrial	Khasi Hills	Khasi & Naga Hills, Sikkim	K. & J. Hills (1000 m)
Gastrodia exilis Terrestrial	Khasi Hills	Khasi Hills	K. & J. Hills (1000 m)
Oberonia parvula Epiphytes	Teesta Valley at Guri Bathaw, Sikkim	Sikkim, Assam	Nongpoh, (*ca* 1000 m) K. & J. Hills
O. sulcata Epiphytes	Selari forest Kameng Dist., NEFA	Kameng Dist., NEFA	Selari forest Kameng Dist. NEFA
Panisea tricalosa Epiphytes	Assam	Bhutan, Assam & Thailand	Assam
Paphiopedilum insigne Terrestrial	Assam	Khasi Hills Assam	Cherrapunji (1000–1300 m)
P. hirsutissimum Terrestrial	Assam	Khasi Hills & Mizo Hills	Khasi & Mizo Hills
P. fairieanum	E. Bhutan	E. Bhutan & Kameng Dist. NEFA	Kameng

Table I (continued)

Family, genus, species and habitat	Original distribution	Maximum range	Assam locality
Terrestrial			Dist., NEFA
P. spicerianum	Bhutan	Assam	Assam
Terrestrial			
P. venustum	Assam	Trop. Himalaya & Assam	Pynursla K. & J. Hills
Terrestrial			
Pennilabium proboscideum	Between Umran & Umsaw, K. & J. Hills	K. & J. Hills	K. & J. Hills
Epiphytes			
Peristylus parishii	South Andamans, Upper Burma, Sikkim	South Andamans, Upper Burma, Sikkim & Assam	Nongpoh, K. & J. Hills
Epiphytes			
Pholidota imbricata var. *sessilis*	Kohima, Naga Hills	Nagaland and Bhutan	Naga Hills
Epiphytes			
Polystachya flavescens	Java	Africa, South India, Assam	Nongpoh, (1000 m) K. & J. Hills
Epiphytes			
Sarcochilus hystrix	Tenasserim	Tenasserim, Java, Assam	Sairang, K. & J. Hills, Cachar, Lakhimpur
Epiphytes			
Satyrium nepalense	Nepal	Temperate Himalaya, Ceylon, Burma	Shillong (1300 m) K. & J. Hills
Terrestrial			
Taeniophyllum crepidiforme	Sikkim-Himalaya	Sikkim Himalaya and Assam	Nongpoh (1000 m) K. & J. Hills
Epiphytes			
T. khasianum	K. & J. Hills	K. & J. Hills	Shillong, K. & J. Hills
Epiphytes			
Thrixspermum muscaeflorum	Between Umran & Umsaw, K. & J. Hills	Between Umran & Umsaw, K. & J. Hills	Between Umran and Umsaw, K. & J. Hills
Epiphytes			
Zingiberaceae	Jowai-Bodorpur	Jowai-Bodorpur	Jowai-Bordorpur
Hedychium calcaratum	Road, 96 km from Shillong K. & J. Hills	96 km from Shillong, K. & J. Hills	Road, 96 km from Shillong K. & J. Hills
Roadside			
H. dekianum	Khliehriat Road from Jowai, Jorain, Umlung, K. & J. Hills	Khliehriat Road from Jowai, Jorain, Umlung, K. & J. Hills	Khliehriat Road from Jowai, Jorain, Umlung, K. & J. Hills
H. gracillimum	Pynursla, K. & J. Hills	Shillong, Cherrapunji & Pynursla, K. & J. Hills	Shillong, Cherrapunji & Pynursla K. & J. Hills
H. gratum	Khasi Hills	Khasi Hills	Khasi Hills
H. longipedunculatum	Amjee, Subansiri Dist., NEFA	Amjee, Subansiri Dist., NEFA & Naga Hills	Amjee, Subansiri Dist., NEFA & Naga Hills
Epiphytes			
H. marginatum	Kohima, Nagaland	Kohima, Nagaland	Kohima, Nagaland
H. rubrum	Jowai-Bodorpur	Jowai-Bodorpur	Jowai-Bodorpur

Table I (continued)

Family, genus, species and habitat	Original distribution	Maximum range	Assam locality
Roadside	Road, 96 km from Shillong, K. & J. Hills, Assam	Road, 96 km from Shillong, K. & J. Hills, Assam	Road, 96 km from Shillong K. & J. Hills
H. wardii Dioscoreaceae	Delei valley, Lohit Dist., NEFA	Delei valley, Lohit Dist., NEFA	Delei Valley Lohit Dist., K. & J. Hills
Dioscorea laurifolia	Penang	Singapore, Malacca, Western region of Malaya Peninsula & Tirap Dist., NEFA	Pungchow, Tirap Dist., NEFA
D. orbiculata On humid densely forested hill slope		Sumatra, Malaya Peninsula, Perak & Assam	Palin vicinity Subansiri Dist., NEFA
D. scortechinii Taccaceae		Malay Peninsula, Sumatra & Siang Dist., NEFA	Siang Dist., (1200 m) NEFA
*Tacca choudhuriana Araceae	Bandardia, Subansiri Dist., NEFA	Bandardia, Subansiri Dist., NEFA	Bandardia, Subansiri Dist., NEFA
*Lagenandra undulata Submerged perennial herbs	Amjee (1220 m) Subansiri Dist., NEFA	Amjee, Subansiri Dist., NEFA	Amjee, Subansiri Dist., NEFA

* In the above table K. & J. Hills stands for Khasi & Jaintia Hills of Assam.

3. *Phytogeographical Affinities of Assam*

CLARKE (1898), from his analysis of the distribution of Cyperaceæ, chiefly *Carex*, considered the Eastern Himalaya and Assam as distinct subareas. HOOKER, in his botanical divisions of India, included the major part of Assam with the Gangetic Plain, treating the Eastern Himalaya as a separate area by itself and considered the hill areas of Assam including the Shillong Plateau, Patkai, Naga and Manipur Hills as an integral part of Burma. CHATTERJEE (1962), following CLARKE, treats Assam as a distinct area because of its distinctive flora. His Eastern Himalayan region is smaller than either HOOKER's or CLARKE's Eastern Himalaya. A satisfactory phytogeographic analysis is possible only with a full knowledge of the distribution of plants, not only the present, but also the past. RIDLEY (1942) recognizes the fact that no story of plant distribution is complete without a considerable knowledge of Tertiary palaeobotany. It can not also be understood fully without a comprehension of the position and form of land surfaces during that period and the time of the evolution of the flowering plants. The modern Asiatic flora is largely a

228

relict of the Oligocene flora, as represented in Europe, and which probably occupied all tropical lands. Many of the early genera and perhaps orders seem to have disappeared, owing to changes in land surfaces and climate, but some species of that date seem to have persisted, to the present day. LAKHANPAL (1970), in a recent review of the Tertiary floras of India, has attempted to visualize the palaeogeography of India during the early Eocene and Miocene times on the combined evidence of fossil plants and animals. He lists from the Middle Tertiary of Assam nearly twenty fossil woods, referable to living genera of fourteen families of flowering plants: Clusiaceae, Dipterocarpaceae, Elaeocarpaceae, Anacardiaceae, Fabaceae, Combretaceae, Ebenaceae, Sterculiaceae, Burseraceae, Sapindaceae and Lecythidaceae. He also records a fossil fruit of *Nypa* from the Miocene of Garo Hills. He refers to the evidence of palynological studies, for the occurrence of Cyatheaceae, Polypodiaceae, Parkeriaceae, Pinaceae, Schizaeaceae, Podocarpaceae, Potamogetonaceae, Poaceae, Bombacaceae, Rutaceae, Anacardiaceae, Caesalpiniaceae, Ericaceae, Polygonaceae, Euphorbiaceae and Fagaceae in the Miocene of Assam. He believes the profusion of Dipterocarpaceae in the Middle Tertiary of Eastern India would show that after the first phase of upheaval culminating in the Oligocene, there was continuity of land from western Malaysia to Eastern India, by which the Dipterocarpus spread northwards into India. During the Miocene, with the rise of the Himalaya, large areas previously occupied by the Tethys sea were converted into land with numerous water basins. Such moist conditions must have prevailed all along the erstwhile Tethyan region upto Africa, furnishing ideal environments for Dipterocarpus to spread upwards. He concludes that further work is necessary to elucidate a number of interesting problems of the Palaeotropical Tertiary geoflora. We have as yet inadequate data from palaeogeography of Assam.

HOOKER used a unit of ten dominant families in his various botanical provinces, as a measure for comparison on the distribution of these families in relation to their distribution in the whole of India. The ten dominant families of flowering plants for India are: 1. Orchidaceae, 2. Fabaceae, 3. Poaceae, 4. Asteraceae, 5. Rubiaceae, 6. Acanthaceae, 7. Euphorbiaceae, 8. Lamiaceae, 9. Cyperaceae and 10. Scrophulariaceae. In comparison with this, for Assam the first ten families of flowering plants are 1. Orchidaceae (550 spp.), 2. Poaceae (435 spp.), 3. Fabaceae (317 spp.), 4. Asteraceae (218 spp.), 5. Cyperaceae (182 spp.), 6. Euphorbiaceae (165 spp.), 7. Rubiaceae (156 spp.), 8. Lamiaceae (101 spp.), 9. Acanthaceae (90 spp.), and 10. Zingiberaceae (75 spp.). The single largest genus is *Dendrobium*, with 62 spp., followed by *Habenaria* 44 spp. and *Impatiens* 42 spp. While it is not within the scope of this review to take up a detailed family-wise analysis, some of the characteristic families and their distribution may be mentioned.

Nepenthaceæ, with the single species *Nepenthes khasiana* endemic to the

Plate 20. Dipterocarpus forest in Tirap Dist., NEFA, Assam.

Shillong Plateau, represents the northernmost limit of this family, with a general range of distribution from Madagascar to Malaysia. The Podostemaceae, with liverwort-like plants, are represented by three genera. *Zeylanidium*, found in the R. Dirang of Kameng District, NEFA, also represents the northernmost distribution of the family in India. The Styracaceae and the Hamamelidaceae occur in the Eastern Himalaya and the Khasi Hills. The occurrence of *Alniphyllum fortunei* and *Huodendron biaristatum* in Subansiri is of considerable significance. The hitherto known distribution of these two species is in case of the former Yunnan, Amoy and W. Hupeh in China, and in the case of latter northeast Burma and Tonkin (Sastry 1967). *Munronia pinnata* has a discontinuous distribution; it occurs in Sikkim and the Khasi Hills, and also in the Western Ghats in South India. *Coptis teeta* (Ranunculaceae) is endemic to the NEFA Himalaya. The primitive family Magnoliaceae has most of its members only in the Eastern Himalaya and Assam within India. The related family Annonaceae is mostly limited to Assam and the Deccan in the Peninsula. *Berberis* and *Mahonia*, two genera of the Berberidaceæ, occur in the Eastern Himalaya and Shillong Plateau and the species show affinities to the Chinese species. Lardizabalaceae has *Holboellia latifolia* in the Eastern Himalaya and in the Shillong Plateau. *Corydalis* and *Dicentra* of the Fumariaceae both occur in Eastern Himalaya, as well as in Shillong Plateau. *Gynocardia* and *Hydnocarpus* (Flacourtiaeae) have limited distribution in the Eastern Himalaya and the hills of Assam. Dipterocarpaceae

230

with *Vatica, Shorea, Dipterocarpus* are an important family of timber-yielding plants. *Dipterocarpus* has an Indo-Malayan distribution, extending in our area in Patkoi, Naga and the Manipur Hills and adjacent Burma.

Among the Balsaminaceae *Impatiens* is one of the largest genera in Assam; it is also the largest genus for India. Six species *I. radicans, I. bella, I. fimbriata, I. acuminata, I. porrecta* and *I. paludosa* are endemic to the Shillong Plateau. *Hydroceras*, another member, is rare and is found in Kamrup. Rosaceae are well represented, with *Pyrus, Prunus, Photinia, Eriobotrya, Pygeum, Rubus* and *Potentilla*, many of which are common to the Assam Hills and the Eastern Himalaya and some common to the Western Himalaya also.

Vacciniaceae and Ericaceae are very well represented both in the Eastern Himalaya and Assam. There are nearly seventy species, of which sixty-four are endemic to Assam-Burma. *Agapetes* and the closely related *Pentapterygium* are quite common and are mostly epiphytic. Of the Ericaceae, *Rhododendron* is the most prominent element, with a profusion of species in the Eastern Himalaya, a few extending to the Shillong Plateau, Naga Hills and Manipur. Nearly 90% of the species are endemic to the area, though a few show distinct Chinese affinities. *R. arizelum R. peramoenum* and *R. stenalum*, hitherto known from W. Yunnan, northeast Upper Burma and Southeast Tibet (3000–3700 m), have been recently discovered in Subansiri (SASTRY & KATAKI 1966). Another new species *R. santapaui*, a pretty epiphytic shrub, has also been described from Subansiri (SASTRY et al. 1969). In the Primulaceae, *Primula* and *Androsace* are well represented in the Himalaya, many being endemic. *Bryocarpum* is confined to the Eastern Himalaya, the adjoining areas of Burma and China. Amongst Asclepiadaceae, *Hoya* is well represented in Assam and *Dischidia rafflesiana* is an interesting member occurring in the warm tropical forests at lower elevations.

Gesneriaceae are well represented, particularly in the Shillong Plateau and Eastern Himalaya, with many of them having ornamental flowers and foliage. Acanthaceae are a fairly well represented family. *Phlogacanthus* is one of the more attractive undershrubs in the hills. Lamiaceae are also well represented with *Leucas* in the plains, *Pogostemon* and several other aromatic herbs and undershrubs in the hills. Euphorbiaceae are another family found at fairly high altitudes, *Croton, Emblica officinalis, Mallotus philippinensis, Macaranga* being some of the frequent elements.

Thirty-four families of Monocotyledons are found in Assam. As already indicated, the Orchidaceae are the largest family not only amongst the Monocotyledons but among the families of flowering plants as a whole. HOOKER (1854) had already observed this peculiarity for the Khasi Hills, and a recent study of the orchids of Khasi Hills has shown that this family includes 75 genera and 265 species. Some of the large genera are *Dendrobium* (62 spp.) *Habenaria* (44 spp.) *Eria, Bulbophyllum* (33 spp.), *Liparis* (32 spp.) and *Coelogyne* (29 spp.). The biological and ornamental importance of this family is well known. An appreciable percentage of cultivated forms of orchids in European gardens have come from the wilds of Assam. The depletion of these valuable plants has now become perceptible and the National Orchidarium has recently been established at Shillong to conserve Assam's orchids and scientifically exploit them. Amongst the species seriously threatened by extermination may be mentioned the celebrated blue vanda, *Vanda coerulea* and several species of lady's slippers *Paphiopedilum. Apostasia*, with two species, now placed under a distinct family, is also limited to Assam-Burma. Poaceae and Cyperaceae are other large families. Zingiberaceae rank as one of the overall ten largest families, with eighteen genera and seventy-five species followed by Liliaceæ with its twenty-one genera and fifty-five species. Arecaceae include seventeen genera and thirty-eight species.

Zingiberaceae have several endemics; thirteen species of *Hedychium* are endemic, other members are *Mantisia wengeri, Globba marantia, Boesenbergia rubrolutea, Amomum pauciflorum* and *Zingiber intermedium*. The Arecaceae of Assam are *Areca nagensis, Pinanga griffithii, Didymosperma nana, D. gracilis, Caryota obtusa, Livistona jenkinsiana, Calamus khasianus, C. kingianus, Plectocomia khasyana, P. assamica* and *Zalacca secunda*. Pandanaceæ are represented by *Pandanus*, with six species. Burmanniaceae, Cannaceae, Musaceae, Taccaceae, Dioscoreaceae, Xyridaceae, Flagellariaceae, Typhaceae, Sparganiaceae, Triuridaceae, Najadaceae, Aponogetonaceae, Potamogetanaceae and Eriocaulaceae have all one genus each; Marantaceae, Stemonaceae, Pontederiaceae, Juncaceae, Lemnaceae, have two genera each; Iridaceae, Amaryllidaceae, Hypoxidaceae and Alismataceae each have three genera, Hydrocharitaceae have five and Commelinaceae eight genera with thirty-five species.

The gymnosperms are well represented by the Conifers. Cycadaceae are represented by *Cycas pectinata*, mostly in the plains of Assam, particularly Kamrup and Nowgong. The Gnetaceae include *Gnetum gnemon* and *G. montanum*, the former with a limited distribution at the edge of the Khasi Hills and in the foothills of Kameng District or border of Darrang District and Sibsagar District, and latter in the hills of the Shillong Plateau and on the Eastern Himalaya. *Ephedra* also occurs in the alpine zone of the Aka Hills in Kameng District, NEFA. Amongst the Conifers are *Abies, Cephalotaxus, Cupressus, Juniperus, Larix, Picea, Pinus, Podocarpus, Taxus* and *Tsuga*. Except *Podocarpus*, all the other genera are Himalayan and form characteristic associations. *Taxus baccata* and *Podocarpus nerifolius* occur in the Khasi Hills. *Pinus insularis* is the characteristic pine of the Khasi, Naga and Mizo (Lushai) Hills and north Burma.

PANIGRAHI (1960) lists three-hundred and fifty species of pteridophytes; Pteridaceae with the largest number, eighteen genera and fifty-nine species. *Helminthostachys zeylanica* occurs in Kamrup and on the Khasi Hills. *Osmunda cinnamomea* has been recorded from the Aka and Khasi Hills. *Dipteris wallichii* is endemic to northeast India and the Blechnoid

Plate 21. Dipteris wallichii near Jowai, Jaintia Hills, Assam.

Brainea insignis is distributed in Khasi Hills, S. China and Mindoro. The tree ferns *Cyathea*, with eight species, are very prominent due to their large size and huge spreading fronds. Equally noticeable is the other large fern *Angiopteris evecta*. *Lygodium flexuosum*, *L. japonicum* and *L. scandens* are all elegant twiners, with pretty dissected foliage. We do not have much data on the fern-allies. *Psilotum nudum* has been noted to occur in Siang District, NEFA. *Selaginella* with thirty species and *Lycopodium* with eleven species are common. *Equisetum*, with two or three species, is also frequent along sandy banks and amidst grass.

A passing reference may be made here on the rôle of man in modifying the vegetation by introduction of exotic plants. In the lower elevations of NEFA, *Mikania micrantha*, which must probably have been introduced into these parts from Tropical America during World War II, has now spread over vast areas, choking other vegetation and forming a troublesome weed. *Tagetes minuta* another Asteraceae, has also become a noxious weed in large areas of the Aka Hills. *Acanthospermum hispidum* has similarly spread in many places in the Tenga Valley. *Eupatorium odoratum* is also a comparatively recently introduced weed that forms now a conspicuous feature of the landscape, even in many remote areas of the Eastern Himalaya and Assam. In the plains of Assam, the numerous waterways and ponds, choked with *Eichhornia crassipes*, are common sights. *Lantana* and *Croton bonplandianum* have completely become a native element of the vegetation in many areas of the Assam plains. In the hills scores of exotic ornamental plants have been introduced; some of them have escaped from gardens and have become gradually naturalized.

233

Plate 22. Engelhardtia spicata from Mowsmai Forest, Khasi & Jaintia Hills, Assam.

234

Table II. Trees of the Khasi Hills Climax (After BOR, 1942)

Species	Altitudinal limits in metres	
	Lower	Upper
Quercus fenestrata	1700	2000
Q. dealbata	1800	2000
Q. lancaefolia	1000	1700
Daphniphyllum himalayana	1000	3300
Michelia doltsopa	1000	2000
Ficus nemoralis	1000	2800
Exbucklandia populnea	1000	2800
Cinnamomum impressinervum	1300	2300
Machilus odoratissima	1000	2300
Engelhardtia spicata	500	2000
Elaeocarpus braceanum	1500	2000
E. lanceaefolium	2000	2500

Table III. Trees of the Naga Hills climax (After BOR, 1942)

Species	Altitudinal limits in metres	
	Lower	Upper
Quercus lamellosa	2800	3000
Q. xylocarpa	1800	3000
Q. pachyphylla	1800	3000
Alcimandra cathcartii		
(Michelia cathcartii)	1800	2000
Michelia doltsopa	1000	2800
Exbucklandia populnea	Plains	2800
Castanopsis tribuloides	-do-	2800
Ficus nemoralis	1900	2800
Evodia fraxinifolia	1300	2800
Acer campbelii	2000	3500
Cinnamomum impressinervum	1300	2300

BOR (1942) quotes CLARKE's letter to HOOKER, in which CLARKE was struck by the marked difference at comparable altitudes of the flora of the Naga Hills from that of the Khasi Hills, hardly 160 km away. There is a far greater resemblance of the Naga Hill flora to the flora of Sikkim, in particular to that of Darjeeling, nearly 800 km away to northwest, with the wide Brahmaputra valley between. He concludes from a comparison of the geology and the flora of the two hills that the very great majority of the species, which comprise both these hills, are Eastern Himalayan and a good many are common to both the Khasi and Naga Hills. CLARKE was right in concluding that the Naga Hills flora is more Himalayan than Khasian.

The trees of the Khasi Hills climax forest are near the upper limit of their range, while those of the Naga Hills are near the lower limit (see

235

Table II & III). The difference in facies is due to the Naga Hills being of a higher altitude, with the well known effect of the higher altitude plants descending lower down in nothern aspects. The low-altitude area between the Naga and Khasi Hills effectively serves as barrier to the high altitude species. He concludes that on the whole the flora of the Eastern Himalaya, the Naga and the Khasi Hills is essentially Indo-Malayan, with a strong admixture of Chinese elements. BOR refers to the importance of the ancient nature of the Khasi Hills, and the later uplift of the Eastern Himalaya and the coming of the glacial age in Pleistocene times. During the glacial epoch, the preglacial flora of the Himalaya spread to the warmer south. The retreat of the ice was followed by the return of an Indo-Malayan flora, the advance of which was helped by the hill country in the south and southeast. This should explain the general resemblance in the floras of the Eastern Himalaya, Naga and Khasi Hills. Apart from a comparison of the high altitude woody vegetation of these areas, Bor lists several Chinese plants, common to these areas, and also notes the presence of *Pinus insularis* in all these areas (but not Eastern Himalaya) and concludes how close is the connection between the vegetation of the mountain regions of Assam, both north and south of the Brahmaputra and that of Burma and to a lesser degree, that of China; the reason for this similarity being the geological history of the area.

In a review of the flora of Manipur, with particular reference to endemism, DEB (1958) remarks that the number of endemic plants for Assam is reduced by about sixty-nine species and that of the Sikkim Himalaya by seven. These discoveries indicate, on the other hand, a greater phytogeographical affinity of Manipur with the Khasi Hills than with the Sikkim Himalaya. He agrees with KINGDON-WARD's view that the position of Manipur in the middle of glaciated mountains and astride one of the glacial escape routes was peculiarly favourable for receiving contributions of flora from all directions. He concludes that Manipur forms a phytogeographical part of Assam, with a very high percentage of Indo-Malayan species and an admixture of some Sikkim Himalayan, Burmese, Siamese and Chinese species. In a detailed study of the flora of the northern face of Khasi Hills, JOSEPH (1969) recently found some Indo-Malayan, Burmese and Sikkim species, which necessitate modification of earlier ideas on the range of distribution.

Reference may be made here to the recent attempt by TAKHTAJAN (1969) to localize the cradle of flowering plants from an analysis of the distribution of primitive plants. In table IV, adopted from TAKHTAJAN, are brought together most of the primitive species of plants, occurring in the northeastern parts of India, especially the Eastern Himalaya, Assam and Burma. It is remarkable that none of these species occurs in any other parts of India. This fact would seem to suggest the phytogeographical distinctiveness of Assam within India. It is, however, not possible, in the present state of our knowledge, to conclude whether the flora of Assam is

Table IV. Primitive flowering plants occurring in Northeastern India and Burma (From
Takhtajan, 1969)

Species	Distribution
1. Magnoliaceae	
Magnolia	Assam, Burma through Indo-China to Malayan Archipelago
Magnolia griffithii	Assam and Burma
Magnolia pealiana ⎱ *Magnolia gustavii* ⎰	Assam
Manglietia	Assam, E. Himalaya and South China through Thailand and Indo-China to Java
Euptelea	E. Himalaya, China and Japan
2. Tetracentraceae	
Tetracentron	E. Himalaya, Upper Burma and Southwest China
3. Menispermaceae	
Pycnarrhena	Assam, E. Himalaya to northwest Australia
Haematocarpus	Assam, E. Himalaya, West Malaysia and New Guinea
Aspidocarya	Assam, E. Himalaya and Southeast Asia
4. Lardizabalaceae	
Decaisnea	Eastern Himalaya and West China
Holboellia	Assam, E. Himalaya, China and Brikin
Stauntonia	Assam, South China, Taiwan, Laos, Vietnam, Korea and Japan
Parvatia	Assam, E. Bengal, South and West China
5. Hamamelidaceae	
Exbucklandia	E. Himalaya, Assam to Sumatra
Distylium	Himalaya, Assam, China, Taiwan, Korea and Japan
Altinglia	Assam, Japan and China to Java and Sumatra
6. Piperaceae	
Houttuynia	Assam, Himalaya to China, Japan, Thailand and Indo-China, Taiwan
7. Myricaceae	
Myrica esculenta	Assam, China, Korea and Japan
8. Betulaceae	
Alnus	Himalaya, Assam and China
Betula	Himalaya and East Asia

truly indigenous but at present marked by large intrusive elements or its affinities are with China, Burma and Malaya.

4. Burma

Burma is separated from India by a mountainous barrier. The important differences in the physical features of Burma from those of India are that its mountains and rivers have a north-south trend and the coasts are different from those of India. The Arakan and the Tenasserim coasts are rocky and are fringed with islands. The Arakan and Pegu Yomas are young fold mountains, with young little-folded soft rocks between in the valley of the Chindwin and Irrawaddy. The mountainous eastern Shan Plateau and the southern continuation into Tenasserim consist of older and hard rocks.

In the north, Burma is a tangle of mountains, which are related to the Tibetan mass and diverge gradually to the south. The Chin Hills are the western highlands border, between the 22nd and the 24th north parallels. To the south are the Arakan and the Pakokku Hills. The southern spur that skirts the Bay of Bengal is known as the Arakan Yoma. To the east of the R. Irrawaddy there is a succession of mountain chains and plateaus, separating the valley of that river from the rocky trough of the R. Salween. Beginning from the north, the eastern Kachin Hills extend southwards and southeastwards to the northern Shan States and the Ruby Mines. The Shan Plateau stretches across the country. The northern Shan State is a much broken group of hills, but the ridges tend to become north-south in the south. Near Toungoo, the Shan Hills give place to the Karen Hills.

An imaginary line drawn along the western bank of the R. Irrawaddy, as far south as Mandalay, and thence further southwards along the foot of the Shan Plateau to the Sittang Valley, roughly divides Burma into two distinct geological halves. To the west of this line are the Chin Hills and the Arakan Yoma of sandstones, shales and Cretaceous limestones and Tertiary formations and to the east are the Archaean formations. The Pegu Yoma of shale and sandstones is more recent that the Arakan.

The general course of the rivers of Burma is north-south. The R. Irrawaddy, which divides Burma into two halves, is formed by the confluence of the R. Maika and N'maika, about 50 km north of Myitkyina, and flows for about 1500 km through rocky defiles, broad level plains and narrow tidal creeks. The R. Salween (Namkong) flows into the Gulf of Martaban. The R. Sittang is fed by affluents from the Karen Hills and also flows into the Gulf of Martaban.

Burma has an equally rich but perhaps a more varied flora than Assam but as remarked by HOOKER (1854), the region is also the least known botanically. He relied almost entirely on KURZ's Forest Flora of British Burma (1877) for his phytogeographical notes on Burma. Some of the early plant collectors in the difficult terrain of Burma were POTTINGER, COLLETT, PARISH and LOBB, and later came TROUP, HANDEL-MAZZET-TI, BARRINGTON, BISWAS, MERRILL, STAMP and KINGDON-WARD. While KINGDON-WARD (1960) described mainly the vegetation of northern Burma, the best general account happens to be that of STAMP (1925).

HOOKER (1854) treated Burma as a distinct botanical province, including within it Assam, Garo, Patkai, Naga, Khasi, Manipur, Cachar and Sylhet Hills, Chittagong, Tippera, Arakan, Pegu and Tenasserim, together with the Shan and other states bordering China and Thailand. CLARKE (1898) recognized, however, two phytogeographical areas in Burma, viz. Ava and Pegu. He also considered Assam as a distinct botanical area. CHATTERJEE's (1940) Upper Burma and Lower Burma are identical with the Ava and Pegu regions of CLARKE.

KURZ (1877) has divided the forests of Burma into i. evergreen forests and ii. deciduous forests, with four subdivisions under each. Under the evergreen forests, he recognized 1. littoral, 2. swamp, 3. tropical and 4. hill forest. Under the deciduous forests be included 1. open, 2. dry, 3. mixed and 4. dune forests. KURZ also separately described the bamboo jungles and secondary vegetation of areas, deserted after cultivation. HOOKER adopted this classification and KURZ's account of Burma forests in his sketch of the vegetation of Burma and analysis of its flora and phytogeography. The vegetation of Burma is classified by STAMP (1925) as 1. mountain vegetation, above 900 m, 2. lowland vegetation below 900 m and 3. grassland and fallowland vegetation. The mountain vegetation is conspicuously similar to that of the Eastern Himalaya, especially in North Burma. The forest abounds in *Quercus, Castanea, Ternstroemia, Exbucklandia* and *Rhododendron*, with undergrowth of *Pteris* and sprinkling of tall grass and bamboo. The high elevation, combined with cool climate, contributes to the development of temperate species. Depending upon the altitude and exposure, these hill forests exhibit distinctive species composition. In many places on the hills, the broad-leaved forest is often wholly replaced by pine forest, composed of two species, viz. *Pinus insularis* in North Burma and *P. merkusii* farther south.

At lower elevations, along the coast, occurs the littoral vegetation, not only along the sea coast but also inwards along the estuaries of the large rivers. This vegetation largely resembles the Sunderban of Bengal. The main factor influencing the aspect of this vegetation type is the salt water. Mangrove vegetation, consisting chiefly of *Rhizophora, Bruguiera, Sonneratia* and *Aegiceras* with smaller trees of *Lumnitzera, Kandelia* and *Ceriops*, dominate the scene. Further inwards and away from the influence of the tidal waves, *Sonneratia* and *Avicennia* prevail, with *Thespesia, Hibiscus, Pongamia, Excaecaria, Antidesma, Erythrina* and *Dalbergia* with *Cerbera* and *Cordia; Phoenix paludosa* also occurs. There is a dense undergrowth of *Clerodendrum* and *Acanthus*. Amongst climbers, several species of *Derris* may be noted. Other noticeable elements, forming gregarious patches, are the palm *Nipa* and *Pandanus*.

The swamp forests occur along river courses and occupy lowlands in alluvial plains. The common trees are *Anogeissus, Mangifera, Xanthophyllum, Memecylon, Elaeocarpus, Symplocos* and *Eugenia* but mostly dwarfed. Many kinds of bushy shrubs like *Capparis* and *Crataeva* are present. Among climbers *Jasminum, Combretum* and *Derris* are frequent. Except for sedges, grasses and some aroids, the herbaceous growth is scanty. The trees support several epiphytes, chiefly orchids and ferns.

The most characteristic vegetation is, however, the tropical evergreen forest, densely covering the shady valleys and shady slopes of the warm humid hilly country. These are seen at their best from Martaban to

Tenasserim. The kinds of trees composing this forest are innumerable and many of them attain gigantic proportions. Some of these gaint species belong to *Tetrameles, Sterculia, Parkia, Albizia, Xylia, Artocarpus, Pterocarpus, Dipterocarpus* and *Duabanga*.

Among the less tall trees may be included *Ficus, Bursera, Kurrimia, Semecarpus, Adenanthera, Lagerstroemia* and *Podocarpus*. Some of these, although found in evergreen forests, are leaf-shedders. A host of still smaller trees like *Garcinia, Dalbergia, Hydnocarpus, Baccaurea, Micromelum* and *Turpinia* are also present in these dense forests. Intertwining with these trees and often enveloping their crowns and dangling from them, there are several woody climbers, including numerous rattan palms. There are huge bamboo clumps of *Dendrocalamus* and *Bambusa*. Palms and screwpine are common, the latter often forming impenetrable thickets near the water edge. Among palms may be mentioned *Arenga, Caryota, Licuala*, and *Zalacca*. The Zingiberaceae herbs are also abundant in moist shady places. These forests are also characterized by heavy epiphytic growth, practically every tree being loaded with orchids, aroids and ferns. About seven hundred species of orchids are known. The herbaceous growth is however poor, the ferns forming the most noticeable elements.

The deciduous forest contains a lesser profusion of species. According as the trees shed their leaves due to cold or to dryness, the forests may be distinguished as winter-deciduous and summer-deciduous forests. The forests on the hills experience a cold season. The most important timber trees of Burma occur in the dry or summer-deciduous forest and hence are of considerable interest. These timber trees are mainly *Dipterocarpus* and *Tectona*, and depending upon their dominance they have given their names to the respective forests. The principal components of these deciduous forests at lower elevations, where they occupy lateritic soil, are *Dillenia, Shorea, Walsura, Buchanania, Diospyros, Emblica, Gardenia, Eugenia, Zizyphus* and *Flacourtia*. The 'eng' or *Dipterocarpus* is the characteristic tree of these forests, which are therefore also called eng-forests. The stemless palm *Phoenix acaulis* is also present, as well as the palm-like *Cycas siamensis*. Both *Dendrocalamus* and *Bambusa* occur. Climbers are also prominent, chiefly the prickly *Calamus*. The shrubby and herbaceous growth is also plentiful, and particularly striking when the herbs, chiefly Acanthaceae, Lamiaceae, Balsaminaceae and Asteraceae are all in their varied bloom. Epiphytic growth of ferns and orchids is also very noticeable. Asclepiadaceæ, including *Hoya*, form part of this epiphytic growth.

The eng-forests on the hills are similar to those in the plains in that *Dipterocarpus* is the dominant element, but the other associated trees are rather different. There are two species of *Dipterocarpus*, viz. *D. tuberculatus* and *D. obtusifolius*. Other trees mixed with these are *Engelhardtia, Quercus, Schima* and *Anneslea*. Some of the other less common elements here are *Dillenia, Rhus, Callicarpa* and *Vernonia*.

In the lower areas, some mixed forests without any eng or Dipterocarp,

240

contain *Terminalia, Dalbergia* and *Strychnos*. In Prome there are distinctive dry deciduous forests, chiefly on calcareous sandstone, and often intermixed with or merging into the Dipterocarp forest. The trees have characteristic dwarf, spreading canopies and are widely separated from each other. The chief components of these forests are *Acacia, Albizia, Melia, Chikrassia, Diospyros, Ulmus, Hymenodictyon, Strychnos, Ehretia, Rhus, Morinda* and *Emblica*. Here and there an occasional *Dipterocarpus* also occurs. The shrubby layer is marked by prickly and thorny species like *Euphorbia* and *Zizyphus*. The palms and bamboos are of the same kind as in Dipterocarp forests. In these, often *Acacia catechu* becomes dominant, forming almost pure forests in Prome District.

Along banks of the large rivers there is a characteristic vegetation, principally of tall, coarse, gregarious grass *Saccharum, Phragmites, Imperata, Arundinaria, Typha*, etc. Higher up on the bank there are stunted trees of *Streblus, Butea, Nauclea* and *Ficus*, with other trees like *Dalbergia, Lagerstroemia, Albizia* and *Zizyphus*. A bamboo also occurs in this.

Another distinctive type is the dune vegetation of sand-clay deposits on sea beach, amidst rivers, on large sand banks and sand islands, characterized by *Pongamia, Erythrina, Bombax, Hibiscus, Terminalia, Eugenia, Calophyllum* and *Barringtonia*. *Cycas rumphii* also forms gregarious growth. Shrubs and climbers are not infrequent. The floor is covered by the creeping grass *Ischaemum* and by *Ipomoea*.

Savannahs or grasslands, practically devoid of any woody vegetation, occur along rivers and occupy inundated areas. The grasses are all stiff coarse kinds like *Arundinaria, Saccharum, Polytoca* and *Arundo*. Sometimes lesser grasses like *Imperata* and *Eragrostis* also cover extensive areas. In some spots the wild sugarcane *Saccharum spontaneum*, with *Andropogon*, may be the chief feature. Due to the prevalent practice of clearing large areas of forest for cultivation and then abandoning it, a characteristic vegetation called 'Poonzohas' succeeds these abandoned grounds. These areas are usually occupied by herbaceous growth of Asteraceae, Malvaceae and Lamiaceae, with numerous grasses and sedges. In some places, particularly in the vicinity of bamboo clumps, the area may be invaded by bamboo. The abandoned cleared areas rarely succeed to the original forest that existed prior to human intervention.

Subsequent to the explorations of KINGDON-WARD, there has been no recent attempt to study the rich forests of Burma. HOOKER's ranking of the ten dominant families of flowering plants more or less still holds good. These are 1. Orchidaceae, 2. Fabaceae, 3. Poaceae, 4. Rubiaceae, 5. Euphorbiaceae, 6. Acanthaceae, 7. Cyperaceae, 8. Asteraceae, 9. Zingiberaceae and 10. Urticaceae.

CHATTERJEE (1940) listed 1071 species of Dicotyledons as endemic to Burma (see table V). He recognized two main outside influences manifest in the Burmese flora; a Chinese one from the northeast, chiefly endemic, with temperate and alpine plants; and a Malaysian influence from the

Table V. Some endemic genera of plants from Assam-Burma (After CHATTERJEE, 1940)

Genera	Distribution
Flacourtiaceae	
1. *Gynocardia*	E. Himalaya, Assam, Burma, Chittagong
Sterculiaceae	
2. *Mansonia*	Burma
Tiliaceae	
3. *Plagiopteron*	Lower Burma
Linaceae	
4. *Anisadenia*	C. & E. Himalaya, Khasi Hills & extreme S. China
Sapindaceae	
5. *Zollingeria*	Lower Burma
Anacardiaceae	
6. *Drimycarpus*	E. Himalaya, Khasi Hills
Fabaceae	
7. *Neocollettia*	Burma
8. *Dicrema*	Burma
9. *Phyllodium*	Burma
10. *Mastersia*	E. Himalaya
Araliaceae	
11. *Tupidanatus*	Khasi Hills
Rubiaceae	
12. *Polyura*	E. Himalaya, Khasi Hills
13. *Parophiorrhiza*	Khasi Hills
14. *Carlemannia*	E. Himalaya, Khasi Hills
15. *Silvianthus*	Khasi Hills
16. *Pentapterygium*	E. Himalaya, Khasi Hills
Pyrolaceae	
17. *Chilotheca*	Khasi Hills
Primulaceae	
18. *Bryocarpum*	E. Himalaya
Myrsinaceae	
19. *Sadiria*	E. Himalaya, Khasi Hills
20. *Antistrophe*	Khasi Hills, S. India
21. *Hymenandra*	Assam
22. *Amblyanthus*	Khasi Hills
23. *Amblyanthopsis*	E. Himalaya, Assam
Styracaceae	
24. *Parastyrax*	Upper Burma
Asclepiadaceae	
25. *Pentabothra*	Assam
26. *Adelostemma*	Burma
27. *Lygisma*	Burma
Convolvulaceae	
28. *Bluikworthia*	Burma
Scrophulariaceae	
29. *Bythophyton*	Khasi Hills
30. *Hemiphragma*	E. Himalaya, Khasi Hills, Burma, S. China
Gesneriaceae	
31. *Tetraphyllum*	Assam & Chittagong
32. *Trisepalum*	Lower Burma

242

Table V (continued)

Genera	Distribution
33. *Phylloboea*	Lower Burma
34. *Leptoboea*	E. Himalaya, Khasi Hills
Acanthaceae	
35. *Ophiorrhizophyllon*	Lower Burma
36. *Phlogacanthus*	Himalaya, Assam, Burma
37. *Cystacanthus*	Burma
38. *Asystasiella*	Khasi Hills
39. *Philacanthus*	N. Assam
40. *Odontonemella*	Khasi Hills
41. *Sphinctacanthus*	Assam
Lamiaceae	
42. *Craniotome*	E. & W. Himalaya, Khasi Hills
43. *Notochaete*	E. Himalaya, Burma
Amaranthaceae	
44. *Stilbanthus*	E. Himalaya, Khasi Hills
Rafflesiaceae	
45. *Sapria*	E. Himalaya & Manipur
Lauraceae	
46. *Purakayasthea*	Khasi Hills
47. *Dodecadenia*	Himalaya, Assam, Burma
Santalaceae	
48. *Phacellaria*	Manipur, S. Burma
Euphorbiaceae	
49. *Platystigma*	Assam

southeast, bringing in a more tropical flora. KINGDON-WARD (1960) has shown that three distinct floral regions have contributed to the bulk of Burmese flora: Indo-Malayan or more correctly the Malayan from the south, the eastern Asiatic from the east, the Sino-Himalayan from the north and northwest. There are also some cosmopolitan elements. The complex phytogeography is correlated with the wide range of altitude, the meridional orientation of the mountains and the geological past of the land. KINGDON-WARD (1960) has also attempted an outline of the probable past history of the flora. He envisages a vast area in the Himalayan and Sino-Malayan mountains, as well as West China and Tibet that was covered with ice in the last glaciation. The fluctuations of glaciation had the effect of isolating the Indo-Malayan subregion from Central Assam, though it did not prevent the mixing of the east Asian and Malayan elements in the southeast. Another pronounced effect was the southward extension of the alpine flora, thus leading to the mingling of the alpines with subalpines, subalpines with warm-temperate and the warm-temperate with the tropical elements of the flora.

The endemics, derived mostly from Indo-Chinese and South Chinese stock and the tropical Malayan elements appear to be nearly equally strong in Burma.

REFERENCES

Anonymous, 1903. An account of the Province of Assam and its administration, Shillong.

Biswas, K. 1941. The flora of the Aka hills. *Indian For. Rec.*, (NS) (Bot.), 3(1): 1–62.

Biswas, K. 1943. Systematic and Taxonomic studies of the Flora of India & Burma. *Proc. 30th Indian Sci. Congress, Pres. Address*, 101–152.

Bor, N. L. 1938. A sketch of the vegetation of the Aka Hills, Assam, a synecological study. *Indian For. Rec.*, (NS) (Bot.), 1(4): i–ix, 103–221.

Bor, N. L. 1942. Some remarks on the geology and the flora of the Naga and Khasi Hills. 150th *Aniv. Vol. Roy. Bot. Gard.*, Calcutta, 129–135.

Bor, N. L. 1942a. The relict vegetation of the Shillong Plateau/Assam. *Indian For. Rec.*, 3(6): 152–195.

Burkill, I. H. 1925. The Botany of the Abor Expedition. *Rec. bot. Surv. India*, 10(1): 1–154 and 10(2): 155–420.

Burkill, I. H. 1965. Chapters on the History of Botany. India, Delhi.

Carter, H. G. 1921. Useful Plants of Lakhimpur, Assam. *Rec. bot. Surv. India*, 6(9).

Champion, H. G. 1936. A preliminary survey of the forest types of India and Burma. *Indian For. Rec.*, (NS) (Bot.), 1(1): 9, 286, 8.

Chatterjee, D. 1940. Studies on the endemic flora of India and Burma. *Roy. Asiat. Soc. Bengal*, (n.s.)

Chatterjee, D. 1962. Floristic Patterns of Indian vegetation. *Proc. Summer School Botany*, Darjeeling (Maheswari Ed.) 32–42.

Clarke, C. B. 1889. On the plants of Kohima and Munneypore. *J. Linn. Soc. London*, 25: 1–107.

Clarke, C. B. 1898. Subareas of British India illustrated by the detailed distribution of the Cyperaceae in that Empire. *J. Linn. Soc. London*, 34: 1–146.

Das, A. 1942. Floristics of Assam – A Preliminary Sketch. *150th Aniv. Vol. Roy. Bot. Gard.* Calcutta, 131–147.

Das, B. N. & S. Rajkhowa, 1968. Woodlands of Assam. *Indian Forester*, 94(2): 137–146.

Das, H. P. 1970. Geography of Assam. New Delhi.

De, R. N. 1923. Assam to Burma across the Hills. *Indian Forester*, 49: 529–539.

Deb, D. B. 1957. Studies on the flora of Manipur. *Bull. bot. Soc. Bengal*, 11(1): 15–24.

Deb, D. B. 1957. *Aldrovanda vasculosa* L. from Manipur. *Curr. Sci.*, 26: 229.

Deb, D. B. 1958. Endemism and outside influence on the flora of Manipur. *J. Bombay nat. Hist. Soc.*, 55(2): 312–317.

Deb, D. B. 1960. Forest types studies in Manipur. *Indian Forester*, 86(2): 94–111.

Deb, D. B. 1961. Monocotyledonous plants of Manipur Territory. *Bull. bot. Surv. India*, 3: 115–138.

Deb, D. B. 1963. Bibliographical review on the botanical studies in Tripura. *Bull. bot. Surv. India*, 5(1): 49–58.

Fischer, C. E. C. 1938. The Flora of Lushai hills. *Rec. bot. Surv. India*, 12(2): 75–161.

Griffith, W. 1847. Journals of Travels in Assam, Burma, Bhutan, Afghanistan & the neighbouring countries. Calcutta.

Hooker, J. D. 1854. Himalayan Journals, 2 vols. London.

Hooker, J. D. & T. Thomson, 1855. Flora indica, London.

Hooker, J. D. 1872–97. The Flora of British India, 7 vols. London.

Hooker, J. D. 1906. A sketch of the flora of British India. London.

Joseph, J. 1969. Flora of Nongpoh and vicinity (D. Phil. Thesis, Gauhati University, 1969).

Joseph, J. & S. Chowdhury, 1966. *Oberonia sulcata* Jos. *et* Chowd. a new orchid from Kameng Frontier District, NEFA, Assam. *J. Bombay nat. Hist. Soc.*, 63(1): 54–56.

Joseph, J. & S. N. Yoganarasimhan, 1967. *Corybas purpureus*. A new species of orchid from United Khasi and Jaintia Hills, Assam. *Indian Forester*, 95(12): 815–817.

Joseph, J. & S. N. Yoganarasimhan, 1967. *Taeniophyllum khasianum*. A new species of orchid from United Khasi and Jaintia Hills, Assam. *J. Indian bot. Soc.*, 46(1): 109–111.

Kanjilal *et al.* 1934–1940. Flora of Assam. 5 vols. Shillong.

Kar, S. K. & G. Panigrahi, 1963. The Rubiaceae in Assam & NEFA. *Bull. bot. Surv. India*, 5: 227–237.

Kataki, S. K. & G. Panigrahi, 1964. Ranunculaceae in Assam and NEFA. *Indian Forester*, 90: 394–400.

Kurz, S. 1877. Forest Flora of Burma, 2 vols. Calcutta.

Lakhanpal, R. N. 1970. Tertiary floras of India and their bearing on the historical geology of the Region. *Taxon*, 19(5): 675–694.

Legris, P. 1963. La Vegetation de l'Inde ecologie et flore. Pondichery.

Maheshwari, P. *et al.* 1965. Flora. *The Gazetteer of India*, 1: 163–229.

Naik, V. N. & G. Panigrahi, 1961. A Botanical tour to Subansiri Frontier Division NEFA. *Bull. bot. Surv. India*, 3: 361–388.

Panigrahi, G. 1960. Pteridophytes of Eastern India. *Bull. bot. Surv. India*, 2: 309–314.

Panigrahi, G. & J. Joseph 1966. A Botanical tour to Tirap Frontier Division, NEFA. *Bull. bot. Surv. India*, 8: 142–157.

Puri, G. S. 1960. Indian Forest Ecology. 2 vols. New Delhi.

Rao, A. S. 1969. The Vegetation of the Khasi and Jaintia Hills. *Proc. Pre-Congress Symposium – 21st International geographical Congress India*, Gauhati, 1968–1969.

Rao, A. S. 1970. A Sketch of the Flora of North-Eastern India – particularly Assam and Meghalaya. Souvenir in connection with 40th session of the Central Board of Irrigation and Power. Shillong 85–90.

Rao, A. S. & J. Joseph, 1967. *Rhynchoglossum lazulinum* – A new species of Gesneriaceae. *Bull. bot. Surv. India*, 9: 280–282.

Rao, A. S. & J. Joseph, 1968. *Pennilabium proboscideum* A. S. Rao & Joseph J. – A new orchid species from K. & J. Hills, Assam, with incidental first record of the genus for India. *Bull. bot. Surv. India*, 10(2): 231–233.

Rao, A. S. & J. Joseph, 1971. *Thrixspermum muscaeflorum* – A new orchid species from K. & J. Hills, Assam. *Bull. bot. Surv. India*, 11. (1–2): 204–205 (1969).

Rao, A. S. & L. C. Rabha, 1966. Contribution to the Botany of Kamrup District (Southern part) Assam. *Bull. bot. Surv. India*, 8(3 & 4): 296–303.

Rao, A. S. & D. M. Verma, 1971. Notes on *Hedychium* Koenig including three new species from Khasi and Jaintia Hills, Assam. *Bull. bot. Surv. India*, 11: 120–128 (1969).

Rao, R. S. & J. Joseph, 1965. Observations on the Flora of Siang Division NEFA. *Bull. bot. Surv. India*, 7: 138–161.

Rao, R. S. & G. Panigrahi, 1961. Distribution of vegetational types and their dominant species in Eastern India. *J. Indian bot. Soc.*, 40.

Ridley, H. N. 1942. Distribution areas of Indian Floras. *150th Aniv. Vol. Roy. Bot. Gard. Calcutta*, 49–52.

Robinson, W. 1841. A descriptive account of Assam. London.

Rowntree, J. E. 1953. An introduction to the vegetation of the Assam valley. *Indian For. Rec.*, (NS) 9(1): 1–87.

Sahni, K. C. 1969. A contribution to the flora of Kameng and Subansiri District, NEFA. *Indian Forester*, 96(5): 330–352.

SASTRY, A. R. K. 1967. *Lagenandra undulata* - A new species of Araceae. *Bull. bot. Surv. India*, 9(1–4): 294–296.

SASTRY, A. R. K. 1967. *Alniphyllum* Mats. and *Huodendron* Rehd. – Two additional generic records to the Indian Styracaceae. *Bull. bot. surv. India*, 9(1–4): 297–298.

SASTRY, A. R. K. 1967. *Merriliopanax cordifolia* – A new species of Araliaceae from India. *Blumea*, 15.

SASTRY, A. R. K. & S. K. KATAKI, 1966. On some species of *Rhododendron* from Subansiri District, North-East Frontier Agency. *Indian Forester*, 93(4): 264–265.

SASTRY, A. R. K. *et al.* 1969. *Rhododendron santapaui* sp. nov. from Subansiri District NEFA in India. *J. Bombay nat. Hist. Soc.*, 65(3): 744–747.

SCHWEINFURTH, U. 1957. Die horizontale und vertikale Verebreitung der Vegetation in Himalaya. Bonn. 372.

STAMP, L. D. 1925. The vegetation of Burma. Calcutta.

SUBBA RAO, G. V. 1963. A new species of *Polygonum* from Assam. *Bull. bot. Surv. India*, 5: 257.

TAKHTAJAN, A. 1969. Flowering Plants. Origin and dispersal (Tr. Jeffery) Edinburgh.

TURRILL, W. B. 1953. Pioneer Plant Geography. Martinus Nijhoff. The Hague.

WADIA, D. N. 1957. Geology of India. London.

WARD, F. KINGDON, 1960. Pilgrimage for Plants. London. (This includes a full list of all WARD's publications).

WATT, G. 1890. Forests of Manipur. *Indian Forester*, 14: 291, 339 & 387.

X. VEGETATION AND PHYTOGEOGRAPHY OF THE HIMALAYA

by

M. A. RAU

1. Introduction

Our knowledge of the vegetation of the Himalaya is derived mainly from the numerous botanical explorations during the past one hundred and fifty years. The observations of members of some of the mountaineering expeditions have also added materially to our knowledge, not only of the prevailing vegetation but also other aspects like climate, topography, soils, glaciers, etc., with which the plant life is closely correlated. In spite of these many adventurous incursions into the inner ranges, there are still numerous gaps in our knowledge and many remote valleys are, as yet, botanically not very well known. It is proposed to give in this chapter a broad survey of the vegetational types found in the various sectors of this vast mountain system and also attempt a discussion of the phytogeographical affinities of its flora.

The Himalayan System, extending over 2400 km nearly east-west and consisting of complex topographical features, naturally presents a wide variety of climate and soils and consequently supports a remarkable assemblage of vegetation types. At the northwestern limit, the mountain reaches a latitude as far north as 35° 50′ N, which is an area of very scanty rainfall compared to the southeastern region lying at 27° N, where the outer ranges receive the full force of the monsoon rains. In its vertical aspect, the mountain system shows a wide altitudinal range between the tropical foothills and the highest limits of alpine vegetation around 6000 m. Between these extremes many kinds of vegetational complexes are seen, depending on the climate, topography and edaphic conditions. Some of these are briefly considered here.

2. The Northwest Himalaya

2.1. KASHMIR

In the extremely arid, northwestern sector, the vegetation in the neighbourhood of the Nanga Parbat has been adequately described by TROLL (1939). A subtropical semi-desert type of vegetation is seen at the foot of the mountain, in which the prominent floristic elements are, *Capparis decidua*, *C. spinosa*, species of *Calotropis*, *Ephedra* and *Pistacia* and the grasses

Cymbopogon, Enneapogon and *Stipa*. On the northern exposition, an *Artemisia*-steppe type of vegetation occurs, in which *Artemisia maritima* is dominant along with the Chenopodiaceæ *Eurotia* and *Kochia*. Here also occur some larger woody elements, like species of *Berberis, Colutea* and *Sophora*. *Rosa webbiana* is occasionally seen. This type of vegetation is seen up to 3000 m on the northern exposition, but on the south exposition and in the Rupan Valley, east of the Nanga Parbat, it may reach a higher altitude, almost up to the alpine zone around 4200 m. A steppe type of forest, including species of *Artemisia, Hippophaë, Juniperus, Lonicera, Salix* and others, is also seen on the southern slopes. In the altitudinal range, 3000–3600 m, where the conditions are favourable, the characteristic west Himalayan conifer forest develops, in which the constituent members are, *Abies pindrow, A. spectabilis, Cedrus deodara, Picea smithiana, Pinus wallichiana* and *Taxus wallichiana* with varying mixture of *Juniperus* spp. In drier localities, *Pinus gerardiana* is seen along with *Juniperus* spp. The distribution of *Pinus gerardiana* is, however, irregular. At higher altitudes, 3800–3900 m and locally up to 4150 m, we find a subalpine forest of *Betula utilis* with *Abies, Juniperus, Pinus wallichiana, Sorbus* and others. In the alpine zone, the exposition has a marked effect. Willows and *Rhododendron anthopogon* are not developed on the south exposition. Here one finds only species of *Juniperus*. The characteristic herbs of the alpine zone include species of *Aconitum, Aquilegia, Aster, Astragalus, Corydalis, Draba, Leontopodium, Lloydia, Potentilla, Primula, Saxifraga, Sibbaldia, Trollius* and others. There are also some alpine rushes, sedges and grasses. In the Deosai Plains towards the Karakoram, where the mean altitude is about 4000 m, an *Artemisia-Tanacetum*-dominated vegetation occurs. Further east, in Ladakh, which lies to the north of the Great Himalayan Range, where also extremely cold and dry conditions prevail, the nature of the terrain neither permits the development of any forest nor does it show the usual zonation of vegetation. There are practically no trees; only in sheltered places and near streams, stunted trees of *Juniperus* are found. The prevailing character of the vegetation is the conspicuous cushion-habit of the shrubby plants, which are thus adapted to withstand the cold dry winds and blizzards. *Caragana pygmaea* is the most prominent plant of this group (KASHYAP, 1925). *Acantholimon lycopodioides*, representing a genus extensively distributed in Middle Asian Highlands, is met with commonly in various parts of Ladakh. *Thylacospermum rupifragum* (Caryophyllaceae) is another shrub, forming hemispherical mounds. It is found in Ladakh, Rupshu and further east in the interior of Tehri-Garhwal Himalaya. There are also species of *Artemisia, Astragalus, Eurotia, Lonicera, Lotus, Oxytropis, Polygonum, Sophora* and others in the region. Woolly herbs of *Saussurea* and the mat-forming, *Hippophaë, Arenaria, Myricaria* and some alpine sedges (*Carex* spp.) are among the other plants seen in Ladakh.

North of Leh, in the Indus Valley, towards Skardu, the same character of vegetation is seen. *Capparis spinosa* is a typical member here along with species of *Artemisia*, *Echinops* and several representatives of the Boraginaceae, Chenopodiaceae and Cruciferae. *Microula tibetica* is a rare Boraginaceous herb seen in Ladakh; interestingly, this species is also seen in north Kumaon at 4800 m.

South of Leh, *Juniperus* spp. are seen at 2700–3000 m. In Rupshu, around the brackish-water lake of Tso Morari, the characteristic plants are again of the alpine steppe, with *Caragana pygmaea* and species of *Artemisia*, *Eurotia*, *Oxytropis*, *Potentilla*, *Stipa* and the larger woody, *Myricaria*. During July-August, appear many alpine herbs like species of *Aster*, *Delphinium*, *Gentiana*, *Leontopodium*, *Primula*, *Sedum* and *Thymus*. One of the very interesting plants occurring in Rupshu and also further east in Hanle is *Glaux maritima* of the Primulaceae. It appears again in the Lonakh Valley of Sikkim. This is a characteristic plant of the coastal and inland salt marshes of the north temperate and arctic regions. *Corydalis crassissima*, most unusual among the fumitories, is known from Ladakh and other localities in Sind and Masjid Valleys, as well as further northwest in Gilgit and Baltistan. It has thick, bluish grey leaves and inflated fruits and generally occurs in stony rubble.

In the Valley of Kashmir, bounded by the Great Himalayan Range on the north and northeast, by the Pir Panjal in the south and southwest and on the west by a prominent spur of the Great Himalayan Range, the forests are dominated by *Pinus wallichiana* and *Abies pindrow*. *Cedrus deodara* is more or less absent on the northern slopes of the Pir Panjal and *Pinus roxburghii* is conspicuous by its absence in the Valley forests, as also *Rhododendron arboreum*, oaks and their usual associates (SHERSINGH, 1929) In some parts of Kashmir, as for example, on the south-facing slopes of Sindh and Lidder Valleys, as well as at the foot of the Pir Panjal on its north face, a characteristic type of vegetation is seen, in which certain shrubs like, *Parrotiopsis jacquemontiana*, *Rosa webbiana*, *Indigofera gerardiana* and species of *Berberis*, *Cotoneaster* and *Viburnum* are prominent. This type of vegetation is designated the Kashmir scrub by TROLL (1939). *Parrotiopsis jacquemontiana* is also prominent in some degraded forests of *Pinus wallichiana*, where the pine has suffered severe damage due to the attack of the minute Loranthaceous parasite, *Arceuthobium minutissimum*.

The Pir Panjal Range, with the highest elevation around 4500 m, shows the altitudinal zonation of vegetation types that is generally seen in other sectors of the Himalaya. The composition of the vegetation is, however, different on the north and south facing slopes. On the mountain slopes facing the plains of Punjab, the foothill zone presents the thorn-scrub type of vegetation, with species of *Acacia* and *Zizyphus* among others. At slightly higher elevations, subtropical, is seen dry evergreen forest, with *Olea cuspidata*, *Dodonea viscosa*, *Pistacia integerrima*, *Punica granatum* and others. *Pinus roxburghii* occurs at altitudes of 900–1700 m, and above this comes the oak-conifer mixed type of species of *Quercus* in association with *Rhododendron arboreum*, along with *Picea smithiana*, *Cedrus deodara* and *Abies pindrow*. Species of *Acer* and *Aesculus* are also found. At 3000 m, *Betula utilis*, often associated with *Abies*, *Juniperus* and *Rhododendron campanulatum*, appears. The higher elevations on the Range present the usual subalpine and alpine elements.

2.2. HIMACHAL PRADESH

The Chenab sector between the R. Chenab and R. Ravi, presents again, at the foot of the mountains, a thornscrub forest of different species of *Acacia* and *Zizyphus*, followed by forests of *Pinus roxburghii*. At higher altitudes, mixed oak-rhododendron forests, with *Acer* and *Aesculus*, are seen. The coniferous forests of the temperate zone include *Pinus wallichiana*, *Picea smithiana*, *Abies pindrow*, *A. spectabilis* and *Cedrus deodara*. *Quercus semecarpifolia* may occur in pure formation and it reaches the timberline in the Dhaula Dhar Range as scattered, stunted trees. The subalpine zone with *Betula utilis* and willows also shows many herbaceous members. The flora further north beyond the Sach Pass (4200 m) is of the characteristic arid, steppe type. In the Brahmaur Valley of Chamba may be seen the oak-conifer mixed forests, with species of *Quercus*, *Cedrus deodara* and *Pinus wallichiana*. On the interior drier expositions, *Pinus gerardiana* occurs in association with *Quercus ilex*. *Betula utilis* with *Rhododendron campanulatum* and *Juniperus* spp. reach up to 4200 m. The alpine herbaceous flora consists of diverse species of *Aconitum*, *Corydalis*, *Delphinium*, *Gentiana*, *Meconopsis*, *Myosotis*, *Primula* and others.

In the Beas Valley in the Kulu Himalaya, in the subtropical zone, are found *Olea cuspidata*, *Punica granatum*, *Zanthoxylum armatum* and others. Forests of *Pinus roxburghii* are also found here. On the drier slopes, the arborescent species of *Euphorbia* may be found. In the temperate zone, mixed conifer forests, with *Abies pindrow* and *A. spectabilis*, show distinct zonation. Blue-pine, spruce and the deodar may also occur in these forests. *Quercus semecarpifolia* is generally found with these conifers.

Approaching the Rohtang Pass (4000 m) from the south, we find, *Betula utilis* with *Rhododendron campanulatum* in the subalpine zone and along the Pass itself, many colourful alpine herbs appear during the summer months. *Primula rosea*, *Saxifraga flagellaris* and *Lagotis cashmeriana* are particularly prominent among a host of others. Later in the year, extensive patches of the colourful *Aconitum violaceum* attract attention.

Beyond the Rohtang Pass lies the district of Lahul and Spiti. The flora in the interior of this district is of the steppe type. Several species of *Astragalus* and *Oxytropis*, along with species of *Artemisia*, *Caragana*, *Ephedra* and the stunted *Hippophaë* (*H. rhamnoides* ssp. *turkestanica* and *H. tibetana*) are found. Rigid, mat-forming species of *Arenaria* are also common. The extreme cushion-form is seen in some spinescent clumps of *Astragalus strobiliferus* and *Arenaria perlevis*. Along the R. Chandra in the Lahul Valley, on some dry slopes, excellent populations of the stately *Eremurus himalaicus* are found. Another characteristic herb on the drier slopes is the spinescent, yellow-flowered, *Morina coulteriana*. It is of considerable biogeographical importance to remark that the genera *Eremurus* and *Morina* are very widely distributed and represented by several species in Middle Asia. The dry slopes in Lahul are also clothed with the shrubby junipers, *Berberis jaeschkeana* and the thistle *Cousinia thomsonii*.

Many temperate and subalpine herbs appear along glacial streams, the most characteristic among them in Lahul being *Caltha palustris* and *Pedicularis punctata*. *Iris kumaonensis* is also abundant on the slopes. One very noticeable feature in the Chandra Valley of Lahul is the profuse development of the blue-flowered Boraginaceous herbs, the genera represented being, *Arnebia*, *Cynoglossum*, *Eritrichium*, *Lappula*, *Lindelofia*, *Myosotis* and others.

The Sutlej Valley system forms another important sector, where excellent coniferous forests are found. *Cedrus deodara* forms pure forests and occasionally, as for example, north of Simla, it is mixed with *Picea smithiana*, *Pinus wallichiana*, *Abies pindrow* and some species of *Quercus*. The dry, sunny spurs are covered with forests of *Pinus roxburghii*, which may reach an altitude of 2100 m in the region. On the dry, southern slopes, on approaching Simla from the plains we find, *Euphorbia royleana* as a conspicuous feature of the landscape. On some northern slopes, mixed forests of evergreens consisting of *Euonymus*, *Ilex*, *Litsea*, *Machilus* and others, with bushes of *Lonicera*, *Rhamnus* and *Viburnum*, are found. On limestone outcrops are frequently seen pure forests of *Cupressus torulosa*.

In the interior of the Sutlej Valley, in the territory lying between Spiti and the Tehri-Garhwal Himalaya, extensive forests are found in the Bushahr Division. The upper Bushahr forests lie entirely in the Sutlej basin. The character and composition of the forests vary greatly, depending on the topography, rainfall and altitude. The effect of monsoon is progressively reduced as one proceeds towards the inner passes. In the monsoon zone, along the right bank, a great belt of *Quercus semecarpifolia* occurs just below the open alpine pasture lands. Below the oaks, there is a middle zone of *Pinus wallichiana* and scattered *Cedrus deodara*, with *Pinus roxburghii* appearing lower down. On the left bank of the river, in cooler situations, the higher forests are of *Abies pindrow* and *Picea smithiana*, with *Cedrus deodara* and *Pinus wallichiana* occurring below. In the lower forest belt, *Pinus roxburghii* fades out towards the inner ranges, where its place is taken by *Pinus gerardiana*. This pine is associated with *Quercus ilex*. With the decrease in rainfall towards the interior, the blue-pine, spruce and fir also diminish and only the deodar is left as the sole survivor among the large conifers (GORRIE, 1929). Further interior, close to the Tibetan border, extremely arid conditions prevail and the *Artemisia*-steppe type of vegetation is alone seen on the hill slopes. Along with *Artemisia* are also seen species of *Ephedra* (*E. intermedia* var. *tibetica*), *Capparis*, *Caragana*,

Colutea and others. In the alpine zone, the grassy slopes present a number of characteristic herbs like representative species of *Androsace, Anemone, Gentiana, Pedicularis, Saussurea, Saxifraga, Sedum* and others, as well as some dwarf willows and *Astragalus*.

3. *The Western Himalaya*

The sector of the Himalaya between the Sutlej and the Kali Valleys, bordering Nepal, includes the river systems of Tons, Yamuna, Bhagirathi, Alaknanda and Gori and the major peaks of Bandar Punch, Kamet, Nanda Devi, Nilkanth, Trishul, Panch Chuli and others. This area has received considerable attention from botanical investigators. This sector, which is also known as the Kumaon Himalaya, comprises the Tehri-Garhwal, Garhwal and Kumaon Divisions. The famed shrines of Jamnotri, Gangotri, Kedarnath and Badrinath are located in this part of the Himalaya. STRACHEY and WINTERBOTTOM made extensive collections of plants in this region, during the middle of the nineteenth century.

The submontane region in the sector is predominantly forested with *Shorea robusta*. *Shorea* forests are seen on the Siwaliks and on the slopes of the lesser Himalaya up to an altitude of 1000 m. Some fresh-water valley type of swamp-forests also occur in the submontane region, with *Bischofia javanica, Salix tetrasperma* and *Pyrus pashia* as the chief components, with occasional cane brakes. *Carallia brachiata* is an interesting member of such swamp-forests in Dehra Dun. In the submontane region are also seen mixed deciduous forests, populated predominantly with *Lagerstroemia parviflora, Dalbergia sissoo, Anogeissus latifolia, Terminalia* spp., and others.

On the outer ranges, *Pinus roxburghii* forms excellent forests at altitudes above 1200 m, often reaching an altitude of more than 2000 m on some spurs. These pine forests are, however, restricted to exposed, dry situations, southfacing slopes, crests of spurs and well drained areas. At its upper limit, the pine may occur in association with *Quercus incana* and *Rhododendron arboreum*. The oak-rhododendron forests are developed at altitudes above 1500 m, where there is sufficient soil moisture. These forests are particularly well developed in cooler habitats, northern exposures and on sheltered slopes. *Lyonia ovalifolia* is an invariable associate in such forests. These forests are densely populated and rich in epiphytes, especially a large variety of orchids, ferns and aroids. In some exceedingly favourable habitats, as for example, in the Mandakini Valley of north Garhwal, the growth of epiphytic ferns is most luxuriant and one finds such interesting plants like *Botrychium virginianum*, growing vigorously on tree bark, along with ferns. In these forests, lianaceous climbers, especially. *Holboellia latifolia, Schizandra grandiflora* and *Vitis* spp., are common. During the rainy season, the development of several genera of the Gesneriaceae, *Chirita, Corallodiscus, Didymocarpus, Lysionotus* and *Platystemma* with their colourful flowers, is a conspicuous feature in some of the valleys. At

252

Plate 23. A fresh-water swamp-forest in the sub-montane region with *Bischofia javanica.*

Plate 24. *Shorea robusta* forest at the foot of the outer ranges in West Himalaya (Photo by G. N. MADHWAL).

Plate 25. Pinus roxburghii in the Western Himalaya (1700 m) (Photo by M. A. RAU).

slightly higher elevations, 2200–2800 m, *Quercus incana* is sometimes replaced by *Q. floribunda*, which has darker foliage and a dense canopy, giving a characteristic physiognomy to the forests. This type of forest occupies an intermediate range between *Q. incana* and *Q. semecarpifolia* forests and

254

is generally found on the higher hills, south of the great snowy peaks. The associates of *Q. floribunda* are the same as those seen in *Q. incana* forests. Clouds hang on these dense forests for a longer time during the monsoon and this results in a more luxuriant development of the epiphytic and ground flora. At altitudes above 2800 m, another characteristic oak-conifer forest occurs, in which the dominant elements are *Quercus semecarpifolia*, *Abies pindrow* along with *Rhododendron arboreum*, *Taxus wallichiana*, *Euonymus*, *Pyrus*, *Viburnum* and others. Forests of *Cedrus deodara* also occur in the altitude range, 2000–3000 m. These are, however, restricted to the inner, drier regions. The deodar forests often have admixture of other species. On the limestone spurs of the outer ranges are also found pure forests of *Cupressus torulosa*.

Betula utilis and *Abies spectabilis*, either pure or mixed in various proportions, constitute the highest forests in Tehri-Garhwal as in west Himalaya in general, extending from 3000 to 4000 m. alt. In many localities, the *Quercus semecarpifolia-Abies pindrow* forests, lying south of the snows, as well as *Cedrus deodara* forests north of them, merge into the above type of forest. Excellent birch forests are found on the way to Gangotri. Near Jumnotri, a typical *Betula-Abies* forest can be seen. At higher altitudes, the birch is associated with *Rhododendron campanulatum*. In such forests, species of *Ribes*, *Salix* and *Sorbus* are frequently seen. In the undergrowth, species of *Trillium*, *Smilacina* and *Clintonia* are conspicuous.

Above the tree limit, only scattered bushes of *Juniperus communis* and *J. wallichiana* occur along with other low shrubs like species of *Cotoneaster*, *Salix*, on the south-exposed sites in the larger valleys. The subalpine forests of *Betula-Rhododendron campanulatum* as well as the shrubby, *Rhododendron anthopogon*, *R. lepidotum* and species of *Cotoneaster*, *Lonicera*, *Ribes*, *Rosa* and others are better developed on the north-exposed slopes. In moist localities, luxuriant meadows of herbaceous perennials are seen with the usual components representing the genera *Anemone*, *Epilobium*, *Geranium*, *Gentiana*, *Polygonum* and others. Along streams and water courses, are particularly to be seen species of *Aletris*, *Caltha*, *Pedicularis*, *Polygonum*, *Potentilla*, *Ranunculus*, etc.

At still higher altitudes, till the upper limit of vegetation is reached, the high-alpine flora is seen. The conditions being extremely difficult, the distribution and luxurance of the vegetation are dependent on favourable location as well as on the availability of adequate moisture. Scattered bushes of *Rhododendron anthopogon* and *Berberis* spp., are found up to 4800 m alt. *Cassiope fastigiata* forms heath-like clumps and on exposed rock faces, many of the herbs present a rosette or cushion habit. Among such herbs are *Paraquilegia anemonoides* and several species of *Androsace*, *Saxifraga* and *Sedum*. *Bergenia stracheyi*, with large fleshy leaves which turn red during autumn season, is conspicuous amidst rocks in this zone.

Many interesting herbaceous members, among them some endemics are found at the extreme limits of vegetation. The curious woolly species

Plate 26. Quercus incana (in a degraded stage due to biotic interference) on limestone slope.

Plate 27. Quercus floribunda forest with a dense canopy in the cloud-forest zone (Photo by M. A. RAU).

256

Plate 28. The Himalayan birch, *Betula utilis* in north Garhwal Himalaya at 3300 m (Photo by M. A. Rau).

of *Saussurea* are widely distributed on the morainic slopes. In many localities in north Tehri-Garhwal and Garhwal extremely dry conditions prevail and, under such conditions, it is not uncommon to find the

257

Plate 29. Corydalis crassissima in Kashmir (3600 m) (Photo by G. N. MADHWAL).

Plate 30. An alpine scrub (Photo by G. N. MADHWAL).

258

Plate 31. Saussurea obvallata in Garhwal (Photo by M. A. Rau).

Plate 32. Acantholimon lycopodioides (Photo by M. A. Rau).

259

Plate 33. Saussurea gossypiphora in Garhwal (Photo by M. A. RAU).

prevalent plants restricted to species of *Caragana*, *Ephedra* (*E. gerardiana* and *E. saxatilis*), *Berberis* and *Juniperus*. *Thylacospermum rupifragum*, which occurs in Ladakh, is found here also. The vegetation of the arid tracts in this sector is similar to that seen in Northwest Himalaya and the Trans-Himalayan territory.

The richness of the herbaceous alpine flora in some of the interior valleys of this sector is visibly demonstrated in the Bhyundhar Valley of north Garhwal, which has acquired fame as the Valley of Flowers (SMYTHE, 1932, 1938). This Valley lies in an area of some excellent Himalayan landscape, with the well known mountain peaks of Ganesh Parbat, Hathi Parbat, Kamet, Rataban and others located in the neighbourhood. It may be of interest to record here that the highest altitude known for a flowering plant in western Himalaya is 6300 m on Mt. Kamet, where a specimen of *Christolea himalayensis* was gathered by Gurdial Singh.

Along the Gori Valley in eastern Kumaon some excellent oak-rhododendron forests are found. A rich assemblage of epiphytic orchids is seen here and many of the Eastern Himalayan orchid species reach their extreme western limit of distribution here.

4. The Central Himalaya

The Central or the Nepal Himalaya may be conveniently divided into the western, middle and eastern sectors. In west Nepal, i.e., the territory to the east of R. Kali and including the Karnali Gandaki and its elaborate river system, the vegetational zonation is similar to that seen towards the west of Kali in eastern Kumaon. *Cedrus deodara* reaches its easternmost distribution here. There exist good deodar forests at the head of the Karnali Valley (COLLIER, 1924). This sector has been explored in recent years by POLUNIN (1950) and WILLIAMS (1953), particularly around Jumla. *Rhododendron barbatum*, which occurs in the eastern Kumaon, is also seen here. Another interesting species, *R. lowndesii*, has been described from this region. The lower reaches of the Valley are very hot and except for occasional patches of subtropical forest and thickets of palms the vegetation is xerophytic and entire hill sides are in some places covered by *Euphorbia royleana*. On the ridges of the outer ranges, forests of *Quercus incana* and *Rhododendron arboreum* are seen. In the interior, the hill sides with a northerly aspect are forested with *Pinus wallichiana* and some *Cedrus deodara*. *Picea smithiana* and *Tsuga dumosa* are the other conifer forest members. *Cupressus torulosa* is found at 2500–2800 m altitude.

On the rather dry hill sides around Jumla, *Stellera chamaejasme* is found in abundance. In the inner ranges, at high altitudes, stunted bushes of the Nepal juniper, *Juniperus squamata*, *Caragana versicolor* and *Lonicera rupicola* occur and at still higher altitudes, the vegetation consists only of characteristic cushion and scree plants like dwarf willows, *Potentilla*, *Saxifraga* and *Androsace*. This sector extends to 83° EL, which is practically the limit of the Western Himalaya. WILLIAMS (1953) recorded the occurrence of *Hydrobryum griffithii* (Podostemaceae) in this sector, a most unusual occurrence in this part of the Himalaya.

The flora of Nepal, in the region lying between longitudes 83° 30' and 85° 10', which includes the valley systems of Marsyandi, Kali Gandaki and Buri Gandaki and the mountain massifs of Manaslu, Ganesh Himal and Thaple Himal, has been recently studied by the Japanese botanists (NAKAO, 1955). At the foot of the mountains, the *Shorea*-zone is seen up to 1000 m. There are also *Pandanus, Phoenix, Ficus* spp. and *Bombax*. Above this *Shorea*-belt comes a zone of *Castanopsis* forest. *Castanopsis indica* is well developed and it is ecologically associated with *Lithocarpus* and *Engelhardtia*. The *Castanopsis* forests extend up to 2100 m. In the altitude range, 1300–2500 m are also seen the mixed evergreen forests, which possess abundant epiphytic orchids, ferns and aroids. Tree rhododendrons begin at 1500 m and reach up to the cold temperate zone amidst *Abies* at 3300 m. In the mixed evergreen forests are generally seen some Lauraceous species, oaks, *Alnus, Myrica, Prunus, Pyrus* and *Symplocos*. In lower ranges of this mixed forest are seen scattered trees of *Schima wallichii* and in the upper belt, *Quercus semecarpifolia*.

Plate 34. Euphorbia royleana on the dry, south-facing slope (Photo by M. A. RAU).

Plate 35. A spruce-fir forest in Kulu Himalaya (Photo by M. A. RAU).

Plate 36. Profuse growth of epiphytic orchids and ferns on an oak (Photo by M. A. Rau).

In the zone of evergreen oak forests, several species of oaks, viz., *Quercus semecarpifolia, Q. glauca, Q. lamellosa* and *Q. floribunda* are seen. The coniferous forests grow poorly on the southern side. A broad belt of oak forest is seen along the lower limit of the conifer forest. The relative width of these forests is reversed on the northern side (Nakao, 1955). In some places, pure *Rhododendron arboreum* is developed, but sometimes mixed with *Magnolia campbellii*. On the northern side of the Great Himalaya, a forest of *Tsuga* and *Picea* is well developed. *Tsuga dumosa* is the dominant species, with *Picea smithiana* and *Taxus wallichiana* as associates. Species of *Acer* also occur in such forests. There are also open forests of *Pinus* and *Juniperus* at 3000–3800 m in the arid lands on the northern flanks of Dhaulagiri and Annapurna Himal, composed mostly of *Pinus wallichiana* and partly tree *Juniperus*, with the shrubby *J. wallichiana* and *J. communis* as undergrowth.

The upper part of the conifer forest has *Abies* associated with *Betula utilis*. At the higher limit, *Betula* may form a pure forest. The Himalayan larch, *Larix griffithiana*, the only deciduous conifer of the Himalaya, is a prominent member in this zone. The larch appears only east of the Buri Gandaki. *Rhododendron barbatum* is rich in density, with species of *Prunus*,

263

Sorbus, Spiroea and *Viburnum* and the herbaceous, *Androsace, Gentiana, Pedicularis, Potentilla, Primula* and *Saxifraga* spp.

In the alpine zone, the lower part has bushes of *Rhododendron anthopogon* and *R. setosum* with *Juniperus squamata*. Continuous plant associations diminish at 4500–6000 m and only the colourful alpine herbs are seen in favourable situations during the summer months. In the interior, at very high altitudes, bushes of *Caragana, Berberis, Artemisia, Lonicera, Spiraea* and others are seen in arid regions. In the grassy alpine zone on the northern flank of the Annapurna Himal, at 4000–4500 m, NAKAO (1955) has described a unique association of grass with a single species, *Helictotrichon virescens* dominating. The conditions here are similar to those of Middle Asian Highlands.

North of Kathmandu, there are forests of *Pinus roxburghii*. In lowlands, there are giant tropical bamboos, species of *Dendrocalamus* and *Arundinaria*, along with *Euphorbia royleana*. *Shorea robusta* occurs in mixed stands with *Pinus roxburghii*. The subtropical landscape with *Engelhardtia*, scattered trees of *Ficus* and dendroid *Euphorbia* along with the exotic *Agave* and giant bamboos is seen up to altitude of 1470 m on the southern slopes (NUMATA 1967). Passing over 2000 m in altitude, *Rhododendron arboreum* is met with and evergreen *Castanopsis-Quercus* forests are also seen. *Abies spectabilis* and *Tsuga dumosa* forests are seen above 2500 m with *Abies* occupying higher ranges, often in pure formation, up to 3900 m. In these forests are also found *Rhododendron* and *Arundinaria*. Secondary forests of *Pinus wallichiana* also occur in this zone. The blue-pine stops to grow above 2800 m and *Abies spectabilis* grows very well above this altitude. *Rhododendron barbatum* is an important member of the shrubby layer in such forests. *Juniperus recurva* along with *Rhododendron anthopogon* and *R. nivale* are predominant above the forest zone at 3900 m. In the alpine zone are seen many cushion plants of the genera, *Androsace, Arenaria, Draba, Saxifraga*, etc. In the Mt. Everest region, some species of *Leontopodium, Sedum* and *Arenaria perlevis* have been recorded from altitudes of 6000 m.

5. The Eastern Himalaya

In the Eastern Himalaya, the region lying between 87° 15' and 89° EL and 26° 30' and 27° 45' NL and including the Singalila Range as well as the territory between the R. Tamur and Teesta in Darjeeling and Sikkim, have been recently explored by Japanese botanists. KANAI (1966) has described the forest types of this region. The northern fringe of the Gangetic Plain and the Himalayan foothills up to an altitude of 700–900 m are covered with a tropical rain-green deciduous forest, dominated by *Shorea robusta, Adina, Dalbergia, Dillenia, Bauhinia, Anogeissus, Litsea, Phoebe, Lagerstroemia, Terminalia* and others. This region has the highest rainfall, as the monsoon strikes in full force the outer ranges. Along the foothills the rainfall may be as high as 5000 mm in the year. Depending on the

264

extent of rainfall, the forests show evergreen or deciduous composition. *Shorea* does not occur in regions with very heavy rainfall, where only *Eugenia*-Lauraceous forests occur. *Shorea-Terminalia, Shorea-Stereospermum-Garuga* and *Schima-Bauhinia* are the other forest types seen in the region. At elevations of 1500–1700 m, mixed broad-leaved forests occur in which the dominant species are *Castanopsis indica* and *Schima wallichii*. Other trees including species of *Ficus, Engelhardtia, Euonymus, Michelia* and *Quercus* also occur. The upper part of this forest overlaps the lower part of the next evergreen oak zone between 1700 and 2000 m. On steep rocky, south-faced slopes, a sparse forest of *Quercus incana* is present. The evergreen oak forests occur at 2500 to 2800 m. In these forests, the oaks are associated with various Lauraceous species, as well as with *Rhododendron arboreum, Lyonia ovalifolia*, species of *Acer, Symplocos* and others. *Castanopsis* predominates up to an elevation of 2000–2300 m and *Quercus* above it to 2500–2800 m. Mixed rhododendron-conifer forests are found above 2500 m and such forests reach the timberline. The chief *Rhododendron* species in the lower belt are, *R. arboreum* var. *campbellii, R. barbatum, R. grande, R. thomsonii*, whereas in the higher zone, *R. campanulatum, R. lanatum* and *R. wightii* dominate up to the timber-line. *R. anthopogon, R. elaeagnoides* and *R. setosum* are found in open places, especially on the margin of the timber-line. *Betula utilis, Gaultheria* and *Sorbus* spp. are seen throughout this *Rhododendron* zone.

In the evergreen and *Rhododendron* forest zones, occasionally, a temperate deciduous forest consisting of *Acer, Betula, Prunus, Schefflera, Sorbus, Viburnum* and others is seen to occur on the northern and eastern slopes. *Abies spectabilis* and *Tsuga dumosa* are the only two species composing the conifer forests, the former occupying the higher zone. Rhododendrons generally occur in the second storey of such forests. The *Rhododendron* forest marks the timber line at a height of about 4000 m and the region up to 5000 m is occupied by alpine meadows of prostrate *Juniperus squamata, Androsace, Arenaria, Cassiope, Saussurea* and others. *Rhododendron* forest is abundant at the top of Mt. Singalila (KANAI, 1966).

A *Daphniphyllum*-forest with *D. himalayense* associated with *Rhododendron grande, R. barbatum* and *R. arboreum* var. *campbelliae* has been described by the Japanese botanists as occurring on the top of the Bhandukay Bhajang at 3100 m. The Japanese botanists, who visited this area in 1963, also made a remarkable discovery on Bhandukay Bhajang of the vesselless Dicotyledon, *Tetracentron sinense*. This find, the first in the Himalaya, appeared to be a new variety and as such has been described as var. *himalense* by HARA and KANAI (1966).

At Siling Tzokupa (3800 m), a purely grass community, with the components *Arundinella birmanica, Calamogrostis emodensis, Cymbopogon stracheyi, Danthonia* sp. and *Helictotrichon virescens*, occurs on the south-faced, steep rocky slope.

5.1. Sikkim

The botany of Sikkim is very well known. Hooker's classical account of his travels in Sikkim (Hooker 1891) provides an excellent picture of the main aspects of the vegetation of the territory. In eastern Sikkim, the area around the ridges lying between the Cho-La and Tankha-La, is one of the wettest in the Himalaya, being exposed to heavy monsoon rains. The moist tropical and subtropical forests in the altitude range, 600–1500 m, comprise of the tree species belonging to *Schima*, *Eugenia*, *Duabanga*, *Engelhardtia*, *Castanopsis* and others with numerous climbers and epiphytes. In the temperate zone are seen oak and conifer forests. As everywhere, in temperate and alpine Sikkim, there are extensive forests of *Rhododendron*. The Magnolias are very prominent in the temperate zone as are the oaks, laurels, maples, birches and alder. The conifers are chiefly confined to a belt lying between 2700 and 3600 m elevation. The most prominent among the conifers is *Abies spectabilis*, which is also the most gregarious. *Larix griffithiana*, *Tsuga dumosa* and the Sikkim spruce, *Picea spinulosa*, along with *Taxus wallichiana*, are the conifers seen in this zone. The dwarf junipers ascend high in the alpine zone. The rhododendrons form a great part of the forest between 2500 and 3600 m, showing a gradual change from the tree habit at lower elevations to a bushy habit at 3600 m. In the alpine zone, they form a heath of prostrate forms less than 0.6 m tall. In the alpine zone, *Meconopsis* is plentiful. Straggling *Arenaria*, *Stellaria* and *Mandragora* are common. *Iris* and *Lloydia* are seen everywhere.

The vegetation of northern Sikkim was described by Smith (1913) and others, besides the classical account of Hooker. The forest above Song is noted for its orchids and in the hot valleys, the Gesneriaceae are prominent. In many localities, *Rhododendron hodgsonii* presents a very dense growth. On entering the valley at Zemu, one observes that the slopes are steep and the valley itself appears dark and thickly wooded: a mixed forest, with the coniferous genera *Abies*, *Picea*, *Larix*, *Tsuga* and *Juniperus*, in which are also found numerous species of *Berberis*, *Euonymus*, *Ilex*, *Pyrus*, *Ribes*, *Rubus*, *Spiraea*, *Viburnum* and others. The rhododendrons and conifers tend to prevail, and at altitudes of 3000 to 3300 m the forest is chiefly composed of them. Above this altitude, the character of the vegetation in the Zemu Valley begins to change. The tall rhododendrons disappear and only some species of intermediate size, like *R. campanulatum* and *R. wightii*, are seen. The conifers also begin to thin off, but herbaceous species become more common. At altitudes of 3300 to 4300 m, up to the base of the Zemu Glacier, small shrubs prevail. The floor of the upper valley is covered with the straggling *Berberis*, *Cotoneaster*, *Lonicera*, *Pyrus*, *Ribes* and rhododendrons among the shrubs, and numerous alpine herbs representing the genera *Angelica*, *Astragalus*, *Corydalis*, *Epilobium*, *Heracleum*, *Potentilla*, *Primula*, *Rheum*, *Saussurea*, *Sedum* and many others. At still higher altitudes of 4300 to 5400 m, the prevailing genera are *Androsace*,

Anemone, Carex, Cassiope, Cortia, Corydalis, Diapensia, Diplarche, Gentiana, Juncus, Lagotis, Pedicularis, Picrorhiza, Poa, Primula and others.

The upper valley, protected by the giant ridge towards Kanchenjunga, is comparatively dry and the northern slopes present a greater variety in its flora than the southern slopes. The valley as a whole shows a transition between the vegetation types of the wet forest of the lower valley and the dry, bleak slopes to the north.

In the Llonakh Valley, the vegetation of the open flats is sparse, trees and shrubs almost disappear here and instead dwarf, gnarled junipers are common. *Rhododendron anthopogon* and *R. lepidotum* may occur along with species of *Berberis, Lonicera, Potentilla arbuscula* and others. Creeping willows and *Hippophaë* are also met with. The mat-forming, rigid *Arenaria* is the most striking feature. In moist localities, herbs representing the genera *Caltha, Pedicularis, Primula, Ranunculus, Saxifraga* and others are seen whereas, on drier flats, the prevailing genera are *Arabis, Astragalus, Delphinium, Guldenstaedtia, Lepidium, Stellaria* and others. On the screes, sheltered from wind by huge boulders, *Anemone, Callianthemum, Draba, Polygonum, Saxifraga, Sedum* and others occur, while on the higher cliffs, *Meconopsis horridula, Braya, Draba, Cochlearia, Potentilla, Primula*, woolly *Saussurea, Thlaspi* and others are found. *Glaux maritima* has again been collected from this part of the Himalaya.

The Llonakh climate and vegetation have greater affinity to the northern Tibetan highlands than to the rest of Sikkim (SMITH and CAVE, 1911).

5.2. BHUTAN

Bhutan was for a long time botanically a terra incognita. In recent years, however, many botanical explorations have been carried out and the main features of the vegetation, as well as the floristic composition, are now fairly known. The territory includes many river valleys, the most important of them being the Manas Valley. The mountain ranges stretch mostly north-south. The flora of western Bhutan is very similar to that of the adjoining Sikkim and the Chumbi Valley. In general it may be said that Bhutan forms a transition between the floras of Sikkim and west China (COOPER, 1942).

As in the neighbouring sectors, the outer spurs receive very heavy rainfall and are as a consequence densely forested with tropical and sub-tropical genera. The density of the vegetation in the valleys, however, depends on the topography and the amount of rainfall received. The moisture-laden winds reach the interior of the valleys, which are disposed in a north-south direction and such valleys bear a luxuriant vegetation. The laterally disposed valleys, which receive comparatively less rainfall, have bare slopes and support moist deciduous forests. *Shorea* and *Pinus* may or may not be present. *Shorea*, when present, dominates the vegetation and occurs along with *Bauhinia, Careya arborea, Dillenia, Melastoma, Syzygium* and others.

In the tropical evergreen forests along the river banks, the dominant genera are *Dillenia, Duabanga, Hydnocarpus* and *Talauma*, with numerous climbers, epiphytes and orchids. *Hodgsonia* is a most conspicuous climber in these forests. In the *Quercus-Rhododendron-Schima* forests, the dominant oak is *Quercus griffithii* associated with *Castanopsis* and *Engelhardtia. Pinus roxburghii* reaches a lower limit at about 600 m. On the dry slopes of lower valleys, *Euphorbia royleana* is commonly met with.

In the interior, wherever there is sufficient moisture, a temperate flora, in which *Rhododendron, Acer* and *Betula* are prominent, is developed. A conspicuous feature of such forests is the profuse development of mosses and lichens on the tree bark. Terrestrial and epiphytic orchids are also abundant in the *Quercus-Rhododendron* forests. In some ranges, the forest is dominated by magnolias, oaks, *Strobilanthes* and others. The climatic conditions are somewhat different in eastern Bhutan and the region lying north of the Khasi Hills is screened from the full effect of the monsoon winds up to an altitude of 2000 m, with the result that the lower ranges present a much drier aspect.

In the inner valleys are also found the characteristic eastern Himalayan, temperate conifer forests, consisting of *Abies spectabilis (A. densa), Larix griffithiana, Picea spinulosa* and *Tsuga dumosa. Larix griffithiana* may occur in association with *Pinus wallichiana* or may form a pure forest above it with an undergrowth of rhododendrons.

In the alpine zone, the vegetation varies between the moist alpine type on the south and the dry steppe type on the northern aspects. *Primula sikkimensis* is abundant in the moist alpine zone. An unusual find in the alpine zone of Bhutan is the remarkable *Lobelia*, which was collected by COOPER (1942) in a remote area at an elevation of 4000 m.

5.3. ASSAM AND NORTH-EAST FRONTIER AGENCY

This sector of the Himalaya includes the Abor, Dafla, Mikir, and Mishmi Hills and the valleys belonging to the rivers Subansiri, Dihang (Siang) and Dibang (Sikang). In the tropical zone, at elevations of about 900 m, are seen the evergreen and semi-evergreen forests, in which species of *Ficus, Sterculia, Syzygium, Terminalia*, along with *Duabanga grandiflora*, are the main tree elements. The palm *Caryota* and the screw-pine *Pandanus* may also occur in forests of this zone. In the subtropical zone are found mixed forests of *Ficus-Castanopsis Callicarpa* on the lower ridges and *Schima-Castanopsis-Engelhardtia-Saurauja* association on the higher ridges. *Rhododendron-Lyonia* forests occur on the drier aspects of hills, and in the deep river valleys *Albizia, Morus* and some bamboos are seen.

In the temperate region, mixed forests of *Acer, Betula, Juglans, Magnolia, Michelia, Quercus, Rhododendron* and others with bamboos characterize the hill tops and valleys. At higher altitudes, the temperate forests have a different composition, with the dominance attained by *Rhododendron*, *Pyrus* and *Tsuga* among the trees. The temperate conifer forests are mostly of *Pinus wallichiana*, associated with *Rhododendron*, *Quercus* spp. and *Lyonia*. Epiphytes are very conspicuous in this zone, among which the orchids deserve particular mention.

The subalpine vegetation includes the tall trees of *Abies spectabilis* (*A. densa*), with shrubby and bushy rhododendrons, junipers, *Berberis*, *Cotoneaster*, *Salix* and others. The herbaceous elements particularly find this a very congenial area for their development and are represented by many species of *Aconitum*, *Anemone*, *Pedicularis*, *Potentilla* and *Primula*. The subalpine zone merges into the alpine zone, where at altitudes above 4500 m and up to 5500 m, an association of dwarfed shrubs like, *Rhododendron anthopogon*, *R. nivale*, *Ephedra* and others is seen. Cushion-forming herbs like *Arenaria*, *Saxifraga*, and *Sedum* are also seen, as well as the woolly *Saussurea* and the large-leaved *Rheum*. At the source of the R. Subansiri, KINGDON-WARD (1960) saw the greatest multitude of *Primula* he had ever seen or imagined.

The Abor Hills, representing the territory between the R. Subansiri and Dihang, have received the careful attention of explorers like BURKILL (1924–1925). The outer Abor Hills are uniformly humid and seasonally, very wet. In the lower belt up to 1100 m, a *Terminalia* forest, with big climbers of *Mezoneurum*, *Vitis* and others, is seen. Forests of the lower slopes are very conspicuous by the large leaves of *Musa*. *Pandanus* and *Alsophila* are also seen in the lower valleys. On the northern slopes of outer hills, pure forests of *Vatica* are found. Pines are, however, absent in Aborland. There are several species of *Quercus* and *Castanopsis*, as also *Betula* and *Engelhardtia*. Excessive moisture of Aborland favours the growth of epiphytic ferns and mosses, *Quercus*, in particular, harbouring a number of them. At the top of the hills, *Rhododendron*, *Vaccinium*, *Daphne*, *Eonymus* and others are found.

A most interesting find in the Abor Hills is *Psilotum nudum*; collected also in the Siang Frontier Division of the Northeastern-Frontier Agency.

6. Phytogeographical Affinities

The foregoing account gives a brief survey of the types of vegetation and their main floristic composition along the length of the Himalaya. It is obvious that there is a wide difference between the climatic conditions of the extreme northwest and the extreme southeast of the vast mountain system. The dry conditions prevailing towards the western ranges, particularly in the interior, have favoured the influx of elements from the western and Middle Asian mountains. The *Artemisia*-dominated steppes of the extreme northwest include many genera, which have a wide distribution in the Middle Asian Highlands. Among them are, particularly, the

269

Chenopodiaceae like *Axyris, Eurotia, Kochia*. Other characteristic Middle Asian plants like, *Juniperus semiglobosa* and *J. turkestanica*, have been recorded from the Nanga Parbat region. Several species of *Astragalus, Hippophaë rhamnoides* ssp. *turkestanica* and *Acantholimon lycopodioides* may also be mentioned in this regard. *Eremurus, Ferula* and *Prangos*, characteristic genera of Middle Asia, have their representatives in this sector of the Himalaya. The Himalayan cedar, *Cedrus deodara*, with a distribution extending from Afghanistan in the west, reaches its easternmost station in west Nepal at 82°50′ EL. This may also be taken as the limit of the West Himalayan botanical province. HOOKER (1906), in his Sketch of the Flora of British India recognized the Western and Eastern Himalaya as two distinct botanical provinces, their eastern and western limits corresponding to the borders of Nepal. The flora of Nepal was very little known in his days. In recent decades, several botanical expeditions have adequately explored Nepal and further east and a more reliable and realistic assessment of the phytogeographical affinities of the Himalayan Flora is now possible. Although KITAMURA (1955) is of the opinion that there is no abrupt change between the west and east Himalaya, the Western Himalayan botanical province is now considered to extend to 83°–84° EL (STEARN, 1960). *According to STEARN (1960), the zone of transition between the western and eastern provinces lies in the area between 80° and 84° EL, where climatic factors 'presumably limit the capacity of the plants suited to one provenance to compete with those of the other'. The species of west and Middle Asian mountains, suited to comparatively dry conditions, extend along the upper region of the Himalaya from Afghanistan to west Nepal. Species originating from the moist areas of high mountains of western China extend as far west in eastern Kumaon bordering Nepal. Based on a close study of the distributional pattern of many high Himalayan plants, STEARN (1960) has provisionally recognized ten important types of ranges of high mountain and alpine Himalayan species. We may consider briefly some of these here. Some species of Western, Middle and Northern Asia are seen in the extreme Western Himalaya. They show varying extent of penetration, some reaching only Kashmir and the others extending the length of western Himalaya. *Thylacospermum rupifragum, Lamium rhomboideum* and *Physochlaina praealta* are among such species. The next type of distributional pattern includes species, which are confined to Northwestern Himalaya, extending in some cases, to Afghanistan in the west. As has already been pointed out, *Cedrus deodara* shows this type of distributional pattern. *Paeonia emodi* is also confined to Northwestern Himalaya, as also *Christolea himalayensis*, several species of *Epilobium* (RAVEN, 1962), *Primula floribunda* and *P. rosea*. There are then the species of western China, which are distributed all along the Himalaya, reaching Kashmir on the west. *Aletris pauciflora, Anemone rupicola* and

---

* This transition in case of animals lies about 78°EL; see Chapter XXI – M. S. MANI.

A. vitifolia are among them. Others originating in western China extend from Yunnan along the Eastern Himalaya, some even reaching Kumaon in the Western Himalaya. *Primula sikkimensis, Magnolia campbellii* are examples for this type of distributional pattern. Some species are confined only to Eastern Himalaya. *Circaeaster agrestis* may be cited as an example for yet another type of distributional pattern, where the species extend from northwestern China (Kansu) across Tibet to the Himalaya. This curious little plant has now been collected as far west as the Mandakini Valley in Garhwal Himalaya.

It thus appears from a general study of the pattern of distribution of the various species that the present day Himalayan flora is related to the flora of western and northwestern China in its eastern sectors and on the western side with the floras of the Western, Middle and North Asian mountains. It has been stated that the Himalaya has served primarily as a 'route of emigration and colonization from the east and northwest, secondarily of endemic development' (STEARN, 1960). There are particularly some areas, where the topography and climatic conditions are such that there are many possibilities for the influx of extraneous elements. As an example, the peculiar set-up of Aborland has been graphically described by BURKILL (1925) in the following words, 'Aborland is where the two earth systems meet, a veritable node in phytogeography. If we could make the world colder from tomorrow, the plants northward of Aborland may wander in; if we could make it hotter, the Malaysian vegetation from the south might enter; if by either change we could let a new group of plants into west Himalaya, it might advance eastwards until Aborland is attained'.

At the same time we must also consider that the topographical features of the mountain ridges, slopes and valleys and in some cases even vegetational barriers have been responsible for the isolation of certain species. As for example, in the flora of Sikkim, many instances of such isolation of species are known (SMITH, 1913) and in some cases the broad belt of rhododendrons are considered to have a rôle in keeping the areas distinct, leading to the isolation of certain species. SMITH (1913) has cited several plants as examples of such isolations, *Calathodes, Meconopsis bella, Cathcartia lyrata, Geranium refractum. Senecio chola, Saussurea laneana, Primula elwesiana, P. wattii* and *Swertia burkilliana*. We have similar instances in other sectors of the Himalaya. In Western Himalaya, HOOKER described a Scrophulariaceous plant from the collection of STRACHEY and WINTERBOTTOM, as a new genus under the name, *Falconeria* (which is a nom. nud.); this is now regarded as a *Wulfenia* (*W. himalaica*). This plant, which occurs in a remote sheltered pass*, was not again collected for more than one hundred years and it was only recently that it was rediscovered in the same locality (RAO, 1961) and, as far as is known, this is the only locality

* Madhari Pass, 2438 m: Kumaon Himalaya. – M. S. MANI.

271

so far recorded for its occurrence. In dealing with such endemic species, one has to be cautious because, in the past, many cases of such endemism have turned out to be the result of inadequate collecting. To cite another instance, towards the close of the nineteenth century, two very rare terrestrial orchids were described from the Lachan Valley in Sikkim, *Didiciea cunninghamii* and *Listera longicaulis* (KING and PANTLING, 1898). They were collected together from a fir forest undergrowth. Till very recently this was the only locality known for these two orchids. It is rather strange that the same two orchids should have appeared in a very similar environment in the forests of north Garhwal, nearly 1000 km further west (RAU and BHATTACHARYYA, 1966). It is possible that further intensive collection in similar habitats elsewhere in Kumaon and Nepal may reveal their presence in other intermediate stations. Incidentally, it may be mentioned that several epiphytic orchids of the subtropical and temperate zones, stated to be purely East Himalayan in distribution, have been recently recorded from localities in Kumaon of the West Himalaya. *Rhododendron nivale* has been collected for the first time from the alpine zone of Tehri-Garhwal in the West Himalaya. Many interesting species, not previously recorded from the respective areas, are now being discovered through intensive exploration.

We may now consider in brief the occurrence of Himalayan plants on the other mountainous regions of India and southeast Asia. JAIN (1967) has recently enumerated such plants occurring on Mt. Abu, Parasnath, Pachmarhi and the Western Ghats. Some of these elements reach only Mt. Abu and are not found further south. There are others which reach the Western Ghats and hills of South India, as well as the hills in Ceylon. *Anemone rivularis* is an example for this type of montane distribution, where some Himalayan species are also found on the hills of Ceylon. Another species of *Anemone*, *A. vitifolia*, extends from the Himalayan region to south China, Formosa and Luzon in the Philippines at altitudes between 1400 and 2400 m. According to VAN STEENIS (1934), *A. vitifolia* has followed the Formosa-Luzon migratory track. The other migratory track recognized by him, viz., the Sumatra-track has been followed by some other Himalayan plants. *Sarcococca saligna*, very widely distributed in the temperate zone of Northwestern Himalaya, has a distribution extending from Afghanistan and Himalaya to China, Formosa, Sumatra, Java and the Lesser Soenda Islands; it is also found in Ceylon. According to VAN STEENIS (1934–1936), the genus *Sarcococca*, represented by only one species, appears to have migrated into Malaysia, independently along both the Formosa-Luzon as well as the Sumatran-tracks.

The occurrence of *Primula prolifera*, the only *Primula* recorded from Malaysian mountains, is of considerable phytogeographic interest. This species, also occurring in the Khasya Mountains and in Eastern Himalaya, belongs to the *Candelabra*-section of the genus, which has its main centre of distribution in Himalaya and west China. It is visualized that *P.*

prolifera must have reached in some former period the island of Java along the Sumatra migratory track. Another instance of a discontinuous distribution of a *Primula* is of *P. sheriffiae*, which was known only from the foothills of Bhutan, but which has been later collected in Manipur Hills, nearly 500 km away (KINGDON-WARD, 1960).

We may also mention the occurrence of a remarkable *Lobelia* in a remote area at high elevation in Bhutan. This was discovered by COOPER (1942) and named as *Lobelia nubigena* by ANTHONY (1936). This has a form resembling some of the giant *Lobelia* of the mountains of East Africa and is unlike any other *Lobelia* at present known from Asia.

The evolution of the Himalayan flora is a subject of continuing interest and discussion. The Chinese mountains, being much older in age, have had considerable influence on the Himalayan flora, and many plants from these mountains have spread westwards to the younger Himalaya. During the Tertiary Period, a common flora must have covered the whole of East Asia, including Himalaya, China and Japan (HARA, 1966). During the subsequent epochs, when great changes took place in the topography and climate of the region, the separation of the floras must have taken place. The climatic fluctuations in the north expressed themselves in the form of intermittent glacial and warm periods (interglacial). During such warmer intermissions, a great many plants and animals must have thrived and many of them must have vanished during each glaciation. Large scale migrations and exchange of floristic elements must have taken place with each advance and recession of the ice sheets. This would explain the presence in the present day high altitude flora of the Himalaya of diverse elements derived from various directions. These successive changes in climate and topography during the Pleistocene not only brought in many new elements to the Himalayan flora from north, west and east but also disturbed the existing elements, driving them out from the old habitats. Fossil evidence has indicated that the Himalayan larch was distributed in the Northwestern Himalaya, but at present it has only a distribution extending eastwards of the Buri Gandaki in Nepal. Such periodic disturbances have resulted in the isolation and disjunct distribution of several genera and species. Writing about the cedar, HOOKER (1862) expressed the opinion that the emergence of three distinct races or subspecies of the Algerian, Lebanese and Himalayan cedars must have been due to the isolation and extinction of transitional forms in intermediate localities of what may have been once a continuous belt of forest of the cedar.

In the Himalaya, the glaciation, however, did not affect the foothills, with the result that the vegetation of the lower belt was not affected. Migration of floras, survival of the relicts, evolution of new species by an intermixing of different floras and acclimatisation of species from the lower altitudes must have all had a rôle in determining the present day composition and distribution of the Himalayan flora of high altitudes.

273

A recent study of the *Saxifraga flagellaris*-complex by HULTEN (1964) is very instructive in this connection. This species, with its curious surculi, has a wide distribution in the world, being found in the high arctic as well as at altitudes of nearly 5000 m in the Himalaya. Several subspecies are recognized, based on their disjunct distribution of the present time. It is visualized that these plants occurred before the Pleistocene glaciation and some populations must have survived the glacial period in unglaciated parts of Alaska and eastern Siberia as well as on the mountains of Asia. The subsequent changes in climate and topography of the regions must have resulted in the selection and survival of altered forms. The subspecies now recognized from the Caucasus, the Northwest Himalaya, Pamir, Tien Shan and in southwestern China are all traced to such an evolutionary origin.

Another group of highly specialized *Saxifraga*, belonging to the *Kabschia*-section of the genus, is represented by as many as fifty-four species in the Sino-Himalayan mountain ranges. Some of them are very local in their distribution and as many as twenty-four of them are known only from single collections. These *Saxifraga* are all confined to exposed rocky habitats, often forming very dense cushions. SMITH (1958), who recently revised this section, believes that their present distribution indicates that they must have flourished during the upheaval of the Himalaya and when the upheaval subsided they must have suffered as they are unable to compete with other plants under more fertile conditions.

An interesting approach to the study of the evolution of the Himalayan flora has been made by JANAKI AMMAL (1960), by study of the chromosome complex of certain genera, which originally formed part of the flora of Asia, before the Himalaya attained the present height. In the genus *Magnolia*, for example, she studied the cytology of some species of China and Tibet, which are considered as being very closely related to certain fossil *Magnolia*. These species proved to be diploids and they also present certain taxonomic characters, which are generally considered as more primitive. Some of the more advanced deciduous species, found in north China, Japan and Korea, were also found to be diploids, but the species occurring in Nepal, Sikkim, upper Burma, Yunnan and Szechuan, including the well known, *M. campbellii* are all hexaploids (JANAKI AMMAL, 1952). She concludes, therefore, that polyploidy in Asian *Magnolia* is restricted to the deciduous Sino-Japanese types, which have migrated along the Himalaya into India.

JANAKI AMMAL (1952) has also observed the high polyploidy in other genera like *Camellia, Lonicera, Rhododendron, Viburnum*, etc., in regions close to the glaciers of Eastern Himalaya, where the bending of the Himalaya is seen. She considers this as a region of active speciation today. The position in the genus *Rhododendron* is particularly interesting because polyploidy is confined to those species distributed in Eastern Himalaya (JANAKI AMMAL, 1950). According to her, this group of hardy mountain plants has adapted itself to life on high altitudes by polyploidy. The highest alpine members show increased chromosome numbers and this is correlated with their smaller size and late-flowering, as adaptations to the environmental conditions. In the northwestern corner of the Himalaya also, a similar region of active speciation is visualised by her. Some of the tetraploid *Artemisia* have been recently discovered there.

In spite of the fact that the vegetation types and their floristic composition in the Himalayan System are generally well known, there is still considerable need for intensive studies for a better understanding of the ecological, phytogeographical and evolutionary processes, which are at work in regard to the localized and isolated populations. There is also need for further cytogeographical studies on various Himalayan genera, so that the situation in regard to the genetic composition and evolution of the high altitude flora of the Himalaya can be better understood.

REFERENCES

ANTHONY, T. 1936. A remarkable alpine *Lobelia* from Bhutan. *Notes Roy. bot. Garden Edinburgh*, 19: 175–176.

AYMONIN, G. G. & R. K. GUPTA, 1955. Études sur les formations végétales et leur succession altitudinale dans les principaux massifs du 'Système alpine' occidental. Essai de comparaison avec l'Himalaya. *Adansonia* (NS) 5: 49–94.

BAEHNI, C., C. E. B. BONNER & S. VAUTIER, 1952. Plantes récoltées par le Dr. Wyss-Dunant au cours de l'Expedition Suisse à l'Himalaya en 1949. *Candollea* 13: 213–236 (see also *Candollea*, 15–19 for other accounts of Swiss expeditions to Nepal in 1952 and 1954).

BALAKRISHNAN, N. P. & S. CHOWDHURY, 1966. Notes on Orchids of Bhutan – I. *Epigenium* Gagnep. and *Katherinea* Hawkes. *Bull. bot. Surv. India*, 8: 312–318.

BALAKRISHNAN, N. P. & S. CHOWDHURY, 1967. Notes on Orchids of Bhutan – II. Some new or imperfectly known species. *Bull. bot. Surv. India*, 9: 88–94.

BANERJI, M. L. 1952. Observations on the distribution of Gymnosperms in Eastern Nepal. *J. Bombay nat. Hist. Soc.*, 51: 156–159.

BANERJI, M. L. 1952. Some noteworthy plants from East Nepal. *J. Indian bot. Soc.*, 31: 152–153.

BANERJI, M. L. 1953. Plants from East Nepal, Part 1. *J. Bombay nat. Hist. Soc.*, 51: 407–423.

BANERJI, M. L. 1955. Botanical Explorations in East Nepal. *J. Bombay nat. Hist. Soc.*, 55: 243–268.

BANERJI, M. L. 1963. Outline of Nepal Phytogeography. *Vegetation*, 11: 288–296.

BANERJI, M. L. 1964. Some salient features of East Nepal Vegetation. *Candollea*, 19: 215–219.

BANERJI, M. L. 1966. Rhododendrons in Nepal. *J. Bombay nat. Hist. Soc.*, 63: 18–31.

BHATT, D. D. 1964. Plant Collection in Nepal. *Madrono*, 17: 145–152.

BHATTACHARYYA, U. C. 1964. *Circaeaster agrestis* Maxim. (Circaeasteraceae) A new record from North Garhwal Himalaya. *Bull. bot. Surv. India*, 6: 297–298.

Biswas, K. 1933. The Distribution of Wild Conifers in the Indian Empire. *J. Indian bot. Soc.*, 12: 24–47.

Biswas, K. 1967. Plants of Darjeeling and Sikkim Himalaya. Calcutta.

Brandis, D. 1883. On the distribution of forests in India. *Indian For.*, 9: 173–189; 221–223.

Burkill, I. H. 1910. Notes from a journey to Nepal. *Rec. bot. Surv. India*, 4: 59–140.

Burkill, I. H. 1924–5. The Botany of the Abor Expedition. *Rec. bot. Surv. India*, 10: 1–420.

Burkill, I. H. 1965. Chapters on the History of Botany in India. Calcutta.

Champion, H. G. 1920. Geology and Forest distribution. *Indian For.*, 46: 152–154.

Champion, H. G. 1923. The influence of the hand of man on the distribution of forest types in the Kumaon Himalaya. *Indian For.*, 49: 116–136.

Champion, H. G. 1936. A preliminary survey of the Forest Types of India and Burma. *Indian For. Rec.* (NS) *Silviculture*, 1.

Chandra, R. 1949. A trip to Bara Banghal in Kangra Dist. *Indian For.*, 75: 501–504.

Collett, H. 1921. Flora Simlensis. Calcutta.

Collier, J. V. 1924. The eastern limit of the natural distribution of deodar. *Indian For.*, 50: 108–109.

Cooper, R. E. 1942. A Plant Collector in Bhutan. *Scott. geogr. Mag.*, 58: 9–15.

Cowan, J. M. 1929. The Forests of Kalimpong, an ecological account. *Rec. bot. Surv. India*, 12: 1–74.

Dang, Hari 1961. A natural sanctuary in the Himalaya: Nanda Devi and the Rishi-ganga Basin. *J. Bombay nat. Hist. Soc.*, 58: 707–714.

Deb, D. B., G. Sen-Gupta & K. C. Mallick, 1968. A Contribution to the Flora of Bhutan. *Bull. Bot. Soc. Bengal*, 22: 169–217.

Dey, A. C., M. R. Unniyal & V. Shankar, 1968. Flora of the Bhillanganga Valley of the erstwhile Tehri-Gahrwal State. *J. Bombay nat. Hist. Soc.*, 65: 384–407.

Dudgeon, W. 1923. Succession of epiphytes in the *Quercus incana* forest at Landour, Western Himalayas. Preliminary Note. *J. Indian bot. Soc.*, 3: 270–272.

Dudgeon, W. & L. A. Kenoyer, 1925. The Ecology of Tehri-Garhwal: A Contribution to the Ecology of the Western Himalaya. *J. Indian bot. Soc.*, 4: 233–284.

Duthie, J. F. 1893. Report on a botanical tour in Kashmir. *Rec. bot. Surv. India*, (I) 1: 1–18.

Duthie, J. F. 1894. Report on a botanical tour in Kashmir. *Rec. bot. Surv. India*, (I) 3: 25–47.

Duthie, J. F. 1906. Catalogue of the Plants of Kumaon and of the adjacent portion of Garhwal and Tibet based on the collections of Strachey and Winterbottom during the years 1846–1849. London.

Duthie, J. F. 1906. The Orchids of North Western Himalaya. Calcutta.

Fawcett, W. E. 1930. A short account of the Kulu Forest Division. *Indian For.*, 56: 335–339.

Gamble, J. S. 1875. Darjeeling Forests. *Indian For.*, 1: 73–99.

Gammie, G. A. 1893. Botanical exploration of Sikkim, Thibet Frontier. *Kew Bull.*, 1893: 297–314.

Gammie, G. A. 1894. Report on a botanical tour in Sikkim, 1892. *Rec. bot. Surv. India*, (I), 2: 1–24.

Ghildyal, B. N. 1956. A botanical trip to the Valley of Flowers. *J. Bombay nat. Hist. Soc.*, 54: 365–386.

Good, R. 1953. The Geography of Flowering Plants. London.

Gorrie, R. M. 1929. A short description of Upper Bashahr Forest Division. *Indian For.*, 55: 534–539.

Gorrie, R. M. 1931. Notes on *Pinus gerardiana*. *Indian For.*, 57: 211–215.

Gorrie, R. M. 1933. The Sutlej Deodar, its Ecology and timber production. *Indian For. Rec. (Silv.)*, 17: 1–140.

Gupta, A. C. 1963. Annual Precipitation and Vegetation of the dry temperate Coniferous Region of the Himalaya. *J. Indian bot. Soc.*, 42: 313–318.

276

GUPTA, M. 1952. Artemisia in Garhwal. *Indian For.*, 78: 423.

GUPTA, R. K. 1955. Botanical Explorations in the Bhillangana Valley of erstwhile Tehri-Garhwal State. *J. Bombay nat. Hist. Soc.*, 53: 581–594.

GUPTA, R. K. 1957. Botanical Explorations in the erstwhile Tehri-Garhwal State II. *J. Bombay nat. Hist. Soc.*, 54: 878–886.

GUPTA, R. K. 1962. Botanical Explorations in the erstwhile Tehri-Garhwal State III. *J. Bombay nat. Hist. Soc.*, 59: 486–512.

HANDEL-MAZZETTI, H. 1929–1936. *Symbolae Sinicae*, 7: 1–1450.

HARA, H. ed. 1963. Spring Flora of Sikkim Himalaya. Osaka.

HARA, H. ed. 1966. The Flora of Eastern Himalaya. Tokyo.

HOOKER, J. D. 1862. On the cedars of Lebanon, Taurus, Algeria and India. *Nat. Hist. Rev.*, 2: 11–18.

HOOKER, J. D. 1891. Himalayan Journals. London.

HOOKER, J. D. 1872–1897. Flora of British India, Vols. 1–7. London.

HOOKER, J. D. 1906. A Sketch of the Flora of British India. Oxford.

HOPKINS, G. M. 1930. Chakrata Forest Division. *Indian For.*, 56: 250–253.

HULTEN, E. 1964. The *Saxifraga flagellaris* Complex. *Svensk bot. Tidskr.*, 58: 81–104.

JACKSON, A. B. 1966. A Handbook of Coniferae and Ginkgoaceae. 4th ed. Revised by S. G. HARRISON. London.

JAIN, S. K. 1956. On a botanical trip to Nainital. *Indian For.*, 82: 22–38.

JAIN, S. K. 1967. Phytogeographic considerations on the flora of Mt. Abu. *Bull. Bot. Surv. India*, 9: 68–78.

JAIN, S. K. & R. C. BHARADWAJA, 1951. On a botanical trip to the Parbatti Valley. *Indian For.*, 75: 302–315.

JANAKI AMMAL, E. K. 1950. Polyploidy in the genus *Rhododendron*. *Rhododendron Yearb.*, 1950: 92–96.

JANAKI AMMAL, E. K. 1952. Chromosome relationships in cultivated species of *Camellia*. *Amer. Camellia Yearb.*, 1952: 106–114.

JANAKI AMMAL, E. K. 1960. The Effect of the Himalayan Uplift on the Genetic Composition of the Flora of Asia. *J. Indian bot. Soc.*, 39: 327–334.

KANAI, H. 1966. Phytogeography of Eastern Himalaya with special reference to the Relationship between Himalaya and Japan. (In: HARA, H. ed. The Flora of Eastern Himalaya. Tokyo. 13–38).

KAPOOR, L. D., R. N. CHOPRA & I. C. CHOPRA, 1951. Survey of Economic Vegetable Products of Jammu and Kashmir, I. Sindh Forest Division. *J. Bombay nat. Hist. Soc.*, 50: 101–127.

KAPOOR, S. L. 1968. Material for a flora of the Doda District of Jammu and Kashmir State. *Bull. bot. Surv. India*, 10: 28–49.

KASHYAP, SHIVRAM 1921. Notes on the distribution of Liverworts of Western Himalaya, Ladakh and Kashmir. *J. Indian bot. Soc.*, 2: 80–83.

KASHYAP, SHIVRAM 1925. The Vegetation of Western Himalayas and Western Tibet in relation to their Climate. *J. Indian bot. Soc.* 4: 327–334.

KASHYAP, SHIVRAM 1932. Some Aspects of the Alpine Vegetation of the Himalaya and Tibet. *Proc. 19th Indian Sci. Congr.* Bangalore, 13–53.

KENOYER, L. A. 1921. Forest Formations and Successions of the Sat Tal Valley, Kumaon Himalayas. *J. Indian bot. Soc.*, 2: 236–258.

KIHARA, H. ed. 1955. Fauna and Flora of Nepal Himalaya. Vol. 1. Kyoto.

KING, G. & R. PANTLING, 1898. The Orchids of the Sikkim Himalaya. *Ann. Roy. bot. Garden Calcutta*, 8: 1–342, tt. 1–448.

KINGDON-WARD, F. 1926. The Riddle of the Tsangpo Gorges. London.

KINGDON-WARD, F. 1930. Plant Hunting on the Edge of the World. London.

KINGDON-WARD, F. 1940. Botanical and geographical exploration in the Assam Himalaya. *Geogr. J. London*, 96: 1–13.

KINGDON-WARD, F. 1960. Pilgrimage for Plants. London (see for full bibliography of Kingdon-Ward's works).

KITAMURA, S. 1955. Flowering Plants and Ferns (In: KIHARA, H. ed. Fauna and Flora of Nepal Himalaya, 1: 73–77. (Kyoto).

KITAMURA, S. 1964. Plants of West Pakistan and Afghanistan. Kyoto.

KLOTZ, G. 1963. The Cotoneasters of the *C. nitidus* Jacques Group. *Bull. bot. Surv. India*, 5: 207–214.

MARQUAND, C. V. B. 1929. The botanical collections made by Capt. KINGDON-WARD in the Eastern Himalayas and Tibet in 1924–25. *J. Linn. Soc.*, 48: 149–229.

MATHEW, K. M. 1966. A preliminary list of plants from Kurseong. *Bull. bot. Surv. India*, 8: 158–168.

MEHRA, P. N. & K. K. DHIR, 1968. Ferns and Fern-allies of Dalhousie Hills. *Bull. Bot. Surv. India* 10: 296–308.

MODDIE, A. D. 1959. A high walk in the Central Himalaya. *Him. J.*, 22: 146–152.

MOHAN, N. P. 1933. Ecology of *Pinus longifolia* in Kangra and Hoshiarpur Forest Division. *Indian For.*, 59: 812–816.

MURRAY, W. H. 1951. The Scottish Himalayan Expedition. London.

NAIR, N. C. 1964. On a botanical tour to Lahul and Spiti (Punjab Himalaya). *Bull. bot. Surv. India*, 6: 219–235.

NAKAO, S. 1955. Ecological Notes (In: KIHARA, H. ed. Fauna and Flora of Nepal Himalaya, 1: 278–290. (Kyoto).

NAKAO, S. 1964. Living Himalayan Flowers. Tokyo.

NUMATA, M. 1965. Ecological Study and Mountaineering on Mt. Numbur in Eastern Nepal, 1963, Chiba, Japan.

NUMATA, M. 1966. Vegetation and Conservation in Eastern Nepal. *J. Coll. Arts & Sci. Chiba Univ.*, 4: 559–569.

NUMATA, M. 1967. Notes on a Botanical Trip in Eastern Nepal, I. *J. Coll. Arts & Sci. Chiba Univ.*, 5: 57–74.

OSMASTON, A. E. 1922. Notes on the Forest Communities of the Garhwal Himalaya. *J. Ecol.*, 10: 129–167.

OSMASTON, A. E. 1927. A Forest Flora for Kumaon. Allahabad.

OSMASTON, A. E. 1931. Notes on *Pinus gerardiana*. *Indian For.*, 57: 351–352.

OSMASTON, F. C. 1935. An Expedition into Sikkim. *Indian For.*, 61: 424–434; 487–499.

PANDE, B. D. 1962. Some aspects of the vegetation of Nepal. *Bull. bot. Surv. India*, 4: 137–140.

PANDE, S. K. 1936. Studies in Indian Liverworts. A Review. *J. Indian bot. Soc.*, 15: 221–240.

PANIGRAHI, G. & V. N. NAIK, 1961. A botanical tour to Subansiri Forest Division (N.E.F.A.) *Bull. Bot. Surv. India*, 3: 361–388.

PARKER, R. N. 1930. Botanical Notes on some Plants of the Kali Valley. *Indian For.*, 56: 105–108.

PARKER, R. N. 1942. The Ecological Status of the Himalayan Fir Forests *(Abies pindrow & Picea smithiana) 150th Anniv. Vol. Roy. Bot. Garden, Calcutta*, 125–128.

PENNELL, F. W. 1943. The Scrophulariaceae of the Western Himalayas. Philadelphia.

POLUNIN, N. 1950. Introduction to Plant Geography. London.

POLUNIN, O. 1950. An Expedition to Nepal. *J. Roy. Hort. Soc.*, 75: 302–315.

POLUNIN, O. 1950. Plant Hunting in Nepal Himalayas. *Geogr. Mag.*, 23: 132–147.

POLUNIN, O. 1956–7. A Kashmir Journey. *Gardn. Chron.*, 140: 546–547; 628–629; 141: 66–67.

PURI, G. S. 1952. The distribution of conifers in Kulu Himalayas with special reference to Geology. *Indian For.*, 76: 144–153.

PURI, G. S. 1960. Indian Forest Ecology, Vols. 1–2. New Delhi (see for extensive bibliography).

RAIZADA, M. B. & K. C. SAHNI, 1957. Vegetation types in the Kumaon Himalaya with special reference to the Panch Chulhi Area. *J. Indian bot. Soc.*, 36: 599–600.

RAIZADA, M. B. & K. C. SAHNI, 1960. Living Indian Gymnosperms. Part 1. (Cycadales, Ginkgoales and Coniferales.) *Indian For. Rec.*, (NS) 5: 73–150.

Rao, Rolla S. 1963. A botanical tour in Sikkim State, Eastern Himalayas. *Bull. bot. Surv. India*, 5: 165–205.

Rao, Rolla S. 1963. The Indian Cho-Oyu Expedition 1958: Observations of a Botanist Member. *J. Bombay nat. Hist. Soc.*, 60: 400–409.

Rao, Rolla S. & J. Joseph, 1965. Observations on the Flora of Siang Frontier Division, North-East Frontier Agency (NEFA) *Bull. bot. Surv. India* 7: 138–161.

Rao, Rolla S. & G. Panigrahi, 1961. Distribution of the Vegetational Types and their dominant species in Eastern India. *J. Indian bot. Soc.*, 40: 274–285.

Rao, T. A. 1959. Report on a Botanical Tour to Milam Glaciers. *Bull. bot. Surv. India*, 1: 97–120.

Rao, T. A. 1960. A Botanical Tour to Pindari Glaciers and Kumaon Hill Stations. *Bull. bot. Surv. India*, 2: 61–94.

Rao, T. A. 1960. Further Contributions to the Flora of Jammu & Kashmir. *Bull. bot. Surv. India*, 6: 47–57.

Rao, T. A. 1961. A Botanical Tour in Kashmir State. *Rec. bot. Surv. India*, 18: 1–67.

Rao, T. A. 1961. An imperfectly known Endemic Taxon of Kumaon Himalayas – *Falconeria himalaica* Hook. f. (= *Wulfenia himalaica*) (Hook. f.) Pennell. *Bull. bot. Surv. India*, 3: 79–82.

Rao, T. A. 1964. Observations on the Vegetation of Eastern Kumaon bordering the Nepal Frontier. *Bull. bot. Surv. India*, 6: 47–57.

Rau, M. A. 1960. On a Collection of Plants from Lahul. *Bull. bot. Surv. India*, 2: 45–56.

Rau, M. A. 1961. Flowering Plants and Ferns of North Garhwal, Uttar Pradesh, India. *Bull. bot. Surv. India*, 3: 215–251.

Rau, M. A. 1963. Illustrations of West Himalayan Flowering Plants. Calcutta.

Rau, M. A. 1963. The Vegetation around Jumnotri in Tehri-Garhwal, U. P. *Bull. bot. Surv. India*, 5: 277–280.

Rau, M. A. 1964. A Visit to the Valley of Flowers and Lake Hemkund in North Garhwal, U.P. *Bull. bot. Surv. India*, 6: 169–171.

Rau, M. A. 1966. Recent Finds of Some Rare Angiosperms in Western Himalaya and their Phytogeographic Significance. *Proc. Indian Sci. Congr.* Chandigarh (Symposium).

Rau, M. A. & U. C. Bhattacharyya, 1966. New Records of three rare orchids for Western Himalaya. *Bull. bot. Surv. India*, 8: 93–94.

Raven, P. H. 1962. The Genus *Epilobium* in the Himalayan region. *Bull. Brit. Mus. (Nat. Hist.)* 2: 325–382.

Roy Chowdhury, K. C. 1951. Sikkim – the Country and its Forests. *Indian For.*, 77: 676–683.

Royle, J. F. 1839–1840. Illustrations of the Botany and other branches of the Natural History of the Himalayan mountains and of the Flora of Cashmere. London.

Sahni, K. C. & M. B. Raizada, 1955. Observations on the Vegetation of Panch Chulhi. *Indian For.*, 81: 300–317.

Sastry, A. R. K. & H. Deka, 1967. *Nertera sinensis* Hemsl. A new find from Subansiri Dist. N.E.F.A. India. *Bull. bot. Surv. India*, 9: 285–286.

Schmid, E. 1938. Contributions to our knowledge of flora and vegetation in the Central Himalayas. *J. Indian bot. Soc.*, 17: 269–278.

Schweinfurth, U. 1957. Die horizontale und vertikale Verbreitung der Vegetation im Himalaya. Bonn. (see for complete bibliography up to 1956).

Sen, G. C. 1963. The epiphytic flowering plants of Darjeeling Hills other than orchids. *Bull. bot. Surv. India*, 5: 111–115.

Shebbeare, E. O. 1934. The Conifers of Sikkim Himalaya and adjoining country. *Indian For.*, 60: 710–713.

Singh, Gurdial 1955. Three months in Upper Garhwal and adjacent Tibet. *Him. J.*, 19: 3–17.

Singh, Sher 1929. The effect of climate on the conifers in Kashmir. *Indian For.*, 55: 189–203.

Smith, H. 1958. *Saxifraga* of the Himalaya I. Section *Kabschia*. *Bull. Brit. Mus. (Nat. Hist.)* 2: 83–129.

Smith, H. 1960. *Saxifraga* of the Himalaya II. Some new species. *Bull. Brit. Mus. (Nat. Hist.)* 2: 227–260.

Smith, W. W. 1913. The alpine and sub-alpine vegetation of south-east Sikkim. *Rec. bot. Surv. India*, 4: 323–431.

Smith, H. & G. H. Cave, 1911. The Vegetation of the Zemu and Llonakh Valleys of Sikkim. *Rec. bot. Surv. India*, 4: 148–258.

Smythe, F. S. 1932. Kamet Conquered. London.

Smythe, F. S. 1938. The Valley of Flowers. London.

Smythies, E. A. 1919. Geology and Forest Distribution. *Indian For.*, 45: 239–243; 46: 319–320 (1920).

Soest, J. L. van 1961. New species of *Taraxacum* from the Himalayan region. *Bull. Brit. Mus. (Nat. Hist.)* 2: 261–273.

Srinivasan, K. S. 1959. Report on a Botanical Tour to Bomdi-La, NEFA (May 1955). *Rec. bot. Surv. India*, 17: 1–38.

Stearn, W. T. 1960. *Allium* and *Milula* in the Central and Eastern Himalaya. *Bull. Brit. Mus. (Nat. Hist.)* 2: 161–191.

Steenis, C. G. G. J. van 1934–1936. On the Origin of the Malaysian Mountain Flora. *Bull. Jardin Bot. Buitenzorg*, (3) 13: 135–262; 289–417.

Stewart, R. R. 1967. The Grasses of Kashmir. *Bull. Bot. Surv. India*, 9: 114–133.

Stewart, R. R. 1967. The Cyperaceae of Kashmir – A Check List. *Bull. bot. Surv. India*, 9: 152–162.

Suri, P. N. 1933. The ecology and silviculture of Himalayan spruce and silver fir. *Indian For.*, 59: 532–550.

Swan, L. W. 1961. The Ecology of the High Himalayas. *Scient. Amer.*, Oct. 68–78.

Troll, C. 1967. Die klimatische und vegetations-geographische Gliederung des Himalaya-Systems. *Khumbu Himal.*, 1: 353–448 (see also literature cited here).

Turner, J. E. C. 1929. West Almora Division U.P. *Indian For.*, 55: 578–586.

Turrill, W. B. 1953. Pioneer Plant Geography. Martinus Nijhoff, The Hague.

'Vagrant' 1887. A high forest of *Quercus dilatata*. *Indian For.*, 13: 124–125.

Vishnu-Mittre, 1963. The Ice Ages and the Evolutionary History of the Indian Gymnosperms. *J. Indian bot. Soc.*, 42: 301–308.

Vohra, J. N. & B. M. Wadhwa, 1963. *Andraea rupestris* Hedw. A new record from Western Himalayas. *Bull. Bot. Surv. India*, 5: 149.

Wadia, D. N. 1957. Geology of India. London.

Walia, M. K. & S. N. Tiku, 1964. Contribution to the Flora of Kashmir-Lolab Valley. *Bull. bot. Surv. India*, 6: 141–149.

Wendelbo, Per 1966. A new species of *Corydalis* Sect. *Oocapnos* from Afghanistan. *Bot. Notiser*, 119: 243–248.

Williams, L. H. J. 1953. The 1952 Expedition to Western Nepal. *J. Roy. Hort. Soc.*, 78: 323–337.

Wiltshire, E. P. 1953. Narrative of a trek and of natural history observations in Kashmir in May-June. *J. Bombay nat. Hist. Soc.*, 51: 825–838.

Wynter-Blyth, M. A. 1951–2. A Naturalist in the North-West Himalaya. *J. Bombay nat. Hist. Soc.*, 50: 344–354; 559–572; 51: 393–406.

XI. THE TRIBAL MAN IN INDIA: A STUDY IN THE ECOLOGY OF THE PRIMITIVE COMMUNITIES

by

PARMANAND LAL

1. Introduction

India, with a land frontier of 15,200 km and a coastline of 5,700 km and an area of about 3,267,500 km², ranks seventh amongst the countries of the world, and has a population of 439 millions (according to the Census 1961). It is the home of diverse types of man, from the most primitive to the modern. While at present it is often difficult to decide where the caste Hindu ends and where the tribal begins, the distinction may, however, be traced back to very early times, indeed to the times of the first entry of the Aryans into India from the northwest. The distinction between the Aryans and the original inhabitants (the *aboriginal*) of the land was fundamental: the one was a cultivator and the other was a non-cultivator. Though the first settlements of the Aryans were in the Trans-Indus area, that barrier (in Sanskrit *Sindhu** = divider or barrier) was soon crossed and the settlements reached as far east as the Sutlej and further east. The aboriginals in these areas were steadily pushed back or they were subjugated. Some of them had also adopted in part the Aryan mode of life, but still the main distinction between the original inhabitants and new-comers was agriculture. The main occupation of the Aryans of the Indus (Hindus), ancient as well as modern, is agriculture, to which we find constant reference in the Rig Veda. Those who practised agriculture were civilized races, and came into interminable conflict with the non-agricultural original inhabitants, when the forests were cleared for cultivation. There are, for example, numerous allusions to wars between the Aryans and the aboriginals, who were considered inferior and as slaves (dasyu). We read: 'O Ye two Asvins! You have displayed your glory by teaching the *Arya* to cultivate with the plough and to sow corn and by giving him rains for the production of his food and by destroying the *Dasyu* by your thunderbolt' (Rig Veda, 1, 117: 21; Max Müller translation) 'Indra, who is invoked by many, and is accompanied by his fleet companions, has destroyed by his thunderbolt the *Dasyus* and *Simyus* and he has distributed the fields to his white-complexioned friends' (I, 100, 8). 'Indra with his weapon, the thunderbolt, and in his vigour,

---

* It is this river that also gave the name Hindu (the letter *H* often being substituted for *S*) for the peoples who settled there and eventually *India* for the country itself.

destroyed the towns of the *Dasyus'* (I, 103, 3). The next hymn refers to the aboriginal robbers, who dwelt on the banks of the four small streams called Sifa, Anjasi, Kulisi and Virapanti, as issuing forth from their forest-fastness and harassing the civilized Aryans. 'Kuyava gets scent of the wealth of others and appropriates it. He lives in water and pollutes it. His two wives bathe in the stream; may they be drowned in the depths of the Sifa river!' (I, 104, 3). It is also evident from the numerous hymns of the Rig Veda that the Aryans speak most uncomplimentarily about the shouts and yells of the aboriginal barbarians. In other places the aboriginals are described as scarcely human: 'We are surrounded on all sides by *Dasyu* tribes. They do not perform sacrifices; they do not believe in anything; their rites are different; they are not men! O destroyer of foes, kill them. Destroy the *dasa* race!' (X, 22, 8). In the face of ruthless onslaught by the superior newcomers, the aboriginal man retreated further and further east and south into the interior of his forest fastness. It was, however, only a matter of time before the Aryan colonists had completely driven him out of his area and had spread their agriculture throughout the Indo-Gangetic Plains. The early Hindus wrested, therefore, the fertile tracts from the age-old homes of the primitive communities, who were the original inhabitants of the country. It must not be supposed that the aboriginals gave up their birthright without a struggle. Retreating before the civilized organization of the new-comers in open field, the aboriginals however hung around in forests, near every Aryan settlement or village, harassed them in diverse ways, waylaid them and robbed them of their cattle. It was by ceaseless fighting that the early Aryan colonists protected their newly conquered land, gradually extended their agriculture, built new villages and all the time despised the aboriginals with genuine hatred, killed numbers of them when they could. The endless battles fought by the Kshatriya princes to protect the sacrificial rites of the ancient rishis, of which we may read repeatedly in the Ramayana and Mahabharata, were all directed against the aboriginals – it was they who are the *asuras* of these epics. The aboriginals were eventually either exterminated or they retreated, before the ever-advancing Aryan civilization, to those refugial forest-fastness, which their descendents – the primitive communities or tribes, as they are here called inhabit today. Some of the weaker aboriginals preferred subjugation to extermination and exile and gradually came to be assimilated into the Aryan community, but remained outside the primary four *varnas* (professional-castes) Brahmana, Kshatriya, Vaisya and Sudra. They were known to and recognized by the *sutrakaras* and Manu, the law-giver, as race-castes, quite distinct from the professional-castes of the Hindus. Some of them even accepted co-existence and collaboration with the new-comers. We read, for example, of the *vanaras*, who are really *vana naras** or forest folk, assisting Rama. Bali was a great king of aboriginals, but his brother Sugriva thirsted for his kingdom and his wife. He befriended Rama, who killed Bali and helped Sugriva to win the kingdom and the widow. Sugriva became an ally of Rama and marched with his army to fight Rama's war in Lanka. It must not, however, be assumed that it was only the aboriginals who adopted the culture, beliefs and professions of the new-comers. The Aryan Hindus also took over countless aboriginal gods and goddesses, the aboriginal modes of worship (which is now universally practised by the Hindus and *not* the Vedic method), marriage and other customs and sanctified them as part of the Hindu traditions! The worship of Kali in Bengal, of Ayyapan, Ayyanar, Mariamma or Sitladevi, etc. by caste Hindus, is one of the numerous aboriginal bequests. We have here an interaction between the endemic and intrusive races, closely paralleled in the history of the fauna.

The descendents of the aboriginals, isolated in widely separated refugial areas, are now called 'tribes'. The term 'tribe' was first used by the British rulers in 1872 for the 'primitive community' of man, supposed to be outside the 'castes' of the natives. Though quite erroneous and a source of great harm to these men, the expression has unfortunately come into general use and has even found a place in the statute books of the land. In the context of this book, the tribal man is thus essentially vanishing relict of early man in India, confined at present to scattered and comparatively small and fast diminishing refugial areas, to which he has retreated under increasing pressure of civilization. He represents now less than 7 % of the total population of India. The ecology of different tribes is of considerable interest in elucidating fundamental problems in biogeography of India. This chapter presents a broad outline of the salient features of the general ecology of the principal tribes, the recent changes in their status and organization, the factors in his disappearance and allied topics. This is largely a summary of my recent investigations and incorporates unpublished data. The life and culture of the civilized races of men do not always reflect the effects of the environment, under which they live. The civilized man has, by virtue of his superior knowledge, science and technology, very profoundly modified his environment to suit his special requirements or has even often created his own particular environment. In the case of the primitive peoples, however, a close correlation between their cultural attainments, habits and their habitats is readily observed. Particularly in India, the basic economy of these peoples depends wholly for its continuance on favourable physical environment. The houses, utensils, tools and every other material equipment, the food, practices, beliefs and indeed nearly every aspect of the tribal man's life are directly influenced by his physical environment – an environment that has been very profoundly altered and greatly impover-

---

* They were considered to be subhuman when compared to the civilized men and thus arose the unfortunate conception of their being apes in English translations of the Ramayana. – M. S. MANI.

283

ished by the activities of the civilized man in India. The geography and ecology of the primitive man in India reveal, therefore, a remarkable parallelism with the fundamentals of the geography and ecology of its flora and fauna.

India was probably one of the cradles of mankind. For several decades hardly a year has passed without the discovery of some exciting new evidence of early man or of manlike apes to strengthen this conclusion. Almost complete human skeletons and various parts of human bodies from the microlithic beds recently discovered in Gujarat, and many other findings, however, lie beyond the time horizon of the present text, which begins with the end of the Palaeolithic period. From that time to the present India has been inhabited by representatives of three races viz. the Dravidian, the Indo-Aryan and the Mongolian. The first two mixed, in varying proportions in different states, with each other and with the Mongolian elements, while the third was largely confined to the northeast frontier and Assam.

The racial factors do not, in themselves, help in explaining the development or distribution of cultures in India, particularly because earlier assumptions of inherent racial differences in the capacity to create and maintain culture have been shown to be untenable. We also recognize that the anthropometric and somatological criteria, by which the three races have been distinguished, are themselves becoming increasingly suspects, as physical anthropology comes to lay its main emphasis upon genetic factors. Unfortunately, genetically precise data are not yet available in sufficient quantity to render much assistance. Though we consequently neither regard race as relevant to culture nor consider traditional typologies as particularly respectable from the scientific point, we nevertheless insist that the older anthropometric and somatological criteria still serve a useful function, as aids in historical reconstruction. They show enough uniformity over limited area, enough stability over time, and enough persistence in mixture, to provide the archaeologist and the ethnologist with an additional tool for tracing important culture historical movements in the past. We thus offer no apologies for three-fold division presented herewith. It would be the vainest of labours to attempt a description of the salient physical and cultural characteristics of even the main groups in these pages.

The populations of India exhibit, in varying degree, characteristics from the four major stocks of mankind: Negroid, Australoid, Mongoloid and Caucasoid. Of earlier peoples, almost the only known skeletal remains of much significance are those of the Indus valley civilization: these show very close affinities with those of Pre-Sargonic Mesopotamia (Al-Ubaid and Kish). The numerous Megalithic remains of the Peninsula undoubtedly hold vital evidence on the peopling of India; their scientific exploration is but beginning.

The earliest of existing groups are the Negritos (Negroids of small

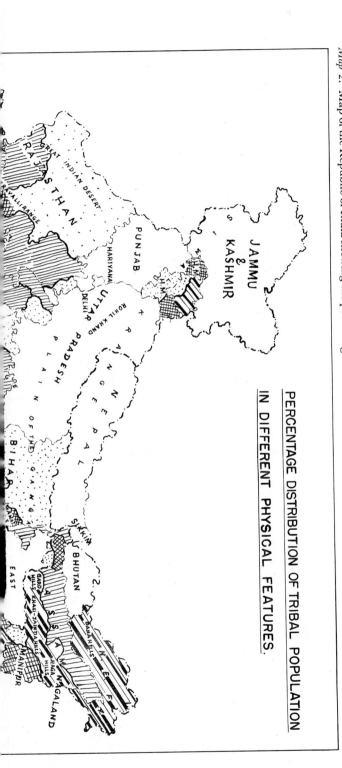

Map 2. Map of the Republic of India showing the percentage distribution of the tribal population in relation to different physical features.

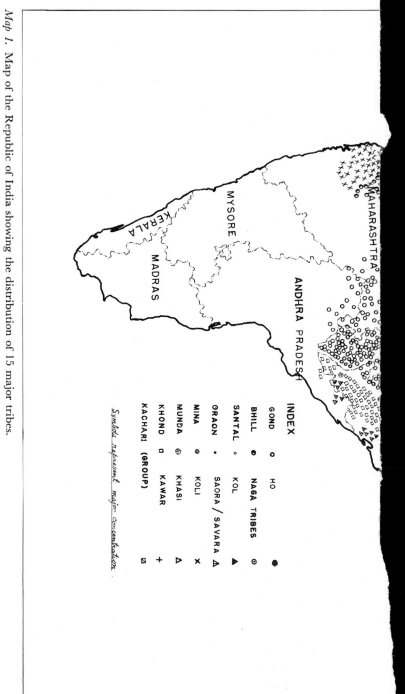

Map 1. Map of the Republic of India showing the distribution of 15 major tribes.

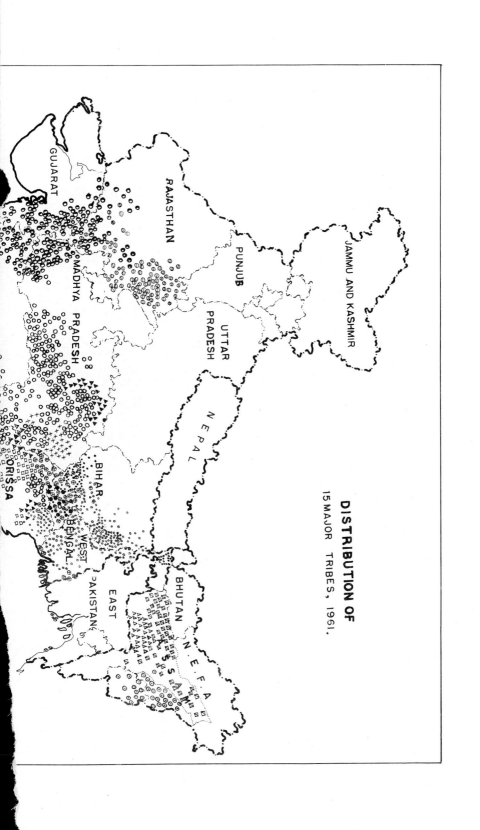

DISTRIBUTION OF
15 MAJOR TRIBES, 1961.

JAMMU AND KASHMIR

RAJASTHAN

PUNJUB

GUJARAT

MADHYA PRADESH

UTTAR PRADESH

NEPAL

ORISSA

BIHAR

WEST BENGAL

EAST PAKISTAN

BHUTAN

N.E.F.A

ASSAM

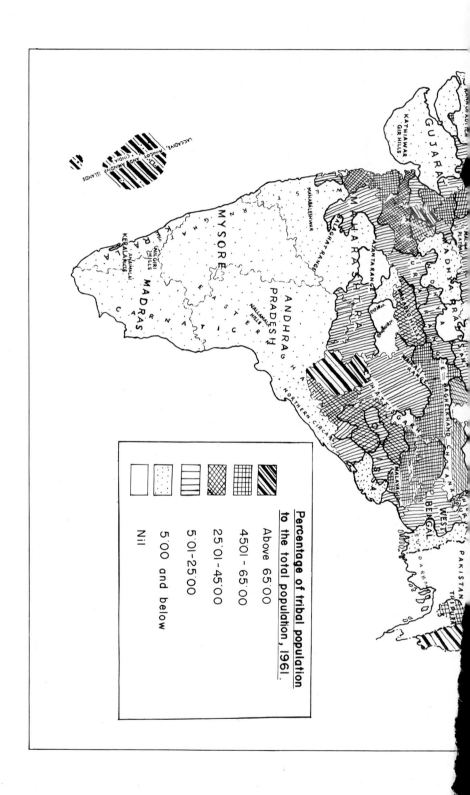

Percentage of tribal population to the total population, 1961.

Above 65·00

45·01 - 65·00

25·01 - 45·00

5·01 - 25·00

5·00 and below

Nil

stature), of whom the Andaman Islanders are good examples (GUHA, 1929). The *Kadars* of Cochin, like the Andamanese still hunters and gatherers, also show some Negrito characteristic and traces at least of Negrito physical types have been reported from the Rajmahal Hills.

Far more significant are the evidences of Australoid stock, which appear in the tribal populations of the South and Central India (e.g. *Mundas, Santhals*). In varying mixtures this is the underlying strain in very much of the Hindu population, especially of lower or exterior castes, south of the Narmada-Chota-Nagpur line.

The tribal peoples of the north are essentially dissimilar, and as might be expected, show marked mongoloid characteristics. They occupy a broad band of Himalayan and Sub-Himalayan country, from Kashmir to Darjeeling. In the hills on either side of the Assam valley a dolichocephalic Mongoloid type is dominant. The Burmese are more brachycephalic. The Assam valley itself has an interesting fusion of Mongoloids (the Shan Ahoms, who were the mediaeval rulers) with Palaeo-Mediterraneans, bearers of Hindu culture.

The populations which show the most marked evidences of these three major stocks (Negroid, Australoid and Mongoloid) are mainly tribal, though of course these elements are neither wholly confined to the tribes nor are they represented in all tribes.

The correlation between ethnic stock and language is rather not strong. However, linguistic relationships provide by far the most dependable evidence of historical connections. If two peoples speak related languages, however much they may differ in race or in culture and however remote their geographical location, either both have descended from a single ancestral society or the ancestors of one have at some time had such intimate contact with a group thus related to the other that they abandoned their own language and adopted that of their neighbours.

The tribes living in India may be classified under three main family of languages.

a. The Australo-Asiatic linguistic branch, under which comes the Munda speeches of central and eastern India, Nicobarese and the Santhali.

b. The Dravidian linguistic group spoken by the tribes of central and South India.

c. The Tibeto-Burman family of languages spoken by tribes living in the Himalaya and in Assam.

Most of the tribal communities are isolated in refugial niches on hills like the Assam Hills, the foothills of the Himalaya, the Central Plateau and the Southern Block (Maps 1, 2). The principal tribes of the Assam Hills are 1. *Garo*, 2. *Khasi*, 3. *Mikiri*, 4. *Kachari*, 5. *Mizo*, 6. *Kuki*, 7. *Naga* and 8. *Abor*, after whom the various hills are named. The *Kuki* tribes include a number of groups of primitive communities like *Purum, Vaiphei, Aimol, Anal, Kom, Thado*, etc. The *Mizo* (of the Mizo or Lushai Hills) are likewise divided into many groups like *Mizo Ralte, Hmar, Pauri, Paihte,*

Reang, Mara (Lakher) Chakma, Panek, Bong, Pang, etc., who came largely from Burma in the seventeenth century. The *Kachari* is an inhabitant of the Cachar Hill and Mikir Hill; the *Mikiri* also live on the Mikir Hill. The *Nagas* are divided into numerous groups like *Angami, Aos, Chakhesang, Chang, Khienmungam, Konyak, Lotha, Phom, Rengma, Sangtam, Sema, Yimchung* and *Zeliang.* The *Abor* or *Adi, Apatami, Dafla, Galong, Khampti, Khowa, Mishmi, Momba, Miri* and *Shingo* are the principal primitive communities of the North East Frontier Agency (NEFA). The *Khasi Synteng (Pnar), War, Bhoi* and *Iyngam* are the tribal peoples of the Khasi and Jaintia Hills. The *Garo* and *Koche* inhabit the Garo Hill. The foothills of the Himalaya are the home of the *Bhotiya.* The Central Plateau is the refuge of the *Munda, Kol, Santhal, Bhumiya, Oraon, Kharia, Juang, Savara, Khond, Gond, Korkup* and the *Bhil.* The *Gonds* are the most numerous of the primitive peoples of India and inhabit an extensive area of the Peninsula; they have figured in Ramayana and the country occupied by them is Gondava-vana (Sanskrit: forest of the Gondava), from which the expression Gondwanaland is derived. The Southern Block is the home of the *Chenchu, Kadar, Paniyan, Urali, Kota, Badaga, Toda* and the *Kurumban.*

2. Tribal Society

Primitive society has generally achieved some kind of adjustment between its material needs and the potentialities of its environment. Four factors largely underlie this adjustment: 1. the size of the social group, 2. the material needs of the group, 3. the resources available and 4. the degree of skill with which the resources are tapped and exploited. What the material needs of a group should be is neither primarily a function of the resources available, nor is it always determined by the size of the social groups. These material needs, the method of acquiring them and the necessary adjustments differ greatly from society to society.

A survey of the different tribal men of the entire range of economic life in India naturally lies outside the scope of the present volume, which attempts to cover intensively only the subsistence economy, i.e. the major types of food acquisition. The various occupations followed by tribes in India are: collection of wild fruits, berries and tubers from the forest, rearing and collection of cocoons, *sawai* grass, etc., gathering honey, fibres and making strings and ropes, manufacture of catechu, crude sugar, pottery, spinning and weaving, lumbering and selling of fire wood and charcoal from the forests, hunting, fishing, raising cattle, *jhuming* of forests for crude agriculture, terraced forming, settled agriculture, mining and labour in factories and plantations.

The tribes in the northeast India are settled agriculturists, living on

* *Panicum miliare,* cultivated for the edible grain, but often run wild; straw is good fodder for cattle. – M. S. MANI.

terraced fields, while in Central India shifting cultivation is the prevalent form of food production. In South India the economic life of tribes is based mainly on the collection of forest produce. Shifting cultivation is, however, a common feature in all the zones. Most of the tribes practise hunting, fishing and resort to minor cottage industries and other subsidiary occupations.

Dependent hunters, who do not practise agriculture, but live on the outskirts of villages and come into the markets to sell jungle produce, include the *Yandi, Chenchu, Kurumban* and some other smaller tribes of Andhra, Madras and Kerala. Among all the large tribes there are sections, which live almost entirely on jungle produce before the autumn crop is harvested.

A variety of handicrafts is practised by the tribes. Among the Assam tribes the most widely practised craft is the manufacture of cloth from cotton, dyed with indigenous vegetables. In the North East Frontier Agency weaving, basketry and wooden images are made. The *Maria Gond* of the Madhya Pradesh are occupied in distilling spirit from the forest products. The *Sawara, Konds* and *Gonds* take to cow-herding, metal working, weaving, cane work and pottery. The *Agarias* of Central India are smelters of iron. The *Tharus*, living in the foothills of the Himalaya, north of the middle Gangetic Plain, make furniture, household utensils, crude weapons, etc. The *Irulas* of Madras do bamboo work and the *Bhotiyas* are expert in spinning and weaving wool.

The tribes living in the vicinity of collieries in Bihar, Orissa and West Bengal work in the local coal mines. In Central India the bulk of the mining labour is derived from the *Gond, Mawasipuds* and *Mahars*. The *Santhals* are good pick-miners and coal cutters. The Tata Iron and Steel Company employs, for example, over 1,700 unskilled labour from the tribes. The manganese mines depend for about 50% of their labour on the tribals. The *Santhals* and *Kols* are largely employed in the iron ore industries. The mica industry of Bihar employs about 2,50,000 tribals. Plantations provide another opportunity for tribal employment. Over half a million adult workers and the same number of children are employed in tea plantations of Assam. The tribals are also employed in the collection of diverse forest produce, like fruits, bark, dyes, leaves for *bidi**, lac, gum, resins, wax, fodder, charcoal wood, drug plants, etc.

Among gleaners and hunters, the quest for food leads them from forest to forest and the search for roots and berries makes them wanderers, with no permanent or stable organization for production. The social organization of the hunting groups has been built up by the needs of economic life and by the cooperation of individuals in the food quest, and as such

---

* *bidi* is a kind of cigarette, with low quality tobacco rolled inside dry leaves of trees like *Butea frondosa, Diospyros tomentosa*, etc. – M. S. MANI.

cooperation is rather sporadic and intermittent, we find their settlements scattered and the economic organization is also less integrated.

The hill *Kharias*, who are confined to the inhospitable hill fastness of Mayurbhanj in Orrisa, Dalbhum (Singhbhum) and Barabhum (Manbhum) in Bihar, have not been much disturbed by contacts with outsiders. Their country does not afford much scope for easy life and they are constantly faced with the problem of shortage of food supply. The iron ores in their hills provide them with material for the tools and implements, required for hunting, fishing, lumbering, etc. The hill *Kharias* practise primitive *jhum* cultivations and also collect honey, fruits, edible tubers, etc. Rice is, however, their staple food. The hill *Kharias*, who live near prosperous villages situated at the foot of the hills, may secure work as day labourers. The size of the Kharia settlements differs according to their cultural stage. The hill *Kharias* live in groups of five or ten families, in huts scattered over the hill sides at distances of a hundred metre or more. The more advanced *Dhelki Kharias* live in regular villages, with sacred groves, dancing arenas and village burial grounds. The *Christian Kharia* villages are neat and more compact, with better houses. The hill *Kharias* and also the *Dhelkis* build dormitories, but the Christian villages have abandoned the practise. The *Kharias* do not eat raw meat and beef is unpopular with all sections of the tribes. Salt is very popular with them and they take plenty of it with their food and leaves, roots and flowers also form a large part in their food. They brew a kind of rice beer, which they consume in copious quantities.

The *Kukis*, living in the Lushai Hills of Assam, provide another example of human adjustment to special habitats. The *Kukis* are known by various clan names. Those of the North Cachar Hills are called *Biete Kukis* and *Khelma Kukis*. To the north of the Lushai Range in the forest-clad hills dwell the *Darlungs*. The Lushai chiefs rule over the country between the R. Karnafuli and its main tributory. Their most northerly villages are found on the borders of the Silchar District. The Lushai culture has shown some 'levelling influence' in absorbing most of the *Kuki* clans. It has shown also how even remote *Kuki* clans have not escaped their cultural influence. The *Kuki* villages consist of tiny settlements in the jungles, of four or five huts, built of bamboo and cane. The *Kukis* are by temperament nomadic and their peculiar vagabond strain leads to villages splitting into hamlets and the latter subdividing till, as in Manipur Hills, there are single houses in the midst of a dense jungle, several kilometres from the next habitation. This vagabond strain also manifests itself in the custom by which each son of a chief, as he attains marriageable age, is provided with a wife at his father's village and sent forth to a village of his own. Henceforth he rules as an independent chief and his success or failure depends on his own talents for ruling. The *Kuki* clans, like many other primitive groups in India and elsewhere, are self-sufficient in all the details of their economic requirements. Both the

288

Lushai and *Kukis* have learnt the use of fire-arms, but a century ago their only weapons consisted of bows and arrows. The forests inhabited by the *Kukis* are thickly clothed with bamboo. The nomad *Kuki* builds light bamboo houses, but where the *Kukis* live a settled life, they construct large solidly built houses, 15–18 m long, 2.5–3.5 m wide and 2–3 m high. The *Kukis* make baskets, mats, tobacco pipes, etc. from bamboo stem. They also practise *jhum* cultivation. Even where the *Kukis* live on the hills they have not taken to terraced cultivation, for which they say they must know the appropriate rituals and sacrifices. Kukis have not also learnt wet cultivation of rice.

A number of tribes of *Gond* extraction are found in the Bastar District of Central India. They are the *Marias, Murias, Parjas, Bhatras* and the *Gaddabas*. They live in villages, which are not self-sufficient. There is usually a family of blacksmiths in the village or several villages may have one such, which supplies the small needs of the people. The artisan elements in the population were probably recruited from the tribal substratum. For example, some *Muria* who was skilled in iron smelting and was adept in making iron implements, may have been allowed to ply the trade of iron-smith and his descendents have taken to this occupation and form today this functional group. Again among the *Saoras* there are a few occupational groups such as the *Arisis*, who weave cloth, and the *Kundals*, who make baskets for tribe. All tribes and groups in Bastar take to fishing as a diversion, but its adoption as a permanent occupation by the *Kurukhs*, widened their social distance from the *Marias*, from whom they are evidently recruited. These *Kurukhs* are indispensable to the social economy of *Maria* country, as they barter their catch in river and tanks for grain at customary rates. Similarly the *Rawats* of Bastar have taken to tending of cattles. The main occupation of the people is agriculture and lumbering, but the wild tribes are still accustomed to their nomadic life in forests and supplement their gleanings by crude cultivation. Agriculture is both by *jhuming* and terracing.

3. Tribal Villages

The tribal villages in India assume different shapes, depending partly on the peculiar social organization and partly on ecological conditions of the village sites. The *Santhal* villages are generally found in dense jungles and the houses are built on either side of one long street. Almost to every house is attached a pigsty or a dove-cot, while bullock or buffalo-sheds are distributed throughout the village. Every village has a *manjithan*, a spot where *manji* the headman meets the villagers and where perhaps some of the older famous *manjis* were buried. The *Manji* is both a civil and moral authority.

The *Munda* village is primarily centred round members of a single exogamous sept. Sometimes the *Munda* villages have been grouped to-

Plate 37. A Gond family (Andhra)*.

gether into a larger unit called *Parha* by the *Mundas*. The *Mundas* call the headman of their village *munda;* he exercises civil power over the villages, assisted by a *Parhan*, who performs religious functions. The *Parha* or *patti-panchayat* (village council primarily of five elders) is kept under the

* All photographs pl. 37–51 in this chapter are published through the courtesy of the Director, Anthropological Survey of India, Indian Museum, Calcutta.

guidance of the most influential of the headmen of the twelve villages.

The socio-administrative side organization of the *Hos* is better structured than in case of the *Mundas* in certain matters. The *Hos* of Kolhan are divided into twenty-four *Pirs* or *Pharganas*, which is linguistically the same thing as the *Parha*. They call their village *hatu* and the headmen of their village, *munda*.

The *Gond* village is often a collection of scattered hamlets, from two to twelve, each one being called *tola* or *khera*. The headman is generally known as the *mandal*. Other castes found in the *Gond* village are the *ahir*, *bari*, *lohar* and some *buka*. In *Gond* villages there is evidence of social stratification in form of *Raj Gonds* or the aristocracy, the *Dhur Gonds* (literally dust Gonds) or the common peasantry.

The *Bhils* live in scattered hamlets, called *phala*, each one consisting of a few huts, which are often less than kilometre apart. The scattered hamlets are grouped together and the unit-group of such hamlets is called *Pal*. The *Bhils*, who have this organization, are known as *Palia Bhils*, in contra-distinction to the other more backward *Kalia Bhils*. Three other castes are associated with the *Bhils* in their village organization. The *balais* and *chamars* are required to do the *Bhils* village work and the *gachhas* or sweepers to clean it. The headman of a village or a *pal* is called *gammaiti*.

4. Tribal Family and Marriage Types

The family is the standard social unit of the tribal peoples and it is found in some form or other at almost all levels of cultural development. The form of familial grouping has naturally varied from time to time and different types of families have been observed in different societies, in point of time and space. Among the tribals we find matriarchal and patriarchal families, and polyandrous and polygynous families; there are also families resulting from voluntary and involuntary monogamy, from group and tribal marriages.

In most parts of India, among the primitive tribes, marriage is a relatively simple affair, in which the couple decide to settle down as man and wife, often without the aid of elaborate ceremony. In most tribes considerable freedom is enjoyed by young men and women in finding their partners by mutual choice, and even where marriage is arranged by parents, the young persons concerned are consulted before the final ceremony. The *Kukis* of Assam allow probationary marriage, where the young man is permitted to live with his would-be wife in the latter's house for weeks and even months. The *Bhils* own two endogamous groups among them, the pure and the impure. Although considerable inter-mixture has taken place, the pure *Bhils* generally restrict their marriages among themselves. The *Parjas* or the *Dhruvas* of Bastar, till recently used to confine the marriageable girls of the village in an underground cell, where young men desirous of matrimony were to join them at night and

291

make their choice. The *Ho* and cognate tribes of *Munda* ethnic stock must pay a heavy bride-price.

Marriage in tribal society is often neither a sacrament nor is it indissoluble in life, as is ordinarily the case among the Hindus. Divorce and mutual separation are freely allowed for reasons of incompetence, cruelty and adultery. Adultery is, however, punishable by the social code of most primitive tribes, and it is the responsibility of the tribal or clan *panchayat* to see that the offence is not frequent.

Where women are dominant and choose their partners in marriage, as among the *Tharus*, adultery is not infrequent. Premarital licence is not frowned upon in such tribal society and in those tribes where late marriage is customary, virginity is not also an essential condition for marriage. Among the *Munda* tribes, girls and boys are allowed to mix freely and marriage may not take place even after they pass their teens. Where the bachelors and maidens of the tribe are housed together, as among the *Gonds* of Central India, sex training is imparted in traditional ways. Many tribes have introduced child marriage, partly as a claim to higher social status and partly as a measure of restricting premarital licence. The *Munda, Ho, Bhils* have, for example, popularized child marriage to restrict such licence.

All the tribes of *Munda* descent in Chota-Nagpur and elsewhere have to pay bride-price at marriage; the *Hos* pay both in cattle and cash, but the *Mundas* and *Santhals* only in cash.

There are some other features in the family life of tribal communities, to which attention should be drawn here. The young men and women may choose their own mates, but this can also be the responsibility of parents. When a bride is selected, a compensation has to be paid to her parents, as they are on the point of losing a working hand in the family. Among some tribes, the bride-price may be very high as, for example, among the *Galong* of North East Frontier Agency in Assam, and may entail very considerable amount of hard labour before one can successfully accumulate the necessary funds. This difficulty is often overcome by three ways. The bridegroom may elope with his chosen bride, in the hope of securing the approval of the elders, or he may serve in the house of his prospective father-in-law as a labourer and thus, in course of time, earn his right to the hand of the daughter. Reciprocal marriage is the third method in which the sister of the bridegroom is married to the brother of the bride; in such cases the dues may be largely written off against each other.

Next to the family comes the clan. The clan is composed of a number of families, often bearing a common designation, and which believe that they have all sprung from a common ancestor. Marriage is usually forbidden among members of the same clan. Among some tribes, the custom is to regard certain others as friendly or related clans; and no marriage takes place between the two. Clan organization regulates marriages, and also ensures cooperation between members when economic assistance is needed. Among the *Juangs* of the highlands of Keonjhar in Orissa, who practise shifting cultivation, the villages are usually inhabited by members of a single clan. When they adopt the more advanced technique of plough cultivation, changes naturally begin to take place. BOSE (S. BOSE, 1961) observed that in a village in Dhenkanel in Orissa, the layout of the new village is after the model of the linear, single-street Orya village, which is quite different from the loose, irregular agglomeration of *Juang* village, as in *Gonasika*. Many clans come to live together, to insure most economical use of both cultivable and homestead lands.

5. *Tribal Demography*

The tribal population has fluctuated widely in numerical strength in different periods. The decennial fluctuation of the tribal population, as percentage of the total general population, is given in Table I.

Table I. Fluctuations in tribal populations between 1881 and 1961

Year	Tribal population	Percentage in the total population of India
1881	6,426,511	2.58
1891	9,112,018	3.23
1901	8,584,148	2.92
1911	10,295,165	3.28
1921	9,775,000	3.09
1931	8,280,000	2.36
1941	—	—
1951	20,000,000	5.60
1961	30,000,000	6.80

The numerical strength of the tribes ranges from a few hundred to more than two millions. Some of the tribes seem to have increased in number, in recent times, but others have apparently declined considerably. The following tables show dominant trends of tribal demography:

Table II. Analysis of populations of some important tribes

Name of tribes	1921	1931	1941	1951	1961
Chenchu	12,402	10,342	12,898	—	17,866
Kota	1,204	1,121	952	—	922
Toda	640	597	630	879	716
Nayadi	301	296	250	—	—
Mavillar	1,737	1,341	—	—	—
Gadabba	53,770	48,154	74,813	54,454	66,907
Malpaharias	38,972	37,437	40,498	374	61,129
Bhoksa	7,628	7,618	274	—	—
Badaga	45,821	43,075	56,047	67,286	—
Naga Tribes	1,47,262	1,39,965	2,80,370	—	3,46,129
Angami Naga	51,730	49,237	52,080	28,678	632
Lhotas	18,309	18,238	19,374	22,402	—
Andamanese	786	460	—	27	19

Table III. Analysis of recent fluctuations in population of some tribes

Tribes	1911	1921	1931	1941	1951	1961
Asur	3,716	2,245	2,024	4,564	—	5,819
Bhil	1,067,792	1,795,808	2,013,177	2,248,152	—	2,609,701
Birhore	2,299	1,810	2,350	2,755	—	3,346
Gond	—	2,902,592	3,069,069	3,201,004	—	3,991,767
Ho	420,179	440,174	523,184	383,737	—	499,144
Juang	12,823	10,454	15,024	17,032	—	21,890
Katkari	91,841	81,202	88,336	69,170	—	*N.A.
Kachari	—	207,266	345,248	428,733	—	236,936
Kharia	133,657	124,521	146,037	167,669	—	224,781
Khond	750,289	698,668	741,078	744,904	—	819,702
Khorwa	200,077	185,553	237,847	205,638	—	66,109
Munda	558,200	559,662	658,450	706,869	—	1,019,098
Oraon	835,994	842,902	1,021,355	1,122,926	—	1,444,554
Santhal	2,078,035	2,189,511	2,508,789	2,732,266	—	3,154,107
Tharu	63,629	61,751	64,403	61,366	—	N.A.

* Not available.

It may be seen from the above table that the most important tribes in India, according to their numerical strength, are the *Gond, Santhal, Bhil, Oraon, Khond* and *Munda*. The decline of the aboriginal population in 1911 may be attributed to the following factors: 1. Tribal areas were exposed to malaria. 2. Absorption of the tribes into Hinduism in the Assam plains and in the north Cachar Hills. 3. Spread of Christianity among the tribes in Lushai, Khasi & Jaintia Hills, Central India, Travancore and Cochin. (the Christians and Hindus are not included among 'tribes' by the census enumerators). 4. Acculturation processes due to contacts with non-tribals in: a. Existence of mines and minerals in the tribal areas of Bihar, Orissa and West Bengal. b. Emigration of tribal labour to mines and factories in Assam and West Bengal. c. Network of communication in tribal areas. d. Activities of the Christian missionaries. e. Visits of administrators, scholars and military personnel in tribal areas.

6. Regional Distribution of Tribal Populations

According to the Census Report of 1961, the population of all tribes throughout the Republic of India stood at 30.13 millions, representing approximately 6.86% of the total population of the country. The tribes are distributed unevenly in the land and wide regional differences in the density of the tribal populations are, therefore, observed. Nearly 65% of the total tribal population of the whole Republic is concentrated, for example, in the Eastern Himalaya and other Assam Hills. The bulk of the rest is concentrated in the hills and plateau of Central India, but particularly Bihar-Orissa. South India and the Bay Islands are inhabited only by a small fraction of the tribes.

7. Tribal Government

Tribal life in India, as elsewhere, is characterized by an absence of a hierarchy of economic organization. The absence of a well assessed division of labour in primitive society does not favour the development of hereditary skill or technique, which leads to the formation of artisan classes, or guilds, so that spontaneous cooperation in domestic and economic life becomes essential. The dormitory affords, therefore, the training ground for educating the children of the village in all matters relating to the social and economic life of the tribes, so that they may participate in all activity of social or economic order.

The institution of a common village dormitory is found among most of the aboriginal tribes of the Chota-Nagpur Plateau, viz. the *Munda, Ho, Oraons* and the *Kharias*. It is also found among the *Gonds* and the *Bhuiyas*. Most of the *Naga* tribes in Assam, the *Aos, Memis, Lhotas, Angamis, Sema, Chang* and the *Konyak* tribes of the Naga Hills and the *Kukis* possess this institution. Some of the tribes in North East Frontier Agency also possess such an institution, especially among the tribes in the upper parts of North East Frontier Agency.

The tribes in Central India have also such organizations. The *Mundas* and the *Hos* call it *gitiora*, the *Oraons* call it *jonkerpa* and the *Gonds* call it *ghotul* (ELWIN, 1947). All young bachelors of a *Munda tola* or village have a fixed common dormitory in the house of a *Munda* neighbour, who may have a hut to spare for the purpose, while the unmarried girls of the village sleep together in the house of a childless *Munda* couple or in the house of *Munda* widow. The 'ghotul' institution appears to have developed

Plate 38. Maria drummers, in ceremonial dress, capped by bison-horn head-gear.

to perfection in certain *Muria* villages, where it has effectively superseded tribal or clan organizations.

In Bastar, the *Murias* have a regular organization; their captain is called *sirdar* and the master of the ceremonies, the *kotwar*, while there are other officials bearing the designations of state officers. The *Oraons* also have a similar organization and the captain, known as *dangar mahato* has an acknowledged position among the village officials. The roots of political organization are to be traced to these dormitories, which are characterized by group solidarity and discipline.

The *Muria ghotul* has a hierarchy of functionaries. The names of the officials are often borrowed from the titles of zamidari (landlords) or state servants. Married people are not allowed in the *ghotul*, but special consideration may be shown to the widows and widowers who want to share the ghotul. The ghotul organization has a tremendous effect on the social life of the tribes concerned.

The dormitory is thus a group organization. Its origin may perhaps be traced to the campings, where the ablest hunters of the community took their shelter for purposes of defence and protection of the weaker members, but in course of time other traits have slowly been woven round it and the elaborate ghotul of the *Murias* is the result. With a settled life and a better control of food supply, predatory excursions of neighbouring groups for women or for cattle become rare, but the economy of accommodation in the house helps to maintain this communal organization, as the members find it a convenient place where not only to sleep in but also a venue for their communal activities. The dormitory institution ensures tribal endogamy, by controlling the movements of women within the tribe's specific area and prohibiting social intercourse between men and women belonging to different tribes.

Since India became independent in 1947, it has become the policy of the Government of India not to hustle the tribal people into a faster pace of political change than they themselves wish. Schedules five and six of the Constitution guarantee to the tribal areas of Assam a far greater measure of autonomy than that enjoyed by other citizens of India. Under the sixth schedule, a large measure of autonomy has been vested in district and regional councils in the regulation of their economic and social life. The laws of inheritance, the appointment of chiefs, the regulation of marriage, control over shifting cultivation and water resources for agricultural purposes, all have been placed under the authority of the two councils named above and of which membership is largely by election. The *panchayati-raj* scheme has also been introduced in tribal areas since 1959.

8. Interrelations with others

The mode of living of the tribal communities and their relationship with their non-tribal neighbours have passed through a series of profound and complex changes. On the basis of these historical changes, the tribes of India may be divided into three classes: 1. The tribes like the *Raj Gonds* and others who have successfully fought the battle against the new-comers and are recognized as members of fairly high status within the Hindu society; 2. the large mass of other tribes that has been more or less completely and gradually absorbed as integral parts of the Hindu society and has come into close contact with the Hindus; and 3. the hill sections, which have exhibited the greatest power of resistance to the alien cultures that have pressed upon their border. The second class has suffered moral depression and decay as a result of their contacts with the non-tribals, from which the third class has largely been free (ELWIN). The causes of this depression as far as contact with Hindus in pre-British days is concerned are mainly two, viz. the loss of their land and the causal and transitory nature of their contact with the Hindu religion. Under such circumstances the aboriginal became keenly aware of the inherent weakness and drawbacks of his customs and came to be ashamed of his own faith, but had no chance to learn another, and the decay of religion was the inevitable result.

We know that the *Santhals* were formerly wandering from place to place in the hope of finding suitable land, from where they hoped they would not be ejected. We have, for example, the record of the family of a *Munda* chieftain, who had turned Hindu and introduced Hindu families into the villages of Chota-Nagpur from the latter part of the seventeenth century onwards. The Hindu families soon began to acquire more and more land, steadily dispossessing the *Mundas* and the *Oraons*. In all these areas the respective tribes were the earlier settlers, particularly so in Chota-Nagpur, where they had reclaimed the land from the jungles.

In Central India, for some generations before the Maratha Conquest of the area in the middle of the eighteenth century, the Hindu colonists had been steadily ousting *Gond* villagers from more fertile tracts, so that at the time of the coming of the Marathas, the Hindus formed the bulk of the population of the plains and also held most of the responsible offices of government. Though there can be no room for doubt that a number of the aboriginal tribes had lost their lands to Hindus, it must be made clear that this loss of land was largely an incident of tribal conquest or a result of the favour of ruling families of the aboriginal stock, and only to a small extent the consequence of deliberate exproprietory tactics of the Hindus.

Most administrators and observers have hoped that improvements in some aspects of the character of the tribals could come about only through their assimilation in the Hindu Society. We shall now examine some of these significant changes by reviewing the history of a few representative tribes.

Some of the tribes had still largely maintained their independent and traditional ways of life at the beginning of the British rule in India, so that they defied the Hindus of the plains, until the British arms brought them under control and opened their country partially to the influence of civilization. The *Hos* of Kolhan were so determined that in order to preserve their area for themselves, they not only refused to allow any non-*Hos* to settle amidst them, but also prevented the Hindu pilgrims, on their journey to the shrine Jagannath, from passing through their tract. The *Paharias*, who were in contact with the plains Hindus, habitually indulged in plundering the Hindus from time to time. The *Santhals* were settled by the Government in the Daman-i-Koh of the *Santhal* Pharganas about 1836. The land was fertile and was long coveted by the Hindu inhabitants of the plains, but their desire had remained unfulfilled owing to the marauding habits of the *Paharias* of hills. The *Santhal* had, however, very rudimentary notions about the value of money and consequently many of them fell victims to unscrupulous traders and money lenders. Again when the work on railways offered itself at their doors and many of the *Santhals* being bound for agricultural service to their landlords, found that they were not free to take advantage of it, they rose in rebellion in 1855. The rising quelled, the *Santhals* were pacified by the creation of a

Plate 39. Santhal women selling pottery in weekly market in neighbouring non-tribal villages.

299

new district called the Santhal Pharganas, which was to be administered by the Santhal tribal organization itself. This arrangement brought back prosperity to the *Santhal*, and he could also profit by the railway employment and was much in demand in tea-gardens, farther away from his home. Unlike the *Paharia*, he has shown willingness to take advantage of all opportunities for selling his labour. The working of huge coal fields at Girideh, Jharria and Ranigunge offered him facilities near home. Work on the coal fields suited him best, as it could be made to fit in with his off-season unemployment in the agricultural operations. Christian missionary activity, starting its work in Bankura in 1864, enabled the *Santhal* to educate himself to such an extent that many *Santhals* have gone forth into the world as clerks, assessors and accountants.

The tribes, so far dealt with, are typical of the peoples whose relations with the Hindus are positively known to have been unhappy and unpleasant. There were also tribes, whose chiefs voluntarily adopted Hindu customs, manners and outlook, and also introduced Hindu landlords by way of bestowing favour. The tribal peoples attitude to the Hindus as a whole was not thus hostile in the beginning; however, later on when their interests clashed with those of the new settlers, they rose against them soon after the British acquired a sort of effective control over the tracts inhabited by them. Principally they are the *Mundas* and *Oraons* of Chota-Nagpur.

A significant change has taken place in the life of the tribal people after India attained independence. This has firstly been due to the planned and comparatively rapid industrial development, and secondly to the acquisition of new political rights through adult franchise. Both have very profoundly altered the relationship of the tribal peoples with their neighbours.

Since Indian independence, there has also been an unparalleled extension of roads all over India, and motor vehicles are now in extensive use even in remote villages. This situation has brought the tribal people into frequent contacts with the urban people and urban ways, and even with members of other tribes in a way that never happened before. In the border areas of the North East Frontier Agency, Nagaland and Himachal Pradesh the tribal people often also come into contact with the Indian Army.

In parts of Bihar, Orissa, Madhya Pradesh and Andhra, the scene is, however, quite different. These states have large reserves of minerals and hydro-electric power, so that mines and industries have sprung up in large numbers during the last two decades in places like Ranchi, Ruarkela Bhilai and Bailadila. Even before the independence of India, industries had been established in the areas predominantly inhabited by tribal communities, as in Jamshedpur, or in the copper belt of Singhbhum. Large contingents of workmen were brought from the Punjab, Andhra Pradesh, eastern Madhya Pradesh and Tamil Nadu, since the local folk

Plate 40. A young woman of the Kuri-Kandh tribe from Koraput (Orissa) bathing her children outside the hut.

were slow in taking to new types of work. It, however, brought discontentment amongst them and the relation of the tribals with them suffered considerably.

Another source of dissatisfaction and tension among the tribal peoples has been the forest departments of the government. With the increase of population and because everyone sought security in agriculture, the forests of the country have nearly all but vanished, but the forest department restricts the wild and unplanned extension of cultivation within the confines of the forests. It fences off new forest plantations in order to prevent the depredations by cattle, goats and sheep, and thus comes into conflict with the immediate short-term interests of the tribal peoples. The tribal peoples feel that this is an encroachment on their traditional rights to use forest land, and they often defy the barriers set up by the forest department, while their animals are encouraged to break through fences, and thus seriously interfere with the growth of new plantations.

Contacts of tribes with their neighbours are no longer slow or on a small scale, but swift and massive and in consequence there is an increasing political consciousness. The immediate result is that the tribal peoples have begun to develop the idea of 'homeland'. Numerous tribal communities react as if their very existence is threatened and some of them have even tried to reaffirm their separate identities and a desire for a kind of nationalistic unification has thus arisen. This attempt at nationalistic unification may be looked upon as a step in the preparation of tribes for taking a due share in the economic and political development that is taking place in the country.

9. Cultural-Ecological Adaptations and Changes

The tribes of India, which number about 29.8 millions and constitute 7 % of the total general population of the country, may be classified into the following socio-cultural units, each of which adapts in various ways to the ecological conditions for its sustainance. The simplest ways would be to arrange the tribes into categories, based on the manner in which they primarily make their living. On basis of their economic life the tribes may be classified into hunting, fishing and gathering tribes; shifting cultivators, peasants, artisans and caste and nomadic groups (BAINBRIDGE, 1907).

9.1. HUNTERS, FISHERS AND GATHERERS

Many tribes on the mainland of India live by hunting, fishing and gathering, though these activities do not form part of their subsistence economy; they are supplementary to their other sources of economy. The aboriginal inhabitants of the Andaman and Nicobar Islands, depend, however, entirely upon these for their livelihood. These islands are

Plate 41. Chulikata hunters from the Mishmi Hills (Assam).

inhabited by a number of tribal groups, which are greatly and nearly completely isolated from each other, and with whom they do not even have any trade relations. Due to their total isolation, the *Onges* of the Little Andaman do not understand the language of *Jarawas* of the Great Andaman or the inhabitants of the North Sentinel Island, though all of which quite are closeby. Each of these groups satisfies all its needs completely with the help of local resources and exercises considerable ingenuity

303

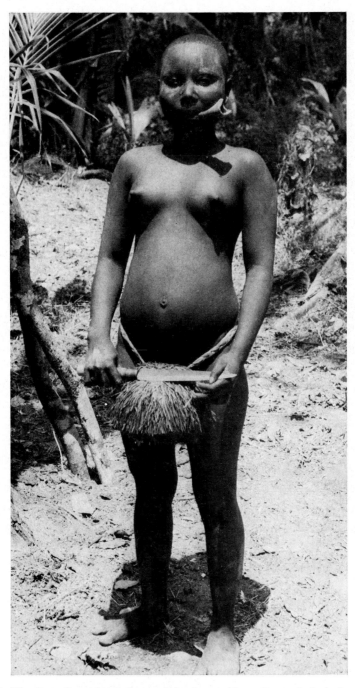

Plate 42. The Onge from the Andaman Islands, smoking pipe and clothed sparingly in grass-rope dress and holding her handy knife.

304

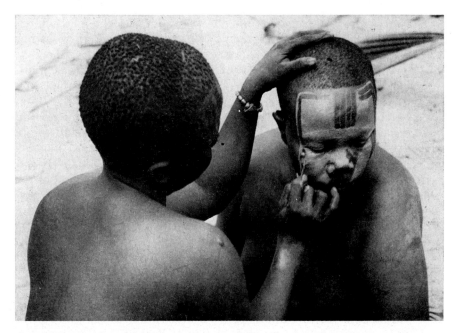

Plate 43. Open-air beauty parlour of the Onge from the Andamans, where the beauty specialist is painting the face of a customer.

in maintaining themselves on these islands. Technologically they are, however, very poorly equipped.

The Andaman Islanders are small-statured, dark coloured, kinky-haired people, but with beautifully proportioned bodies. They belong to the Negrito race and are physically akin to the *Semang* of the Malaya Peninsula and East Sumatra, and *Aeta* of northernmost Philippine Islands, namely Luzon. On the whole the Andamanese live either near the coast, where they depend on fishing or in the forest, where they depend on hunting for their existence.

The Andaman Islands, which together with the Nicobar Islands form a Union Territory of India, lie along the eastern part of the Bay of Bengal, and hence are also collectively called Bay Islands. Their total area is about 6500 km². The largest island, Great Andaman, is nearly 190 km long, but its breadth is nowhere more than 30 km. The Little Andaman, the only other large island, is about 42 km long and 25 km wide. All the rest are merely outlying islets. The estimated native population, before the effects of European diseases were felt, was 5500 (SERVICE, 1963).

The climate of the islands is tropical, warm and moist, with little variation in annual temperature. The greater part of the relatively high annual precipitation (about 350 cm) falls during the monsoon season, from May through part of November. The rest of the year is quite dry.

305

Fresh-water streams are rare and not also large, except in the Great Nicobar Island, and the rain water drains into large interior swamps.

Our knowledge of the Andamanese is mainly due to the labours of E. H. MAN, a British government official in the islands from 1869 until 1880 and A. R. RADCLIFFE-BROWN, who studied the Andamanese from 1906 to 1908. By 1906, European diseases had reduced the native population to 27% of what it had been during MAN's period of residence and this reduction apparently affected the local village organization and kinship nomenclature also.

About 1900 the aboriginal population of the Andaman Islands constituted an important proportion of the civilized population; the aboriginal population was the 10.5% of the total civilized population. The aboriginal population has since then declined, as a result of natural and man-made causes and in 1961 it constituted only 0.4% of the total civilized population.

The Andaman forests are devoid of large game, but pigs are plentiful, and their meat and fat are greatly relished by the Andamanese. Some years ago, the government of India introduced the spotted dear in a few islands, where they have multiplied greatly. It is however, curious that the Andamanese do not hunt these animals at all. When two *Jarawa* young men, arrested and kept in detention at Port Blair some years ago, were offered several kinds of meat, they smelt and rejected them all, but when pork was offered, they went into excitement.

The Andamanese catch very little fish with nets, but they use bows and arrows and spears for this purpose. There are coral reefs around some of the islands, where the water is shallow and crystal-clear, so that it is comparatively easy to spot fish and turtles from canoes. Turtle's eggs can also be collected easily from some of the lonely islands. Shellfish of various kind and crabs are also gathered as food. It is, however, interesting that, so far as the *Onges* are concerned, they do not shoot birds for meat, although the bird life of the island is not poor. It has been suggested that they do not do so for fear of losing their arrows in the thick vegetation that covers the island.

Though the arrows of the *Jarawas*, sometimes used against outsiders, are found to be tipped by some kind of hard wood, they appreciate the use of iron; and try to steal bits of this metal from the cottages of those, who have settled down as peasants in the jungle. It is said that formerly they used to collect iron from the wreckages of ships cast upon the coast.

It is remarkable that the food, which the Andamanese eat by simple boiling, is never seasoned with salt. If meat cooked with salt is offered to them, they reject it forthwith. Honey is one of their favourite foods, and from January to March they spend considerable time in gathering honey from wild hives in the forest. There is some kind of leaf called tongee, the juice of which is mixed with saliva and besmeared over the body, and the smell prevents the bees from stinging the men when collecting honey.

306

An enquiry about the food habits of the *Onge* (S. Bose, 1964) reveals that the average intake of food is about a kilo per man per day. Proteins, consisting of pig's meat, fish, turtle's eggs, crabs and bivalves, constitute 76 %; carbohydrates consisting of root crops and tubers 22.6 % and fruits and honey, 1.4 %. When food is abundant from the hunt, the *Onges* even consume 3 kilos in a day, but they may also go without food for two days or more in succession, if none is available. In this respect they are somewhat like the larger carnivores of the forest of Chota-Nagpur.

Another interesting feature of their food habits may be mentioned here. In one of the villages studied, the number of men and women during the period under investigation, fluctuated between 16 and 60, in the second between 41 and 102. It all depended upon how many came to share the feast. There was apparently no quarrel over who should join and how the food was to be divided. Everything seemed to belong to everybody, and one could eat as much as one needed.

It seems also extra-ordinary that the Andamanese should be one of the very few peoples in the world who use fire, but do not know how to produce it. They have, therefore, to tend the fire very carefully in a country that is subject to a rainfall of 375 cm and where they live in huts thatched with leaves and grass.

The Andamanese practically have no pottery, but use containers hollowed out of wood. They also make fine baskets and now the government started presenting them with iron buckets, aluminium vessels and tin containers. Axes are also left as presents for the *Jarawas;* and although they are very timid and suspicious and make themselves scarce when an outsider approaches them, there is abundant proof that they make use of these presents very effectively in cutting down big trees.

Bose (N. K. Bose, 1971) gives an interesting account of a *Jarawa* settlement, which he visited quite recently. The *Jarawa* men promptly disappeared as his boat approached it through the coral reefs. Bose found that between two trees, there was fairly long strip of cane on which about a dozen sting rays had been hung up to dry. The intestines had been removed, and the fish strung through the eyes. The fish had become quite hard by drying, and the line was stretched from north to south so that the sun's rays could beat upon them in full.

About the social organization of the Andamanese, very little is known, except what has been recorded about the dwindling tribe of the Great Andamanese, studied by A. R. Radcliffe-Brown (1948). The Andamanese are divided into various sub-divisions, based on differences in language. It is also known that many of the small local groups combine in hunting and festivity; and there is also no bar to marriage between neighbouring but distinguishable local groups. The family, consisting of husband, wife and children, forms the most important social unit; but there is no trace of a clan. A few of the local groups may be said to constitute a tribe subject to their elders, but not with any well-defined political authority.

The *Onges* of the Little Andaman are divided into various septs or local groups within clearly defined hunting grounds. Such a local group of about ten families, headed by a chief, builds its permanent encampment, a large communal hut, roughly circular in form and thatched with palm leaves. The communal huts are found in two locations on the island; firstly the communal huts near the sea coast and secondly the communal huts of the interior forest regions. The huts are connected with each other by path running through the jungles. Each communal hut is located by the side of some sweet-water stream. They also make temporary sheds when they go out on hunting expeditions. An average temporary shed is composed of ten to twelve huts, which may sometimes even increase to twenty.

The *Onges* hardly wear any clothes, except those which are now being given to them as gifts by others. A woman's garment consists of a waist-band and an ornamental tuft of vegetable fibres, suspended in front. They are very fond of decorating their bodies with ornamental and geometrical designs, painted grey, yellow or red, with coloured earth, mixed with pig fat or spittle.

BOSE (S. BOSE, 1964) has calculated the land-man ratio among this hunting and gathering tribe of the Little Andaman. The area of this island is 1420 km². In 1964 the population of the *Onge* was found to be 132, with perhaps an addition of ten or fifteen more, who might have been overlooked while they were out hunting in other parts of the island. If the total is taken as 150, then the density of the *Onge* per km² is 9.4; or roughly every person has at his command 4.7 km². It was also found that the land, which can support a man with food in the forested interior for nine months, would be sufficient for him for twelve months, if he lives on the coast, for the sea offers him a greater store of food than the forest, provided he is skilled enough to utilize it.

The Nicobarese have Mongoloid features, and may be divided into two major groups, namely the *Nicobarese* and the *Shompen*. The latter are more primitive than the *Nicobarese* and are found in the interior regions of the Great Nicobar Island. According to my own observations in the Great Nicobar Island during 1966, the average height of the male *Nicobarese* is 160.7 cm and that of the *Shompen* male is 159.1 cm. Technically the *Nicobarese* may termed as 'below medium' and the *Shompen* as 'short'. The *Shompens* are 'medium-headed' and the Nicobarese are highly 'broad-headed', but both have medium nose. The hair form among both the groups is mostly straight, although some have flat wavy hair. The skin colour of *Shompen* is light brown and that of the *Nicobarese* is somewhat fairer.

The Nicobar Islands are quite hilly and densely forested. The climate is more uniform than in the Andamans. It has copious rainfall throughout

Plate 44. A Shompen woman, man and child from the Great Nicobar Island.

Plate 45. A Typical Nicobarese village from the Nancowry Island.

the year, except scanty rains during February and March. The mean annual rainfall recorded on Great Nicobar is 3226.8 mm. The mean annual temperature recorded for Great Nicobar is 26.7 °C. April, being the hottest month, records a temperature of 28.1 °C. The forest wealth of Nicobar Islands abounds in a variety of hard and soft wood, bamboo, areca palm and cocoanut trees. There are no large mammals in the forest, except wild pigs and deers, which are hunted by the Nicobarese. Monkeys, with black face and long tail, are also found in the forests. There is a large variety of birds but curiously they are not hunted by the *Nicobarese*. The rivers are full of fishes and crocodiles. The sea is an inexhaustible source of fish, octopus and shells, which are regularly collected by the Nicobarese. The Nicobarese practise no agriculture but collect the fruits of *Pandanus*, (large orange-coloured fruit, of the size of jackfruit) and use them as food. It is boiled and its pulp is taken out to be eaten. The cocoanuts are another source of food. Fish and octopus are caught by spears made from the wood of the areca palm. They also collect areca nuts, cane strips and honey from the forest and sell or barter them mostly with Gujarati traders.

The *Nicobarese* build their huts with wood, cane and bamboo, covered with leaves. The traditional dress of *Nicobarese* male is a *lagoti* with a tail-like end and mostly they remain bare bodied. The women wear *lungi* made

310

from printed or plain coloured cloth. They generally remain bare bodied, but in front of outsiders they cover themselves with blouse or tuck the upper part of the *lungi* under their armpit. The *Nicobarese* make nice cane baskets and canoes. They are very fond of chewing betel nuts and betel leaves.

The *Nicobarese* are neither patrilineal nor matrilineal, but their rule of descent is bilateral. Households are based on joint families and in the majority of cases the daughters remain with their parents after marriage and their husbands come to live with them. They are monogamous but separation is not uncommon and is executed without a fuss. More than half of the population of Nicobar Islands are Christians, but in the Great Nicobar Island the people still follow the traditional religion of animism. Both the groups bury their dead.

The land-utilization survey of a Nicobarese village carried out by me in 1966 reveals the following facts: The approximate area of Pulo-Babi village is 129.5 hectares, out of which 15.6% land was under *Pandanus* gardens and 25% under cocoanut gardens. The rest was under homestead, bush and forest growth. The total population of the village was 38 persons. I calculated the daily production and consumption pattern of the village population for a period of thirty days. It revealed that the *Pandanus* garden of the village can support twenty-five persons in perpetuity, while the cocoanut gardens can support twenty persons. So the total carrying capacity of the village land was 45 persons, if they entirely depend on pandanus and cocoanut gardens. The village already had a population of 38 persons and a number of domesticated animals like pigs, fowls, dogs etc., which also depend for their food on the men. This leads to shortage of food, which has therefore to be supplemented by the products of the sea.

9.2. SHIFTING CULTIVATORS

Assam, Nagaland, Manipur and Tripura enjoy a warm and humid climate and the rainfall is well over 250 cm annually. The whole of the land is covered with a thick mantle of vegetation. In the midst of these warm, rain-drenched forests, there live a number of tribal communities, which depend principally upon a rather simple and primitive form of cultivation. The same method is also in vogue among some of the tribes of Orissa and Central India, and outside India it is also practised extensively in northern Burma, Sumatra, Borneo, New Guinea, as well as in parts of the African continent*.

Among the Indian tribals, a village community controls a certain measure of land comprising mountains and valleys, and brings a small part of it every year under cultivation. Ploughs and cattle are not

---

* Shifting cultivation is common also among the South American aboriginals (Kuikuru Indians) of Central Brazil. Note added in proof – M. S. MANI.

employed, but axes and bill hooks and digging sticks are the only implements used for the purpose. After winter, a portion of the hill-side or jungle is marked off, cleared by lopping off the undergrowth and branches of trees, which are left to dry in the sun for some time. Shortly before the rains set in, the dry leaves and bushes are set on fire. Farmers take care that the fire does not spread into the forest, but are not always successful, so extensive damage to the forest results. When the fire dies down, the ashes are lightly spread over the ground where necessary. The fire kills the weeds and insects, and the ashes fertilize the ground. Then the former goes to the field with a digging stick or bill-hook in hand, makes a hole in the ground, sows a few seeds, and covers them over with earth by pressing it down with his toes. As the rains come, the seeds begin to sprout, and the harvest is gathered as each crop ripens. In Nagaland or in North East Frontier Agency the land may be used for only one season or two; while in more crowded areas like Orissa, it may be used for three seasons and then left as a fallow for a number of years for regeneration. The period of recovery may vary from three or four to ten years, depending upon the needs of the farmer and the pressure of population in the locality.

This practice of shifting the area of cultivation is known under many names. In Assam it is known as *jhum* or *jum*, in Orissa as *podu*, *dahi* or *kamana*, and *penda* in Madhya Pradesh, etc. It is best described as slash and burn or swidden cultivation or shifting cultivation. An important feature of this cultivation is that those who practise this form of cultivation do not themselves move from place to place to form new settlements. What they do is, every family goes on adding a fresh patch of forest every year, while a patch which has been used several times is abandoned. The villages remain, however, in the same place, generation after generation.

The crops which are grown in these fields vary very much from place to place, depending on climatic and soil conditions. In North East Frontier Agency upland paddy, maize, millets and job's tears are grown in abundance in the *jhum*-fields. In North Cachar and Mikir Hills District of Assam, cotton, papaya and vegetables of many kinds form the principal crops. Among the *Juang* of Keonjhar District in Orissa, the first year's crop, which is grown in a *podu*-field for sale, consists of *Sesamum*. It is purchased largely by the neighbouring peasants for extracting edible oil. *Sesamum* is followed in the second year by upland paddy, and then come millets of one kind or another. The second or third years's crop is generally kept for home consumption. In the hills and plateaus of Western Palamau District in Bihar, a variety of pulse named *ram-arhar* is grown copiously in fields cleared by axe and fire. This has a ready market among the neighbouring peasantry.

It is thus interesting that, in all the areas mentioned above, the people who practise shifting cultivation do not wholly use the produce of their land for their own consumption. The cotton of Mikir or Naga Hills, the pulses of Palamau, the vegetables and cotton of Dimsa Kachari or of

312

Riang of Tripura are all meant for outside sale. With the money thus earned, the tribal people buy their other necessities, like cloth or iron, tobacco, salt, sugar and tea. The poorer crops, like coarse paddy or millets, grown in the partially exhausted soil, are largely used by the farmer and his family. In this way much of the shifting cultivation, which is still practised, is bound up with the economy of the outside market. This market supplies the requirements of the peasant populations of both tribal and non-tribal origin, which live nearby, and which pay for goods and services in cash. Unlike the hunting and gathering of the Andamanese, it has come to be an ancillary to a larger peasant economy and has lost its independent status. Yet wherever possible, the tribal communities continue to practise it, because in many of the hill sides this is the only practicable method of land utilization. When we consider the thin population of the area, it is not possible to convert the hill slopes into terraced fields for growing wetland paddy.

At this stage, it would also be useful to refer to some practices in vogue among the *Adis* of Siang District (NEFA), *Juang* tribes (Orissa) and the inhabitants of Mizo or Naga Hills in Assam. Among the *Adis* of Siang (NEFA), particularly, *Gallong* and *Minyong* tribes, the village community controls the *jhum*-land of its own. The entire *jhum*-land of a village is known as *patat*, which may consists of several hillocks around the village. The *patat* is subdivided into blocks owned by individual households. It often happens that one household may possess *jhum*-plots on several hills. Every year the village council meets and decides which hillock is to be opened up for *jhum*-field. After a decision is taken, the village elders go to that hillock and cut down the trees, while the women and boys cut the bushes. The cut trees and bushes are then left to dry, burnt and the ashes are mixed with the soil. Seeds are sown at the appropriate time by digging stick. The cultivation in this block of land continues for two or three years, depending upon the fertility of the soil and the density of population. It is then abandoned for some years. Thus every year two blocks of *jhum*-land are utilized: an old plot is put under cultivation and a new plot is opened up. The *jhum*-field thus moves in a circle, till after a rotation of 10–11 years the first plot is reopened. Sometimes the households of a village form two or three parties and open three *jhum*-fields on three different hills. It has also been noticed that a particular household may not cultivate the entire plot of *jhum*-land, earmarked for it, but may sublet a portion of it to some resident of the village and charge a rent on it. The tenant cultivator takes all the produce of the field, but he does not own the plot of land.

Among the *Savaras* of southern Orissa, a piece of land is placed under *podu* for one year or so. Then the villagers combine and erect small stone walls on the hill slope and convert the whole of the latter into series of terraces. Water from the neighbouring hill streams is carefully diverted to these fields for irrigation; and thus, with great care and ingenuity, the

Table IV. Summary of the carrying capacity of different tribal areas (S. Bose, 1967)

Area and village	Population per km²	Carrying capacity per km²
Mizo Hills		
Mampui	2.5	16.4
Sairap	7.4	12.4
Bastar District		
Batar Batar	4.0	6.3
Gundakote	1.5	5.2
Kondakote	1.8	4.0
Dandrawada	4.7	6.4
Keonjhar District		
Raidiha	4.3	9.1
Kadalibadi	31.0	10.5
Hatisila	1.7	8.5

Savaras eventually turn into irrigated paddy fields what was opened up originally as a *podu*-field.

A detailed survey, undertaken by the Anthropological Survey of India in 1961–63, determined the carrying capacity of land under shifting cultivation (N. K. Bose, 1971). About 328 km² of land in the Mizo District of Assam, Keonjhar in Orissa and the Abujhmar Plateau in Bastar in Madhya Pradesh were subjected to detailed survey. The soil and rainfall vary conspicuously in these areas; but the technique of cultivation is not very different and can be looked upon as fairly uniform. The soil in Mizo Hills is the result of the disintegration of friable Tertiary sedimentary rocks. In Keonjhar and Bastar, it is largely derived from Archaean gneisses and schists, interspersed with veins of quartzite and igneous rocks. The rainfall differs much in the three areas of investigation; it is 325 cm in the Mizo Hills and between 125 and 150 cm in Bastar and Keonjhar.

Only the cereals grown in the *jhum* fields are included here and supplementary sources like meat and eggs, fruits and vegetables are omitted, because of the inadequacy of records. Secondly, during the investigation, the Mizos (Lushais) had just collected their harvest and the food was also plentiful. The consumption rate was thus found to be as high as 3,500 calories per adult per day. This was taken as a uniform standard, from which to calculate the carying capacity. Children under twelve years were treated as half adults.

It is evident from an examination of the table that, although all the tribes in question live by shifting cultivation, the problems with which they are faced are very different.

314

In Mampui, for instance, there is an abundance of land, while in Sairap the land which can actually be brought under slash-and-burn cultivation is small and the forest is thin, the inhabitants have to supplement their earnings by working as labourers in road building.

In Bastar also, the margin between land and population is narrow. The *Gonds* of Abujhmar continue, however, to live in their old ways, since their territory is far away from roads and markets, and as they have nothing else to fall back upon, except their own ingenuity in extracting food from the soil.

In marked contrast, the *Juangs* of Keonjhar are in a sad plight. Where the carrying capacity is no more than 9.5 per km^2, they have to feed 26.1 adults, on an average, from the same area. They are, therefore, constantly faced by acute shortage of food and have naturally to resort to diverse measures for survival. The first step is, of course, supplementing their stocks of food by gathering leaves, fruits and tubers from the jungle. In some seasons, they depend largely upon wild mangoes and jack fruits for their diet. The stones of mangoes are also ripped open and the thick cotyledons are dried and pulverized into flour, which may be mixed with other stuff while cooking. Jack fruits in season are also most welcome; besides the flesh, the seeds are also good food. The second step is that, when food in a village becomes scarce, some families move off, climb to higher and to the more inaccessible hill slopes, where they found new settlements, in which they continue their accustomed way of life. A village named Panasanesa threw up, for example, a colony on the upper reaches of a neighbouring mountain, which came to be known as Upara Panasanesa (Panasanesa of the heights), in contrast to the original one, which came to be designated as Tala Panasansea or to be (the Lower Panasanasea).

Such favourable new sites are not, however, always readily available, and on account of their height and rocky character, they can hardly support all the people, faced with shortage in their home village. The situation is then met with in several ways: they hire themselves out as agricultural labourers in fields owned by others, who may not belong to their own tribe; they supplement their meagre earnings by collecting and selling jungle produce like firewood, honey, leaves of *kendhu* tree (*Diospyros tomentosa*) for the manufacture of country cigarettes; they may also give up shifting cultivation altogether and take to use of the plough and bullocks. This last step requires some amount of initial capital outlay, which most *Juangs* can rarely afford.

In the District of Dhenkanal in Orissa a large number of *Juangs* have taken to the plough cultivation, but in the uplands of Gonasika in Keonjhar, they still cling to the old practice. The small settlements of the *Juangs* in the uplands of Keonjhar are mostly inhabited by members of a single clan. When, however, they come down to the valleys of Dhenkanal and adopt the farming methods of their neighbours, the villages grow more populous, as land cannot be wasted and new problems arise.

Marked and rapid changes are taking place even in the tribal communities, which still cling to the shifting cultivation. Hardly any community in either north-eastern India or Central India lives now entirely by shifting cultivation. It has largely become an auxiliary means of support for those, who have been changing over to the plough cultivation or working as labourers wherever jobs are available. The produce of the *jhum*-fields is often meant for the market. The more efficient productive system of the peasants in the plains may be said to gradually, but inexorably swallow up the less efficient productive system of many tribes in India. The social system of the Hindus has also begun to exercise its influence over the tribal peoples.

In the villages of Dhenkanal and Keonjhar, where the *Juangs* have given up shifting cultivation and taken to the plough, there has arisen a new urge to be regarded by others as one of the peasant castes. They have even set up a caste panchayat, on the model of the trading *tailikavaishya* or *teli* (oil monger) and other castes. Meetings are held in the same manner, resolutions regarding internal reforms passed and the government's attention is drawn to the need for establishing schools and for the promotion of economic development.

Affiliation of the tribal communities with the productive system of the Hindu rural communities has thus led to the gradual assimilation of some sections of the *Juangs*, practically into a new Hindu caste. It is also interesting to observe that all the reported changes have taken place without any effort on the part of the Hindus to proselytize; it is the tribal community moving within the Hindu circle. The source of the R. Baitarni of Gonasika in Keonjhar is held sacred by the *Juangs*. The Hindus of Orissa also consider this as a place of pilgrimage, so that at the religious level, a bridge has thus been built up, just as at the economic level the *Juangs* have been deeply affected by the productive system. A chain reaction was thus started, which began at one point and gradually engulfed other aspects of tribal life and culture. Inter-personal and inter-communal relations were recast in a new way, so that the *Juangs*, inspite of their distinctive identity, have come to be observed as an integral part of the local social structure.

The tribal events elsewhere are, however, very different. The tribes of North East Frontier Agency, for example, were not subject to the same kind of contacts with the peasants of the Brahmaputra Valley as the *Juangs* were with those of the Baitarani. Moreover, the contacts of the *Juangs* have extended over many generations, while those of the *Adis* of North East Frontier Agency or of the *Nagas* of Nagaland with others is very recent and rather sudden.

9.3. Peasants, artisans and castes

Having thus explained how as a result of contact of the tribals with the civilized peoples, the former are gradually changing and adopting the productive system of the latter, we shall now consider some of the results of such a process. There is no doubt that the main stimulus for the tribes in this process is the greater promise of food, which the more advanced methods of the non-tribals hold. It is precisely for this reason that the centuries-old system of shifting cultivation is being progressively given up in favour permanent cultivation in terraced fields.

Terraced cultivation is being practised by tribes in the Himalaya, particularly in the western and central regions. Even in North East Frontier Agency among the *Apatani* tribes, fields are terraced and ingenuously irrigated by diverting the hill streams. The *Apatani*, like the *Newars* of Nepal, use only the hoe but no plough or draft animals for cultivation. In the mountainous regions of the Himachal Pradesh, terraced forming is, however, carried on with plough and bullocks. In some of these areas, there are no specialized castes of artisans; in others such castes are present. In the Jaunsar-Bawar region of the Dehra Dun District (U.P.) the population is distributed in three groups, which occupy different altitudinal zones. They are the high-caste group, intermediate caste and low-caste groups. The high castes occupy the main and higher parts of the hill, while the low-caste groups are given shelter on the lower slopes. The intermediate castes live here and there within the main cluster of huts.

As all the tribal communities are rapidly changing as a result of their contacts with the rest of the population, it is evident how the *Juangs* of Orissa, the *Gonds* of Madhya Pradesh or the *Santhals* of Bihar and Bengal have eventually come to be assimilated within the orbit of the peasant civilization of the Hindus, and finally classified as cultivators agricultural labourers and workers and certain other primary types of occupations.

In the census of 1961, 11.59% of the scheduled tribes were classified as cultivators, who owned some land, 10.58% were agricultural labourers, who owned no land and 11.08% were engaged in the primary occupations of mining, quarrying, foresting, gardening, fishing, hunting and rearing of livestock. The *Santhals* of Bihar, Orissa and Bengal, the *Mundas* and *Oraons* of Bihar, and the *Gonds* of Central India have thus largely given up their adherence to the relatively primitive forms of production, and have taken to work which affiliates them with the more prosperous communities living in the neighbourhood. These tribes are thus no longer self-contained, as the primitive fishing-and-gathering peoples of the Andaman Islands continue to be even today.

Contacts of the tribal peoples with the outside world have also naturally led to extensive changes in other aspects of the tribal life. The languages spoken by *Mundas, Santhals, Oraons* and *Gonds* have assimilated numerous words from Hindi, Bengali and Oriya, as they have taken to the use of

317

Plate 46. The Munda men with their oil-press in the Munda village.

things and processes from the more prosperous communities living nearby. Apart from language, their social customs and religious beliefs have also come to be very profoundly influenced.

The *Oraons* (Roy, 1915) live in the eastern part of the Ranchi District (Bihar) and the adjacent districts of Orissa in the south and Madhya Pradesh in the west. The *Oraons* are mostly confined to their own villages, or they also live mixed with other tribes like the *Mundas*, *Kharias* or *Bhumyas* in joint villages. They have, however, their own system of communal organization, presided over by hereditary secular or religious officials. Bachelors have a separate dormitory of their own, called the *dhumkuria*, near which is the dancing ground, where men and women gather in the evenings for dance and recreation.

With the passage of time and frequent and closer contacts with the Hindu peasantry, they have begun to be ashamed of the custom of men and women dancing together, so that the youth organization of *dhumkuria* has come to be frowned upon. The puritanistic Hindus look askance at such license and freedom, characteristic of the *Oraon* life. Besides, the Hindus are more prosperous, while the *Oraons* are comparatively poor.

318

Plate 47. Kinaur beauty in full bridal regalia.

Plate 48. Kinaur grandmother and grand-daughter from Kalpa in the Simla Hills.

Among the *Adis* of North East Frontier Agency, who have come into contact with the Hindus of the Brahmaputra plains, the institution of youth dormitories is also gradually becoming defunct.

According to Bose (N. K. Bose, 1968, 1971) the *Oraons* have, during the last fifty or sixty years, been subjected to a number of puritanic social and religious movements, the ultimate sources of which have been either the Hindu or the Christian. The most important among these is known as the *tana bhagat* movement. Under its influence, the *Oraons* of western Ranchi have given up drinking wine and eating meat. Word also went round that all land belonged to god, and it was for god to provide the daily bread of his children! Some of them went to the extreme of giving up the practice of agriculture, freed their cattle and retired into the forest in a spirit of absolute resignation to god! The result has been naturally tragic for all concerned.

There were a number of similar revivalistic movements also among the *Oraons*, connected with the worship of Siva or the acceptance of the

320

teachings of saints like Kabir. Several sections of the *Oraons* split off from their traditional customs and came as near Hinduism as they could. They were eventually assimilated into the Hindu society, not through caste, but through the doors thrown open to all without reservation by various forms of the *bhakti* movements (sect of the Hindus who emphasize the importance of devotion, prayer and surrender to god).

Developments of a comparable nature also took place among the *Mundas* of Bundu and Tamar in Ranchi (Bihar), which are near the centres of influence of the *Vaishnava* cult in the adjoining district of *Manbhum*. These Mundas distinguish themselves from the socalled impure *Mundas* of central and western Ranchi. Their religious practices and even food habits have also been profoundly influenced by followers of the bhakti cult. Having given up eating meat, the need for oil or fat was newly felt. As no oil presser or *teli* is, however, present in some of their villages, they have themselves started pressing oil by manual power. Although their ploughs are drawn by bullocks, they do not harness the animals to the oil-presses, which they have set up, lest they be equated with the *teli*, (oil presser) who occupies a lower rank than the peasant castes! The *Mundas* reckon themselves as one among the peasant castes of the locality. Many tribes in India have thus thrown off sections, which have become regular *jatis* or castes of the Hindus (N. K. Bose, 1941).

The people of the scheduled areas of Kinaur (Himachal Pradesh), for instance, are clearly grouped into castes. Some are farmers, some are silversmiths, carpenters, blacksmiths, or leather workers, like any other caste in the valley below; yet they are still treated as tribes by politicians, in the interest of votes.

9.4. Transhumant herders and terrace cultivators

In the Western and Northwestern Himalaya in general and Himachal Pradesh, Jammu & Kashmir and U.P. Himalaya in particular, there are a number of tribal groups, who live by terrace cultivation at lower altitudes and move with their flocks of cattle to higher altitudes in search of pastures during summer. We have, for example, the *Gujjars* of Kashmir Valley, *Gaddis* and *Lahaulis* of Himachal Pradesh and *Jaunsaris* of Jaunsar and Bawar (Dehra Dun).

The *Gaddis*, whose numerical strength according to 1961 census is 51,369, are mainly concentrated in the District of Chamba, and Mandi in Himachal Pradesh. They reside chiefly on the slopes of the Dhauladhar Range and its spurs at elevations between 1066 and 2285 m. The *Gaddis* make their houses on the snowy range and many of them have homes on both sides of the range. They regularly undertake journeys from one side to the other side of the mountain range through high passes, when they are free from snow in summer. In the valley the *Gaddi* settlements form a dendritic pattern, as they are chiefly concentrated along the river banks.

Such settlements may be noticed in the Chamba, Kulu and Kangra valleys. The villages are in the midst of terraced rice fields and orchards.

The environment of the *Gaddi* is characterized by high relief, snow-capped summits, deeply dissected topography, antecedent drainage, complex geological structure, temperate flora and fauna. Popular legends ascribe the origin and migration of *Gaddis* from the Indian plains, to escape persecution by the Muslim rulers. Their early home was *Gaddheran*, a place in the upper reaches of the R. Ravi.

Plate 50. A Gaddi family in their picturesque dress, with the background of the mountains their home in the Himalaya.

The *Gaddis* build their houses of stones, wood and thatch. The houses are two storied and do not always face the Valley; the availability of the sunlight determines site and arrangement of the room, etc.

More than 90% of the environment of the *Gaddi* is unfit for cultivation, owing to the mountainous character of the terrain. It is readily reflected in the semi-nomadic and semi-pastoral economy of the people. They make small terrace-fields, use primitive implements and the yields are generally poor. The chief crops produced are maize, hill millets, potatoes and

323

pulses. In the higher regions buckwheat (*Polygonum*) is also grown. After the harvest and before the winter sets in, the *Gaddis* move to the lower elevations, leaving behind the old and infirm with enough fuel and provisions to last through the winter season. In the valley towns the *Gaddis* work as sawyers, and their women work as domestic servants. The *Gaddis* stay only four months in a year in their dwellings at high altitude. In the rainy season, they go to high pastures in Lahaul, Pangi and Spiti (Himachal Pradesh). In winter they descend to the valleys of Kangra, Kulu and Chamba. Thus they change their habitat and economy with the seasons.

The Lahaulis, whose numerical strength according to 1961 census is 12,106, live in the Lahaul, with an area of 4566.6 km², of which only 14.24 km² are cultivated. Thus the population pressure on cultivated land is 865 per km². Some investigations among the Lahaulis in 1968 show certain interesting patterns of ecological adaptations.

The Lahaulis are composed of the *Badhs*, *Sanglas*, *Sippis* and *Lohars*. The *Badhs* and the *Sanglas* are 'scheduled' tribes; the *Sippis* and the *Lohars* are 'scheduled' castes. The bulk of the population lives in hamlets, which are unevenly distributed all over the valley. The Pattan valley, with seventy-two hamlets accommodates 5,697 people, while 3,322 and 3,087 people live in forty-four and forty-three hamlets in the Bhaga and the Chandra Valleys respectively.

Agriculture is the main source of livelihood in Lahaul, which is supplemented by seasonal migration for work in the towns of the lower valley. Climate is the limiting factor in the growth of the crops. In the Bhaga and Chandra Valleys only one crop is grown, while two crops are grown in Pattan Valley. Fields are terraced along the river valleys, where the tributaries form fluvioglacial fans. Irrigation is practised by means of the channels. The agricultural implements are primitive. The plough is usually drawn by *churu* (a hybrid between the Indian cattle and Tibetan yak). The chief crops grown are buckwheat, barley, wheat and potato. Kuth (*Saussurea lappa*), a kind of aromatic medicinal herb, is the important cash crop and in the past there was considerable trade with Tibet for this plant*.

Besides agriculture, the Lahaulis practise wool spinning and weaving, mule driving and also work as labourers in government projects. There are about 21,000 native sheep in the area, out of which four hundred are merino sheep. Wool and wool products are locally consumed. Besides there are 10,360 horses, 1005 mules and 304 donkeys in possession of the Lahaulis.

The *Gadia Lohar* or the blacksmiths of Rajasthan, Gujarat and Madhya Pradesh are another nomadic caste of artisans, who may also be of tribal origin. They claim that their original home was in Chittor in Rajasthan,

---

* Considered to be an aphrodisiac – M. S. MANI.

from where they took to a roving life after its fall at the hands of the Mogul rulers of Delhi. They move from place to place with their heavy and richly-decorated bullock-carts, camp in one place as long as work is available, and when there is no work, move over to another place. They are generally called *gadia lohar*, because they move about in their carts or *gadis*.

The next example of adaptation to a complex and specialized organisation of production is furnished by cattle-keeping or shepherd tribes, like the *Rabari* of Gujarat, the *Gaddi* of Himachal Pradesh or the *Toda* of Tamil Nadu. The *Rabaris* are included in the list of scheduled tribes in Gujarat. They are a semi-nomadic cattlerearing people. They live in small conical huts, called *kubha*, unlike those of their neighbours, but resembling in some ways the leaf huts of the *Birhors*.

The *Todas* of Tamil Nadu are a distinctive group of buffalo-breeders, who have preserved their identity in a very remarkable manner. They live in the hills of Ootacamund and their houses are half-barrel-shaped in design. Their entire religious worship centres round the buffalo and they have continued to maintain their custom of polyandry. The *Todas* are now tied up with the economy of their neighbours.

The case of the *Banjaras* or *Lambadis* of Gujarat, Rajasthan, Maharashtra, Madhya Pradesh and Andhra Pradesh is also of great interest. These ethnic groups were originally carriers of merchandise with the aid of their pack-animals from one part of the country to another. They wear very colourful dresses and ivory ornaments and move about from place to place even now. Probably it was the coming of the British rule, when communication were developed all over the country, that the *Banjaras* lost a large part of their trade, and took to criminal activities.

9.5. NOMADIC GROUPS

In the early pages it has been shown how the tribal groups have changed under the ever increasing pressure of the more efficient productive organization built by the Hindus. It has also been indicated how the majority of tribes became farmers or farm-labourers, while some others turned to such arts and crafts or services as fitted into the local, regional scheme. There is, however, another kind of adaptation, which we may now consider. Some tribes, instead of settling down, became nomadic or itinerant artisans, economically related and generally subservient to the settled communities of peasants and artisans (N. K. BOSE, 1968).

Let us begin our observations with the *Birhors* of Bihar and Orissa. This tribe lives on the Plateau of Chota-Nagpur and in the neighbouring Orissa districts of Mayurbhanj and Keonjhar. In the latter place, they are known as the *Makarakhiya Kulha* or people who eat monkey flesh. Indeed, they had once earned the reputation of hunting monkeys in

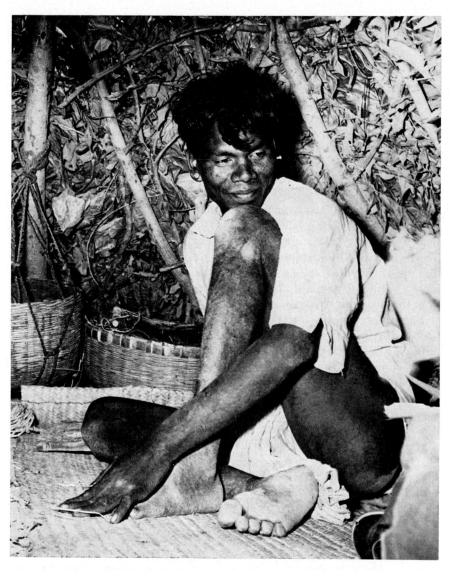

Plate 51. A Birhor youth inside his hut.

particular, either for meat or skin, which is supposed to be good for making drums. They were even supposed to be cannibals; but there is not a shred of evidence to prove that they ever indulged in this practice. The word *Birhor* really means men of the forest. They live at present in small communities, in temporary settlements on the fringe of forests, generally not far from peasant villages belonging to tribal or non-tribal folk. Their settlements consist of from half a dozen to a score or more of

326

thatched huts, arranged in an irregular circle, in places where they choose to settle down for a few months. The huts, called *kumba* are conical in shape, made carefully of boughs, twigs and leaves. They are about 2.4–3 m in diameter at the base and nearly 2.1 m high within. The open space, cleared between the huts is used by different families for cooking. The women set up ovens there and also scrub and clean and the men work on ropes.

BOSE (N. K. BOSE, 1971) refers to an interesting observation made some forty years ago. The *Birhors* changed their habitation thrice in a year. In summer they sought the shade of big trees, in winter some comparatively open and sunny space was preferred, while during the rains the huts were laid on some high ground from which water drained off quickly. An American anthropologist has recently shown that one particular group of *Birhors* changed their place even more than half a dozen times in the year. It does not take long for a group of *Birhors* to build such settlements.

Birhor's economy is based on hunting and trapping small game, and the collection of the bark of a wild creeper called chop (*Bauhinia vahlii*), which is turned into excellent cordage. The ropes are in great demand among neighbouring peasants, who barter them for paddy or millets. The small game, like rabbits or jungle fowl and also medicinal herbs are sold to villagers for cash.

The forests of Hazaribagh (Bihar) were once full of hyenas, leopards and tigers. As a result of large-scale deforestation and shooting of wild game, the population of these carnivores has substantially decreased. Consequently at least several known groups of *Birhors* have now began to keep goats, which thrive well in the forests. These are also sold in the weakly markets and the *Birhors* have substantially been able to add to their income in this way. One indirect result of this has been that the size of the particular settlements of *Birhors* has grown. Now more families can live together than before and their frequency of movement has also decreased considerably.

The *Birhors* of Hazaribagh and of Orissa have economic relationship to the local peasant population. To all intents and purposes, they have become a *jati* or caste, which lives within easy reach of villages, yet within the forest. They specialize in the production of certain commodities, which are needed by the peasant folk who live nearby. Some sections of *Birhors* have started settling and adopted agriculture, but they continue to marry only among members of their own tribe, and have thus developed one more characteristic of caste, namely, endogamy.

The religious beliefs and rites of the *Birhors* are not very different from those of other *Mundari* groups. In Raman's group BOSE (N. K. BOSE, 1971) observed an interesting development. This group had become a little more efficient than some other groups because they come more frequently to towns and markets to sell small game or medicinal herb or for the purchase of cloth, ornaments and so on. One women in that group had even started lending money to her own tribesmen and had thus taken the place of moneylenders from the market to whom formerly, those in need used to go. Besides in the Raman's group of *Birhors* some Hindu ideas of purity and some of their gods and godesses have been absorbed. They have also started looking upon themselves as one of the regular Hindu castes.

Just as the Birhors have specialized in the production of certain commodities, there have been others also who have followed an approximately similar course. The members of the *Pentia (Bhoi)* tribe of Orissa are blacksmiths by profession. They are apparently the same as the *Suras* of Palamau in Bihar and of the *Agariya* of Madhya Pradesh. They use double bellows, worked by foot and have a tradition that formerly they used to smelt iron from ores. Today they are just blacksmiths who are, however, looked upon as lower in rank than these who use bellows worked by hand.

REFERENCES

BAINBRIDGE, R. B. 1907. The Saora Paharia of the Rajmahal. *Mem. Asiatic Soc. Bengal*, 2(4).

BISWAS, P. C. 1957. The Santal. Delhi: Adim Jati Sewa Sangh.

BOSE, N. K. 1937. The geographical background of Indian culture. In: Cultural Heritages of India, vol. 1. Calcutta: Ramakrishna Mission Institute of Culture.

BOSE, N. K. 1941. The Hindu Method of tribal absorption. *Soc. & Cult.*

BOSE, N. K. 1956. Culture zones of India. *Geogr. Res. India*, 18(4).

BOSE, N. K. 1966–1967. Report of the Commissioner for Scheduled Castes and Tribes. New Delhi: Govt. India Publ.

BOSE, N. K. 1968. Some aspects of nomadism in India. *Proc. Internat. Geogr. Seminar*, pp. 93–98.

BOSE, N. K. 1971. Tribal Life in India. New Delhi: National Book Trust of India.

BOSE, S. 1961. Land use survey in a Juang village. *Man in India*, 41: 112–183.

BOSE, S. 1962. Land and People of Dhauli Ganga Valley. *Man in India*, 42: 292–304.

BOSE, S. 1964. Economy of the Onge of Little Andaman. *Man in India*, 44: 298–310.

BOSE, S. 1967. Carrying capacity of land under shifting cultivation. *Asiatic Soc. Bengal Mon. Serv.*, 12.

CHATTERJEE, A. N. & T. DAS, 1928. The Hos of Saraikela. Calcutta.

DALTON, E. T. 1872. Descriptive Ethnology of Bengal. Calcutta.

DAS, T. C. 1905. The Wild Kharias of Manbhum. Calcutta.

DUTTA-MAJUMDAR. 1955. The Santhal. Calcutta.

ELWIN, V. 1939. The Baiga. London: John Murray.

ELWIN, V. 1942. The Agaria. London: Oxford University Press.

ELWIN, V. 1947. The Muria and their ghotul. London: Oxford University Press.

ELWIN, V. 1950. Bondo Highlanders. London: Oxford University Press.

ELWIN, V. 1959. The Art of the North East Frontier Agency, Shillong: 16.

ELWIN, V. 1959. A philosophy for NEFA. Shillong. 2nd ed.

ELWIN, V. Loss of nerve: a comparative study of the result of contact of peoples in the aboriginal areas of Bastar State and the Central Provinces in India. New Delhi.

FURER-HAIMENDORF, C. VON, 1946. Culture types in the Assam Himalaya. *Indian geogr. J.*, 21: 1–9.

FURER-HAIMENDORF, C. VON, 1950. The Chenchus The Tribal, Hyderabad, Pres. Address Anthropology Section, Indian Sci. Congr. 1950, *Proc. Indian Sci. Congr.* 1950.

FURER-HAIMENDORF, C. VON, 1955. The Himalayan Barbary. London.

GHUREY, G. S. 1932. Caste and Race in India.

GHUREY, G. S. 1945. The Aborigines socalled and their future. Gokhale Inst. Politics & Econ.

GRIGSON, W. V. 1938. Maria Gonds of Bastar. Oxford University Press.

GUHA, B. S. 1929. Negrito racial strain in India. *Nature*, 1928, May 19.

HODSON, T. C. 1911. Primitive Culture of India Vol. The Naga Tribes of Manipur.

HUTTON, J. H. 1921. Sema Nagas.

HUTTON, J. H. Angami Nagas.

HUTTON, J. H. 1938. The primitive Philosophy of Life.

HUTTON, J. H. 1946. Caste in India.

MAJUMDAR, D. N. 1937. A Tribe in Transition. London.

MAJUMDAR, D. N. 1944. The fortunes of Primitive Tribes.

MAJUMDAR, D. N. 1948. The Matrix of Indian Culture.

PANT, S. D. 1935. The Social Economy of the Himalaya. London: George Allan & Unwin.

RADCLIFFE-Brown, A. R. 1948. The Andaman Islanders. Illinois: Glencoe.

ROY, S. C. 1912. The Mundas and their country.

ROY, S. C. 1915. The Oraons.

ROY, S. C. 1925. The Birhors.

ROY, S. C. 1928. Oraon religion and custom.

ROY, S. C. & R. C. ROY, 1937. The Kharias, vols. 2.

RUSSELL, R. V. & HIRA LAL, 1916. The Tribes and Castes of Central Provinces of India.

SAHAY, K. N. 1963. Impact of Christianity on the Oraon of three villages of Chota-Nagpur (Ph.D. Dissertation Ranchi University).

SARKAR, S. S. 1938. The Maler of the Rajmahal Hills. Calcutta.

SCHMIDST, P. W. 1926. The Primitive Races of Mankind.

SEN, B. K. & J. Sen, 1935. Notes on Birhor. *Man in India*, 35 (3).

SERVICE, E. R. 1963. Profiles in Ethnology: A revision of a profile of primitive culture. New York: Harper & Roa.

SINHA, S. 1958. Tribal cultures of Peninsular India, as a dimension of little tradition in the study of Indian civilization. *J. Amer. Folklore*, 17.

THURNWALD, R. 1932. Economics of Primitive Communities.

THURSTON, S. R. 1906. Ethnographic Notes on South India.

VIDYARTHI, L. P. 1958. Cultural types in Tribal Bihar. *J. Soc. Res.*, 1(1).

VIDYARTHI, L. P. 1963. The Maler: Nature-Man-Spirit-complex in a Hill Tribe of Bihar. Calcutta: Bookland Pvt. Ltd.

XII. SOME EXAMPLES OF RECENT FAUNAL IMPOVERISHMENT AND REGRESSION

by

A. K. MUKHERJEE

1. Introduction

This chapter deals with a brief account of the pronounced impoverishment and regression, which have occurred among the land vertebrates of the Indian fauna within the last century.

The impoverishment of the Indian fauna is the result of disappearance of many species by extinction, extermination and isolation. Regression has been brought about by progressive shrinking of the size of the distributional range of species.

The rich and highly diversified fauna of India is today composed, in nearly all groups, largely of vanishing elements. The most striking character of this fauna is the very recent regressive changes. The distributional ranges of most species of our land vertebrates have shrunk to small fractions of their size about fifty or hundred years ago. Species that were very abundant over much wider areas are now severely restricted to localized refugial areas or have totally vanished from the land within the last two or three decades. The vanished and vanishing elements of our fauna include interesting extinct as well as exterminated species, geographical relicts and isolates (see Chapters XIX & XXIV). Former continuous distribution has, in the majority of cases, given place to more or less pronounced discontinuity that is not often older than a hundred years. While it is not possible to discuss here all or even the major extinct animals, brief mention of some of the more important vertebrate fossils of the Siwalik should serve to amply illustrate the main theme of this chapter. Fossils of *Hipparion* are common from the Pliocene. Deltaic deposits of *Equus*, *Elephas* and *Bos* are abundant in the early Pleistocene or late Pliocene. In the Middle Pleistocene of the Narmada Valley are found *Equus nomadicus*, *Hippopotamus*, *Cervus*, etc. *Mastodon* and as many as eleven species of *Elephas* flourished during those times; only one species of these elephants survives today. These remains are associated with fossils of bison, buffalo, ox, giraffe, hippopotamus, chimpanzee, rhinoceros, *Sivatherium*, sabre-toothed tiger and hunting leopard, all of which have now completely vanished from our fauna. While the giraffe, hippopotamus and chimpanzee are at present confined to Africa only, it is thus evident that their range formerly included India also. Even during the late Pleistocene and perhaps hardly ten thousand to eight thousand years ago, the rhino, wild buffalo and other species, now known only as Siwalik fossils, existed in extensive swamps that spread in many parts of the Indo-Gangetic Plains.

Not only have the Pliocene and nearly all the Pleistocene vertebrates completely disappeared from the Indian fauna within the past ten or eight thousand years, but the greatest majority of those that differentiated in Recent times and were abundant until the middle of the last century, are either rapidly disappearing or have already vanished.

Some of the more striking examples of these vanishing species are listed below, with brief notes on the changes in their geographical ranges that have occurred, particularly within fifty or hundred years. Some of the major factors which have contributed to this faunal impoverishment and regression are briefly outlined. The species discussed below serve to emphasize the fundamental but generally overlooked fact that we deal with the phenomena of extermination rather than extinction and migration in India.

2. Some Vanishing Species of Reptiles

Among the numerous reptiles that have become sparse and whose distributional range has suffered marked regressive changes, *Gavialis gangeticus* and *Crocodilus palustris* undoubtedly represent the most striking examples. While nearly every other lesser known species of chelonians, lizards and snakes have been reduced to the level of geographical relicts rather gradually, the regressive changes of these two species have occurred largely within the last fifty years.

Gavialis gangeticus ranged formerly from the R. Indus across the whole of the Gangetic Plains of north India to the R. Brahmaputra, R. Kaladan in the Arakan Yoma (Burma) and in R. Mahanadi in the Peninsula. It is a shy animal that prefers larger rivers, with extensive sand and gravel banks and islets for basking in the sun. It breeds during May-June and lays clusters of about fifty eggs in sand. The young hatch shortly before the outbreak of the monsoon rains and migrate to the large river system. Some years ago, the young were caught in large numbers by fishermen on moonlight nights, especially after the rains, but now they are rarely if ever found in those areas. At present it is found sparsely and in isolated and relatively inaccessible places in the R. Ganga and its larger feeders in the Nepal-Terai, Bihar and Bengal. Its occurrence in the Brahmaputra and Mahanadi is now doubtful. Even in the Indus the only report by Manton in recent years has not been sufficiently authenticated.

Crocodilus palustris, which may attain a size of four metres, is the commonest Indian crocodile. It is essentially an inhabitant of marshes and small pools, but it also occurs in rivers that course through dense and shady forests. It is a rapacious predator that even turns cannibalistic on the smaller ones. It ventures out on the ground in dry weather in search of tanks and ponds and under conditions of severe drought, it buries itself in the mud to aestivate. Several dozen eggs are deposited in a crude nest in sand and the mother may be seen brooding over the cluster. The newly hatched baby crocodiles fall prey to the male crocodile, fishes, otters and to storks.

Both *Gavialis gangeticus* and *Crocodilus palustris* have been mercilessly hunted in India since long. Their skins find a ready market abroad for the manufacture of handbags, shoes, belts and other fancy articles. The flesh and fat are consumed by the primitive communities of man. As the estuarine crocodile has been practically decimated by ruthless hunting, the attention of hunters has fallen on these two species within the past fifty years.

Crocodilus palustris would undoubtedly have been totally exterminated many years ago, but for the fact that this species has somehow acquired a religious significance, at least in certain parts of India. This crocodile has, in consequence, been artificially introduced by man in totally arid areas and flourishing colonies, like those in the Pushkar Lake near Ajmer in Rajasthan and the *muggar peer* near Karachi (Pakistan), exist even today. It was a fashion in former times to maintain crocodiles in a semi-domesticated condition, in moats around fortresses as a means of defence, but with the modernization of warfare, the moats and with them the crocodiles too have totally vanished.

Among the major factors, which have contributed to the regression of the range of these two species, massive commercial hunting and extensive destruction of their natural habitats by river-taming, irrigation projects, draining off of marshes and rapid urbanization must be considered the most important. The destruction of natural habitats by man must indeed be considered as the most dominant determining factor in these species.

3. Some Vanishing Species of Birds

Rhodonessa caryophyllacea

It was resident of the savannah-forest of the foothills of the Siwalik, especially the thick cover reeds and tall grasses, growing in marshes. It was solitary and shy and occurred in pairs during the breeding season. Its food comprised aquatic organisms, vegetable as well as animal matter. The call is a whizzy-whistle, like that of a mullard but soft, sometimes with two syllables 'wugh-ah'. The female has a low quack. Nesting was observed in April and egg laying in June and July. Nests were generally built in tufts of tall grass and *Andropogon*. The egg-clutch varied from 5–10 in number. The bird was found to breed in north Bengal (Malda District) and in North Bihar (Bhagalpur and Darbhanga Districts).

The present distribution of the species is not precisely known, but it is believed to have been exterminated and no authentic record of it is

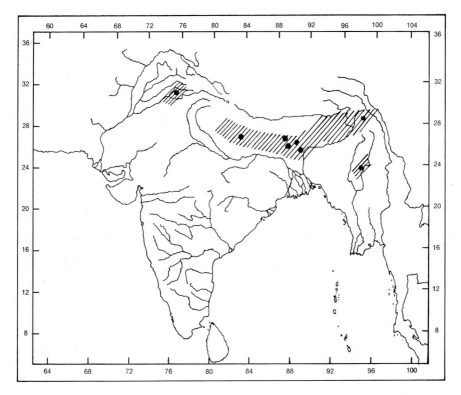

Fig. 26. Past (striped area) and the present (black area) distribution of *Rhodonessa caryophyllacea.*

available since 1935. Over fifty years ago, however, it ranged in north and eastern India, from the upper Gangetic Plain to Assam, Burma, through the foothills of the Himalaya and its adjoining forest tracts in the plains. The centre of its restricted distribution was the terai of north Bihar. Skins have been obtained as far northwest as Rupar (Punjab) and as far south as Pulicat Lake (Madras); its distribution was always localized even in the past.

There is no recent information of the availability of the bird from anywhere in its former range during the last thirty-five years. The last record was by C. M. INGLIS in 1935 from Darbhanga District (Bihar). In captivity it thrived till 1945 in Sir DAVID EZRA's aviary in Calcutta (ALI, 1960). HOULTON (1948) relates that he shot it in 1948 in Manbhum (Bihar). It is also reported that the Choudhuris of Simri Bahktyarpur shot one specimen on a marsh not far from Barauni (Bihar) in 1948, but this information has remained unconfirmed. The present distribution represents 0.63% of its former range if at all the species still survives.

333

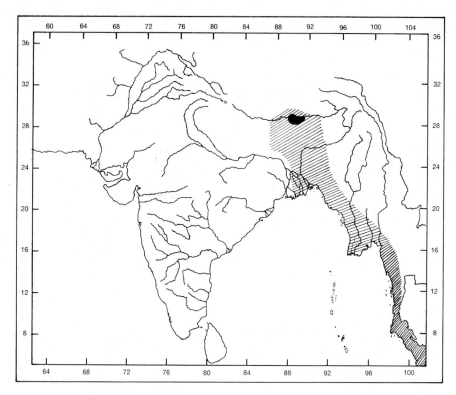

Fig. 27. Past (striped area) and the present (black area) distribution of *Cairina scutulata*.

Cairina scutulata

This is a large duck, which is easy to identify by its white patches on its wings. It resembles comb-duck in size and in having spotted white and black head and neck, but differs in having the lower parts chestnut-brown instead of white and in lacking a comb. It is a resident species that inhabits practically inaccessible dense swampy forest areas, which are studded with pools and sluggish creeks. It is generally found in pairs and sometimes in small parties of four and five. During the day it avoids the heat of the sun and remains in shade of trees, perching on a branch or swimming. It is typically nocturnal, although it may utilize the early morning hours or cloudy days in feeding on aquatic organisms as well as standing cereal crop. It is an expert diver and a fisher. An unmistakable long-drawn honk is emitted by the bird from time to time while feeding. It prefers holes and hollows of trees for nesting. It breeds from May to August.

The range of this species is from Assam to the Malaya Peninsula

334

through Burma and further southeast to the Greater Sunda Island. Within the Indian limits, it ranges from the western districts of Assam to the eastern boundary of north Bengal and also extends south to Cachar. It also occurs in the Lohit Frontier Division, Mishmi Hills, Manipur and in the Nagaland.

Over sixty years ago, this species was quite common in Assam, specially in the eastern parts. Within the last forty years it has become extremely rare. Recent reports of stray individuals are a pair from Tezu and Brahamakund, Lohit Frontier Division in 1947; three birds from Dum Duma area in 1958 and two pairs from the Ranga reserve forest in 1958. It has been known to breed in captivity (in Holland). In 1969 some six live specimens have been flown from Dum Duma (Assam) to Britain for conservation breeding, in specially protected conditions at the International Wild Life Reserve Centre. The specimens were obtained by Mr. M. J. S. MACKENZIE, in 1959 from northeast Assam. The present distributional area is only 1.5% of that at the beginning of this century.

Ophrysia superciliosa

This species, popularly called mountain quail, is related to the bloodpheasants (*Ethaginis*) in appearance, and to the spurfowls (*Galloperdix*) in its habits. It is rather larger than the common grey quail.

It occurs in grassland and bush, at elevations of 1520–2130 m on the Western Himalaya (Mussoorie and Naini Tal). Its flight is slow, heavy and short. The bird has been reported in coveys of six to ten individuals, in pairs or even single. The quail-like note is peculiar, unlike that of any other bird. It is a resident bird. This species was first described by J. E. GRAY in 1845 and the last collection was by G. CARWITHEN in 1876. Within thirty years less than dozen specimens were collected as follows: KNOWSLEY collection, 1846, India, 2 examples; KENNETH MACKINNEN, 1865, Budraj and Beneg, 2000 m Mussoorie, 2 examples; CAPT. HUTTON and PETY 1867, Jerepani, 1875 m Mussoorie, 5 examples; Major G. CARWITHEN 1876, Sherkadanda, 2200 m Naini Tal, one example. No recent information is available of the occurrence of the bird, although RIPLEY (1952) recently reported that a specimen was shot in the eastern Kumaon. The area in which it occurs at present is only about 2% of the former size of its range.

Choriotis nigriceps

This is the largest Indian game-bird, reminiscent of a young ostrich, having a heavy body, long neck and long bare running legs. The plumage is dull brown above, finely vermiculated and white below. It has a height little over a metre and wing span about 2.5 m. It is essentially an inhabitant of wide, open, dry, scrubby plains and waste, broken undulating

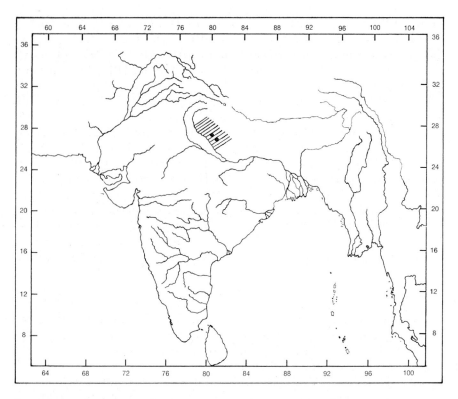

Fig. 28. Past (striped area) and the present (black area) distribution of *Ophrysia superciliosa*.

lands of the Rajasthan, West Punjab and Gujarat, where it is a permanent resident now. Formerly it occurred over a much wider area in the Peninsular India, as far south as the Malabar Coast and Ceylon and eastwards to West Bengal till the early part of the present century. It prefers to live among thorny bushes, tall grass and in cultivated patches. It is normally solitary, but flocks of usually three or four and sometimes as many as twenty-five to thirty individuals are also met with. It is omnivorous and feeds on all types of animal food, specially arthropods, lizards, and vegetable food such as wild berries, grass seeds, cultivated grains, etc. It breeds from July to September in grass fields or open waste, where there are shrubs and grass covers. Usually a single olive-brown egg is laid. The female does all the incubation and takes care of the young ones.

As it is a large and spectacular bird and as its flesh was greatly relished, this magnificent bird has now nearly been exterminated. Since it frequents open country, where the pressure of human population and agriculture is also high, the species has been pushed back by man. The regressive effect has been greatly accentuated by the fact that only a single egg is laid in a

336

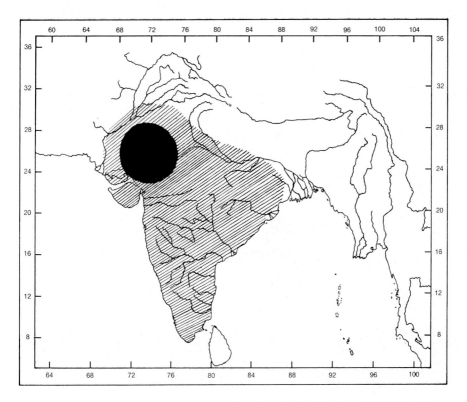

Fig. 29. Past (striped area) and the present (black area) distribution of *Choriotis nigriceps*.

year. The large drove described in the past are seldom seen now, but stray individuals may still be met with in Rajasthan, Gujarat and Bombay. No specimens have been recorded from Deccan since 1924, although flocks of twenty or thirty individuals were not once an uncommon sight there. The present distributional range of this great Indian bustard is hardly 1.7 % of its former size.

Cursorius bitorquatus

This is a light brown lapwing-like bird, with two white bands across the upper and lower breast and a small and straight bill. Its upper plumage is light brown, breast and flanks chestnut and lower abdomen grey. It has a prominent broad white supercilium running from lores to nape. The legs are long with no hind toe, signifying adaptation for running.

It occurred in thinly forested areas and scrubs or in deciduous bush-jungles of the Eastern Ghats, from the Godavari Valley in the north to Madras in the south. JERDON (1864) in 1848 discovered it in Nellore and

337

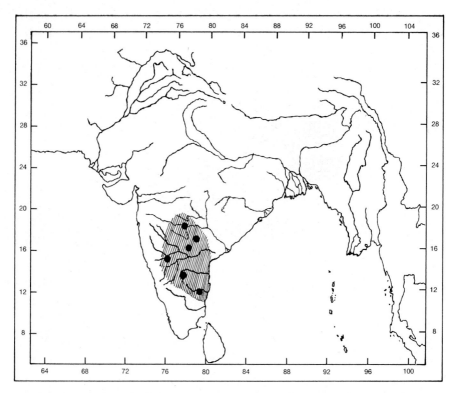

Fig. 30. Past (striped area) and the present (black area) distribution of *Cursorius bitorquatus*.

Cuddapah in the south and BLANFORD in 1898 obtained it in Sironcha, Bhadrachalam (north of the R. Godavari). Since then it was reported from Borgumpad in Hyderabad District and in Madras. The bird was last seen by H. CAMPBELL in 1900 near Anantapur, after which there has been no report. The last specimen collected was in 1871.

Recent surveys of the former Hyderabad State and the Eastern Ghats did not reveal even a single specimen and no sportsman has reported about it since 1900; it seems to have been totally exterminated. The size of the range, if at all the species still exists, is no more than 0.6% of the former.

4. Some Vanishing Mammals

Equus hemionus khur

This species is the wild ass of the Rann of Cutch and Tibet, characterized by short erect dark mane, continuing as a dark brown stripe along the back to the root of the tufted brown tail. The shoulders, saddle and sides

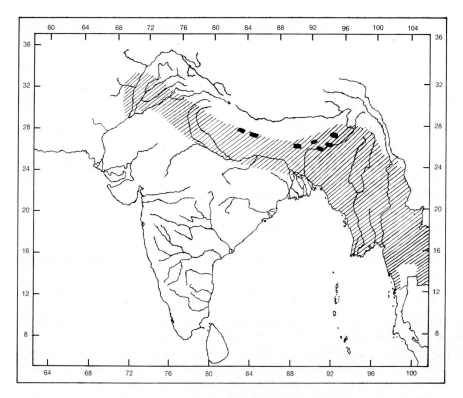

Fig. 31. Past (striped area) and the present distribution (black area) of *Rhinoceros unicornis*.

of the rump are fawn-coloured. The muzzle, legs and under parts are white. Its ears are short like those of zebra. There are callosities on the fore legs only. It stands 110–120 cm at shoulder.

The wild ass is generally found in small and large parties, of a pair to about three dozens, grazing throughout night in grass-covered expanses. In the Rann of Cutch, the hillocks of the dry weather turn into small islets, covered with scent grass and scrubs known as *bets*. The wild ass swiftly moves about here in search of food from place to place and thus covers long distances, maintaining an average speed of 50 km per hour. It often raids crops at night. Rutting starts in June–July, when the sociability of the wild ass ceases. The mare separates from the troop with a stallion, which fights viciously for her possession, the combatants rearing up on their hind legs. When the combat is over, the stallion takes to galloping around his harem, throwing himself on the ground, rolling over on his back and braying. Mating takes place during the rainy season. The period of gestation is eleven months. It never interbreeds or mixes with local domestic donkeys.

About a hundred and fifty years ago, the Indian wild ass used to roam in thousands in northwest India, Pakistan and southeast Iran. It has since been exterminated from Iran and perhaps a few stragglers may be found in the Thar Desert. The onager race has always been esteemed for sport and food. It has long been hunted by certain tribals for food. This animal was speared or shot by the Baluchis on horse back. Some were also trapped in pitfalls. It has become extinct in practically all over its range, except for the pocket in the eastern Rann. GEE estimated in 1962 that about 860 wild asses existed in India and about ten strayed into Pakistan (GEE, 1964). A number of wild asses died in 1960 and some in 1961 from *surra* disease. The wild ass now occurs in an area that is about 2.6 % of its former distributional range.

Rhinoceros unicornis

The one-horned rhinoceros, the largest of the three species of rhinoceros found in Asia, is a huge ungainly creature, with a blackish-grey hide, formed into characteristic folds or shields and devoid of hair except on the tail and ears. At the sides it is studded with convex tubercles. It possesses a horn on the snout, sharper and longer in the female but blunt and short-ened in the male by frequent combat. A full grown rhino may attain a length of 3.9 m and height of 1.8 m, with the horn about 30 cm in length. The great one-horned rhinoceros is confined to the grassland and jungle areas of the foothills of the Himalaya (Central Nepal) and to isolated areas of the plains of West Bengal and Assam. It prefers swamps and open savannah, covered with the tall elephant-grass, but is also found in wooded forests along the low hills and river valleys. It is essentially a grazing animal that prefers young grass shoots, which grow up after the tall elephant grasses are burnt. It may also feed on reeds and cultivated crops. It is a slow, solitary animal and is strictly territorial. It has special places for dropping excreta, always tending to use a fixed mud wallow and moving along regular trails. The animal is neither fierce nor does it charge at sight, except when the female is with her calf. The Indian tiger seems to avoid the rhino, a fact which the deer and the buffalo take advantage of and graze in company of the rhino, for protection. The rhino is helpless in quicksands and shallow pits, and if caught it utters deafening cries and dies of exhaustion. Breeding is practically all the year round but in Assam, it generally mates in the spring (March–April). Generally a single calf is born about October. The life-span of the animal is about 50–70 years.

The great one-horned rhinoceros has been known to have once been extremely common and wide-spread throughout the Indo-Gangetic Plains (RAO, 1957) and the neighbouring countries, but by the end of the 18th century it had completely disappeared from most of its range, except Nepal, Bengal and Assam. In 1904 about a dozen rhinos alone remained in Kaziranga (Assam) and fewer in Bengal. Enforcement of protective

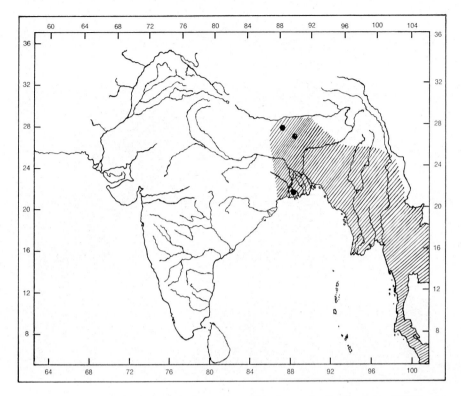

Fig. 32. Past (striped area) and the present (black area) distribution of *Rhinoceros sondaicus.*

measures helped it to re-establish itself in Nepal, West Bengal and Assam. According to reliable estimates, there are three hundred individuals in Nepal, forty-five in West Bengal and three hundred and thirty in Assam. The animal exists at present only in a few sanctuaries, of which Kaziranga has the largest number. The species may soon be wiped out totally if a serious contagious fatal disease, carried by the grazing domestic cattle, spreads to the rhino. The distributional range of the one-horned rhino is at present about 0.97 % of its former vast size.

Rhinoceros is commercially a very valuable creature. Though every part of its body has a market, the horn is the most highly prized, because of its supposed aphrodisiac property. In Europe, during the middle ages, its horn was believed to have peculiar medicinal virtues. A cup made of the horn is still believed to render poison innocuous in China, so that a single horn often fetches Rs. 1500–2000. In Nepal the flesh and the blood of the rhino are considered highly acceptable as food by all classes. The blood is used in religious ceremonies and the urine is said to have antiseptic properties. From the hide war-shields and other kinds of articles are made.

341

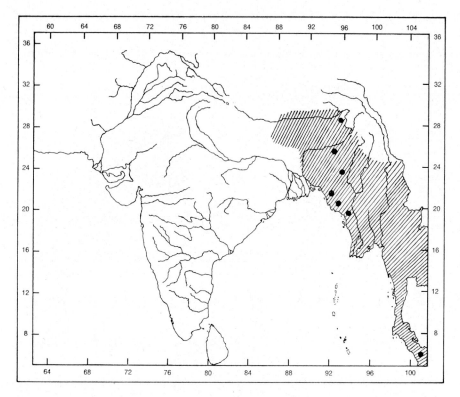

Fig. 33. Past (striped area) and the present (black area) distribution of *Rhinoceros sumatrensis.*

Senseless destruction of the animal was highly fashionable till 1958 in Nepal.

Another cause of the shrinking range of the rhino is the rapid expansion of cultivation in the rhino-territory in recent years. The introduction and vast development of the tea plantations in India have not only driven them out of many of their holds, but also served as incentive for the wanton destruction of the animal.

Rhinoceros sondaicus

This is the smaller one-horned rhinoceros, very similar in appearance to the great Indian rhinoceros, but slightly smaller. Its skin is dusky-grey like the other Indian rhino, but the skinfold in front of the shoulder continues right across the back. There are no tubercles in the skin, which is divided by cracks into small, polygonal, scale-like discs throughout. A median horn on the snout is present only in male.

In earlier times this species was widely distributed, from northern

342

India and southern China through southeast Asia to Sumatra and Java. Today it is, however, confined to Java, though a small number may have taken refuge in inaccessible parts of the Malaya Peninsula and in Sumatra. Even in Java, where it was quite common about fifty years ago and was seen throughout the island, it is now restricted to one small game reserve. The Javan rhinoceros inhabits forests rather than grasslands, hill tracts upto an elevation of 2330 m, but dense forests in lowlands are also included in its territory. It browses on the leaves of all kinds of forest trees and shrubs, and this habit may possibly have helped the animal to extend its range through the great forest tracts in northeast India, Malaysia and southeast China.

The last specimen from Assam was reported from Manipur in 1874, and from the swamps of the Sundarban in Bengal in 1870. It was also reported from Chittagong in 1864. It was not uncommon in Sikkim Terai and Assam till the middle of the nineteenth century, but it was hunted beyond its thriving limit by the rhino-horn collectors. Serious attempts to preserve the species were made from 1921 in the Udjung Kulong Game Reserve, West Java and it is estimated that there may be no more than two dozens to eight animals surviving at present. The present range of the smaller one-horned rhino amounts to 0.5 % of its former size.

Rhinoceros sumatrensis

This is the two-horned Asiatic rhinoceros, earthy-brown to almost black and covered with black or brown hair. It has a single pair of lower front teeth instead of two pairs as in the other Asiatic species; it is the smallest of all the five existing rhinoceros of the world, and attains a length of about 250 cm and a height of 110–135 cm. Its front horn rarely exceeds 30 cm in length.

The Sumatran rhinoceros prefers well-wooded forests, ascending upto an elevation of 1100 m. It loves shade and vicinity of water and bathes in streams at night and in hot part of the day. It has a habit of wallowing in mud, like the buffalo and pig. Its tracks, leading off from the wallows, appear like large tunnels hollowed out through forests. The wallows are usually visited singly or sometimes in pairs, the cow and the bull together. It descends to lower country during the monsoon and in winter. It is usually shy and timid but can also be tamed easily.

The original distribution of the two-horned rhinoceros was similar to that of the Javan rhinoceros and extended from eastern India through Burma to Sumatra and Borneo. It is now found in small numbers in scattered patches in Central and Lower Burma (in the Myitkina District, the Arakan and Pegu Yomas, Katha District and Lower Tenasserim). It is also found in small numbers in the Malaya Peninsula, Sumatra and Borneo.

The Sumatran rhinoceros probably no longer exists in any part of India or China, but about eighty years ago it was known to be sparse in Assam, Bhutan and Northern Bengal. There is a possibility of small numbers occurring in remote parts of the Lushai Hills and the Chittagong Hill tracts. It is still believed to inhabit remote forest areas in Burma, Thailand, Cambodia, Laos, Vietnam, Malaya, Sumatra and Borneo, and the number may be very small even in these areas. In Burma there were about forty in 1959, in scattered pockets. It has not only been exploited by poachers for its horn and flesh but the military operations during the second World War, wiped it by the wanton destruction of the defenceless animal, when the soldiers operated through the dense forests where the rhino lived. The present distributional range is about 1.2 % of the past size of the range.

Bos grunniens mutus

The wild yak is a massive, short-legged, blackish-brown bison-like animal with drooping head and high-humped shoulder. Its coat consists of long, coarse hair, in shaggy fringes that hang from its flanks, shoulders and thighs. An adult bull stands about two metres at the shoulder and may weigh 600 kg. The smooth black horns may be a metre in length. To tide over the severe winter the yak has a dense underfur, soft and closely matted, which gives additional warmth. In spring, this underfur comes away in great masses. The domesticated yak is smaller, and has patches of white on the chest and tail, sometimes reddish-brown or black with the horn less developed, but the fur is much developed.

It inhabits the high, desolate and rugged snow-covered mountains and valleys of Tibet and adjoining western Indian borders (Changchen-mo Valley in Ladak, Sutlej Valley and Kangri-Bingi Pass of the Kumaon Hills) at elevations of 3300–4500 m.

The wild yak lives in small herds of two to five, but the cows, calves and young bulls gather in herds, of two thousand. Its food consists of tufts of grass, shrubs, salt encrusted earth and frozen snow, when water is not available; it can withstand starvation for days together. The herds wander for considerable distances, feeding during the early mornings and evenings and sleeping on the steep slopes during the day. Saddles, saddle-girth, bridle, reins, whips, boots and numerous other articles are made out of yak-hide. Its wool is used for rope and clothing and its tail is used as a flywhisk in religious ceremonies. Yak meat and fat are eaten and its heart and blood are used by the Mongols for medicinal purposes. The species has disappeared from most of its regular tracts and perhaps it is now restricted in very remote and inaccessible areas. The wild yak has been practically replaced by its domesticated cousin as a beast of burden for riding, for milk, meat and wool for the nomadic tribes.

Capra falconeri falconeri

This is the largest of the wild goats of the Pamir and its radiating

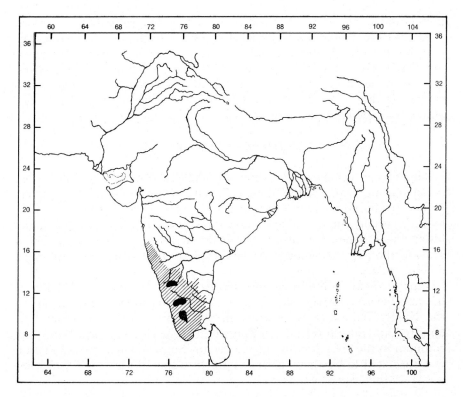

Fig. 34. Past (striped area) and the present (black area) distribution of *Hemitragus hylocrius.*

mountain ranges; its name markhor is derived from the Persian meaning a 'snake-eater'. It is a long silky-haired, thick-coated large and heavy animal, standing 90–100 cm at the shoulder. The magnificent horns of the male spread out in a heavy cork-screw and measure about 165 cm. The females are half the size of the male, and have short-twisted horns. The older males have a long black beard, which covers their throats throughout the year and also a shaggy grey mane. There are four distinct subspecies in the Himalaya, Kashmir, Hindukush and Sulaiman Ranges. The nominate race is represented within the Indian limits.

Markhor has as its natural predators the snow leopard and Asiatic wild dog. It has also been extensively hunted by man. About fifty years ago it used to roam about in large herds along the tributaries of R. Astor and Harmosh Nullahs, and on the Pir Panjal. Although the animal is gifted with incredible agility in climbing the most difficult and dangerous cliffs in order to avoid the predators, the use of modern long-range telescopic-sight fire-arms has decimated this species. Poaching by the nomads, and hunting by modern sportsmen and the increasing human

345

population in the land of markhor are also other contributory factors.

In the Pir Panjal and Great Himalayan Ranges of Kashmir, the markhor inhabits dense pine and birch forests. In the Sulaiman Range it lives on barren slopes. It occupies the most difficult and precipitous ground along the margin between the deep forest and the high snow-capped peaks. In winter it descends to lower valleys. Its present range amounts to about 2 % of its former size.

Hemitragus hylocrius

This species occurs at elevations of 1000–1200 m on the Nilgiri and Anamalai and parts of the Western Ghats in South India. It is closely related to the Himalayan form, but is larger and has also single pair of teats. It usually inhabits crags above forest level, but also descends to the lower grassy slopes. Herds of half a dozen to a dozen graze on patches of grass in the early morning and evening. It is alert and climbs the most difficult ridges.

It pays a heavy toll to predators, like leopard, tiger and the wild dog. It has also been excessively shot by man. According to the census taken by the Nilgiri Wild Life Association in 1963, there are only about four hundred animals on the Nilgiri Plateau. The species was, however, quite common in the later part of the nineteenth century, when its population was estimated to be about 1500. Its present range is about 6.9 % of what it was about hundred years ago.

Cervus elaphus hanglu

This is certainly one of our most spectacular animals, related to and similar in size and appearance to the European red-deer and the American wapti. It is little smaller and less robust than *Cervus unicolor* and bears magnificent 10 to 16 pointed, spreading antlers, with the browtine curved upward. Its tail is short, less than one-third of the head length. In the breeding season it acquires long shaggy fur in foreneck. The winter coat is light brown, fading to dingy white on lips, chin, underparts and buttocks; upper surface of the tail black, with a white rump-patch which does not extend much above the tail and is divided by the broad median stripe extending down to the base of the tail. In summer its coat becomes lighter, the hinds show traces of spotting on flanks and back. Fawns are spotted. A full grown stag at withers is 110–130 cm. Horns in adult are 100 cm in length.

It inhabits the densely wooded mountain slopes, in the summer in the high ranges about 3000 m, after shedding its antlers in March and April. By October it leaves the snow-bound heights and comes down to the valleys between 1550–2480 m to feed on the sprouting grass and budding larches, wandering a great deal from one glade to another. In summer it is found with its harems of hinds, composed of 10–20 individuals or more. The hinds and young ones live in a family group. Towards the end of September the stags commence to call and by that time the new antlers are hardened enough to challenge. Pairing takes place in October after which the stag deserts the hinds.

It was widespread in Kashmir about 150 years ago, but at present, it is restricted to the northern and the adjacent eastern valleys of Kashmir.

Until 1947, this species was in no danger of extinction; about 2000 individuals were believed to exist in the 'Kashmir Valley Preserve', the property of the former Maharaja of Kashmir and was strictly protected. Since then military activities, extension of cultivation and the use of guns in crop protection and the activities of poachers have reduced the numbers to 250 in 1954. A sanctuary for its protection was established at Dichigam near Srinagar, in an area of 54 km², at an elevation of about 2000 m. A census carried out in 1957–58 recorded their numbers as 550. Unfortunately, however, in the course of the last decade the number has once again fallen to 150 in the Dichigam Sanctuary. A sheep farm established in the Sanctuary is a major menace and today there over 1300 sheep are competing with the deer for food.

Cervus elaphus wallichi

This is similar to the foregoing species, but larger and heavier. It has longer and massive five pointed antlers, the browntine of which is less constantly longer than the second and closer to the burr. The terminal fork is placed at right angles to the axis of the head. The stag has a rufous-brown coat above, with large light rump patch. The stag stands at 140–150 cm at the shoulder and the antler measures 152 cm. It inhabited the rhododendron forests of the Chumbi Valley and some of the adjacent valleys of Bhutan, northeastern Nepal and Tibet, at elevation of 3000–4000 m. It is really a rare animal within our limits. It was first described from Nepal in 1823. LYDEKKER thought that the type came from the Mansarowar Lake. There is no authentic report of it during the past one hundred years and in any case the present distributional range does not exceed 1.6 % of its former size.

Cervus duvauceli

This is the swamp deer, which is a little smaller than the common Indian sambar (Cervus unicolor). It has splendid antlers with much variation in form. The number of points on the horns may vary from 10–14. Average horns measure 75 cm round the curve. The colour of the coat is generally brown in winter and chestnut in summer. The hinds are lighter in colour than the stags; the fawns are white-spotted. The stag has a mane. It is exclusively confined within the Indian limits and is

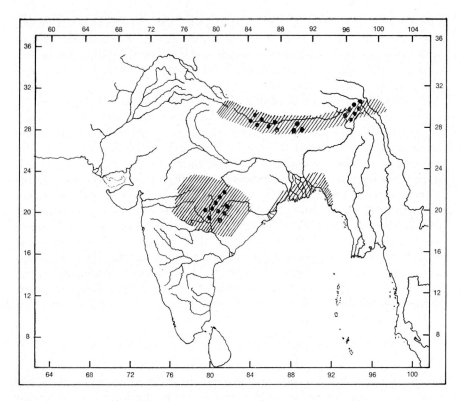

Fig. 35. Past (striped area) and the present (black area) distribution of *Cervus duvauceli*.

represented by two races, the northern and eastern race, (*Cervus duvauceli duvauceli*) and the Central Indian race (*Cervus duvauceli branderi*). It lives in marshy tracts of the Terai and the Duars, from northern parts of the Upper Gangetic Plain to Assam, eastern Sundarban and Central India. In the Terai it lives in bogs and swamps and seldom out of water. In Assam it prefers dry thatch land in proximity of water and sometimes far away. In Central India it lives mostly on dry ground and is less dependent on water. Its range in Central India covers more or less the *Shorea robusta* forest tracts. In swampy areas in northern and eastern India the hooves of the deer are more or less splayed out to give it a greater support in soft ground, whereas those that are found in the hard open ground have smaller well knit hooves.

Freshly growing grass attracts the swamp deer, which feeds mostly by day, resting at noon. They are highly gregarious; one to four dozen of one sex are usually noticed to move about separately. Its rutting season in northern India is November–December, in eastern India April and May and in Central India December–January. In Central India the stags

348

have been observed to retain horns till April. In Assam the stags have horns in velvet in March–April.

With the reduction in the area of *Shorea robusta* forests and with the swampy areas being reclaimed for agriculture, its population has decreased to a considerable extent. In Pachmarhi (Central India) it was extremely common in the early part of the present century, but it has now been completely exterminated. In Central India it is now protected in the Kanha National Park and some flocks exist in small numbers of isolated patches. About one hundred animals of the race *Cervus duvauceli branderi* survive today. In the Terai and Duars, in sanctuaries along the Himalayan foothills and adjoining plains its population has increased slightly, because of protective measures and it is fairly common and easily seen in the Kaziranga Sanctuary.

Formerly it was also known to occur in the Sundarban swamps, but has now definitely disappeared from the western parts of Sundarban, and its status in the eastern part of the range is not very satisfactory. In the Indus swampy plains, Bahawalpur, Rohri and the upper Sind it existed till the end of the 19th century (FINN, 1929), but seems to have now been exterminated from there. The size of its range represents about 4.4% of that at the beginning of the present century.

Cervus eldi eldi

This species is slightly smaller than the preceding and differs from all other deers in having a distinctive bow-shaped antler, like a prostrate C. Its winter coat is brown, changing to chestnut or nearly black; the female is fawn-coloured, and the young are brown and spotted. This gregarious species is an inhabitant of floating swamps of tall reeds and grasses and other hydrophytes, which grow on a mat of humus in the Loktak Lake in the Manipur Valley. It seems to avoid hills, hard ground and heavy forest. It feeds on wild rice, grass and other marsh-plants and sometimes raids crops.

The species ranges from Manipur, southern Assam to Thailand, Annam and Hainan through Burma and possibly to Malaya Peninsula. The race *Cervus eldi eldi* is confined to Manipur, *Cervus eldi thamin* is found in Burma and Thailand and *Cervus eldi siamensis* extends from Thailand to Hainan. The Manipur race is readily distinguished from the other two in having hairless, hard, horny hind pasterns. It was once found in all the swamps of the Manipur Valley, but is now confined to only the southern portion of the Logtak Lake (Keibul Lamjao), within an area of approximately 26 km². Hardly one hundred animals were known to exist in 1960. It was believed to be almost extinct in 1950, but since 1956, protection has helped it to multiply. The animal was hunted excessively by the local people for meat and for its horns, in the Far East Asia for medicinal purposes, even after it was declared as a protected species. It now

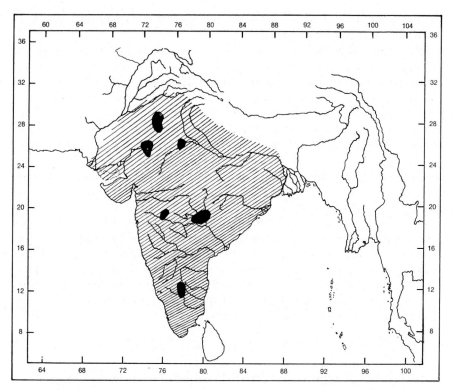

Fig. 36. Past (striped area) and the present (black area) distribution of *Antelope cervi-capra*.

occurs in an area that is hardly 4% of its former distributional range.

Moschus moschiferus moschiferus

This is the well known musk-deer, an inhabitant of the forests of the Himalaya. It is a primitive form, combining some characters of antelopes and of the deer. It has no face-glands but possesses a gall-bladder, which no deer possesses. The male has no antlers but the doe grows a pair of curved tusks of about 3–5 cm in length, thrusting from under its upper lip. The most interesting feature of the animal is the globular musk gland in the male, located beneath the skin of the abdomen. The musk gland helps the hind to seek out the male in the breeding season. Besides the musk gland, it has also a caudal gland, which is also said to play a part in breeding. It has a peculiar tail, which is completely buried in the long hairs of the anal region and is for the most part naked but the tip has a tuft. It has thick hind limbs, with the help of which it bounds. The colour of the coat is sepia-brown to golden-red; hairs of the coat are coarse thick and brittle.

350

The musk deer inhabits the dry temperate mountain forests of southern China and Tibet and the Himalaya from Kashmir to Sikkim, at elevations between 2500–3810 m, but generally in rhododendron forests. At times it may come to lower levels and remain in thick covers. In habit it is like the hare, unsocial and found singly or in pairs, concealing itself in a self-scraped out shallow, feeding on grass, lichen, leaves, flowers, etc., at dusk and dawn. It pairs during the severest period of cold in December and January and the young are born in June. Usually a single fawn is born but twins are not unusual. It is a prolific breeder, since the young breed again within a year.

The commercial importance of the musk is well known. Its odorous secretion has been long recognized as one of the best natural fixatives for perfumes. The musk hunters annual toll approaches 100,000 animals (STREET, 1961). Though musk is produced by the male only, it is difficult to distinguish a male from a female, so that many females are shot. The musk-deer now occurs in an area that is only about one-fourth of its former range.

Antelope cervicapra

This is the blackbuck, the male of which has spiralled-horns, blackish-brown above, white below and a pronounced white ring round each eye. The doe and young buck are yellowish-fawn above and white below. Old bucks become gradually blacker with age. The spiral of horns develop during the second year. Fully developed horns at the end of the third year are about 46 cm. A well grown male stands about 80 cm and at the shoulder 175 cm.

The blackbuck is one of the fastest creatures in the world, perhaps the only racing competitor is the cheeta, which is extinct in India today. This antelope was the main target-prey of the cheeta and with the deterioration in the population numbers of buck, the population of the cheeta seems also to have been seriously affected. This antelope leaps high into the air as it runs and relies on its great speed to escape from its enemies. It has a keen eye-sight. It thrives in small and large flocks, varying from a dozen to hundred in numbers, in grasslands. It feeds on grass as well as cereal crops. Its enemies are the tiger, panther, lion, wild dog and the wolf.

The blackbuck is exclusively an Indian species and occurs in the plains, and avoids hilly and mountainous terrain and forests. It is more common in the northwest and Deccan than in other parts of India.

Blackbuck herds of 50–100 were common sight practically all over the plains of India about a hundred years ago. Conditions have definitely deteriorated in recent years and such congregations are now never seen. It is ruthlessly hunted by man, since it does not keep itself confined to reserve or protected areas but moves out and is, therefore, hunted in jeep

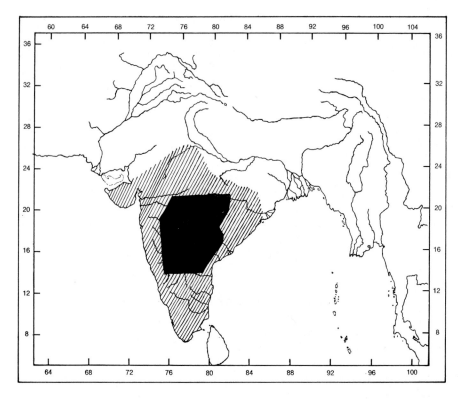

Fig. 37. Past (striped area) and the present (black area) distribution of *Tetracerus quadricornis.*

with modern long-range fire-arms. The size of its present range is now 4.6% of the former.

Tetracerus quadricornis

The four-horned antelope is the only four-horned animal in the world and is also peculiar to India. It is about 600 cm high, with a narrow muzzle. The males have two pairs of short straight horns and the females are hornless. The posterior horns are 8–10 cm and the anterior are 1–2.5 cm long. It is distinguishable from other antelopes by the presence of a pair of well developed glands between the false hooves of the hind legs. It is dull red-brown above and white below, with dark stripes down the legs. Its coat is short and coarse.

The four-horned antelope inhabits thin forests of bamboo-mixed jungles in undulating and hilly terrain. It behaves more as a deer than an antelope, preferring open forests and water to drink, and, therefore, resorts to the forest edge.

This antelope is found almost throughout the Peninsular India but is absent in the Malabar Coast. About a hundred years ago it was said to be a common animal throughout its range of distribution. In Central India, STERNDALE (FINN, 1929) reported it to be plentiful in the early part of this century. It has become scarce everywhere, as a result of ruthless destruction, specially in waterholes.

Sus salvanius

This is the smallest pig in the world and is the diminutive form of the wild boar. Its upper tusks are short, the snout is shorter than in the large wild boar, the ears and tail are naked and short. Its coat is coarse and scanty and the teats are in three pairs instead of four as in the wild boar. It possesses a fourth toe and bears forty teeth in its short jaw. The adult has a blackish-brown colour and the young is striped brown with the underparts white. The average height is about 30 cm and length from the muzzle to tail is 66 cm, the tail is hardly 3 cm.

The pigmy hog occurs in tall grass in jungles of *Shorea robusta* forests of the Himalayan Terai and the Duars. There is no report about this little hog for the last fifty years and it may have been totally exterminated.

Panthera leo persica

The Asiatic lion is distinguishable by its pale yellowish-brown colour, shaggy mane of the male and tufted tail in both the sexes, tall tuft and blackish outside of the ears. It has a flatter skull. The female is maneless and the young are invariably spotted.

The Asiatic lion, within historic times, roamed all the Middle East and seemed to have reached even southeastern Europe, where it was exterminated between A.D. 80–100. About 150 years ago it used to range over most of north India, except the easternmost parts, and as far south as the Narmada river. It is practically extinct from Asia, except in the Gir Forests in Kathiawar in India.

The Gir Forest, now the only home in India, is an area of about 1280 km² of rugged open country, with mixed deciduous and thorny scrub of stunted trees of *Ficus, Butea, Eugenia*, etc. and patches of bamboo, together with an undergrowth of thorny bushes of dense *Acacia, Euphorbia, Zizyphus* and grass. It preys on any herbivorous animal, but mostly it depends on stray domestic cattle that graze in the forested areas.

In the early part of the nineteenth century the lion was reported to be a common game animal. There are records that about the middle of the last century 300 lions were killed by Colonel AUCLAND SMITH. KINNEAR (1920) states that lions were formerly found in Sind, Bahawalpur and the Punjab, and became extinct in Harjana in 1842 and in Sind in 1810. In Palamau District (Bihar) the last lion was shot in 1814, and in 1832 in

Baroda. It was fairly common in Central India. By 1890 it was already on the verge of extinction and by 1908 barely a dozen were left in the Gir Forest. This precarious status of the Asiatic lion was saved by the endeavours of the Nawab of Junagarh. Thanks to his effort, the census of lions in 1950 revealed 240 and again in 1963 some lions were reported to have been destroyed by the villagers. A fresh count in the year revealed the population to stand at 280. It had been the royal game from time immemorial, till LORD CURZON declined to shoot it and urged its protection in the early part of the present century. Most of the normal prey of the lion have become scarce due to hunting pressure in the area of the lions. As a result of the spread of human settlement and the rapid progress of cultivation in the Gir Forest area, the territory of the lion has shrunk to considerable extent and the lion now commits considerable havoc amongst the cattle, which are brought into the Gir Forest for grazing. The professional graziers, locally known as the *maldharies*, therefore, bear a grudge against the lion and do not hesitate to entice it by tie-ups. CRADELL (1935) reports that by this method twenty animals were slain within the three seasons ending 1934. Besides this unauthorized shooting, they have also from time to time beers poisoned by dozens.

In view of the fact that the congregation of lions in a single area involves the risk of contagious feline disease wiping out the whole population, the Wild Life Board suggested in 1952 the setting up of an alternative home. Within its former range of distribution, a site known as Chandraprabha Sanctuary, 67 km from Varanasi, was selected and a lion and two lionesses were released in 1957. They are said to have multiplied, perhaps now about a dozen.

Panthera tigris tigris

The tiger is found in a variety of habitats, from the snowline and in cold coniferous forests of the Himalaya to the tropical dense forests and savannahs and marshes, including the tidal swamps; in the latter habitat it leads an amphibious life. In Peninsular India it is found in open dry grassy plains and in mixed jungles, preferring well shaded ravines and nullahs. In the rain-forests of eastern India its favourite haunts are the bamboo and grassy glades, impenetrable canebrakes, etc.

The tiger, although Asiatic, is widely distributed south of a line drawn from R. Euphrates along the southern shores of the Caspian and Aral by Lake Baikal to the sea of Okhotask. The southern limits reach the Malaysian islands. In the west it is limited to Turkish Georgia and to the east, the Sakhalin Island. It is in reality a species of the temperate north, which has only invaded the warmer climates relatively recently. The tiger is believed to have entered India from northern Asia after the last ice age through China and northeastern areas. There are eight recognizable

354

races of the species. The nominate race *Panthera tigris tigris* is found practically throughout the Indian subregion, except Ceylon, in Rajasthan deserts and the desiccated zones of Punjab and Kutch and the snow-covered terrain of the higher Himalaya.

The tiger had a wide distribution in India about 2500 years ago. It was abundant in the Indus reed beds, where the last one was shot in 1886. Except the desert parts of Rajasthan, Lower Sind and Kutch, it was found practically in every situation. By 1850 it had already been exterminated in many parts, specially western and northwestern India and greatly reduced in other parts (RAO, 1957). There were possibly 4,000 tigers about fifty years ago (GEE, 1964). Since 1930, there has, however, been a steady contraction of tiger jungles. In South India they have almost been exterminated by poachers and armed villagers. In 1948, the tiger population was estimated at 20,000–25,000, but in 1958 about 4000. GEE (1964) and PERRY (1957) state that rather more than 10% of the present population is being killed every year and it seems that its number has diminished to less than 3000 (SESHADRI, 1970). A new menace has appeared in the growing tourist demand for the tiger-skins; for example, over a thousand skins were sold in Delhi in 1967–68. As the tiger has always been considered as one of the best of all sporting trophies, overshooting and shrinkage in the tiger forests have diminished their numbers very rapidly.

Panthera unica

The snow leopard is smaller than the panther and has a pale ashy-brown coat, marked with black rosettes, with exceptionally long fur. It has a short face and head, bearing solid black spots. Its total length is 200–230 cm. Its tail is longer than in the panther.

The snow leopard is found throughout the Himalaya and other connected ranges, from Kashmir to Sikkim, near the snowline. It is not of course a permanent resident among the snows of the highest peaks, but it readily migrates up and down between 2000–4000 m, depending on climatic differences. Its haunts are rather inaccessible, but wherever wild sheep, goat and Himalayan thar, the goat-antelope and other small mammals are available, the snow leopard is found to prey upon them. It is nocturnal, but stalks prey with agility.

About a hundred years ago the snow leopard was as common as the leopards of the plains, but due to the great demand of its attractive coat, this beautiful leopard has been excessively hunted for fashionable ladies' dresses, handbags, gloves, cushion covers, etc. It has been mercilessly persecuted and killed to such an extent that it is now extremely rare even in Kashmir.

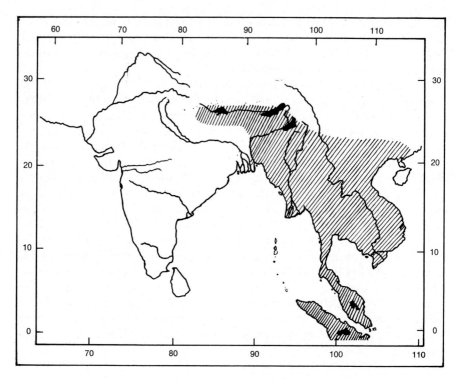

Fig. 38. Past (striped area) and the present (black area) distribution of *Felis temmincki*.

Felis temmincki

The golden-cat, as its name signifies, is rich golden-brown and head marked with white, black and grey stripes. The underside is paler with black markings. The ears are black. It is about a metre long, excluding the tail, which is little less than half of its body length.

It inhabits dense forests and is found from the foothills of the Himalaya in Nepal and Assam, and through Burma to western China and as far south as Sumatra. It lives among rocks and preys upon sheep, goat and even buffalo calves, as well as birds and various other small mammals. It is now rather rare, and in the recent years the animal collectors are finding it difficult to procure it even for zoos. Its range now is only 3.3 % of the size of the former.

Acinonyx jubatus venaticus

The cheetah is a dog-like cat, which is the fastest four-legged animal. It is a powerful sprinter and is a long-legged, lanky, short headed cat, with a tail exceeding half the body. Its coat varies from tawny to

356

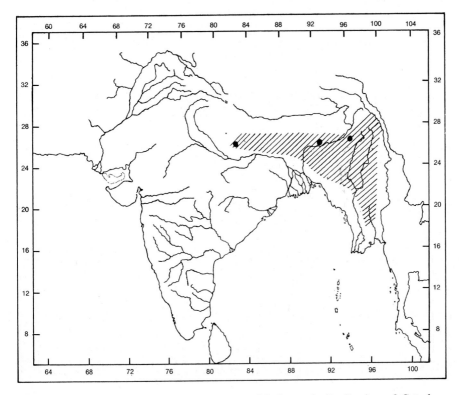

Fig. 39. Past (striped area) and the present (black area) distribution of *Caprolagus hispidus.*

pale buff and is heavily patterned with closely set solid black spots.

Nearly a hundred years ago, the cheeta ranged from the Northwest India to Bengal and Deccan in Peninsular India. Relatively recent records from India are from Meleghat (Central India) three examples in 1890, Wanoi (Central India) a single example in 1895, Rajkot a female and four cubs in 1894, and Mirzapur District (Middle Gangetic Plain) 2 examples in 1918–1919. FINN (1929) mentions that four examples were captured from Rewa in 1925. SESHADRI (1970) states that ARTHUR LOTHIAN shot one cheetah in Talcher (Orissa) in 1932. TALBOT (1960) speaks of a definite report in 1951, when three were shot in one night in Hyderabad. KIRKPATRICK's report (1952) of a cheeta in 1952 in Chitoor District has since proved to be a leopard. As late as 1947, there were reports of hunting with cheetas in the princely states. In Kolhapur there were thirty-five animals in the cheetah house. The Indian race extended through Baluchistan to West Asia and as far west as Palestine and Arabia. The race is extinct throughout its range for all practical purposes; its present range is hardly 2 % of its former.

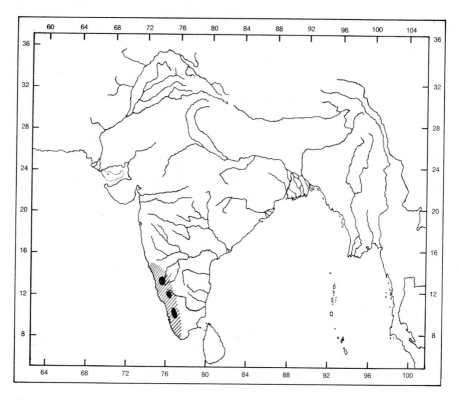

Fig. 40. Past (striped area) and the present (black area) distribution of *Macaca silenus*.

The easily tamed cheetah has been used for coursing for several thousand years. It had been very popular with princes and noblemen in India and in other countries. About a thousand cheetahs were maintained by the Emperor Akbar in 16th century and his successors also followed his trait.

Caprolagus hispidus

The hispid hare is in general appearance very much like a rabbit, but its coarse bristly dark-brown or rusty coat, short tail and hind legs and small eyes, readily distinguish it from other hares.

It inhabits grass jungles. Like rabbits it burrows but is not social. It is said to live on roots and barks. The hispid hare occurs along the foothills of the Himalaya in the Terai and Duars and as far south as Dacca in East Bengal.

It was reported to be common throughout its range about hundred years ago. In the early part of the present century, SHEBBEARE frequently found it in Assam. Since then it has been sporadically reported, the last

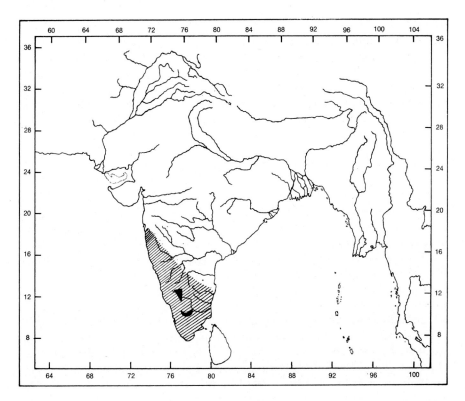

Fig. 41. Past (striped area) and the present (black area) distribution of *Presbytis johni*.

record being in 1951. In recent years attempts have been made to collect it from Assam, which is supposed to be its real home, but so far no authentic report of having found it there has been received.

Macaca silenus

The lion-tailed macaque is a medium-sized animal, with head and body about 50–60 cm and the short tail of about 25–38 cm terminating in a tuft. It is unique dark grey or brownish-grey, and is distinguished from other macaques by the ruff growing from temples and cheeks.

Its range extends along the Western Ghats, from North Kanara to Kerala, the northern limit is about 14° NL. It mainly inhabits evergreen and semi-evergreen forests, with the trees of 20 m or more in height, at elevations of 610–1070 m.

SUGIYAMA (1968) estimated in 1967 that the population of the lion-tailed macaque is not less than one thousand. As the hair is rather silky, black or very dark brown, it is a hunted for fur trade, and also for its meat

359

by certain primitive communities of Nayakanmar of the Nilgiri. Persecution by man has made them extremely shy even in their natural
habitat and has driven them to high forests. The population of the species
has been considerably reduced during the last fifty years. These are more
or less now restricted to the Nilgiri, Anamalai and Cardamom Hills and
the vicinity of the Periyar Lake. No information of its occurrence north
of 11 ° 30″ is available. If rigid protection is not extended, the lion-tailed
macaque will become extinct before the end of the century.

Presbytis johni

The adult Nilgiri langur is glossy-black or blackish-brown, with a
yellowish-brown head, but the young are jet black. The head hair is
longer than body hair but not radiating. The tail is almost the size of its
head and body together. It is almost as large as the common langur.

This langur is found in the South Indian Hill ranges, from Coorg to
Cape Kumarin, the Nilgiri, Anamalai, Palni and adjacent ranges. It is an
inhabitant of dense forests and prefers the sholas (dense evergreen forest
stretches, following water courses on hill slopes), at elevations between
900 and 2000 m. Small troupes of five to ten individuals may be seen to
move about, foraging not only in forests but also in belts of woodland and
gardens. Its food comprises fruits, leaves, flowers, etc. It prefers the cool
of morning and evening for feeding. The Nilgiri langur has been persecuted
for its fine fur. The primitive tribal man has systematically hunted it for
food and medicine; every organ including, blood is used by him. It has
thus completely disappeared from several areas within the past fifty years,
so that its range is now only 1.8 % of its former.

5. Impoverishment and Regression

The preceding examples are perhaps sufficient evidence of the pronounced impoverishment of the vertebrate fauna of India. The composition
of this fauna was formerly more complex than at present. Though it is
difficult to present a satisfactory quantitative estimate of the numbers of
genera and species that have totally or nearly wholly disappeared, it is
nevertheless obvious that the intensity of impoverishment is high. The
impoverishment set in nearly five thousand years ago, about the time
of the Mahabharata War, but continued rather gradually and only came
to be abruptly accelerated within the last half a century, so that it is now
irreversibly high. Concomitant with this qualitative and quantitative
impoverishment, there has also been a progressive reduction in the
distributional range of nearly all species, leading to a high degree of
general faunal regression. The degree of faunal regression indeed greatly
surpasses that of faunal impoverishment. The range of many species,
formerly continuous and extensive, is now highly disjunct, broken up in

small isolated patches and less than even a hundredth part of its former size. The rate of regression of different species has naturally been different, but the mean faunal regression in India as a whole is very large, almost 90%. The following table summarizes the coefficient of regression of twenty-one of the illustrative species discussed above, expressed as percentage shrinkage of the range during the past one century:

$$\frac{(R-r)\ 100}{R} = Q$$

Where Q = Coefficient of faunal regression; R = Size of the former distributional range in km^2; r = The present size of the same in km^2.

Table showing the regression of the distributional range of some important land Vertebrates in India

No.	Name	Coefficient of Regression	Variation from the mean
1.	Rhodonessa caryophyllacea	99.4	+10.4
2.	Cairina scutulata	98.5	+ 9.5
3.	Ophrysia superciliosa	97.9	+ 8.9
4.	Choriotis nigriceps	98.3	+ 9.3
5.	Cursorius bitorquatus	99.4	+10.4
6.	Equus hemionus khur	97.4	+ 8.4
7.	Rhinoceros sondaicus	99.5	+10.5
8.	Rhinoceros unicornis	99.1	+10.1
9.	Rhinoceros sumatrensis	98.8	+ 9.8
10.	Capra falconeri falconeri	98.0	+ 9.0
11.	Hemitragus hylocrius	94.1	+ 5.1
12.	Cervus elaphus hanglu	98.4	+ 9.4
13.	Cervus duvauceli	95.6	+ 6.6
14.	Cervus eldi eldi	96.0	+ 7.0
15.	Moschus moschiferus moschiferus	74.0	—25.0
16.	Antelope cervicapra	95.4	+ 6.4
17.	Tetracerus quadricornis	65.0	—26.0
18.	Felis temmincki	96.7	+ 7.7
19.	Acionyx jubatus venaticus	98.0	+ 9.0
20.	Macaca silenus	66.0	—23.0
21.	Presbytis johni	98.2	+ 9.2

Mean = 89%, Standard deviation = 12.2.

Reference to the distribution maps of these species shows that in nearly every case the reduction of the distributional range has occurred consistently large in certain parts of India, more than the others. Even a casual observer cannot fail to note that these areas are precisely the parts of India, where human interference with the natural environment has also been the maximum. It is also of importance to observe that the species discussed here are of some economic interest to man. He has no doubt pursued these species from time immemorial, but the destructive effects of his interest have become pronounced only with recent advances in his standard of living.

6. Major Factors determining Faunal Impoverishment and Regression

While it is by no means easy to understand the major factors that underlie the extinction of the Pre-Pleistocene and Pleistocene elements of our fauna, the causes that have led to the disappearance of so many species within the past one hundred years are, however, readily described. The rôle of physiographic and climatic factors is on the whole extremely insignificant, at least in so far as the impoverishment and regression that has taken place within the past one hundred years or even less. Some of these factors are themselves the result of others that have contributed at the same time to impoverishment and regression. The dominant determining factors arise from human activities; considerable numbers of our species have vanished from large parts of India, because man has exterminated them, others are also vanishing, because he is exterminating them and the distributional ranges of diverse species and indeed of the whole faunal complex have regressed because man has extensively destroyed their natural habitats and has occupied them himself; he has transformed the entire ecology of the land into massive human ecology, which has no place for the dominant members of the flora and fauna (see Chapter V). Man has been in India since at least the last ten thousand years, but these irreversible destructive processes have only set in abruptly within a century. In order to understand this paradoxical situation, we may briefly outline here some of the events of the preceding century.

In ancient times forests and the animals of the forests were protected as essential parts of religious practice. Nature was worshipped as manifestation of God and preservation of Nature was an integral part of human responsibility in India; it was his *dharma* or sacred duty and responsibility. Large-scale destruction of animals and deforestation were not only unknown in ancient India but considered illegal and acts wholly unbecoming of Arya or the civilized man. The *ashrams* (hermitages) of the Brahmanas were the *sanctum sanctorum* for wild life, where plants as well as animals flourished undisturbed. All temples were protected places and as a rule large trees enveloped the area. The hunting of any animal and the felling of trees in the environs of a temple were completely prohibited and these measures had rigorous religious sanction. Fishing in the river near a temple or in the temple tank was unthought of; pious people indeed fed the fishes and birds in temples, as a part of their ritual of worship. The original inhabitants took refuge from the advancing Aryans in the forests and depended completely on forest products and killed and trapped animals for subsistence, but were not responsible for the total destruction of forests. Indian mythology, art and literature are bound up intimately with its wild life, as testified by a great number of sacred animal representatives like the elephant-headed *Ganesha*, rhino-headed* *Varahavatar*, the turtle-shaped *Kurmavatar*, the monkey-faced *Hanuman*, the lion-headed *Narasinghavatar*, (Nagas) snake-worship and (garuda) eagle worship. With the passage of time, there was an increasing pressure on wild life, till forest management laws were framed and enforced by the Maurya kings. The *Arthasastra* of Kautilya (1924) enumerates, for example, eight forest divisions, (VISHNUDHARMOTTARA) of which the *gajavanas* (elephant forest) the dense forests, which sheltered elephants, were regarded as very important, since the elephant was the most important factor that determined the rise and fall of ancient empires. These forests also sheltered the panther, tiger, lion, rhinoceros, wild buffalo, gaur, yak, stag, blackbuck, iguana, crocodile and numerous other animals. Regular game laws were enforced during the reign of Chandra Gupta, the Maurya. There was a regular forest organization, headed by the *kupyadhyaksha* or superintendent of the forests and forest products. Reserve forests were exclusively meant for the rulers and were well guarded and rendered inaccessible by deep ditches on all sides and provided with a single entrance. Afforestation was also considered essential. Wild animals and forest products were strictly protected, and only the king had the privilege of hunting, once a year with his courtiers and members of his family. Killing an elephant was punished with death. MEGASTHANES (MACCRINDLE, 1926) has left a graphic account of the royal chase of Chandra Gupta, the Maurya, but ASOKA (B.C. 296–227) abolished the practice of his grandfather. He made the game laws even more elaborate and stringent. He mentioned them as *dharma-niyamas* (duty of moral discipline) on the principle of *ahimsa* (not molesting and not causing harm by thought or deed to any being), which not only prohibited the killing of certain species of birds, fishes and wild animals throughout the empire, but also protected them as** *pradishtābhāyanām abhayavanavasinām chamrigapasupakshimatasyānām* in wild life sanctuary. ASOKA in the stone pillar Edict V prohibited the killing, for the purpose of eating, of birds like parrot, mynah, red-crested pochard, brahmini duck, swan, crane, stork, vulture, peacock, mammals like bats, porcupine, squirrel, stag, rhinoceros, etc. Killing of Primates and Carnivores on specific days of the lunar years, totalling 72 days in a year, was also

---

* The *varaha* of the puranas is really *ekashringi varaha* or the single-horned *Rhino* and not boar as has been widely considered; *varaha* included many animals that uprooted by digging underground tubers like rhino, pig, boar. Aard vark is derived also from the Sanskrit root varaha – M. S. MANI.

Indeed nearly every animal and forest tree was an object of worship – Nature as a manifestation of the all pervading *Parabrahman* or Supreme Being.

** Establishment of a refuge for protected forest denizens and deers, birds, beasts, fishes, etc.

prohibited. The forest organization under the Guptas (A.D. 400–600) was due to Sukracharya, who greatly improved the forest administration by dividing forests into ranges, beats and blocks and strictly enforced the laws for preservation of forests and wild game. The king alone had the right to kill, but for him also the hunting of peacock and other species of birds was prohibited. Royal elephant corps and camel corps were allowed to be camped inside the forest, but not the infantry or a cavalry. The king also used to lead regular afforestation campaigns. The animals depicted on the Sanchi Stupa, the Ajanta Cave frescos and Khajuraho temples are eloquant testimony of the care for the fauna during that period.

The condition of our forests and wild life started deteriorating from about the end of the Hindu period in Indian history and was accelerated with the beginning of the Muslim rule. Wild life protection and forest conservation were completely neglected till the Moguls came to power in A.D. 1526. Some records of hunts of the late 14th century and early 16th century are revealing. According to *Zafarnamah*, Timur once killed several rhinoceroses with sword and spears on the frontier of Kashmir. The *Tariki-Mubarak-shahi* states: 'In the month of Zi-i-kada of the same year (ca 1387) he (prince Muhammad Khan) went to mountains of Sirmor (west of Yamuna) and spent two months in hunting the rhinoceros and the elk' (Ettinghausen, 1950). Babur relates in his memoirs (Beveridge, 1922) how he went to a rhinoceros hunt on the Sawati (ca 1519) and set fire to bushes to drive out the animals and finally killed a calf. The hunting incidents clearly show that rhinoceros existed in northwest India till the early part of 16th century. Further he (Babur) has often mentioned the presence of the rhino in different parts of northern India. 'There are number of them (rhino) in the jungles of Peshawar and Hashnajar as well as between the rivers Sind and Behreh in the jungles. In Hindusthan too they abound in the bank of Saryu (Gogra)' (Leyden, 1921). Babur also mentions that when he visited Chunar on March 24 of A.H. 935 (A.D. 1528) and was proceeding to Benares, he halted at an intermediate station. He notes in his diary 'At this station a man said that in an island close to the edge of camp he had seen lion and rhino, etc. (Jarrett).

In the mediaeval period, hunting was the most fashionable method of amusement and recreation among the rulers and their satellites. Elephant, lion, tiger, buffalo, wild goat and blackbuck were the mostly hunted animals. Akbar (A.D. 1556–1605) introduced a special kind of hunting called *gamargha* – hunt, which became very popular with the Moughal kings (Chopra, 1963). Every successive emperor and the associated nobles took a lively interest in this sport (Alvi & Rahman, 1968). He maintained about a thousand hunting leopards (Harper, 1948), a species, which, as already indicated, has now disappeared from India. During his time elephants were found throughout the Indo-Gangetic Plains and the Peninsula and their herds sometimes amounted to a thousand elephants (ABU-I-FAZL, 1590). ABDUL FAZL mentions rhinoceros in the Sambal Sarkar of Delhi Suba during the Akbar's region (ABU-I-FAZL, 1590).

Emperor JAHANGIR (1605–1627) was a great naturalist, who hunted male tigers and lions only. In his time lion hunting was the exclusive prerogative of the king and elephant hunting required special permission, which was sparingly granted to only professional hunters. He had made successful efforts to breed hunting leopard in captivity.

Till the middle of the 18th century the natural environment of India remained more or less undisturbed. The establishment of the East India Company (1701 A.D.) provided a happy hunting ground for the colonists. The company's officers regularly carried home trophies of lion, tiger or panther-head and skins, rhinoceros horns, elephant tusk and feet for waste-paper basket. The lion, which was not uncommon in northern India about 200 years ago, was shot without any restriction. KINNEAR (1920) states that during the Sepoy Mutiny in India (1857) Col. GEORGE ACLAND SMITH killed upwards of 300 lions, of which 50 were in the Delhi District alone. By 1870, tigers had been exterminated in many parts of India. This animal, once abundant in the Indus reeds, was shot out of existence there by 1886 (PERRY, 1964). The one-horned rhinoceros that was not uncommon in the Gangetic Plains in the 18th century became alarmingly rare everywhere by the end of the 19th century due to the demand for its horn and other products. WILLIAM FINCH, during his journeys in 1608–11 to different parts of India, describes that Ayodhya was a great centre for sale of products made from rhino horns. 'Here is great trade, and such abundance of Indian asse-horns (Rhino) that they make here of bucklers and diverse sorts of drinking cups' (FOSTER). The Rajmahal Hills were inhabited by the species till 1850 (LYDEKKER, 1900) and in 1876 big game in the Malda District (Bengal) included rhinoceros, though very rare. Since the beginning of the 19th century ruthless destruction, not only of the wild life but also of the natural environment, has been on the increase. A typical example may be cited about the environmental changes in southern as well as northern West Bengal during the last two hundred years. The tidal mangrove forests, known as the Sundarban, then extended from the sea-face to north of Calcutta and the extensive areas of the districts of Midnapore, Hooghly, Murshidabad and Nadia were covered with dense forests, specially along the rivers, which served as game reserves of the princes and the nobles. Large and small game such as elephant, wild buffalo, gaur, pig, panther, tiger, antelope, deer, crocodile, etc. were everywhere in plenty. These forests had been completely cleared by the end of the 18th century, following the foundation of Calcutta by JOB CHARNOCK in 1690. Rapid industrialization in and around that city turned the boggy and swampy areas into habitable human colonies. The thick forests of the Sundarban that flourished within a few kilometres north of Calcutta till the end of the 19th century was thrown open for cultivation and human settlement. This resulted in shrinkage of forest area from 4096 km² to 2320 km². The ecology of the reclaimed area was completely altered by deforestation, levelling for processing into arable land and introduction of plants from other areas, which were never in existence there. It would be disappointing to note that the Javan rhino, *Rhinoceros sondaicus*, and the wild buffalo, *Bubulcus bubalis*, have completely disappeared from these forests. The last record of this rhinoceros from this area is based on a specimen collected in 1870 and the buffalo was known to exist even in 1885. The swamp deer *Cervus duvauceli*, the barking deer *Muntiacus muntiak* and the fishing cat *Felis viverina* that existed in those swamp-islands are no more found in these parts. The estuarine crocodile *Crocodilus porosus*, which was quite common fifty years ago, is becoming scarcer day by day.

The intensity of the human factor in the disappearance of so many animals from the Indian fauna has rather abruptly increased during the past twenty years. As indicated in Chapter V, the cult of civil disobedience, propagated by political leaders during the freedom struggle and regularly practised by diverse parties since then, has had the result of generating a wholesome contempt for all laws, including forest and game protection laws, so that killing of wild life flourishes completely unchecked, and particularly because there is a very stimulating export market for such goods. It is impossible to even roughly estimate the quantity of prohibited forest and animal products that is being regularly smuggled out of India, year after year. The result of the utter lack of regard for our national wealth, combined with the introduction of highly sophisticated modern weapons, improvements in tracking and quick transport, is that the amount of habitat and faunal impoverishment during the recent twenty years equals, if it does not actually surpass, the sum-total of the entire past history of man in India. There has, in addition to the tendency for ignoring law and authority, also been a pronounced haste during the past two decades to completely break away from centuries-old traditions, religious beliefs, taboos and customs, so that the sanctity of temples and places of pilgrimage is also now openly mocked at. Idols from the temples and irreplaceable works of art from ancient centres of pilgrimages are stolen and exported on a massive scale to the hungry American market. These places that in former times guaranteed absolute safety to diverse animals like birds, deer, monkey, snakes, bats, etc., do not now even offer them retreat from the hunter.

The factors leading to disappearance and depletion of the fauna are not only due to direct human interference with environment, but also partly due to some topographical changes of courses of rivers, (see Chapter II) which have cut off sweet-water flow into the estuarine rivers, resulting in the increase in the salinity of the Sundarban channels. The flora has also been affected to a considerable extent and some important trees like *Heritiera* sp. and *Nepa fruticans* have been greatly reduced in number and the saline-resisting stunted mangrove trees have outnumbered the less salt-resisting plants. Another example of recent change in the general ecology is the Salt Lakes just east of modern Calcutta. It was formerly fed by brackish-water channels, but within the past one hundred years the feeding channels have become silted up and the lakes have become almost salt-free and are now being utilized for fresh-water fisheries. These lakes, which support a huge number of water-birds, are being at present reclaimed to establish a satellite township for accommodating the over growing population of Calcutta.

The greater part of the southwest region of Midnapore District was covered with forest, continuous with the Orissa Hill and Chota-Nagpur forests. Through these forests a regular wildlife traffic flowed from the western parts of the Sundarban to the Peninsular India even a century ago, but the forests have all disppeared.

To summarize it may be concluded that the impoverishment and regression of our fauna are recent occurrences, wholly brought about by indiscriminate killing and destruction of natural habitats by man.

REFERENCES

Abu-i-Fazl Alam. 1590. Ain-i-Akbari.

Ali, S. 1960. The pink-headed duck. *Wild-fowl Trust 11th Ann. Rep.*, 1958–1959. p. 58.

Alvi, M. A. & A. Rahman, 1968. Jahangir the Naturalist. New Delhi.

Arthasastra (Ed. Sama Sastry, R) Mysore: 1924.

Beveridge, Annette S. 1922. Babur-nama in English (Memoirs of Babur). Translation from the original Turkish Text of Zahirudin Muhammad Babur Padshah Ghazi. London: 1.

Cadell, P. 1935. The preservation of wild life in India. No. 5 The Indian lion. *J. Bombay nat. Hist. Soc.*, 37(4): 165–166.

Chopra, P. N. 1963. Some aspects of society and culture during the Mughal age (1526–1707). Agra: pp. 68–72.

Ettinghausen, B. 1950. Studies in Muslim iconography. *Smithsonian Institution Free Gallery of Art Occ. Papers*, 1(3) (3993): 45.

Finn, F. 1929. Sterndale's Mammalia of India. Calcutta.

Foster, W. Early Travels in India 1583–1610. Oxford.

Gee, E. P. 1964. Wild Life of India. London.

Harper, F. 1948. Extinct and vanishing Mammals of the World. New York.

Houlton, J. 1948. Bihar the heart of India.

Jarrett, H. S. In: J. N. Sircar, Ain-i-Akbari. Calcutta: 2: 285.

Jerdon, T. C. 1864. Birds of India. Calcutta: vol. 3: 629.

Kinnear, N. B. 1920. The past and the present distribution of the lion in southeastern Asia. *J. Bombay nat. Hist. Soc.*, 27(1): 37–39.

Kirkpatrick, K. M. 1952. A record of the cheetah in Chittoor District. *J. Bombay nat. Hist. Soc.*, 50: 931.

Leyden, J. 1921. Memoirs of Zahir-ud-Din Mohammed Babur. London.

Lydekker, R. 1900. The great and small game of India, Burma and Tibet. London.

MacCrindle, J. M. 1926. Ancient India as described by Megasthenes and Arrian. Megasthenes Fragments xxv, xxvi, xxvii. Calcutta edition.

Perry, R. 1964. The World of Tiger. London.

Rao, H. S. 1957. History of our knowledge of the Indian Fauna through the ages. *J. Bombay nat. Hist. Soc.*, 54: 251–280.

Ripley, S. D. 1952. Vanishing and extinct bird species of India. *J. Bombay nat. Hist. Soc.*, 50: 903.

Seshadri, B. 1970. The Twilight of India's Wild Life. Oxford University Press.

Street, P. 1961. Vanishing Animals. London.

Sugiyama, Y. 1968. The ecology of the lion-tailed macaque *Macaca silenus* (Linnaeus) – A pilot Study. *J. Bombay nat. Hist. Soc.*, 65(2): 283.

Talbot, L. M. 1960. A look at the threatened species. *Oryx*, 5: 255.

Vishnudharmottara Purana, 1. 251, 22–37.

XIII. THE ECOLOGY OF VERTEBRATES OF THE INDIAN DESERT

by

ISHWAR PRAKASH

1. Introduction

The Indian Desert, situated on the eastern-most boundary of the Saharo-Rajasthan Desert, is of recent origin and offers considerable scope for ecological studies of fundamental and economic importance. Rao (1957) records evidence on the hunting of rhinoceros (which inhabits humid regions) in the Indus Valley by the Mugals as late as in 1519. At present, however, the prevailing climatic as well as the habitat conditions in this desert can only support predominantly xerophile and xerobiont fauna. Inspite of the interesting zoögeographical and evolutionary importance of the Indian Desert, adequate attention has not been paid so far to the zoology of the region. For instance, the Mammal Survey of India, Burma and Ceylon, conducted by the Bombay Natural History Society, did not include the Indian Desert. Realizing this lacuna in our knowledge of the fauna of the Indian desert, Dr. Daya Krishna, formerly Professor of Zoology at the Jaswant College, Jodhpur, moved the UNESCO to finance a project on the 'Ecological studies of vertebrates of the Indian Desert'. The author had the privilege to be associated with it from 1953 to 1955. These ecological studies were carried out mainly on reptiles and mammals and their findings were incorporated in university dissertations (Dave, 1961; Prakash, 1957).

The Central Arid Zone Research Institute was established in Jodhpur by the Ministry of Food and Agriculture, Government of India. The author collaborated in the research activities as the head of the Animal Ecology Section of the Central Arid Zone Research Institute, Jodhpur, since its inception.

While writing this chapter, the author had to depend very largely on the results of his own work. An attempt is also made to bring together here all the existing information on the arid regions of the Punjab, Haryana and Northern Gujarat, which collectively constitute the present Indian Desert. The work done in the adjoining desert areas in Pakistan has also been included, wherever possible. A special reference, in this context, may also be made to Minton's (1966) monograph on the Herpetology of Pakistan. With regard to the Rajasthan Desert, the work of Daya Krishna, who initiated the establishment of a School of Ecology of Desert Vertebrates, and those of Dr. P. K. Ghosh, Dr. K. C.

369

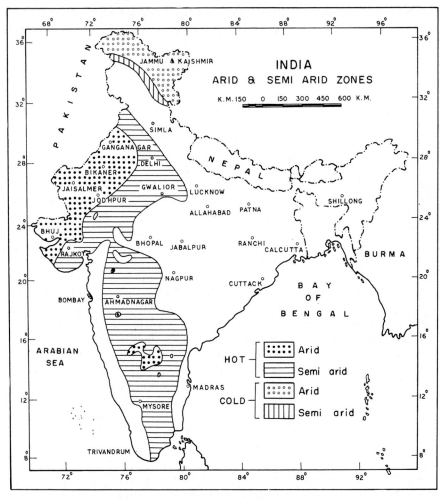

Fig. 42. Sketch map of the Republic of India, Nepal, Bangladesh, Burma and Ceylon, showing the semi-arid and arid areas; this chapter deals with the Vertebrata from the hot arid areas (Modified from KRISHNAN).

DAVE, MESSRS A. P. JAIN, B. D. RANA and I. K. SHARMA, as the associates of the author, have also been included.

This chapter deals with the ecology of vertebrate fauna of Indian Desert and comprises lists of fishes and amphibians. The distribution of reptiles in various types of habitats occurring in the desert, their food, breeding season and zoögeography are dealt with. The enumeration of birds is restricted to game birds and to those which are considered to be of economic importance. A detailed account of mammals is also included.

Their distribution in different habitats, relationships of the rodents with vegetation types, their food, certain reproduction aspects, population characteristics, behavioural and physiological adaptations to the desert conditions and zoögeography are also discussed.

2. The Indian Desert

Position and Extent: The Indian Desert lies on the northwestern boundary of the Republic of India and merges with the desert areas of Pakistan (Fig. 42). Prior to 1947 the deserts of both the countries were known under a common name, the 'Great Indian Desert'. The Indian Desert lies between 21° 25' and 30° 30' NL and 67° and 75° 25' EL. The main part of the Indian Desert lies mainly in the western and northeastern regions of Rajasthan (196 150 km²). The remaining part lies in some districts of Haryana (12 840 km²) and Punjab (14 510 km²), to the north Rajasthan and in some districts of Gujarat (62180 km²) in the southwest. The total area of the northwestern desert is 285 680 km² (A. KRISHNAN, 1969). Western Rajasthan (Fig. 43), like the Deccan Plateau, has resisted the orogenic forces, but has been subjected to marine transgressions, particularly in the area of Jodhpur and Bikaner, during the Jurassic, Cretaceous and Eocene periods. The area apparently became dryland during the Miocene and Pliocene, when the sea, occupying what is now western Rajasthan, gradually receded (M. S. KRISHNAN, 1952). WADIA (1953) concludes, however, that western Rajasthan began to get gradually dry only after the Pleistocene and last glacial period. The monsoon climate probably came to be established when the Himalaya had risen high enough to obstruct the southwesterly winds and cause the precipitation of moisture on their southern flanks (M. S. KRISHNAN, 1952).

On the basis of the findings of Chalcolithic period on kiln-burnt bricks, system of drains, and various other characteristics of the Indus civilisation, archaeologists also support the view that the Indian arid zone is of very recent origin (A. GHOSH, 1952; BHARADWAJ, 1961). A good river system (see Chapter II) (Saraswati and Ghaggar) seems to have existed in the desert region of India even during the Vedic period (A. GHOSH, 1952). Subsequent deterioration of the climate in the Indian Desert is indicated by the disappearance of the two classical river-systems and depression of the underground watertable to 80–120 m (WADIA, 1960). Considering the geological history and archaeological evidence, KRISHNAN (M. S. KRISHNAN, 1952) and WADIA (1960) conclude that the Indian Desert is not older than 5000–10,000 years. AHMED (1969), however, does not agree with this view, but feels that it may be actually much older.

Geology: The major geological features of the Indian Desert are concealed under the aeolian sand deposits. According to WADIA (1953), the aeolian sand deposits are due to long and continued aridity. Sand

Fig. 43. Sketch-map of the Rajasthan State, showing the administrative districts of the desert area to the west of the Aravalli Range.

deposition is also aided by drifting, caused by the southwest monsoon, which blows the material derived from atmospheric weathering of rocky outcrops with considerable force (BHARADWAJ, 1961). The topography of the Indian Desert is dominated by the Aravalli Ranges on its eastern border, which consist largely of tightly folded and highly metamorphosed Archaean rocks. The igneous rocks are represented by Erinpura, Jalor and Siwana granites, and Malani rhyolite at Jodhpur. The Vindhyan sandstone at Jodhpur is regarded to be different from the sandstones of Pokran-Jaisalmer region. The latter sandstones are dark with conglomerate base (M. S. KRISHNAN, 1952). The Pokran-Jaisalmer group of sandstones lie over the Malani rhyolite, which shows evidence of glaciation (BLANFORD, 1877). The Jurassic sandstone is exposed near Jaisalmer and the Cretaceous sandstone, found near Barmer, is of estuarine character and contains fossils of fruits. The Eocene *Nummulites* and *Assilina* limestones and shales, are found in Bikaner and Jaisalmer (M. S. KRISH-

Plate 52. Sand dune habitat in the Indian Desert.

NAN, 1952). In the Bikaner region, Palana lignite and Fuller's earth deposits are encountered. In the northeastern part, at Jhunjhunu, the Aravalli Ranges are represented by the broken ridges of sandstone.

Soil: The soil of the Rajasthan State has been extensively investigated (MEHTA & SHARMA, 1967; ROY & SEN, 1968) and grouped into the following seven categories (ROY & SEN, 1968): 1. desert soils confined to interdunal areas in the districts of Ganganagar, Bikaner, Churu, Jhunjhunu, Barmer, Jaisalmer, Jodhpur and Jalor; 2. dune sands, extensively distributed in the entire western Rajasthan, longitudinal, transverse, and *barkhans;* longitudinal sand dunes developed where the wind is strong and are confined to the southern parts of the Indian Desert; their direction is parallel to the prevailing southwest winds; the transverse sand dunes predominantly occur in the eastern and northern regions of the Indian Desert; the longitudinal axis of these dunes is transverse to the wind direction; barkhan type of dunes are common in the Central Desert; soils over the sand dunes are very deep fine sand deposits; 3. red desertic soils are pale brown to dark brown, almost loose and well drained. The texture varies from sandy loam to sandy clay loam. These are distributed in Nagour, Jodhpur, Jalor, Pali, Barmer, Churu and Jhunjhunu districts; 4. sierozems are more or less like red desertic soils, but are sandy clay loams and have fairly rapid permeability; these are met

373

Plate 53. Habitat in the Indian Desert, sandy plain, monsoon vegetation, with sheep grazing.

with in the Nagour and Pali districts; 5. saline soils are dark grey to pale brown heavy soils with water table very close to the surface. The soils have salinity of various degrees; these are distributed in *ranns* (saline depressions) of Barmer, Jaisalmer, Bikaner, Nagour districts; 6. lithosols and regosols are shallow, light textured, fairly drained soils of sloping hillsides, found on rocky outcrops throughout the desert; 7. red and yellow soils are found along the foothills in the desert. The soils along with the vegetation types form an influencing factor regulating the number of rodents and also their distribution in the desert region.

Climate: The Indian Desert is characterized by high atmospheric temperatures and low and erratic rainfall. More than 90% of the annual precipitation occurs during the monsoon months of July-September. The annual rainfall in the desert varies from 88 to 425 mm (PRAMANIK & HARIHARAN, 1952). The number of rainy days (day on which 10 cents or more rain occurs) varies from 2.7 to 28.5 per annum. The desert

374

Plate 54. Gravel Plain.

exhibits great extremes of temperatures; the lowest and highest temperatures recorded are —4.4 °C and 50.5 °C. The average annual minimum
and maximum temperatures vary, however, from 16.4 ° to 20.2 °C and
32.2 ° to 34.6 °C respectively. The mean annual relative humidity of air
varies from 26 to 66 %. Winds are strong and for most of the year the
direction is southwest to northwest. The maximum velocity of wind that
can be expected is about 140 km/hour. Reference may be made to
PRAMANIK & HARIHARAN (1952) for details of the climate in the desert
region. The reader will also find further useful information on the climatic
conditions of these areas in Chapters IV and V.

Habitats and Vegetation of the Desert: Four chief habitat types, based on
typical land forms, can be recognized in the Indian Desert.

Aquatic habitat: Perennial lakes, ponds and *nadis* (temporary rain water

Plate 55. Rocky habitat in the Indian Desert.

collections). In addition to some aquatic vegetation, a luxuriant, oasis-type of vegetation is found around them. The lakes also support considerable man-made gardens around them. These provide an ideal habitat to a large number of vertebrates.

Sandy habitat: The plains have thick deposits of older alluvial sand, but have a calcareous zone at various depths varying from 45–150 cm. *Prosopis cineraria, Capparis decidua* and *Zizyphus nummularia* are the main wooded species of plants. *Acacia nilotica, A. leucophloea* and *Salvadora persica* are also found in areas of relatively higher rainfall. *Prosopis juliflora,* an exotic, has thoroughly established itself in a wide region in the Indian Desert. Other common plants are *Leptadenia pyrotechnica, Capparis decidua, Calligonum polygonoides, Crotalaria burhia, Aerva tomentosa, Fagonia cretica, Indigofera* spp., *Tephrosia purpurea, Calotropis procera, Panicum antidotale, P. turgidum, Lasiurus sindicus, Cenchrus ciliaris, C. biflorus, C. setigerus, Dichanthium annulatum, Aristida* spp., etc. (BLATTER & HALLBERG, 1918–1921).

Rocky habitat: The hills are sparsely clothed with vegetation due to the scanty soil layer. The characteristic plant of this habitat is the shrub *Euphorbia caducifolia.* The common trees of rocky region are *Anogeissus*

Plate 56. Ruderal habitat showing a nadi or a water-pond on the right.

pendula, Acacia senegal, Commiphora mukul, Maytenus emarginatus and *Grewia tenax.* The shrub flora is dominated by *Zizyphus nummularia, Barleria acanthoides, Lepidagathis trinervis,* and *Capparis decidua.* The grass cover consists of *Cenchrus* sp., and *Sehima nervosum.* Near the hills extensive gravelly plains also occur.

Ruderal habitat: The *Village Complex* occurs on any land form, depending upon the availability of water and forage for the livestock population. The typical plant species of this habitat are *Azadirachta indica, Prosopis cineraria, Tamarindus indica, Ficus* spp. and *Salvadora oleoides,* which are mostly planted. *Calotropis procera* is probably the most characteristic plant of the habitat. The ruderal habitat is very well inhabited by vertebrates. This may be due to the maximum availability of shelter, food and water under the influence of man and his livestock; it is fundamentally a man-made habitat.

3. *Fishes

It appears unnatural to speak of fishes in a desert, but due to presence of large perennial lakes throughout the desert, a number of species of

---

* The nomenclature followed in this text is after BERG (1940) for fishes, MINTON (1966) for amphibians and reptiles, RIPLEY (1961) for birds, ELLERMAN & MORRISON-SCOTT (1951) for mammals other than rodents and ELLERMAN (1961) for rodents.

fishes are known from this region (ADAMS, 1899; DATTA-GUPTA et al., 1961; DHAWAN, 1969; HORA & MATHUR, 1952; KRISHNA & MENON, 1958; B. B. L. MATHUR, 1954; D. S. MATHUR & YAZDANI, 1968, 1970; YAZDANI & BHARGAVA, 1969). HORA & MATHUR (1952) reported the occurrence of *Labeo nigripinnis* in the Aravalli Ranges, which form the southeastern border of the Indian Desert. This species of fish is also known only from the Sind hills. These two hilly areas are at present separated by a large stretch of desert country. This discontinuous distribution is explained by the authors by the fact that the Kirthar Range of Sind and the Aravallis were in the past connected by a hilly link, which has since been smothered by desert sand. The actual period of migration of *Labeo nigripinnis* from Kirthar to Aravalli Range is supposed to be during the last Glacial Period, which ended about 7,500–10,000 years ago.

HORA & MATHUR (1952) have also recorded from the Aravalli Hills other species like *Oxygaster clupeoides*, *Tor khurdee*, *Puntius amphibia*, *Garra mullya* and *Noemacheilus denisonii*, which are otherwise known only from the Peninsular India. During the late Himalayan orogenic movements, the northern and northwestern parts of the once extensive Aravalli ranges sank, with the result that there was down-warping of the range northwards, so that the aquatic fauna of the south may have had a chance to be transferred to the Aravalli Range (HORA & MATHUR, 1952).

As there are no perennial rivers, all the species are exclusively found in large lakes and ponds, most of which also dry completely during years of drought, resulting in total loss of the fish fauna. In the absence of detailed ecological data, only a list of fishes reported from the desert region is given here. The common food fishes are: *Labeo rohita*, *Labeo dero*, *Catla catla*, *Cirrhinus mrigala*, *Clarias batrachus*, *Heteropneustes fossilis*, *Wallago attu*, *Ompok bimaculatus*, *Channa marulius*, *Channa punctata* and *Mastocembelus armatus*. Other smaller fishes are *Puntius ticto*, *P. sophore*, *P. sarana*, *P. amphibius*, *P. vittatus*, *Oxygaster bacaila*, *Esomus danrica*, *Rasbora daniconius*, *Danio devorio*, *Osteobrama cotio*, *Aspidoparia morar*, *Labeo boggut*, *Mystus bleekeri*, *Lepidocephalus guntia*, *Aphanius dispar*, *Aplocheilus panchax*, *Mastocembelus pancalus*, *Aplocheilus blochii*, *Barilius barna*, *Channa gachua* and *Amblypharyngodon mola*. MATHUR & YAZDANI (1970) have recently described *Noemacheilus rajasthanicus* from Rajasthan.

4. Amphibians

The amphibian fauna of the desert region is restricted to one species of toad and five species of frogs. The Anderson's toad, *Bufo andersonii* is widely distributed along the water courses and temporary ponds. It is, however, plentiful during monsoon season even in urban areas. The Indian bull-frog, *Rana tigrina*, is most common in ditches, marshes and tanks. During the day it spends the time in small crevices on the banks. It is more active during the night in the warmer season, but is found to be active during the day in winter. Its diet consists of a variety of aquatic insects, *Galeodes*, scorpion, land crab, *Uromastix*, metre-long ratsnake and mouse (PRAKASH, 1963).

The other desert amphibians namely the Indian cricket frog *Rana limnocharis*, skittering frog *Rana cyanophlyctis*, Indian burrowing frog *Rana breviceps* and the marshy toad *Microhyla ornata* are found near water sources. Unfortunately, however, their ecology in the Indian desert has not been adequately studied.

5. Reptiles

The Reptilia are represented in the desert by one species of Loricata and two of Testudines, eighteen species and one subspecies of Sauria and eighteen species and seven subspecies of Serpentes (DAVE, 1961).

5.1. ECOLOGICAL DISTRIBUTION

DAVE (1961) recognizes five ecosystems with respect to the distribution of reptiles in the Rajasthan desert, viz. aquatic, rocky, psammophile, soil and ruderal. I have, however, preferred the term habitats instead of ecosystems and have called psammophile as the sandy habitat and 'soil' as the gravely habitat in conformity with the account of mammals.

The Aquatic Habitat: In perennial lakes and in the Jawai Dam catchment, *Crocodilus palustris* and the freshwater turtle *Geoclemys hamiltoni* are found. The crocodile is, however, fast vanishing from the desert region, due to the gradual desiccation of lakes from years of continuous drought. The turtle is also commonly found in perennial ponds.

The Rocky Habitat: *Alsophylax tuberculatus* takes shelter under bushes of *Euphorbia caducifolia*, *Acacia* spp., *Capparis decidua*, and *Zizyphus nummularia*. At times it is also found under stones. The keeled rock-gecko *Cyrtodactylus scaber*, the Persian gecko *Hemidactylus persicus* and the fat-tailed lizard *Eublepharis macularius* are crevice-dwellers, but also occur under stones. Among the agamid lizards, *Calotes versicolor* and *Agama agilis* are usually found perched on shrubs and trees. The skink *Mabuya dissimilis* occurs in the vicinity of piles of stones or in crevices in stone pavements near lakes. MINTON (1966), who reported it in damp grasslands in Pakistan, found that it readily enters water. *Varanus bengalensis* is also met with in this habitat. On the foothills near Erinpura, I have observed *Testudo elegans* and collected an *Eumeces taeniolatus* on a hill at about 450 m elevation near Jalor. The common snakes which occur on rocks are *Coluber ventrimaculatus*, *C. arenarius*, *Lytorhynchus paradoxus*, *Bungarus caeruleus caeruleus*, and *Naja naja naja*.

The Sandy Habitat: Almost all the lizards found in the rocky habitat inhabit also the sandy habitat, usually in association of the bushes *Calotropis procera*, *Capparis decidua*, *Calligonum polygonoides*, *Zizyphus nummularia*, etc. In addition, *Mabuya macularia*, *M. aurata* and *Eumeces taeniolatus* are also found in association of desert grasses and shrubs. *E. taeniolatus* also abounds in marshy clayey soils near irrigation wells. The most common reptile of this habitat is, however, the lacertid, *Acanthodactylus cantoris cantoris*, which is found on undulating sand dunes and interdunal sandy plains. It is a close associate of the skink, *Ophiomorus tridactylus*, commonly known as the sand-fish due to its wriggling mode of locomotion under the loose soil, which looks similar to swimming in water (RATHOR, 1969). *Varanus griseus* is quite commonly found. The two sand boas, *Eryx conicus* and *Eryx johni* prefer sandy plains, where they mostly occupy rodent burrows. Other snakes collected from this habitat are *Ptyas mucosus*, *Coluber ventrimaculatus*, *Coluber rhodorachis*, *Coluber arenarius*, *Sphalerosophis diadema*, *Contia persica*, *Lytorhynchus paradoxus*, *Telescopus rhinopoma*, *Bungarus caeruleus sindanus* and *Echis carinatus*.

The Gravelly Habitat: In addition to the species *Hemidactylus persicus*, *Calotes versicolor*, *Agama agilis*, *Varanus bengalensis* and *V. griseus*, the most common lizard is *Uromastix hardwickii*, which abounds in depressions and saline patches. The snakes *Ptyas mucosus*, *Psammophis condanarus*, *S. diadema*, *Bungarus caeruleus caeruleus*, *B. caeruleus sindanus* and *Echis carinatus* also occur in this habitat.

The Ruderal Habitat: *Hemidactylus brooki* and *H. flaviviridis* are the common geckos, found in ruined buildings and in inhabited houses. *Calotes versicolor* is plentiful in the hedges of fields and gardens. The blind burrowing snakes, *Typhlops braminus* and *Leptotyphlops macrorhynchus*, are invariably met with in the village complex, but MINTON (1966) collected the latter species at an ancient archaeological site, about 1.5 m below the ground level, in Pakistan. The other snakes, which commonly occur in villages and towns in the Indian Desert, are *Eryx conicus*, *Eryx johni*, *Ptyas mucosus*, *S. diadema*, *Bungarus caeruleus* sbspp. and *Naja naja*.

5.2. FOOD

In absence of first-hand information on the food of these reptiles, I have largely drawn my account from the excellent observations of MINTON (1966) in Pakistan, a region which is not very different from the Indian Desert.

The spotted pond turtle, *Geoclemys hamiltoni* generally feeds on snails, but at times undigested algae are also found in the stomach, though MINTON (1966) regards that algae are swallowed only accidentally, as his captive turtle fed upon meat, snails and insects, but refused vegetable food. The faecal matter of the two Indian star-tortoises, collected by me at Erinpura, consisted mostly of grasses. It also feeds upon a wide variety of fruits, leaves and flowers, but shows no interest in animal food (MINTON, 1966). It appears that the pond turtle is mostly carnivorous, while the terrestrial tortoise is a vegetarian.

Fish and turtle appear to be the main food of the *Crocodylus palustris*, but it is reported to attack livestock and even human beings.

380

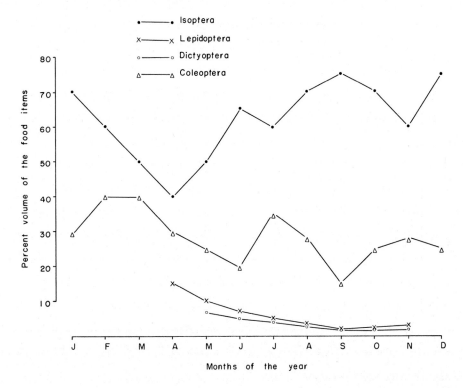

Fig. 44. Monthly fluctuations in various food items of the Indian sand skink, *Ophiomorus tridactylus*.

Most of the lizards thrive upon grasshoppers, crickets, beetles, dragon-flies, antlions, butterflies and termites. Spiders are also found in the stomachs of *Calotes versicolor* and *Mabuya dissimilis*, whereas scorpions constitute a part of the food of *Eublepharis macularius*. *Uromastix hardwickii* appears to be herbivorous (DAVE, 1961). A wide variety of food found in the stomach of *Varanus bengalensis* includes the musk shrew, striped palm squirrel, snake, *Calotes*, beetles, locust, downy feathers of bird, fish, crabs and crayfish (MINTON, 1966).

RATHOR (1969), who studied the food of *Ophiomorus tridactylus* by analysing the monthly samples of stomach contents of freshly captured lizards (Fig. 44), found termites to form its major food all over the year, varying from 50 to 75 % of the total food by volume. Fluctuations in the occurrence of Coleoptera in the stomachs range from 15 to 40 %, the peak being during February to April. Lepidoptera occur during April to June and Dictyoptera and Orthoptera from May to November, and March to November respectively. The author experimentally provided a wide variety of food to captive lizards and concluded that this lizard is primarily insectivorous.

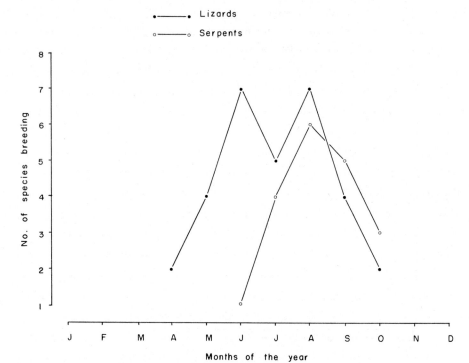

Fig. 45. Monthly fluctuations in the breeding of lizards and snakes in the Indian Desert.

Contrary to the general belief that snakes mostly feed upon rodents, the stomachs of most of the species examined by MINTON (1966) included the lizards *Acanthodactylus*, *H. flaviviridis*, *Cyrtodactylus*, *Hemidactylus* sp., *M. dissimilis*, *A. cantoris*, *A. agilis*, *M. macularia*, *H. persicus*. *Rana tigrina*, toad, finchlark and centipede were also found in the stomach contents of *Ptyas mucosus*, *Bungarus caeruleus*, *Psammophis schokari* and *Echis carinatus* respectively. Musk-shrews were found in the stomachs of *Eryx johni*, bats and musk-shrew in *Coluber ventrimaculatus*, gerbil in *Sphalerosophis arenarius* and palm squirrel and mice in *Echis carinatus*. MINTON's (1966) findings reveal that mammals were eaten only by a few species of snakes, which are supposed to be the main predators of rodents. Further work will probably throw some light on the predator-prey relationship of snakes and rodents, and on the magnitude of the rôle the former play in acting as natural control of rodents.

5.3. BREEDING

In his comprehensive monograph on the herpetology of Pakistan, MINTON (1966) has mentioned the months of egg laying and hatching for most of the reptiles. Reliable information about the breeding of certain reptiles of this region is also available in DAVE (1961). I have depended mostly on these two sources for the information on breeding. Eggs of most of the lizards (seven) hatch during June and August, whereas the eggs of peak number of serpent species (six) hatch during August (Fig. 45). From these data it appears that the hatching of peak number of lizard species occurs two months earlier than that of snakes. It is rather difficult to explain this difference in hatching periods, particularly when the above accounts are based only on a few species. Moreover, the information which is available for any species is also based on study of only a few specimens. Further extensive work on the reproduction of reptiles in the desert region is desirable. It is, however, sufficiently clear that the main hatching period of all the reptiles starts in June and continues upto September. This period corresponds also to the period of maximum precipitation in the desert region. This type of correlation has also been reported (PRAKASH, 1960) in mammals, which have maximum litter during the rainy reason, when the living conditions are optimum and survival rate of the young may be higher than in any other month.

5.4. ZOOGEOGRAPHY

On the basis of distributional records mentioned by SMITH (1931, 1935) and MINTON (1966), the reptiles of the Indian Desert fall under two major groups viz. Palaeotropical and the socalled Indo-Malayan (Oriental). Among the Palaeotropical species we may recognize 1. the species which have wide distribution in the Palaeotropical region; 2. Saharo-Oriental or species which are distributed from the Saharan region to the Indian; 3. Saharo-Rajasthani, species distributed from the Sahara to Rajasthan desert; 4. Irano-Oriental, species distributed from Iran to the Indian; 5. Irano-Rajasthani, species distributed from Iran to Rajasthan desert; and 6. Endemic, restricted to Great Indian Desert (but with Saharan affinities). Among the Oriental we have 1. the species with distribution in the Indo-Malayan region and 2. species distributed in the Indian inclusive of the Great Indian Desert.

This analysis (Table I) shows that 69.2% of the reptiles have Palaearctic affinities, so that most of the herpetofauna of the Indian desert is an extension of the Saharan elements. Out of this, 7.7% species and 1 subspecies are endemic to the Great Indian Desert. Only 31.8% of the reptiles are of Indo-Malayan affinities, out of which 23% are distributed in the Indian subregion. This shows that among the reptiles inhabiting the Indian Desert there is an admixture of both the western and eastern elements (see Chapter XVII).

6. Birds

The birds of the desert region have attracted the attention of many naturalists (ADAM, 1873, 1874; ADAMS, 1899; BISWAS, 1947; BUTLER, 1875, 1876; HOLMES & WRIGHT, 1968–1969; HUME, 1878; RANA, 1969; TICEHURST, 1922–1924; WHISTLER, 1938). Unfortunately, however, almost all the authors have, except for brief field notes, merely listed species but practically nothing about their ecology and zoögeography. A notable exception is, however, the recent study of RIPLEY (1961), who considered the Indian avifauna from a zoögeographical standpoint. His analysis of 176 species, endemic to the Indian Subregion, shows that 62 % of them have Indo-Chinese affinities, whereas 17 % are Palaearctic and 17 % Ethiopian. Evidently, the influence of the Indo-Chinese faunal centre in the Indian avifauna is maximum, as is apparent in the birds found along Himalaya, Peninsula and Ceylon. RIPLEY (1961) adds, however, that the Ethiopian influence is conspicuous in the open dry plain area, the western deserts, the dry parts of Gangetic Plains, Deccan Plateau and the dry areas of the Peninsula. The birds enumerated in the following account are mostly either Ethiopian or Palaearctic, suggesting that the avian fauna of the Indian Desert is more western than elsewhere.

I have restricted my account of birds of the desert to only some of the species, which have economic importance. The painted partridge, *Francolinus pictus pallidus*, is distributed in the southern parts of the desert, near irrigation channels and open grasslands. The species is common in Gujarat and it is quite probable that the partridge might have migrated into western Rajasthan from Gujarat, along with riverine and irrigation systems. SHARMA (1965) analysed its crop contents and found that in winter it feeds mostly upon the seeds of *Tephrosia purpurea*, *T. uniflora*, *Panicum antidotale*, *Zizyphus nummularia*, *Citrullus* sp., *Brachiaria ramosa*, *Cucumis callosus*, *Cyperus rotundus*, etc. These plant species consumed are not common in the natural grasslands and represent only 0.5 % of the total natural vegetation. The most common species of plants do not seem to be consumed by the painted partridge. He concluded, therefore, that the bird is selective in its feeding habits. The large ant, *Monomorium indicum* and ladybird beetles are also included in its dietary. Five to eight eggs are laid in a depression in the ground, covered by grass or crops, from May to October (DHARMAKUMARSINHJI, 1954). The grey partridge, *Francolinus pondicerianus*, is commonly found throughout the Indian Desert, particularly near villages, where large number of livestock is present. The francolin is especially attracted to cowdung containing undigested seeds. The termites, found under dried dung, also form substantial feed for the partridge. FARUQUI et al. (1960) and BUMP & BUMP (1964), who studied the crop contents, report that out of 54 crops examined, only 23 contained plant material, one only insects, and 30 both plant and animal parts. In all, 33 species of plants and 7 orders of insects were identified. Other items found in the crop contents are coal, baked brick pieces, grit and snail shells. Seeds of a variety of weeds constituted the bulk of the plant material. Insect food was taken abundantly in the summer, an observation true for desert gerbil also (PRAKASH, 1962), showing a high preference for ants and termites. Beetles were also found in large quantities. The francolin is considered useful to the agriculturist, as it rids the former of the weed seeds and the insects.

DHARMAKUMARSINHJI (1954) found its clutch size to be 5–10, the usual number being 7. Egg laying is observed from February to June, but the principal breeding months are March to May. Old pairs breed after the rains and during winter. The common quail *Coturnix coturnix* is a Palaearctic species and is resident as well as winter visitor. My observations are that it stays in the Indian Desert throughout the year, but the resident population is augmented during winter by migratory *Coturnix* from Middle Asia. It occurs in grasslands and in the cultivated areas. Flocks of five to fifteen are quite common in the southeastern desert, but pairs are found throughout the desert, except in the very arid parts of Jaisalmer and Bikaner. MUKHERJEE (1963) analysed its food in western Rajasthan and found the seeds of *Andropogon contortus, Brachiaria ramosa, Cenchrus montanus, Indigofera cordifolia, I. linifolia, Panicum antidotale, Tephrosia purpurea, T. tenuis, P. strigosa, Pennisetum typhoideum, Sorghum vulgare, Phaseolus radiatus;* arachnids and insects (*Chrotogonus* spp., *Brachytripes* sp., *Microtermes* sp. and *Dorylus* sp.) in their crops. The breeding season lasts from February to October (SALIMALI, 1961) and 6 to 14 eggs are laid. DHARMAKUMAR-SHINHJI (1954) observed, however, that it breeds from June to September in northern Gujarat. Other species of the quail, reported from the Indian Desert, are the black-breasted quail *Coturnix coromandelica*, rock bush quail *Perdicula argoondah meinertzhageni*, little bustard-quail *Turnix sylvatica* and button quail, *Turnix tanki* (WHISTLER, 1938). The great Indian bustard, *Choriotes nigriceps*, is a large and heavy bird, which prefers to live alone or in small groups on open grasslands and is found in most of the arid western India (PRAKASH & GHOSH, 1963). During the monsoon period it migrates only locally. Due to persecution as a rare trophy, and partly due to the transformation of grasslands into cultivated areas, this magnificent bird has become rare and is threatened with extinction, though totally protected by law. However, in the remote parts of the desert, where they are free from molestation, we have seen these birds in appreciable numbers. It feeds upon grasshoppers, locusts, beetles, small snakes, lizard, grass seeds, fruits and food grains (DHARMAKUMAR-SINHJI, 1954). It breeds from March to September (SALIMALI, 1961). We observed a single bustard egg during June near Pokran (PRAKASH & GHOSH, 1964).

The houbara, *Chlamydotis undulata macqueeni*, is distributed from the Canary Islands, North Africa, Middle East to Middle Asia, but the subspecies *C. undulata macqueeni* is found in Pakistan and is a winter visitor in the Indian Desert. The bird is found in small groups and is well camouflaged against the ground. It inhabits sandy plains, having large numbers of *Zizyphus nummularia*, the fruits of which are its main food.

DHARMAKUMARSINHJI (1954) mentions that it feeds upon insects, lizards and fruit of *Capparis decidua*. The bird breeds in Baluchistan and Mekran. It is much hunted for food purposes.

The lesser florican, *Sypheotides indica*, occurs throughout the Indian Subregion, except Assam; in the southeastern Indian Desert it arrives only during the monsoon season. DHARMAKUMARSINHJI (1954) believes that it migrates from the Narmada and Tapti Valleys to northeast, into Gujarat and parts of the desert. It feeds upon grasshoppers, beetles especially the blister beetle, termites, seeds, plant-shoots, etc. The breeding season lasts from June to October. It displays a queer courtship behaviour; it jumps up about 2–3 m in the air and at the same time emits a short croak-like sound. This courtship display makes it vulnerable to sportsmen and the bird is shot during the breeding season. The meat is of good quality.

The common crane, *Grus grus*, is a Palaearctic breeding species that winters in the Mediterranean, North Africa and China but the subspecies *G.g. lilfordi* migrates in winter to the Indian Desert also. Flocks of 20–50 may be observed near lakes, in the rocky habitat, from the third week of October to March. It feeds upon insects, especially on locusts and grasshoppers, green shoots, groundnut and grains (DHARMAKUMARSINHJI, 1954). While migrating, the flocks keep to a V-shaped formation.

The common Indian sandgrouse, *Pterocles exustus erlangeri* is found in western India, especially Punjab, Kutch and northern Gujarat (RIPLEY, 1961), but the nominal form has a wide range from North Africa to India. It inhabits ploughed open fields, barren gravel plains and areas with short grass. The bird is generally found in pairs and small flocks, but larger flocks of 200–4000 birds have also been observed by me during summer near water sources. The sandgrouse takes water daily, primarily two hours after sunrise. The flocks fly to nearby *nadis* (rain water catchment pond) and after drinking water, shift to the grazing grounds. CHRISTENSEN (1962) observed coveys of three thousand birds coming to water tank near Pokran. He trapped a large number of them near water tanks, taking advantage of their regular drinking habit. In periods of severe drought, it is a remarkable sight to observe great flocks arrive from considerable distances in the few collections of water. In periods of drought FARUQUI et al. (1960) studied its food habits at different seasons of the year. Insects were not found in the forty-seven crops examined, although HUME & MARSHALL (1880) recorded two insects in the crop of *P. exustus*. The food of the bird is apparently restricted to wild leguminous seeds. The common Indian sandgrouse breeds chiefly from January to May (SALIMALI, 1961). I have seen eggs of the bird during May and June in the desert region, usually three eggs are laid.

Considering its importance as sport, a large number was introduced from the Thar Desert to Nevada, U.S.A. during 1960–1961. The birds from both the 1960 and 1961 liberations apparently disappeared from

Pahrump Valley within two months after each release. Later on, two birds were shot during February 1962 in Navoja, Sonora, Mexico, suggesting a southward migration to 27° NL, which is similar to their native habitat in India (CHRISTENSEN, 1963). The imperial sandgrouse, *Pterocles orientalis*, is a Palaeotropical bird and a winter visitor to the desert region. It arrives in the Western Indian Desert in flocks of 20 to 500. Its habits are like those of the foregoing species. Both the species of sandgrouses are hunted in large numbers for the table.

The other sandgrouses reported (WHISTLER, 1938) from the Indian Desert are the large pintail sandgrouse *Pterocles alchata*, the spotted sandgrouse *P. senegallus*, and the painted sandgrouse *P. indicus indicus*. These species are, however, not so abundant as the former two sandgrouses.

The perennial lakes and other large ephemeral ones are inhabited by a variety of birds in which a hunter is interested. Among the resident ones are the spotbill duck *Anas poecilorhyncha* and the cotton teal *Nattapus coromandelianus*. The Indian Desert also receives some winter visitors, the white-fronted goose *Anser albifrons*, the wigeon *Anas penelope*, the garganey *Anas querquedula*, red-crested pochard *Netta rufina*, and tufted duck *Aythya fuligula*.

MUKHERJEE (1962) has discussed the economic importance of Indian birds from the point of view of products of feathers, especially of *Pavo cristatus*. This is abundant in the desert region, particularly near human settlements. It is considered sacred and the erstwhile princely states gave the bird complete protection by law.* The peafowl moults yearly and the old feathers are dropped. These are picked up by people and sold in large numbers. A systematic and extensive study is needed on the peafowl as practically very little is known about its ecology and biology. Recently, SHARMA (1965, 1969), did some work on the peafowls in the Rajasthan Desert. He examined over hundred nests during March–June and found that the clutch size varies from 3 to 10. He (1969, 1970) further studied the breeding of twelve other species of birds around Jodhpur. Ten species of birds lay eggs during summer whereas two, the kite *Milvus migrans* and the vulture *Gyps bengalensis* during winter.

The most common among the vultures are *Gyps bengalensis* and the white scavenger vulture, *Nephron percnopterus*, but the king vulture *Torgos calvus* also feeds on carcass of animals in the desert. The jungle crow *Corvus macrorhynchus* and the raven *Corvus corax*, the house crow *Corvus splendens* and the pariah kite *Milvus migrans* also feed upon dead animals. Other Raptores, *Accipiter badius*, peregrine falcon *Falco peregrinus*, merlin *Falco*

* As indicated in Chapter XII, with the abolition of the princes, since the Indian independence, the strict enforcement of these laws has also disappeared and the ideas of sacredness of the peafowl have been conveniently set aside to earn dollars by export of the peacock feathers — M. S. MANI.

columbarius, kestrel *Falco tinnunculus*, collared scops owl *Otus bakkamoena*, great horned owl *Bubo bubo*, mottled wood owl *Strix ocellata*, and owlet *Athena brama* commonly feed upon the desert rodents, but their depradation scarcely seems to influence the rodent populations in the desert, which is much higher than that of the predatory birds.

Among the insectivorous birds, special mention is made of the rosy pastor or rose-coloured starling *Sturnus roseus* and of the starling *Sturnus vulgaris*. These Palaearctic birds winter in India. We may watch very large flocks or 'clouds' of these birds, moving from their roosting places to feeding grounds and vice versa. In the desert these birds feed upon insects, particularly the locusts.

HUSAIN & BHALLA (1939) recorded the food habits of certain birds in the Lyallpur region. Recently RANA conducted extensive investigations on the food of the desert birds, and examined the crops and gizzards of babblers, doves, parakeet, bulbul and many others to find out their relationship to agriculture. He (RANA, 1970) found that annually the jungle babbler *Turdoides striatus* and the common babbler *Turdoides caudatus* feed chiefly on millet grains up to 74.4 and 36.2% of total food, respectively. Rest of the food is composed of wheat, sorghum, grass seeds and leafy matter. It was also found that *T. striatus* feeds on insects mainly during summer, but *T. caudatus* feeds on insects all the year round, and in summer on large amounts of grass blades. His studies point out that *T. striatus* is more important to the agriculturist than *T. caudatus*.

RANA (unpublished paper) found that the parakeet, *Psittacula krameri* feeds mainly on millet grains, but seeds of *Albizia lebbek* are also found in its food contents. The red-vented bulbul *Pycnonotus cafer* feeds upon grass blades and fruits of *Zizyphus nummularia*, *Azadirachta indica*, *Grewia asiatica*, spiders and larger ants. The ring dove, *Streptopelia decaocto* and the little brown dove, *S. senegalensis* depend mainly on millet grains.

JAIN & PRAKASH (unpublished paper) estimated that the birds, mostly house sparrow, damage the grains of *Pennisetum typhoideum* to the extent of 8 to 10% before harvest.

7. Mammals

Out of fifty-one species so far recorded from the Indian Desert, three belong to Insectivora, eleven to Chiroptera, two to Primates, thirteen to Carnivora, one to Pholidota, four to Artiodactyla, one to Lagomorpha and sixteen to the Rodentia. Recent information on their taxonomy and distribution is available in PRAKASH (1959, 1961, 1963, 1964). MOORE & TATE (1965), GUPTA & AGARWAL (1966), BISWAS & TIWARI (1966), AGARWAL (1962, 1967), BISWAS & GHOSE (1968), PRAKASH & JAIN (1967).

7.1. Ecological distribution

The desert species occur in all the three major types of habitat and the aquatic habitat is partially occupied by only one species, the soft-furred otter, *Lutra perspicillata sindica*. The otters are reported, and once observed by me, at the Sardarsamand Lake, about 61 km east of Jodhpur, but due to frequent droughts leading to complete drying of the lake, the otters are no more seen at this place.

The sandy habitat is extensive and includes sand dunes, sand hills, sandy plains and artificial mines (PRAKASH, 1957, 1964). The sand dune habitat is usually unstable and is subject to effects of strong winds. The movement of the sand is at its maximum from April to June. During the monsoon, however, scanty vegetation, comprising *Aristida adscensionis*, *Indigofera cordifolia*, *Citrullus colocynthis*, *Cyperus rotundus*, *Tephrosia purpurea*, etc. establishes itself on the sand dunes and may partially stabilize them. All these annuals, however, dry up by October–November. During and soon after the rainy season, when the sand dunes become stationary and consolidated, the desert gerbil *Meriones hurrianae* digs its burrows, which are usually associated with the creeper, *Citrullus colocynthis*, as they feed on the seeds of this cucurbit. Burrows of the hairy-footed gerbil, *Gerbillus gleadowi*, are frequently observed on the sand dunes. Hedgehogs, foxes, hares and gazelles also occasionally visit this sub-habitat, mainly for feeding upon the beetles and tender vegetation.

The sand hills are permanent, consolidated sand dunes, initially formed by drifting sand but stabilized by growth of perennial vegetation, like *Calligonum polygonoides*, *Capparis decidua*, *Zizyphus nummularia*, *Aerva tomentosa*, *Panicum turgidum*, etc. The desert gerbil *M. hurrianae* and Indian gerbil *Tatera indica* are the chief inhabitants of the sand hills, but *Gerbillus gleadowi*, the two hedgehogs, the foxes *Vulpes v. pusilla* and *V. bengalensis* and cats *Felis chaus prateri*, *Felis libyca ornata*, the two species of mongoose *Herpestes edwardsi ferrugenius* and *H. auropunctatus pallipes* are also commonly found.

The sandy plains are the chief characteristics of the desert and cover the greater part of the desert region. The vegetation here consists of grasses, bushes and trees (BLATTER & HALLBERG, 1918–1921). Hedgehogs, gerbils, foxes, cats, mongoose, pangolin, artiodactyles and several species of rodents inhabit the sandy plains.

The artificial mines are long, though not deep caverns, excavated for the Fuller's earth, near Bikaner; they are inhabited by the bats *Rhinolophus lepidus lepidus* and *Rhinopoma hardwickei* (PRAKASH, 1961).

The Rocky habitat: The southeastern part (Sirohi and Pali District) of the Desert is rocky; rocky outcrops are also scattered throughout the Desert. The vegetation in this habitat is typically represented by *Sehima nervosum*, *Panicum* spp., *Aristida* sp. and *Cenchrus biflorus* among grasses. The shrubs and trees are represented by *Euphorbia caducifolia*, *Acacia senegal*, *Grewia tenax*,

Capparis decidua, etc. The Cutch rock-rat *Rattus cutchicus* is found in appreciable numbers in fissures and crevices in the rocks. On the slopes, where some sand accumulates, the porcupine *Hystrix indica indica* makes its burrows. Hares take refuge inside the huge bushes of *Euphorbia caducifolia* on the hill slopes. These bushes also provide shelter to the spiny mouse *Mus cervicolor* and *Mus platythrix*. The langur *Presbytis entellus entellus* and northern palm squirrel *Funambulus pannanti* dwell on the trees. The ruddy mongoose *Herpestes smithi* and the ratel *Mellivora capensis* have usually been observed on the foot of the hill.

The large natural caves are inhabited by the wolf *Canis lupus pallipes;* the jackal *Canis aureus aureus*, *Hyaena hyaena*, and by the panther *Panthera pardus*. The man-made caverns, worked for excavating lime and other useful earth, are the chief resort of bats *Rhinopoma kinneari*, *Rhinopoma hardwickei hardwickei*, *Taphozous perforatus perforatus*, *Rhinolophus lepidus lepidus*. If these caverns have crevices in the ceiling, *Taphozous kachhensis kachhensis* occupy them. *Herpestes* spp. are associated with bats and are found in these caverns.

Due to the scarcity of water in the desert, all the rain water is collected from the entire catchment area in large lakes near the rocky outcrops, princes had formerly established good gardens around the lakes, which serve as favoured abodes of mammals. Such lake areas would approximate to oasis of other deserts. The fruit bat *Pteropus giganteus giganteus* prefers to live on large *Ficus* trees and the langur visits the garden from nearby rocks for feeding. Squirrels become prolific. The wild boar *Sus scrofa cristatus*, once very common in rocky plains, is also found in small numbers near the Sardarsamand Lake. The black buck *Antelope cervicapra rajputanae*, a fast vanishing species, is common at certain places, where protection is given.

The Ruderal habitat: This term was used by BLATTER & HALLBERG (1918–1921) for the ecological association of plants near human settlement. TABER et al. (1967), however, have termed it as a village complex. It is an important habitat with respect to the mammal distribution also (PRAKASH, 1957, 1963, 1964). It provides mainly four types of sub-habitats. The inhabited houses and vicinity, the ruins of buildings and abandoned forts, drainage nullahs and cultivated fields.

In inhabited houses the house rat, *Rattus rattus rufescens*, the common house mouse, *Mus musculus bactrianus* and the house shrew, *Suncus murinus sindensis* are found. The Pipistrelle, *Pipistrellus mimus*, inhabits crevices in the verandahs. In the vicinity of houses in Bikaner we also find large numbers of *Tatera indica indica*, but in Jodhpur and in surrounding villages these are found in backyards of houses and in the thorn fencings. The Indian gerbil sometimes enters houses and comes into contact with *Rattus rattus*. The former species is found to be natural reservoir of plague infection (BALTAZARD & BAHMANYAR, 1960). The domestic rat acts as a liaison rodent between man and field rodents. In the backyards, in thorn and mud-fences we find a variety of mammalian fauna, comprising the squirrel *Funambulus pennanti* on the trees, *Tatera indica, Suncus murinus sindensis, Gerbillus gleadowi, Meriones hurrianae, Rattus meltada pallidior* and *Golunda ellioti gujarati*.

The ruins of buildings are scattered throughout the desert region and forts and fortresses of the erstwhile Maharajas are also not uncommon. *Herpestes auropunctatus pallipes* and in dark places, the bats, *Rhinopoma kinneari* and *Taphozous perforatus perforatus* are quite common. *Megaderma lyra lyra, Rhinopoma hardwickei, Taphozous kachhensis* and *Pipistrellus mimus* are also found in ruins and forts.

Drainage channels are common in the vicinity of towns and villages, particularly near hillocks. *Rhinolophus kinneari, R.h. hardwickei, Rhinolophus l. lepidus, T.p. peforatus* and *Rousettus arabicus* are found in this type of habitat (PRAKASH, 1961).

The cultivated fields are invaded by the artiodactyles; special mention is made of the blue bull, *Boselaphus tragocamelus* (PRAKASH, 1959). Herds of 5–10 animals visit the fields at night and ravage standing crops. The black buck and gazelle, which do considerable damage, are considered sacred and protected from killing by people of the Vishnoi caste, illustrating an outstanding situation where the threatened wildlife is conserved on sentimental and religious grounds. The fields are the favourite resorts of the rodents *Tatera indica indica, Rattus meltada pallidior, Mus booduga*, and *Nesokia indica. Meriones hurrianae* is found on the borders of the crop fields from where it freely feeds on the crops. *F. pennanti* occupies the trees scattered in the fields.

Recently we studied* the ecological distribution of various rodent species. Eleven trapping stations were selected and trapping was carried out during winter of 1968–1969. At each station trapping was done in sandy plain, gravel plain, rocky and in ruderal habitats. In each habitat two trap lines were run for seventy-two hours. Each trap line consisted of thirty snap traps, fixed at 10 m interval. Vegetation composition was studied to see its relationship with rodent distribution.

Table II indicates that in sandy and ruderal habitats occur a large number of rodent species, whereas the gravel habitat is the least favoured one. This observation is further confirmed by the number of rodents collected from each habitat. In all the administrative districts of the Rajasthan Desert 168 rodents, belonging to various species, were collected from the sandy habitat, 167 from ruderal, 81 from rocky and 36 from the gravel habitat.

Some rodents are found in only one habitat, for example, *Gerbillus dasyurus indus* only from the sandy habitat, *R. cutchicus* and *M. cervicolor* from the rocky habitat, and *Mus booduga, M. musculus* and *Rattus rattus*

---

* Detailed report will be published by Dr. W. Junk b.v., The Hague.

Table II. Percentage distribution of rodents in different habitats in the Indian Desert

Species	Habitats			
	Sandy	Gravel	Rocky	Ruderal
F. pennanti	14.2	—	35.8	50.0
G.d. indus	100.0	—	—	—
G. gleadowi	56.0	—	—	44.0
T.i. indica	28.8	10.0	3.6	57.6
M. hurrianae	60.0	17.0	—	23.0
R.c. cutchicus	—	—	100.0	—
R.m. pallidior	37.0	1.6	5.1	56.0
R. gleadowi	66.6	33.3	—	—
M.m. bactrianus	—	—	—	100.0
M.b. booduga	—	—	—	100.0
M. cervicolor ssp.	—	—	100.0	—
M.c. phillipsi	—	—	100.0	—
M.p. sadhu	28.0	13.5	53.3	13.3
G.e. gujarati	25.0	12.5	—	62.5

from the ruderal habitat only. Other species are, however, found in more than one habitat (Table II).

Studies on the association of the rodents and vegetation indicate that a single plant species cannot be regarded as indicator of a rodent species, since most of the rodents occur in various plant communities in the desert (GUPTA & PRAKASH, 1969). Some broad associations may, however, be established. *Mus cervicolor phillipsi*, the spiny-furred mouse, is always collected in traps situated on rocky slopes under bushes of *Euphorbia caducifolia*. *Rattus cutchicus cutchicus* is always found on hilly outcrops having the dominant grass, *Sehima nervosum*; *Rattus meltada pallidior*, *Rattus gleadowi* and *Mus platythrix sadhu* also occur in gravely plains, where the *Eleusine compressa* community occurs. Well drained sandy soils, where *C. setigerus*, *C. ciliaris* and *Lasiurus sindicus* are the common grasses, are inhabited by the four gerbils. In sandy loam, with good moisture-holding capacity which can sustain *Dichanthium* type of vegetation, *Tatera indica indica*, *Rattus meltada pallidior* and *Golunda ellioti* are found. The only mammal which inhabits the sand dunes more or less permanently is the hairy-footed gerbil, *Gerbillus gleadowi*. *Panicum antidotale* is the main grass growing on sand dunes. Its fragments were found in the gerbil burrows. Its burrows are of simple nature, with two to three openings, a few bolt-runs, a long arm, and a resting pouch (PRAKASH & PUROHIT, 1967). Once, on a drifting sand dune, on which their burrows were situated, marked *G. gleadowi* were released at dusk. Next morning the burrow openings were clearly visible, but as drifting sand deposited in the burrow openings, they were completely plugged by afternoon. In the evening *G. gleadowi* conveniently dug their way out.

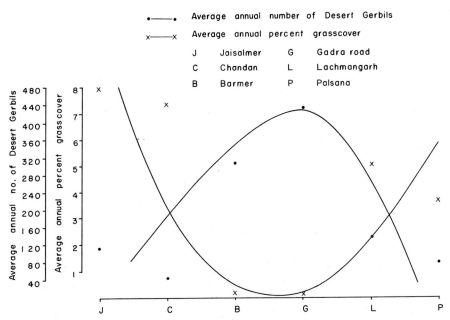

Fig. 46. Graph showing the relation between the mean annual grass-cover and the mean annual numbers of the desert gerbil, *Meriones hurrianae.*

In the course of another study (PRAKASH et al., 1971), we observed that the number of the desert gerbil, *Meriones hurrianae*, was higher in plots in which the frequency of occurrence of *Dactyloctenium sindicum, Aristida adscensionis, Lasiurus sindicus, Perotis hordeiformis* and *Digitaria marginata* was more. On the other hand the gerbil density was low in habitats where the grasses, *Cenchrus biflorus* and *Erianthus munja* are present (PRAKASH, 1964). An inverse relationship (Fig. 46) between the grass-cover and population density of *M. hurrianae* was also found, i.e. the higher gerbil number is associated with a lower density of grass-cover (PRAKASH et al, 1971). This is paradoxical, as it may apparently be expected that rodent density should be more in habitats with good cover of grasses, which are their chief food. This paradox can, however, be explained on the basis of their burrowing habits. The gerbils do not seem to be capable of readily digging in soils having a large amount of anasto-mosing and fibrous grass root, the dense root system making the soils more compact. Clayey and compact soils are not also preferred by them (PRAKASH et al., 1971).

7.2. FOOD

The hedgehogs and shrews are primarily insectivorous mammals, but the former feed also on appreciable quantities of scorpions, toads, lizards,

Table III. Showing the percentage stomach contents (PRAKASH, 1959) of *Tatera indica indica* during 1954

	Jan.	Feb.	Mar.	Apr.	May	June	July	Aug.	Sept.	Oct.	Nov.	Dec.
No. of stomachs examined	14	15	15	20	11	9	21	17	11	7	6	8
Seeds	40	30	25	15	15	10	0	10	10	20	25	25
Stems and rhizomes	15	20	30	30	25	25	20	20	25	25	25	25
Leaves and flowers	25	20	20	25	20	20	30	25	20	20	25	30
Insects	10	15	15	20	30	35	40	35	30	25	15	10
Miscellaneous	10	15	10	10	10	10	10	10	15	10	10	10

eggs of ground birds, (partridges, sandgrouse, quail) and small mammals (KRISHNA & PRAKASH, 1955, 1956, 1960; PRAKASH, 1956). On the basis of the study of stomach contents, KRISHNA & PRAKASH (1960) found that their food is composed of insects 53.6%, amphibians 15.8%, egg shells 10.4%, reptiles 8.4%, other arthropods 7.4% and mammals 4.1% respectively. The hedgehog also shows cannibalistic propensities (PRAKASH, 1953, 1954).

The bats *Pteropus giganteus giganteus* and *Rousettus arabicus* are fruit-eating chiropterans and rest of them are insectivorous. The Indian false vampire bat, however, feeds upon other food items also. House lizards, house sparrow and several species of bats were observed being eaten by the false vampire at Jodhpur (PRAKASH, 1959). Their stomach contents included bony pieces of amphibians, bones and scales of fishes. In addition to flesh, the foxes and cats, in the desert region, consume quantities of the berries of *Zizyphus nummularia* and seeds of *Cucumis callosus*. It is quite surprising to see vegetable matter in the stomachs of these carnivores. The stomachs of foxes, cats and mongoose contain fragments of scropion, hare, grey partridge, common pigeon, sandgrouse and other birds (PRAKASH, 1959).

The food of artiodactyles has not been studied in detail, but stray observations indicate that they primarily feed upon grasses and crops like gram, wheat, millet and other vegetables, which are grown in the desert, particularly where irrigation facilities, mostly from wells, are available. In periods of drought, when grass is not readily available, the Indian gazelle has been observed to feed upon *Capparis decidua, Crotalaria burhia, Aerva tomentosa* and *Calligonum polygonoides*, which are usually avoided during the years of normal rainfall.

The food of certain rodents has been studied in detail (PRAKASH, 1957, 1959, 1962, 1964, 1968; PRAKASH et al., 1969; PRAKASH & KUMBKARNI, 1962; PRAKASH et al., 1967). The palm squirrel, *Funambulus pennanti*,

Table IV. Showing the percentage stomach contents of *Meriones hurrianae* (PRAKASH, 1959) during 1954

	Jan.	Feb.	Mar.	Apr.	May	June	July	Aug.	Sept.	Oct.	Nov.	Dec.
No. of Stomachs	17	14	19	9	9	4	11	17	25	13	18	16
Seeds	60	50	40	30	20	10	20	20	20	30	50	60
Stems and rhizomes	10	15	25	35	40	45	30	20	25	20	15	10
Leaves and flowers	25	25	15	10	5	15	25	35	30	40	30	30
Insects	0	0	0	5	15	15	15	10	10	0	0	0
Miscellaneous	5	10	10	20	20	15	10	15	10	10	5	0

feeds mostly on the fruits of *Prosopis cineraria, Acacia senegal, Grewia tenax, Azadirachta indica,* etc. In the ruderal habitat, the squirrel is mostly dependent on man for its food. It also feeds in orchards on fruit and saplings of vegetables, thus causing severe damage.

The gerbils, *T. indica indica* and *M. hurrianae,* the two most common rodents in the Indian Desert, feed upon grasses, shrubs and tree species. The stomach contents of these gerbils were examined throughout the year (PRAKASH, 1962). The study revealed that both the gerbils feed on maximum quantities of seeds during the winter (Tables III and IV). The rhizomes and stems are found upto 40 % of the total food in the stomachs of *M. hurrianae* during summer, but these food items constitute lower proportion (20 %) of the total monthly contents during rest of the year. In *Tatera indica indica* this food item fluctuates from 15 to 30 % of the total stomach content all the year round. During the rainy and post-monsoon seasons, their chief diet is constituted by leaves and flowers. *M. hurrianae,* a phytophagous rodent, also feeds upon insects during the summer and they fluctuate from 5 to 15 % of the monthly food. *Tatera indica indica* feeds on insects throughout the year. It is interesting to note that the feeding habits of *T. indica indica* and *T. indica cuvieri* (PRASAD, 1954) are almost similar throughout the year, inspite of the fact that both the subspecies are widely separated geographically.

The monsoon season foods of the desert gerbil, *M. hurrianae* were further studied in nature by comparing the frequency of occurrence of different plant species in a community with that of freshly cut and unconsumed parts of plants lying near gerbil burrow openings (PRAKASH, 1968). The study indicated that the gerbil also prefers the same species of grasses, which constitute the chief fodder for livestock. In the study region (80 hectares) the rodent population was 477 per hectare. This population is sufficient to depradate the total yearly grass produce of the region, leaving nothing for the livestock to graze upon. The average annual

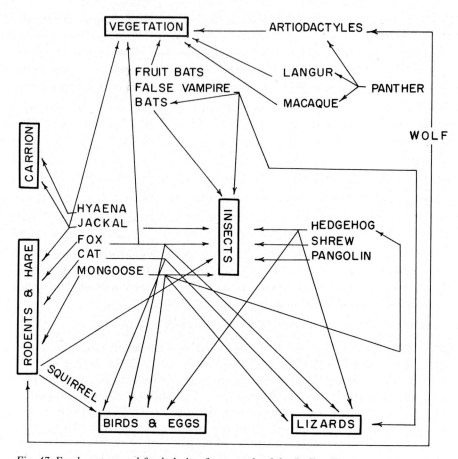

Fig. 47. Food centres and food chain of mammals of the Indian Desert.

population of *M. hurrianae* in the Indian Desert fluctuates from 7.4 to 523 per hectare (PRAKASH et al., 1971).

Food Centres and Food Chain: On the basis of the study of natural food of mammals, found in the Indian desert, it is possible to recognize certain food centres, following KASHKAROV & KURBOTOV (1930). Six food centres are recognized (PRAKASH, 1964): vegetation, insects and scorpion, lizards, birds and eggs, rodents and hares and carrion. Simplified food-chains (Fig. 47) indicate that vegetation and insects are the busiest food centres in the Indian Desert.

Shortage of food for the bad season is characteristic of the mammals of the steppe and the predesert (BODENHEIMER, 1957). It was accordingly expected that storing habit should be found in several mammalian species in the Indian Desert, particularly among the rodents. A thorough examination of dozens of burrows each of the several species of carnivores and rodents revealed that none of them, except the India gerbil, *Tatera indica indica*, stores food. The hoards of this gerbil were also found to be very small, compared with those of *Bandicota bengalensis* (PARRACK, 1969). It is not surprising that most of the mammals do not hoard food for bad season in the Indian Desert, mainly because they appear to be well adapted to the changing availability of food in nature.

7.3. REPRODUCTION

Oestrous Cycle: There is very scanty information on the oestrous cycle of desert mammals. GHOSH & TANEJA (1968) worked out the oestrous cycle of *Tatera indica indica* and *Meriones hurrianae*. The mean interval between vaginal cornifications, with standard deviation, is 3.825 ± 0.359 days for the former species and 6.22 ± 0.687 days for the latter. The maximum probable duration for cornification was found to be one day for both the species. *T. indica* may, therefore, be said to have an average oestrous cycle of 4.82 days, while the cycle in *M. hurrianae* may repeat every 7.22 days. The authors observed that period of oestrous of *T. indica* is in conformity with the average duration of 4.0 to 4.8 days reported by ASDELL (1946) for various species of *Rattus*.

Gestation period: Authentic information on the gestation period is available for only a few species of the desert mammals. ANANDKUMAR (1965) reported a gestation period of 123 days in *Rhinopoma kinneari*. A gestation period of 140–150 and 150–160 days was found in *Megaderma lyra lyra* and *Pteropus giganteus giganteus* respectively (GOPALAKRISHNA, 1969). BANERJI (1955, 1957) reports a gestation period of 42 days in the squirrel, *F. pannanti*. A female porcupine, *Hystrix indica* littered in the Bikaner Zoo on 3.6.1963 and again on 20.9.1963, the difference between the two deliveries was of 109 days. Although calling the period between two deliveries as the gestation period is not always correct, particularly when we do not know about the occurrence of super-foetation or post-parturition oestrous in the animal, yet this record is the only information on Indian crested porcupine, by which a rough estimate of the gestation period could be made (PRAKASH, 1968). ZUCKERMAN (1953) mentions that for the north African species of porcupines the gestation takes 63 or 112 days (ASDELL 1946, KENNETH 1947). The gestation period in *Tatera i. indica* was found to be 27 to 30 days; average 28.22 days (PRAKASH et al., 1971), that of *M. hurrianae* 28–30 days; average 29 days (PRAKASH, 1964) that of *Rattus rattus* 21.5–22 days (LONG & EVANS, 1922) and that of *Mus musculus* 19–20 days (PARKES, 1926).

Litter Size: Information on the litter size of some mammalian species is given in Table V. Most of the data are based on observations in the Indian Desert, but some information has also been taken from other publications.

Extensive studies on the monthly distribution of litters of various sizes

Table V. Litter size and littering season of some mammals of the Indian Desert

Species	Litter size	Littering season	Source of information
H.a. collaris	1–6	July-September	PRAKASH 1954
P.m. micropus	1–5	August	GUPTA & SHARMA
P.g. giganteus	1	April-September	1961
R. kinneari	1	May-July, Oct.-Nov.	PRAKASH 1960
R.h. hardwickei	1	June-July	PRAKASH 1960
M.l. lyra	1	May	PRAKASH 1960
	(Twin in only one case)*		RAMASWAMY & ANAND KUMAR 1963
S. heathi	1	August	PRAKASH 1960
P.m. glaucillus	1	September	PRAKASH 1960
M.m. mulatta	1	March-May, Sept.-Oct.	PRAKASH 1958, 1962
P.e. entellus	1	January-May, October	PRAKASH 1958, 1962
M. crassicaudata	1	November	PRAKASH 1960
C.l. pallipes	3–8	October-December	BLANFORD 1888–91
C.a. aureus	about 4	—	BLANFORD 1888–91
V.v. pusilla	3	January	PRAKASH 1960
V. bengalensis	4	February-April	BLANFORD 1888–91
H. hyaena	3–4	—	BLANFORD 1888–91
F.c. prateri	3–4	Twice a year	BLANFORD 1888–91
P. pardus	2–4	February-March	BLANFORD 1888–91
S.s. cristatus	8	August	PRAKASH 1960
B. tragocamelus	1 (occasionally 2)	April, July-Aug.	Unpublished record
A.c. rajputanae	1	August-September	PRAKASH 1960
G.g. bennetti	1 (occasionally 2)	April, July-Sept.	PRAKASH 1960
L.n. dayanus	1–4	All the year round	PRAKASH & TANEJA 1969
H.i. indica	1–3	March-Oct. (in zoo)	PRAKASH 1968
F. pennanti	1–4 (5)	March-September (all the year round)	PUROHIT et al. 1966 AGARWAL 1965
T.i. indica	1–9	All the year round	JAIN 1970
M. hurrianae	1–9	All the year round	PRAKASH 1964
G.d. indus	2–3	April, June, December	PRAKASH & JAIN 1971
G. gleadowi	2–4	June, November, Jan.	PRAKASH & PUROHIT 1967
R.r. rufescens	1–9	April-September	PRAKASH 1960
R.c. cutchicus	2	April-September	PRAKASH 1971
R.m. pallidior	2–3(4–6)	March-September	BINDRA & SAGAR 1968
R. gleadowi	2–3	August-September	PRAKASH 1971
M.m. bactrianus	4–8	All the year round	BLANFORD 1888–91
M.p. sadhu	3	October	PRAKASH 1971

in some mammals, like *Funambulus pennanti* (PUROHIT et al., 1966), *Tatera indica* (JAIN, 1970), *Meriones hurrianae* (PRAKASH, 1964), *Lepus nigricollis dayanus* (PRAKASH & TANEJA, 1969), have shown that the litters having

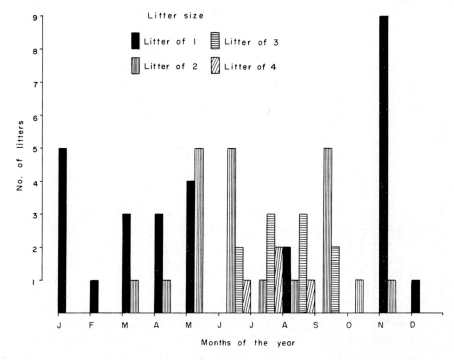

Fig. 48. Monthly distribution of various litter sizes of the Indian Desert hare, *Lepus nigricollis dayanus.*

large number of young occur during the rainy season. Fig. 48 shows that the larger litters of the desert hare, *Lepus nigricollis dayanus* carrying three and four young, are encountered only during the monsoon period. SPENCER (1896) has reported an analogous situation that in Central Australia the small mouse-like marsupial, *Sminthopsis crassicaudata,* usually delivers ten young at a time. In seasons of scarcity the litter size is, however, restricted to 4–5. The delivery of large litters during the monsoon months in the Indian Desert is readily correlated with the optimum living conditions which prevail during and soon after the rainy season, when ample green food is also available and the climate is relatively moderate. It was also observed that the young desert hare, *Lepus nigricollis dayanus* born in rainy season has a higher survival rate than in the first half of the year (PRAKASH & TANEJA, 1969).

Breeding Season: Extensive data on this aspect of reproduction are wanting, but I have tried to compile the available information from our published work and it has been supplemented by our observational records during the last two decades. The information with respect to some species is not, however, complete and some data have been incorporated from literature. From table V it is evident that most of the mammalian species litter between April and September, while the number of species

399

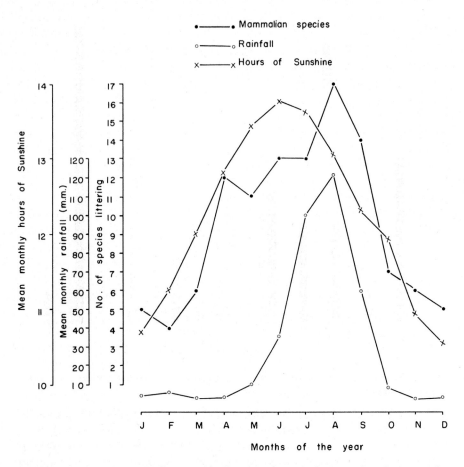

Fig. 49. Number of mammalian species, littering through the year, in relation to the rainfall and day-length.

giving birth to young ones is low during winter season. The peak in littering activity of the desert mammals occurs during August (Fig. 49). Those species, which breed all the year round, show maximum prevalence of pregnancy during rainy season (Fig. 50). These peaks coincide with the higher nutritional status of green food, moderate temperatures, and higher relative humidity of air during the rainy season. During this period, the health of the mother remains better to sustain stress of lactation and hence the survival of young ones is superior. Moreover, no sooner the young are weaned, than not only ample food is available but it is of high calorific value. BODENHEIMER (1957) also observed that gazelles of northeastern Africa give birth one month later than the peaks of rains. In the Indian Desert also largest number of mammals litter one month (August) after the initial month (July) of rainy season (Fig. 49).

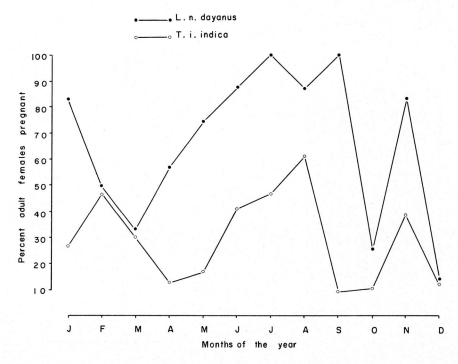

Fig. 50. Prevalence of pregnancy in the Indian Desert hare, *Lepus nigricollis dayanus* and the Indian gerbil, *Tatera indica indica.*

It is interesting to observe that minimum births occur during the winter season and not during the summer, when the conditions are largely unfavourable. This observation may indicate that length of the day is one of the important factors influencing the reproduction pattern, as it can be seen from Fig. 49 that the trend of the day length and that of the number of species littering run almost parallel during the year (greater the day length larger the number of species littering).

7.4. Population characteristics

Yearly fluctuations: Data are available for only a few mammalian species of the Indian Desert, like the desert hare, *L. nigricollis dayanus* (Prakash & Taneja, 1969), the northern palm squirrel, *Funambulus pennanti* (Prakash & Kametkar, 1969), the Indian gerbil, *Tatera indica indica* (Jain, 1971) and Indian desert gerbil, *Meriones hurrianae* (Prakash et al., 1970). The population trend of the palm squirrel, *F. pennanti* (Fig. 51) shows a gradual decrease from April to October. It could not, however, be ascertained whether this fast decrease in squirrel numbers was due to mortality or due to emigration from the experimental area. The seasonal

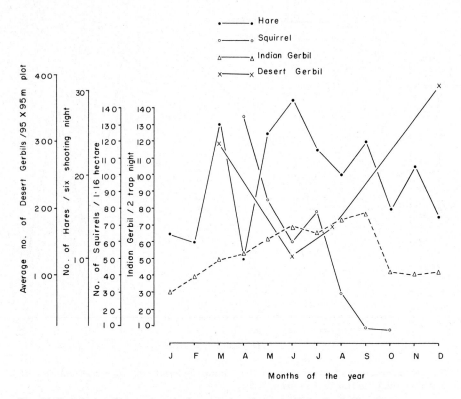

Fig. 51. Population fluctuations in four species of mammals in the Indian Desert.

fluctuations in the density of the desert gerbil, *M. hurrianae* indicate a build up in winter, which continues till spring but their numbers decrease during summer. The peak in desert gerbil numbers during winter is attributed to their higher reproductive rate during the rainy season (PRAKASH et al., 1971). The data presented (Fig. 51) on the monthly numbers of the desert hare and Indian gerbil are based on the monthly catches by shooting hares during six nights in a month and live trapping of *Tatera* on two nights per month. Although the data do not pertain to their density, they do reflect a monthly fluctuation in their numbers. Both the species were collected in greater numbers during summer and monsoon. The fluctuations in their numbers show an entirely different trend from that of *M. hurrianae*. It is not, however, worthwhile to compare these trends, as the census methods for these species were different and hence a generalization is not possible. It is, however, observed that in general, the numbers are low during summer, but tend to increase after the rainy season, which is their chief breeding period.

Age structure: The age structure in a population is also known with respect to the four species discussed above. These mammals breed all the

Table VI. Ratio of males: females of some bats and rodents of the Indian Desert

Species	Sex ratio
Pteropus g. giganteus	2.00:1.0
Rhinopoma kinneari	1.09:1.0
Rhinopoma h. hardwickei	1.60:1.0
Taphozous p. perforatus	1.30:1.0
Taphozous k. kachhensis	1.00:2.5
Megaderma l. lyra	1.00:2.2
Rhinolophus l. lepidus	1.00:2.0
F. pennanti	3.30:1.0
G.d. indus	1.00:1.5
G. gleadowi	1.00:1.5
T.i. indica	1.00:1.4
M. hurrianae	1.00:1.4
Rattus c. cutchicus	1.10:1.0
R.m. pallidior	1.05:1.0
R. gleadowi	1.00:2.0
Mus m. bactrianus	3.00:1.0
M.b. booduga	2.00:1.0
M.c. phillipsi	1.00:1.0
M.p. sadhu	2.66:1.0
Golunda e. gujerati	1.00:1.1

year round, with the peak during the monsoon and the young ones are found throughout the year, but it is observed that the survival rate of the young born in the later half of the year is much higher than those born earlier. Evidently the young ones are not capable of withstanding the extreme conditions during the summer. This is mainly true of the squirrel and the hare. One of them being arboreal and the other takes shelter under bushes or thickets during the hot, sunny day. The two gerbils are decidedly better off than the squirrel or hare, as they pass hottest time of the day in their burrows, which are cooler than the outside environment (PRAKASH et al., 1965). It was also observed that the older squirrels perished during the drought period (PRAKASH & KAMETKAR, 1969).

Sex ratio: An appreciable preponderance of males over females is observed in certain bats and rodents, *P.g. giganteus*, *F. pennanti* and three species of *Mus* (Table VI). Females of *T.k. kachhensis*, *M.l.lyra*, *R.l. lepidus* and *T.i. indica* outnumbered the males. Besides the sex ratios observed in these random collections, Table VII indicates the monthly distribution of numbers of male and female of four mammal species. It is evident that the annual sex ratios in the desert hare, palm squirrel, desert and Indian gerbil do not differ significantly from the 50:50 ratio. PRAKASH (1971), however, reported that the males *M. hurrianae* were significantly ($P < 0.01$) less in number in the yearly sample of 1963–64,

403

Table VII. Monthly fluctuations of percentage of males in four species of mammals of the Indian Desert

	J	F	M	A	M	J	J	A	S	O	N	D
Desert hare (PRAKASH & TANEJA 1969)	50.0	66.6	50.0	30.0	48.0	62.0	66.6	47.3	62.5	50.0	38.1	40.0
Palm squirrel (PRAKASH & KAMETKAR 1969)	—	—	—	73.8	78.2	42.1	73.3	71.4	50.0	33.2	—	—
Indian gerbil (JAIN 1970)	46.0	48.0	60.7	50.9	50.7	51.4	63.4	48.6	35.8	47.6	40.0	51.0
Indian desert gerbil (PRAKASH 1962)	45.8	47.0	43.1	55.5	33.3	50.0	45.4	58.8	52.9	69.2	55.5	50.0
Indian desert gerbil (PRAKASH 1971)	6.2	42.2	20.2	15.3	50.0	50.0	75.0	53.3	50.2	—	—	29.3

being only 35.5% of the total. This significant difference was attributed to the probable higher rate of mortality in male desert gerbils, particularly in the sub-adult age class, as they are subjected to hostile encounters with adult males at the time when the former reach the age of sexual maturity. FITZWATER & PRAKASH (1969) observed that in the peak breeding season the adult male desert gerbil increases its home range to twice its usual territory. It is quite possible that due to the aggressive interactions of the dominant males, the sub-adult males succumb to fights. GHOSH (personal communication) mentions that male desert gerbils are less adaptable to xeric conditions than the female desert gerbils. This inadaptability may also add to lowering of the numbers of male desert gerbils.

Contrary to this we (PUROHIT et al., 1966) found that the male-female ratio in the new born *Funambulus pennanti* was 1.1:1, in sub-adult individuals 1.83:1 and in the adult squirrels 2.28:1. These ratios indicate that male and female squirrels are born almost in equal numbers, but the proportion of males gradually increases, indicating a higher mortality rate of female squirrels. The monthly samples of the male and female desert hare and Indian gerbil do not show any significant deviation from the normal 1:1 ratio.

Movements: The blackbucks, gazelles, and bluebulls have been observed to move long distances between their feeding grounds, lakes and ponds for drinking water and back to shelter places. Although no systematic study was undertaken, it was estimated that near Pali two combined herds of these three artiodactyles move about 20 to 25 km in 24 hours. Some studies have, however, been conducted on the home ranges of the palm squirrel (PRAKASH et al., 1965), Indian gerbil (PRAKASH & RANA, 1970) and Indian desert gerbil (FITZWATER & PRAKASH, 1966).

404

The average home range of adult male palm squirrel (0.21 $\pm$ 0.073 hectares) does not differ significantly from that (0.15 $\pm$ 0.034 hectare) of adult female squirrel, but the average greatest distance between capture points of the adult male squirrel (65.61 $\pm$ 4.80 metres) significantly, $P < .05$, differs from that of sub-adult male (41.71 $\pm$ 10.93 m), the adult female 46.87 $\pm$ 5.40 m) and the sub-adult female (43.95 $\pm$ 1.85 m). This significant difference may be due to the promiscuous mating habits of squirrels. The female mates with more than one male; several males surround and compete with each other for the receptive female. Thus the adult male squirrels wander a lot more than the female and other age classes. Likewise the home ranges of the male and female Indian gerbil (1875.0 m² and 1912.5 m²) and Indian desert gerbil (88.7 $\pm$ 14.3 m² and 154.7 $\pm$ 24.6 m²) do not differ significantly between sexes of each species. The males of latter species, however, doubled (159.9 $\pm$ 44.3 m²) their range of movement during mating season. As compared to *Tatera indica*, the range of movement of *M. hurrianae* is significantly low.

Aestivo-Hibernation: In deserts which are exposed to cold winters, certain mammals hibernate, in the true physiological sense (fattening, lower metabolism, temperature, pulse, respiration, etc.). Although precise studies are still wanting on these aspects, from my observations during the last twenty years in the Indian Desert, I have watched almost all mammal species on cold nights during winter in nature and have found no evidence of hibernation among them, except in the bats. During a very cold spell the long-eared hedgehog, *Hemiechinus auritus collaris*, does not come out of its burrows for the nightly sojourns. This continued for about fifteen days. It is not possible to say if the hedgehogs were truly hibernating or were in a state of semi-hibernation or were simply passing through a period of torpor. BUXTON (1923) believed that in southern Iraq *E. auritus*, and porcupines hibernate for a few months, but BODENHEIMER (1957) considers this statement with caution. I have seen the Indian crested porcupine, *Hystrix indica indica*, fully active at night during winter near Jodhpur. Certain bats (Microchiroptera) *R. kinneari*, *R. hardwickei*, *T.p. perforatus* and *T.k. kachhensis* do hibernate and locally migrate to warmer environment, from tunnels to deep wells (PRAKASH, 1961). They accumulate sufficient fat in their thighs before entering hibernation. The pipistrelle are, however, active on cold nights.

It has been observed that hedgehogs in Madagascar (HILZHEIMER, 1913), prairie dog, *Cynomys fulvus*, in Middle Asia (KACHKAROV & KOROVINE, 1942) and the northern ground squirrel, *Citellus columbianus*, in North America (SHAW, 1925) aestivate for a few months in the dry season. On the basis of our field observations it can be said that none of the species of mammals aestivate in the Indian Desert. HEIM DE BALSAC (1936) also stated that no case of aestivation in mammals of the Sahara is known to him but he feels that it is a lacuna in our knowledge.

8. The Vanishing Desert Wildlife*

Two large carnivores, the cheetah *Acinonyx jubatus* and the lion *Panthera leo persica*, which were commonly distributed in the southeastern parts of the Desert, have totally vanished from this region and the caracal, *Felis caracal*, has become extremely rare (PRAKASH, 1958). Other mammalian species fast vanishing in the Indian Desert are the otter *Lutra perspicillata sindica*, the Indian wolf *Canis l. pallipes*, the wild boar *Sus scrofa cristatus*, the black buck *Antilope c. rajputanae* and Indian gazelle *Gazella g. bennetti*.

9. Behavioural and Physiological Adaptations to the Xeric Environment

The desert mammals and other biota are faced with the problem of survival under conditions, mainly characterized by high and low temperatures during summer and winter respectively, large fluctuations in diurnal temperatures, paucity of food and water and strong winds blowing dust for about eight months during a year. Most of the desert animals are behaviourally as well as physiologically adapted to face these harsh conditions.

By far the majority (82.1 %) of the mammals reported from the Indian Desert are nocturnal, and thus avoid the extreme heat of the day. The diurnal primate, *Presbytis entellus*, which is usually found in rocky habitats, stays on tree tops and under shady branches during the day. The artiodactyles, except the Indian gazelle, *Gazella gazella bennetti*, inhabit thickets of bushes and trees near the lakes and small ponds. By this habitat selection these larger animals reduce exposure to the sun. Moreover, in summer they are more active during the early morning and late evening, when the temperature is not so high. During the daytime, they rest under shady trees, avoiding direct exposure. The diurnal rodents are *Funambulus pennanti*, *Meriones hurrianae*, and *Golunda ellioti*. The squirrel *F. pennanti* is arboreal and a nest dwelling form, and the bush rat *Golunda* stays in bushes, or under the dense thorn hedges in the Indian Desert. During winter, the desert gerbil, *M. hurrianae*, another diurnal rodent, ventures out of its extensive burrow system late in the morning and is active outside the burrow almost throughout the day and ceases its surface activities at dusk. By maintaining such a diurnal activity, it avoids the chilly mornings and cold evenings. During summer, however, it breaks its activity pattern into two parts. It is actively foraging early in the morning and ceases its activity before it becomes too hot. It ventures out again late in the evening for a few hours till dusk. By modifying its activity pattern during the hot season in the desert, it avoids the unfavourable environmental conditions. During the hotter parts of the day, most of its time is spent inside its extensive burrow system (AGARWAL,

* See Chapter XII.

406

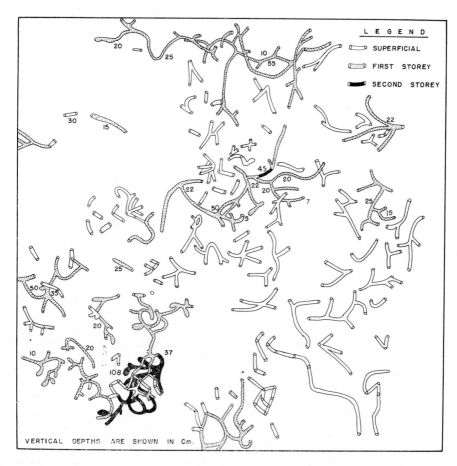

Fig. 52. Surface view of the excavated burrow system of the desert gerbil, *Meriones hurrianae*, in a 15 × 15 m plot (After FITZWATER and PRAKASH.)

1965; FITZWATER & PRAKASH, 1969; GANGULI & KAUL, 1962; PETTER, 1961; WAGLE, 1927) (Fig. 52), which is comfortably cool throughout the year (Fig. 53, 54), when compared with the outside environment (PRAKASH et al., 1965). Table VIII clearly indicates that at the time of the maximum temperature epoch of the soil surface, the burrows are cooler than the soil surface by 19.2° to 20.9 °C, except during the monsoon when this difference is 11.4 °C. It was observed that during summer the desert gerbil retires to its burrow for short periods, in between its spells of surface activities. It is quite possible that it allows some hyperthermia to develop during surface activities and the excess heat is then intermittently unloaded in the cooler surroundings of the burrow. It is, therefore, evident that the fossorial mammals are at an advantageous situation, in fact, are not usually exposed to the true xeric climate, as the

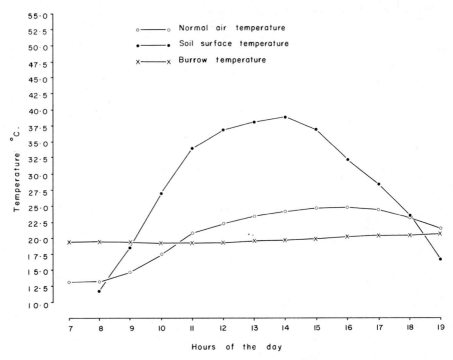

Fig. 53. Mean monthly temperature regime inside the burrow of the desert gerbil, *Meriones hurrianae*, during winter, in the Indian Desert.

microclimate inside their burrows remains almost constant throughout the year. This is probably the reason why 56.8% of the mammals inhabiting the Indian Desert are fossorial in habit.

The desert mammals are also adapted to the constant fluctuations in the availability and scarcity of food. The desert gerbil feeds on seeds during winter, on rhizomes and stems during summer and on green leaves, flowers and stems during the monsoon. During summer this usually phytophagous rodent feeds on insects, the bodies of which contain a high level of water, which helps the gerbil to meet its water requirement (PRAKASH, 1964). Thus by changing to various dietary items during various seasons, the desert gerbil not only adapts to the availability of food but also tides over the intermittent periods of water scarcity. It is quite likely that desert gerbils accumulate neutral body fat to act as efficient water store in the body, as 106 parts of water can be obtained from 100 parts of fat by oxidation. This is evidenced by the work of GHOSH et al. (1962). A group of desert gerbils were maintained on a water-restricted diet, and the food given to them was oven-dried. After 60 days of water deprivation, the body fat increased to almost double its initial quantity (Table IX).

The increase in total body fat presumably serves as a water reservoir

Table VIII. The air and burrow temperatures °C at the maximum temperature epoch of the soil surface during different seasons (PRAKASH et al., 1965)

	Winter 2 pm	Hot weather 2 pm	Monsoon 1 pm	Post-monsoon 12 noon & 1 pm
Soil surface	39.0	55.5	45.8	49.3
Air	24.1	38.6	32.4	31.6
Burrow temperature averaged for the four depths	19.8	34.7	34.4	28.4
Difference between soil surface and average burrow temperature	19.2	20.8	11.4	20.9
Difference between soil surface and air temperature	14.9	16.9	13.4	17.7

and it is another part of the adaptation syndrome developed by the rodent to fight chronic water stress. Table IX also indicates that during the water restriction period the total body water of the rodent is decreased, which probably results in a decrease of plasma volume, as evidenced by the haemo-concentration (Table X) seen in this animal during this period.

The above findings clearly prove that the desert gerbil, *Meriones hurrianae* is physiologically adapted to the drought conditions, as it is able to survive by tolerating a significantly high level of haemo-concentration by excreting concentrated urine.

Morphologically also the desert mammals are adapted to xeric environment. Among rodents a distinct difference is found in the length of the two pairs of legs (LAVAUDEN 1926), the hind limbs being longer than the fore. Their tail is also usually longer than the head and body length. The longer hind limbs keep the body away from the soil surface and help them in adopting a hopping mode of locomotion, which is also assisted by the long tail which acts as a fifth leg in the true jumping rodents (BODENHEIMER 1957). The hind limbs and tail are longer than fore limbs and head and body respectively in the true desert elements of the Indian desert viz., *Meriones hurrianae*, *Tatera indica indica*, *Gerbillus gleadowi* and *Gerbillus dasyurus indus* – all gerbils.

The hypertrophy of the bullae tympanicae is a common phenomenon in desert mammals (BODENHEIMER 1957, DE BALSAC 1936, PETTER 1961). An evaluation of this aspect (PRAKASH 1959) showed that the tympanic bullae of the Indian desert mammals are also hypertrophied. The large bullae possibly act as resonators, facilitating the perception of

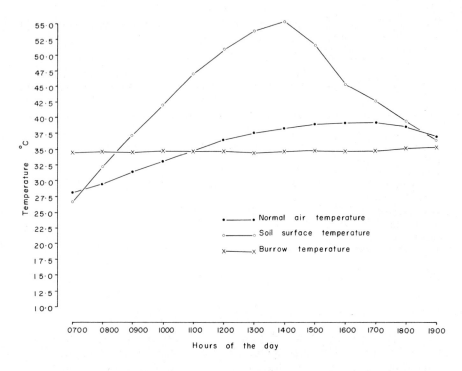

Fig. 54. Mean monthly temperature regime inside the burrow of the desert gerbil, *Meriones hurrianae*, during summer in the Indian Desert.

soil vibrations and acting as amplifiers (BODENHEIMER 1957, DE BALSAC 1936). FITZWATER and PRAKASH (1969) observed that the sound of wing beats of predatory birds are quickly perceived by the desert gerbil, *M. hurrianae* ducking in the burrows, but interestingly, the sound of wing-beats of babblers, pigeons etc., which are non-predatory, never disturb them. The gerbil, thus, shows a specialization to differentiate between the wing-beat sounds of harmless and predatory birds. BODENHEIMER (1957), quoting ZAVATTARI (1938), mentions that in the fennec *Vulpes zerdo*, which has large bullae as well as ears, the former are used for the perception of soil vibrations and the large ears enable the perception of flying or fluttering birds which are its main prey. However, even though the ears of *M. hurrianae* are comparatively very small, the way they are able to differentiate between the fluttering sounds of wings of predatory and unharmful birds is a matter that needs further research.

The observation that a high salt and a high protein diet to the water-deprived desert gerbil, *M. hurrianae* increases the concentration of chlorides and urea in the urine, shows that the gerbil kidney is highly efficient in filtering out a large excess of the salt and nitrogenous metabolites

410

Table IX. Effect of prolonged dehydration on body composition of the desert gerbil (values are mean ± standard error). Results are expressed in gm per 100 gm wet weight (GHOSH et al. 1962)

Condition of animals	Total body water	Total body fat	Total body cholesterol	Total body phospholipid
Freshly captured	70.17 ± 2.11	4.04 ± 0.65	0.158 ± 0.021	0.359 ± 0.047
Dry-fed for 60 days	59.99 ± 3.8	11.62 ± 3.62	0.19 ± 0.018	0.343 ± 0.051
	$P < 0.001$	$P < 0.005$	NS	NS

Table X. Red blood cells, white blood cells and haemoglobin content of the desert gerbil blood (values are mean ± standard error (GHOSH et al. 1962)

Condition of animals	Red blood cells (millions per cu mm)	White blood cells (thousands per cu mm)	Haemoglobin (gm per 100 ml)	Average corpuscular haemoglobin content (in picograms)
Freshly captured	2.352 ± 0.381	4.91 ± 0.62	12.66 ± 1.09	54
Dry fed for 60 days	7.29 ± 0.852	6.62 ± 0.33	16.83 ± 0.28	23
	$P < 0.001$	$P < 0.025$	$P < 0.01$	

Table XI. 24 hours urinary excretion levels of chloride, total electrolytes in urea, total nitrogen in desert gerbil values are mean ± standard error (GHOSH et al., 1962).

Condition of animals	Chloride (mg)	Total electrolytes (mg)	Urea (mg)	Total nitrogen (mg)
Freshly captured	11.03 ± 0.37	75 ± 4.7	93 ± 6.5	82 ± 2.8
Dry-fed for 60 days	24.3 ± 1.95	120 ± 3.61	168 ± 6.5	153 ± 3.15
	$P < 0.001$	$P < 0.001$	$P < 0.001$	$P < 0.001$

(Table XI), even when the body is subjected to severe water stress (GHOSH & GAUR 1966, GHOSH et al., 1962).

This reveals that the mechanism of excreting urine of a very high osmotic ceiling is apparently aimed at conserving maximum body water under xeric environment. It is achieved primarily by increased re-absorption of urinary water during its passage through the loops of Henle,

411

Table XIII. Palaearctic elements in different mammalian orders of the Indian Desert

Mammal orders	No. of species occurring in the Indian Desert	No. of Palaearctic species	No. of Oriental species	Palaearctic %
Insectivora	3	3	0	100
Chiroptera	11	4	7	36
Primates	2	0	2	00
Pholidota	1	0	1	00
Carnivora	13	9	4	69
Artiodactyla	4	2	2	50
Lagomorpha	1	0	1	00
Rodentia	16	7	9	44
Total	51	25	26	49

which are relatively longer in desert mammals (SPERBER 1944). BOR-THOLOMEW and DAWSON (1968) and SCHMIDT-NIELSEN (1964) have dealt with these aspects of desert mammals in great detail.

10. Zoögeography

The zoögeography of mammals, particularly of the rodents, has been discussed in earlier communications (PRAKASH 1963). Further information is incorporated here, including the reports of occurrence of additional species from the desert region (AGARWAL 1967, BISWAS & GHOSH, 1968, MOORE 1960, PRAKASH & JAIN 1967, PRAKASH et al., 1970). The species are grouped (Table XII) according to their distribution, mostly taken from ELLERMAN & MORRISON SCOTT (1951) and ELLERMAN (1961), under various zoögeographical categories, as defined in the case of the reptiles of this region. In this table the area of origin of each species is given in parenthesis; (P) indicates Palaearctic origin, (O) Oriental origin and (SR) Sind-Rajasthani origin. According to the origin of various species, it is evident that all insectivores and majority of carnivores and rodents inhabiting the Indian Desert have Palaearctic origin (Table XIII). The Primates, Pholidota and Lagomorpha are purely Oriental in their origin. Two species and fifteen subspecies of rodents are endemic to the Great Indian Desert. On an overall basis 49% of the total of mammal species are of Palaearctic origin and 51% are Oriental elements. This shows that the Indian Desert presents an admixture of the Saharan and Oriental

Note. – 2 species of rodents, as per their present geographical status, are endemic to the Great Indian Desert; 17 subspecies are endemic to Great Indian Desert

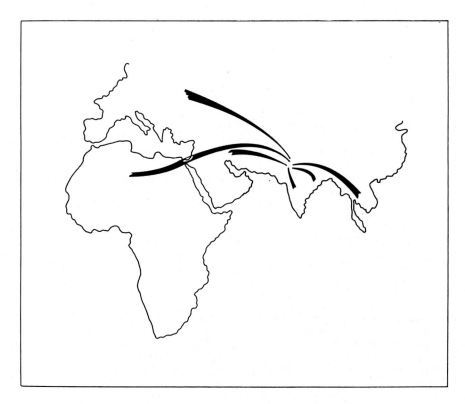

Fig. 55. Possible routes of the intrusive mammals to the Indian Desert.

faunas. Some species (Saharo-Rajasthani and Irano-Rajasthani) have their eastern limits in this dersert, whereas most of the Oriental species have their western limits in this desert.

From the above, it can be concluded that, as aridity appeared in the region of Great Indian Desert, some Saharan and Iranian species extended their ranges into this newly formed desert. The original fauna either migrated or have mostly become adapted to the new environment. Certain species of the Oriental Region also advanced from the east into the desert. It is also likely that certain mammals, *Herpestes smithi*, *Rattus cutchicus*, *Mus cervicolor phillipsi* and *Golunda ellioti* entered the desert through the Gujarat border. Considering the zoögeography of animals of the Indian Desert, the intrusion of mammalian fauna into the desert seems to have occurred through the following routes (Fig. 55): Sahara to the Indian Desert and/or further, Iran to the Indian Desert and/or further, from Indo-Chinese-Malayan region to Indian Desert through north India, from the Indian Deccan through the Gujarat border into the Indian Desert.

413

REFERENCES

ADAM, R. M. 1873. Notes on the birds of the Sambhar Lake and its vicinity. *Str. Feath.*, 1: 361–404.

ADAM, R. M. 1874. Additional notes on the birds of the Sambhar Lake and its vicinity. *Str. Feath.*, 2: 337–341; 456–466.

ADAMS, A. 1899. Western Rajputana States. London, Taylor & Francis.

AGARWAL, V. C. 1962. Taxonomic study of skull of Oriental rodents in relation to ecology. *Rec. Indian Mus.*, 60: 125–326.

AGARWAL, V. C. 1965. Observations on habits of five-striped squirrel, *Funambulus pennanti*, in Rajasthan. *J. Bengal nat. Hist. Soc.*, 34: 76–83.

AGARWAL, V. C. 1965. Field observations on the biology and ecology of the desert gerbil, *Meriones hurrianae* (Rodentia, Muridae) in western India. *J. zool. Soc. India*, 17: 125–135.

AGARWAL, V. C. 1967. New mammal records from Rajasthan. *Labdev J. Sci. & Tech.*, 5: 342–344.

AHMED, E. 1969. Origin and geomorphology of the Thar Desert. *Ann. Arid Zone*, 8: 171–180.

ANANDKUMAR, T. C. 1965. Reproduction in the rat-tailed bat, *Rhinopoma kinneari*. *J. Zool.*, 147: 147–155.

ASDELL, S. A. 1946. *Patterns of Mammalian Reproduction*. New York, Constable.

BALTAZARD, M. & M. BAHMANYAR, 1960. Recherches sur la peste en Inde. *Bull. World Hlth. Org.*, 23: 169–215.

BANERJI, A. 1955. The family life of a five-striped squirrel, *Funambulus pennanti* Wroughton. *J. Bombay nat. Hist. Soc.*, 53: 261–265.

BANERJI, A. 1957. Further observations on the family life of the five-striped squirrel, *Funambulus pennanti* Wroughton. *Ibid.*, 54: 336–343.

BARTHOLOMEW, G. A. and W. R. DAWSON, 1968. Temperature regulation in desert mammals. Chapter VIII: 396–423 in *Desert Biology*. I. (Ed. G. W. BROWN) New York, Academic Press.

BERG, L. S. 1940. Classification of fishes both Recent and Fossil. *Trav. Inst. Zool. Acad. Sci. U.S.S.R.*, 5: 87–517.

BHARADWAJ, O. P. 1961. The arid zone of India and Pakistan. in *A History of land use in arid regions* (Ed. I. DUDLEY STAMP) Arid Zone Research – 17: 143–174 Paris, UNESCO.

BINDRA, O. S. and P. SAGAR, 1968. Breeding habits of the field rat, *Millardia meltada* (GRAY). *J. Bombay nat. Hist. Soc.*, 65: 477–481.

BISWAS, B. 1947. On a collection of birds from Rajputana. *Rec. Indian Mus.*, 45: 245–265.

BISWAS, B. and K. K. TIWARI, 1966. Taxonomy and distribution of common Indian rodents. *Indian Rodent Symposium:* 1–45, Calcutta, The John's Hopkins University CMRT and USAID.

BISWAS, B. and R. K. GHOSE, 1968. New records of mammals from Rajasthan, India. *J. Bombay nat. Hist. Soc.*, 65: 481–482.

BLANFORD, W. T. 1877. Geological notes on the Great Indian Desert between Sind and Rajputana. *Rec. geol. Surv. India*, 10: 10–21.

BLANFORD, W. T. 1888–1891. *The Fauna of British India, including Ceylon and Burma. Mammalia* (Vols. 1 & 2). London: Taylor & Francis.

BLATTER, E. and F. HALLBERG, 1918–1921. The flora of Indian Desert. *J. Bombay nat. Hist. Soc.*, 26: 218–246, 525–551, 811–818; 27: 40–47, 270–279, 506–519.

BODENHEIMER, F. S. 1957. The ecology of mammals in arid zones. *in Human and Animal Ecology*. Reviews of research. Arid Zone Research, 8: 100–137, Paris, UNESCO.

BROWN, G. W. 1968. Desert Biology, Vol. I New York: Academic Press, i–xvii, 1–635.

Bump, G. and G. Bump, 1964. A study and review of the Black and Grey francolin. *Bull. Wildl.*, 81, Washington D.C.

Butler, E. A. 1875. Notes on avifauna of Mount Aboo and northern Gujarat. *Str. Feath.*, 3: 437–500.

Butler, E. A. 1876. Notes on avifauna of Mount Aboo and northern Gujarat. *Str. Feath.*, 4: 1–41.

Buxton, P. A. 1923. Animal life in deserts. London: Arnold.

Christensen, G. C. 1962. Use of clap net for capturing Indian sandgrouse. *J. Wildl. Mgmt.*, 26: 399–402.

Christensen, G. C. 1963. Sandgrouse released in Nevada found in Mexico. *Condor*, 65: 67–68.

Datta-Gupta, A. K., P. K. B. Menon, C. K. G. Nair and C. R. Das, 1961. An annotated list of fishes of Rajasthan. *Proc. Rajasthan Acad. Sci.*, 8: 129–134.

Dave, K. C. 1961. Contribution to the systematics, distribution and ecology of the reptiles of the desert of Rajasthan, with special reference to the ecology of lizards. Doctoral Thesis. University of Rajasthan, Jaipur.

Day, F. 1875–1888. The fishes of India. Vols. I & II, London (Reprinted, 1958).

Dharmakumarsinhji, R. S. 1954. Birds of Saurashtra, India. Dil Bahar, Bhavnagar.

Dhawan, S. 1969. Fish fauna of Udaipur lakes. *J. Bombay nat. Hist. Soc.*, 66: 190–194.

Ellerman, J. R. 1961. The Fauna of India including Pakistan, Burma and Ceylon, *Mammalia*. 3(1 & 2). Delhi: Manager of publ., Govt. of India.

Ellerman, J. R. and T. C. S. Morrison Scott, 1951. Checklist of Indian and Palaearctic mammals. London: Brit. Mus. Nat. Hist.

Faruqui, S. A., G. Bump, P. C. Nanda and G. C. Christensen, 1960. A study of the seasonal foods of the Black Francolin (*Francolinus francolinus* Linnaeus), the Grey Francolin (*F. pondicerianus* Gmelin), and the Common Sandgrouse (*Pterocles exustus* Temminck) in India and Pakistan. *J. Bombay nat. Hist. Soc.*, 57: 354–361.

Fitzwater, W. D. and Ishwar Prakash, 1966. Handbook of Vertebrate pest control. CAZRI, Jodhpur (Mimeo): 1–111.

Fitzwater, W. D. and Ishwar Prakash, 1969. Burrows, behaviour and home range of the Indian desert gerbil, *Meriones hurrianae* Jerdon. *Mammalia*, 33: 598–606.

Ganguli, B. N. and R. N. Kaul, 1962. Preliminary study on the behaviour and control of Indian desert gerbille (*Meriones hurrianae*). *Indian Forester*, 88: 297–304.

Ghosh, A. 1952. The Rajputana desert – its archaeological aspects. in 'Symposium on Rajputana desert'. *Bull. Nat. Instt. India*, 1: 37–42.

Ghosh, P. K. and B. S. Gaur, 1966. A comparative study of salt tolerance and water requirements in Desert Rodents, *Meriones hurrianae* and *Gerbillus gleadowi*. *Indian J. Exptl. Biol.*, 4: 228–230.

Ghosh, P. K., K. G. Purohit and Ishwar Prakash, 1962. Studies on the effects of prolonged water deprivation on the Indian desert gerbil, *Meriones hurrianae*. Proc. Symp. Environmental Physiology and Psychology in Arid conditions, Lucknow & Paris: UNESCO.

Ghosh, P. K. and G. C. Taneja, 1968. Oestrous cycle in the desert rodents, *Tatera indica* and *Meriones hurrianae*. *Indian J. Exptl. Biol.*, 6: 54–55.

Gopalakrishna, A. 1969. Gestation period in some Indian bats. *J. Bombay nat. Hist. Soc.*, 66: 317–322.

Gupta, B. B. and H. L. Sharma, 1961. Birth and early development of Indian hedgehogs. *J. Mamm.*, 42: 398–399.

Gupta, P. D. and V. C. Agarwal, 1966. Distribution of Indian Hairy-footed Gerbil, *Gerbillus gleadowi*. *Sci. Cult.*, 32: 470–471.

Gupta, R. K. and Ishwar Prakash, 1969. Management of range resources in the Indian Arid Zone with reference to rodent control. Australian Arid Zone Conf.

Heim de Balsac, H. 1936. Biogeographie des mammifères et des oiseaux de l'Afrique du Nord. *Bull. Biol. de la France et de la Belgique*. Paris (Supplement 21).

Hilzheimer, M. 1913. Handbuch der Biologie der Wirbeltiere. Stuttgart: Enke.

415

HOLMES, D. A. and J. O. WRIGHT, 1968–69. The birds of Sind: A review. *J. Bombay nat. Hist. Soc.*, 65: 533–556, and 66: 1–30.

HORA, S. L. and B. B. L. MATHUR, 1952. On certain palaeogeographical features of Rajasthan as evidenced by the distribution of Fishes. 'Proc. Symp. Rajputana Desert'. *Bull. Natl. Instt. Sci. India*, 1: 32–36.

HUME, A. O. 1878. The birds of a drought. *Str. Feath.*, 7: 52–68.

HUME, A. O. and C. H. T. MARSHALL, 1880. The Game Birds of India, Burma and Ceylon. 2.

HUSAIN, M. F. and H. R. BHALLA, 1939. Some birds of Lyallpur and their food. *J. Bombay nat. Hist. Soc.*, 39: 831.

JAIN, A. P. 1970. Body weights, sex ratio, age structure and some aspects of reproduction in the Indian Gerbil, *Tatera indica indica* Hardwicke in the Rajasthan desert. *Mammalia* 34(3):415–432.

KACHKAROV, D. N. and E. P. KOROVIN, 1942. La vie dans les deserts. Paris, Payot.

KASHKAROV, D. and V. KURBOTOV, 1930. Preliminary ecological survey of the verte-brates of Central Karakum Desert in Western Turkestan. *Ecology*, 11: 35–60.

KENNETH, J. H. 1947. Gestation periods. Edinburgh: Imp. Bureau of Animal Breeding and Genetics.

KRISHNA, D. and C. B. MENON, 1958. A note on the fishes of Jodhpur (Rajasthan). *Vijnana Parishad Anusandhan Patrika*, 1: 207–209. (In Hindi with English Abstract).

KRISHNA, D. and ISHWAR PRAKASH, 1955. Hedgehogs of the desert of Rajasthan. Pt. 1. Distribution and fossorial habits. *J. Bombay nat. Hist. Soc.*, 53: 38–43.

KRISHNA, D. and ISHWAR PRAKASH, 1956. Hedgehogs of the desert of Rajasthan. Pt. 2. Food and feeding habits. *J. Bombay nat. Hist. Soc.*, 53: 362–366.

KRISHNA, D. and ISHWAR PRAKASH, 1960. Hedgehogs of the desert of Rajasthan. Pt. 3. Food in nature. *Proc. Raj. Acad. Sci.*, 7: 60–62.

KRISHNAN, A. 1969. Some aspects of water management for crop production in arid and semi-arid zone of India. *Ann. Arid Zone*, 8: 1–17.

KRISHNAN, M. S. 1952. Geological history of Rajasthan and its relation to present day conditions. Proc. Symp. Rajputana Desert. *Bull. Natl. Instt. Sci. India*, 1: 19–31.

LAVAUDEN, L. 1926. Les Vertebrés du Sahara. Tunis, Guénard.

LONG, J. A. and H. M. EVANS, 1922. The oestrous cycle in the rat and its associated phenomenon. *Mem. Univ. Calif.*, 6.

MATHUR, B. B. L. 1954. Notes on fishes from Rajasthan, India. *Rec. Indian Mus.*, (1952): 105–110.

MATHUR, D. S. and G. M. YAZDANI, 1968. Occurrence of *Aplocheilus blochii* (Arnold) in Rajasthan. *Labdev J. Sci. and Tech.*, 6: 77.

MATHUR, D. S. and G. M. YAZDANI, 1970. *Noemacheilus rajasthanicus*, a new species of loach from Rajasthan, India. *J. zool. Soc. India*, (In press).

MAYHEW, W. W. 1968. Biology of desert amphibians and reptiles. Chapter 6: 196–356, in Desert Biology. I. (Ed. G. W. BROWN). New York: Academic Press.

MEHTA, K. M. and V. C. SHARMA, 1967. Soils of Rajasthan. Seminar on Soil Workshop. Hissar.

MINTON, S. A. 1966. A contribution to the herpetology of West Pakistan. *Bull. Amer. Mus. nat. Hist.*, 134(2): 27–184.

MOORE, J. C. 1960. Squirrel geography of the Indian Subregion. *Syst. Zool.*, 9: 1–17.

MOORE, J. C. and G. H. H. TATE, 1965. A study of the diurnal squirrels, Sciurinae, of the Indian and Indo-Chinese subregions. *Fieldiana* (Zoology) 48: 1–351.

MUKHERJEE, A. K. 1962. Some economic products of Indian birds. *Sci. & Cult.*, 28: 306–312.

MUKHERJEE, A. K. 1963. An analysis of the food of the Grey Quail in Western Rajasthan (India). *Pavo. Indian J. Ornith.*, 1: 31–34.

PARKES, A. S. 1926. Observations on the oestrous cycle of the albino mouse. *Proc. Roy. Soc.*, (B) 100: 151.

PARRACK, D. W. 1969. A note on the loss of food to the Lesser Bandicoot Rat, *Bandicota bengalensis. Curr. Sci.*, 38: 93–94.

PETTER, F. 1961. Répartition géographique et écologie des rongeurs désertiques (du Sahara occidental à l'Iran Oriental). *Mammalia*, 25 (No. Special): 1–222.

PRAKASH, ISHWAR, 1953. Additional to recorded food items of the Bull frog *(Rana tigrina). J. Bombay nat. Hist. Soc.*, 51: 750–751.

PRAKASH, ISHWAR, 1953. Cannibalism in hedgehogs. *J. Bombay nat. Hist. Soc.*, 51: 730.

PRAKASH, ISHWAR, 1954. Note on desert hedgehog (*Hemiechinus auritus collaris* Gray). *J. Bombay nat. Hist. Soc.*, 52: 921–922.

PRAKASH, ISHWAR, 1956. Studies on the ecology of the desert hedgehogs. *Proc. Raj. Acad. Sci.*, 6: 24–30.

PRAKASH, ISHWAR, 1957. A survey and ecological studies of the mammals of the desert of Rajasthan with special reference to food and feeding habits of certain insectivores and rodents. Doctoral thesis. University of Rajasthan, Jaipur.

PRAKASH, ISHWAR, 1958. The breeding season in the monkey, *Macaca mulatta* (Zimmerman). *J. Bombay nat. Hist. Soc.*, 55: 154.

PRAKASH, ISHWAR, 1958. Extinct and vanishing mammals from the Desert of Rajasthan and the problem of their preservation. *Indian Forester*, 84: 642–645.

PRAKASH, ISHWAR, 1959. Food of the Indian False Vampire. *J. Mamm.*, 40: 545–547.

PRAKASH, ISHWAR, 1959. Destruction of vegetation by desert animals in Rajasthan. *Indian Forester*, 88: 251–253.

PRAKASH, ISHWAR, 1959. Food of some Indian desert mammals. *J. Biol. Sci.*, 2: 100–109.

PRAKASH, ISHWAR, 1959. Food of certain insectivores and rodents in captivity. *Univ. Raj. Stud.*, (B): 1–48.

PRAKASH, ISHWAR, 1959. Hypertrophy of bullae tympanicae in the desert mammals. *Sci. & Cult.*, 24: 580–582.

PRAKASH, ISHWAR, 1959. Checklist of the mammals of the Rajasthan desert. *Univ. Rajasthan Stud.*, (B) 4: 30–56.

PRAKASH, ISHWAR, 1960. Breeding of mammals in Rajasthan desert, India. *J. Mamm.*, 41: 386–389.

PRAKASH, ISHWAR, 1961. Taxonomic and biological observations on the bats of the Rajasthan desert. *Rec. Indian Mus.*, 59: 149–170.

PRAKASH, ISHWAR, 1962. Group organisation, sexual behaviour and breeding season of certain Indian monkeys. *Jap. J. Ecol.*, 12(3): 83–86.

PRAKASH, ISHWAR, 1962. Ecology of the gerbils of the Rajasthan desert, India. *Mammalia*, 26: 311–331.

PRAKASH, ISHWAR, 1963. Taxonomical and ecological account of the mammals of Rajasthan Desert. *Ann. Arid Zone*, 1: 142–162.

PRAKASH, ISHWAR, 1963. Zoogeography and evolution of the mammalian fauna of Rajasthan desert, India. *Mammalia*, 27, 342–351.

PRAKASH, ISHWAR, 1964. Ecology of the Indian desert gerbil, *Meriones hurrianae* Jerdon. Proc. Symp. Problems of Indian Arid zones. UNESCO & Min. Edu. Jodhpur: 305–310.

PRAKASH, ISHWAR, 1964. Taxonomical and ecological account of the mammals of Rajasthan desert. *Ann. Arid Zone*, 2: 150–161.

PRAKASH, ISHWAR, 1964. Ecotoxicology and control of the Indian desert gerbille, *Meriones hurrianae* (Jerdon). Pt. II. Breeding season, litter size and post-natal development. *J. Bombay nat. Hist. Soc.*, 61: 142–149.

PRAKASH, ISHWAR, 1966. A review of the studies of the food of gerbils. Indian Rodent Symposium, Calcutta: 147–151. The John's Hopkins University CMRT and USAID.

PRAKASH, ISHWAR, 1968. Ecotoxicology and Control of Indian desert Gerbille, *Meriones hurrianae* Jerdon. Pt. V. Food preference in the field during monsoon. *J. Bombay nat. Hist. Soc.*, 65: 581–589.

PRAKASH, ISHWAR, 1968. Biology of the rodents of Rajasthan desert. Proc. Symp. 'Natural Resources of Rajasthan' (In press).

417

PRAKASH, ISHWAR, 1970. Rodents in rural areas. *Participant J.*, 4(8): 16–18.

PRAKASH, ISHWAR, 1971. Breeding season and litter size of Indian desert rodents. *Zeitsch. Zool.*, 58(4): 441–454.

PRAKASH, ISHWAR, 1971. Eco-toxicology and Control of Indian Desert Gerbil, *Meriones hurrianae* (Jerdon). Pt. VIII. Body weight, sex ratio and age structure in the population. *J. Bombay nat. Hist. Soc.*, 68(3): 717–725.

PRAKASH, ISHWAR, W. D. FITZWATER and A. P. JAIN, 1969. Toxic chemicals and baits for the control of two Gerbils, *Meriones hurrianae* Jerdon and *Tatera indica* Hardwicke. *J. Bombay nat. Hist. Soc.*, 66(3): 500–509.

PRAKASH, ISHWAR and P. K. GHOSH, 1963. The Great Indian Bustard in Rajasthan desert. *Newsletter for Birdwatchers*, 3: 1.

PRAKASH, ISHWAR and P. K. GHOSH, 1964. The Great Indian Bustard breeding in Rajasthan desert. *Newsletter for Birdwatchers*, 4: 1.

PRAKASH, ISHWAR and A. P. JAIN, 1967. Occurrence of *Rattus meltada* and *Gerbillus dasyurus* in the Rajasthan desert. *Ann. Arid Zone*, 6: 235.

PRAKASH, ISHWAR and A. P. JAIN, 1971. Some observations on Wagner's Gerbil, *Gerbillus dasyurus indus* (Thomas). *Mammalia*, 35(4): 614–628.

PRAKASH, ISHWAR, A. P. JAIN and K. G. PUROHIT, 1971. A note on the breeding and post-natal development of the Indian Gerbil, *Tatera indica indica* Hardwicke, in the Rajasthan desert. *Säugetierk. Mittel.*, 19(4): 375–380.

PRAKASH, ISHWAR, A. P. JAIN and B. D. RANA, 1971. New records of rodents from the Rajasthan desert. *J. Bombay nat. Hist. Soc.*, 68(2): 447–450.

PRAKASH, ISHWAR and L. R. KAMETKAR, 1969. Body weight, sex and age factor in population of the Northern Palm Squirrel, *Funambulus pennanti* Wroughton. *Ibid.*, 66: 99–115.

PRAKASH, ISHWAR, L. R. KAMETKAR and K. G. PUROHIT, 1968. Home range and territoriality of the Northern Palm Squirrel, *Funambulus pennanti* Wroughton. *Mammalia*, 32: 603–611.

PRAKASH, ISHWAR and C. G. KUMBKARNI, 1962. Eco-toxicology and control of Indian Desert Gerbil, *Meriones hurrianae* (Jerdon). Pt. I. Feeding behaviour, Energy requirements, and selection of bait. *J. Bombay nat. Hist. Soc.*, 59: 800–806.

PRAKASH, ISHWAR, C. G. KUMBAKARNI and A. KRISHNAN, 1965. Eco-toxicology and Control of the Indian desert Gerbille, *Meriones hurrianae* Jerdon. Pt. III. Burrow temperature. *J. Bombay nat. Hist. Soc.* 61: 237–244.

PRAKASH, ISHWAR and K. G. PUROHIT, 1967. Some observations on the Hairy-footed gerbille, *Gerbillus gleadowi* Murray, in the Rajasthan desert. *J. Bombay nat. Hist. Soc.*, 63: 431–434.

PRAKASH, ISHWAR, K. G. PUROHIT and L. R. KAMETKAR, 1967. Intake of seeds of grasses, shrub and tree species by three species of gerbils in Rajasthan desert. *Indian Forester*, 93: 801–805.

PRAKASH, ISHWAR and B. D. RANA, 1970. A study of field population of rodents in the Indian desert. *Zeitsch. Zool.*, 57(2): 129–136.

PRAKASH, ISHWAR and G. C. TANEJA, 1969. Reproduction biology of the Indian Desert Hare, *Lepus nigricollis dayanus* Blanford. *Mammalia*, 33: 102–117.

PRAKASH, ISHWAR, G. C. TANEJA and K. G. PUROHIT, 1971. Eco-toxicology and control of the Indian desert gerbille, *Meriones hurrianae* (Jerdon). Pt. VII. Relative numbers in relation to ecological factors. *J. Bombay nat. Hist. Soc.*, 68(1): 86–93.

PRAMANIK, S. K. and P. S. HARIHARAN, 1952. The climate of Rajasthan. Proc. Symp. Rajputana Desert. *Bull. Natl. Inst. Sci. India*, 1: 167–178.

PRASAD, M. R. N. 1954. Food of Indian gerbil, *Tatera indica cuvieri* Waterhouse. *J. Bombay nat. Hist. Soc.*, 52: 321–325.

PUROHIT, K. G., L. R. KAMETKAR and ISHWAR PRAKASH, 1966. Reproduction biology and post-natal development in the Northern Palm Squirrel *Funambulus pennanti* Wroughton. *Mammalia*, 30: 538–546.

418

RAMASWAMI, L. S. and T. C. ANAND KUMAR, 1963. Differential implantation of twin blastocysts in *Megaderma* (Microchiroptera). *Experientia*, 19: 641.

RANA, B. D. 1969. Some observations on the birds at Jawai Dam. *Newsletter for Bird-watchers*, 9(10): 1–2.

RANA, B. D. 1970. Winter food of the common Babbler *(Turdoides caudatus)* in Rajasthan. *Indian Forester*, 96: 153–155.

RANA, B. D. 1970. Some observation on the food of Jungle Babbler, *Turdoides striatus* and Common Babbler, *Turdoides caudatus* in Rajasthan desert. *Ibid.*, (In press).

RAO, H. S. 1957. History of our knowledge of the Indian fauna through the ages. *J. Bombay nat. Hist. Soc.*, 54: 251–280.

RATHOR, M. S. 1969. Fossorial and nocturnal adaptations of the Indian sand lizard, *Ophiomorus streeti* Anderson & Levington. *Jap. J. Ecol.*, 19: 67–69.

RATHOR, M. S. 1969. Food and feeding habits of the Indian sand skink, *Ophiomorus trodactylus* (Blyth) Boulenger. *J. Bombay nat. Hist. Soc.*, 66: 186–190.

ROY, B. B. and A. K. SEN, 1968. Soil map of Rajasthan. *Ann. Arid Zone*, 7: 1–14.

RIPLEY, S. D. 1961. A synopsis of the birds of India and Pakistan. Bombay: Bombay Natural History Society.

SALIM ALI, 1961. The Book of Indian birds. Bombay: Bombay Natural History Society.

SCHMIDT-NIELSEN, K. 1964. Terrestrial animals in dry heat: Desert rodents. Chapter 32, pp. 493–507, in Handbook of Physiology – Adaptation to Environment. (Ed. D. B. DILL). Washington DC. Am. Physiol. Soc.

SCHMIDT-NIELSEN, K. 1964. Desert Animals. London and New York: Oxford Univ. Press.

SHARMA, INDRA KUMAR, 1965. Some observations on Peafowl at Jodhpur. *Peacock*, 2: 26–34.

SHARMA, I. K. 1969. Habitat et Comportment due Paon *(Pavo cristatus)*. *Alauda*, 37: 219–223.

SHARMA, INDRA KUMAR, 1969. Breeding of Indian White-Backed Vulture at Jodhpur. *Ostrich*, 10.

SHARMA, INDRA KUMAR, 1970. Breeding of some common birds of semi-arid Jodhpur. *Ecology* (In press).

SHARMA, S. C. 1965. Winter food of the Painted Partridge, *Francolinus pictus* (Jardine & Selby) in Rajasthan. *J. Bombay nat. Hist. Soc.*, 61: 686–688.

SHAW, W. T. 1925. Duration of the aestivation and hibernation of the Columbian Ground Squirrel *(Citellus columbianus)*. *Ecology*, 6: 75–81.

SMITH, M. A. 1931. The fauna of British India including Ceylon and Burma. Reptilia and Amphibia. Vol. I. *Loricata, Testudines*. London: Taylor & Francis.

SMITH, M. A. 1935. The fauna of British India including Ceylon and Burma. Reptilia and Amphibia. Vol. 2. *Sauria*. London: Taylor & Francis.

SMITH, M. A. 1943. The fauna of British India including Ceylon and Burma. Reptilia and Amphibia. Vol. 3. *Serpentes*, London: Taylor & Francis.

SPENCER, B. 1896. Report on the work of the Horn Scientific expedition to Central Australia. II. Zoology, *Mammalia*. London, 1–52.

SPERBER, I. 1944. Studies on the mammalian Kidney. *Zoologiska bidrag fron Uppsala*, 22: 249–431.

TABER, R. D., A. N. SHERI and M. S. AHMED, 1967. Mammals of the Lyallpur region, West Pakistan. *J. Mamm.*, 48: 392–407.

TICEHURST, C. B. 1922–24. The birds of Sind, I–VIII. *Ibis*, 4–6.

WADIA, D. N. 1953. Geology of India. 3rd Ed. London: MacMillan.

WADIA, D. N. 1960. The post-glacial desiccation of Central Asia: Evolution of the arid zone of Asia. Natl. Inst. *Sci.*, India, Monogr., 10: 1–25.

WAGLE, P. V. 1927. The Rice rats of lower Sind and their control. *J. Bombay nat. Hist. Soc.*, 321: 330–333.

WHISTLER, H. 1938. The ornithological survey of Jodhpur State. *J. Bombay nat. Hist. Soc.*, 40: 213–235.

Yazdani, G. M. and R. N. Bhargava, 1969. On a new record of minnow, *Aphanius dispar* (Ruppell) from Rajasthan. *Labdev J. Sci. & Tech.*, 7: 332–333.

Zahavi, A. and J. Wahrman, 1957. The cytotaxonomy, ecology and evolution of the gerbils and jirds of Israel (Rodentia: Gerbillinae). *Mammalia*, 21: 341–380.

Zavattari, E. 1938. Un problema di biologia sahariana: l'ipertrofia delle bulle timpaniche dei Mammiferi. *Atti Accad. gioenia.*, 3: 1–8.

Zavattari, E. 1938. Essai d'une interpretation physiologique de l'hypertrophie des bulles tympaniques des mammifères sahariens. *Mammalia*, 2: 173–176.

Zuckerman, S. 1953. The breeding seasons of mammals in captivity. *Proc. Zool. Soc. London*, 122: 859.

XIV. ECOLOGY AND BIOGEOGRAPHY OF THE TERMITES OF INDIA

by

P. K. SEN-SARMA

1. Introduction

This chapter summarizes the salient results of two decades' studies by the author on the general field ecology, distributional peculiarities and faunistic affinities of the termites of the Indian region. Although the termites are perhaps better known than most other groups of insects in this country, there are nevertheless large gaps in our knowledge and many important areas in the country have not been satisfactorily explored so far. In the present state of our knowledge any account of the ecology and biogeography of these insects is therefore bound to be rather sketchy. It is, however, hoped that this chapter will serve to focuss attention on some of the outstanding problems.

2. Some Aspects of the Ecology of Termites in India

The ecology of the termites in India is largely determined by the geological history and the predominantly monsoon-climate of the region and the extensive deforestation within historical times. While it is not possible to give here a comprehensive account of the complex ecology of termites, some aspects may be presented in their relation to biogeographical problems.

2.1. TERMITES AND SOIL

As is well known, a great many species of termites are subterranean insects, the majority of which feed on wood, humus, lichen, moss, etc. The effects on the soil of the activities of subterranean termites present a number of interesting features, but in India we really know very little of the changes in the soil. The large quantities of the subsoil brought to the surface in the course of the burrowing and mound-building activities of termites contribute evidently to the maintenance of soil fertility. Further, the burrowing activity seems to increase the rate of percolation of rain water and of the aeration of both the top and subsoils. This beneficial effect is rather pronounced in areas rich in lateritic soils (see Chapter II). Fallen branches, trees, dried stems of shrubs and herbs, grasses and fallen leaves devoured by the termites disintegrate in these soils. The humus-feeding species are active in the top soils, which are, therefore, depleted of organic matter. The rapid removal of the organic litter from the surface of the soil can thus be a serious economic problem in areas deficient in humus. The destruction of the grass mulch by termites serves to keep the soil temperatures low and moisture content high, but these activities often pose serious problems of humus formation.

According to JOACHIM & KANDIAH (1940), the soil of the mounds of *Odontotermes redemanni* from Ceylon is heavier in texture and poorer in organic matter than the soil of the surrounding ground. The replaceable bases are highly variable, though generally lower than in the surrounding soil. PENDLETON (1941) analysed the mound soil of termites in Thailand (the species not indicated) and found a higher calcium carbonate content in the basal parts of the mound, higher plant nutrient content and better moisture relationship than in the soil of the surrounding ground. On the basis of the analysis of the mound soils of three species of *Macrotermes* from Africa, HESSE (1953) concluded that the mounds are constructed exclusively of the subsoil, which is not at all altered in its chemical status by the termites. He has also observed that comparison of the mound soils with that of the surrounding ground shows considerable significant differences. Even the results of chemical analysis of the soil of the deserted termite mounds are known to be different from those of the soil of mounds inhabited by termites. MUKERJI & MITRA (1949) did not find any differences in the pH values of the mound soils of *Odontotermes redemanni* and of the soil of the surrounding ground in Bengal. The author did not also find any marked differences in the pH values of the mound soil of *Odontotermes obesus* from that of the surrounding ground at Dehra Dun. The mounds of *Trinervitermes biformis* have, however, entirely different relations with the surrounding soil. The inside of the mounds in this case is highly reticulate in structure and is constructed perhaps from the excreta of the termites, with the addition of some soil to the outer layers. The high lignin content, together with a fair quantity of undigested cellulose, are characteristic of these mounds (SEN-SARMA & MATHUR, 1961).

2.2. TERMITES AND LIVING PLANTS

The close association of termites with living plants, especially crops, horticultural and forest trees, cocoanut, coffee and tea plantations, grass, etc. has attracted considerable attention, primarily because of the heavy damage caused. According to FLETCHER (1912), the loss to crops alone by termite attack amounts to Rs. 280 000 000 annually. It is not, however, always known precisely whether the termites are the primary pests or the damage by termites is secondary to other predisposing factors.

The crops commonly damaged by termites are wheat, millets, pulses, cotton, spices, vegetables, etc. The common species of termites causing serious damage to these crops include *Coptotermes heimi, Odontotermes obesus, Microtermes obesi* and *Trinervitermes biformis*. Of these, *Microtermes obesi* is a serious pest of wheat and destroys the roots of small plants, resulting in the yellowing of the leaf blades and ultimate death of the affected plant. A loss of 6–25 % of the wheat crop has been reported by HUSAIN (1935). *Trinervitermes biformis* is a serious pest of a number of crops like cotton, wheat, eggplant, groundnut, etc. in the Maharashtra State (PATEL & PATEL, 1954). *Odontotermes obesus* has been reported to damage groundnut crops by generally restricting their attention to the ripening pods underground and causing them to break up during harvesting.

The sugarcane crop is damaged by a number of species of termites in India. Broadly speaking, there are three periods when sugarcane is prone to serious damage by termites. The first stage of attack takes place during the pre-monsoon period, when the seed setts are planted out and during the post-monsoon period when the foraging termites destroy the eye-buds, causing failure of germination. The second period of attack takes place when the crop is nearing maturity and in the third stage, the termites take advantage of the damage caused by rodents, stem-boring Lepidoptera larvae, etc. to gain access to the soft inner layer. The most serious damage is, however, caused to the newly planted setts and nearly 40–60 % of the eye-buds may be destroyed. The damage by termites is usually very serious when the monsoon rains are delayed, under conditions of poor irrigation and severe drought. According to AGGARWAL (1955), a loss of 2.5 % in the tonnage of the cane and 4.5 % of the sugar output is caused in Bihar, mainly due to the attack of *Microtermes obesi*. In addition to this species, the others known to damage the sugarcane crop in India are *Coptotermes heimi, Eremotermes paradoxalis, Odontotermes assmuthi, Odontotermes obesus, Odontotermes wallonensis, Trinervitermes biformis* and *Trinervitermes heimi* (AGGARWAL, 1970).

The damage to cocoanut palm by termites is restricted primarily to the seedlings in the nursery and to the young palms in the plantations. *Odontotermes obesus* is an important pest in cocoanut nurseries and plantations in Kerala, Andhra, Madras and Mysore (KRISHNAMOORTY & RAMASUBBIAH, 1962; NIRULA et al., 1953). The first signs of the attack by termites are revealed by the wilting of the central shoot. In Ceylon *Coptotermes ceylonicus* and *Odontotermes redemanni* are reported to be destructive to the cocoanut plantations (JEPSON, 1931).

The tea-bush is often attacked by local subterranean termites, giving rise to damage in a variety of ways not only of the tea plant but also of the shade trees (KAPUR, 1958). Conditions of drought and soil erosion seem to greatly accelerate the frequency and intensity of attack by termites, particularly in old tea gardens. *Odontotermes obesus, Odontotermes parvidens* and some other species of *Odontotermes* and *Microcerotermes* are pests of tea bushes in India and Pakistan. In Ceylon *Postelectrotermes militaris, Neotermes greeni, Coptotermes ceylonicus, Odontotermes horni* and

Odontotermes redemanni are reported to attack tea bushes (HARRIS, 1961).

AGGARWAL (1964) reports that *Coptotermes* and *Glyptotermes* damage the rubber tree in India. *Coptotermes ceylonicus* attacks rubber in Ceylon. Furthermore, diverse species of *Odontotermes* and *Microtermes* also feed on moribund and dead trees.

According to DUTT (1962) the jute plant in Bengal is occasionally attacked by *Microtermes obesi*. Another fibre crop *Crotolaria juncea*, grown extensively throughout India, is reported to be damaged by *Odontotermes obesus* in the Uttar Pradesh. The incidence of termite attack on this crop is high during November and afterwards, when the soil moisture is considerably low. *Agave sislana*, also an important fibre plant, is damaged by diverse species of *Odontotermes* in Orissa, where the affected plants gradually wither away and die prematurely within two or three years (TRIPATHI, 1970).

Newly planted setts of the grapevine are often attacked by termites that cut hollow the entire vine and kill the tender sprouting shoots in South India (AYYAR, 1940). In the Maharashtra State, *Trinervitermes biformis* commonly attacks *Mangifera indica, Citrus, Psidium guyava, Punica granatum, Achras sapota* and other fruit trees (PATEL & PATEL, 1954). KUSHAWAHA (1964) reports damage by *Odontotermes obesus* to the seedlings of *Mangifera indica, Psidium guyava* and *Punica granatum* from Udaipur in the Rajasthan. Dead and semi-rotten parts of standing trees of *Mangifera indica* are attacked by *Neotermes bosei, Neotermes mangiferae, Neotermes megaoculatus*, etc. in different parts of India (CHATTERJEE, 1970; ROONWAL & SEN-SARMA, 1955).

While a number of species of termites feed almost exclusively on grasses, the true harvester termite *Anacanthotermes macrocephalus* from the arid tracts of Sind, Afghanistan and Rajasthan is a foraging and grass-gathering species that causes considerable damage to pastures in these areas. The nasute harvester termite *Trinervitermes biformis*, also occurring in semi-arid parts of India, is equally destructive to pasture (SEN-SARMA & MATHUR, 1961).

In forest nurseries and plantations the roots of trees are damaged by different species of termites, particularly when the plants are about one to three years old. The valuable and quick-growing *Eucalyptus* is perhaps the worst sufferer in recent years in such situations, especially in arid areas, by the attack of *Anacanthotermes macrocephalus, Microcerotermes minor, Odontotermes feae, Odontotermes obesus* and *Odontotermes parvidens* (CHATTERJEE, 1970).

2.3. RELATION BETWEEN ATMOSPHERIC TEMPERATURE AND HUMIDITY AND TERMITES

The atmospheric temperature and humidity are two dominant environmental factors that influence termites very profoundly. The termites

424

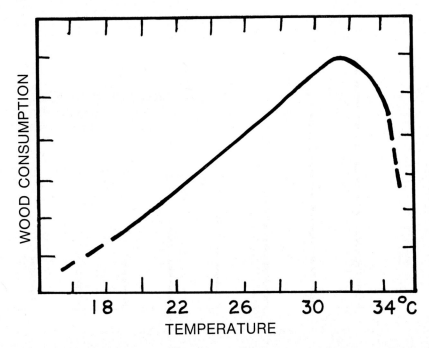

Fig. 56. Optimum temperature requirements of *Heterotermes indicola* (After BECKER).

seem to be extremely sensitive to changes in the atmospheric temperature and the distribution of individual species is very greatly influenced by temperature. Many species descend deep into the soil in dry and hot localities, so as to escape the extreme heat of the midday, though active foraging is known in the early morning hours. The termites that attack buildings similarly seem to be concentrated in locations, which are damp or moist. The distribution of certain species is influenced equally by the type of soil and by the soil moisture. Water-logged areas or parts where the subsoil water-tabel is high are generally avoided by the mound-building species of termites. The depth of the soil, at which the termites remain active, depends not only on the moisture content of the soil but also on the depth of the subsoil water. Every kind of aeration leads to desiccation, unless the moisture content of the surrounding area is at saturation level or the termites have immediate access to water. The ability of different termites to survive in unsaturated air depends, how-ever, on the possibility of obtaining drinking water. To some extent, the water is also provided as a metabolic end product of the break down of cellulose and other carbohydrates. Several genera of termites do not thus require any additional source of water, as the relative humidity within their nests is constantly maintained near saturation point (BECKER, 1969).

The primitive termite *Archotermopsis wroughtoni* is capable of with-

425

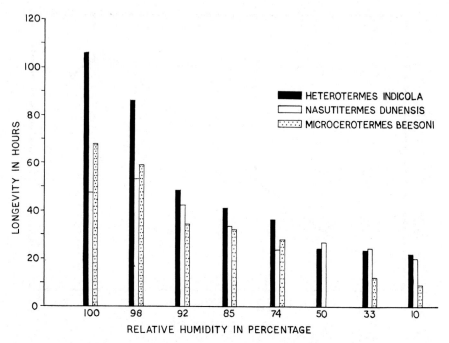

Fig. 57. Survival time of workers of *Heterotermes indicola*, *Nasutitermes dunensis* and *Micro-cerotermes beesoni* under varying conditions of relative humidities when kept isolated.

standing a wide range of temperature variations. In the areas frequented by *A. wroughtoni*, the summer temperature often rises to 37.7 °C during June and during December–February falls almost to the freezing point. The termites remain active throughout all these months (IMMS, 1919). Since *Archotermopsis wroughtoni* does not build any terminaria, the insulation of its nest against temperature fluctuations seems difficult to understand. The temperature in the interior of the mounds of *Odonto-termes obesus* varies somewhat during different seasons, although the outside temperatures show much higher fluctuations. Inside the mound (CHEEMA et al., 1962) the temperature in the areas, where the fungus combs are lodged, is always higher than in the surrounding parts.

Under laboratory conditions (SEN-SARMA & CHATTERJEE, 1965) *Neotermes bosei* prefers a temperature of 28 °C. The optimum temperature requirements of *Heterotermes indicola* (BECKER, 1962a) seem to lie between 30 and 32 °C ($\pm$ 1) (Fig. 56). *Microtermes beesoni* is cultured in the laboratory best at a temperature of 29 °C $\pm$ 1. (SEN-SARMA & CHATTERJEE, 1968).

Comparatively more information on the water relations and humidity tolerance of Indian termites is available than in case of temperature. Most species prefer high humidities, practically near the saturation point. While the species of Kalotermitidae are able to tolerate low humidities for

426

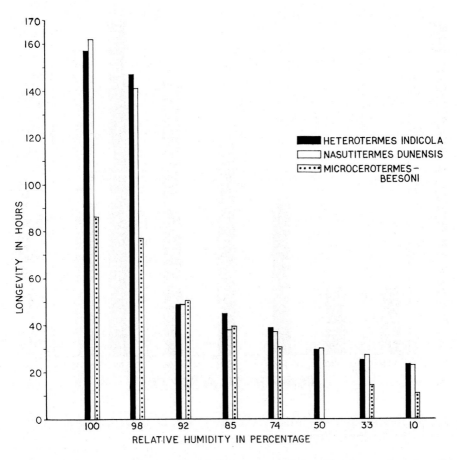

Fig. 58. Survival time of workers of *Heterotermes indicola, Nasutitermes dunensis* and *Microcerotermes beesoni* under varying conditions of relative humidities, when kept in groups.

a considerable time, the species of other families seem to be lacking in this low humidity tolerance. Among the Kalotermitidae *Neotermes* requires as high a relative humidity as 98%, but *Bifiditermes beesoni* and *Crypto-termes bengalensis* survive best at 92% RH. At a relative humidity of 98%, there is quick mortality of termites, due perhaps to the water poisoning, described by Buxton (1932). This effect is readily observed when the metabolic water production is in such copious quantities that it cannot be eliminated by evaporation, so that the water content rises above normal level. The survival times under different percentages of relative humidities differ not only in different species, but even in the same species it varies from caste to caste (Fig. 57, 58) (Sen-Sarma & Chatterjee, 1966a; Sen-Sarma, 1969). From figs. 57 and 58 it is evident that the survival

427

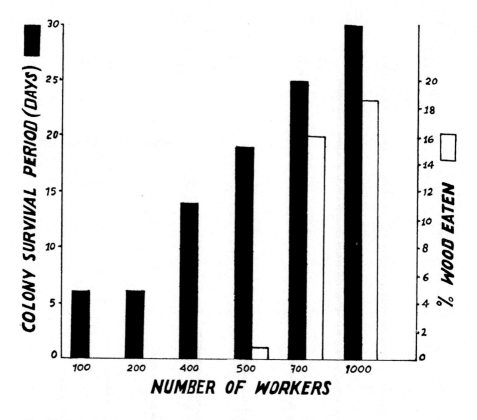

Fig. 59. Relation between the size of the colony, survival period and the quantity of wood eaten by *Microcerotermes beesoni.*

time is greater in grouped termites than in isolated and single individuals. GRASSE & CHAUVIN (1944) have suggested that the survival period is higher in groups of social insects, probably as a result of effects of sensory stimuli. SEN-SARMA & CHATTERJEE (1968) found, for example, that *Microtermes beesoni* in the laboratory does not feed on wood when the number of individuals in the group is less than five hundred workers; a minimum of at least one thousand workers seems to be the optimal size under such conditions (Fig. 59). Investigations with radio-active isotopes (Fig. 60) have demonstrated that the intense trophallactic exchange accounts for the higher survival time in grouped individuals of termites (ALIBERT, 1959, SEN-SARMA & KLOFT, 1965). PENCE (1956) believes, however, that among the grouped individuals the reduction of exposed surface due to the huddling-together of numerous individuals should account for the higher survival time in groups than in isolated and single individuals.

The intensity of response to humidity fluctuations among termites is

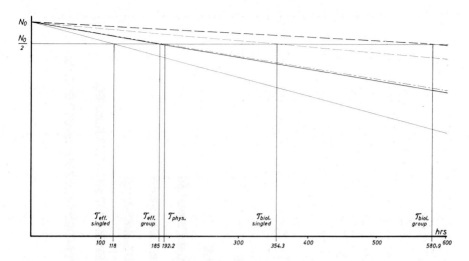

Fig. 60. Graphical representation of trophalactic exchange of radio-active I^{131} in grouped (thickline) and single (thin line) termites based on measurements of biological half-life of the radio-active I^{131} from the following formula

$$\text{Tbiol.} = \frac{1}{\dfrac{1}{\text{T effect}} - \dfrac{1}{\text{T Phys.}}}$$

Abbreviations: *Tbiol* (- - -) = biological half life of a radio active substance due to biological decay; *T-effec.* (- - -) = effective half-life which is defined as the total effective decrease in the impulse rate of a radio-active substance (in time unit) due both to biological and physical decay of a radio-active substance; *Tphys.* (-.-.-) = physical half-life of a radio-active substance due to physical decay.

Lower rate of decrease in impulse rate of radio-active I^{131} in grouped termites is due to exchange and circulation of the radio-active substance among the members of the group (After SEN-SARMA & KLOFT.).

influenced by the conditions of humidity to which they had previously been subjected. The workers of *Microcerotermes beesoni*, preconditioned in a dry atmosphere (5% RH) even for 30 minutes, exhibit much quicker orientation to high humidity than those preconditioned in a humid atmosphere (100% RH) for the same duration, in a humidity gradient apparatus (Fig. 61–64). The humidity preferendum, which is defined as the humidity to which an animal moves if given its choice of a humidity gradient, seems to be about 90–95% RH (SEN-SARMA & CHATTERJEE, 1966b). The water content of the soil depends, however, on its water-holding capacity, which varies according to the type of soil; it is very low in case of pure sand, but high in case of the humus-soil. The optimum water-content of sand is about 2–4% and of the humus-soil 15–25%. In the humus-soil the optimal feeding and survival occurs in a broad range,

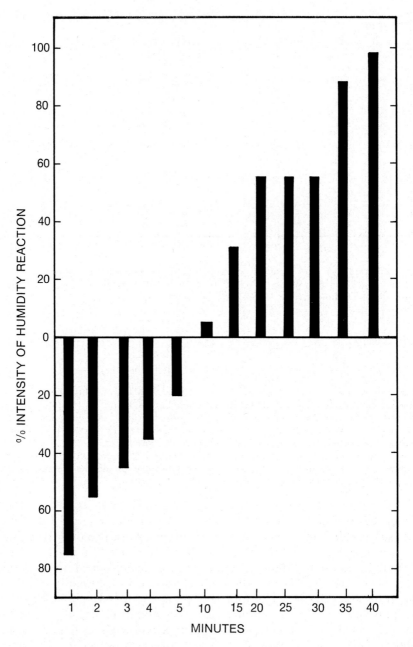

Fig. 61. Intensity of humidity reactions expressed as the excess percentage ratio of termites in the humid zone (80–95% r.h.) of dry-preconditioned workers of *Microcerotermes beesoni* to humidity gradient. (After SEN-SARMA & CHATTERJEE).

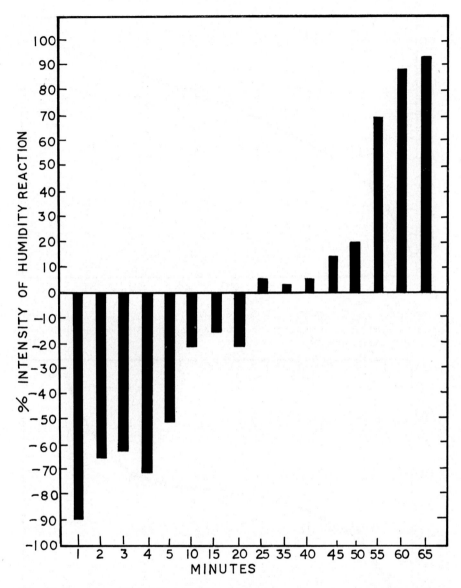

Fig. 62. Intensity of humidity reactions expressed as the excess percentage ratio of termites in the humid zone (80–95% r.h.) of wet-preconditioned workers of *Microcerotermes beesoni* to humidity gradient. (After SEN-SARMA & CHATTERJEE).

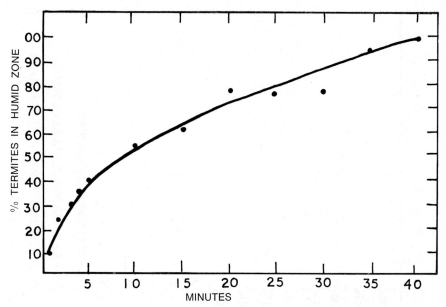

Fig. 63. Percentage of dry-preconditioned workers of *Microcerotermes beesoni* in the humid zone (80–95% r.h.) of a humidity gradient apparatus at various intervals of time (After SEN-SARMA & CHATTERJEE).

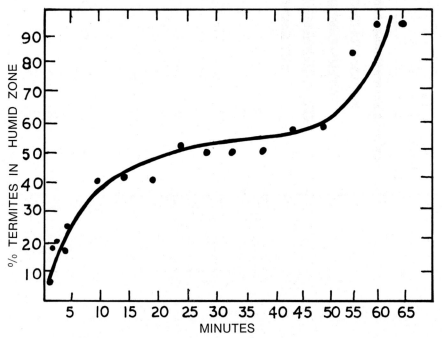

Fig. 64. Percentage of wet-preconditioned workers of *Microcerotermes beesoni* in the humid zone (80–95% r.h.) of a humidity gradient apparatus at various intervals of time (After SEN-SARMA & CHATTERJEE).

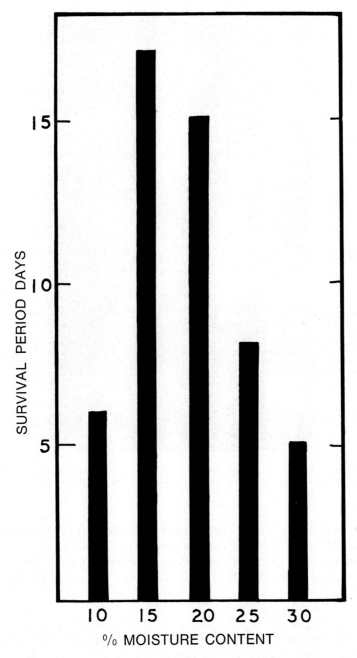

Fig. 65. Colony survival period in days, of the workers of *Microcerotermes beesoni* under different moisture-contents of the soil, used as culture medium (After Sen-Sarma & Chatterjee).

Plate 57. Neotermes bosei in natural habitat, nesting in a standing tree of *Acer oblongum*
(In the background are palm trees in the Botanical Garden, New Forest,
Dehra Dun).

but the same is not true in case of sand which has a poor water-holding
capacity. About 20–30% water content of the soil seems to be preferred by
Heterotermes indicola. The rate of feeding is higher in case of wood with a
higher moisture content. *Heterotermes indicola* prefers wood having about
80–100% moisture content. It has further been demonstrated (BECKER,
1965) experimentally, by using a vertical glass tube (about 120 cm long),
that notwithstanding low soil and wood moisture, *Heterotermes indicola* feeds
mostly near the surface of the soil, thus indicating a strong acrotropic
behaviour. The observations of HARDY (1970) and of SEN-SARMA
et al. (1968) have shown that a soil with a moisture content of 15–20%
seems to be preferred in the genus *Microcerotermes*, as is evident from the
values of the survival periods (Fig. 65).

2.4. Gravity response and vertical distribution of termites

For subterranean species gravity must be considered as one of the most important environmental factors. While the subterranean termites live in soil, the drywood species inhabit wooden structures like trees, rafters, etc. without maintaining any connection with the soil. On the basis of experimental observations, Sen-Sarma et al. (1968) found that the workers of the subterranean *Microcerotermes beesoni* show negatively geotactic response and the foraging pseudoworkers of the drywood termites *Neotermes bosei* show positive geotactic response.

The depth to which subterranean termites penetrate has an important bearing on preventive measures by soil poisoning. Except for the work by Hoon (1962), no extensive study has, however, been made of this problem. The top soil in the area studied by Hoon comprises reddish-brown or yellowish-brown earth and is about 30 cm or more thick. Beneath this is a layer of harder moorum-stratum, which overlies decomposed and sound rock. It has been found that the top soil is sufficiently penetrated by the termites up to a depth of generally 30 cm. At places where the top soil consists, however, of deep alluvial deposits, the termite tracks penetrate as deep as 300 cm. In these studies there is, however, no mention of depth of the subsoil water, which determines the depth to which the termite galleries can penetrate. As the area of the study is markedly a semi-arid tract, with low annual rainfall, the subsoil water must be expected to be relatively very deep. A number of observers have also noticed that overground activities of termites are very much restricted in dry summer months, a period when the termites penetrate much deeper into the soil in search of moisture.

2.5. Population of termite colonies

Mukerji & Mitra (1949) found that the ratio of soldier-worker-nymphs in the fungus combs of *Odontotermes redemanni* is 2.42 % soldiers, 14.25 % workers and 81.33 % immature forms, but did not count the total number of individuals in the mound. Gupta (1952) estimated the total number of individuals in a colony of *Odontotermes obesus*. According to him, the total population of a mound, comprising soldiers, workers and immature forms, ranges between 4548 to 90,961 individuals, depending upon the size of the mound in the non-mound-building months (except during and immediately after monsoon rains); he did not however study the possibility of a correlation between the population and mound sizes. He largely confirmed the relatively high population figure of the percentage of workers and soldiers in the fungus combs, reported earlier by Mukerji & Mitra (1949). Roonwal (1954) reported the worker-soldier ratio to be 97.8 %:2.2 % in *Odontotermes parvidens* found in a dead tree. The same author reports a much higher

435

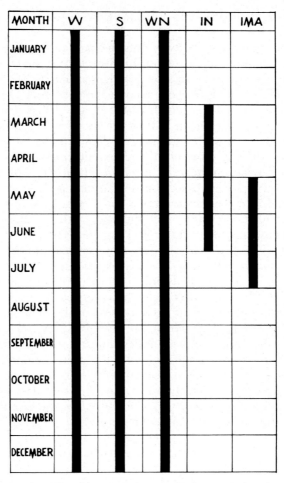

MONTH	W	S	WN	IN	IMA
JANUARY					
FEBRUARY					
MARCH					
APRIL					
MAY					
JUNE					
JULY					
AUGUST					
SEPTEMBER					
OCTOBER					
NOVEMBER					
DECEMBER					

Fig. 66. Occurrence of different castes and stages of *Microcerotermes beesoni* through the year (After SEN-SARMA & MISHRA). W, Worker; S, Soldier; WN, Worker-Soldier nymph; IN, imago nymph; IMA, imago.

ratio of 33 % soldiers and 67 % of workers in *Coptotermes heimi*. It is known that in *Dicuspiditermes incola* there is roughly one soldier for 80 workers (BUGNION, 1915; ESCHERICH, 1911).

Seasonal fluctuations in the nest population of *Microcerotermes beesoni*, a carton-nest building termite of North India, have recently been reported (SEN-SARMA & MISHRA, 1969). This species is a denizen of *Shorea robusta* forest in Dehra Dun and its environs (MATHUR & SEN-SARMA, 1960). Population estimates, made volumetrically, were checked by actual counts of small samples. The total nest population, which does not of course include the foraging individuals outside at the time, varies from 7000 to 45 000 individuals, depending on the season and size of the colony.

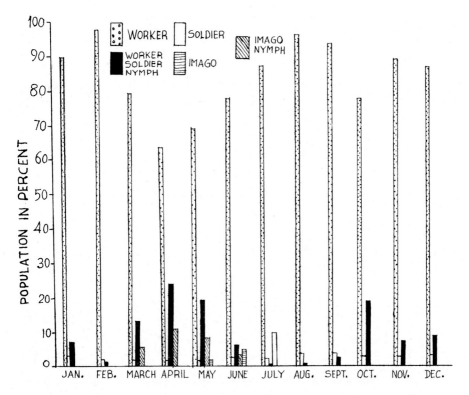

Fig. 67. Percentage fluctuations of population in different castes and stages in the nests of *Microcerotermes beesoni* through year (After SEN-SARMA & MISHRA).

Workers, soldiers and immature forms of worker-soldiers are found throughout the year. The nymphs of alates are present in the nest from March to June and the alates from May to July (Fig. 66). The workers constitute nearly 63.43 % to 98.08 % of the total population. The soldier population is very low and fluctuates between 0.55 % to 3.8 % (Fig. 67). The population of immature forms of the worker-soldier is inversely proportionate to the residual worker population (Fig. 68). The highest population density of immature forms of worker-soldier is recorded in April, when the worker and soldier population is the lowest in the nest. It is explained that on account of the presence of highest density of the alate-nymphs during this month a very active foraging is necessary for the maintenance of the improved nutrition for the development of the young stages to maturity and a decline of worker and soldier population takes place due to predation by ants and other natural enemies. A regression study has not shown statistically significant correlation between the weight of the nest and the number of the workers in the nest (Fig. 69), because of the foraging activity. It is, therefore, to be concluded from

437

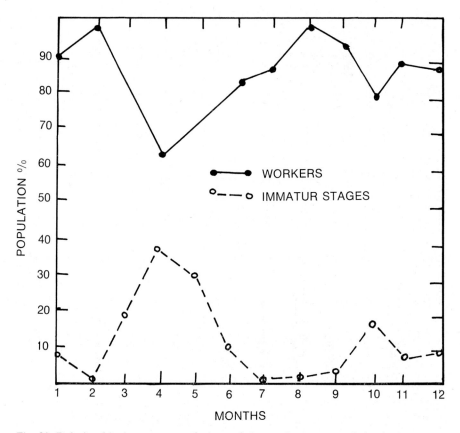

Fig. 68. Relationship between populations of the worker caste and the immature or young stages in *Microcerotermes beesoni* through the year (After SEN-SARMA & MISHRA).

these observations that the polymorphic populations of termites are not only qualitative but also quantitative. Each neuter caste represents a rather constant proportion of the total population. This proportion is, however, greatly influenced by the longevity of various individuals, as well as by the differential accidental mortality. The percentages tend, however, to be readjusted by factors which determine the differentiation of the various castes. Each species of termite seems, therefore, to possess its characteristic constants of composition.

2.6. INQUILINISM

Inquilinism seems to be fairly common among the termites of India. Different species of termites frequently and regularly share the nest with other species and certain termite species are guests more often than others. The association of different species may be obligatory, common, occasion-

438

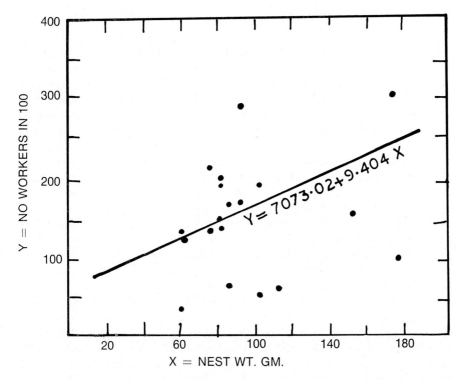

Fig. 69. Relationship between the weight of nests and number of workers in *Micro-cerotermes beesoni.* (After Sen-Sarma & Mishra).

al and exceptional. The numerous cases reported in the literature are, however, not always well documented to enable us to define the precise relationships. An intimate relationship seems, however, to exist between *Microtermes obesi* and some species *Odontotermes* (like for example *Odonto-termes feae, Odontotermes microdentatus* and *Odontotermes obesus*) that share the same abode. *Microtermes obesi* generally builds its nests in the walls of the mound of *Odontotermes.* The soldiers and workers forage, not infrequently, accompanied by *Odontotermes obesus.* Even the adults of *Microtermes obesi* have been observed in the mounds of *Odontotermes obesus.* Diverse other species of *Microtermes* are found either in the mounds of or within the close proximity of the nests of *Macrotermes, Odontotermes,* etc. For example, *Microtermes incertoides* occurs in vicinity of *Macrotermes gilvus* in Indo-China, *Microtermes insperatus* in the mounds of *Macrotermes* and *Odontotermes* in Indonesia, *Microtermes pakistanicus* occurs in the nest of *Macrotermes carbonarius* (Annandale, 1924; Assmuth, 1915; Bathellier, 1927; Beeson, 1941; Mathur & Sen-Sarma, 1962; Roonwal & Sen-Sarma, 1960). *Termes laticornis* from Vietnam is primarily found in the mounds of *Macrotermes gilvus,* in which it makes very narrow and intricate galleries, but the galleries of the two species remain separate from each other (BATHELLIER, 1927). Among other examples of common associations of diverse species of termites, mention may be made of *Dicuspiditermes incola* in the mounds *Odontotermes ceylonicus* and *Odontotermes redemanni* and *Hypotermes obscuriceps* in India and Ceylon. The galleries and chambers are made throughout the mound of the host species. Similarly, *Pericapritermes ceylonicus* is often found in the mounds of *Odontotermes redemanni* and *Hypotermes obscuriceps* in Ceylon (BUGNION, 1915; ESCHERICH, 1911). *Odontotermes parvidens* is frequently found in the field in association with *Odontotermes feae*, *Odontotermes obesus*, *Microtermes* sp., *Coptotermes* sp., *Nasutitermes* sp., etc. (BEESON, 1941; MATHUR & SEN-SARMA, 1959, 1962; ROONWAL & CHHOTANI, 1962a). Two rather exceptional cases of inquilinism may be mentioned here. The carton nest of *Microcerotermes bugnioni* has been reported to occur inside the mound of *Odontotermes redemanni* in Ceylon (BUGNION, 1915; BUGNION & POPOFF, 1912). In the second case two sexual pairs of *Schedorrhinotermes longirostris* were recorded by John (1925) from the nest of *Termes propinquus* in Sumatra. As many as eight species of *Heterotermes*, *Eurytermes*, *Synhamitermes*, *Discuspiditermes*, *Odontotermes* and *Ceylonitermes* have been found in the mounds of *Hypotermes obscuriceps* in Ceylon (ESCHERICH, 1911). ROONWAL (1954) describes the example of *Coptotermes heimi* sharing the same food as *Odontotermes redemanni* in a dead but standing trunk of *Boswellia serrata* in India. The core was infested by *Coptotermes*, which filled the hollowed-out central portion with its characteristic spongy mass of nest material. The outer portion (sapwood) was heavily infested by *Odontotermes parvidens*, the excavations of which did not, however, penetrate deeper than a centimetre below the surface. Dead stumps or fence-posts of *Shorea robusta* are often found simultaneously infested by *Microcerotermes beesoni*, *Odontotermes obesus*, *Microtermes obesi* and *Nasutitermes dunensis* in Jhajra Forest, Dehra Dun. The author found that the galleries of different species remain separate from each other, thus avoiding chances of mutual antagonism. It is apparent that the inquiline species, sharing the abode of the host species, is perhaps incapable, on its own, of creating the environmental conditions, but adjusts itself to those created by the host species. Inquilinism also illustrates interspecific tolerance among different species of termites.

2.7. TERMITES AND FUNGI

The association of termites and different fungi, though extensively investigated elsewhere, has been neglected in India by workers. Some groups of termites seem to be obligatorily associated with fungi more than others. The termites belonging to Termopsidae and Rhinotermitidae usually attack wood that has undergone partial decay under the action of fungi. The primitive species *Archotermopsis wroughtoni* is known to always attack and build nest inside stumps and fallen logs, in which the process of wood decay has already set in. According to Imms (1919), this species never attacks or nests in sound wood, a fact that has been amply confirmed by my own extensive observations. The species can, however, be made to attack sound wood by forced feeding tests under experimental conditions. Similarly the Rhinotermitid genera are often found in the field attacking rotten wood in preference to healthy ones. On the other hand the species of the Kalotermitidae live inside and readily attack fungus-free wood in the field. Hendee (1933) has, however, recorded as many as seventeen genera of fungi from the colonies of the Neotropical *Incisitermes minor*. The brown-rot and white-rot fungi cause decay in many kinds of wood. The former fungus is known to break down the harmful extractives in the wood and thus prepare it for feeding by termites. The common brown-rot fungi attacking wood in India are *Polyporus versicolor, Polyporus sanguineus, Polyporus abietinus, Polyporus palustris, Lenzites trabea, Lenzites striata, Pobia monticola*, etc. The white-rot fungi seem to break down the lignin of the wood, thus providing extra assimilable carbohydrates for the termites. Thus many species of termites are attracted to decayed wood, attacked either by the brown-rot or the white-rot fungi. The precise relationships of different species and of the wood-rotting fungi are not, however, clearly understood. There does not appear to be evidence of a symbiotic relation between the termite species and the wood-rotting fungi, though there is little doubt that the termites are nutritionally dependent on the wood-rotting fungi (Lund, 1963).

König (1779) was the first to give an account of the sponge-like fungus gardens of Macrotermitinae, built from vegetable residues and supporting a mycelium, with white nodules of conidia and conidiospores. He believed that the conidia might be used as food for the newly hatched young termites. Petch (1913) has given a general resumé of the fungi found associated with termites from the world. The conidia form a white mealy growth on the surface of the combs, milky-white, shiny and about a millimetre in diameter. The white spheres are named by Berkeley (1847) as *Aegeriltha duthei* and by Cifferi (1935) as *Termitosphaeria* and are in reality the conidial stage of the agaric *Collybia albuminosa* that develops inside termite mounds (Cheo, 1942). A number of workers like Doflein (1906), Petch (1906, 1913), Escherich (1909, 1911), Bugnion (1915), Hegh (1922), Annandale (1923, 1924), Bose (1923), Bathelier (1927), Heim (1940), Cheo (1942), Grasse (1944), Bakshi (1951), Dass et al. (1962), and Batra & Batra (1966) have studied these fungi. Heim placed all the agarics that develop from termitaria under the genus *Termitomyces*, which he differentiated into two subgenera: *Praetermitomyces* with a single species *Termitomyces (Praetermitomyces) microcarpum* and the subgenus *Eutermitomyces* for the remaining species that fructify at the tips of long pseudorhiza growing out from the fungus combs through the mounds or through the soil above the termite nests. Of these only two

441

species *Termitomyces (Eutermitomyces) albuminosa* and *Termitomyces (Euter-mitomyces) eurhiza* are known so far from the Indo-Malayan Region. The former species of fungus has been recorded from the termitaria of *Odonto-termes redemanni* (PETCH, 1906), *Odontotermes (Hypotermes) obscuriceps* (PETCH, 1906), *Odontotermes horni* (BATHELLIER, 1927), *Odontotermes sundaicus* (KEMNER, 1934), *Microtermes insperatus* (KEMNER, 1934) and *Odonto-termes obesus* (BOSE, 1923). The latter species occurs in the termitaria of *Odontotermes gurdaspurensis* (BATRA & BATRA, 1966). The genus *Termito-myces* is apparently restricted largely to *Odontotermes* in India and according to SANDS (1969) the single record by KEMNER (1934) of *Termitomyces (Eutermitomyces) albuminosa* from the fungus-combs of *Microtermes insperatus* must be a case of wrong identification.

All species of *Termitomyces* fructify early during the rainy season and discharge the basidiospores on the surface of the soil, where they are readily gathered by foraging workers, thus bringing about reinoculation of new combs. BATRA & BATRA (1966) have, however, doubted this method of reinoculation, as they claim to have observed conidia in the gut of the swarming termites and also in crevices in their integument. The observations of SANDS (1960) and of LÜSCHER (1951) that the combs, newly built in cultures reared in the laboratory, remain entirely sterile should, however, prove that the alate founders do not carry the fungal inoculum of viable spores. That the basidiospores are host specific for the reinoculation of sterile combs has been experimentally proved by SANDS (1960). His attempts to inoculate the sterile combs of *Ancistrotermes guineensis* by means of the combs from a related species *Ancistrotermes crucifer* failed, but on the other hand when the laboratory colonies were supplied with the combs from another nest of the same species, the sterile combs became readily reinoculated.

In addition to *Termitomyces*, the spores of a number of other fungi have also been recorded from the fungus garden and from the soil in the neighbourhood of the fungus garden. DAS et al. (1962) have recorded, for example, the following organisms from the mounds of *Odontotermes obesus*: From the fungus garden: The fungi *Aspergillus flavus, Aspergillus oryzae, Aspergillus ustus, Alternaria* sp., *Curvularia* sp., Dematiaceae (non-sporulating), *Fusarium equiseti, Fusarium* sp., *Monilia* sp. *Penicillium* sp., *Paecilomyces* sp. *Rhizopus* sp., *Rhodotorulla* sp., *Xylaria* sp. and *Actinomyces* spp. and the anaerobic sulphate-reducing Bacteria *Desulphovibrio* sp. In the soil in the vicinity of the fungus garden: The fungi *Aspergillus oryzae, Aspergillus* sp., *Diplodia* sp., Dematiaceae, *Fusarium* sp., *Penicillium* sp., *Rhodotorula* sp. and *Syncephalastrum* sp. and anaerobic sulphate-reducing *Desulphovibrio orientalis* bacterium. Similarly BATRA & BATRA (1966) have also reported the occurrence of a number of microörganisms in the mounds of *Odontotermes gurdaspurensis*.

When a termite mound is deserted by the termites or becomes partially exposed, the fruit bodies of a species of *Xylaria* make their appearance.

442

If an active comb, having spheres, is placed in the laboratory, numerous long, white (turning black subsequently) stalks arising from a grey coloured mycelium and bearing cylindrical stromata at the tips, develop. *Xylaria nigripes* is the common species in the nest of *Odontotermes redemanni*, *Odontotermes obesus* and *Odontotermes gurdaspurensis* (BAKSHI, 1951; BATRA & BATRA, 1966; BOSE, 1923; MUKERJI & MITRA, 1949). *Xylaria* species are saprophytic. GRASSE (1937) found that the mycelium of *Termitomyces* is often mixed with the finer hyphae of *Xylaria* in termite fungus-combs. The growth of the fungus *Xylaria* brings about the death of the termite workers. BATRA & BATRA (1966) have also reported the finding of another sparophytic fungus *Neoskofitzia termitum* in the mounds of *Odonto-termes gurdaspurensis* from India. *Padoxan* sp. grows from small mounds of *Trinervitermes germinatus* in the Nigerian savannah. According to the observation of SANDS (1969), *Padoxan* sp. occurs in places, where the grass-feeding and harvesting genera of Nasutitermitinae are found and its possible host-genus in India, Ceylon and Malaysia seems to be *Trinervitermes*. There is, however, no definite record of the genus in the Indo-Malayan region.

The problem of how the fungi other than *Termitomyces* are prevented from growing in the mounds does not seem to have been satisfactorily solved so far. Earlier authors believed that the termites actively weed the fungus combs to keep down alien fungi (PETCH, 1906). This belief seems to be untenable from the observation that *Xylaria* may grow in partially deserted combs, while the workers, soldiers and nymphs wander about in the tunnels, until they gradually become immobilized and die in a profuse growth of the mycelia and stromata of *Xylaria*. Lack of power of weeding in the termites does not seem to account for these events. BATRA & BATRA (1966) have recently demonstrated that not only the secretions of the soldiers but even the soil that has been manipulated by the workers with saliva exert a strong fungistatic effect on genera of fungi other than *Termitomyces* in the mounds of *Odontotermes gurdaspurensis*. SEN-SARMA first suggested that the high CO_2-content inside the mound might perhaps inhibit the growth of *Xylaria*, and this has since been confirmed by the observations of BATRA & BATRA (1966). The high CO_2-content seems to inhibit the growth of many other foreign fungi that may have been carried inside the mound as contamination by foraging termite workers.

The precise rôle of the fungus *Termitomyces* in the life of the termite colony is not understood at present. A number of workers like KÖNIG (1779), PETCH (1906, 1913), ANNANDALE (1923, 1924) and DASS et al. (1962) have regarded the fungus combs in the nests of Macrotermitinae as possible source of food for the termites. HEIM (1948) believed that the fungus garden is merely a part of the architecture and the fungi growing on the combs act as commensals to be expelled when overgrown. GRASSE (1944), LÜSCHER (1951) and DASS et al. (1962), however, observed that the conidia were occasionally eaten by the termites in small quantities

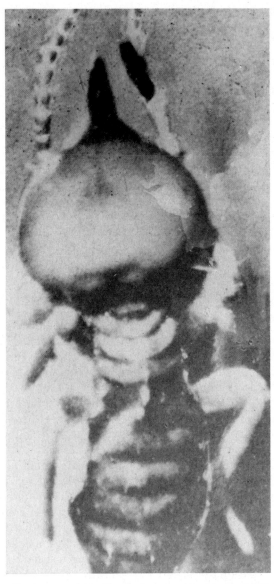

Plate 58. Fungus disease caused by *Termitaria* sp. on the fore leg of *Nasutitermes fletcheri*, (After SNYDER).

and this observation was interpreted as showing that the fungi act as a source of necessary vitamins for the termites. The masses of eggs of the termites found in the fungus combs led KÖNIG (1779), ANNANDALE (1923, 1924), ESCHERICH (1911) and MUKERJI & ROY-CHOUDHRY (1943) and MUKERJI & MITRA (1949) to the belief that the fungus

444

combs serve as egg depositories and nurseries for the termites. KALSHOVEN (1936) observed that the fungus combs of *Macrotermes, Odontotermes* and *Microtermes* were eaten away from below by the termites, when the combs have attained a certain size. The writer has himself observed on a number of occasions that overgrown fungus combs in the nests of *Odontotermes obesus* and *Odontotermes microdentatus* are concave from below and often a new comb may be constructed in the space created by the concavity. The concavity is perhaps the result of the fungus being eaten away by the termites. LÜSCHER (1951) and GEYER (1951) believe, on the other hand, that the fungus combs may generate heat as a result of fermentative processes and thus assist in the maintenance of constant temperatures inside the mound. GHIDINI (1938) considers, however, that the fungus garden serves as humidity regulator.

Reference may be made here to *Termitaria*, a genus of fungus that is pathogenic to termites. This fungus forms grey or brown flat plates, with a dark brown or even black rim, closely applied to any part of the body of a termite. The plate-like mature form of the fungus consists of columnar cells that arise from a flat, primary thallus-like brown structure and forms a firm substratum, adhering to the insect cuticle (THAXTER, 1920). The fungus penetrates the insect cuticle by a kind of rooting pedicle (FEYTAUD & DIENZEIDE, 1927), invades the adipose tissue and profoundly alters it. The infection of *Nasutitermes luzonicus* by *Termitaria coronata* is described by COLLA (1929). SNYDER (1933) has figured the infection of the leg of *Nasutitermes fletcheri* from India by *Termitaria* sp. ALSTON (1947) has described the attack of *Coptotermes curvignathus* from Malaya by the parasitic fungus *Entomophthora* sp., resulting in 100% mortality of the termites within forty-eight hours. *Aspergillus flavus* is known to be a pathogenic parasitic fungus, attacking the kings and queens *Odontotermes obesus* in South India (SANNASI, 1969a). The fungal infection is associated with melanosis of the insect cuticle (SANNASI, 1969b & c). The fungus *Aspergillus flavus* occurs very commonly in soil and has also been found in the fungus garden inside termite mounds. It is not, however, clearly understood how the fungus fails to be parasitic on the termites in Nature, eventhough the mouthparts of the workers of *Odontotermes latericus* are often loaded with the spores (SELLSCHOP, 1965).

2.8. NATURAL ENEMIES OF TERMITES

Although early writers like LEFROY & HOWLETT (1909) and FLETCHER (1914) have remarked that termites pay a heavy toll to lizards, frogs, birds, etc., it was MATHUR (1962) who first gave an extensive list of the natural enemies of termites in India.

The ants are perhaps the most implacable enemies of termites in India; a number of species *Acantholepis frauenfeldi, Camponotus compressus, Crematogaster* sp., *Lobopelta diminuta, Monomorium destructor, Monomorium indicum,*

Plate 59. A unusual case of mound formation indicated by the arrow by *Odontotermes wallonensis* at the top of a house in Shimoga (Mysore).

Myrmecocystus satipes and *Solenopsis geminata* are predators on termites like *Coptotermes*, *Dicuspiditermes*, *Heterotermes*, *Microcerotermes*, *Microtermes* and *Odontotermes*.

The dragonfly *Pantala flavescens* is also known to predate on the swarming adults of *Odontotermes assmuthi*, *Odontotermes feae*, *Odontotermes obesus* and *Odontotermes parvidens* at Dehra Dun. The robberflies *Machimus rufipes* and *Prochamus duvaucelli* are predators on swarms of *Odontotermes obesus* and *Coptotermes heimi* at Dehra Dun. The muscoid fly *Ochromyia jejuna* preys upon the swarming *Coptotermes*, *Odontotermes* (GREEN, 1906; LEFROY & HOWLETT, 1909; MATHUR, 1962). Cockroaches have been reported to devour the winged adults, as they swarm out of the nest (FLETCHER, 1914).

Rana tigrina, *Rana breviceps* and *Bufo melanostictus* devour termites. Most lizards are also predators on the winged and other forms of termites. The termites constitute an important item of food of the common Indian birds like *Corvus splendens*, *Corvus macrorhynchus*, *Dicrurus macrocercus*, *Molpostes cafer*, *Acridotheres tristis tristis*, *Acridotheres ginginianus*, *Hirundo rustica* and *Milvus migrans*. Among the other insectivorous birds that take a heavy tol of termites are *Upupa upupa orientalis*, *Gallus bankiva murghi*,

446

Plate 60. Assemblage of mounds of *Odontotermes wallonensis* in Shimoga (Mysore), The soil is red and lateritic.

Gallus sonneratii, Coturnix coturnix, Francolinus francolinus, Francolinus pondicerianus, Turdoides somervillei, Saxicoloides fulicata, Uroloncha punctulata and *Athene brama*. Among the termitovorous mammals, the scaly-anteater *Manis crassicaudata* and the sloth-bear *Melursus ursinus* are perhaps the most important. Both these predators break open the mounds of termites and devour not only the adults but also the nymphs and eggs. Insectivorous bats and *Vulpes bengalensis* have also been recorded as eating swarms of the termites. The stomach contents of rats contain quantities of termites (Roonwal, 1949).

2.9. Termitophile associates

The most important termitophilous insects in India are the Staphylinidae (Coleoptera) and Phoridae (Diptera). The distribution and phylogeny of the termitophiles are mirror images of those of the termites (Seevers, 1957). The major specializations, associated with termitophily, are physogastry, lumuloid body outline and presence of exudatory appendages. The termites lick the physogastric abdomen and exudatory appendages of the termitophile insects, which in return obtain shelter and food from the termites.

447

The important termitophilous species from India include the Staphylinid beetles *Termitodiscus heimi* from the mounds of *Odontotermes obesus* and *Odontotermes wallonensis*, *Termitodiscus echerichi* and *Termitodiscus butteli* from the mounds of *Odontotermes redemanni*, *Odontotermes ceylonicus* and *Hypotermes obscuriceps* in Ceylon, *Doryloxenus* and *Termozyrus* in the mounds of different species of *Odontotermes* and of *Hypotermes obscuriceps* and *Trinervitermes biformis* in India and Ceylon. *Zyrus (Rhyncodonia) termiticolus* is associated with *Coptotermes gestroi* and *Zyrus (Rhyncodonia) feae* with *Odontotermes feae* in Burma. *Zyrus (Myrmedonia) punctissima* occurs in the mounds of *Odontotermes horni* in Ceylon. *Odontoxenus termitophilus* is associated with *Odontotermes obesus* in India. The Phorid *Termitoxenia peradeniyae* has been recorded from the nest of *Odontotermes taprobanes* from Ceylon. Collembola are also found as termitophilous species (BUGNION, 1915; CAMERON, 1932; JOHN, 1925).

3. Limiting Factors in Distribution of Termites

The termites are strictly tropical insects, with extremely narrowly specialized food habits, bound down to cellulose, and highly capable of creating their own specific environment. Their soft bodies, social habits and weak flight are other important features that govern their dispersal and distribution. These conditions largely explain the fact that extremely few species are really cosmopolitan in their distribution. Their narrow specialization also precludes adaptative radiations for diverse food habits. A high degree of phylogenetic correlation exists with their geographical distribution and all the living families of termites apparently dispersed to the major geographical tropical areas of the world by the late Mesozoic Era and subsequent differentation seems to have occurred locally during the Tertiary times (KRISHNA, 1970). One of the most important conditions that determine their wide dispersal is the restriction of dispersal to the brief swarming periods, even during which the flight range is not great, so that even a small body of water constitutes a major barrier.

The principal ecological factors that influence the dispersal of termites in India are the monsoon rainfall pattern (mean annual rainfall, mean number of rainy days), atmospheric temperature, atmospheric humidity, vegetation, altitude, soil type, natural enemies and other associated organisms. Of these the vegetation and soil types are perhaps more directly important than the others.

Although the influence of vegetation type on the distributional pattern of termites is generally recognized, we really know very little about the precise relations of different species. The problem is also complicated by the fact that the vegetation type is in its turn influenced by the soil and monsoon rainfall patterns. We may state, however, that certain well known termites like *Anacanthotermes* and *Trinervitermes* are restricted to open vegetation of grassland savannah, but others like *Microcerotermes*

Plate 61. Arboreal carton-nest of *Nasutitermes beckeri* in the rain-forest of Kerala (After BECKER).

and *Nasutitermes* are characteristics of forest vegetation. A forest seems to constitute an effective barrier to *Trinervitermes*, although most moist rain forests have relatively larger number of genera of termites. *Eurytermes* may be said to be confined to the deciduous forests of the central and Peninsular areas of India and to Ceylon (ROONWAL & CHHOTANI, 1966b). *Procapritermes* seems to be characteristic of tropical rain forest areas. *Archotermopsis wroughtoni* occurs generally in the Sub-Himalayan tracts,

449

at elevations of 1290–2600 m above mean sea-level in India, Pakistan and Afghanistan (HARRIS, 1967). *Anacanthotermes* is distributed discontinuously in arid areas. *Stylotermes* seems to be primarily a hill form. For example, *Stylotermes faveolus* occurs in the Kulu Valley (1200–1350 m), *Stylotermes bengalensis* occurs from Darjeeling to Garhwal and *Stylotermes chakratensis* in Chakrata and *Stylotermes fletcheri* occurs in the Shevroy Hills of South India. The lowest altitude at which the genus has so far been recorded is 680 m at Dehra Dun. *Psammotermes rajasthanicus* is typical of sandy areas of Western Rajasthan, and occurs under bark of trees, in the wood-work of buildings and under cowdung and stones in the open country (ROONWAL & BOSE, 1962, 1964). *Heterotermes indicola* is found throughout northern India up to elevations of 2135 m in the Himalaya, in West Pakistan and in parts of Afghanistan.

4. Distribution of some important Termites from India

The termites of India comprise predominantly Oriental elements (65 %). About 22 % of the genera known so far from India are endemic; the subfamily Nasutitermitinae has perhaps the largest number of endemic genera. The Ethiopian elements come next in order of importance, but there are also distinct Australian and Neotropical facies in the termites of India. The discontinuous distribution of certain genera like *Synhamitermes* in South America and in India and of *Speculitermes* in South America, Africa and India, must be explained on the basis of continental drift hypothesis. EMERSON (1955) and KRISHNA (1970) have, however, attempted to explain such examples on the ideas of the now defunct land-bridge theory. A land-bridge with tropical conditions is believed to have existed across the Bering Strait to account for the occurrence of South American termites in Australasia. They also believe that some genera may have reached India from Africa through west Asia. It must, however, be recognized that the distribution of termites in India presents a number of still unsolved and complex problems.

4.1. KALOTERMITIDAE

Of about two dozen living and fossil genera of this family, only *Postelectrotermes*, *Neotermes*, *Kalotermes*, *Glyptotermes*, *Bifiditermes*, *Procryptotermes* and *Cryptotermes* are so far known from India. None of these genera is endemic and no genus has also been reported so far from Burma, but the family is well represented in the Australian, Ethiopian, Malagasy, Papuan, Palaearctic, Nearctic Neotropical regions. Three species of *Postelectrotermes* occur in widely separated areas: *P. bhimi* occurs in Kerala, *P. militaris* in Ceylon (1000–1375 m), and *P. pishinensis* is reported from Pakistan. *Neotermes* is known from the mainland of India, as well as the Andaman Islands and from Ceylon. *Kalotermes* is represented by the single species *Kalotermes jepsoni* in Ceylon at elevations of 900–1200 m. *Glyptotermes* is known from north and South India and Ceylon. *Bifiditermes* with the single species *Bifiditermes beesoni* is known from northwestern parts of the Gangetic Plains and from the Punjab and Pakistan. The genus *Procryptotermes* has been recently discovered from India (ROONWAL & CHHOTANI, 1963). *Cryptotermes* occurs in India and Ceylon, and seems to be restricted to coastal tracts. There is, however, a recent report of *Cryptotermes* attacking dead and dry trees in central India, parts of the Gangetic Plains and in Bangalore. *Cryptotermes domesticus* is one of the few widely distributed species known to occur in South India, Ceylon, Borneo, Java, Sumatra, Malaya, Thailand, Vietnam, Hainan, Taiwan, Japan, Central America and in a number of islands of the north and south Pacific Ocean (GAY, 1967; HARRIS, 1968). *Cryptotermes dudleyi* is known from Africa, Asia, Australia, New Guinea, Central America, the Caribbean Islands and South America; in India it occurs in Assam, East Bengal, Sundarbans, Andaman Islands and is also known from Ceylon. *Cryptotermes bengalensis* is endemic in the Sundarbans. *Cryptotermes havilandi* has recently been reported from the Andamans (ROONWAL & BOSE, 1970).

4.2. TERMOPSIDAE

The Termopsidae are a small family of primitive, damp-wood termites, divided into three subfamilies, Termopsinae, Stolotermitinae and Porotermitinae. Only the first named subfamily occurs within our limits and is represented by two genera *Archotermopsis* and *Hodotermopsis;* the only other genus *Zootermopsis* is Nearctic. *Archotermopsis wroughtoni*, the only living species of the genus, occurs at an elevation of 850–2890 m (mostly above 1200 m) in Kumaon to Kashmir in the Sub-Himalayan tracts, Hazara in Pakistan and Kabul (Afghanistan). It is found under the bark and inside dead and decaying Conifer trees and never in broad-leaved trees. EMERSON (1955) believes that this family probably differentiated in temperature Eurasia in early Mesozoic and reached North America later.

4.3. HODOTERMITIDAE

The Hodotermitidae or the socalled harvester termites forage in the open and store bits of grass in their nest. Of the three genera *Hodotermes*, *Microhodotermes* and *Anacanthotermes*, known so far, only the last named occurs within our limits (Fig. 70). The genus seems to have differentiated in Africa from *Hodotermes*-like stock and reached India. Within India, the genus is discontinuously distributed in the Rajasthan and in South. We have three species, *Anacanthotermes macrocephalus*, *Anacanthotermes viarum* and *Anacanthotermes rugifrons* in India. The first named species extends from Afghanistan, Karachi to Rajasthan (WEIDNER, 1960; ROONWAL

451

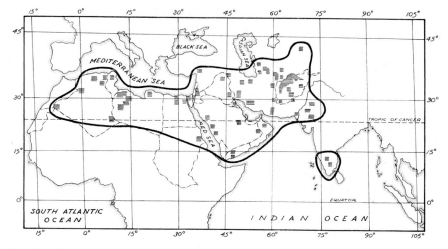

Fig. 70. Map showing the geographical distribution of *Anacanthotermes*.

& Bose, 1964) and the remaining two species are found in South India.

4.4. Rhinotermitidae

All the six subfamilies Coptotermitinae, Heterotermitinae, Psammo-termitinae, Termitogetoninae, Stylotermitinae and Rhinotermitinae, occur in Indian region. The Coptotermitinae, with the single genus *Coptotermes*, are distributed all over the tropics and is perhaps the most primitive subfamily in the family. *Coptotermes ceylonicus* occurs in the Western Ghats, Malabar Coast, Mandapam, Krusadai Islands, Ceylon and Indo-China (doubtful) (Becker, 1962b, c; Ghanamuthu, 1947; Roonwal & Chhotani, 1962b). *Coptotermes gestroi* occurs in Assam, Burma and Malaya. *Coptotermes heimi* is common throughout India and Pakistan (Mathur & Sen-Sarma, 1959; Roonwal & Bose, 1970). It is a polyphagous termite that attacks over thirty-five different species of plants (Roonwal, 1970). *Coptotermes travians* occurs in Assam, Bengal, Orissa, East Pakistan, Burma, Malaya, Sumatra, Java and Borneo (Roonwal & Chhotani, 1962b; Roonwal & Maiti, 1966).

Of the two genera *Heterotermes* and *Reticulitermes* of the Heterotermitinae, the former is the more primitive and occurs in all the tropical and sub-tropical regions (excluding the Palaearctic). Emerson (1955) believes that the genus *Heterotermes* has perhaps had an obscure origin in the tropics, possibly before the Cretaceous time and before the drift of Australia from the Indo-Malayan areas. *Heterotermes ceylonicus* occurs in Ceylon. *Heterotermes indicola* is primarily tropical in its distribution, but has extensively intruded into subtropical and warm temperate areas of the Sub-Himalayan regions, up to elevations of 1980 m. It meets the southern

452

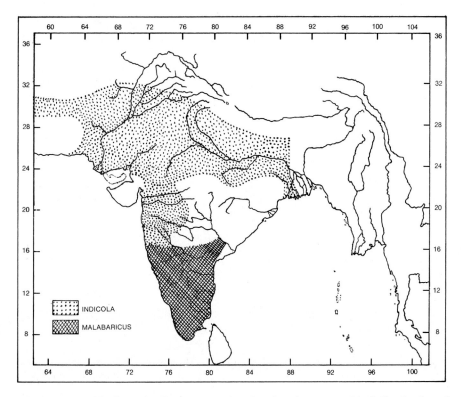

Fig. 71. Map of India and adjacent countries, showing the geographical distribution of
Heterotermes indicola and *H. malabaricus*. (Modified from BECKER).

species *Heterotermes malabaricus* at about the 15–20 NL and extends also to
West Pakistan and parts of Afghanistan (BECKER, 1962b, & c; MATHUR
& SEN-SARMA, 1959; WEIDNER, 1960). *Heterotermes malabaricus* is widely
distributed in the Peninsula (Fig. 71). *Reticulitermes* is the only Holarctic
genus, but its centre of origin seems to be obscure (EMERSON, 1955).
Reticulitermes chinensis and *Reticulitermes saraswati* are reported from Assam;
the former species occurs also in China and Indo-China (HARRIS, 1968;
ROONWAL & CHHOTANI, 1962a).

The Psammotermitinae are represented by two genera *Glossotermes*
(with a single South American species) and *Psammotermes*, with seven
species, one of which occurs in Madagascar and the others are found in
the arid areas of Africa, West Asia, Arabia and northwestern India
(Fig. 72). *Psammotermes* seems to have originated in Africa and entered
India perhaps through Miocene grasslands across West Asia. We have a
single species *Psammotermes rajasthanicus* from Rajasthan (ROONWAL &
BOSE, 1962).

The small subfamily Termitogetoninae, with a single genus *Termitogeton*

453

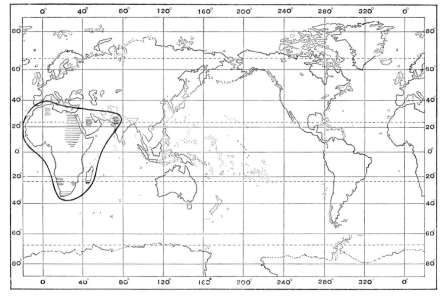

Fig. 72. World map showing the geographical distribution of *Psammotermes*. (After ROONWAL, CHHOTANI & BOSE).

is Oriental. *Termitogeton planus* occurs in Boreno and *Termitogeton um-bilicatus* is widely distributed in Ceylon up to elevations of 1370 m and is a moisture-loving species that lives in damp decaying tree trunks, crevices of softwood trees, etc. (BUGNION, 1914, 1915; ROONWAL, 1970).

The subfamily Stylotermitinae (considered by some as a family) is represented by the living genus *Stylotermes*, distributed in diverse climatic conditions in India and China. We know about half a dozen species from India. *Stylotermes bengalensis* extends from Darjeeling to Garhwal in the Sub-Himalayan areas, *Stylotermes chakratensis* in Chakrata also in the Sub-Himalayan Garhwal tract at an elevation of 2100 m, *Stylotermes fletcheri* at an elevation of about 1085 m in the Shevroy Hills in South India and *Stylotermes faveolus* occurs in the Kulu Valley of the Northwest Himalaya at an elevation of 1190–1280 m.

The Rhinotermitinae are represented by three genera *Prorhinotermes*, *Parrhinotermes* and *Schedorrhinotermes*. *Prorhinotermes* occurs both in the Old and New World and is represented by three species within the Indian region. *Prorhinotermes flavus* is recorded from the Andaman Islands and from Ceylon, *Prorhinotermes shiva* and another species also from the Andaman Islands. *Parrhinotermes*, extending from the Australian to the Indo-Malayan areas, is represented by *Parrhinotermes khasii* in Assam. The genus *Schedorrhinotermes* occurs in the Malayan, Papuan, Australian and Ethiopian areas and within our limits has so far been reported from only the Nicobar Islands.

454

4.5. Termitidae

This is the largest family and comprises nearly three-fourths of all the species of the world. Four subfamilies Amitermitinae, Termitinae, Macrotermitinae and Nasutitermitinae are generally recognized; all these subfamilies are known from the Indian region.

Amitermitinae are the most primitive subfamily, with *Protohamitermes* as the most primitive genus, represented by the single species *Protohamitermes globiceps* from Borneo. The number of endemic genera in the Malayan subregion is large. Within the Indian area we know the genera *Eurytermes, Doonitermes, Speculitermes, Euhamitermes, Amitermes, Synhamitermes, Globitermes, Eremotermes* and *Microcerotermes*. The genera *Eurytermes* and *Doonitermes* are endemic; the former is confined to the deciduous forest areas of Central India and South India and Ceylon. Of the four species, three occur within India and one is known from Ceylon (Fig. 73). *Doonitermes* is monotypic from the Doon Valley. *Speculitermes* is known from Thailand, Burma, India and Ceylon and from the Neotropical areas (Fig. 74). The soldiers of this species were first found in India by Roonwal & Chhotani (1960, 1966a) and later from Thailand (Ahmed, 1965) and Ceylon (Krishna, 1970). As the soldier caste is absent in the Neotropical region, Krishna is of the view that the Oriental species must be quite distinct from the Neotropical species. We have seven species of the genus in the Indian region, with some subspecies in the Peninsula, Ceylon, western India, Assam and Burma. *Speculitermes cyclops* is widely distributed from western India through central India, eastern India and Burma. *Speculitermes sinhalensis* occurs in the Peninsula and in Ceylon. *Speculitermes triangularis* has so far been recorded only from Dehra Dun. *Speculitermes decanensis* is confined to the Peninsula. The genus *Euhamitermes* is represented by one species in Malaya and five species in India, especially the Peninsula and central India. *Amitermes* is a tropicopolitan genus with the largest number of species in the Ethiopian Region; the Australian and Neotropical Regions come next in importance. The Indo-Malayan has only four species. The primitive relatives of this genus are Indo-Malayan and the occurrence of larger numbers of species in outside areas is interpreted by Emerson (1955) as indicative of higher rate of speciation in response to favourable ecological conditions than in the centre of differentiation and dispersal. We have in India *Amitermes belli* and *Amitermes paradentatus*, found commonly in Rajasthan and West Pakistan; the report of the genus from Calcutta by Becker (1962b, c) seems to be doubtful. The distribution of the small genus *Synhamitermes* presents a number of interesting features, particularly its discontinuous distribution. It occurs in the Neotropical Region and in India and Ceylon (Fig. 75). It seems that this genus arose in the Indian region and is supposed to have spread through the Bering Strait to the New World. *Synhamitermes ceylonicus* and *Synhamitermes colombensis* occur in Ceylon and *Synhamitermes*

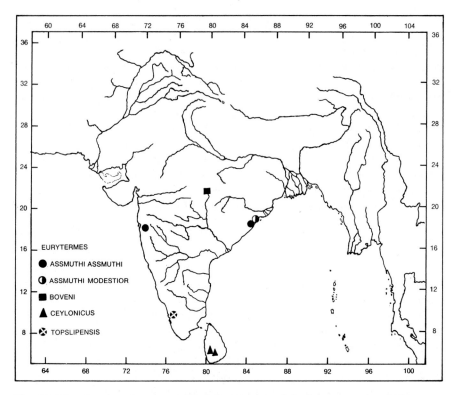

Fig. 73. Map showing the geographical distribution of the known species of *Eurytermes*. (After Roonwal & Chhotani).

quadriceps extends from Goa southwards to Kerala and eastwards to Central India and westwards to Rajasthan; it is also reported from Tripura in the east. The genus *Globitermes* is wholly Oriental; *Globitermes sulphureus* is known from Lower Burma and extends eastwards to Thailand, Cambodia and Indo-China, but does not occur within India, Pakistan and Ceylon (Fig. 76). *Eremotermes* is represented by seven species, of which one is from North Africa, one from the Middle East and five others from India, where they occur in the arid, desert tracts as well as wetter areas (Dehra Dun and the Peninsula). *Eremotermes dehraduni* occurs at elevations of 750 m in the Sub-Himalayan tract, *Eremotermes fletcheri* is reported from South India and Baluchistan, *Eremotermes madrasicus* from Madras, *Eremotermes neoparadoxalis* is confined to the arid western areas and to West Pakistan and *Eremotermes paradoxalis* is widely distributed in India and West Pakistan. *Microcerotermes* is widely distributed, except in the Nearctic Region, but there are perhaps more species in the Ethiopian than elsewhere. The genus may have differentiated in Africa during the Cretaceous times and spread across the Oriental Region to the New World

456

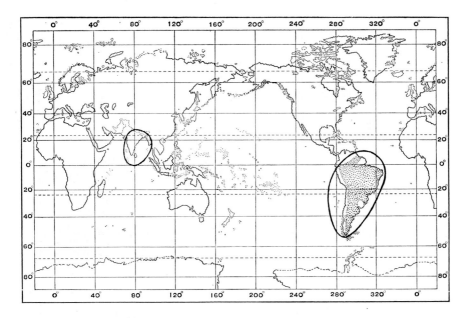

Fig. 74. World map showing the geographical distribution of *Speculitermes.* (After ROONWAL, CHHOTANI & BOSE).

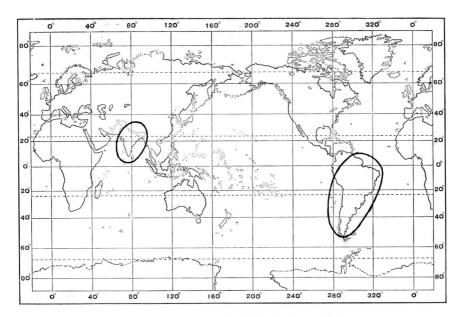

Fig. 75. World map showing the geographical distribution of *Synhamitermes.*

457

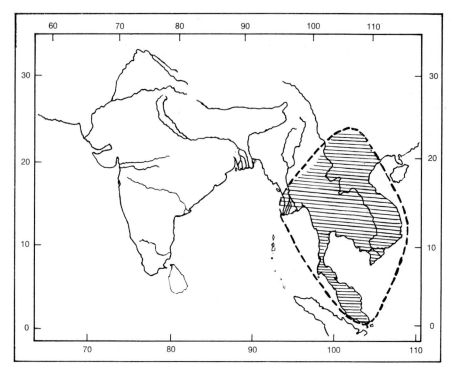

Fig. 76. Map showing the geographical distribution of *Globitermes sulphureus.*

(Krishna, 1970). *Microcerotermes annandalei* occurs in Bihar and Orissa; *Microcerotermes baluchistanicus* in the arid tracts of Baluchistan; *Microcerotermes cameroni* in the Peninsula (coastal area); *Microcerotermes crassus* in Burma, Thailand, Hainan and Indo-China; *Microcerotermes cylindriceps* is endemic in Ceylon; *Microcerotermes heimi* is widely distributed in the Peninsula, Pakistan and Ceylon and *Microcerotermes tenuignathus* is found in the arid tracts of Gujarat, Deccan and Rajasthan and West Pakistan (Ahmed, 1955; Becker, 1962b, c; Prasad et al., 1967).

The subfamily Termitinae is rather poorly represented in India. Of the fifty-seven genera known so far from the world, hardly seven occur within the limits our region, representing only 12.28 %. Over 34 genera are endemic in the Ethiopian region, including the most primitive genus *Hoplognathotermes.* It is, therefore, generally believed that this subfamily differentiated from Amitermitinae in tropical Africa during Cretaceous times (Emerson, 1955). The genera found in India are *Angulitermes, Homallotermes, Discuspiditermes*, Pericapritermes*, Procapritermes, Labioca-*

---

* The Indian species of these genera were mostly included by Snyder (1949) under *Capritermes,* now monotypic genus from Madagascar (Krishna, 1968).

458

*pritermes**, and *Microcapritermes. Angulitermes* is the most primitive among the Indian genera and it is also known from the Ethiopian, Palaearctic and Oriental Regions. We have in India nine species; *Angulitermes acutus* occurs in the Peninsula; *Angulitermes akhorisainensis* in the Sub-Himalayan area, *Angulitermes ceylonicus* in Ceylon; *Angulitermes dehraensis* in the Sub-Himalayan area; *Angulitermes fletcheri* in the Peninsula; *Angulitermes husaini* in West Pakistan; *Angulitermes paanensis* in Lower Burma; *Angulitermes obtusus* in the Peninsula and *Angulitermes resimus* in Upper Burma. *Homallotermes* is exclusively Oriental, with but two species. *Homallotermes foraminifer* is found in Malaya Peninsula and Borneo and *Homallotermes pilosus* occurs in South India. *Dicuspiditermes* is an Oriental genus found in Thailand, Sumatra, Malaya, Burma, India, Pakistan and Ceylon. We find in the Peninsula *Dicuspiditermes fletcheri, Dicuspiditermes fontanellus, Dicuspiditermes gravelyi, Dicuspiditermes incola, Dicuspiditermes parname*; Central India *Dicuspiditermes abbreviatus, Dicuspiditermes obtusus, Dicuspiditermes incola cornutella* and *Dicuspiditermes punjabensis;* in Ceylon *Dicuspiditermes hutsoni* and *Dicuspiditermes incola;* Burma *Dicuspiditermes laetus* and in West Pakistan *Dicuspiditermes incola. Pericapritermes*, from the Ethiopian, Oriental and Papuan regions, is represented in India by five species: *Pericapritermes assamensis* in Assam, *Pericapritermes ceylonicus* in Ceylon, *Pericapritermes dunensis* in the Sub-Himalayan tracts, *Pericapritermes latignathus durga* and *Pericapritermes tetraphilus* in the Chittagong Hills and Burma. *Procapritermes, Labiocapritermes* and *Microcapritermes* are wholly Oriental. The first named genus occurs in Borneo, China, Taiwan, Java, Malaya, Sumatra, Thailand and India. In India the three species are *Procapritermes fontanellus* from South India, *Procapritermes goanicus* from Goa (also thus South India) and *Procapritermes tikadari* from Assam. *Labiocapritermes* is a monotypic genus from Kerala. *Microcapritermes* is known from Borneo, Thailand, and Burma and is at present unknown in India and Ceylon.

The subfamily Macrotermitinae includes the fungus-growing termites of the Ethiopian, Malagasy and Oriental Regions. Of the twelve genera known at present, seven are endemic in Tropical Africa, including the most primitive *Acanthotermes* and *Pseudacanthotermes*. Two genera *Hypotermes* and *Euscaiotermes* are Oriental. *Macrotermes, Odontotermes* and *Microtermes* occur both in the Ethiopian and Oriental Regions. *Odontotermes* has given rise to *Hypotermes* and *Euscaiotermes* in the Oriental Region. Of about forty-five species of *Macrotermes*, only ten occur within the Oriental and five in the India, viz. *Macrotermes annandalei* in Lower Burma and also Thailand and China; *Macrotermes estherae* from the Peninsula and Ceylon; *Macrotermes gilvus* and *Macrotermes serrulatus* from Burma and *Macrotermes serrulatus hopini* from Assam.

Odontotermes (Fig. 77) represents the most dominant genus of termites of India and includes over thirty-eight species and subspecies, the distribution of which is summarized in Table I. *Hypotermes* is represented by three species: *Hypotermes nongpriangi* occurs in Bangladesh, Assam and the

459

Table I. Distribution of the Indian species of Odontotermes

Sl. No.	Species	Andaman Islands	Nicobar Islands	Ceylon	Peninsular India Zone below 20°N Latitude	Central India Zone	Arid Zone	Eastern Zone (Including Bangladesh)	Himalayan Region	W. India Zone (including Pakistan)	Burma	Rest of Oriental Region	Elsewhere
1	2	3	4	5	6	7	8	9	10	11	12	13	14
1.	Odontotermes almorensis	—	—	—	—	—	—	—	+	—	—	—	—
2.	Odontotermes anamallensis	—	—	—	+	—	—	—	—	—	—	—	—
3.	Odontotermes assmuthi	—	—	—	+	—	—	—	—	—	—	—	—
4.	Odontotermes bellahunisensis bellahunisensis	—	—	+	—	—	+	—	—	—	—	—	—
5.	Odontotermes bellahunisensis guptai	—	—	—	—	—	—	+	—	+	—	—	—
6.	Odontotermes bhagvati	—	—	—	—	+	—	—	—	—	—	—	—
7.	Odontotermes brunneus brunneus	—	—	—	—	+	+	—	—	+	—	—	—
8.	Odontotermes brunneus kushwahai	—	—	+	—	—	—	—	—	—	—	—	—
9.	Odontotermes ceylonicus	—	—	+	—	—	—	+	—	—	—	—	—
10.	Odontotermes dehraduni	—	—	—	—	+	—	—	+	—	—	—	—
11.	Odontotermes distans	—	—	—	+	—	—	+	+	—	—	—	—
12.	Odontotermes escherichi	—	—	+	+	+	+	—	—	—	—	—	—
13.	Odontotermes feae	—	—	—	+	—	—	+	+	—	—	—	—
14.	Odontotermes feaeoides	—	—	—	+	—	—	—	—	+	+	+	—
15.	Odontotermes flavomaculatus	—	—	—	+	—	—	—	—	—	—	—	—
16.	Odontotermes formosanus	—	—	—	—	—	—	+	—	—	+	+	—
17.	Odontotermes gravelyi	—	—	—	—	—	—	+	—	—	—	—	—
18.	Odontotermes giriensis	—	—	—	+	—	—	—	—	—	—	—	—
19.	Odontotermes gurdaspurensis	—	—	—	—	—	—	—	—	+	—	—	—
20.	Odontotermes hainanensis	—	—	—	—	—	—	—	—	—	+	+	—
21.	Odontotermes horai	—	—	—	—	—	—	+	—	—	—	—	—

Table I (continued)

Sl. No.	Species	Andaman Islands	Nicobar Islands	Ceylon	Peninsular India Zone below 20°N Latitude	Central India Zone	Arid Zone	Eastern Zone (Including Bangladesh)	Himalayan Region	W. India Zone (including Pakistan)	Burma	Rest of Oriental Region	Elsewhere
1	2	3	4	5	6	7	8	9	10	11	12	13	14
22.	*Odontotermes horni*	—	—	+	+	—	—	—	+	—	—	—	+
23.	*Odontotermes horni* var. *hutsoni*	—	—	+	—	—	—	—	—	—	—	—	—
24.	*Odontotermes horni* var. *minor*	—	—	+	—	—	—	—	—	—	—	—	—
25.	*Odontotermes kapuri*	—	—	—	—	—	—	+	—	—	—	—	—
26.	*Odontotermes koenigi*	—	—	+	—	—	—	—	—	—	—	—	—
27.	*Odontotermes kulkarni*	+	+	+	—	—	—	—	—	—	—	—	—
28.	*Odontotermes latigula*	—	—	—	—	—	—	—	—	—	—	—	—
29.	*Odontotermes lokanandi*	—	—	—	+	—	—	—	+	+	—	—	—
30.	*Odontotermes malabaricus*	—	—	—	+	—	—	—	+	+	—	—	—
31.	*Odontotermes meturensis*	—	—	—	+	+	—	—	—	—	—	—	—
32.	*Odontotermes mathadi*	—	—	—	+	—	—	—	+	—	—	—	—
33.	*Odontotermes microdentatus*	—	—	—	—	+	—	+	—	—	—	—	—
34.	*Odontotermes mirganjensis*	—	—	—	+	+	+	+	+	+	—	—	—
35.	*Odontotermes obesus*	—	—	—	—	—	—	+	—	+	+	—	—
36.	*Odontotermes oblongatus*	—	—	—	+	+	+	—	+	+	—	—	—
37.	*Odontotermes paralatigula*	—	—	—	—	—	—	+	—	+	+	—	—
38.	*Odontotermes parvidens*	—	—	—	+	—	+	—	—	+	—	—	—
39.	*Odontotermes preliminaris*	—	—	+	—	—	+	—	+	—	—	—	—
40.	*Odontotermes redemanni*	—	+	+	+	—	—	—	—	—	—	—	—
41.	*Odontotermes taprobanes*	—	+	+	—	—	—	—	—	—	—	—	—
42.	*Odontotermes wallonensis*	—	—	+	+	—	—	—	+	—	—	—	—

461

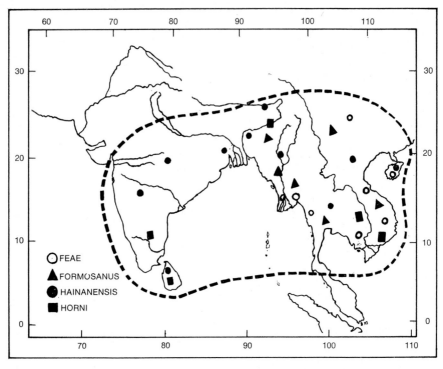

Fig. 77. Map showing the geographical distribution of *Odontotermes feae, O. formosanus, O. hainanensis* and *O. horni.*

Northeast Frontier Agency; *Hypotermes obscuriceps* occurs in North Bengal and Assam and *Hypotermes xenotermitis* in Assam, Bangladesh and Burma and Thailand. *Euscaiotermes* is endemic, with a single species from the Barkuda Island in the Chilka Lake (Orissa). *Microtermes* is perhaps the most highly specialized genus of the subfamily and includes all subterranean termites, with about nine species in India. The most widely distributed species is *Microtermes obesi* (Fig. 78); *Microtermes obesi curvignathus* occurs throughout India and Pakistan (excluding the Andaman and Nicobar Islands), Ceylon, Burma, Thailand, Indo-China and Cambodia. *Microtermes globicola* is widely distributed in the Peninsula and Ceylon. *Microtermes imphalensis* occurs in Manipur (Assam). *Microtermes incertoides* occurs in West Pakistan, Peninsula, Indo-China (BATHELLIER, 1927; ROONWAL, 1970). *Microtermes macronotus* is endemic in Ceylon and *Microtermes mycophagus* occurs in the arid and semi-arid northwestern India and West Pakistan. *Microtermes pakistanicus* is known from Assam, Tripura, Bangladesh, Burma and Malaya. *Microtermes sindensis* is confined to Sind and *Microtermes unicolor* occurs in the Sub-Himalyan tracts (Dehra Dun).

462

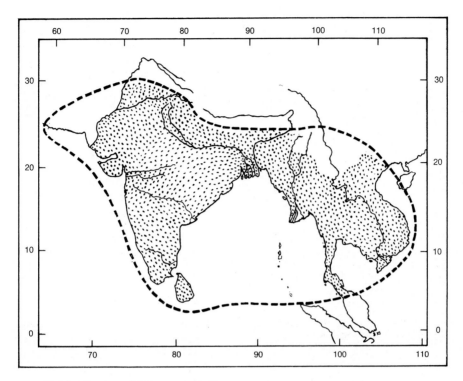

Fig. 78. Distribution of *Microtermes obesi.*

The Nasutitermitinae appear to have evolved from an extinct stock, from which also arose the Macrotermitinae (AHMED, 1950; SEN-SARMA, 1968). The most primitive members of the subfamily, with rudimentary nasute and developed mandibles in soldiers, are found in the Neotropical Region. The most primitive genera with vestigial soldier mandibles and well developed nasute '*Hirtitermes* and *Longipeditermes*' are Oriental. Of about sixty genera from the world, only eleven (of which six are endemic) occur in India: *Nasutitermes, Ampoulitermes, Indograllatotermes, Fletcheritermes, Bulbitermes, Ceylonitermes, Hospitalitermes, Emersonitermes, Ceylonitermellus, Aciculitermes* and *Trinervitermes.* The genus *Nasutitermes* is represented by about twenty-one species in India, Ceylon, Andaman and Nicobar Islands, Pakistan but not Burma (Fig. 79). The distribution of these species is summarized in Table II. The genus extends also to the Australian, Papuan, Oriental, Ethiopian, Malagasy and Neotropical Regions. *Ampoulitermes* (with one species), *Indograllatotermes* (two species) *Fletcheritermes* (one species) and *Emersonitermes* (one species) are endemic in the Peninsula. *Ceylonitermes* (one species) and *Ceylonitermellus* (two species) are endemic in Ceylon. *Aciculitermes* is represented by a single species in Burma. *Bulbitermes* is represented by a single species in Burma, Thailand

463

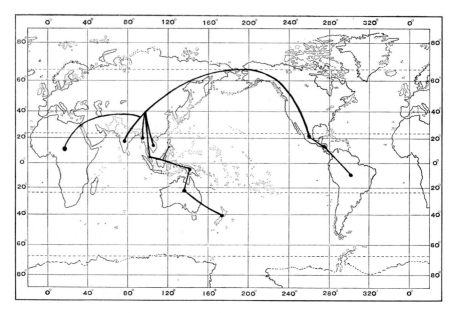

Fig. 79. Map showing the probable routes of dispersal and distribution of the genus *Nasutitermes*. (After SEN-SARMA).

and Indo-China (HARRIS, 1968); the record of *Bulbitermes singaporensis* from Ceylon by SNYDER (1949) needs further confirmation. *Hospitalitermes* is represented by six species: *Hospitalitermes ataramensis* occurs in Burma, Thailand, Indo-China and Cambodia; *Hospitalitermes blairi* occurs in the Andaman Islands; *Hospitalitermes burmanicus* in Burma; *Hospitalitermes jepsoni* in Burma, Cambodia and Thailand; *Hospitalitermes madrasi* in the Peninsula and *Hospitalitermes monoceros* is endemic in Ceylon.

The genus *Trinervitermes* occurs in savannahs of the Ethiopian and Oriental Regions. Of about sixty-four species known so far, only seven occur in the Oriental Region. We have in India five species; none is known from Burma. *Trinervitermes biformis* occurs in India, Pakistan and Ceylon. *Trinervitermes fletcheri* occurs in the Peninsula and *Trinervitermes indicus* in Central India (thus also in the Peninsula), *Trinervitermes nigrirostris* also from the Peninsula and *Trinervitermes rubidus* is endemic in Ceylon.

4.6. INDOTERMITIDAE

The family Indotermitidae (ROONWAL & SEN-SARMA, 1960) comprises the single genus *Indotermes*, with three species. *Indotermes isodentatus* occurs in South China; *Indotermes maymensis* in Burma and *Indotermes thailandis* in Thailand (Fig. 80).

Table II. Distribution of Indian species of the genus Nasutitermes

Sl. No.	Species	Andaman Islands	Nicobar Islands	Ceylon	Peninsular India Zone below 20°N Latitude	Central India Zone	Arid Zone	Eastern Zone (Including Bangladesh)	Himalayan Region	W. India (Zone including Pakistan)	Burma	Rest of Oriental Region	Elsewhere
1	2	3	4	5	6	7	8	9	10	11	12	13	14
1.	*Nasutitermes anamalaiensis*				+								
2.	*Nasutitermes beckeri*				+								
3.	*Nasutitermes brunneus*				+								
4.	*Nasutitermes ceylonicus*			+									
5.	*Nasutitermes cherraensis*				+			+					
6.	*Nasutitermes crassicornis*				+				+				
7.	*Nasutitermes dunensis*				+								
8.	*Nasutitermes emersoni*							+					
9.	*Nasutitermes fletcheri*				+								
10.	*Nasutitermes gardneri*				+			+					
11.	*Nasutitermes garoensis*							+					
12.	*Nasutitermes horni*			+				+					
13.	*Nasutitermes indicola*							+					
14.	*Nasutitermes jalpaigurensis*							+					
15.	*Nasutitermes kali*	+											
16.	*Nasutitermes krishna*			+									
17.	*Nasutitermes lacustris*		+										
18.	*Nasutitermes matangensiformis*												
19.	*Nasutitermes moratus*					+							
20.	*Nasutitermes oculatus*			+									
21.	*Nasutitermes thanensis*							+					

465

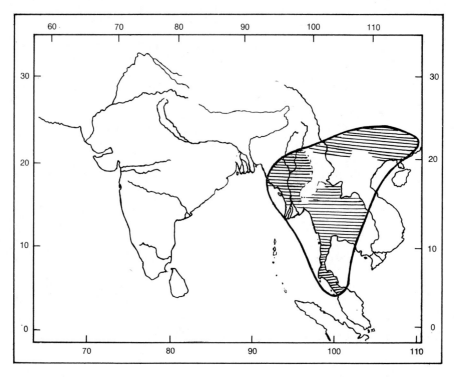

Fig. 80. Map of India and adjacent countries showing the geographical distribution of *Indotermes.*

REFERENCES

AGGARWAL, S. B. D. 1955. Control of sugarcane termites 1946–1953. *J. econ. ent.,* Menasha, 48: 533–537.

AGGARWAL, S. B. D. 1964. *Termites. Entomology in India:* 361–383. New Delhi. *(Ent. Soc. India).*

AGGARWAL, R. A. 1970. Problems of termites of sugarcane in India. *Meeting Termitologist India,* New Delhi; 1–12.

AHMED, M. 1950. The phylogeny of termite genera based on imago-worker mandibles. *Bull. Amer. Mus. Nat. Hist.,* 95: 37–86.

AHMED, M. 1955. Termites of West Pakistan. *Biologia,* Lahore, 1: 202–264.

AHMED, M. 1965. Termites (Isoptera) of Thailand. *Bull. Amer. Mus. Nat. Hist.,* 131: 1–113.

ALIBERT, J. 1959. Les échanges trophallactiques chez le Termite a cou jaune *Calotermes flavicollis* (Fabr.) études à l'aide du phosphore radioactif. *C.R. Acad. Sci.* Paris, 248: 1040–1042.

ALSTON, R. A. 1947. A fungus parasite on *Coptotermes curvignathus* Holmg. *Nature,* 160: 120.

ANNANDALE, N. 1923. The fauna of an Island in the Chilka lake. The habits of the termites of Barkuda Island. *Rec. Indian Mus.* Calcutta, 25: 233–252.

ANNANDALE, N. 1924. Termite mounds. *J. Bombay Nat. Hist. Soc.*, 30: 25–35.

ASSMUTH, J. 1915. Indian Wood destroying white ants (Second contribution). *J. Bombay nat. Hist. Soc.* 23: 690–694.

AYYAR, T. V. R. 1940. *Hand book of entomology for South India.* XVIII + 528 (Supdt. Govt. Press, Madras).

BAKSHI, B. K. 1951. Fungi in the nest of *Odontotermes obesus. Indian Phytopathol.* 4: 1–4.

BATHELLIER, J. 1927. Contribution à L'étude Systématique et biologique des termites de l'Indo-China. *Faune Colonies France* 1: 125–365.

BATRA, L. E. & S. W. T. BATRA, 1966. Fungus growing termites of tropical India and associated fungi. *J. Kansas. Entomol. Soc.* 39: 725–738.

BECKER, G. 1962a. Laboratoriumsprüfung von Holz und Holzschutzmitteln mit der sudasiatischen Termiten, *Heterotermes indicola* Wasmann. *Holz. Roh-Werkstoff,* 20: 476–486.

BECKER, G. 1962b. *Report to the Govt. of India on termite investigation. Report No. 1592* (Expanded Programme of Technical Assistance).

BECKER, G. 1962c. Beiträge zur Kenntnis der geographischen Verbreitung und wirtschaftlichen Bedeutung von Termiten in Indien. *Z. angew. Ent.,* 50: 143–165.

BECKER, G. 1965. Feuchtigkeitseinfluss auf Nahrungswahl und Verbrauch einiger Termiten-Arten. *Insectes Sociaux.,* 12: 151–184.

BECKER, G. 1969. Rearing of termites and testing methods used in the laboratory. In 'Biology of termites' Vol. I. (Ed. KRISHNA, K. & WEESNER, F. M.) 351–385 (Academic Press, New York).

BEESON, C. F. C. 1941. The ecology and control of the forest insects of India and the neighbouring countries. Dehra Dun: Vasant Press, ii + 1007.

BERKELEY, M. J. 1847. Decades of fungi – Ceylon fungi: Hooker's. *London J. Bot.* 6: 479–574.

BOSE, S. R. 1923. The fungi cultivated by the termites of Barkuda Island. *Rec. Indian Mus.* Calcutta., 25(2): 253–258.

BUGNION, E. & N. POPOFF, 1912. Anatomie de la reine et du roi termite *(Termes Redemanni obscuriceps* et *horni). Mem. Soc. Zool. France,* 25: 210–232.

BUGNION, E. 1914. Le *Termitogeton umblicatus* Hag. (Ceylon) (Corrodentia, Termitidae). *Ann. Soc. Ent. France* 83: 39–47.

BUGNION, E. 1915. La biologie des termites de Ceylon. *Bull. Mus. Hist. Nat.,* Paris, 20: 170–232.

BUXTON, P. A. 1932. Terrestrial insects and the humidity of the environment. *Biol. Rev.,* 7: 275–320.

CAMERON, M. 1932. The fauna of British India, including Ceylon and Burma. Staphylinidae Vol. III & IV(2) London: Sidgwick & Jackson.

CHATTERJEE, P. N. 1970. The rôle of termites in Indian Forestry. *Meet. Termitologist, New Delhi,* 1970-3. (in press).

CHEEMA, P. S., S. R. DAS, H. K. DAYAL, T. KOSHI, K. L. MAHESWARI, S. S. NIGAM & S. K. RANGANATHAN, 1962. Temperature and humidity in fungus garden of the mound building termite *Odontotermes obesus* (Rambur). *Termites in the Humid Tropics Proc. of the New Delhi Symp.* 1960 UNESCO, 145–149.

CHEO, C. C. 1942. A study of *Collybia albuminosa* (Berk) Petch. the termite growing fungus, in its connection with *Aegeritha duthiei* [*Termitosphaeria duthiei* (Berk. Ciferri)]. *Sci. Red. Acad. Sin.,* 1: 243–248.

CIFERRI, R. 1935. Sulla posizione sistematica dell'*Aegeritha duthiei*', fungo dell'ambrosia' dei termitii. *Att. Ist Botn. Univ. Paisa,* 6: 229–246.

COLLA, S. 1929. Su alcuni funghi parassiti delle Termiti. *Boll. Lab. Zool. Portici,* 22: 39–48.

DAS, S. R., K. L. MAHESHWARI, S. S. NIGAM, P. K. SHUKLA & R. N. TANDON, 1962. Micro-organisms from the fungus gardens of the termite *Odontotermes obesus* (Rambur). *Proc. Intern. Symp. on Termites in the Humid Tropics, New Delhi* 1960: 163–166.

467

DOFLEIN, F. 1906. *Die pilzzüchtenden Termiten*. In Teubner (Ed. by) Ostasienfahrt Erlet. Naturf. in China, Japan. u. Ceylon, 454–473.

DUTT, N. 1962. Preliminary observations on the incidence of termites attacking jute. *Proc. Intern. Symp. on Termites in Humid Tropics, New Delhi*, 1960: 217–218.

EMERSON, A. E. 1955. Geographical origins and dispersions of termite genera. *Fieldiana (Zool.)* 37: 465–21.

ESCHERICH, K. 1909. Die pilzzüchtenden Termiten. *Biol. Cbl. Leipzig*, 21(1): 16–27. 1 pl.

ESCHERICH, K. 1911. *Termitenleben auf Ceylon*. Neue Studien zur Soziologie der Tiere zugleich ein Kapital Kalonialer Forstentomologie, 32+263. pp.

FEYTAUD, J. & R. DIENZEIDE, 1927. Sur un champignon parasite du *Reticulitermes lucifugus* Rossi. *Rev. Zool. Agric. Appl. Bordeaux*, 26(11): 161–163 (Also *C.R. Acad. Sci. Paris*, 185: 671–672).

FLETCHER, T. B. 1912. Termites or white ants. *Agric. J. India* 7(3): 219–239.

FLETCHER, T. B. 1914. Some South Indian insects. Madras: Govt. Press, pp. 565.

GAY, F. J. 1967. A world review of introduced species of termites. *Bull. Commonwealth Sci. and Ind. Res. Organ Melbourne, Australia*. 286: 1–88.

GEYER, J. W. 1951. A comparision between the temperatures in termite supplementary fungus garden and in the soil at equal depths. *J. Ent. Soc. A. Afr.* Pretoria, 14(1): 36–43.

GHIDINI, G. M. 1938. La presumibile funzione delle spugne legnose nei nidi dei Metatermitida. *Riv. Biol. Colon.*, Rome, 1: 261–267.

GHANAMUTHU, C. P. 1947. The occurrence of termites at Krusadai Island. *Curr. Sci.*, 16: 154–155.

GRASSE, P. P. 1937. Ecologie animale et microclimate. *Rev. Sci. Assoc. Trane Advance Sci.*, 16: 383–390.

GRASSE, P. P. 1944. Recherches sur la biologie des termites champignonnistes (Macrotermitinae). *Ann. Soc. Nat. Zool. Biol. Animale*, (II)6: 97–171.

GRASSE, P. P. & R. CHAUVIN, 1944. L'effect de groupe et la survie des neutres dans les sociétés d'Insectes. *Rev. Scien.*, 82: 461–464.

GREEN, E. E. 1906. Note (Flies (*Bengalia*) hunting winged termites at night). *Spolia Zeylanica*, 2: 220.

GUPTA, S. D. 1952. Ecological studies of termites. Part I. – Population of the mound building termites, *Odontotermes obesus* (Rambur). (Isoptera: Fam. Termitidae). *Proc. Nat. Inst. Sci. India.*, (B), 19(5): 697–704.

HARRIS, W. V. 1961. *Termites: their recognition and control*. London: Longmans-Green. pp. 187.

HARRIS, W. V. 1967. Beiträge zur Kenntnis der Fauna Afghanistans (Sammlergebnisse von O. JAKEZ 1963–64, D. POVOLNY 1965, D. POVOLNY & FR. TENORA 1966. J. SIMEK 1965–66) Isoptera. *Acta Musei Moraviae Sci. Nat.*, 52 (Suppl.): 211–215.

HARRIS, W. V. 1968. Isoptera from Vietnam, Combodia and Thailand. *Opuscula Entomol.*, 33: 143–154.

HEGH, E. 1922. *Les Termites* Partie Général Description, Distribution Géographique, Classification. Biologie, vie social, Alimentation Constructions Rapports avec le monde Extérieur. – 4+756: (L. Desmet Vertenc.).

HEIM, R. 1940. Les champignonnières des termites et les grands champignons d'Afrique tropicale. *Rev. Bot. Appl.*, 20: 121–127.

HEIM, R. Nouvelles reussites culturales sur les Termitomyces. *C.R. Acad. Sci. Paris*, 226: 1488–1491.

HENDEE, E. C. 1933. The association of the termites *Kalotermes minor. Reticulitermes hesperus* and *Zootermopsis angusticollis* with fungi. *Univ. California Publ. Zool.*, 39(5): 111–134, 1 fig.

HESSE, P. R. 1953. A chemical and physical study of the soils of termite mounds in Africa. Ph.D. thesis, Lond. Univ. (Referred by HARRIS. 1961).

HOON, R. C. 1962. The incidence of white ants (termites) in the region of the Hirakud Dam Project. *Proc. Internat. Symp. on Termites in the Humid Tropics, New Delhi* (UNESCO). 141–149.

HRDY, I. 1970. Laboratory colonies of *Microcerotermes cameroni* Snyder (Isoptera) *J. Indian Acad. Wood. Sci.*, Bangalore, 1(1).

HUSSAIN, M. A. 1935. Pests of wheat crop in India. *Proc. World's Grain Exhib. & Conf.*, 2: 562–564.

IMMS, A. D. 1919. On the structure and biology of *Archotermopsis* together with description of new species of intestinal Protozoa and general observations on the Isoptera. *Phils. Trans. Roy. Soc. London*, (B) 209: 75–180.

JEPSON, F. P. 1931. The termites which attack living plants in Ceylon. Rutherford's planters' notebook. (9th Ed.). 579–596.

JOHN, O. 1925. Termiten von Ceylon, der malayischen Halbinsel, Sumatra, Java und den Aru Inseln. *Treubia*, 6: 360–419.

JOACHIM, A. W. R. & S. KANDIAH, 1940. A comparison of soils from termite mound and adjacent land. *Trop. Agric.*, 95: 333–338.

KALSHOVEN, L. G. E. 1936. Onze Kennis van de Javaansche Termieten. *Handel Ned. Ind. Natuurw. Cong.*, 7: 427–435.

KAPUR, A. P. 1958. A report reviewing entomological problems in the humid tropical regions of South Asian. *Prob. Humid Trop. Reg.* UNESCO, Paris 63–85.

KEMNER, N. A. 1934. Systematische und biologische Studien über die Termiten Javas und Celebes. *Kgl. Svenska Vetenskaps Akad. Handl.*, (3) 13: 1–241.

KÖNIG, J. G. (1779) Naturgeschichte der sogenannte weissen Ameisen. *Beschr. Berlin. Ges. Naturf. Freunde*, 4: 1–28. (English translation & comments by T. B. FLETCHER in Proc. 4th Entomol. Meeting Pusa, 1921. 312–333, Calcutta 1921).

KRISHNA, K. 1968. Phylogeny and generic reclassification of the *Capritermes* complex (Isoptera: Termitidae: Termitinae). *Bull. Am. Mus. Nat. Hist.*, 138: 294–304.

KRISHNA, K. 1970. Taxonomy, Phylogeny and distribution of termites. In 'Biology of Termites. Vol. II New York & London: Academic Press, pp. 127–152.

KRISHNAMOORTY, C. & K. RAMASUBBIAH, 1962. Termites affecting cultivated crops in Andhra Pradesh and their control: retrospect and prospect. *Proc. Internat. Symposium on Termites in the Humid Tropics, New Delhi.* Oct. 4.12.1960. (UNESCO), 6: 243–245.

KUSHWAHA, K. S. 1964. A note on infestation of termites (Insecta: Isoptera) around Udaipur (Rajasthan) (India.). *Univ. Udaipur Res. Stud.*, 2(Spec. No.): 105–107.

LEFROY, H. M. & F. M. HOWLETT, 1909. Indian insect Life, pp. 115–121, 268–272, 277, 1 pl. 3 figs. Calcutta, London: W. Thacker & Co.

LUND, A. E. 1963. Subterranean termites and fungi. Theoretical interactions. *Pest Control* 31: 78.

LÜSCHER, M. 1951. Significance of fungus gardens in termite nests. *Nature*, London, 167: 34–35.

MATHUR, R. N. 1962. Enemies of termites (Whiteants). *Proc. international Symp. Termites in the Humid Tropics*, New Delhi. (UNESCO), 4: 137–139.

MATHUR, R. N. & P. K. SEN-SARMA, 1959. Notes on the habits and biology of Dehra Dun termites, Part I. *J. Timber Dryers & Preserv. Assoc. India*, 5(3): 3–9.

MATHUR, R. N. & P. K. SEN-SARMA, 1960. Notes on the habits and biology of Dehra Dun termites, Pt. II. *J. Timber Dryers & Preserv. Assoc. India*, 6(2): 23–27.

MATHUR, R. N. & P. K. SEN-SARMA, 1962. Notes on the habits and biology of Dehra Dun termites. Pt. III. *J. Timber Dryers & Preserv. Assoc. India*, 8(1): 1–8.

MUKERJI, D. & P. K. MITRA, 1949. Ecology of the mound building termites, *Odontotermes redemanni* (Washm.) in relation to measures of control. *Proc. zool. Soc. Bengal*, Calcutta, 2(1): 9–26.

MUKERJI, D. & S. RAYCHODHARI, 1943. Structure, function and origin of the exudate organs in the abdomen of the physogastric queen of the termite, *Termes redemanni* Wasm. *Indian J. Ent.*, 4(2): 173–199 (1942).

NIRULA, K. K., J. ANTONY & K. P. V. MENON, 1953. Some investigations on the control of termites. *Indian Cocoanut J.*, 7: 26–34.

PATEL, G. A. & H. K. PATEL, 1954. Seasonal incidence of termite injury in the Northern parts of the Bombay State. *Indian J. Ent.*, 15(4): 376–378.

469

Pence, R. J. 1956. The tolerance of drywood termite *Kalotermes minor* Hagen to desiccation. *J. econ. ent.*, 49(4): 553–559.

Pendleton, P. L. 1941. Some results of termite activity in Thailand soil. *Thai. Soc. Bull.*, 3: 29–53.

Petch, T. 1906. The fungi of certain termite nests. *Termes redemanni* Wasm. and *T. obscuriceps* Wasm. *Ann. Roy. Bot. Gardens, Peradeniya*, 3(2): 185–270.

Petch, T. 1913. Termites and fungi: a resumé. *Ann. Roy. Bot. Gardens, Peradeniya*, 5(5): 303–341.

Prashad, B. & P. K. Sen-Sarma, 1960. Revision of the termite genus *Hospitalitermes* Holmg. (Isoptera: Termitidae: Nasutitermitinae) from the Indian region. Monograph No. 10–29, 1–32; 9 figs.

Prashad, B. & P. K. 1960. Review of the genus *Trinervitermes* Holmgren from the Indian Region (Isoptera: Termitidae: Nasutitermitinae). *Indian For. Bull. (N.S.), Ent.*, Delhi, 248: 1–17.

Prashad, B., Thapa, R. S. & P. K. Sen-Sarma, 1967. Revision of Indian species of the genus *Microcerotermes* Silvestri (Isoptera: Termitidae: Amitermitinae). *Indian For. Bull.*, 246: 1–56.

Roonwal, M. L. 1949. Systematics, ecology and bionomics of mammals studied in connection with tsutsugamushi disease (scrub tymphus) in the Assam-Burma War Theatre during 1945. *Trans. nat. Inst. Sci. India*, 3: 67–122, 92–95; termites as food for rats).

Roonwal, M. L. 1954. Biology and ecology or Oriental termites (Isoptera). No. 2. On ecological adjustment in nature between two species of termites, *Coptotermes heimi* (Wasm.) and *Odontotermes parvidens* (Wasm.) in Madhya Pradesh, India. *J. Bombay nat. Hist. Soc.*, 52: 463–467.

Roonwal, M. L. 1970. Termites of the oriental region. In 'Biology of termites' Vol. II ed. K. Krishna & F. M. Weesner New York & London: Acad. Press. pp. 315–391.

Roonwal, M. L. & G. Bose, 1962. An African genus *Psammotermes* in Indian termite fauna with fuller description of *P. rajasthanicus* from Rajasthan, India. *Rec. Indian Mus.*, 58: 151–158.

Roonwal, M. L. & G. Bose, 1964. Termite fauna of Rajasthan, India. *Zoologia*, Stuttgart, 40: 1–58.

Roonwal, M. L. & G. Bose, 1970. Taxonomy and Zoogeography of the termite fauna of Andaman and Nicobar Islands, Indian Ocean. *Rec. Zool. Surv. India*, 62 (1964): 2–170, 4 pls.

Roonwal, M. L. & O. B. Chhotani, 1960. Soldier caste found in the termite genus *Speculitermes. Sci. & Cult.*, Calcutta, 26: 143–144.

Roonwal, M. L. & O. B. Chhotani, 1962a. Termite fauna of Assam Region Eastern India. *Proc. Nat. Inst. Sci. India*, (B) 28: 281–406.

Roonwal, M. L. & O. B. Chhotani, 1962b. Indian species of termites genus *Coptotermes. Ent. Mong. Indian ounc. Agric. Res.*, No. 2 New Delhi.

Roonwal, M. L. & O. B. Chhotani, 1963. Discovery of termites genus *Procryptotermes* (Isoptera-Kaloter mitidae) from Indo-Malayan region with a new species from India. *Biol. Zbl.* 82(3): 265–273.

Roonwal, M. L. & O. B. Chhotani, 1966a. Soldier and other castes in termite genus *Speculitermes* and phylogeny of *Anoplotermes – Speculitermes* complex. *Biol. Zbl.* 85: 183–210.

Roonwal, M. L. & O. B. Chhotani, 1966b. Revision of the termite genus *Eurytermes* (Termitidae: Amitermitinae). *Proc. Nat. Inst. Sci. India*, (B) 31: 81–113.

Roonwal, M. L. & P. K. Maiti, 1966. Termites from Indonesia including West Irian. *Treubia*, 27: 63–140.

Roonwal, M. L. & P. K. Sen-Sarma, 1955. Biology and ecology of Oriental termites (Isoptera). No. 3 Some observations on *Neotermes gardeneri* (Snyder) (Family: Kalotermitidae). *J. Bombay nat. Hist. Soc.*, 53: 234–239.

470

Roonwal, M. L. & P. K. Sen-Sarma, 1960. Contribution to the systematics of Oriental termites. *Entomol. Monograph* No. 2 (I.C.A.R. New Delhi) 1–407.

Roonwal, M. L., O. B. Chhotani & G. Bose, 1962. Some recent zoogeographical findings in Indian termites. *Proc. Internat. Symposium on Termites in the Humid Tropics*, New Delhi, Oct. 4–12, 1960 (UNESCO) 1: 51–54. 3 figs.

Sands, W. A. 1960. The initiations of fungus comb construction in laboratory colonies of *Ancistrotermes guineensis* (Silvestri). *Insectes Sociaux*, 7: 251–259.

Sands, W. A. 1969. The association of termites and fungi. In 'Biology of termites' Ed. Krishna, K. & Weesner, F. G. Vol. I New York: Academic Press. pp. 495–524.

Sannasi, A. 1969a. Apparent infections of queens and drones of the mound building termite Odontotermes obesus by *Aspergillus flavus J. Invertebrate Path.*, 10: 434–535.

Sannasi, A. 1969b. Studies of an insect mycosis. I. Histopathology of the integument of the infected queen of the mound building termites *Odontotermes obesus. J. Invert. Path.*, 13: 4–10.

Sannasi, A. 1969c. Studies of an insect mycosis. II. Biochemica lchanges in the blood of the queen of the mound building termite *Odontotermes obesus* accompanying fungal infection. *J. Invert. Path.*, 13: 11–14.

Seevers, C. H. 1957. A monograph on the termitophilous Staphylinidae (Coleoptera). *Fieldiana. Zool.* 40: 1–334.

Sellschop, J. P. F. 1965. Field observations on conditions conducive to the contamination of ground nuts with mould *Aspergillus flavus. S. African Med. J. (Suppl. S. African J. Natur.)*, 774–776.

Sen-Sarma, P. K. 1968. Phylogenetic relationship of the termite genera of the subfamily Nasutitermitinae (Isoptera: Termitidae). *Oriental Insects*, 2: 1–34.

Sen-Sarma, P. K. 1969. Effect of relative humidity on the longevity of starving workers and soldiers of *Heterotermes indicola* (Wasm.) *VI Cong. Intern. Union Studies Social Insects*, Bern. 263–265.

Sen-Sarma, P. K. & P. N. Chatterjee, 1965. Studies on the natural resistance of timbers to termites. IV. Qualitative and quantitative estimations of resistance of sixteen species of Indian wood against *Neotermes bosei* Snyder based on laboratory tests. *Indian Forester*, 91: 805–813.

Sen-Sarma, P. K. & P. N. Chatterjee, 1966a. Humidity behaviour of termites. I. Effect of relative humidity on the longevity of workers of *Microcerotermes beesoni* Snyder (Insecta: Isoptera: Termitidae) under starvation condition. *Indian For. Bull., (Ent.)* 55: 1–6.

Sen-Sarma, P. K. & P. N. Chatterjee, 1966b. The effect of preconditioning on the humidity reactions of workers of *Microcerotermes beesoni* Snyder (Isoptera: Termitidae). *Insectes Sociaux*, 13: 267–276.

Sen-Sarma, P. K. & P. N. Chatterjee, 1968. Studies on the natural resistance of timbers to termite attack. V. Laboratory evaluation of the resistance of three species of Indian wood to *Microcerotermes beesoni* Snyder (Termitidae: Amitermitinae). *Indian Forester*, 94: 694–704 + 3 flag leaves.

Sen-Sarma, P. K. & B. K. Gupta, 1968. Responses of termites (Insecta: Isoptera) to gravity. *Curr. Sci.*, 37: 501–502.

Sen-Sarma, P. K. & W. Kloft, 1965. Trophallaxis in pseudoworkers of *Kalotermes flavicollis* (Fabr.) (Insecta: Isoptera: Kalotermitidae) using Radioactive. 131 *Proc. Zool. Soc.*, Calcutta, 18: 41–46.

Sen-Sarma, P. K. & R. N. Mathur, 1961. *Trinervitermes biformis* (Wasmann). a mound building termite in South India. *Indian Forester*, 87: 252, 1 pl.

Sen-Sarma, P. K. & S. C. Mishra, 1969. Seasonal variations of nest population in *Microcerotermes beeson* Snyder. *Proc. nat. Inst. Sci. India*, (B) 35: 361–367.

Snyder, T. E. 1933. New termites from India. *Proc. U.S. Nat. Mus.*, 82: 1–15.

Snyder, T. E. 1949. Catalog of the termites (Isoptera) of the world. *Smithsonian Misc. Coll.*, 112: 1–490.

471

THAXTER, R. 1920. Second note on certain peculiar fungus parasites of living insects. *Bot. Gaz.*, 69(1): 3–9, pl. 2. figs. 13–17.

TRIPATHI, R. L. 1970. Termite problem in jute and other allied fibre crops. *Meet. Termitologists of India*, 1970, 2. (in press).

WEIDNER, H. 1960. Die Termiten von Afganistan. Iran and Irak (Isoptera). (Contribution al 'entude de la fauna d'Afganistan 29). *Abhandl. ver. naturw. ver. Hamburg*, (NF) 4: 43–70.

XV. THE BIOGEOGRAPHY OF INDIAN
BUTTERFLIES

by

J. D. HOLLOWAY

1. Introduction

In HOLLOWAY (1969) a new application of cluster analysis was described for the study of the geographical distribution of taxa. The method enabled the derivation of *faunal centres* by classifying genera numerically, both according to their geographical distribution and according to where their component species were concentrated. This method was exemplified by analysis of a sample of the butterfly genera to be found in India. The results were related to what is known of the continental drift of the part of India, south of the Plain of the Ganges. This survey will be reviewed here and the discussion of the results will be expanded. An analysis of the distribution of Indian butterfly species (mainly from the genera used in the survey of centres) is also presented, using the *faunal element* method described in HOLLOWAY & JARDINE (1968). The distribution of species from the various generic centres amongst these elements and the ecological preferences of the species in each element will be described and discussed.

The data used in both these surveys were extracted from revisions of genera or descriptions of the fauna of the Indian region in the literature. The generic revisions were based mainly on material that had been accumulated in museums at a time when detailed locality and habitat recording was rare. Hence the data that can be gleaned from such works can only be handled on a rather broad geographical basis, such geography being political as much as ecological or physical. Thus the butterfly species faunal elements derived in this survey might be expected to contain forms with a variety of habitat preferences. In fact it turns out that this is of rare occurrence and most of the elements derived are also of ecological significance. It is to be hoped that future collectors will employ a more ecological approach and the value of this is demonstrated in HOLLOWAY (1970) where the biogeography of the moth fauna of Mt. Kinabalu is analysed from a large series of samples collected with light traps on an altitude transect. A series of such transects made along the length of the Himalayan Ranges and in a series of localities in South India and neighbouring countries would be of immense value and their analysis would make this survey pale into insignificance.

2. The faunal centres of Indian butterfly genera

Data on 123 genera of butterfly were used to derive the faunal centres, consisting of genera with both coincident geographic distribution and coincident geographical concentrations of species. Where possible, the data for these genera were taken from recent revisions, viz. EVANS (1937, 1949, 1957), GABRIEL (1943), HIGGINS (1941), HOWARTH (1957), TALBOT (1928–1937), and ZEUNER (1943). The remaining distributions were taken from SEITZ (1909–1927). The centres derived are illustrated in figures 81–84. The numerals in each province indicate the average of the percentage of the total number of species of each genus found in that province, i.e. the hypothetical 'typical' genus of each centre is illustrated. The provinces are picked out in lines of varying thickness and brokenness in order to display each centre more graphically. They indicate major changes of species-concentration rather than denoting fixed percentage intervals.

The provinces have been delimited as realistically as possible, given the crudity of the data. Where possible, physiogeographical and ecological boundaries have been used, for example the African provinces follow CARCASSON's (1964) analysis of the African butterfly fauna. The provinces used, apart from the inclusion of African provinces, differ from those used for the study of species distributions (listed later) in the following respects: provinces 1 and 3 (of the species study – Turkey and the Middle East) are taken together; provinces 6 and 7 are arranged so that Turkestan is included with the Pamirs, with the Caucasus treated separately; 9 and 10 (Baluchistan and Sind) are taken together; Nepal, the United Provinces (now Uttar Pradesh) and Bengal 12, 13 & 19 are treated as one North Indian province; 15 and 16 (Western Ghats and Deccan) are treated together as South India; Sikkim and Assam 18 & 20 are combined; 21 and 24 (Szechwan and Yunnan) form a West Chinese province; Burma south of Assam 27 & 28 is taken as a single province; Japan is treated separately from Amur; the lowlands of Cambodia are split from Thailand 29 and included with the more mountainous Annam 25 and Thailand is included with 30. These differences in treatment in the two studies become apparent on comparison of relevant figures (e.g. 81 and 89).

The centres derived in HOLLOWAY (1969) were as follows:

1. (Fig. 81). Seventy-two genera centred in the Oriental Region with subcentres as follows:

1a. (Fig. 81). Nine genera centred on Sundaland (*Elymnias, Cirrochroa + Smerina, Narathura, Cyrestis + Chersonesia, Ancistroides, Plastingia, Euploea, Troides, Oriens*).

1b. (Fig. 82). Seventeen genera centred broadly from Sundaland to Assam (*Horaga, Appias, Notocrypta, Hasora, Potanthus, Caltoris, Coladenia, Bibasis, Flos, Matapa, Choaspes, Eriboea, Neptis + Rahinda, Pantoporia*

474

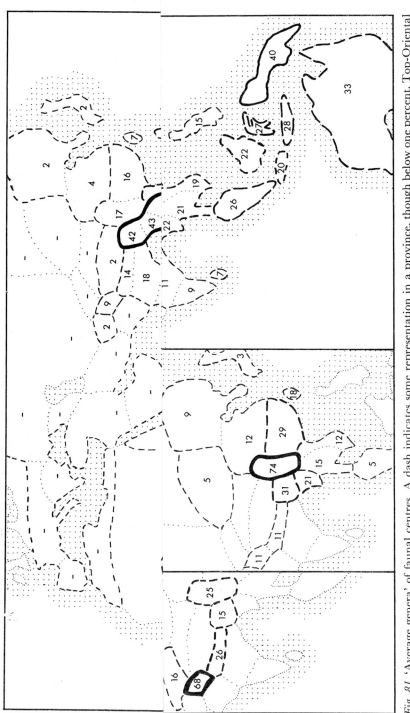

Fig. 81. 'Average genera' of faunal centres. A dash indicates some representation in a province, though below one percent. Top-Oriental centres, 1; centre left – Centre 1a; bottom left – Centre 1c; centre right – Centre 6; bottom centre – Centre 1e; Bottom right – Centre 1d.

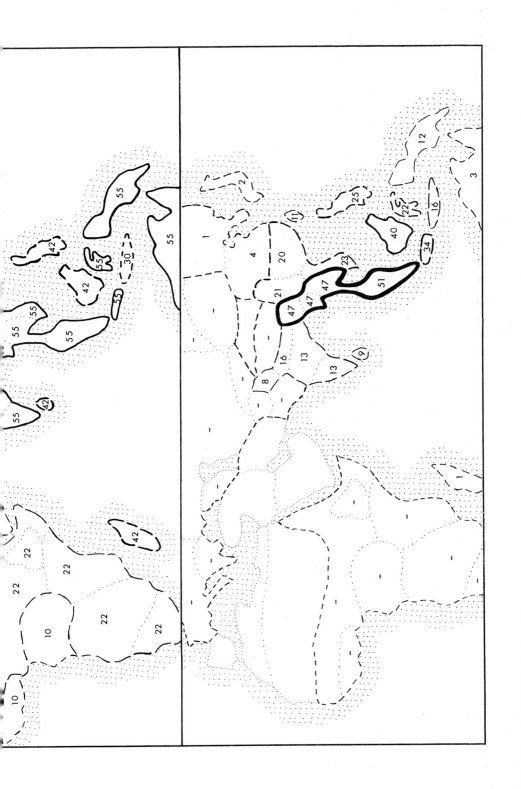

WILTSHIRE	GROSS	HOLLOWAY
Palaearctic Pacific	{ Himalayan-Chinese	1e, 9
	{ Chino-Japanese	5
Angaran	Mongolian	2b
Caucaso-Iranian	Turkestan	2a
Atlantic-Mediterranean	Mediterranean	4

and deserts broke it up into several separate centres, illustrated in the
survey. These centres, and their correspondence to those recognized by
GROSS (1961) and WILTSHIRE (1962), are listed in Table I (from
HOLLOWAY, (1969)).

These Palaearctic centres are of varying ecology, but their common
past history may be illustrated by butterfly groups, such as the papilionid
subfamily Parnassiinae (as defined by MUNROE & EHRLICH, (1960))
which have development in each centre at the generic level but form a
group at a higher taxonomic level with a distribution that might be
representative of a widespread, ancestral, temperate taxon. The Parnassiinae are thought by MUNROE & EHRLICH to be an early offshoot
from the Papilionidae (ZEUNER, 1961) described fossil Parnassiinae
from the early Tertiary of Europe. The subfamily is taxonomically
isolated and can be assumed fairly strongly to be monophyletic. It is
made up to the following genera (the centre of the numerical survey to
which they are referable is indicated by the appropriate number):
Thais 4; *Doritis* 4; *Hypermnestra* 2a or 6; *Parnassius* 2a; *Armandia* 1e;
Serecinus 9; *Luehdorfia* 5. Their distributions are mapped in Fig. 85. Several
of these genera impinge on the Indian Region and their species are
included in the faunal element survey. The first three genera are characteristic of Mediterranean and semi-arid habitats, *Parnassius* is almost
entirely restricted to high altitude and latitude, commonly above the
treeline, and the other three genera are subtropical and montane.

The area to the north of the temperate centres is still largely unsaturated,
an effect augmented by the recent glaciations, but it is largely colonised
by forms from the more mountainous temperate centres (e.g. by *Parnassius* and *Colias* from Tibet and the mountains of Turkestan) and by
forms from the less heavily glaciated Mongolian 2b and Nearctic centres
(MANI, 1962, 1968). Interchange between the Mongolian and Nearctic
centres via the Bering Straits (ROSS, 1953) has probably been more or
less continuous throughout the Tertiary and some genera in the survey
(e.g. *Oeneis* of 2b) are perhaps better referred to the Nearctic.

Consideration of these events is important to the study of the butterflies
of the Indian region as the majority of the centres, both temperate and
tropical, surround India to the north. Their origins have a direct bearing

on the distribution of their component species amongst the faunal elements to be described in the next section.

Recent discoveries in the investigation of sea-floor spreading* have demonstrated that, during the Tertiary, the southern part of India drifted steadily northwards through the Indian Ocean, at first rapidly (14 cm per year) and then, in the Oligocene, the rate slowed to its present magnitude (8 cm per year). The evidence for this, together with corroboration from palaeomagnetic evidence, is reviewed in HOLLOWAY (1969). During this drift phase the greater part of West Pakistan and the Ganges Plain up to and including the Siwalik foothills of the Himalaya were submerged and accumulating deposits of sediment under shallow marine conditions. These regions did not become land to any great extent until the Pliocene (WADIA, 1953). The combination of these facts indicates that southern India was isolated from mainland Asia by an ocean gap of approximately 2000 km at the beginning of the Miocene and 700 km at the beginning of the Pliocene.

If continental drift had not occurred, the stratigraphic evidence would indicate that South India existed as a large island closely offshore from mainland Asia (with a 200–400 km water gap) until the late Pliocene. Such an island would probably have borne a similar biogeographic relationship to mainland Asia as Madagascar does to Africa (PAULIAN 1958), as the relative ecological conditions are not all that different. There would have been a high degree of independent development in the flora and fauna of such an island (both initially derived from the mainland), with the radiation of endemic genera and species. Interchange with mainland Asia after unification in the Pliocene might be expected to be approximately equal, bearing in mind the relative areas involved (DARLINGTON, 1957; MACARTHUR & WILSON, 1967), perhaps comparable to the exchange of tropical biotas between New Guinea and N.E. Australia (GOOD, 1960).

The drift evidence suggests a much greater isolation of the South Indian landmass up to the Pliocene. Hence the development of a rich, endemic biota of Asian derivation would probably not commence until the early Pliocene. Evidence of such development would be swamped by further mass invasion by mainland forms, following unification of the two landmasses. During the more isolated phase previous to the Pliocene South India would probably only have received those forms characteristic of oceanic islands, ephemeral species of subclimax vegetation that are highly dispersive (CORBET, 1941). Such forms have been known to undergo adaptive radiation on isolated islands, but are not good competitors when faced with forms from evolutionary centres in large areas (CARLQUIST, 1966). Therefore the survival of such a biota, if it occurred in South India, would be unlikely after unification with the mainland.

* Since going to press MCKENZIE & SCLATER (1971) have published a precise account of the plate tectonics of the Indian Ocean.

However, the relationships between the butterflies of India and the Malagasy Region discussed below are of great interest in this context.

The pattern revealed by the faunal centre survey is far more concordant with the situation predicted for the drift alternative. Specific endemism is low in South India (see next section) and there is no generic endemism or evidence of generic centres. The hesperid genus *Thoressa*, clustering at a high level in the 'Oriental' centre, has three species endemic to the Western Ghats and one to Ceylon, but has its main centre in Assam. It may well be evidence of minor Pliocene radiation before unification but needs to be investigated further. Otherwise all endemic species are closely related to southeast Asian forms, speciation perhaps being due to ecological isolation of wet-forest forms as discussed for species element 1(f) in the next section. The faunal centre survey presents a picture of relatively recent invasion of South India from several centres in mainland Asia and from one in Africa. The ecological segregation of these centres will be discussed in the next section.

An outstanding question is whether India had any considerable biota during the drift phase. A well developed podocarp flora, allied to that of Africa and other fragments of Gondwanaland, occurred in South India in the Jurassic (COUPER, 1960). Post-Jurassic sediments in South India are few and mainly coastal but, at present, have not yielded any clue as to the nature of the land biota*. Indications that any elements of the present flora and fauna of India are remnants of such a biota are slight. In HOLLOWAY (1969) it was stated that most other faunal and floral groups show comparable distribution patterns to the butterflies. The uropeltine burrowing snakes are a large group, restricted to South India and Ceylon, but are generally related to Asian forms. It is possible that these snakes could have evolved in India, following colonisation before unification with Asia, but the mode of dispersal to the drifting landmass would then be problematic. The species of brevicipitid frog of a subfamily found only in India and central Africa (DARLINGTON, 1957) could be construed as a relict distribution from the predrift period, but this is not the only possible explanation as will be seen below. There are several endemic genera of plants, found in the southern part of India, largely of Asian affinity (HOOKER, 1906; GOOD, 1953). It was suggested in HOLLOWAY (1969) that, when the tropical regions of the Asian mainland were contracting during the Tertiary climatic deterioration, their biota would become supersaturated and taxa would be extinguished in adjustment. AXELROD (1960) mentioned such extinction as occurring in the flora, especially of the American element. The increase in area available to such a biota presented by the arrival of the South Indian landmass and by the uplift of land in northern India could well have

---

* ROBINSON (1970) reviewed the literature on Cretaceous land dinosaurs from southern India. These are indicated to have affinities with Madagascan and Patagonian forms.

reversed the process of contraction to a certain extent and conserved some of the taxa in the process of extinction. The endemic plant genera in India could well have arisen through such a process, a hypothesis supported by the fact that the majority of these endemic genera are rare, locally distributed and monotypic, characteristics which often identify taxa in the process of extinction (FISHER, 1930; PRESTON, 1962).

There appears, therefore, to be little, if any, evidence of the survival of elements of the predrift biota in South India. Links between South India and Africa could as well be explained by disjunction of the ranges of tropical taxa widespread between Asia and Africa in the early Tertiary when, as suggested above, tropical habitats in the two areas may have exchanged taxa fairly freely; this disjunction would then be followed by the 'conservation' of the Oriental representative in South India as suggested in the previous paragraph. This would apply only to elements of known poor dispersive ability. Those with good means of dispersal could have developed such patterns of distribution more recently, as indicated in the discussion of Indian Ocean relations below.

What, then, happened to the podocarp flora and its associated fauna*? Three points are worth considering. Firstly, any biota existing in South India during the late Cretaceous and the early Tertiary would probably have been 'south-temperate' in nature; it is possible that the passage of this landmass across the equatorial belt was too rapid for the adaptation of such a biota to tropical conditions, resulting largely in its extinction or restriction to a very small area of suitable habitat**. Secondly, though most of South India was land during the drift period, there was a period of intense vulcanism and laval outpouring that affected the western part of the mass-area of the Deccan Traps (WADIA, 1953). This may have rendered precarious the existence of a biota or again reduced the area available to it. Both these factors would have reduced the area available

---

* Such a fauna probably did not include butterflies as these did not radiate in Africa or Asia until the drift phase had commenced (HOLLOWAY & JARDINE, 1968). The main radiation of the African avifauna is dated at the Miocene (MOREAU, 1952). Therefore the African rain-forest element in the Indian avifauna (RIPLEY, 1955, 1961) was probably derived by dispersal across the Indian Ocean later than this and would, in any case, depend upon the development of rainforest in southern India.
** Very recent (BRIDEN, SMITH & DREWRY, in press and at Geological Society/ Palaeontological Association/Systematics Association Symposium at Cambridge, December, 1971) palaeogeographic reconstructions of the Tethyan region, based on sea-floor spreading evidence, place India joined to Madagascar and Africa (adjacent to Somalia, Kenya and Tanzania) about 30°S with Antarctica joined to the south. Mesozoic and Tertiary marine faunas of southern India are generally Tethyan-tropical in nature (A. A. MEYERHOFF, in litt.) indicating that India enjoyed a tropical or subtropical climate during that period. Hence any stress of adaptation to climatic changes would not be large. Furthermore, N. F. HUGHES (in press and at the above Symposium) suggested that the 'south-temperate' flora existing today is limited to high latitudes by competition from the modern angiosperm flora rather than by climatic factors.

to the biota, if not resulting in its extinction. They would also tend to place a premium on adaptive evolution rather than on evolution for greater general efficiency (characteristic of mature, stable environments), reducing the competitive ability of the biota (CARLQUIST, 1966). Thus, thirdly, the invasive Asian elements in the Pliocene, coming from a large area of development, relatively stable and ecologically mature, may have been of such greater efficiency that they eliminated the endemic biota in competition.

It has been suggested above that, apart from any predrift biota, South India might have developed a biota during the drift phase derived from highly dispersive, ephemeral taxa. This biota would also be at a competitive disadvantage in face of the Pliocene invasion from Asia. There are indications that some butterfly taxa may have spread to Africa via India and Madagascar from the Oriental rain-forest centre. The divergence in Africa and Madagascar of some of these taxa would be consistent with their spread there having occurred before the main invasion of South India by the Oriental taxa. The islands in the Indian Ocean today have a fauna of dispersive, ephemeral species, derived from both Asia and Africa (FLETCHER, 1909). There are elements in their flora that are heavy seeded (E. J. H. CORNER, pers. comm.). Though a hypothesis of long-distance dispersal of these heavy seeded plants in recent time cannot be ruled out, it is unlikely. It is more probable that these plants have evolved from ephemeral species that have dispersed there in the past and have since lost their dispersive powers. After the initial colonisation of a small island, good dispersal is disadvantageous and heavy seeds may be evolved (CARLQUIST, 1966). Such plants may well indicate the sort of flora that South India may have supported during the drift phase.

The butterfly species that appear to have reached Africa from Asia via the Indian Ocean are almost entirely of dispersive genera, or related to dispersive species. In the faunal centre survey the Oriental centre, 1, shows extension to Africa, with slightly stronger representation on Madagascar. The actual genera in the survey that show this are as follows:
1. *Hypolimnas*, a nymphalid genus of very dispersive species that have colonised many isolated islands in the Pacific.
2. *Melanitis*, a satyrid genus of few, widespread species common in subclimax secondary vegetation and areas of habitation.
3. *Atella*, a nymphalid genus, again with species extending well into the Pacific.
4. *Cirrochroa* + *Smerina*. *Cirrochroa* is an entirely Oriental nymphalid genus extending strongly through the East Indies to the Pacific. It is characteristic of secondary growth and forest fringes. *Smerina* is a closely related monotypic genus restricted to Madagascar.
5. *Tagiades* is a hesperid genus whose species fall into the widespread Oriental elements in the species faunal element survey.
6. *Arnetta* is a hesperid genus with the greatest concentration of species

481

in Madagascar, but well represented in southeast Asia and India. Its species are not particularly dispersive. It clusters at a high level into the 'Oriental' centre. It could possibly indicate spread across the Indian Ocean from Madagascar to Asia, but this is tenuous.

7. *Ampittia* is a hesperid genus centred strongly in West China but with forms in East Africa and the Congo rainforest. It would appear to be exceptional in this and requires further investigation. The possibility of undercollecting should not be ruled out.

8. *Parnara* is another hesperid genus of wide distribution, centred in Asia and better represented in Madagascar than in Africa.

9. *Baoris*, *Borbo* and *Pelopidas*, hesperids, are similar to *Parnara* and all have species in the widespread Oriental faunal element, 1.a. *Pelopidas* is very strongly represented in East Africa.

10. *Eurema* is a pierid genus that typifies secondary vegetation. Its most widespread species, *hecabe*, has spread all over Africa. In Madagascar there is a distinct, but closely related species *floricola* (PAULIAN & BERNARDI, 1951) as well as *hecabe* itself. Also in Madagascar and nearby regions of East Africa there is a species related to *E. andersoni*, *E. hapale*. Both *floricola* and *hapale* may have been derived across the Indian Ocean, and their nearest relatives, *hecabe* and *andersoni*, both fall into the widespread lowland species element 1(a)ii. This element will be seen to contain a high proportion of species characteristic of secondary growth. A fuller account of *Eurema* may be found in HOLLOWAY (forthcoming paper).

11. *Catopsilia* is a pierid genus that falls within centre 7, containing widespread genera with no definite centre (*Borbo*, above, is another). Its species are characteristic of secondary vegetation and often migrate in swarms (WILLIAMS, 1930). One species, *florella*, occurs in Africa and extends through the Middle East to India (faunal element 4 below). Three species occur in the Oriental Region and all are widespread. One of these, *pomona*, has a subspecies *(grandidieri)* in Mauritius and Madagascar that tends towards the remaining species, *thauruma*, in markings (but not in genitalia). *Thauruma*, also from Madagascar, is morphologically closest to *pomona*, though its genitalia are markedly modified in some respects. It would appear, therefore, that *pomona* stock has crossed the Indian Ocean to colonise Madagascar twice.

12. *Libythea* (Libytheidae) has a similar distribution and species concentration to *Parnara*. It also contains few, widespread species.

The species and genera showing a distribution pattern across the Indian Ocean are, in summary, generally widespread, mobile (often known migrants) species of ephemeral habitats. Several cases of such spread are of sufficiently long standing for divergence and minor radiation to have occurred in Africa or Madagascar (especially *Cirrochroa* + *Smerina*). Other examples outside the sample in the survey may be found in COR-

BET's (1949) study of the relationships between Indian and Madagascar butterflies. There is much scope for further study of these interesting genera, especially those (e.g. *Arnetta*) shown statistically to be exceptions to the general distributional rule.

4. The faunal elements of Indian butterfly species and their relation to the generic centres

In order to learn more about the way in which the generic centres are made up and the distribution patterns and the ecology of their component taxa, faunal elements were derived for a sample of 436 species of Indian butterfly. These species were, for the large part, from the genera used in the faunal centre survey above. In some cases the literature used, though sufficient to give a good indication of generic distribution, was somewhat inaccurate at the species level and therefore unsatisfactory for a faunal element survey. In addition it was, however, possible to use the data in TALBOT (1939) on all the species of the Pieridae and Papilionidae found in India. Distributional data on the following were also used: species of the majority of the hesperid genera used in the centre survey (EVANS, 1949); species from several groups of the Lycaenidae (*Neozephyrus* (HOWARTH, 1957), the *Arhopala* group (EVANS, 1957), *Anthene* (TITE, 1966), and the tribes Horagini (COWAN, 1966) and Cheritrini (COWAN, 1967)); and species from nymphalid groups (*Melitaea* (HIGGINS, 1941) and the Neptini tribe (ELIOT, 1969)).

Several of the provinces used in HOLLOWAY (1969) were divided up in order to make them more realistic, both topographically and ecologically. African distribution was not taken into account but will be mentioned in the discussion of relevant elements. Forty-nine provinces were recognized as follows: 1. Asia Minor, 2. Arabia, 3. Middle East (Syria, Jordan, Israel, Lebanon, Iraq), 4. Iran, 5. Tibet (excluding southeast), 6. Caucasus and Turkestan, 7. Pamirs and Tien Shan, 8. Hindu Kush (N. Afghanistan, Chitral), 9. Baluchistan and S. Afghanistan, 10. Indus Plain (Pakistan), Thar, Kutch, Sind and S. Punjab, 11. Northwest Himalaya (Karakoram, Gilgit, N. Punjab, Jammu, Kashmir, Kumaon), 12. Nepal, 13. Ganges Plain (United Provinces), 14. Central Hills (Central India, Central Provinces, Orissa), 15. Western Ghats (Bombay, Coorg and W. Madras), 16. Deccan (to east of W. Ghats), 17. Ceylon, 18. Sikkim, Bhutan and Southeast Tibet, 19. Bengal, 20. Assam (including Khasia, Lushai, Chin, Naga and Kachin hill regions), 21. Szechwan, 22. China north of Hupeh and Shanghai, 23. South China (Kweichow, Hunan, Kiangsi, Chekiang, Fukien, Kwangsi and Kwangtung), 24. Yunnan, 25. Annam (including Vietnam, Laos and Hainan), 26. Taiwan, 27. Shan States of Burma, 28. Burma to Rangoon (excluding 20 and 27), 29. Lowland Thailand and Cambodia, 30. Burma (Tenasserim) south of Rangoon and western and peninsular

Thailand, 31. Malaya, 32. Sumatra, 33. Borneo (and Palawan), 34. Java and Bali, 35. Philippines, 36. Celebes, 37. Lesser Sundas (Lombok to Tenimber), 38. Moluccas, 39. New Guinea, 40. Australia, 41. Bismarcks, Solomons and New Hebrides, 42. Altai and Mongolia, 43. Manchuria, Korea, Amur and Japan, 44. Europe south of Brittany, excluding Spain and the Balkans, 45. Balkans, 46. N. Africa, 47. Spain, 48. N. Europe and Russia (excluding 6) to Urals and 49. Siberia.

The distributions of the taxa amongst these provinces were recorded and, through coincidence of distribution, the taxa fell into 351 units to be classified. Dissimilarity coefficients [1 – (number of provinces in which both taxa are found)/(number of provinces occupied by one or both taxa in toto)] were calculated for all pairs of units and single-link cluster analysis performed on these as described in HOLLOWAY & JARDINE (1968). The resultant dendrogram is depicted in Fig. 86 and the clusters (elements) recognized are indicated and labelled. The numbers at the foot of each 'stalk' of the dendrogram refer to the units under analysis, containing the taxa listed under these numbers in the following paragraphs. The various elements are mapped in Fig. 87–94. In these figures the number in each province refers to the number of species from the element concerned that are found in that province. The variations in thickness and brokenness of the lines outlining the elements indicate where the major drops in species number occur, passing out from the centre of the element. They are not at fixed intervals of species number as were the contour lines depicting elements in HOLLOWAY & JARDINE (1968). The elements derived are described below, together with lists of their component taxa and an estimate of their ecological character. The ecological information was collected for as many species as possible from TALBOT (1939), WYNTER-BLYTH (1957), CORBET & PENDLEBURY (1957 – on Malayan species) and from BARLOW, BANKS & HOLLOWAY (1971 – Bornean species).

The main element in the Indian butterfly fauna is the Oriental element (1). This is made up of subelements as follows.

1.a.1. (Fig. 87). This is an element centred on Sundaland, Burma and Assam with some extension to Thailand, Indochina and Sikkim. It is weakly represented in southern China, Bengal and the Wallacean subregion (as defined in HOLLOWAY & JARDINE, 1968) but is not represented in southern India. It is made up mainly of species of wet, lowland forest that are often found drinking at wet places. Two species *(Graphium payeni, Lamproptera meges)* are perhaps more characteristic of altitudes up to 1000 m. This is therefore a predominantly a lowland rainforest element and is perhaps excluded from southern India through inability to adapt to the drier conditions (see 1.f.). It contains the following taxa:

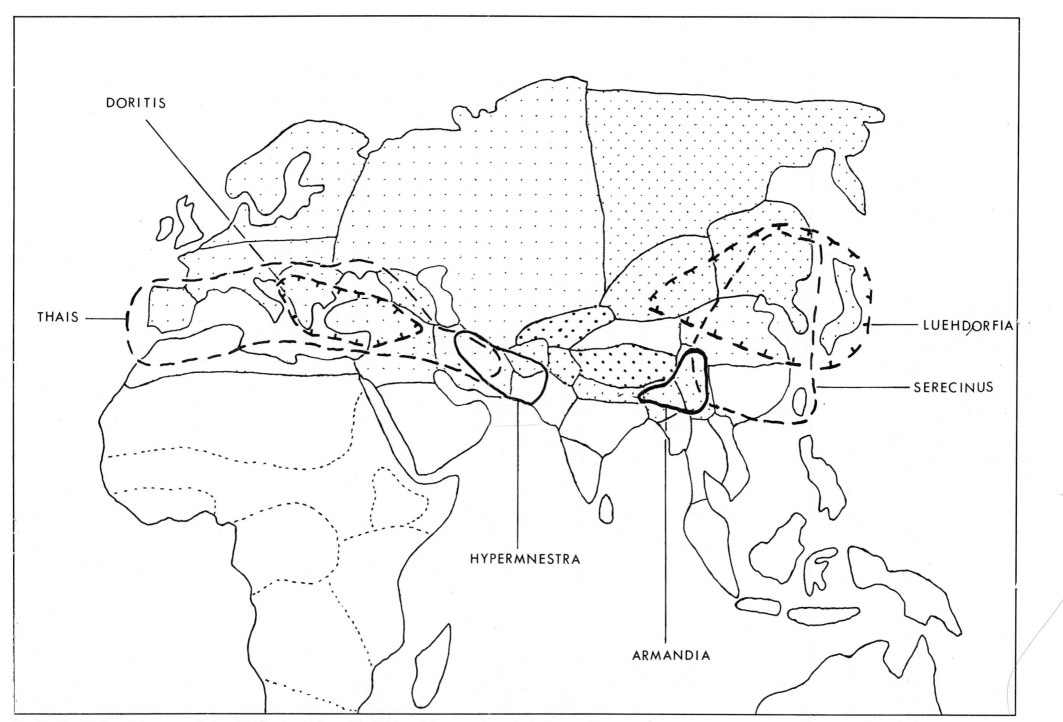

Fig. 85. Distribution of the genera of the subfamily Parnassiinae. Distribution of the genus *Parnassius* is illustrated by stipple that gets heavier toward the generic centre.

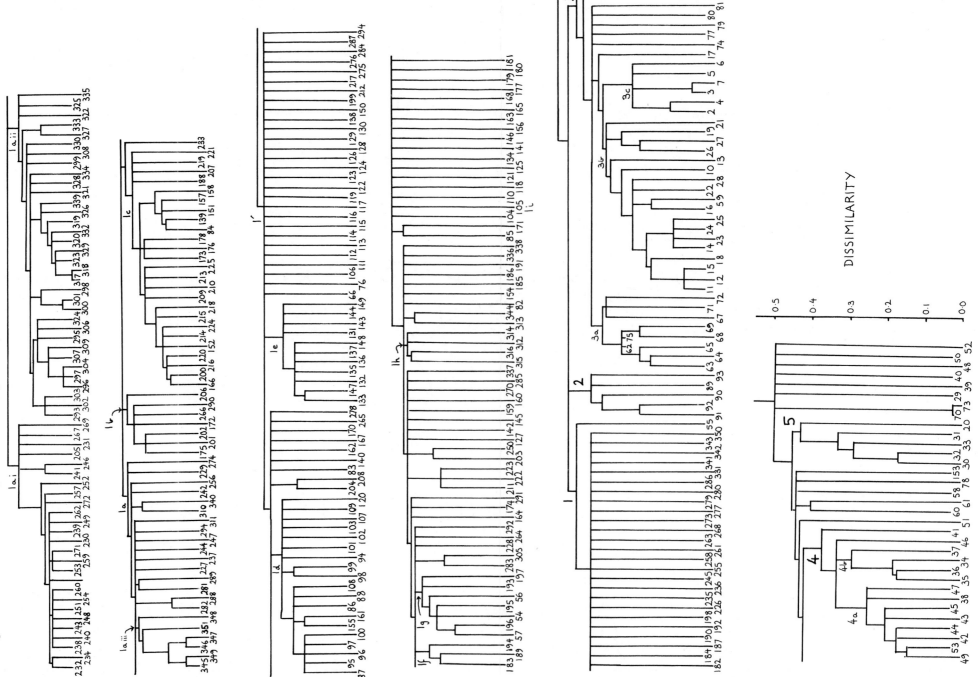

Fig. 86. Dendrogram derived for 351 units of butterfly distribution by single-link cluster analysis. The clusters (elements) recognised are indicated. For convenience the dendrogram has been broken up into several sections. Horizontal lines broken at the left of each section join on to comparable horizontal line ends to the right of the section below. The numbers at the foot of each 'stalk' of the dendrogram indicate the units listed, with the taxa they contain, for each element in the text.

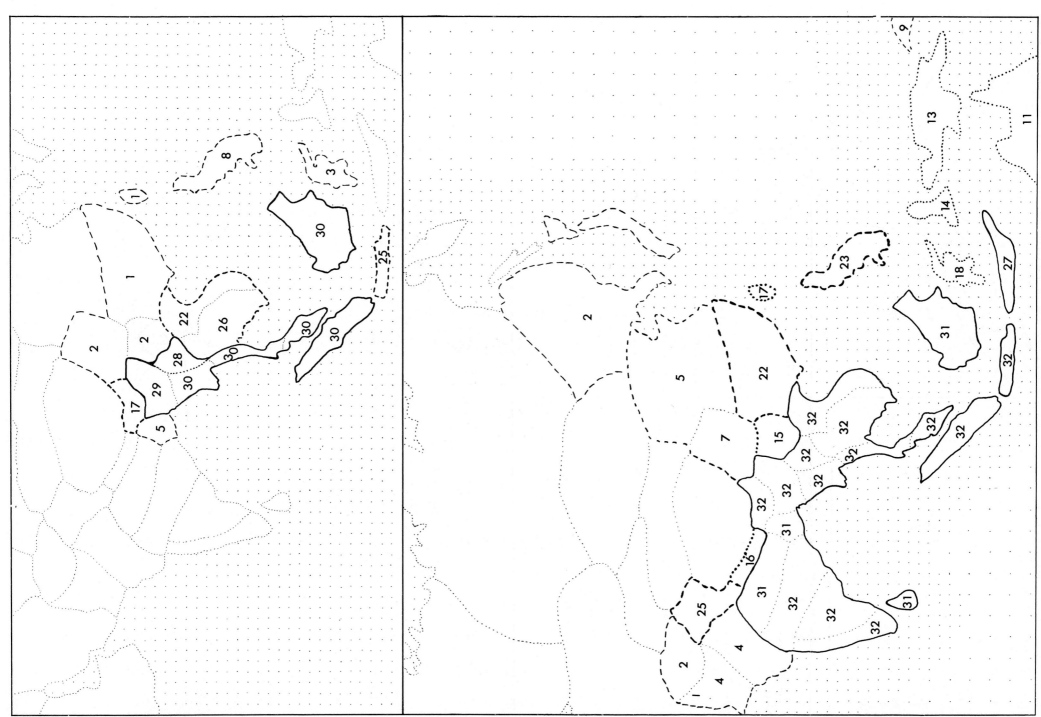

Fig. 87. Faunal elements of Indian butterfly species. top – Element 1(a)i; bottom – Element 1(a)ii.

Fig. 88. Faunal elements of Indian butterfly species. top left – Element 1(i); top right – Element 1(a)iii; lower centre – Element 1(b); upper centre – Element 1(a)iii; lower centre left – Element 1(f); lower centre right – Element 1(d); bottom – Element 1(e).

205 *Celaenorrhinus aurivittata*
230 *Chilasa paradoxa*
    Narathura anarte
    Pantoporia paraka
    Neptis magadha
    Graphium evemon
231 *Neptis miah*
232 *Potanthus trachala*
234 *Telicota linna*
238 *Flos fulgida*
240 *Narathura silhetensis*
241 *Celaenorrhinus putra*
243 *Celaenorrhinus asmara*
246 *Neptis ilira*
248 *Graphium macareus*
    Graphium bathycles
    Graphium payeni
    Tagiades parra

249 *Chilasa slateri*
251 *Hasora schoenherr*
252 *Narathura agrata*
253 *Narathura eumolphus*
254 *Coladenia agni*
257 *Graphium megarus*
    Capila phanaeus
259 *Neptis harita*
260 *Lamproptera meges*
262 *Flos diardi*
267 *Narathura athada*
    Narathura allata
269 *Darpa pteria*
271 *Bibasis harisa*
272 *Halpe zema*

1.a.2. (Fig. 87). An element that is centred more broadly than the previous one, from Sundaland to Indo-China, Assam and all India. It is strongly represented in the Himalaya, southern China and Wallacea, and more weakly in the provinces to the northwest of India, in northern China and areas to the east of Weber's line (Papuan Region). It contains again a number of lowland forest species, but also a high proportion of species that are found in open places and plains as much as the forest. A number of species are those characteristic of secondary vegetation. Several of the pierid and papilionid species are those commonly found drinking in swarms. A number show seasonal dimorphism and several are known migrants. It is an element of marked dispersive and ephemeral character. It contains the following taxa:

293 *Anthene emolus*
295 *Leptosia nina*
    Odontoptilum angulatum
    Hebomoia glaucippe
296 *Appias indra*
297 *Horaga onyx*
298 *Appias lyncida*
299 *Tagiades gana*
300 *Eurema andersoni*
301 *Cheritra freja*
302 *Anthene lycaenina*
303 *Bibasis sena*
304 *Cepora nerissa*
306 *Neptis hylas*
308 *Graphium antiphates*
309 *Graphium doson*
317 *Graphium sarpedon*

318 *Eurema hecabe*
319 *Troides helena*
320 *Tagiades japetus*
321 *Papilio polytes*
322 *Telicota ancilla*
323 *Graphium agamemnon*
324 *Parnara naso*
325 *Hasora taminatus*
326 *Borbo cinnara*
327 *Polydorus aristolochiae*
328 *Papilio helenus*
329 *Appias albina*
330 *Telicota colon*
333 *Ixias pyrene*
334 *Pelopidas matthias*
335 *Badamia exclamationis*
339 *Pelopidas agna*

1.a.3. (Fig. 88). A small element centered on Sundaland, Wallacea, the

Moluccas, Burma, Assam and Sikkim, but with not much representation outside this area. The ecology of its component taxa, listed below, is similar to that of element 1.a.1. – wet lowland forest.

345 *Gandaca harina*
346 *Graphium aristeus*
347 *Graphium eurypylus*

348 *Horaga syrinx*
349 *Appias nero*
351 *Telicota ohara*

1.a. (Fig. 89). This element contains the three previous elements plus the additional taxa listed below. It is centred on Sundaland, Thailand, Burma and Assam. It is strongly represented throughout the Oriental and Papuan Regions but extends only weakly into the Palaearctic Region north of China and west of the N.W. Himalaya. It consists entirely of widespread species of lowland forests and associated open plains and secondary vegetation. The following taxa are additional to the three subelements above:

227 *Oriens gola*
229 *Potanthus mingo*
237 *Narathura hellenore*
242 *Flos apidanus*
244 *Narathura arvina*
247 *Lasippa tiga*
256 *Cephrenes chrysozona*
274 *Phaedyma columella*

281 *Narathura centaurus*
282 *Tagiades litigiosa*
288 *Neptis nata*
289 *Narathura abseus*
294 *Appias libythea*
310 *Papilio pario*
311 *Chilasa clytra*
340 *Potanthus confucius*

1.b. (Fig. 88). An element centred broadly through the mountainous provinces of the Northwest Himalaya, Nepal, Sikkim, Assam, Yunnan, Indo-China, Shan and Burma, with some representation in Bengal, Szechwan, South China and Taiwan, and extending south to Malaya and Sumatra. It contains the following species, predominantly of forest habitats at moderate (1000–2000 m) altitudes:

172 *Polydorus aidoneus*
175 *Chilasa agestor*
201 *Troides aeacus*
202 *Polydorus philoxenus*

206 *Chilasa epycides*
266 *Neptis sankara*
290 *Tagiades menaka*

1.c. (Fig. 90). This element resembles 1.b. in general distribution but is more markedly centred in the Eastern Himalaya (Sikkim, Assam, Shan, Burma) and in Annam. It has proportionately more representation to the south of its centre than 1.b., down to Malaya. As with 1.b. it consists largely of forest species from moderate altitudes. There are a few lowland species (e.g. *Valeria avatar*, *Prioneris thestylis*, *Graphium xenocles*) of forest habitats. *Prioneris thestylis* is interesting in that it is said (TALBOT, 1939) to mimic three species of *Delias*, all of which fall into the montane element

486

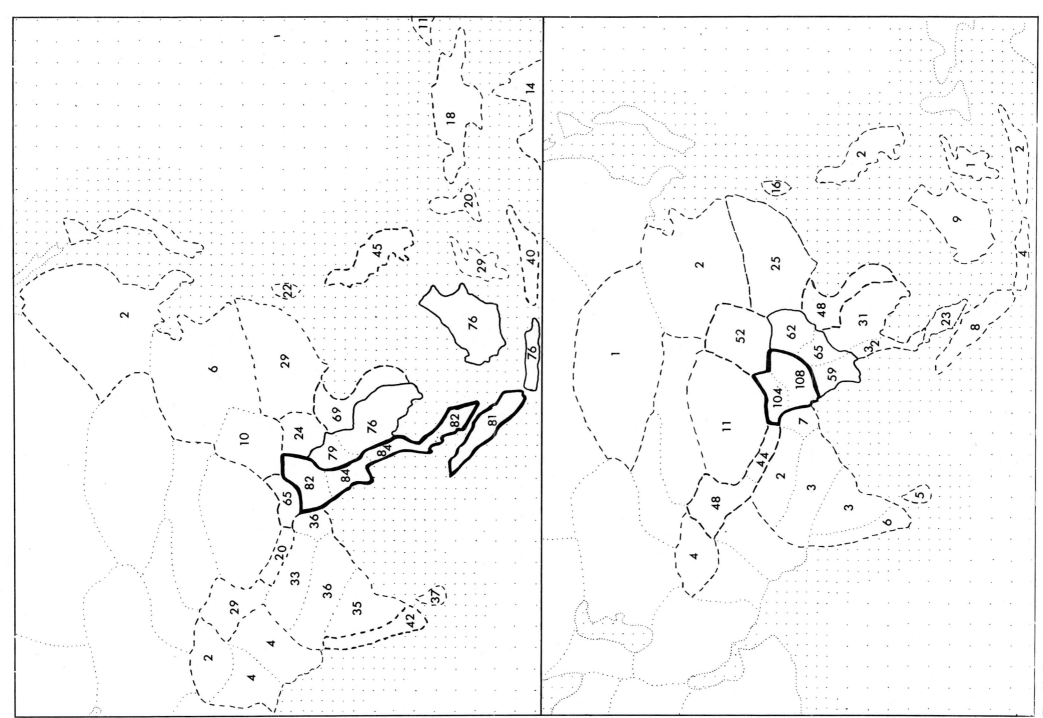

Fig. 89. Faunal elements of Indian butterfly species. top – Element l(a); bottom – Element l'-l(a).

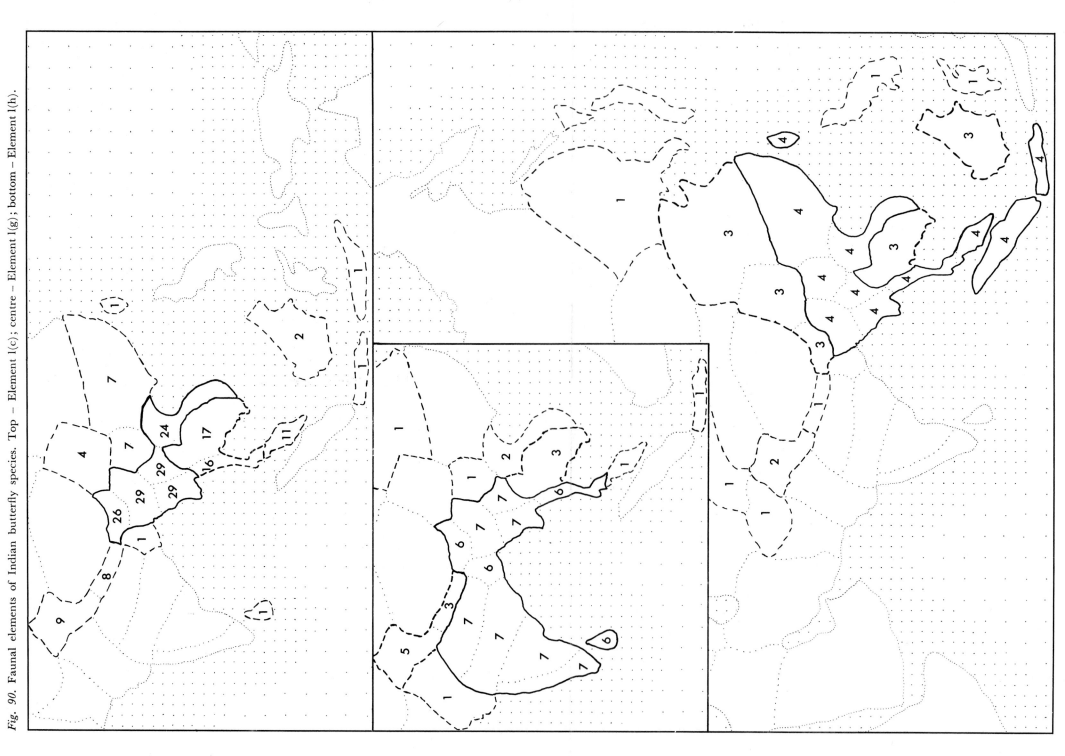

Fig. 90. Faunal elements of Indian butterfly species. Top – Element l(c); centre – Element l(g); bottom – Element l(h).

1'.-1.a. described below and are characteristic of moderate altitudes. The element contains the following taxa:

84	*Narathura rama*		*Halpe arcuata*
139	*Neptis ananta*	215	*Graphium xenocles*
151	*Neptis zaida*		*Thoressa cerata*
152	*Pelopidas assamensis*		*Bibasis amara*
166	*Polydorus varuna*		*Narathura camdeo*
169	*Prioneris thestylis*		*Valeria avatar*
173	*Papilio arcturus*	216	*Prioneris clemanthe*
176	*Papilio rhetenor*	218	*Hasora danda*
178	*Polytremis discreta*	219	*Baoris penicillata*
188	*Dercas verhuelli*	220	*Potanthus rectifascia*
200	*Delias agostina*	221	*Polydorus crassipes*
207	*Papilio castor*	224	*Odina decoratus*
210	*Neptis pseudovikasi*		*Arnetta atkinsoni*
213	*Celaenorrhinus pyrrha*	225	*Halpe hauxwelli*
214	*Graphium agetes*	233	*Potanthus nesta*

1.d. (Fig. 88). This element, like 1.c., is centred in the N.E. Himalayan provinces, but the centre also extends north to Szechwan and to Yunnan. It is moderately represented in the Shan States and has five species in Tibet. In accordance with its more northerly distribution its species are, in general, found at higher elevations (1,500–4000 m. and higher) and are characteristic of more open montane woodland and meadows up to the tree line. The element contains the following taxa:

86	*Hasora anura*	102	*Thoressa hyrie*
87	*Delias lativitta*	103	*Celaenorrhinus tibetana*
88	*Flos chinensis*	107	*Papilio krishna*
94	*Neptis cydippe*		*Sovia lucasii*
	Pantoporia bieti		*Neozephyrus desgodinsii*
95	*Delias berinda*		*Polydorus plutonius*
96	*Graphium glycerion*	108	*Phaedyma aspasia*
97	*Neptis armandia*	109	*Polydorus nevilli*
98	*Dercas lycorias*	120	*Carterocephalus avanti*
99	*Neozephyrus duma*	155	*Neptis nashona*
100	*Pelopidas sinensis*	161	*Celaenorrhinus patula*
101	*Parnassius imperator*		

1.e. (Fig. 88). An element centred along the range of the Himalaya from Kashmir to Sikkim, Assam and Yunnan. It is moderately represented in Szechwan. Its component species, listed below, are of similar ecology to those of element 1.d.

131	*Ochlodes brahma*	136	*Seseria sambara*
	Sovia grahami	137	*Narathura singla*
	Polydorus latreillei	143	*Celaenorrhinus ratna*
132	*Potanthus mara*	144	*Celaenorrhinus pero*
133	*Papilio bootes*	147	*Neozephyrus ataxus*
	Choaspes xanthopogon	148	*Pedesta masuriensis*

Neptis narayana   149 *Celaenorrhinus pulomaya*
135 *Delias sanaca*
    Seseria dohertyi

All the elements above cluster together at a dissimilarity level of .25 (see Figure 86) to form the element 1'. 1.a. clusters with 1.b., 1.c. and 1.d. at the .23 level by one link only and by two more links at level .25. 1.b., 1.c., 1.d. and 1.e. (clustering at the .25 level) cluster with each other by more links than this. The single units coming in at both these levels, e.g. 111–117, are all linked to one or more of 1.b–e. by one link or more, and not at all to 1.a. Hence if one applied B_2 or B_3 (double- or treble-link) cluster analysis to the data as described in JARDINE & SIBSON (1968) and mentioned in HOLLOWAY (1969) it is likely that 1.a. and 1'.-1.a. would cluster out as separate entities at a certain level. Most of the taxa in 1.a. are lowland forms and those of the other elements are characteristic of medium altitudes and hence inclusion of an altitude parameter in the data would tend to enhance the separation of these two elements. The additional units clustering in are almost entirely characteristic of medium altitudes and several are from Palaearctic genera (e.g. *Parnassius* and *Colias*) and are found at very high altitudes. The separation of 1.a. and 1'.-1.a. as distinct elements is supported by the studies of COR-BET (1941) and CORBET & PENDLEBURY (1956) on the ecological and geographical distribution of the Malayan butterfly fauna. These authors recognised three elements in Malaya. The first they called the Indo-chinese element, only just extending south to Malaya from the north. Over half the species of this element are montane and therefore, in distribution and ecological character, this element could be referred to 1'.-1.a. Their second element is of Malaysian species, largely confined to lands on the Sunda Shelf. As these species do not occur in India they were not included in this survey, but the presence of such an element is confirmed in a similar numerical survey in HOLLOWAY (forthcoming paper). The third element in the Malayan fauna is of widespread Oriental species of lowland forest or associated secondary vegetation. This element may be equated with 1.a.

1'.-1.a. is illustrated in Fig. 89, and is largely restricted to the mountainous provinces of the Himalaya and western China. In addition to subelements 1.b., 1.c., 1.d. and 1.e., it contains the following taxa:

  66 *Parnassius hardwickei*
    Narathura dodonaea
  76 *Parnassius acco*
    Taractrocera danna
  83 *Melitaea arcesia*
106 *Celaenorrhinus aspersa*
111 *Pyrgus dejeani*
113 *Pieris extensa*
    Carterocephalus houangty

129 *Graphium gyas*
    Sebastonyma dolopia
    Teinopalpus imperialis
130 *Neozephyrus bhutanensis*
    + *N. triloka*
138 *Graphium cloanthus*
    Graphium eurous
140 *Aporia agathon*
150 *Neptis cartica*

	Aporia harrietae	162 *Lobocla liliana*
	Colias montium	167 *Polydorus dasarada*
114	*Neptis radha*	170 *Papilio polyctor*
115	*Celaenorrhinus plagifera*	199 *Appias lalage*
116	*Cheritrella truncipennis*	204 *Caltoris cahira*
117	*Pedesta pandita*	208 *Papilio chaon*
119	*Satarupa zulla*	212 *Narathura aceta*
122	*Neozephyrus kabrua*	217 *Potanthus lydia*
123	*Neozephyrus disparatus*	265 *Ticherra acte**
124	*Narathura paralea*	275 *Bibasis gomata**
	Narathura alex	276 *Pantoporia sandaka**
126	*Darpa hanria*	278 *Tapena thwaitesii**
128	*Aeromachus kali*	284 *Ampittia dioscorides**
	Celaenorrhinus sumitra	287 *Pantoporia hordonia**
	Flos adriana	294 *Appias libythea**

* These taxa tend towards 1.a. in distribution.

Element 1'. is illustrated in Fig. 91. It shows great similarity to element 1. itself and hence will not be discussed further here. Element 1. has a few more subelements, some of very few taxa and none of any great size.

1.f. (Fig. 88). This is a very small element, but it is important in that it illustrates the tendency for species in the Oriental element in India to be more strongly represented in Ceylon and the Western Ghats where the heavier and somewhat less seasonal rainfall produce an ecology distinct from that of the other southern provinces in India and closer to that of the Oriental generic centres in Burma, Indochina and Sundaland. Its taxa, favouring wet lowland forests, are:

183	*Potanthus pallida*	194	*Neptis jumbah*
189	*Baracus vittatus*		

1.g. (Fig. 90). A small element strongly represented in southern India, and extending to Sikkim, Assam, Burma, Shan and Tenasserim. It consists of taxa of drier lowland forest and of the plains, resembling 1.a.2. The taxa are:

54	*Spialia galba*	195	*Graphium nomius*
56	*Daimio bhagava*	196	*Taractrocera maevius*
57	*Narathura amantes*	197	*Coladenia indrani*
193	*Sarangesa dasahara*		

1.h. (Fig. 90). This is a very small element broadly centred over Sundaland, S.E. Asia and China, with some extension west through the Himalaya. Its taxa are characteristic of open montane forest, similar to element 2 below. The taxa are:

312	*Parnara guttatus*	315	*Delias aglaia*
314	*Narathura bazalus*	316	*Lamproptera curius*

489

1.i.   (Fig. 88). This element is made up of one unit, 105, of 11 taxa restricted to the Sikkim and Assam provinces. As this unit occupies only two provinces it cannot cluster at any level lower than 0.33, whereas in character it resembles element 1.c. This is a disadvantage of using data where distribution is crudely recorded on a presence/absence basis with regard to rather loosely defined geographical provinces. Taxa with restricted distributions can never cluster below rather high levels. This situation is largely eliminated if transect data are used as in Holloway (1970). The taxa of this element are similar in ecology to those of element 1'.-1.a. and are:

105 *Capila zennara*   *Capila jayadeva*
     Celaenorrhinus morena   *Celaenorrhinus zea*
     Celaenorrhinus badia   *Neozephyrus assamicus*
     Neozephyrus kirbariensis   *Potanthus sita*
     Halpe kumara   *Halpe knyvetti*
     Armandia lidderdalei

All the above elements, together with the taxa listed below, cluster together to form the Oriental element, 1. illustrated in Fig. 91. It bears a close resemblance to the Oriental generic centre (Figure 81) except that it is not so strongly represented in Sundaland. This is largely due to exclusion of species of the Malaysian element mentioned above. Element 1. will be seen in Table II to consist almost entirely of species from this centre. Of note is the strong representation in the Western Ghats and Ceylon, as compared to the rest of southern India. Reasons for this have been indicated in the discussion of element 1.f. Additional taxa clustering into this element are:

 82 *Panchala ganesa*   211 *Potanthus juno*
 85 *Narathura paramuta*   222 *Halpe wantona*
104 *Neptis nemorum*   223 *Narathura belphoebe*
110 *Neozephyrus vittatus*   226 *Aeromachus jhora*
118 *Pieris naganum*   228 *Neptis clinia*
121 *Neptis manasa*   235 *Capila pieridoides*
125 *Bibasis vasutana*   236 *Celaenorrhinus nigricans*
127 *Flos asoka*   245 *Panchala ammonides*
134 *Neptis mahendra*   250 *Aeromachus pygmaeus*
141 *Colias stoliczkana*   255 *Satarupa gopala*
142 *Neptis yerburyii*   258 *Bibasis iluska*
145 *Narathura oenea*   261 *Horaga albimacula*
146 *Colias nina*   263 *Delias belladonna*
156 *Panchala paraganesa*   264 *Celaenorrhinus dhanada*
159 *Polytremis eltola*   268 *Narathura ace*
160 *Flos areste*   270 *Choaspes plateni*
163 *Daimio sinica*   273 *Pithauria stramineipennis*
164 *Daimio phisara*   277 *Pelopidas conjuncta*
165 *Bibasis anadi*   279 *Choaspes benjaminii*
168 *Delias thysbe*   280 *Caprona agama*

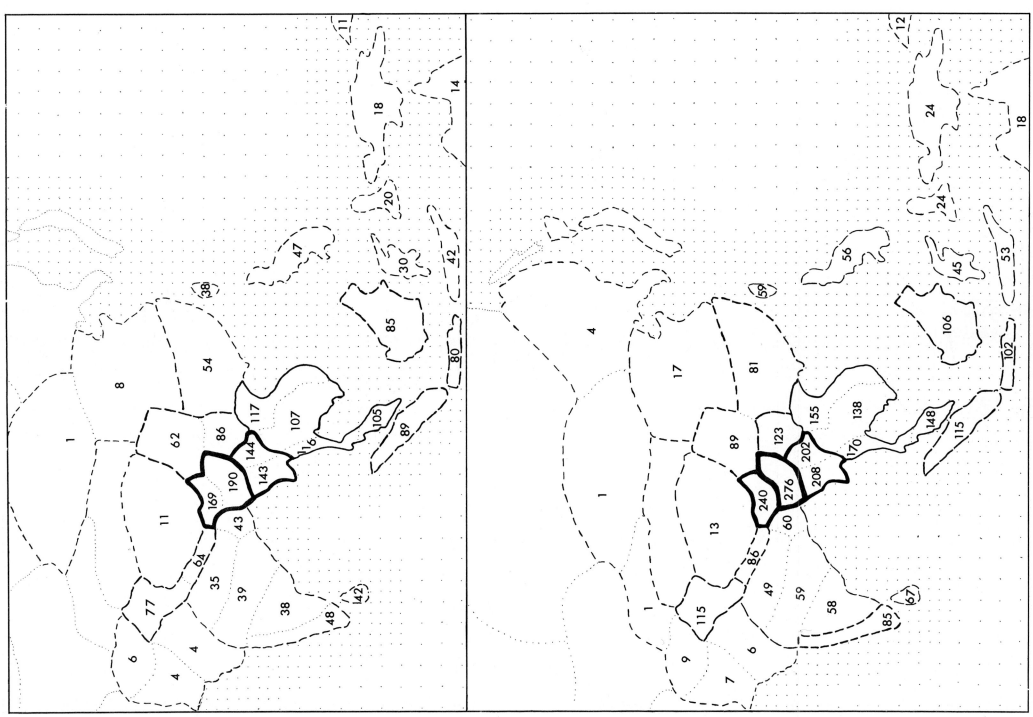

Fig. 91. Faunal elements of Indian butterfly species. Top – Element I'; bottom – Element 1.

Fig. 92. Faunal elements of Indian butterfly species. Top – Elements 3(a); centre – Element 3(b); bottom – Element 4.

171 *Ochlodes siva*
174 *Papilio protenor*
177 *Parnara ganga*
179 *Potanthus pseudomaesa*
180 *Capila pennicillatum*
181 *Pieris canidia*
182 *Caprona alida*
184 *Neptis soma*
185 *Halpe homolea*
186 *Oriens goloides*
187 *Potanthus palnia*
190 *Celaenorrhinus leucocera*
191 *Lasippa viraja*
192 *Pelopidas subochracea*
198 *Taractrocera ceramas*
203 *Halpe porus*

283 *Iambrix salsala*
285 *Delias descombesii*
286 *Caltoris kumara*
291 *Bibasis oedipodea*
292 *Cepora nadina*
305 *Baoris farri*
313 *Polytremis lubricans*
331 *Eurema libythea*
336 *Eurema laeta*
337 *Hasora vitta*
338 *Hasora chromus*
341 *Borbo bevani*
342 *Choaspes hemixanthus*
343 *Potanthus pava*
344 *Coladenia dan*
350 *Appias paulina*

2.   (Fig. 94). This is a small element centred broadly over Assam, China and the Amur province. It may be compared with centre 5 in the survey of genera. It has some representation in the Himalaya and Tibet. One species extends to the Philippines via Taiwan and it is worth noting that this element may represent the source of species exhibiting the 'Luzon track' distribution indicated by VAN STEENIS (1934) for the high altitude flora of the East Indies. The species are characteristic of open, mesic (subtropical) woodland. Where they extend to the Himalaya they are found from moderate to high altitudes. Amur-Chinese species such as these may be far more numerous and such elements might be much larger in a survey including species additional to those impinging on the Indian region. Taxa included are:

89 *Carterocephalus dickmanni*
90 *Gonepteryx mahaguru*
91 *Daimio tethys*

92 *Ochlodes subhyalina*
93 *Papilio xuthus*

3.a.   (Fig. 92). This is the first of three subelements of the Palaearctic element, 3. It is centred on the N.W. Himalaya and in nearby mountainous provinces of the Pamirs, Turkestan and the Chitral. It extends to Tibet and the eastern Himalaya but not to the Mediterranean or northern Palaearctic. Its taxa are all characteristic of alpine meadows at, or above the treeline. Several species are found up to the snow line. The following taxa are included:

62 *Baltia butleri*
63 *Parnassius simo*
64 *Pyrgus alpina*
65 *Parnassius tianschanica*
   Parnassius actius
   Colias eogene

67 *Parnassius jaquemontii*
68 *Parnassius epaphus*
69 *Parnassius delphius*
71 *Parnassius charltonius*
72 *Colias cocandica*
75 *Baltia shawi*

491

3.b.   (Fig. 92). Another subelement of the Palaearctic element, this resembles 3.a. in being centred in the mountainous provinces to the north-west of India but is not so restricted in distribution. It extends strongly to Iran, the Middle East and Asia Minor and is moderately represented in the Mediterranean provinces. The taxa embrace a somewhat broader ecological spectrum than those of the previous subelement, from subalpine, deciduous woodland *(Aporia leucodice)* to semiarid habitats *(Pontia glauconome)*. The former tend to be from the Palaearctic generic centre 2. and the latter from the Mediterranean centre 4. On average the species are characteristic of rocky, sparsely wooded hillsides of moderate altitude and Mediterranean-type ecology. The taxa included are:

10 *Pontia daplidice*	21 *Gonepteryx farinosa*
11 *Euchloë ausonia*	22 *Papilio alexanor*
12 *Gegenes pumilio*	23 *Melitaea trivia*
13 *Gegenes nostrodamas*	24 *Carcharodus floccifera*
14 *Pieris krueperi*	25 *Erynnis marloyi*
15 *Carcharodus alceae*	26 *Eogenes alcides*
16 *Euchloë charltonia*	27 *Hypermnestra helios*
18 *Melitaea persea*	28 *Pontia glauconome*
19 *Spialia geron*	59 *Aporia leucodice*

3.c.   (Fig. 93). This, the last subelement of element 3., consists of very widespread species centred broadly over the whole Palaearctic Region and extending into the Himalaya, China and the mountains of Burma. Its taxa are characterised by an ecology reminiscent of that of the taxa of Oriental element 1.a.2., being widespread, of high ecological amplitude and dispersive. They may be regarded as the 'ephemeral' species of the Palaearctic. Several are migratory. Where they penetrate the Himalaya they are found over a wide range of altitude (1000–5000 m.). The taxa are:

2 *Gonepteryx rhamni*	5 *Pieris napi*
3 *Pieris brassicae*	6 *Pieris rapae*
4 *Papilio machaon*	7 *Colias croceus (fieldi)*

3.   (Fig. 93). The complete Palaearctic element resembles the Palaearctic generic centre, 2., but is centred in the N.W. Himalaya rather than in the Pamirs and Turkestan. This is to be expected as only those species found in provinces of the Indian region were included in the survey. Its species are broadly alpine to Mediterranean and semiarid in ecological preference. Transect data might serve to differentiate the ecological elements more exactly. The following taxa in addition to the three subelements are included:

492

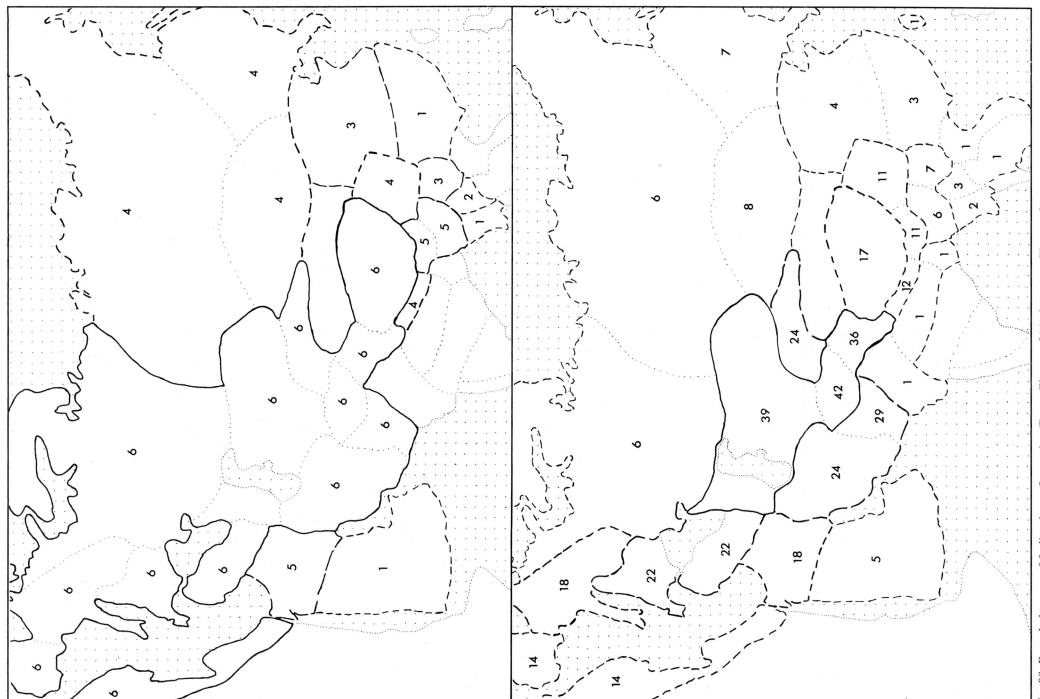

Fig. 93. Faunal elements of Indian butterfly species. Top – Element 3(c); bottom – Element 3.

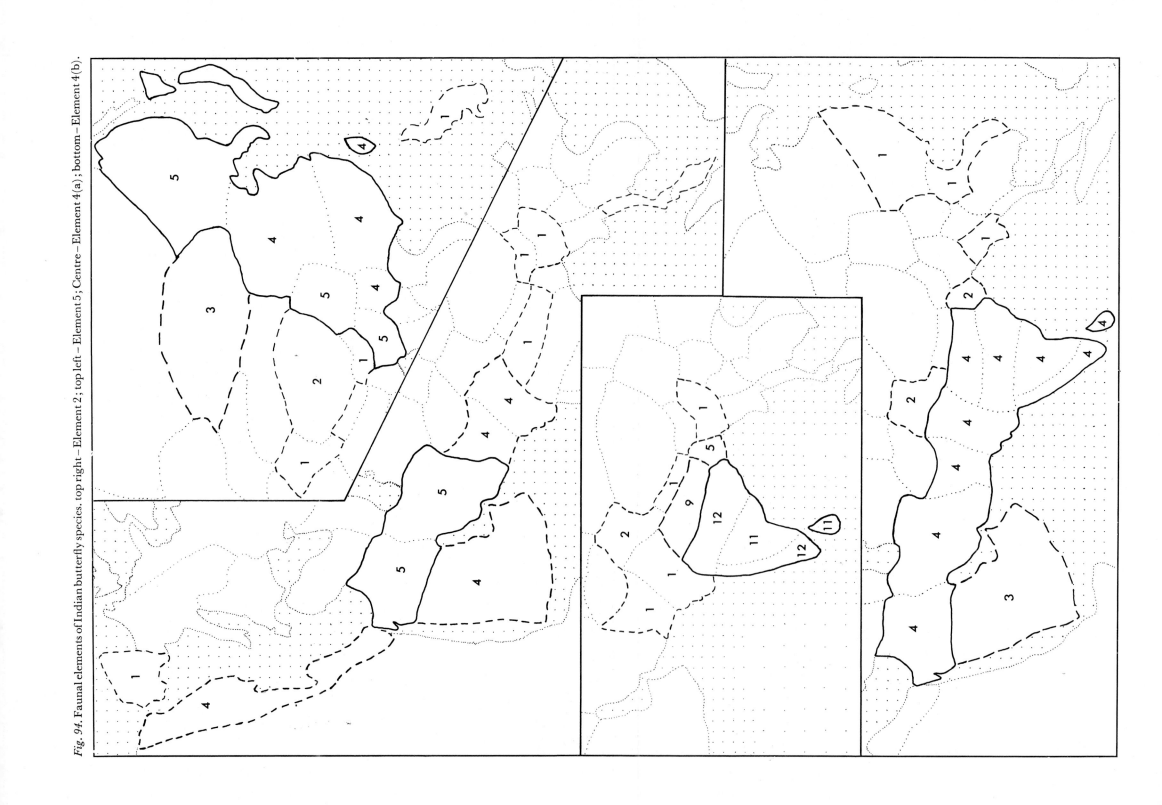

Fig. 94. Faunal elements of Indian butterfly species. top right – Element 2; top left – Element 5; Centre – Element 4(a); bottom – Element 4(b).

1	*Neptis sappho*		77	*Aporia nabellica*
8	*Melitaea didyma*		79	*Melitaea lutko*
9	*Spialia sertorius*		80	*Neozephyrus syla*
17	*Synchloe callidice*		81	*Neozephyrus birupa*
74	*Aporia peloria*			

4.a.   (Fig. 94). This element is centred on the southern provinces of India, strongly represented in the Ganges and Bengal provinces and extending slightly to the north and west. In this latter it differs from the other element centred in southern India, 1.g., which has a strong representation in the provinces of S.E. Asia. Several of the species in 4.a. might fall within 1.g. if the data were more ecological, as they are likewise characteristic of the drier forest regions and plains of India. Such species are derived from the Oriental centre. Other species in the element, such as the two *Colotis*, are found in very dry places of savannah character and might fall into the next element, 4.b., if the data were more ecological. The element contains the following taxa:

38	*Colotis etrida*		47	*Caprona ransonnettii*
42	*Polydorus hector*		49	*Delias eucharis*
43	*Papilio buddha*			*Prioneris sita*
44	*Ixias marianne*			*Papilio polymnestor*
45	*Colotis eucharis*			*Papilio crino*
	Valeria ceylanica		53	*Rathinda amor*

4.b.   (Fig. 94). An element more broadly centred than 4.a., the centre extending from southern India to the provinces of the Middle East via Sind and Baluchistan. It is strongly represented in South Arabia and the species are also found in East Africa and the Sudan. They are species of semi-arid and arid habitats as follows:

34	*Colotis danae*		36	*Colotis fausta*
35	*Colotis calais*		37	*Catopsilia florella*

4.   (Fig. 92). This element is made up of subelements 4.a. and 4.b. with the following additional taxa:

41	*Sarangesa purendra*		46	*Polydorus jophon*

5.   (Fig. 94). This element resembles 4.b. except that it is absent from southern India and its centre extends to North Africa. It is also represented in savannah regions south of the Sahara not included in the survey. Like 4.b. its component taxa are almost entirely those favouring arid habitats. It contains the following taxa:

20	*Euchloë belemia*		32	*Colotis phisadia*
30	*Spialia doris*		33	*Colotis vestalis*
31	*Pelopidas thrax*			

Taxa in the sample endemic to the provinces of the Indian region were noted, though not included in the cluster analysis for similar reasons to those mentioned in the discussion on element 1.i. Those taxa endemic to the N.W. Himalayan province were, like the taxa of element 3.a., primarily taxa of very high altitudes from the Palaearctic generic centre. Similarly, those endemic to Sikkim and those endemic to Assam resemble the taxa of element 1.i., being forms of moderate to high altitudes. The taxa endemic to Ceylon and those found only in the Western Ghats are mainly species favouring thick, wet, evergreen forests. Therefore they probably diverged in these areas in ecological isolation after colonisation by ancestors from regions of similar ecology in S.E. Asia (see 1.f.). The species concerned are:

N.W. Himalaya
Parnassius stoliczkanus   *Colias ladakensis*
Colias leechi   *Potanthus dara*

Sikkim, Bhutan
Colias berylla   *Colias dubia*
Neptis nycteus   *Neozephyrus sikkimensis*
Neozephyrus sandersi

Assam
Polydorus polla   *Papilio elephenor*
Pantoporia assamica   *Neozephyrus tytleri*
Neozephyrus paona   *Neozephyrus khasia*
Neozephyrus intermedius   *Neozephyrus suroia*
Neozephyrus jakamensis   *Caltoris aurociliata*

Western Ghats
Papilio dravidarum   *Papilio liomedon*
Appias wardi   *Narathura alea*
Sovia hyrtacus   *Thoressa astigmata*
Thoressa sitala   *Thoressa evershedi*
Oriens concinna   *Arnetta mercara*

Ceylon
Narathura ormistoni   *Thoressa decorata*
Celaenorrhinus spilothyrus

The following taxa in the sample cluster in at a high level and are not referable to any specific element:

29 *Spialia zebra*   58 *Muschampia staudingeri*
39 *Gomalia elba*   60 *Pontia chloridice*
40 *Narathura atrax*   61 *Colias erate*
48 *Arnetta vindhiana*   70 *Colias marcopolo*
50 *Celaenorrhinus ruficornis*   73 *Pieris deota*
51 *Celaenorrhinus ambaresa*   78 *Synchloe dubernardi*
52 *Caltoris canaraica*   153 *Panchala aberrans*

494

Table II. Distribution of a sample of species of butterfly found in India amongst the centres and elements described in this chapter.

Species Element	Or.	1a	1b	1c	1d	1e	1	2a	2b	2	3	4	5	6	7	8	9
							Generic Centre										
1.a.1	26	6	11		1		23									3	
1.a.2	25	1	10	1	4	1	23								1		
1.a.3	3		2				2										
1.a.	68	12	31	1	6	1	62								1	3	
1.b.	3	1	2				3										
1.c.	18	2	7				15								1		
1.d.	16		7				8	1		1			1			2	
1.e.	13	1	3				6									3	1
1'.-1.a.	66	10	27		1	1	49	6		8			2			8	1
1.f.	3		2				3										
1.g.	4	1	1		1		4				2						
1.h.	3	1	1				3										
1.i.	10		1				5									3	
1.	206	20	82	1	10	2	169	10		12			4		2	17	2
2.	3						1	1		1			1				1
3.a.								9		9				2			
3.b.								5		5	1	6					
3.c.								5		5							
3.	3		1				1	23		23	2	6		2			
4.a.	3						2			2							
4.b.										3					1		
4.	3						2			6					1		
5.	1						1			3							
Endemic																	
N.W.H.	1		1				1	3		3							
W. Gh.	8	3					3										
Ceyl.	3	1					1									1	
Sik.	3		1				1	2		2							
Ass.	6		1				1										

Table II demonstrates how the various generic centres contribute to the various species elements derived. In general the pattern revealed confirms the conclusions of TALBOT (1939, pp. 41–42) and MANI (1962, 1968) concerning the derivation and distribution of the butterfly fauna of India.

Butterfly species from genera of centre 1, the Oriental centre, are found mainly in the Oriental element, 1. Very few are found in the other elements. Similarly the complete 'Oriental' centre, containing centres 1, 5, 7, 8, and 9, is largely restricted to the Oriental element in the distribution of its species. HOLLOWAY (1969) noted that genera centred strongly on the N.E. Himalaya and West China tended to cluster into the Oriental centre loosely at a high level. He suggested that the inclusion of some altitude parameter in the data might separate such Himalayan-centred genera from the more lowland Sundaland-Burmese-centred genera. This hypothesis is supported by the observation in Table II that West Chinese centres, 5 and 9, are more strongly represented in the montane element, 1'.-1.a. Also, element 1'.-1.a. derives fewer species than 1.a. from genera of centre 1., but derives an equal number of species from the 'Oriental' centre, where more of the Himalayan genera have clustered in. But resolution of the situation must await the collection of better data. Then it may be possible to understand better the transition between Oriental and Palaearctic, lowland and montane faunae in East Asia.

Another point worthy of note is the preponderance of the widespread 1.a. species, especially 1.a.2. 'ephemeral' species, among the Asian representatives of genera in centre 1.d., the New Guinea centre. HOLLOWAY & JARDINE (1968) and HOLLOWAY (1969) suggested that these genera had originally colonised the Papuan Region from Asia in the Miocene, and had undergone rapid radiation in the unsaturated conditions there. Such spread from Asia would be expected primarily from the opportunist, dispersive type of species found in element 1.a.

The Palaearctic centre, 2, is represented most strongly in the Palaearctic element, 3, with most species in the high altitude element, 3.a. It contributes several species to the montane Oriental element 1'.-1.a. at high altitudes and has a few species endemic to the N.W. Himalaya and to Sikkim. The centre is thus largely restricted to high altitudes in the mountainous regions of India, but also has species at moderate to low altitudes in the Mediterranean-type semi-arid habitats to the north-west of India.

Species from genera of the African savannah centre, 3., are distributed amongst the elements centred on southern India (1.g. and 4.) and the desert element 5. There are two species in the Palaearctic element, 3., at the semi-arid extremity of its ecological range. All the species are of arid or semi-arid habitats. The Kashmir centre, 6., contributes species only to the high altitude Palaearctic element, 3.a. Species from the Mediterranean centre 4. are found only in Palaearctic element 3.b. which embraces most species of Mediterranean-type habitats.

It would appear that, in general, the faunal elements derived correspond fairly closely to the various generic centres, i.e. there is little intermingling between the species of the various centres, suggesting that such centres have been developing independently from each other for some time. This apparent segregation is open to confirmation by careful studies in the field such as the transect survey suggested earlier.

5. Conclusion

The biogeography of the Indian butterfly fauna can now be summarized. The part of India south of the Ganges has a well authenticated history of continental drift northwards through the Indian Ocean during the Tertiary. Interaction with the biota of mainland Asia appears to be almost entirely recent and overwhelmingly unidirectional. The history of the biota of southern India before this (probably late Pliocene) Asian invasion is obscure, apart from evidence of a Jurassic podocarp flora. The hypothesis of minor colonisation of southern India by 'ephemeral' species prior to the main invasion from Asia needs further investigation. The evidence indicates colonisation of southern India primarily by species from S.E. Asian tropical centres, but with a few species derived from African savannah regions via Arabia and the Middle East. These latter are restricted to arid habitats. The Oriental species in southern India contain a high proportion of widespread species of dry lowland forest and associated plains and ephemeral habitats. Species with more marked wet rain forest preferences are commoner in the wetter regions of the Western Ghats and Ceylon. These regions also support several endemic wet rain forest species that have probably diverged recently from Oriental stock in ecological isolation. Invasion of southern India may not have been continuous, but perhaps broken up into episodes related to factors such as the Pleistocene fluctuations of climate. Investigation of this must await further fieldwork and taxonomic studies on the fauna of southern India. Evidence for such waves of invasion in the East Indies will be presented in HOLLOWAY (forthcoming paper).

In the Himalaya the lower and medium altitudes are predominantly occupied by species from centres in the Oriental Region, more especially from genera with marked centres in the Eastern Himalaya. The higher altitudes at or above the tree line, especially in the Northwest Himalaya, have been colonised mainly by species from the Palaearctic centres of the Pamirs and Turkestan. In the northwest there are also semi-arid and Mediterranean-type habitats at medium altitude containing species from the Palaearctic and Mediterranean centred genera.

These patterns are suggested to be the result of the adaptation of a relatively uniform, early Tertiary, Asian biota (grading from tropical in the south to temperate in the north) to progressive climatic deterioration throughout the Tertiary, combined with the disruption of the ecological regularity of the area by mountain building episodes and associated climatic effects in central Asia, the Middle East and southern Europe. This mountain building was probably a direct result of the interaction with mainland Asia of the northward-moving Indian Ocean tectonic plate and (during the latter part of the Tertiary) the northern continental shelf of the southern Indian land mass.

REFERENCES

AXELROD, D. I., 1960. The evolution of flowering plants. In SOL TAX (ed.), Evolution after Darwin, 1: 227–305. Chicago: Univ. Chicago Press.

BARLOW, H. S., H. J. BANKS & J. D. HOLLOWAY, 1971. A collection of Rhopalocera Mt. Kinabalu, Sabah, Malaysia. *Oriental Insects*, 5:269-296.

CARCASSON, R. H., 1964. A preliminary survey of the zoogeography of African butterflies. *E. Afr. Wildlife J.*, 2: 122–157.

CARLQUIST, S., 1966. The biota of long distance dispersal. 1. Principles of dispersal and evolution. *Q. Rev. Biol.*, 41: 247–270.

CORBET, A. S., 1941. The distribution of butterflies in the Malay Peninsula. *Proc. R. ent. Soc. London (A)* 16: 101–116.

CORBET, A. S., 1949. Observations on the species of Rhopalocera common to Madagascar and the Oriental Region. *Trans. R. ent. Soc. London*, 99: 589–607.

CORBET, A. S. & H. M. PENDLEBURY, 1956. The Butterflies of the Malay Peninsula. Revised edition. London and Edinburgh: Oliver & Boyd.

COUPER, R. A., 1960. Southern hemisphere Mesozoic and Tertiary Podocarpaceae and Fagaceae and their palaeogeographic significance. *Proc. R. Soc. (B)* 152: 491–500.

COWAN, C. F., 1966. Indo-Oriental Horagini (Lepidoptera: Lycaenidae). *Bull. Br. Mus. nat. Hist. (Ent.)*, 18: 103–141.

COWAN, C. F., 1967. The Indo-Oriental tribe Cheritrini (Lepidoptera: Lycaenidae). *Bull. Br. Mus. nat. Hist. (Ent.)*, 20: 78–103.

DARLINGTON, P. J. JR., 1957. Zoogeography: The Geographical Distribution of Animals. New York: John Wiley.

ELIOT, J. N., 1969. An analysis of the Eurasian and Australian Neptini (Lepidoptera: Nymphalidae). *Bull. Br. Mus. nat. Hist. (Ent.)*, Supplement 15.

EVANS, W. H., 1937. A Catalogue of the African Hesperiidae in the British Museum. London: British Museum (Nat. Hist.).

EVANS, W. H., 1949. A Catalogue of the Hesperiidae from Europe, Asia and Australia in the British Museum (Natural History). London: British Museum (Nat. Hist.).

EVANS, W. H., 1957. A revision of the *Arhopala* group of Oriental Lycaenidae (Lepidoptera: Rhopalocera). *Bull. Br. Mus. nat. Hist. (Ent.)*, 5: 85–141.

FISHER, R. A., 1930. The Genetical Theory of Natural Selection. New York: Dover.

FLETCHER, T. B., 1909. Lepidoptera, exclusive of the Tortricidae and Tineidae, with some remarks on their distribution and means of dispersal amongst the islands of the Indian Ocean. *Trans. Linn. Soc. London*, 13: 264–324.

GABRIEL, A. G., 1943. A revision of the genus *Ixias* Hübner (Lepidoptera: Pieridae). *Proc. R. ent. Soc. London*, 12: 55–70.

GOOD, R., 1953. The Geography of the Flowering Plants. 2nd. edition. London, New York, and Toronto: Longmans.

GOOD, R., 1960. On the geographical relationships of the angiosperm flora of New Guinea. *Bull. Br. Mus. nat. Hist. (Bot.)*, 2: 205–226.

GROSS, F. J., 1961. Zur Geschichte und Verbreitung der euroasiatischen Satyriden (Lepidoptera). Verh. dt. zool. Ges. Munster 1960 *Zool. Anz., (Suppl.)*, 24: 513–529.

HIGGINS, L. G., 1941. An illustrated catalogue of the Palaearctic *Melitaea* (Lep: Rhopalocera). *Trans. R. ent. Soc. London*, 91: 175–365.

HOLLOWAY, J. D., 1969. A numerical investigation of the biogeography of the butterfly fauna of India, and its relation to continental drift. *Biol. J. Linn. Soc.*, 1: 373–385.

HOLLOWAY, J. D., 1970. The biogeographical analysis of a transect sample of the moth fauna of Mt. Kinabalu, Sabah, using numerical methods. *Biol. J. Linn. Soc.*, 2:259–286.

HOLLOWAY, J. D. The affinities within four butterfly groups (Lepidoptera: Rhopalocera) in relation to general patterns of butterfly distribution in the Indo-Australian area. *Trans. R. ent. Soc. Lond.* Forthcoming paper.

HOLLOWAY, J. D. & N. JARDINE, 1968. Two approaches to zoogeography: a study based on the distributions of butterflies, birds and bats in the Indo-Australian area. *Proc. Linn. Soc. London*, 179: 153–188.

HOOKER, J. D., 1909. A sketch of the flora of British India. Imperial Gazetteer of India, Vol. 1. London: Oxford University Press.

HOWARTH, T. G., 1957. A revision of the genus *Neozephyrus* Sibatani and Ito (Lepidoptera: Lycaenidae). *Bull. Brit. Mus. nat. Hist. (Ent.)*, 5: 236–285.

JARDINE, N. & R. SIBSON, 1968. The construction of hierarchic and non-hierarchic classifications. *Comput. J.*, 11: 177–184.

MACARTHUR, R. H. & E. O. WILSON, 1967. The Theory of Island Biogeography. Princeton, New Jersey: Princeton University Press.

MANI, M. S., 1962. Introduction to High Altitude Entomology. London: Methuen.

MANI, M. S., 1968. Ecology and Biogeography of High Altitude Insects. The Hague: Dr. W. Junk N.V.

MCKENZIE, D. & J. G. SCLATER, 1971. The evolution of the Indian Ocean since the Late Cretaceous. *Geophys. J. R. astr. Soc.*, 25: 437–528.

MOREAU, R. E., 1952. Africa since the Mesozoic with particular reference to certain biological problems. *Proc. zool. Soc. London*, 121: 869–913.

MUNROE, E. & P. R. EHRLICH, 1960. Harmonization of concepts of higher classification of the Papilionidae. *J. Lepid. Soc.*, 14: 169–175.

PAULIAN, R., 1958. Le peuplement entomologique de Madagascar. Proc. 10th. Int. Congr. Ent. Montreal, 1956, 1: 789–794.

PAULIAN, R. & G. BERNARDI, 1951. Les *Eurema* de la region malgache (Lep: Pieridae). *Naturaliste malgache*, 3: 139–154.

PRESTON, F. W., 1962. The canonical distribution of commonness and rarity. *Ecology*, 43: 185–215, 410–432.

RIPLEY, S. D., 1955. Considerations on the origin of the Indian avifauna. *Bull. natn. Inst. Sci. India*, 7: 269–275.

RIPLEY, S. D., 1961. A Synopsis of the Birds of India and Pakistan. Bombay: Bombay Natural History Society.

ROBINSON, P. L., 1970. The Indian Gondwana formations – a review. First Symposium on Gondwana Stratigraphy, 1967. Published by I.U.G.S.

ROSS, H. H., 1953. On the origin and composition of the nearctic insect fauna. *Evolution, Lancaster, Pa.*, 7: 145–158.

SEITZ, A., (ed.). The Macrolepidoptera of the World. Stuttgart: Alfred Kernen. 1909. Vol. 1. The Palaearctic Butterflies. 1925. Vol. 9. The Indo-Australian Rhopalocera. 1927. Vol. 13. The African Rhopalocera.

STEENIS, C. G. G. J. VAN, 1934–1936. The origins of the Malaysian mountain flora. *Bull. Jard. Bot. Buitenz. (ser. 3)*, 13: 135–262, 289–417; 14: 56–72.

TALBOT, G., 1928–1937. Monograph of the Pierine genus *Delias*. London: John Bale, Danielsson and British Museum (Nat. Hist.).

TALBOT, G., 1939. The Fauna of British India, including Ceylon and Burma. Butterflies. Vol. 1. Second edition. London: Taylor & Francis.

TITE, G. E., 1966. A revision of the genus *Anthene* from the Oriental Region (Lepidoptera: Lycaenidae). *Bull. Br. Mus. nat. Hist. (Ent.)*, 18: 255–275.

WADIA, D. N., 1953. Geology of India. London: Macmillan.

WILLIAMS, C. B., 1930. The Migration of Butterflies. Edinburgh: Oliver & Boyd.

WILTSHIRE, E. P., 1962. Studies in the geography of Lepidoptera VII. Theories of the origin of the west Palaearctic and world faunae. *Entomologist's Rec. J. Var.*, 74: 29–39.

WYNTER-BLYTH, P., 1957. Butterflies of the Indian Region. Bombay: Bombay Natural History Society.

ZEUNER, F. E., 1942. Two new fossil butterflies of family Pieridae. *Ann. Mag. nat. Hist., (11)* 9: 409.

ZEUNER, F. E., 1943. Studies in the systematics of *Troides* Hbn. and its allies. *Trans. zool. Soc. London*, 25: 107–184.

ZEUNER, F. E., 1961. Notes on the evolution of the Rhopalocera (Lep.). Verh. XI Int. Kongr. Ent. Wien, 1960, 1: 310–313.

XVI. SOME ASPECTS OF THE ECOLOGY AND GEOGRAPHY OF DIPTERA

by

SANTOKH SINGH

1. Introduction

The Diptera constitute one of the major orders of insects in India and present a number of complex problems in ecology and distribution. With perhaps the exception of certain groups of medical and veterinary importance, the order has, however, been sadly neglected and the total number of species so far described from within the biogeographical area of India is less than two thousand. Our limited experience in the field with certain families like Chloropidae, Agromyzidae, etc. has demonstrated that this must represent perhaps less than one-twentieth of the species that still await discovery. Our knowledge of the distribution of even the known species is extremely meagre and the life-histories and habits of hardly half a dozen species are adequately known. An attempt at discussing the ecology and geography of the order under these discouraging circumstances must, therefore, appear purposeless; our main object here is on the other hand to draw attention of workers to certain broad trends and thus stimulate further research. We consider in the following pages in more or less general terms certain salient aspects of the ecology, distributional peculiarities and faunistic affinities mainly of Tipuloidea, Mycetophiloidea, Culicidae, Simuliidae, Bibionidae, Blepharoceridae, Deuterophlebiidae, Psychodidae, Stratiomyiidae, Tabanidae, Syrphidae, Calliphoridae, Chloropidae and Agromyzidae.

2. Major Ecological Types

The general ecology of Diptera in India is dominated by two major factors of great fundamental importance, which may be traced back to the salient facts of the geomorphological evolution of the region, viz. 1. the division of the region into the Peninsular and Extra-Peninsular areas (see chapter II) and 2. the characteristic monsoon climate (see chapters IV & V). Both in the Peninsula and in great parts of the Extra-Peninsular areas the ecology of the order is dominated by the monsoon climate, but particularly the pattern of rainfall distribution. It is the abundance and distribution of the rainfall and not the atmospheric temperature that determine the habits, phenology, life-cycles, number of generations, population levels and indeed nearly every other aspect of the ecology of nearly all the families throughout the land. While this fact may be readily observed anywhere in the Peninsula, a little close study will show that likewise in the Extra-Peninsular areas of the Eastern Himalaya, Assam, large areas of the Lower Ganga Plains and of the Terai (see chapter II) the Diptera ecology is essentially monsoon dominated. It is also of particular interest to remark that even among the typical Diptera of the western borderlands, the semi-arid and arid zones and the upper Ganga Plains the fundamental factor in ecology is the monsoon rainfall. It is only in the case of the northern margins of the Middle Ganga Plains and in the Himalaya to the west of the great defile of the R. Sutlej that the atmospheric temperature plays any significant rôle in determining the composition and character of the Diptera communities, the life-cycles and habits of the species and their geographical distribution. The typically monsoon-dominated ecology is modified to some extent by other factors, but particularly the ones that have resulted from recent rapid urbanization of large areas of the country and the increase in industrialization in the east.

A most striking character of the ecology of Diptera, especially of the Peninsula and parts of the Indo-Gangetic Plains is the remarkable cyclic periodism or seasonal succession, reflecting the fundamental influence of the monsoon climate. Even casual observation will show that over large areas, the species complex in nearly all families during the period of the southwest monsoon rainfall is predominantly composed of humid-tropical endemic or humid-tropical eastern intrusive elements, but in the post-monsoon period the species complex in the same areas shows considerable affinity to the Mediterranean-Ethiopian stock. The species complex of the Indo-Gangetic Plains during the monsoon rainfall is, for example, typically Peninsular, but in the dry season or winter of north India, we find that the species which succeed the Peninsular forms are either identical with or are closely related to Mediterranean and the northern Palaearctic species. The Agromyzid genus *Phytomyza* is, for example, always associated with the socalled winter weeds or the temperate elements of the flora in the northern plains and totally disappears in the monsoon season from these plains, though it is common in the outer Himalaya during the summer. There is thus a more or less pronounced phenological alternation of ecological and faunistic types among the Diptera of at least the transitional areas between the Peninsula on the one hand and the Himalaya on the other hand. This seasonal succession of the species complex throws considerable light on the history of the faunal interchanges and the changes in the composition of the Diptera during the Pleistocene and Post-Pleistocene times. This may be interpreted as recapitulating the southward advance of the geographical range of the Palaearctic elements during the periods of glaciation on the Himalaya and of the advance of the humid-tropical Peninsular elements northwards during the Inter-Glacial times.

501

On the basis of their general ecology and the species composition, we may recognize the following major types among the Diptera of India: 1. the monsoon or the wet-season communities, 2. the dry season or the winter communities, 3. the humid-tropical forest elements, 4. the dry tropical elements of the deciduous forest, savannah and semi-arid communities, 5. the temperate zone communities and 6. the synanthropous forms.

2.1. The monsoon communities

The monsoon-breeding forms are also known as the wet-season Diptera, because the adults are found almost exclusively during the months of the southwest monsoon rainfall. Like nearly every other wet-season group of insects, the monsoon Diptera is a complex of multivolent or even continuously breeding species. There is a rapid succession of several, often as many as a dozen overlapping generations during these months. They do not as a rule disappear entirely during the dry season, but only become more or less sparse as adults and the intensity of breeding slows down very considerably. During this time there is also a marked dominance of terrestrial-breeding over the aquatic-breeding species, inspite of the rains mainly because during the rainy months, the water is generally turbid, greatly disturbed and also rapidly drains off the major part of the Peninsula. The bulk of the terrestrial species breeds also in living plant parts in the early weeks, but later after the peak of the rainfall is past we find an increasing proportion of species that breed in decaying organic matter. Although there are considerable numbers of Nematocera like Psychodidae and Mycetophiloidea, the wet-season Diptera are predominantly Brachycera. The character species of the monsoon complex belong typically to Pipunculidae, Chloropidae, some Agromyzidae, Drosophilidae, Asilidae, Syrphidae, Stratiomyiidae, Empididae, Dolichopodidae, Platypezidae, some Trypetidae, Conopidae, Celyphidae, Diopsidae, many Calliphoridae, Tachinidae, Anthomyiidae and other Muscoidea. Though the blood-sucking flies are considerably less abundant during the monsoon rainfall months than in the post-monsoon period, there are nevertheless numerous species of the Phlebotomini, Ceratopogonidae; and some typical wet-season Culicidae (especially the tree-hole breeding forms) are also typical members of the monsoon communities. *Anopheles culicifacies* which breeds, for example, commonly in open irrigation channels, river beds, rainwater puddles and pools, ponds, burrow-pits, etc., is a common monsoon species. *Anopheles minimus* is also characteristic of paddy fields during the monsoon months, but it usually avoids areas flooded with muddy rain water. The most typical monsoon breeding mosquito is perhaps *Anopheles varuna*, which is characteristic of any collection of rain water, even by the roadside during and immediately after the monsoon rains. Other common species associated with the

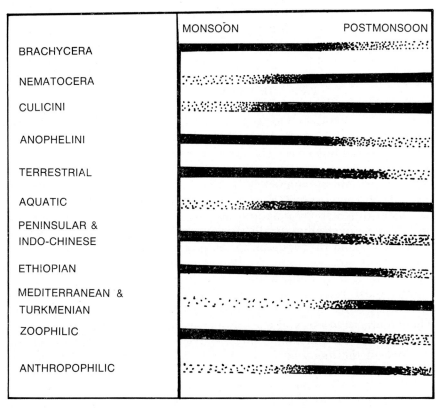

Fig. 95. Diagrammatic representation of the characteristic shift in the dominance of types of Diptera synchronizing with the rhythms of the alternation of the monsoon and post-monsoon periods in India.

monsoon rains include *Anopheles stephensi, Anopheles annularis, Anopheles pallidus, Anopheles jeyporensis, Anopheles subpictus,* etc. It seems curious that among the Culicidae, Anophelini are conspicuously dominant over the Culicini in the monsoon ecosystem. It may also be remarked that the Anophelini, characteristic of the monsoon ecosystem, have largely their faunistic affinities with the east and the northeast tropical countries.

It seems paradoxical that during the months of the monsoon rainfall the aquatic species of Diptera should constitute minor elements over large areas of India, but become dominant only in the post-monsoon months. Heavy and frequent floods, rapid flow and high turbidity of water and other rapidly fluctuating conditions that mark monsoon rain water collections would appear to act rather unfavourably to the breeding of Diptera, so that only strictly eulimnetic rather than the rheatic elements may be observed. Even the eulimnetic forms are relatively sparse, during the monsoon weeks. The greatest bulk of the wet-season species is typically widely distributed in the Peninsula, in the eastern borderlands, the whole of the Indo-Gangetic Plains and in the semi-arid and arid zones (the Indian Desert). They also extend northwards deep on the Siwalik and some of the transverse valleys of the outer Himalayan ranges.

2.2. The winter communities

The species-complex characteristic of the post-monsoon communities represents a most remarkable ecological succession, in which the adults of species breeding extensively during the southwest monsoon rains more or less completely disappear when the rains recede in a given area and are replaced almost completely by those of other species. The latter are typically characteristic of basically different ecosystems. The larvae breed in wholly different habitats and also show different faunistic affinities. Fig. 95 shows diagrammatically the shift in the general dominance among the Diptera from the monsoon to the post-monsoon months.

The most striking fact of taxonomic, ecological and distributional importance is the disappearance of the dominance of Brachycera (that characterized the monsoon communities) and the very pronounced ascendency of the Nematocera during the winter in large areas of India. There is also a curious and marked increase in the total aquatic elements, particularly the Culicini, Chironomidae, Simuliidae, etc., after the cessation of the monsoon rains. Among the few winter Brachycera the dominance now markedly shifts to the Syrphidae, Empididae, Asilidae, etc. We may also observe a marked shift in the dominance, among the Brachycera, to debris-breeding from live-plant breeding species. The few plant-breeding species are now characteristically restricted to the socalled winter-weeds or the temperate species of plants, described in chapter VI. This is readily illustrated by the occurrence of such genera like *Phytomyza*, *Agromyza* and *Cerodontha* of the Agromyzidae, mining in leaves of winter weeds. Even the fruit infesting Trypetidae are now largely of Ethiopian affinity. The faunistic affinities to the humid tropical eastern amphi-theatres and the proportion of endemic elements diminish remarkably and we may also observe in their place an increased population of the Ethiopian, Mediterranean and steppes elements in the Peninsula. The number of blood-sucking flies rises remarkably abruptly when the monsoon rains stop. It may also be of interest to observe that while the sparse blood-sucking flies of the monsoon months are primarily zoophilic, during the post-monsoon weeks the blood-sucking flies are strikingly anthropophilic. These peculiarities are readily illustrated by the domi-nance of *Culex fatigans*, *Armigeres*, *Aëdes*, etc. The succession of the Palae-arctic and temperate elements after the typically Peninsular types is particularly striking during the winter in the Indo-Gangetic Plains of north India, especially in the western parts of the Middle Ganga Plain and in the arid areas of Rajasthan.

504

2.3. The humid-tropical elements

The humid-tropical forest elements constitute the greatest bulk of all Diptera, nearly throughout the whole of India, including great parts of the semi-arid and deciduous forest-covered tracts. This ecological anomaly and occurrence of habitat-fremde species have already been referred to in an earlier chapter; the occurrence of these Diptera in ecologically anomalous areas is strong evidence that formerly these areas were largely humid-tropical in nature and aridity is of relatively recent occurrence in these parts. The humid-tropical Diptera are, therefore, the most widely distributed and ecologically dominant types that have secondarily specialized for a wide variety of other habitats. There is abundant evidence to conclude that the original Diptera fauna of India comprised these elements in great proportion. It is these elements which are also dominant among the intrusive components of our Diptera, particularly the genera and species which differentiated in the Indo-Chinese amphitheatre. In the course of the phenological succession, synchronizing with the rhythm of the monsoon climate, the humid-tropical elements of the Diptera do not completely disappear, but merely show a more or less pronounced fall in relative abundance. This ecological type may be met with practically in all families known so far from India, with perhaps the exception of exclusively Palaearctic genera.

2.4. The dry tropical deciduous forest communities

These Diptera constitute minor elements, which also occur in isolated pockets, markedly disjunctly rather than continuously like the humid-tropical elements. The available evidence would seem to indicate that the species characteristic of only the deciduous forests are relatively recently differentiated offshoots of the primarily humid-tropical forms or they are recent intrusives from the arid-zone stock. They appear in the deciduous forests not during the months of the southwest monsoon (when actually the humid-tropical elements dominate even here), but only after the retreat of the monsoon.

2.5. The temperate-zone communities

The characteristic species of the temperate-zones represent an insignificant proportion of our Diptera and are largely of Pleistocene differentiation. The species belong mostly to the Tipuloidea, Bibionoidea, Chironomidae, Simuliidae, Ceratopogonidae, Tabanidae, Stratiomyiidae, Syrphidae, Dolichopodidae, Empididae, some Agromyzidae like *Cerodontha*, *Agromyza*, *Liriomyza*, etc., Tachinidae, etc. They are generally concentrated in the western borderlands, in the upper reaches of the forest zone of the outer ranges of the Northwest Himalaya, the high altitudes of the inner Himalaya and disjunctly as Pleistocene relics on Mahendragiri, Shevroy, Palni, Anamalai and Cardamom Hills and on the Western Ghats in the Peninsula.

2.6. THE SYNANTHROPOUS COMMUNITIES

The anthropophilic species are secondarily specialized recent derivatives of the primarily zoophilic Diptera, largely as byproducts of rapid urbanization. It is interesting to remark, for example, that the species of the blood-sucking flies that are decidedly anthropophilic in urban areas are, however, wholly or nearly completely zoophilic in rural environment and prefer cattle to man. The pronouncedly synanthropophilous types are dominated by the humid-tropical and temperate forms almost in equal proportions, though there are marked regional and seasonal differences. The eusynanthropophilic species constitute an insignificant minority of the total species of Diptera known at present and their distribution and abundance are influenced largely by the degree of congestion and insanitary conditions prevailing in our urban areas. It is interesting to note that the strictly synanthropous species are remarkably sparse in rural environment.

3. Nematocera

The Tipuloidea (BRUNETTI 1912) are perhaps among the most common Nematocera of India and nearly three hundred species have been described so far. The great majority of the species are typically concentrated in the forest-covered Siwalik and Lesser Himalayan Ranges and on the Western Ghats. The dominant genera include *Ptychoptera, Pseliphora, Tipula, Pachyrhina, Dicranomyia, Geranomyia, Limnobia, Erioptera, Trichocera, Limnophila, Eriocera*, etc. The Himalayan species, especially those that occur below the timberline, seem to have been derived nearly equally from the eastern Asiatic humid-tropical forest amphitheatre and from the temperate Euro-Asiatic or the Palaearctic amphitheatre. The derivatives of the former have apparently extended in bulk westwards more than the latter, which represents essentially the southern fringe of the range of Turkmenian forms, so abundant on the Himalaya above the timberline. Considerable western genera and also a number of species with western affinities are found in the Northwest Himalaya, Sind, the Punjab plains, northern and western parts of the Peninsula and have even infiltrated into Ceylon. Some of these forms have intruded eastwards on the mountains of north Burma across Assam and their outliers have reached Java. The Peninsular Tipulids have often very pronounced affinity to the Himalayan forms and represent no doubt the Pleistocene relics and endemics which differentiated from the western intrusive elements. Exceedingly few species seem to be widely distributed; *Sym-*

plecta punctipennis occurs, for example, throughout Europe and extends through the Himalaya to nearly Assam.

The Mycetophiloidea (BRUNETTI 1912) are likewise known largely from the Himalayan forests. Of about one hundred and fifty species known at present, nearly 115 occur in the Himalaya, from Assam to Kumaon. Only 20 % of these species show western affinity, so that the bulk of the Himalayan Mycetophiloidea are humid-tropical eastern Asiatic elements, which have intruded westwards from the Indo-Chinese subregion. The Peninsular forms are partly of eastern and partly of western affinities. There are besides considerable numbers of monotypic endemic genera, in the Peninsula. The dominant genera are *Platyura, Mycomyia, Macrocera, Sciara, Ceroplatus, Sciophila, Odontopoda, Leia, Myceto-phila, Rhynchosia*, etc. Recent field experience shows that the Peninsula is nearly as rich in Mycetophiloidea as the Himalaya, but the area still remains almost unexplored.

The Bibionoidea (BRUNETTI 1912) are relatively poorly known, but nevertheless seem to be nearly equally abundant in the Himalaya and Peninsula. The family is apparently a relatively recent intrusive from the west or north and its presence in the Peninsula, particularly its concentration on the Western Ghats, is evidently a Pleistocene event.

The Psychodidae are predominantly a humid-tropical eastern group in India and are nearly equally abundant in the Peninsular and Extra-Peninsular areas. The common genera include *Psychoda, Pericoma*, and the remarkable endemic *Brunettia, Parabrunettia* and *Horaiella* and *Tel-matoscopus* with the subgenus *Neotermatoscopus*. *Parabrunettia* (BRUNETTI 1912, HORA 1934) occurs in the east, in Orissa, Tranvancore and Ceylon. The Phlebotominae are on the contrary largely of Mediterranean-Ethiopian descent and are particularly abundant in the Peninsula and in the transitional Indo-Gangetic Plains. They are, however, conspicuously poorly found on the Himalaya. The Ethiopian and Oriental elements of the Phlebotominae converge and intermingle in the north-west plains. The Simuliidae are even less known than Psychodidae, though they are abundant in mountain streams throughout the Himalaya, Assam, Eastern and the Western Ghats. The Chironomidae are perhaps the most neglected of the Nematocerous families, though they are extremely common (BRUNETTI 1912). The family is represented by tropical forms in the Peninsula and by Palaearctic forms like *Diamesa, Brillia, Orthocladius*, etc. on the Himalaya (KAUL 1970, SINGH 1958).

The extremely interesting Blepharoceridae (HORA 1934, KAUL 1971, TONNOIR 1931) are confined to the hill streams in Assam, Himalaya, and the Western Ghats and include species of *Blepharocera, Apistomyia, Philorus, Hammatorhina, Liponeura* and *Horaia; Hammatorhina* is known so far from Ceylon only.

The family Deuterophlebiidae is confined to the cold torrential streams of glacial origin in the Northwest Himalaya; the distribution of this family is discussed by MANI (1968).

The Itonididae (MANI 1934–38, 1942–45, 1964) are widely distributed throughout the region. In the eastern parts the affinities of the species are largely with the Malayan subregion but in the Peninsula there are numerous endemic genera and species. Considerable part of the Peninsular species like *Asphondylia ricini*, *A. sesami*, *A. tephrosiae*, etc. are either identical with or very closely related to Ethiopian forms. Large numbers of Mediterranean like *Rhopalomyia millefolii*, and northern Palaearctic elements like *Geocrypta galii*, *Kiefferia*, etc. are found among them in the Northwest Himalaya. The dominant genera include *Lasioptera*, *Dasineura*, *Oligotrophus*, *Schizomyia*, *Asphondylia*, *Contarinia*, *Hormomyia*, *Lestodiplosis*, etc. Although most of the gallicolous species known so far from India are associated typically with Dicotyledons, there is, however, abundant field evidence to conclude that the Indian gall-forming Itonididae associated with Monocotyledons are not inconsiderable.

The Culicidae (BARRAUD 1934, CHRISTOPHER 1933, ROY & BROWN 1970) have been studied extensively in India, because of their medical importance and nearly three hundred species have been described so far. Nearly one-sixth of the total Culicidae known at present belongs to the Anophelini and the rest are Culicini; some Megarhinini are also known. Ecologically the Culicidae of India fall into major groups, viz. the wet-season complex of species and the dry season complex of species. The wet-season complex is characteristically rich in Indo-Chinese and Malayan elements and the species breed generally during the months of the southwest monsoon in most parts of the Peninsula or in the east coast during the retreating monsoon rainfall weeks. The species are typically concentrated in the northeast areas of the Peninsula, in Bengal, Assam and parts of Bihar. They are rather sparsely distributed in the Upper Ganga Plains. The dry-season complex comprises the winter-breeding species, largely of Ethiopian affinities and often also with more or less pronounced Mediterranean facies. They are abundant in the Indus Plains, the Upper Ganga Plains and large parts of the Peninsula, but a number of them extend as far east as Bengal and Assam. We observe, therefore, a characteristic seasonal oscillation of the tropical and the temperate, Oriental and Mediterranean-Ethiopian elements in the densely populated areas of north India.

A most striking difference that distinguishes the Indian Culicidae from nearly all other families of Nematocera is the fact that it is rich in relatively young, highly variable and unstable endemic subgenera, species and sub-species, which have differentiated locally in the receding lagoons and marshes of the Post-Pleistocene times (in areas formerly covered by the Tethys Sea). We find thus that almost 50% of the species are endemic. Genera like *Uranotaenia*, *Heizmannia*, *Haemogogus*, *Armigeres*, etc. are, for example, extremely rich in endemic species. Some of the genera of

508

Table I. Tropical Asiatic eastern (Indo-Chinese Malayan complex) *Anopheles* species found in India, with their distribution

Species & subspecies	Distribution in India	Distribution outside India
1. *aitkini bengalensis*	Assam-Burma, E. Himalaya, Central & W. Himalaya	Java, Cochin-China, China, Hong Kong
2. *insulaeflorum*	Assam-Burma, E. Himalaya, Central & W. Himalaya, Western Ghats	Java, New Guinea, Lesser Sunda Is.
3. *hyrcanus nigerrimus*	Northwest to E. Himalaya, Assam-Burma	Java, Sumatra, Borneo, Cochin-China
4. *hyrcanus sinensis*	Northwest to E. Himalaya, Assam-Burma	China, Japan, Formosa, Tonkin
5. *barbirostris*	Himalaya, Assam, Burma, Peninsula, Indo-Gangetic Plain	New Guinea, Cochin-China, Lesser Sunda Is.
6. *umbrosus*	Assam, Eastern to Kumaon Himalaya, Andamans Is.	Java, Borneo, Sumatra, Celebes, Tonkin
7. *kocki*	Northeast India	Philippines, Lesser Sunda Is., Borneo, Java
8. *leucosphyrus*	Eastern to Kumaon Himalaya, Peninsula	Assam-Burma, Philippines, Java, Borneo, Lesser Sunda Is.
9. *tessellatus*	Himalaya, Peninsula: Western Ghats, Assam-Burma, northeast end of Peninsula, Andaman Is.	China, Tonkin, Philippines, Thailand
10. *minimus*	Eastern to Kumaon Himalaya, Indo-Gangetic Plain, Eastern Ghats, Western Ghats	Cochin-China, Tonkin, Malaya, Java, Philippines
11. *aconitus*	E. Ghats, W. Ghats, E. Himalaya to Kumaon	Philippines, Cochin-China, Thailand, Malaya, Borneo
12. *vagus*	Himalaya up to Kumaon, Indo-Gangetic Plain, Assam, NE end of the Peninsula	Java, Sumatra, Tonkin
13. *sundaicus*	Himalaya up to Kumaon, Assam, Ganga Plain, Andaman Is.	Java, Sumatra, Celebes, Borneo, Lesser Sunda Is.
14. *maculatus*	Himalaya, Assam, Peninsula	Philippines, S. China, Java, Borneo, Tonkin, Thailand
15. *theobaldi*	Himalaya, Assam, Peninsular northeast	Philippines
16. *karwari*	E. Himalaya to Kumaon, Assam, Peninsula	Philippines, Tonkin, Cochin-China, Thailand, Borneo
17. *jamesi*	E. Himalaya to Kumaon, Peninsula including the northeast	Cochin-China, Tonkin
18. *ramsayi*	E. Himalaya to Kumaon, Peninsula including the northeast, Ceylon	Java, Sumatra, Thailand

509

Table I (continued)

Species & subspecies	Distribution in India	Distribution outside India
19. *annularis*	Himalaya, Assam, Peninsula	Philippines, Formosa, Java, Borneo, Thailand, Sumatra, Tonkin
20. *philippinensis*	E. Himalaya to Kumaon, Peninsula including the northeast	Thailand, Philippines, Java, Sumatra, Cochin-China
21. *pallidus*	E. Himalaya to Kumaon, Ganga Plain, Peninsula	Thailand, Sumatra
22. *minimus*	E. Himalaya to Kumaon, Assam-Burma, Peninsula	Yunnan, Indo-China, Thailand, Malaya, Java, Sumatra

Ethiopian stock like, for example, *Aëdes* have differentiated and diversified into a number of local subgenera *Christophersomyia, Indusius, Diceromyia, Aëdomorphus, Finlaya,* etc. *Culex* has also given rise to a number of subgenera like *Lutzia, Barraudius, Neoculex, Mochothogenes,* etc.

Among the Anophelini almost half the known species are derived from the Oriental stock and the rest comprise Mediterranean, Ethiopian and endemic elements, each in nearly equal proportions. The more important species of *Anopheles*, with Indo-Chinese and Malayan affinities, are listed in table I.

The Mediterranean-Ethiopian affinities are revealed by *Anopheles barianensis, Anopheles lindsayi, A. gigas simlensis, A. gigas baileyi, A. dthali, A. sergenti, A. culicifacies, A. subpictus, A. turkhudi, A. multicolor, A. superpictus, A. moghulensis, A. stephensi, A. splendidus, A. pulcherrimus* and others.

The Culicini of Indo-Chinese, Malayan or Peninsular affinity include *Culex sinensis, Culex whitmorei, C. gelidus, C. mimulus, Lophoceratomyia, Uranotaenia, Armigeres,* etc. The Mediterranean-Ethiopian Culicini, occurring in India, include *Culex mimeticus, C. barraudi, C. theileri, C. vagus, Theobaldia, Aëdes,* etc.

It may be observed that the African *Aëdes*-complex has radiated by way of Iraq and western Asia, across Middle Asia to the east and southeast Asia, diversifying into a remarkable number of locally endemic subgenera and species the whole of the way. Its influx into India has, therefore, been both from the northwest and to some extent from the northeast as well. The subgenera differentiated in the Indo-Chinese area are largely concentrated in the northeastern parts of India and in the Lower Ganga Plain, but the subgenera of the western stock are continuously distributed in the northwest, Upper Ganga Plain and in the Peninsula.

4. Brachycera

The Brachycera (Brunetti 1920, 1923, Senior-White, Aubertin & Smart 1940) are less adequately known than the Nematocera. Except for the Tabanidae and perhaps some other flies of veterinary importance, the group as a whole has been the most neglected of Diptera. Some of the important families we may consider here include Stratiomyiidae, Tabanidae, Bombyliidae, Pipunculidae, Agromyzidae, Syrphidae, Muscoidea especially Calliphoridae, Sarcophagidae, Oestridae, Hypodermatidae and Gastrophilidae.

The Stratiomyiidae are represented by about eighty species belonging to the genera *Odontomyia, Stratiomyia, Ptecticus, Sargus, Chloromyia, Microchrysa, Xylomyia, Pachygaster, Ptilocera, Oxycera*, etc. Though some typical Palaearctic forms like *Lasippa villosa* (with the subspecies *himalayensis* on the Northwest Himalaya), and the Mediterranean *Beris geniculata* (Himalaya) are known, the greatest bulk of the species are strictly Oriental and have a particularly strong Malayan affinity. The family as a whole is widely distributed throughout India.

The Tabanidae (Roy & Brown 1970) of India include *Tabanus striatus, T. hilaris, T. albimedius, T. rubidus, T. speciosus, T. brunnipes, T. ditaeniatus, T. rufiventris* and *T. orientis* as common species. Some of these species are of Ethiopian and Mediterranean affinity and the rest are strictly Oriental. *Tabanus crassus* is one of the largest species occurring in India and is typically a monsoon type. *Tabanus speciosus* occurs throughout the plains. It is of considerable interest to point out that *Tabanus striatus* is also known as the most prevalent horsefly of the Philippines, where it is associated with cattle and buffaloes. The distribution of the genus *Chrysops* is also interesting; in Africa the genus is confined to the rain-forests of the West and Central Africa and in India it is found in the tropical humid forests of Assam. The common species include *Chrysops dispar, C. indiana* and *C. stimulans. Chrysops indiana* occurs in Assam and Bengal and is largely an Indo-Chinese species; it also occurs discontinuously in South India. *Haematopota* is largely Indo-Chinese; *H. indiana, H. annandalei, H. singularis, H. assamensis* and *H. marginata* occur in Assam-Burma, but *H. horalis, H. tessellata* and *H. cingalensis* are found in Ceylon and *H. dissimilis* and *H. bilineata* are found in the Peninsula. *Pongomia longirostris* is restricted to the forest areas of India, especially Assam, Himalaya and the Western Ghats. *Stomoxys calcitrans* is widely distributed in India. The African *Stygeromyia* is represented by *S. maculosa* in India. The Palaearctic *Haematobosca* is sparsely found in north India. Another Palaearctic genus *Haematobia* is represented by *H. sanguisugens* in north India and in the Eastern Himalaya. *Bdellolarynx sanguinolentus* is the only species of that Palaearctic genus to be found in India; it also occurs in East Africa. *Hypoderma crossi* and *H. aeratum* (of the Palaearctic genus *Hypoderma*) occur in Pakistan and the Punjab. *Chrysomyia bezziana*, the

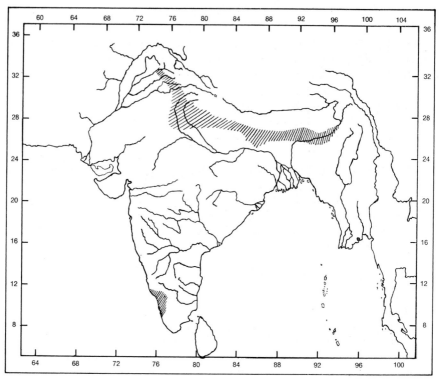

Fig. 96. Distribution of *Agromyza.*

common screw-worm fly of India, is also known from Africa. The cosmopolitan *Hippobosca maculata* is found commonly throughout India and
among the other Indian species we may include *H. capensis* and *H.
camelina.* Most of the blood-sucking zoophilous Brachycerous flies,
associated with domestic animals in India, apparently owe their present
distribution to human agency. *Calliphora erythrocephala* and *C. vomitoria*
are confined to Baluchistan and extend along the Himalaya eastwards
to Sikkim. *Lucilia illustris,* known from North America, Europe, Manchuria and China, occurs on the Himalaya and in Burma within our
limits. *Lucilia papuensis* has an interesting distribution; it is known from
Malabar and Ceylon and is also reported from Malaya, Thailand, Java,
Borneo, Celebes, Amboina, Ternate, Sumatra, New Guinea, South
China, Philippines, Australia and New Hebrides. The distribution of
Cainsa indica includes Ceylon, South India, Orissa Hills in the northeast
corner of the Peninsula, Malaya, Java, Formosa and Celebes. *C. testacea*
occurs in the hills of Ceylon, South India, Malaya and Philippines. The
genus *Bengalia* occurs in the Ganga Plains, eastern India and in Ceylon
and South India and three species extend to Malaya and Formosa and
Thailand.

512

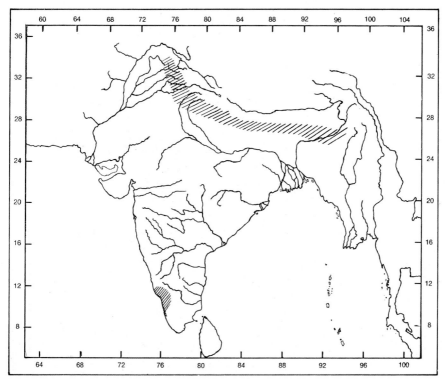

Fig. 97. Distribution of *Phytomyza*.

An intensive survey of the Agromyzidae was carried out recently by the writer nearly throughout India (GARG 1971, IPE M. IPE 1971, IPE M. IPE & BERI 1971, SINGH 1971, SINGH & IPE M. IPE 1971a). Over one hundred and twenty species have been found; most of these species fall under the genera *Agromyza*, *Phytomyza*, *Melanagromyza* (Fig. 98), *Cerodontha*, *Liriomyza*, *Phytagromyza* and *Pseudonapomyza*. *Melanagromyza* has Ethiopian and Malayan distribution and in India nearly 40% of the species of the family so far known belong to this genus. The species are typically monsoon type and the genus is therefore relatively sparsely represented in north. The few species that do occur in the Himalaya are typically restricted to the transverse river valleys. The predominantly Holarctic *Phytomyza* (Fig. 97), *Agromyza* (Fig. 96) and *Cerodontha* are characteristically concentrated along the temperate and subtemperate areas of the Himalaya; some of them are met with as isolates in the High Ranges of the Western Ghats. We have, for example, *Phytomyza nilgiriensis*, *Agromyza sahyadriae* and *Liriomyza flaviola* in the Western Ghats. *Cerodontha* has diversified into the subgenera *Poëmyza* and *Dizygomyza* in the Himalaya and South India, and *Icteromyza* in the Upper Ganga Plains. *Liriomyza* is predominantly confined to the temperate Himalaya. *Pseudonapomyza*,

513

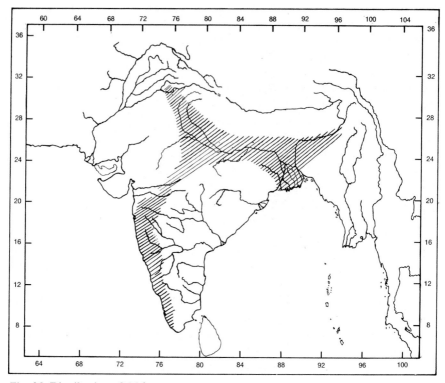

Fig. 98. Distribution of *Melanagromyza.*

a tropical genus, is found almost exclusively in the transitional area of the Indo-Gangetic Plains and in the Peninsula.

5. Conclusion

The Diptera of India present an interesting admixture of autochthonous endemic forms of the Peninsula, Indo-Chinese and Malayan derivatives often distributed discontinuously in the Peninsula and in the eastern borderlands, Palaearctic elements on the Himalaya and found discontinuously as Pleistocene relicts in South India and Mediterranean and Ethiopian forms widely and continuously distributed in the Peninsula and often also occurring as isolates in Assam. The ecology of the groups as a whole is dominated in large parts of India by the monsoon rainfall pattern. The composition, general distribution and the characteristic ecology of the order find their natural explanation in the vast changes, leading up to the uplift of the Himalaya and formation of physical connection of the Peninsula with Asia in Assam as an early phase of the Himalayan uplift. The ecology and biogeography of the order in India strongly reflect, therefore, the events of continental drift.

514

REFERENCES

BARRAUD, P. J. 1934. Fauna Brit. India, Diptera, Culicidae, Megarhini, Culicini, pp. 463, pls. viii.

BERI, S. K. & IPE M. IPE. 1971. Descriptions of two new species of the genera *Phytobia* Lioy and *Phytagromyza* Hendel (Diptera: Agromyzidae). *Oriental Ins.* (Suppl.), 1: 1–8.

BRUNETTI, E. 1912. Fauna Brit. India, Diptera, Nematocera, pp. 581, pls. xii.

BRUNETTI, E. 1920. Fauna Brit. India, Diptera, Brachycera, pp. 401, pls. IV.

BRUNETTI, E. 1923. Fauna Brit. India, Diptera III. Pipunculidae, Syrphidae, Conopidae, Oestridae, pp. 424, figs. 85, pls. VI.

CHRISTOPHER, S. R. 1933. Fauna Brit. India, Diptera IV. Culicidae, Anophelini, pp. 371, pls. III.

GARG, PRAMOD K. 1971. Taxonomic studies on Agromyzidae from Gangetic Basin. Part I. Descriptions of three new species. *Oriental Ins.*, 5(2): 179–188.

HORA, S. L. 1934. Remarks on Tonnoir's theory of evolution of the ventral suckers of Dipterous larvae. *Rec. Indian Mus.*, 35: 283–286.

IPE M. IPE. 1971. Descriptions of six new species of Agromyzidae from India. *Oriental Ins.*, 5(2): 165–178.

IPE M. IPE and S. K. BERI, 1971. Descriptions of two new species of the genus *Lemurimyza* Spencer (Diptera: Agromyzidae). *Oriental Ins.* (Suppl.), 1: 9–16.

KAUL, B. K. 1970. Torrenticole Insects of the Himalaya II. Two new Diamesini (Diptera: Chironomidae) from the northwest Himalaya. *Oriental Ins.*, 4(3): 293–297.

KAUL, B. K. 1971. Torrenticole Insects of the Himalaya V. Description of some new Diptera: Psychodidae and Blepharoceridae *Oriental Ins.*, 5(3): 401–434.

MANI, M. S. 1934–38. Indian Itonididae. *Rec. Indian Mus.*, 36–38, pp. 126.

MANI, M. S. 1942–45. Indian Itonididae. *Indian J. Ent.*, 4–7, pp. 65.

MANI, M. S. 1962. Introduction to high altitude entomology. London: Methuen & Co., pp. 304, figs. 88, pls. xvii.

MANI, M. S. 1964. Ecology of plant galls. The Hague: Dr. W. Junk Publishers, pp. 434, fig. 164.

MANI, M. S. 1968. Ecology and biogeography of high altitude Insects. The Hague: Dr. W. Junk Publishers, pp. 527, figs. 80.

MANI, M. S. and SANTOKH SINGH. 1961. Entomological Survey of the Himalaya. Part XXV. A contribution to our knowledge of the ecology of high altitude insect life of the Northwest Himalaya. *Proc. Zool. Soc. Calcutta*, 14(2): 61–135.

MANI, M. S. and SANTOKH SINGH, 1961–63. Entomological Survey of the Himalaya. Part. XXVI. A contribution to our knowledge of the geography of the high altitude insects of the nival zones from the Northwest Himalaya. *J. Bombay Nat. Hist. soc.*, 58: 387–406, 724–748 (1961): 59: 77–99, 360–381 (1962); 60: 140–172 (1963).

ROY, D. N. & A. W. BROWN, 1970. Entomology, Medical and Veterinary. Bangalore: Bangalore Printing & Publishing Co. pp. 855.

SENIOR-WHITE, R., D. AUBERTIN and J. SMART, 1940. Fauna Brit. India, Diptera VI. Calliphoridae, pp. 288, figs. 152.

SINGH, SANTOKH. 1958. Entomological Survey of Himalaya. Part XXIX. On a collection of nival Chironomidae (Diptera) from Northwest Himalaya. *Proc. Nat. Acad. Sci. India*, B, 28(4): 308–314.

SINGH, SANTOKH. 1961. Entomological Survey of Himalaya. Part XXXII. A note on a larva of an apparently undescribed species of *Deuterophlebia* Edw. (Deuterophlebiidae: Diptera) from the Northwest Himalaya. *Agra Univ. J. Res. (Sci.)*, 10(1): 109–114.

SINGH, SANTOKH. 1971. Agromyzidae exploration in India. School of Entomology, St. John's College, Agra (India), pp. 73 (Pvt. publication).

SINGH, SANTOKH and IPE M. IPE. 1967. A new species of *Phytomyza* from Western Himalaya (Agromyzidae: Diptera) *Oriental Ins.*, 1(1–2): 61–64.

515

SINGH, SANTOKH and S. K. BERI. 1968. Notes on the biology and descriptions of immature stages of *Phytomyza kumaonensis* Singh & Ipe, from Western Himalaya. *Bull. Ent., Delhi*, 9(1): 1–5.

SINGH, SANTOKH and IPE M. IPE. 1968. Descriptions of two new species of Agromyzidae from Northern India. *Oriental Ins.*, 2(1): 89–96.

SINGH, SANTOKH and IPE M. IPE. 1970. Descriptions of two new species of *Phytobia* Lioy from South India. *Oriental Ins.*, 4(1): 59–64.

SINGH, SANTOKH and P. K. GARG. 1970. Descriptions of two new species of Agromyzidae from India. *Oriental Ins.*, 4(4): 427–433.

SINGH, SANTOKH and S. K. BERI. 1971. Studies on the immature stages of Agromyzidae from India. Part I. Notes on the biology and descriptions of immature stages of four species of *Melanagromyza* Hendel. *J. Natural Hist. London*, 5: 241–250.

SINGH, SANTOKH and IPE M. IPE. 1971. Descriptions of two new species of *Melanagromyza* Hendel (Diptera: Agromyzidae). *Oriental Ins.*, 5(2): 223–228.

SINGH, SANTOKH and IPE M. IPE 1971a. A new Agromyzid genus *Indonapomyza* from India. *Oriental Ins.*, 5(4): 571–576.

TONNOIR, A. L. 1931. Notes on Indian Blepharocerid larvae and pupae with remarks on the morphology of Blepharocerid larvae and pupae in general. *Rec. Indian Mus.*, 32: 161–214.

XVII. ECOLOGY AND DISTRIBUTION OF
FRESH-WATER FISHES, AMPHIBIA AND REPTILES

by

K. C. JAYARAM

1. Introduction

This discussion on the ecology and distribution of fishes, amphibians and reptiles is based mainly on the researches of the author during the past two decades. For many years he was closely associated with Hora and Prof. Mani in their biogeographical studies. Considerable work was done on the fresh-water fishes in collaboration with Hora, who more than any one else contributed in recent years to focuss attention on some of the outstanding problems of the biogeography of India. In addition to specializing on the taxonomy of the Bagrid fishes of Asia and Africa, the author has also had numerous opportunities of personally collecting fishes from nearly all over India and of making observations on their ecology, habits and distributional peculiarities; some attention was also paid to the amphibians and reptiles. The extensive collections of the Zoological Survey of India were also available to him for study and reference. Numerous discussions on some of the intriguing problems of the fresh-water fishes were also held with Prof. Mani in recent years. The results of the pioneer investigations of Medlicott and Blanford (1879), Day (1876–78, 1889), Blanford (1901), Smith (1931, 1935, 1943) and Hora (1933–55, Hora & Naïr, 1941) have also been extensively relied upon in this contribution.

I am thankful to the Director, Zoological Survey of India, Calcutta, for facilities for work. I am much indebted to Prof. Dr M. S. Mani for the benefit of many fruitful discussions and for helpful criticisms. I also thank Mr T. S. N. Murthy, Assistant Zoologist, Zoological Survey of India, Madras, who readily provided me with an up-to-date list of the Amphibia.

2. Fish

Day (1889) listed 1418 species of fresh-water and marine fishes, under seventy-two families and three hundred and forty-two genera, from the faunal limits of India. Since then considerable additions have been made to this list, until we know now over 1700 species under six hundred genera, distributed among two hundred families. The vast majority of these species are peripheral or sporadic in their distribution; many enter fresh-waters from the sea in river mouths or sometimes even spend much time

517

far away from the sea. Leaving these out of consideration here, the fresh-water fishes fall into two natural groups, viz. the primary fresh-water species and the secondary species. The primary fishes are as a rule generally confined to the fresh-water habitats, both limnetic and rheatic. The secondary fresh-water fishes on the other hand tolerate varying degrees of salinity and considerable changes in the media. Even among the strictly primary fresh-water fishes, some like the Bagridae, especially *Mystus gulio*, may be more salt-tolerant than some of the secondary fresh-water forms (MARSHALL, 1965; MILLER, 1966; SCHWARTZ, 1964). It is possible to consider about 15 % of the genera known so far from within our limits as strictly primary fresh-water forms. The secondary fresh-water genera amount to no more than 3 %. The marine and the peripheral genera of fish that constitute the bulk of our fish fauna are not considered here.

The distribution of six hundred and eighty-three species of primary fresh-water fishes, belonging to eighty-nine genera under seventeen families is discussed in the following pages. Of the seventeen families, the Cyprinidae are the largest and account for no less than thirty-one genera and three hundred and seventy-three species. The eleven families of Siluroid fishes, with forty-one genera and one hundred and seventy-six species, come next in importance. The other common families are Cobitidae, Homalopteridae, Psilorhynchidae, Indostomidae and Chan-nidae (see Table I).

The Oriental of India has perhaps the largest number of genera, viz. fifty-eight, representing nearly 63 % of the total genera of primary fresh-water fishes known so far from within our faunal limits. It is of considerable interest to observe that the Ostariophyseans dominate, particularly the Siluroid genera. Within the Oriental of India, the Indo-Chinese and Malayan subregions have a larger number of endemic genera, viz. 19 than the Indian subregion, which has only 9 endemic genera. The number of genera common to both these subregions is far more than those restricted to or endemic in either of the subregions and amounts to nearly fifty. These genera are phylogenetically old and well stabilized taxonomically. The genera endemic in the Indian subregion have interestingly fewer species than those endemic in the Indo-Chinese and Malayan subregions and even than those that are common to both.

The following is a list of the secondary fresh-water fishes (though not included in our final analysis):

Notopteridae
1. *Notopterus*
Belonidae
2. *Polyacanthus*
Horaichthyidae
3. *Horaichthys*
Cyprinodontidae
4. *Cyprinodon*
5. *Haplochilus*

Synbranchidae
6. *Amphipnous*
7. *Monopterus*
8. *Synbranchus*
Nandidae
9. *Badis*
10. *Nandus*
11. *Pristolepis*

Cichlidae
  12. *Etropius*
Anabandidae
  13. *Anabas*
  14. *Colisa*

Osphronemidae
  15. *Osphronemus*
Mastacembelidae
  16. *Rhynchobdella*
  17. *Mastocembelus*

The Synbranchid and Labyrinthicine fishes, rather well represented, are hardy forms, with well developed means of dispersal. *Osphronemus* has been widely introduced in the Orient. It may be observed that, as in the case of the primary fresh-water fishes, here also the Oriental of India has eleven genera but the Palaearctic part has only a single endemic genus and eight genera are common to the two major subregions within the Oriental of India. *Etropius* and *Horaichthys* are perhaps the only genera, which are restricted to the Indian subregion and *Monopterus* is the only genus endemic in the Indo-Chinese subregion. Five genera are distributed both in the Palaearctic and Oriental of India. *Notopterus* and *Mastocembelus* that come under this category are African genera, widely distributed and taxonomically stabilized in that continent, but with a few species in the Oriental Region. Two species *Polyacanthus signatus* and *Horaichthys setnai* are interesting in their distribution; the former occurs in Java and Ceylon and the latter is a Cyprinodont endemic in the West Coast of the Peninsula, about 160 km north and south of Bombay. Its closest relative *Tomerus* occurs in South America. A second species of *Polyacanthus* is endemic in the Peninsula.

2.1. Distributional patterns of the primary fresh-water fishes

The genera restricted to the Palaearctic of India include the following four:

Cyprinidae
  1. *Schizopygopsis*
  2. *Schizothorax*
  3. *Ptychobarbus*
  4. *Cyprinion*

Schizopygopsis and *Ptychobarbus* are endemic in the Turkmenian subregion, *Cyprinion* is a Mediterranean genus and *Schizothorax* is common to both the Mediterranean and Turkmenian subregions. Of the eighteen species of *Schizothorax*, known at present, ten are found in the Mediterranean subregion. All the seven species of *Cyprinion*, so far known, occur in the northwest borderlands and extend to southeast Iran. The three species of *Ptychobarbus* and a solitary species of *Schizopygopsis* are endemic in the Northwest Himalaya, mainly in the headwaters of the R. Indus, the tributaries of the R. Yarkand and other rivers in the area. These genera have not so far been recorded from areas east of the Great Defile of the R. Sutlej. They also occur at lower elevations only very rarely and the

519

few isolated reports represent no doubt stray specimens that have been washed down to the lakes in the plains by sporadic flash floods, to which most of the Himalayan rivers are subject. The small streams generally lack macro vegetation, though in some favourable situations some plants do grow (HORA & MUKERJI, 1935). *Schizothorax* and *Ptychobarbus* are generally found in lakes and in large rivers, with backwaters. *Schizopygopsis* is equally at home in lakes, large rivers and even in small rapids rich in algal growth. The various streams and lakes in the Northwest Himalaya, where these fishes occur, are not very widely separated from each other by high ridges. Many species are, therefore, common to different streams and unlike the species in the plains, they are not geographically restricted to specific watersheds. It would seem that these lakes and rivers of the Northwest Himalaya are relicts of a former larger and continuous mass of water.

These genera are undoubtedly very poor representatives of the Palaearctic elements, which have intruded from Middle Asia. It is interesting to observe that *Schizothorax* is known as fossil from the Pleistocene Karewas in Kashmir.

The following genera are common to the Palaearctic area and the Oriental of India; some of them are also Ethiopian:

Cyprinidae
1. *Garra*
2. *Crossocheilus*
3. *Oreinus*
4. *Schizothoracichthys*
5. *Diptychus*
6. *Labeo*
7. *Catla*
8. *Amblypharyngodon*
9. *Puntius*
10. *Tor*
11. *Esomus*
12. *Rasbora*
13. *Aspidoparia*
14. *Barilius*
15. *Danio*
16. *Chela*

17. *Oxygaster*
Cobitidae
18. *Botia*
19. *Lepidocephalichthys*
20. *Noemacheilus*
Bagridae
21. *Rita*
22. *Mystus*
Sisoridae
23. *Bagarius*
24. *Glyptothorax*
25. *Glyptosternum*
Clariidae
26. *Clarias*
Channidae
27. *Channa*

The genera *Oreinus*, *Glyptosternum*, *Schizothoracichthys* and *Diptychus* occur in a more dynamic environment than the four genera that are restricted to the Palaearctic area of India. *Diptychus* is a monotypic genus that occurs in rapid to very rapid streams supporting or also without algal growths (HORA & MUKERJI, 1935). *Glyptosternum* is a Palaearctic genus, apparently derived from the Indo-Chinese *Glyptothorax*-like ancestors. Of the three species known at present, *Glyptosternum akhtari* is confined to the R. Oxus in Afghanistan, *Glyptosternum reticulatum* is found in Ladak, (Leh), Kashmir, Chitral and Afghanistan and *Glyptosternum maculatum* is the only species, which extends to Sikkim in the Eastern Himalaya, far from its main

Table I. Percentage analysis of genera of primary fresh-water fish

Family	Total	Pal. + Ethiop.	Oriental
1. Cyprinidae	35.0	4.5	95.5
2. Cobitidae	11.0	—	100.0
3. Psilorhynchidae	1.1	—	100.0
4. Homalopteridae	4.5	—	100.0
5. Bagridae	4.5	—	100.0
6. Siluridae	3.3	—	100.0
7. Scheilbeidae	9.0	—	100.0
8. Pangasiidae	1.1	—	100.0
9. Amblycipitidae	1.1	—	100.0
10. Akysidae	1.1	—	100.0
11. Sisoridae	20.2	—	100.0
12. Clariidae	2.2	—	100.0
13. Heteropneustidae	1.1	—	100.0
14. Chacidae	1.1	—	100.0
15. Olyridae	1.1	—	100.0
16. Indostomidae	1.1	—	100.0
17. Channidae	1.1	—	100.0

home in Tibet. Of the three species of *Oreinus* so far known, one is endemic in the Mediterranean subregion, one in the Indo-Chinese subregion and only *Oreinus plagiostomus plagiostomus* extends to the Eastern Himalaya from the west, where it also occurs as far as Arabia and eastern Iran. *Oreinus plagiostomus* has diverged into many local forms in the Northwest Himalaya and hybridization is not also uncommon (Jayaram & Majumdar, 1964). *Schizothoracichthys* is a derivative from the Palaearctic *Schizothorax* and only one of the two species known at present extends into the Eastern Himalaya. *Diptychus maculatus*, the monotypic form, has also similarly extended into the Eastern Himalaya.

The Ethiopian elements are represented by ancient, well stabilized genera, occurring in a wide range of ecological niches, but preferably in streams in plains. *Rita* and *Mystus* are known as fossils from the Pliocene of the Siwalik (Lydekker, 1886). *Chrysichthys* from Africa is considered as the ancestor of *Rita*, indicating the Ethiopian origin of the genus. *Mystus* is also similarly related to the African *Porcus* (Jayaram, 1966). It is of considerable biogeographical significance to observe here that the genus *Chrysichthys* is also known from the Pliocene of Siwalik, although the living forms occur only in Africa. The genus *Clarias* is the most widely distributed member of the Clariidae, a family found at present in Africa, Syria, southeast Asia up to the East Indies. Although *Clarias* is known as fossils from the Siwalik, the family is at present absent from Baluchistan, Iran and Arabia. Thirty-three species of *Clarias* are known from Africa (Buolenger, 1911), but only seven are found within the limits of our region. *Clarias batrachus* is found continuously from Sind to the Malaya Archipelago and from Kashmir to Ceylon. The remaining six species are restricted in their distribution to the Indo-Chinese sub-region. *Garra*, *Labeo*, *Puntius* and *Barilius* are also similarly widely distributed in Africa, Iran and the northwest of our area. Forty-nine species of *Labeo* occur in Africa. *Tor* is believed to have differentiated from *Puntius*, which is extensively distributed from Europe, throughout Asia and Africa. DAY (1889) observes that the number of species appears to diminish as we approach the Malaya Archipelago. *Rasbora* occurs in Africa, India, Ceylon and through Burma to the Malaya Archipelago. *Noemacheilus* extends, on the other hand, from the fresh-waters of Europe to Asia. The most generalized species of these genera occur in Africa rather than in the Oriental Region. The Ethiopian elements have apparently spread along the Jacob's Arabian region of distribution.

On the basis of their distributional patterns, the genera occurring within the Oriental of India fall into the following major groups: 1. genera endemic in the Indian subregion, 2. genera endemic in the Indo-Chinese subregion and 3. genera common in both these subregions.

1. The following genera are endemic in the Indian subregion:

Cyprinidae
  1. *Lepidopygopsis*
Homalopteridae
  2. *Bhavania*
  3. *Travancoria*
Cobitidae
  4. *Jerdonia*
  5. *Noemacheilichthys*

Bagridae
  6. *Horabagrus*
Scheilbeidae
  7. *Silonopangasius*
  8. *Neotropius*
Clariidae
  9. *Horaglanis*

A most striking fact is that these genera are truly autochthonous in the Peninsula, where they differentiated from phylogenetically older, widely distributed and well stabilized ancestral forms. Most of them are monotypic and are also restricted at present in their distribution to the Western Ghats.

Lepidopygopsis is known from the Periyar Lake, but its nearest relatives are found in Middle Asia (Fig. 99). HORA (1949) supposed this genus to have differentiated from a *Schizothorax*-like ancestor of the Eastern Himalaya, but it has since been shown that *Schizothorax* is more Palaearctic than Indo-Chinese in origin. *Horabagrus*, with a single species *Horabagrus brachysoma*, is considered to be a derivative of *Pelteobagrus* from South China, and is endemic in Kerala (JAYARAM, 1953, 1968). Similarly *Bhavania* and *Travancoria* are believed to have differentiated from Homalopteroid ancestors, perhaps of Malayan origin (Fig. 100). *Jerdonia* and *Neomacheilichthys* are small loaches confined to the Deccan and believed to be derived from the Malayan *Acanthopsis*-like stock. *Horaglanis* is a degenerate blind Clariid fish that occurs in wells in the Kerala part of the Western Ghats. MENON (1950, 1951) has traced its affinities to the Ethiopian *Clarias*. *Neotropius*, restricted in its distribution to the northern parts of the Western Ghats, is perhaps a derivative

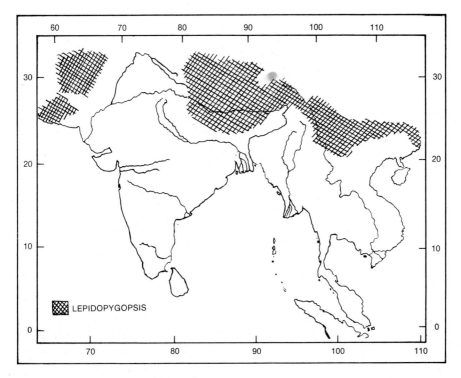

Fig. 99. Distribution of *Lepidopygopsis.*

from *Pseudeutropius. Silonopangasius* is the Deccan-form of *Silonia* (SILAS, 1952), which is found in India and Burma.

The autochthonous elements of the Indian subregion are thus curiously restricted at present to the southern part of the Peninsula and none occurs in Ceylon, probably indicating their relatively recent (post-Pleistocene) origin. These genera, apparently differentiated from widely distributed ancestral forms, are now relicts, though they have diverged to different taxonomic levels. Their ancestral stocks have had their origin from the Indo-Chinese and Malayan areas to a large extent, but sometimes also from the Palaearctic area.

2. Genera endemic in the Indo-Chinese subregion are the following:

Cyprinidae
1. *Labiobarbus*
2. *Semiplotus*
3. *Accrossocheilus*
Cobitidae
4. *Acanthopsis*
5. *Somileptis*
6. *Acanthophthalmus*
7. *Apua*

8. *Aborichthys*
Akysidae
9. *Akysis*
Sisoridae
10. *Pseudecheneis*
11. *Parapseudecheneis*
12. *Coraglanis*
13. *Myersglanis*
14. *Euchiloglanis*

523

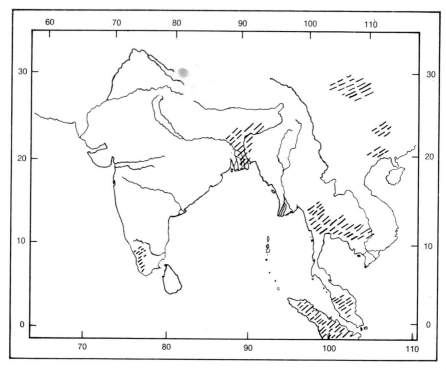

Fig. 100. Discontinuous distribution of the Homalopteridae, a diphyletic family of fresh-water stream-fishes that includes the interesting genera *Bhavania*, *Travancoria* (endemic in the Peninsula) and *Balitora*.

15. *Oreoglanis*	Olyridae
16. *Exostoma*	18. *Olyra*
17. *Conta*	Indostomidae
	19. *Indostomus*

The great majority of these are inhabitants of swift-flowing streams, either of low mountains or of rivers in the foothils of the higher ranges. The Cobitid genera occur in rivers of low hills and mountains, no more than a thousand metres in elevation. The Sisorid genera are inhabitants of torrential streams with rocky bottom and many of them have developed adhesive organs in the thoracic region. The first ray of the pelvic and pectoral fins is completely segmented and pinnate. The lips are expanded with an interrupted labial fold, mainly for the purpose of adhesion to the substratum.

The Sisorids are represented by eight genera. *Euchiloglanis* and *Exostoma* are the only genera with many species. Of the six species of *Euchiloglanis* known so far, three occur in South China and the Eastern Himalaya, and the remaining species are endemic one each in Burma, Cambodia and the Eastern Himalaya. *Coraglanis* and *Parapseudecheneis* are endemic

524

Table II. Percentage analysis of the Oriental genera of the primary fresh-water fishes (Fig. 101)

Family	No.	Indian Endemic	Indo-Chinese endemic	Common to both
1. Cyprinidae	31	3.7	12.1	63.0
2. Cobitidae	10	20.0	50.0	—
3. Psilorhynchidae	1	—	—	100.0
4. Homalopteridae	4	50.0	—	50.0
5. Bagridae	4	50.0	—	50.0
6. Siluridae	3	—	—	100.0
7. Scheilbeidae	8	25.0	—	75.0
8. Pangasiidae	1	—	—	100.0
9. Amblycipitidae	1	—	—	100.0
10. Akysidae	1	—	100.0	—
11. Sisoridae	18	—	50.0	11.1
12. Clariidae	2	50.0	—	50.0
13. Heteropneustidae	1	—	—	100.0
14. Chacidae	1	—	—	100.0
15. Olyridae	1	—	100.0	—
16. Indostomidae	1	—	100.0	—
17. Channidae	1	—	—	100.0

in South China. *Pseudecheneis*, *Myersglanis* and *Conta* are restricted to the Eastern Himalaya and Burma. These Sisorid genera are believed to have become differentiated from the generalized *Glyptothorax*-like ancestors under the relatively more dynamic environment of the areas in which they now occur. *Glyptothorax* is itself an Indo-Chinese genus. These genera appear to have spread from a main Indo-Chinese centre. The differentiation of the phylogenetically younger and specialized genera like *Myersglanis*, *Conta*, *Coraglanis* and *Parapseudecheneis* must also be considered as an integral part of such a radiation. *Accrossocheilus* is known by eight species, of which six are endemic in Thailand. *Accrossocheilus hexagonolepis* is found in Assam and the Eastern Himalaya and the eighth species *Accrossocheilus dukai* has extended up to Malaya in the south. The last named species is believed to have differentiated from the generalized *A. hexagonolepis*. The Olyrid catfish *Olyra* and the Cyprinid *Semiplotus* are endemic in the Eastern Himalaya and Burma. *Accrossocheilus* and *Semiplotus* are considered to have differentiated from *Tor*-like ancestors of the Indo-Chinese stock.

The five Cobitid genera and *Akysis* and *Labiobarbus* are probably Malayan forms, which have spread into the Indo-Chinese area. Of the eight species of *Akysis*, three are endemic in Burma, two in Thailand and one in Malaya. A species each is also found in Thailand and Burma, and Thailand and Sumatra. *Akysis* was known for a long time from only

525

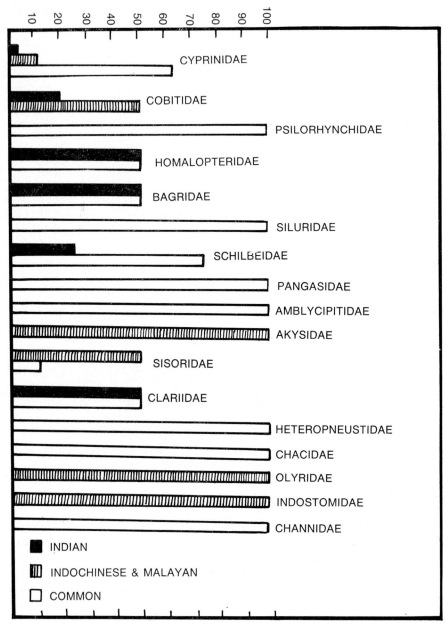

Fig. 101. Percentage analysis of the Oriental genera of primary fresh-water fishes in the Indian region.

526

Java, Sumatra, Borneo and Tenasserim, but it occurs also in Thailand and Burma. *Labiobarbus* is a Thailand genus, with only two species out of nine not occurring in Thailand; five have an extended range, including Malaysia. It may have differentiated from the typically Malayan *Osteocheilus*-like stock. Amongst the five Cobitid genera, only *Aborichthys* and *Acanthophthalmus* are polytypic. *Acanthophthalmus pangia* occurs in the Eastern Himalaya and extends through Assam to Burma southwards. *Acanthophthalmus kuhlii* and *Acanthophthalmus javanicus* extend from Thailand to the East Indies. *Acanthopsis* is not known north of Burma. *Aborichthys elongatus* is confined to the Eastern Himalaya and *Aborichthys kempi* extends further to Burma. *Somileptis gongota* occurs only in the Khasi Hills (Assam). These are small-sized loaches, having a restricted distribution. All of them are phylogenetically young and their rarity and absence from the areas to the north of Burma and their abundance in Malaya are of considerable biogeographical significance. These are mainly Malayan elements, which have radiated northwards and northeastwards into the Indo-Chinese subregion. KREMPF & CHEVEY (1934) have ably discussed the radiation of such Malayan faunas.

3. The following genera are common to the Indian and Indo-Chinese subregions (including in part also the Malayan subregion):

Cyprinidae
   1. *Osteocheilus*
   2. *Cirrhinus*
   3. *Schismatorhynchus*
   4. *Thynnichthys*
   5. *Osteobrama*
   6. *Rohtee*
Psilorhynchidae
   7. *Psilorhynchus*
Homalopteridae
   8. *Homaloptera*
   9. *Balitora*
Bagridae
   10. *Chandramara*
   11. *Batasio*
Siluridae
   12. *Wallago*
   13. *Ompok*
   14. *Silurus*
Scheilbeidae
   15. *Silonia*

   16. *Ailia*
   17. *Pseudeutropius*
   18. *Proeutropius*
   19. *Eutropiichthys*
   20. *Clupisoma*
Pangasiidae
   21. *Pangasius*
Amblycipitdae
   22. *Amblyceps*
Sisoridae
   23. *Sisor*
   24. *Gagata*
   25. *Erethistes*
   26. *Erethistoides*
   27. *Hara*
   28. *Laguvia*
Chacidae
   29. *Chaca*
Heteropneustidae
   30. *Heteropneustes*

Of the thirty genera listed above, it may be seen that the majority are siluroids like Sisoridae and Scheilbeidae. These are generally lethargic bottom-dwellers in relatively sluggish streams, but most of the genera seem to prefer slightly more dynamic environments. *Amblyceps* and *Conta* are mainly found at the Himalayan foothills and occur in hillstreams subject to flash floods. Being bottom-dwellers, and also habituated

527

to stony and pebbly substratum, they are not so directly exposed to the full force of high current velocities as in case of other forms of the upper layers (HORA, 1933). By means of its adhesive organs on the thorax and abdomen, *Conta* anchors itself to rocks and boulders (HORA, 1951). *Erethistes, Erethistoides* and *Hara* occur rather in sluggish and deep waters, overgrown with macrophytes. The Scheilbeid genera occur in both large and small rivers, some of them such as *Silonia* ascending up to nearly the source of the stream (HORA, 1938). The Cyprinid genera generally prefer swift, clear and deep waters. The Homalopteroid and Psilorhynchid genera occur, however, in torrential streams and have thus developed special modifications, especially on their mouthparts and thorax.

Wallago, Chaca, Heteropneustes and *Cirrhinus* are relatively old genera, widely distributed and well stabilized. *Chaca* is restricted in its distribution to north India. *Chandramara, Sisor, Erethistoides* and *Laguvia* occur in the Eastern Himalaya, in streams at the base of the hills and occasionally also descending to the Gangetic Plains. *Hara* and *Amblyceps* have, however, a relatively wider range, and are found in Burma also. *Amblyceps* and *Laguvia* have also been recently discovered to occur in Hoshangabad (HORA & NAIR, 1941) in the northern part of the Peninsula and in the R. Rihand also in the northeast of the Peninsula (HORA, 1949). These are phylogenetically very young genera; they are indeed younger than those found in Yunnan, South China and other areas further east.

Ompok is a Malayan genus, represented in India by *Ompok bimaculatus, Ompok palo*, and *Ompok palda*. Out of the eight species so far known, one is endemic in the Eastern Himalaya and Assam and the others occur in the East Indies and Malaya. The Scheilbeid genera *Silonia, Pangasius, Eutropiichthys* and *Proeutropiichthys* show marked discontinuous distribution. *Silonia silondia* of north India and Burma is replaced by *Silonopangasius* in the R. Cauvery. *Pangasius* is also similarly found in northeast India, Burma, Thailand, Malaya, Java and in the Cauvery of South India. DAVID (1962) has recently described a subspecies of *Pangasius pangasius* from the R. Godavary. *Eutropiichthys vacha* has apparently given rise to *Neotropius goongware* in the Deccan. The genus *Thynnichthys*, represented by *Thynnichthys thynnoides*, occurs in Thailand, Malaya Peninsula and the East Indies: a second species *Thynnichthys sandkhol* is known only from the Godavari and Krishna river systems (Fig. 102).

The discontinuity is wider and still more pronounced in the case of the Homalopterid genera (Fig. 100). *Homaloptera* is known by twenty-five species (SILAS, 1953) and is distributed extensively in Burma, Thailand, Indo-China, Malaya Peninsula and the East Indies, Sumatra, Java and Borneo. A single species *Homaloptera montana* is endemic in the Anamalai Hills in South India. *Balitora* is known by two species; *Balitora brucei brucei* occurs in the Eastern Himalaya and Assam; *Balitora brucei mysorensis* is endemic in Mysore; and *Balitora brucei burmanicus* in Burma and *Balitora brucei melanosoma* in the Burma-Thailand border. The second

528

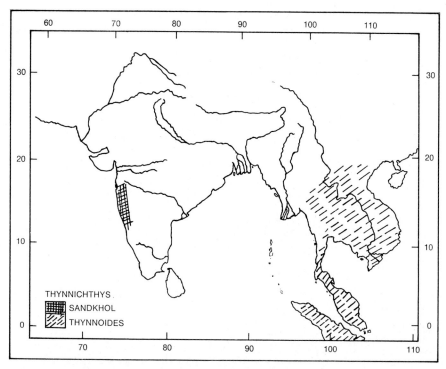

Fig. 102. Discontinuous distribution of the Malayan fresh-water stream-fish *Thynnichthys* in the Peninsula, Indo-China, Malaya, Thailand, Sumatra; the genus extends further east to Borneo also.

species *Balitora maculata* is restricted in its distribution to Darjeeling in the Eastern Himalaya. The distribution of *Gagata* is shown in fig. 103.

Silurus is a very widely distributed, old genus exhibiting a similar pattern of discontinuous distribution (Fig. 104) like *Balitora*, of which twelve species are known (HAIG, 1950) from its entire range. *Silurus cochinchinensis* occurs in Cambodia, Thailand, Malaya Peninsula and Burma. *Silurus berdmorei wynaadensis* is endemic in Mysore, but *Silurus berdmorei berdmorei* is found only in Akyab (Burma). *Silurus goae*, recently described from Goa (HAIG, 1950), is closely related to *Silurus berdmorei*. *Batasio*, of which three species are known, also exhibits a similar discontinuous distribution (HORA & LAW, 1941) (Fig. 105). *Batasio batasio* is found only in the Eastern Himalaya, Assam, Burma and Malaya. *Batasio havmolleri* is endemic in Thailand. *Batasio travancoria* is only known from Travancore in the extreme southwest of the Peninsula. *Osteobrama* has given rise to subspecies in the extremes of its range, for example, *Osteobrama cotio cotio* in the Eastern Himalaya, Burma and the Gangetic Plain; *Osteobrama cotio cumna* in Burma and *Osteobrama cotio peninsularis* in the Western Ghats.

529

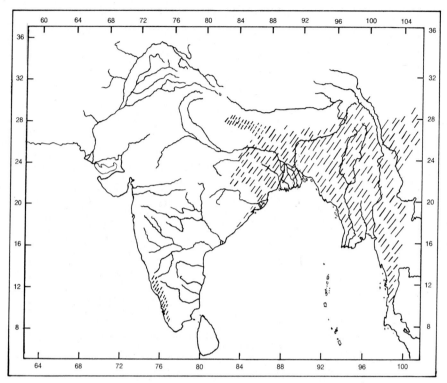

Fig. 103. Distribution of *Gagata* (Sisoridae). This Fresh-water fish genus is represented by *G. gagata*, *G. cenia*, *G. viridescens*, *G. nangra*, *G. itchkeea* and *G. schmidti* in different parts of its range.

The above mentioned patterns of distribution range from simple discontinuity of the undifferentiated populations to discontinuity of populations which have differentiated into subspecific and sometimes even beyond the subspecific levels.

Osteocheilus, *Schismatorhynchus* and *Rohtee* have differentiated subgenerically. The first named genus, with thirty-one species, occurs in South China, Laos, Vietnam, Thailand, Malaya Peninsula and the East Indies (Fig. 106). *Osteocheilus (Osteocheilichthys) thomasi*, *Osteocheilus (Osteocheilichthys) nashii* and *Osteocheilus (Osteocheilichthys) brevidorsalis* are the only species found discontinuously on the Western Ghats. *Schismatorhynchus* is also likewise discontinuously distributed; the genus occurs in Sumatra and Borneo and the subgenus *Nukta* is represented by *Schismatorhynchus (Nukta) nukta* in Deccan. *Rohtee* is restricted to the Peninsula and the subgenus *Mystacoleucus* accommodates all the species from Burma, Thailand and Malaya (Fig. 107).

The Indo-Chinese elements intrusive in the Himalaya are represented by eight genera, which are taxonomically stabilized and contain a number

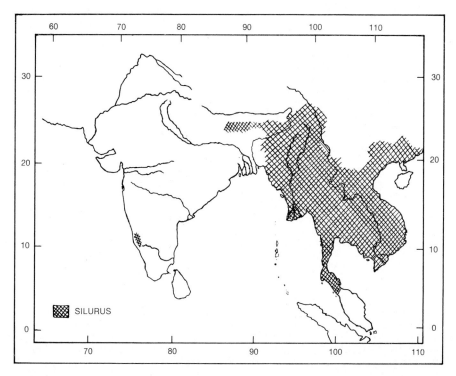

Fig. 104. Discontinuous distribution of the fresh-water fishes of the genus *Silurus*, which have had their differentiation in the Yunnan area or perhaps farther northeast.

of phylogenetically older species in the Indo-Chinese subregion. All of them occur in clear swift streams. *Bagarius bagarius*, described as a living fossil (HORA, 1939), is also known from the Pliocene Siwalik fossils from India (LYDEKKER, 1886) and from other Tertiary beds in Padong in Sumatra. The species occurs throughout India, Burma, Malaya Archipelago and Tonkin. Its occurrence in the Punjab and in Sind is sporadic. *Glyptothorax, Chela, Oxygaster, Esomus, Botia* and *Lepidocephalichthys* are Indo-Chinese genera, with only one or two species extending west in the Himalaya. *Glyptothorax* is known by twenty-four species, of which thirteen occur in the Indo-Chinese subregion and only a single species *Glyptothorax conirostrae* has intruded west into the Northwest Himalaya. This species occurs along the wooded slopes of the Himalaya, South China, Eastern Tibet and also the Western Ghats. *Chela, Oxygaster* and *Esomus* are represented by the widely distributed *Chela cachius, Oxygaster bacaila* and *Esomus danrica. Chela chachius* is, however, absent from Ceylon (SILAS, 1958). *Botia* and *Lepidocephalichthys* are, however, represented by *Botia birdi* and *Botia lohachata* restricted to the Western Himalaya and the adjacent parts of the Gangetic Plains and *Lepidocephalichthys guntia* occurs both in the

531

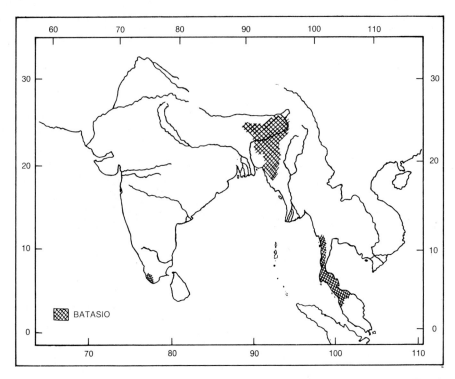

Fig. 105. Discontinuous distribution of two catfishes. The interesting genus *Batasio* (Bagridae) is represented by four species: *B. batasio, B. tengana, B. havmolleri* and *B. travancoria.*

western and eastern parts of the Himalaya and in the Gangetic Plains. *Crossocheilus* is mainly an Indo-Chinese genus, with *Crossocheilus latius* having diverged subspecifically in the Punjab (MUKERJI, 1934).

Amongst the one hundred and six genera of the primary and secondary fresh-water fishes occurring in India, the Ostariophyseans are the dominant forms. The number of endemic genera in any of the subregions is, however, considerably smaller than those that are common to them. No fossils of the present-day genera, earlier than Eocene, are known, although records of Triassic and Jurassic Dipnoans and Teleosts are known from the Maleri (OLDHAM, 1859) and the Kota beds (EGERTON, 1845). *Rita, Mystus, Clarias, Heteropneustes, Bagarius, Schizothorax, Channa, Ambassis* and *Nandus* are the few genera known as Tertiary fossils. All the Siluroid genera are recorded from the Pliocene of the Siwalik (LYDEKKER, 1886). *Schizothorax* is recorded from the Pleistocene of Karewas of Kashmir (HORA, 1937), *Ambassis* from the Eocene of Kohat (HORA, 1937) and a single rather doubtfully referable species of *Nandus* comes from the Eocene of Deccan and Kheri beds (HORA, 1938) in Central India. It may be remarked that none of the modern Cyprinid genera is known as

532

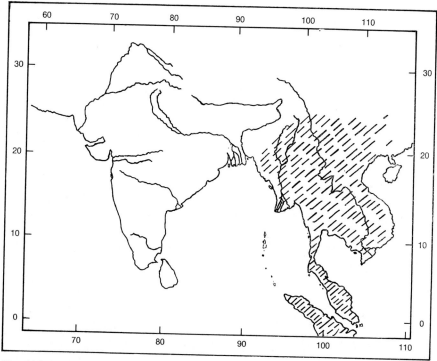

Fig. 106. Distribution of the Cyprinid fish *Osteocheilus*.

fossil. *Mystus* is replaced by *Percus* and *Rita* by *Chrysichthys* in Africa; *Clarias* is prolific in Africa. These facts would seem to show that they are of African descent. The Ethiopian element is, however, small compared to the faunal contribution from south Asia. Of the eighty-nine genera, nineteen are endemic in the Indo-Chinese subregion, thirty are common to the Indian and Indo-Chinese subregions and twenty-seven are common to the Palaearctic and Oriental. The fauna is thus essentially Ostariophysean, dominated by the modern Cypriniform and Siluriform genera (Table I). The Cypriniform fishes have in general become the dominant fresh-water group in any landmass to which they have had access. Wherever they occur naturally in any considerable number they have differentiated into many more species than in other groups. In most places, the species of the small-sized Cyprinoids are the principal forage-fishes, on which the larger predatory species feed. They swarm in great abundance in the lowland rivers and in the highland brooks. Wherever the other larger predators are few or are absent, and occasionally even when these are fairly numerous, cyprinoids themselves have evolved large and important predatory forms, like *Barilius*, *Tor*, *Catla*, *Semiplotus*, *Labeo*, *Cirrhinus*, etc. Though the Siluriform fishes may have accompanied the Cyprinoids in their geographical radiation, the ability of some genera to

533

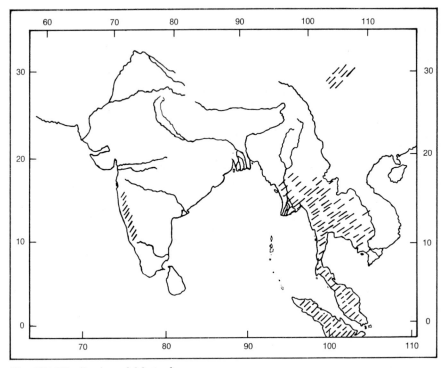

Fig. 107. Distribution of *Mystacoleucus.*

transgress short stretches of sea may have materially facilitated their radiation. The carps attain their maximum diversity only in Southeast Asia; the most generalized types are also found only here. The number of distinct subfamilies recognized is far greater, for example, in China (CHU, 1935) than in any other area of its distribution (MILLER, 1958; MYERS, 1966) in the New World. It is generally considered that the Cypriniform fishes, which have risen in Southeast Asia must have descended from a toothless protocyprinoid characoid stock. The Cypriniformes have blossomed in Eurasia into the largest familiar group of Ostariophyseans* (GREENWOOD et al., 1966; MYERS, 1967).

To summarize, we may observe that the fish fauna is dominated by the Indo-Chinese elements, though small amounts of the Palaearctic, Ethiopian and Malayan elements are also present. It is also important to observe that none of the Peninsular autochthonous genera have penetrated Ceylon. Marked discontinuity of distribution is characteristic of the fishes occurring in the Peninsula.

---

* The origin of Ostariophysi was in Africa, where the Siluroids and Characoids diverged during the Cretaceous; the Cyprinoids arose later (from Characoid-like stock?) and the radiation in oriental area is explained on the basis of continental drift. (Note added in proof – M. S. MANI).

534

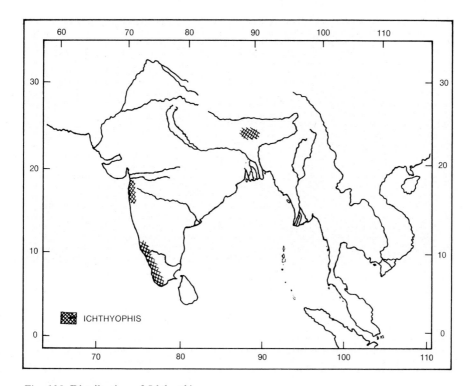

Fig. 108. Distribution of *Ichthyophis*.

3. Amphibia

It is common knowledge that frogs and toads are abundant in the tropics and the salamanders in the temperate regions. As in the case of the fresh-water fishes, discussed above, the distribution of the amphibians is generally governed by their breeding-site preference and the moisture content of a locality. Twenty-four genera and one hundred and eleven species under eight families and three orders occur within our limits. The majority are inhabitants of the evergreen, moist forest of the Western Ghats, with annual rainfall of 300–500 cm. About 41.6% of the genera are endemic in the Western Ghats.

All the five genera and the thirteen species of Gymnophiona are restricted in their distribution to South India, with the exception of *Ichthyophis sikkimensis*, endemic in the Eastern Himalaya (Fig. 108). Only the genus *Herpele* occurs in the eastern area, but all the others are restricted to the Western Ghats. Of the seventeen genera and seventy-five species of Caecilians so far known from the world (NEIDEN, 1913), *Ichthyophis* is the most primitive. All the Indian genera are limbless, long-bodied, with a series of transverse grooves and have a superficial resemblance to earth-

535

Plate 62. Ichthyophis beddomii with egg mass in nature. (Photo by Dr. B. K. Tikader, Deputy Director, Zoological Survey, Poona).

worms. The males are provided with a protrusible copulatory organ. They are seldom seen, since they live buried underground in wet soil. The entire order is probably derived from the Carboniferous Amphibia. It is absent from Madagascar.

The order Urodela is represented by *Tylotriton* (Salamandridae), which is the most primitive. All the salamanders are Eurasian, with the exception of *Triturus* of America. Fossil salamanders are known from the Oligocene, Miocene and Recent formations of Europe. *Tylotriton verrucosus* occurs in the Eastern and Nepal Himalaya and *Tylotriton andersoni* in the Okinawa Island in the Loochoo Archipelago. They are all rough skinned and unlike their other cousins more terrestrial than of montane-brook habitat. *Tylotriton* has also been recently discovered in the Miocene of Switzerland. From the available evidence, it would seem that Europe is the centre of differentiation of the salamanders. The occurrence of a single, primitive genus of the family in the Eastern and Nepal Himalaya and Yunnan is of particular interest.

The order Salientia is the best represented of the amphibians within our limits. *Rana*, with thirty-five species, is mainly Peninsular and four-teen of these species are endemic in the Western Ghats, eleven in the Eastern Himalaya, three in the Northwest Himalaya and one in Nicobar Islands and the remaining ones are more or less widely distributed. The Ranids are primarily Old World inhabitants and *Rana* alone being the single exception to occur in America. Six subfamilies of Raniidae are known, of which four are restricted to Africa and the other two are found in southern Asia. The genus *Staurois*, with two species, is mainly Indo-Chinese with a single species *Staurois afghanus* occurring, besides the

536

Plate 63. Habitat of a tree-frog in the Western Ghats.

Khasi Hills (Assam), in Kangra in the Northwest Himalaya. Outside our limits, the genus comprises a large series of species from the Philippines, Hainan, East Indies, Thailand and Burma. The genera *Micrixalus*, *Nyctibatrachus* and *Nannobatrachus* are restricted to the Western Ghats, with occasional records in the Carnatic Plains. These are small-sized frogs, representing mainly local specializations of the *Rana*-stock.

The Rhacophoridae, a family of small-sized tree-frogs, are represented by two genera. Of the nine species of *Rhacophorus*, known so far from within our limits, four are endemic in the Western Ghats, four in the Eastern Himalaya and Assam and one is widely distributed. *Philautus* is known by ten species, of which eight are endemic in the Western Ghats and two in the Eastern Himalaya (Fig. 109). Both these genera have developed special modifications for arboreal life. *Philautus* is believed to have differentiated from the widely distributed *Polypedates*.

The family Microhylidae is represented by five genera. *Microhyla* has an extensive range in Southeast Asia and the adjoining islands. The monotypic *Kaloula*, though restricted to Bengal, Western and South India in the Indian subregion, has also like *Microhyla*, an extensive range in Southeast Asia and the adjoining islands. *Ramanella*, with three species,

537

Plate 64. Habitat of a tree-frog in the Western Ghats.

is restricted to the Peninsula. It is a small derivative of *Kaloula; Kaloula* and *Microhyla* being the more primitive members of the family. The representatives of this family now occur in China and the United States of America.

The Bufonidae are represented by *Nectophryne*, *Bufo* and *Ansonia*. The first named genus has an interesting discontinuous distribution (Fig. 110); *Nectophryne tuberculosa* occurs in the Western Ghats and *Nectophryne kempi* in the Garo Hills (Assam), but absent in the intervening areas. *Ansonia*, considered a local representative of *Bufo*, is monotypic and endemic in the Brahmagiri Hills of the Mysore Plateau. *Bufo* is known by fifteen species, of which seven are found only in the Western Ghats, two in the Eastern Himalaya and one in Ladak. *Bufo* occurs all over the world, except in New Guinea, Polynesia, Australia and Madagascar. Fossils of *Bufo* are known from Miocene and later formations of Europe.

The family Pelobatidae is known by only *Megalophrys* and *Aelurophryne*, both restricted to the Himalaya. All the three species of *Megalophrys* are restricted to Darjeeling in the Eastern Himalaya and the Khasi Hills in Assam. *Aelurophryne*, with a single species occurring in Kashmir, is considered to be a recent divergent from *Megalophrys*. The genus *Megalophrys*

538

Plate 65. Habitat of a tree-frog in the Western Ghats.

Plate 66. Nannobatrachus.

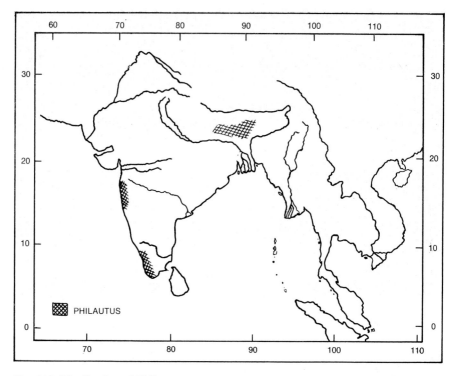

Fig. 109. Distribution of *Philautus*.

is itself known by twenty-five species, distributed across southern Asia and western end of the Indo-Australian Archipelago. This genus, evidently Indo-Chinese in origin, seems to be a recent intrusive in the Himalaya, which would appear to lie on the outermost periphery of its range.

The family Hylidae has only one monotypic genus in the Khasi Hills (Assam). It is considered a specialized Bufonid, which is otherwise found almost all over the world, except for a hiatus in the Indo-Australian Archipelago, Polynesia, Ethiopia and Madagascar. Only one fossil record of *Hyla* from the Miocene of Europe is known. As in the case of *Megalophrys*, *Hyla* is also a recent intrusive element in the Indo-Chinese subregion.

Fossil amphibia so far known from India come from the Jurassic. The Caecilians are unknown as fossils. Of the living forms, *Bufo melanostictus* is recorded from the Pleistocene cave deposits of Karnool (LYDEKKER, 1886). A species of *Rana* is also recorded from the Post-Pleistocene deposits of Bill-Surgam. The existing Indian genera are more autochthonous in the Indian Peninsula, especially in the Western Ghats, than anywhere else. Nearly 50% of the genera are endemic in the Indian sub-

540

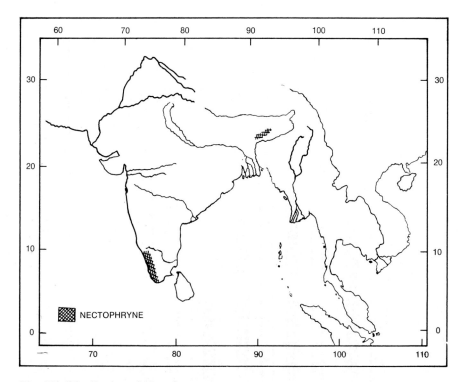

Fig. 110. Distribution of *Nectophryne*.

region, 12.5 % in the Indo-Chinese subregion and 21 % are common to both the subregions. Only a single genus is known from the Northwest Himalaya, in the Turkmenian subregion. (Table V) (Fig. 111). The genera and the families occurring in South India are generalized forms and from the evidence of the fossils and from their phylogenetic relations, it must be concluded that they are typical relicts.

The world distribution of some of the important genera shows that they were formerly widely distributed. Except for Madagascar, Australia and most of Polynesia, *Bufo* occurs nearly everywhere and indeed frogs and toads are common in the tropics and salamanders in the temperate regions. There is, however, a wide hiatus in the Indo-Malayan region in the case of Hyldiae, but the Ranidae are absent from South America (NOBLE, 1954). Thus the geographical limits of each family of Amphibia have their own peculiarities and various groups show different geographical patterns. The present distribution of some groups like the American *Proteus* and *Necturus* and of *Tylotriton* is of considerable interest.

To summarize we may observe that the most primitive genera of some families are found in our limits. The discontinuous distribution of genera is characterized by their occurrence in the Eastern Himalaya and Western

541

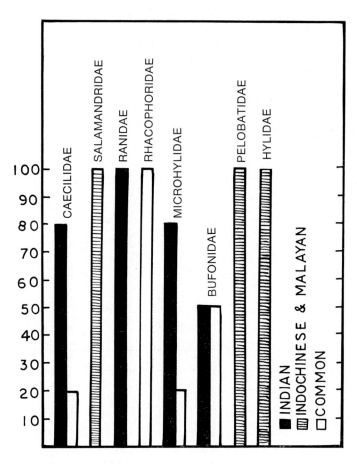

Fig. 111. Percentage analysis of amphibian genera in the Indian region.

Table V. Percentage analysis of amphibian genera. Total No. of genera 24.

Family	Genera	Palae-arctic	Indian (A)	Oriental Indo-Chinese (B)	Common to A & B	Total
1. Caecilidae	28.3	—	80	—	20	100
2. Salamandridae	4.2	—	—	100	—	100
3. Ranidae	28.3	—	100	—	—	60
4. Rhacophoridae	8.3	—	—	—	100	100
5. Microhylidae	28.3	—	80	—	20	100
6. Bufonidae	12.5	—	50	—	50	100
7. Pelobatidae	8.3	—	—	100	—	100
8. Hylidae	4.2	—	—	100	—	100

542

Ghats; the majority occur in the Western Ghats but many are also found in the Eastern Himalaya. The genera are relics of a group that was formerly widely distributed. Some genera have extended within recent times westwards along the Himalaya, which marks the extreme periphery of their range. Generic divergence has also occurred in some cases.

4. Chelonia

The Chelonians have existed perhaps almost unchanged since the Triassic. The fossil *Trionyx*, *Chitra* and *Lissemys*, practically indistinguishable from the present-day forms, have been found in the Pliocene and Pleistocene of Siwalik. *Kachuga* is known from the Pleistocene of Siwalik and of the Narmada Valley. The Trionychidae are relatively old and appear first in the Upper Cretaceous of North America. Living forms of *Trionyx* have, however, a much wider distribution across Asia, Africa and North America. The Chelonians seem as a whole to have attained their maximum development at the end of the Mesozoic and have since then remained a relict group wherever they are found today. The antique and well stabilized nature of the group lends support to this conclusion. Of the fifty-seven species living at present in the Oriental, a high percentage of endemism, viz. 55% is seen in the Indo-Chinese subregion. There is only a single genus *Hardella* endemic in north of the Indian subregion. Of the twelve genera endemic in the Indo-Chinese subregion, nine are monotypic. It is also remarkable that there is no endemic genus in the Palaearctic. The poverty of Chelonians in the Indian subregion is of considerable significance.

The Testudines are represented by six families, twenty-six genera and fifty-seven species. The families Sphargidae and Chelonidae, with four genera and four species, are marine forms and are not considered here. The remaining twenty-two genera, with fifty-three species, fall into four distinct categories, depending on their distributional patterns. Sixteen genera are found in the Oriental and only six are common to the Palaearctic and Oriental. The Testudines are a very old group, dating back to the Permian and many fossils are known from the Triassic, much in same form as we find them today. The greatest development of the group was reached towards the end of the Mesozoic and in the early Tertiary.

Among the sixteen genera occurring in the Oriental, twelve are endemic in the Indo-Chinese subregion, one in the Indian subregion and three are common to the two subregions.

1. The genera endemic in the Indian subregion: *Hardella*, with a single species *Hardella thurgii*, is the only genus endemic in the Indian subregion. This aquatic tortoise, living in slow-flowing and stagnant waters, occurs in the rivers Ganga and Brahmaputra. This represents a relict of a former widely distributed form like *Kachuga* and its range has become restricted

in the Indian subregion relatively recently. It must be remarked here that the type specimen of this species is a skull that is reported to have come from the R. Indus.

2. Genera endemic in the Indo-Chinese subregion:

Platysternidae
1. *Platysternon*
Emydidae
2. *Cyclemys*
3. *Cuora*
4. *Damonia*
5. *Hieremys*
6. *Notochelys*

7. *Siebenrockiella*
8. *Clemmys*
9. *Chinemys*
10. *Ocadia*
Trionychidae
11. *Pelochelys*
12. *Dogania*

The fresh-water tortoises comprise the bulk of our Chelonians. Over twenty genera of Emydidae are known from the world; they are very closely related to the Testudinidae (land tortoises) and are separated by small, but distinct characters. Some genera like *Geoemyda* are terrestrial.

On the basis of their distributional patterns, we may recognize two major groups. Five genera occur in Thailand, Cambodia and radiate to South China, Yunnan, Hainan, Formosa and China in the northeast. *Clemys*, *Chinemys* and *Ocadia* are found mainly in Annam, South China, Formosa and Hainan. *Chinemys* extends through China to Japan. *Platysternon* occurs in south Burma, Thailand, Laos, Vietnam, South China and Hainan. *Pelochelys* the most widely distributed of the fresh-water chelonians, is definitely known to occur in the Indo-Chinese Peninsula, southern China and Hainan, in addition to the Philippines. With the exception of *Clemys*, which has two species, all the other genera are monotypic. In sharp contrast to the above pattern, seven genera *Cyclemys*, *Cuoria*, *Damonia*, *Hieremys*, *Notochelys*, *Siebenrockiella* and *Dogania* radiate southwards and southeast of the main Indo-Chinese centre of differentiation. With the exception of the last mentioned genus, the others are Emydid fresh-water tortoises. *Dogania* is an archaic Trionychid, distributed widely. Of the closely related *Cuora* and *Cyclemys*, the former has radiated to South China. The other five genera are more restricted in the sense that they are mainly found in Thailand, transgressing to the Malaya Peninsula, but only exceptionally to the East Indies.

3. Genera common to the Indian and the Indo-Chinese subregions: *Geoemyda*, *Morenia* and *Batagur* are common to the two subregions. The last two genera are aquatic, herbivorous tortoises; *Morenia* is restricted to Bengal and south Burma, but *Batagur* extends from Bengal through Burma, Cambodia to the Malay Peninsula and Sumatra. *Geoemyda* is, on the other hand, a hill species, almost completely terrestrial and vegetarian. It is widely distributed from throughout India to Japan and the Malaya Archipelago, Central and South America. *Geoemyda trijuga* has differentiated into five local subspecies within its range (Fig. 112). It would appear that these are old forms, mainly of Indo-Chinese origin.

544

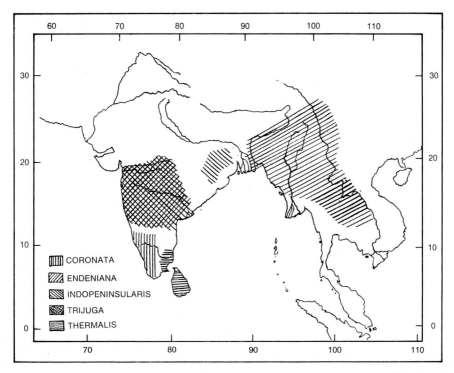

Fig. 112. Distribution of fresh-water Chelonian *Geoemyda trijuga* represented by loeal forms. The marked discontinuity and the presence of local forms, in isolated pockets, may be traced back to former distributional continuity of a single species. Destruction of habitats in the intervening areas had the concomitant result of disappearance of the species; the isolated populations evolved into local forms.

Unlike *Morenia* and *Batagur*, *Geoemyda* is widely distributed and taxonomically well stabilized.

We may also refer here to the following genera that are common to the Oriental and Palaearctic in India:

Emydidae
1. *Geoclemys*
2. *Kachuga*
Testudinidae
3. *Testudo*

Trionychidae
4. *Lissemys*
5. *Chitra*
6. *Trionyx*

Some of these like *Kachuga*, *Lissemys* and *Trionyx* have numerous species and have also diverged subspecifically. *Geoclemys*, *Kachuga* and *Testudo* are also known as Siwalik fossils. Six species of *Kachuga* are known from north India and Burma, all of which are wholly aquatic and herbivorous. *Kachuga tectum* has diverged subspecifically in South India. *Kachuga trivittata* is the only species found in the rivers Irrawady and Salween of

545

Burma; all the other species are mainly found in the rivers Ganga, Brahmaputra and Indus.

The land tortoise *Testudo*, with seven species, is rather widely distributed. *Testudo horsfieldi* is widely distributed from the Caspian and Aral Seas to the northwest corner of India. *Testudo elegans* occurs throughout Central and South India. *Testudo travancorica* occurs only in the Western Ghats up to Coorg. This species is very closely related to *Testudo elongata*, found in northeastern India, Tonkin to Malaya Peninsula. ANNANDALE (1913) considered as a distinct species the form from Chota-Nagpur, but SMITH (1931) believes that these are descendents of an ancestor that once ranged over the whole of India and Indo-China. While *Testudo travancorica* has reached a level at which it may be distinctly separated from *Testudo elongata*, the Chota-Nagpur form *Testudo parallelus* is not considered to have diverged to that level.

The nominate form of *Lissemys punctata* occurs in the R. Indus and Ganga; the race *granosa* occurs on the other hand in the Indian Peninsula, south of the Gangetic Plains. *Lissemys punctata scutata* is a Burmese form.

Chitra is mainly an Indo-Chinese genus that has spread to North India. *Trionyx* has on the other hand spread to the Peninsula also. One of the eight species, *Trionyx sinensis*, has diverged subspecifically in the Indo-Chinese subregion.

The distribution of Testudines shows that they are a group that date from the Pliocene times. The living genera are now almost mostly Indo-Chinese and have radiated north, northeast and south, but are relatively sparsely found in the Peninsula, where they would seem to be recent intrusive elements.

The Chelonians have existed practically unchanged since the Triassic. Fossils of *Trionyx*, *Chitra* and *Lissemys*, indistinguishable from the present-day forms, have been found in the Pliocene and Pleistocene of Siwalik. *Kachuga* is known from the Pleistocene of Siwalik and the Narmada Valley. The Trionychidae are not geologically very old. They appear first in the Upper Cretaceous of North America, but living forms of *Trionyx* have a wider distribution in Asia, Africa and North America. The Chelonians seem to have attained their maximum development at the end of the Mesozoic and have remained as a relict group since then in nearly all parts where they occur at present. The fifty-seven species occurring in the Oriental region are characterized by high endemism, amounting to 55 % in the Indo-Chinese subregion. Only a single genus *Hardella* is endemic in the Indian subregion. Nearly 75 % of the genera occurring in the Indo-Chinese subregion are monotypic.

5. Lacertilia

The Lacertilia comprise more than three hundred genera and about three thousand species from the world. Though they are particularly

Family	Genera %	Palae-arctic Ethio-pian	Indian endemic (A)	Indo-Chinese endemic (B)	Common to A & B	Total Oriental	Palae-arctic + Oriental
1. Geckonidae	32.4	27.3	25.0	25.0	50.0	54.5	18.2
2. Agamidae	28.0	5.3	50.0	42.9	7.1	73.7	21.0
3. Chamaeleonidae	1.5	—	100.0	—	—	100.0	—
4. Scincidae	26.4	22.2	45.5	27.3	2.3	61.1	16.7
5. Dibamidae	1.5	—	—	100.0	—	100.0	—
6. Lacertidae	7.3	20.0	50.0	50.0	—	40.0	40.0
7. Anguidae	1.5	—	—	100.0	—	100.0	—
8. Varanidae	1.5	—	—	—	—	—	100.0

abundant in the tropics, many of them range far into the temperate areas also. They exhibit a wide range of form and habits and both far-reaching simplification and high degree of specialization are met with in the living forms today. SMITH (1935) deals with two hundred and ninety-seven species, under sixty-eight genera of eight families from our limits. The dominant members are the Geckonidae with twenty-two genera, Agamidae with nineteen genera and Scincidae with eighteen genera. The first named family is undoubtedly the most ancient and has also a worldwide distribution. Though no fossils of Geckonids are known, their range is wide in both the north and temperate zones. The Agamidae are, on the other hand, completely an Old World group, though their close relatives the Iguanidae are found only in the New World. Fossils of Iguanidae are known from the Eocene of Europe. Scincidae, though cosmopolitan like the Geckonids, are however most abundant in the Australian region, in the islands of the west Pacific Ocean, in the Oriental Region and in Africa. They are, however, rather poorly represented in America. Lacertidae are known since the Eocene times from Europe. Varanidae are likewise known from the Upper Cretaceous and Eocene of North America and Eurasia.

Analysis of the distributional patterns of sixty-eight genera reveals interesting patterns. Of the sixty-eight genera known so far from within our limits, twelve are endemic in the Palaearctic part of India (this includes some Ethiopian elements also) forty-two are endemic in the Oriental of India and fourteen are common to both. The fauna is thus predominantly Oriental in composition. The purely Palaearctic elements have recently intruded into the Oriental and some of the Palaearctic genera have also spread to the Far East across Middle Asia. Oriental elements too have intruded deep into the fringe of the Palaearctic.

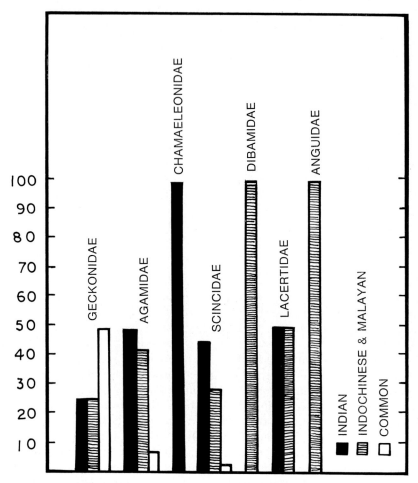

Fig. 113. Percentage analysis of Lacertilian genera in the Indian region.

1. The Mediterranean elements of our Lacertilia comprise the following genera:

Geckonidae
  1. *Teratoscincus*
  2. *Stenodactylus*
  3. *Alsophylax*
  4. *Agamura*
  5. *Pristurus*
  6. *Ptyodactylus*

Scincidae
  7. *Scincus*
  8. *Ophiomorus*
  9. *Chalcides*
Lacertidae
  10. *Eremias*

The following genera are common to the Mediterranean and Turk-menian subregions:

Agamidae
  11. *Phrynocephalus*

Scincidae
  12. *Ablepharus*

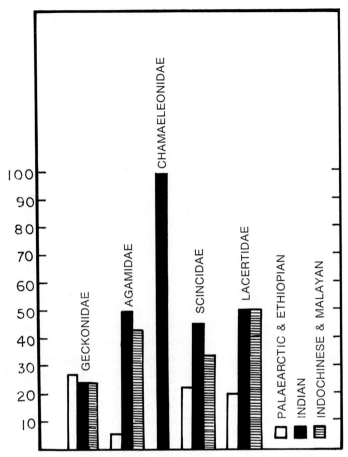

Fig. 114. Percentage analysis of Lacertilian genera in the Indian region.

The following genera are common to the Oriental and Palaearctic of India:

Geckonidae
1. *Gymnodactylus*
2. *Hemidactylus*
3. *Teratolepis*
4. *Eublepharis* *

Agamidae
5. *Japalura*
6. *Calotes*
7. *Agama*
8. *Uromastix*

Scincidae
9. *Mabuya*
10. *Leiolopisma*
11. *Eumeces*

Lacertidae
12. *Acanthodactylus*
13. *Ophiops*

Varanidae
14. *Varanus*

Stenodactylus orientalis, according to BLANFORD (1876), is a nocturnal species that burrows in the desert sand. *Agamura persica* occurs in barren

---

* *Eublepharis* is now included under a separate family Eublepharidae.

549

stony plains and hill slopes (see Chapter XIII). *Pristurus rupestris* is, on the other hand, diurnal and occurs on limestones. The Agamidae are also of diurnal habits and occur chiefly in rocky areas. *Ophiomorus*, with a cuneiform snout, burrows in desert sands (SMITH, 1935). Amongst the twelve genera from the Palaearctic of India, not one is endemic in the Northwest Himalaya. *Phrynocephalus* and *Ablepharus*, common to the Oriental and Palaearctic of India, are mainly Mediterranean in origin, with one species each intrusive in the Northwest Himalaya. The genera *Alsophylax*, *Pristurus* and *Scincus* have twelve, eight and seven species respectively in the extreme west of their range. *Eremias* is represented by forty-five species and numerous subspecies in southeastern Europe, Middle Asia and Africa. One species each of *Pristurus* and *Ptyodactylus* occurs in Sind; the remaining four genera are known from Baluchistan or Afgha- nistan-Baluchistan border. A species of *Stenodactylus* occurs in Sind. *Alsophylax tuberculatus* has also Sind as its range. *Chalcides pentadactylus*, reported from South India, has not been subsequently found and is of doubtful provenance. The genus is mainly an inhabitant of southern Europe, North Africa, Southwest Asia, with fifteen species known so far.

The Scincid genus *Ablepharus* is peculiar in that it occurs in Europe, Africa, Southwest Asia and extends through the East Indies to Australia and Polynesia. SMITH (1935) observes that its range is not in the Oriental Region proper and there is no evidence to show that it occurs in the New World. The genus must be considered to be of polyphyletic origin from independent centres. It is derived perhaps from the more widely dis- tributed *Leiolopisma*. The occurrence of the genus in Polynesia and Australia through the East Indies and in the Palaearctic would seem to suggest parallel evolution of the same characters in a number of unrelated species in the above mentioned parts of its range (see the last chapter). A Palaearctic element, occurring exclusively and abundantly in the main centre of its range, has spread and transgressed the eastern fringe of the Palaearctic. That this transgression was a recent event is evident from absence of differentiation beyond the species level. A case of parallel evolution of some groups of species in two far-flung areas of its range is also undoubtedly observed here. The groups of species have not also diverged beyond the species level.

The wide distribution of *Gymnodactylus* and *Hemidactylus* is attributable perhaps to their being often transported by human agency in ships. Their ability, at least in some cases, to survive long periods without food, may have also facilitated this passive dispersal. The monitor lizard *Varanus* is also widely distributed, but in this case its readiness to take to water, both fresh and brackish, may have greatly facilitated its wide dispersal. *Mabuya* is cosmopolitan and most numerous in the Australian, Oriental and Ethiopian regions.

With the exception of these four genera, most others prefer desert or hilly country. The nocturnal *Eublepharis macularius* hides during the daytime under stones, but *Agama, Eumeces, Acanthodactylus, Ophisops* and *Uromastix* are diurnal lizards that live inside burrows in desert sand. Most of these genera are Palaearctic and have intruded into the Oriental to varying extent. *Teratolepis, Eublepharis, Agama, Uromastix, Eumeces, Acanthodactylus* and *Ophisops* have intruded in the extreme western fringe of the Oriental. *Uromastix hardwickeii* has spread from northwestern parts of India as far east as the western parts of Gangetic Plain. *Acanthodactylus cantoris cantoris* also has similarly spread up to Agra in the Gangetic Plain, in addition to Ambala, Lahore, Ferozpore and other parts of the Punjab. *Agama minor* is the only species that has spread to the Peninsula (Saugor, Ratlam), and the Middle Gangetic Plain (Allahabad). The remaining ten species occur in Afghanistan, Iran, Iraq and Sind. *Ophiops* has likewise some species that occur in the Peninsula, Deccan and Western Ghats, as far south as Wynaad in South India, but otherwise largely Palaearctic.

Eublepharis and *Teratolepis* exhibit a peculiar distribution. Three species of *Eublepharis* are known; *Eublepharis hardwickii* is definitely known to occur in the Gangetic Plain, Chota-Nagpur, Orissa, Bengal and the Peninsular north and east; *Eublepharis macularius* (closely related to the foregoing species) occurs mainly in northwest India, Sind, Punjab, Rajasthan and the northern parts of the Deccan; *Eublepharis lichtenfelderi* occurs on dry rock in the Iles de Norway, Hainan. The species are all thus restricted to arid tracts.

Eumeces is known by about thirty-five species distributed in North and Central America; southeast and southwest Asia and North Africa. Of the five species occurring within our limits, three are also found in the Palaearctic, one in South China, Hong Kong and Hainan and another in South China, Hong Kong, Hainan, Cambodia and Thailand. *Teratolepis* is a monotypic genus, with *Teratolepis fasciata* from Sind and Khasi Hills. As SMITH (1935) observes, if these localities are correctly recorded, the lizard has indeed a most remarkably discontinuous distribution, differing widely in climate and separated by a wide gap of the whole of the Indo-Gangetic Plains.

A third type of distributional pattern is illustrated by the genera *Japalura, Calotes* and *Leiolopisma*. The three genera are found mainly in the Indo-Chinese and Indian subregions, but they have also penetrated the fringe of the Palaearctic. *Japalura* is a mountain form that frequently occurs at fairly high elevations in the Himalaya and Trans-Himalayan areas, southwest and central China, Formosa, the Riu-Kiu Islands, Borneo, Sumatra and the Natunas. Of the twelve species included in our analysis here, two alone are known from the Northwest Himalaya. Similarly only two species of *Leiolopisma* are known from the Northwest Himalaya. *Calotes versicolor* is the single (out of twenty-three) species occurring in Baluchistan, where also it is rather scarce. The main range of the genus is the Indo-Chinese subregion, where the majority of the species are concentrated, but nine species occur in South India of the Indian subregion. These are essentially Oriental elements that have penetrated into the fringe of the Palaearctic.

2. The genera endemic in the Indian subregion comprise the following:

Geckonidae
1. *Callodactylus*
2. *Dravidogecko*
3. *Lophopholis*
Agamidae
4. *Sitana*
5. *Otocryptes*
6. *Cophotis*
7. *Ceratophora*
8. *Lyriocephalus*
9. *Salea*

10. *Psammophilus*
Chamaeleonidae
11. *Chamaeleon*
Scincidae
12. *Ristella*
13. *Barkudia*
14. *Sepsophis*
15. *Chalcidoseps*
16. *Nessia*
Lacertidae
17. *Cabrita*

These seventeen genera are restricted in their distribution to the Peninsula. It is remarkable that ten genera are monotypic, four genera have two species each and only *Nessia* has six species. *Ceratophora* and *Ristella* are known respectively by three and four species. Five are endemic in Ceylon.

Sitana, Psammophilus, Chamaeleon and *Cabrita* are found throughout the Peninsula. The distribution of the Agamids *Sitana* and *Psammophilus*, which are typically inhabitants of arid open country or bare rocky terrain, is interesting. *Sitana ponticeriana* occurs from Ceylon to the foothills of the Himalaya, but not in Sind or in Bengal east of the R. Ganga. SMITH (1935) differentiates two forms of this species, on the basis of size, but intermediate types are not uncommon. The larger form, 70–80 mm long, is confined to area around Bombay and the smaller form, 40–50 mm long, occurs in the rest of the range and in Ceylon. The typical form comes from Pondicherry (South India). *Psammophilus dorsalis* is restricted to South India, south of 16° NL and the closely related *Psammophilus blanfordanus* occurs in the hills of the Chota-Nagpur Plateau, Orissa Hills, Eastern Ghats and south up to Trivandrum, in addition to the northern parts of the Peninsula (Central India). The genus has not, however, spread to the Sub-Himalayan ranges.

Chamaeleon and *Cabrita* represent the Ethiopian elements in the fauna of the Indian Peninsula. The family Chamaeleonidae is extensively distributed in Madagascar and Africa; over eighty species and four genera are known so far from its entire range, but a single genus and species occur in India. The common chamaeleon of Africa is also found in the eastern islands of the Mediterranean; two species occur in south Arabia and Socotra. *Chamaeleon zeylanicus* is the most common species occurring in the wooded districts of the Peninsula and Ceylon, but it is absent from the Himalaya and is sparsely found in Cutch.

552

Cabrita is known only by two species, both of which occur in the Peninsula and Ceylon and the genus is absent in the Himalaya and its vicinity. Like *Chamaeleon*, it is an inhabitant of open dry forest. Its close relative *Acanthodactylus* is widely distributed from southeast Europe, Africa north of the Equator, West Asia to the northwestern parts of West Pakistan.

The genera endemic in Ceylon are *Cophotis, Ceratophora, Lyriocephalus, Chalcidoseps* and *Nessia* (Fig. 121); they are naturally absent elsewhere. All the five genera are mostly arboreal and prefer dense jungle of the hilly parts of the island and occur at elevations of 1200–2100 m above mean sea-level. *Cophotis* is viviparous. The first three Agamid genera are closely related to each other and the two Scincid genera are also related to each other. Excepting *Nessia*, which has diverged into six species, the others are not rich in species. *Chalcidoseps* and *Nessia* are derived from the mainland Peninsular *Barkudia* and *Sepsophis*. These Ceylonese genera are no doubt derivatives from the mainland Peninsular stock, which spread to the island in the late Pleistocene before its separation.

The genera common to South India and to Ceylon are *Lophopholis* and *Otocryptes*, which appear to have penetrated the island prior to the stock that has given rise to the endemic genera. The same species of *Lophopholis* is found in South India and Ceylon. The species of *Otocryptes* occurring in Ceylon is closely related to the one found in the Cardamom Hills in South India.

3. Genera endemic in the Indo-Chinese and Malayan subregions are as follows:

Geckonidae
  1. *Phyllodactylus*
  2. *Ptychozoon*
  3. *Phelsuma*
Agamidae
  4. *Ptycholaemus*
  5. *Goniocephalus*
  6. *Mictopholis*
  7. *Oriocalotes*
  8. *Physignathus*
  9. *Leiolepis*

Scincidae
  10. *Ateuchosaurus*
  11. *Topidophorus*
  12. *Ophioscincus*
Dibamidae
  13. *Dibamus*
Lacertidae
  14. *Takydromus*
Anguidae
  15. *Ophiosaurus*

Most of the genera mentioned above are terrestrial and insectivorous, but some like *Physignathus* and *Tropidophorus* readily take to water and *Ptychozoon* and *Leiolepis* even show powers of gliding. *Phyllodactylus* is very widely distributed and is doubtfully distinct from *Diplodactylus* and is perhaps a genus of convenience rather than one reflecting real phyletic position. Forty-five species are known from Tropical America, Australia, Africa and the Mediterranean islands; three occur in Asia and only one species in Indo-China. *Physignathus* is known by eight or nine species from north Australia and Papuasia, Thailand and Laos. *Physignathus cocincinus*, the only species from the Oriental Region, occurs in the eastern Thailand, Laos, south China up to the West River. The distribution of the genus is paralleled by that of *Ophioscincus*, a genus of degenerate dwarf skinks, of which three occur in Indo-China and the fourth in Australia.

The genera *Leiolepis*, *Ateuchosaurus*, *Tropidophorus* and *Dibamus* have spread north and northeastwards. The first named genus occurs in Sumatra, Malaya Peninsula, south Burma as far as 18° NL, Thailand, South China and Hainan. *Ateuchosaurus* is likewise found in South China, Tonkin and the Riu-Kiu Islands. *Tropidophorus* is, on the other hand, represented by nine species in the Indo-Chinese subregion, of which one occurs in the Harigaj Range of the Sylhet Hills in Assam.

Ptychozoon and *Goniocephalus* are Malayan elements intruding into the Indo-Chinese area. The former is distributed from Malaya, South Burma, southeast and northern Thailand, Nicobar Islands and the Philippines. Of the four species known so far, only one extends to the Philippines. *Goniocephalus* is also represented by four species and is distributed more or less like *Ptychozoon*, but *Goniocephalus armatus* has diverged into subspecies endemic in Malaysia and another in the Peninsular Thailand.

The genera *Ptycholaemus*, *Mictopholis* and *Oriocalotes*, from the Eastern Himalaya and Assam, are represented by their close relatives in South India. Both *Ptycholaemus* and *Mictopholis* are monotypic and occur in the Khasi Hills (Assam) and Dafla Hills (Eastern Himalaya) respectively. They are believed to be derived from genera like *Otocryptes* and *Salea*, endemic in the Western Ghats. *Oriocalotes* is considered as a dwarfed *Calotes*, differing from it in unequal scalation on the dorsum, occasionally covered by tympanum and in the absence of any basal swelling in the tail; it is also monotypic and is endemic in the Khasi Hills.

Another distributional pattern is exemplified by *Takydromus* and *Ophiosaurus*. *Takydromus sexlineatus sexlineatus* occurs in Burma and the East Indies and another subspecies in the Khasi Hills. A second species is endemic in Goalpara (Assam). *Ophiosaurus* has spread from the Eastern Himalaya to western Yunnan, South China and Formosa. These genera have differentiated on the Tertiary mountains of Yunnan-Assam-Burma and spread westwards along the Himalaya.

About fifteen species of *Phelsuma* are known from Madagascar, the Comoro, Seychelles and Muscarene Islands and a single species *Phelsuma andamanense* occurs in the Andaman Islands. This illustrates the Madagascar-Indo-Australian distributional pattern. Some genera have transgressed from the Australian and Polynesian into the Malayan-Indo-Chinese subregions. Indo-Chinese genera have abundantly transgressed and radiated in diverse directions.

Plate 67. Ophisaurus gracilis.

Plate 68. A live specimen of *ophiosaurus* with a cluster of eggs in its natural habitat.

555

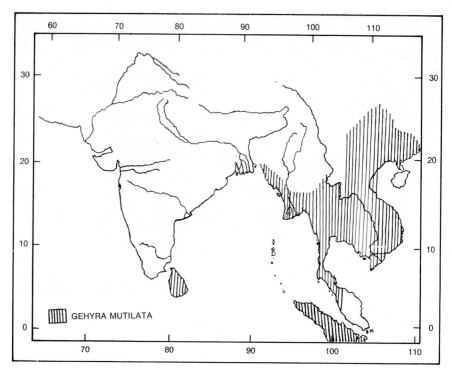

Fig. 115. Distribution of *Gehyra multilata.*

4. The following genera are common to the Indian and Malayan-Indo-Chinese subregions:

Geckonidae
  1. *Cnemaspis*
  2. *Platyurus*
  3. *Gehyra*
  4. *Hemiphyllodactylus*
  5. *Gecko*
  6. *Lepidodactylus*

Agamidae
  7. *Draco*
Scincidae
  8. *Dasia*
  9. *Lygosoma*
 10. *Riopa*

The genera *Gehyra, Hemiphyllodactylus* and *Gecko* have come to be widely distributed, possibly because of their being often passively transported by human agency. Their ability to survive long periods without food may have also greatly facilitated this passive dispersal. *Gehyra* (Fig. 115) and *Hemiphyllodactylus* are conspicuously discontinuously distributed (Fig. 116), but have been left out of the present discussion in view of the possibility of passive human dispersal. *Lepidodactylus* may also have to be similarly excluded from our discussion for the same reason. This is an Indo-Chinese genus that is widely distributed and apparently

556

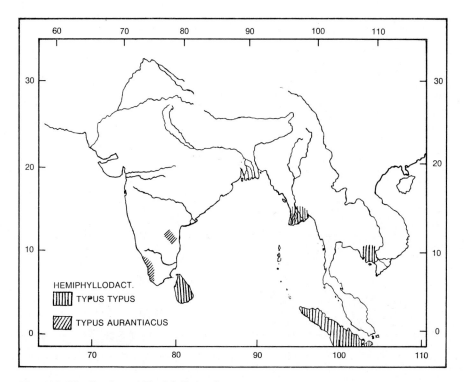

Fig. 116. Distribution of *Hemiphyllodactylus typus*.

common in its main centre. Its record from Ceylon and its absence from the intermediate areas is rather striking, but even in Ceylon, it is relatively scarce (Fig. 117). *Riopa* is widely distributed throughout the two sub-regions (Fig. 118).

The remaining five genera exhibit a remarkable discontinuous distribution. The range of *Lygosoma* is perhaps the least broken (Fig. 119); thirteen species are known so far from the Oriental Region and eight of these species occur in the Eastern Himalaya, Assam, Burma, southwest Yunnan to the Malaya Peninsula and also the Pareshnath Hills in the extreme northeast corner of the Indian Peninsula. One species occurs in the Western Ghats, from South Kanara to Trivandrum and Ceylon and four species are endemic in Ceylon. *Platyurus*, with two species, also exhibits more or less similar distributional pattern. (Fig. 120). *Platyura platyura* occurs in Ceylon, the Eastern Himalaya, Laos, Hong Kong, Formosa and the East Indies. The second species *Platyura craspedotus* occurs in the Malaya Peninsula and Borneo.

The range of *Cnemaspis* and *Dasia* is characterized by a larger discontinuity than the cases mentioned above. Thirteen species of the former genus are known; of these, eleven are restricted to the Western Ghats

557

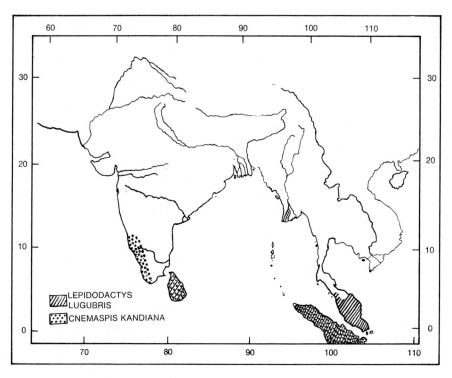

Fig. 117. Distribution of *Lepidodactylus* and *Cnemaspis.*

and the hilly parts of Ceylon and two are confined to the Indo-Chinese
subregion (Fig. 117). It is remarkable that not one of the eleven species
occurring in Ceylon is endemic. *Dasia* is a small genus of eight, more or
less arboreal species, of which three occur in the Oriental. *Dasia olivacea*
occurs in Borneo, Sinkip Island, northeast Sumatra, Malaya Peninsula
and the Philippines. *Dasia subcaerulea* occurs in South India and *Dasia*
haliana is endemic in Ceylon. The South Indian species resembles the
Indo-Chinese species *Dasia olivacea* very closely (Fig. 121).

The genus *Draco* affords us the best example of high degree of discontinuous distribution (Fig. 122). These flying lizards are completely
arboreal and seldom descend voluntarily to the ground. About forty
species of *Draco* are known, but excepting one, the rest are found in the
Indo-Chinese subregion, the East Indies and the Philippines. *Draco*
dussumieri is the only species occurring in the Western Ghats from the
south to Goa in the north. The gap between this species and the Indo-
Chinese form found in Assam, is at least 1600 km. These are typically
Indo-Chinese and Malayan genera which have transgressed across the
Eastern Himalaya and reached South India.

The lizards are only imperfectly known as fossils. Though fossil

558

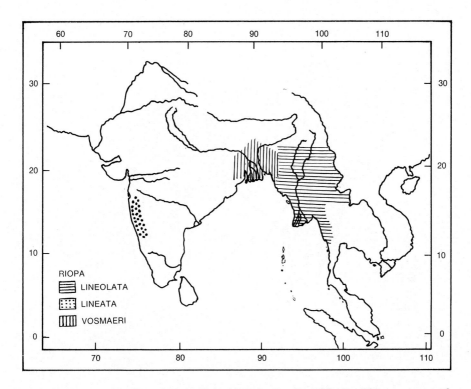

Fig. 118. The distribution of *Riopa* (Scincidae) in our faunal limits. The genus extends from Australia and Polyneisa to South America and Africa. The marked discontinuity that characterizes the distribution of the genus in India is illustrated by three species, which are considered by some authorities to be merely local forms of a single species, which was formerly continuously and more widely distributed, but has since disappeared in the intervening areas. The range of another species *Riopa anguina*, not plotted here, on the northern parts of the Western Ghats, overlaps in part with that of *Riopa lineata*. Two other species, also not plotted here, include *R. albopunctata* from India and *R. bowringi* from Indo-China.

Varanidae are known from the Pliocene of Siwalik, no other living families of lizards from India have so far been discovered as fossils and in any case the group as a whole is not older than the Triassic (BELLAIRS & UNDERWOOD, 1951). Unlike in the case of the Ophidians, the number of lizard genera and families in the Palaearctic is high. Similarly there are not many specialized families in the Indo-Chinese subregion. Nearly 25% of the genera are endemic in the Indian subregion, 23% in the Indo-Chinese subregion, 15% are common to the two subregions and about 21% are common to the Oriental and Palaearctic. The geckos are considered as an ancient group, but no fossils have been found as yet; they are abundant in the warmer parts of the world, particularly in the Australian and Oriental regions. The Agamidae that closely resemble the

559

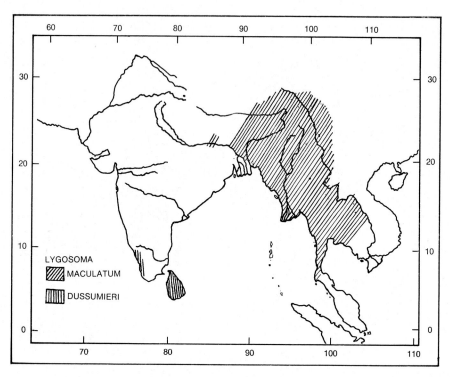

Fig. 119. Discontinuous distribution of *Lygosoma.*

Iguanidae of the New World are largely inhabitants of Asia, but some species extend into southeast Europe, Africa (but not Madagascar), Australia and New Guinea Archipelago (excluding New Zealand). Fossil Iguanidae are known from the Eocene of Europe. The distribution of Agamidae is not discontinuous unlike that of the Iguanidae; the two families resemble each other closely in external and internal characters and are separated only by differences in the dentition. The ancestry of Lacertidae can likewise be traced to the Teidae of America, which appear to be as old as the Cretaceous. The Lacertidae are inhabitants mainly of the Old World and occur in Europe, Asia and Africa, not however in Madagascar and Australian region. They are, however, most abundant in Africa and comparatively rare in the Oriental Region. The Chamaeleonidae, another African family, like the Lacertidae, are a vast assemblage of species, of which majority occur in Africa and Madagascar. The family is related to the Agamidae and CAMP (1923) considers it as having been derived from the highly developed Agamids at the beginning of the Tertiary. The Scincidae are cosmopolitan.

To summarize, we may conclude that a small Palaearctic element has transgressed into the extreme fringe of the Oriental within recent

560

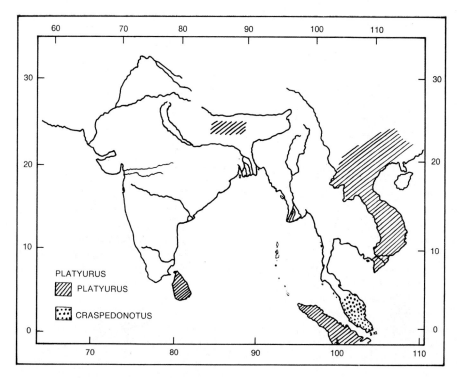

Fig. 120. Distribution of *Platyurus.*

times. Some of the Palaearctic genera have spread along the Central and Middle Asian areas to the Far-East. Some Oriental genera have also intruded into the Palaearctic Region. Many genera of purely Indo-Chinese stock have radiated in diverse directions. A distinct Malayan and Indo-Chinese element is present in the Eastern Himalaya and Peninsula. A peculiar and rather rare Malagasy and Ethiopian element is also found. The Peninsular and Ceylonese fauna are of different ages and have been derived at different times. Subspecific differentiation is evidently more on the Western Ghats than in the Eastern Ghats. Peculiar patterns of discontinuous distribution are exhibited by some genera. Parallel evolution has also occurred extensively in the case of some groups of species in two very widely separated areas of the range.

6. Serpentes

About three hundred genera and over two thousand six hundred species of snakes of eleven families are known so far from the world. Snakes occur in a variety of habitats and in all continents and continental islands and even in some of the most isolated oceanic islands. Though the

561

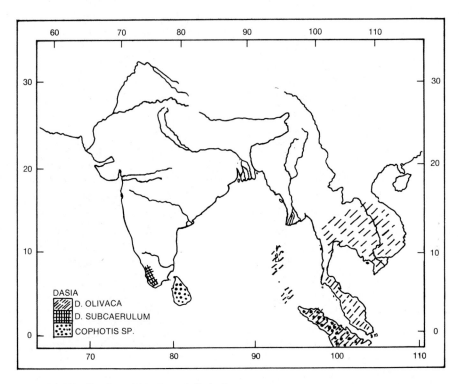

Fig. 121. Distribution of *Dasia* and *Cophotis.*

majority of snakes are tropical forms, many are common in the temperate
areas also (DITMARS, 1927). It is interesting that all the eleven families
of snakes are found within our limits, especially in the Oriental. As
DARLINGTON (1948) observes, the Orient is a cross-road at the dispersal
of snakes, even if not the main centre of their differentiation and evolution.
The principal aquatic snakes are found in or around the Oriental region,
which must perhaps be considered as the amphitheatre of the origin.
The snakes are, however, but poorly represented as fossils, particularly
the genera now living. The Boidae are known from the Eocene of Egypt
and the Elapidae from the Miocene and Pliocene of France. The Colubrids
are known from the Oligocene and the vipers from the Miocene. The
true vipers are confined at present to the Old World, but a vast majority
of Crotalinae occur in both the New and the Old World. The Elapine
snakes are abundant in Australia.

Ninety-one genera and three hundred and eighty-eight species occur
within our limits. Of these, twelve genera and twenty-nine species are
marine snakes of the family Hydrophidae and are not considered here.
Of the remaining seventy-nine genera, the Typhlopidae, Leptotyphlopidae, Boidae, Colubridae and Elapidae are cosmopolitan. The Viperidae

562

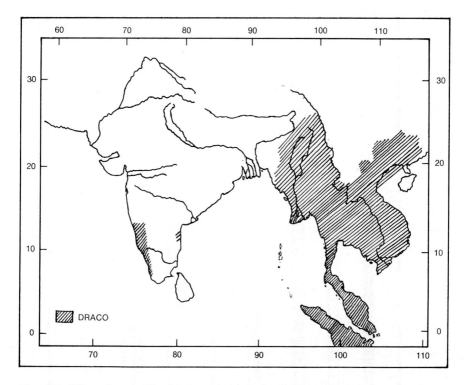

Fig. 122. Discontinuous distribution of the Indo-Chinese *Draco*, a genus of flying lizards, represented by about a dozen species, of which one, *Draco dussumieri*, is a geographical relict in the west and southwest corner of the Peninsula. *Draco norvilli* extends from Indo-China to the Naga Hills in Assam.

are Old World, but the Crotalinae are cosmopolitan. The Uropeltidae and Xenopeltidae are peculiar to the Oriental region. The Anilidae and Dipsadinae inhabit today the Oriental and Neotropical regions. The Dasypeltidae are highly specialized forms that occur in Africa and northern Bengal. The percentage analysis of distribution of genera is summarized in table VII (Fig. 123).

Excluding the Hydrophidae, we find that fifty-five of the seventy-nine genera, representing nearly 70% of the total genera known occur within the Oriental of India. Only six genera occur within the Palaearctic and eighteen genera, or about 23%, are common to the Oriental and Palaearctic of India.

Of the six genera occurring within the Palaearctic of India, five are endemic in the Mediterranean subregion and the genus *Contia* is common to the Mediterranean and Turkmenian subregions. This is a genus of dwarfed and degenerate snakes, probably derived from the Indo-Chinese *Liopeltis*, from which except for the presence of the apical pit,

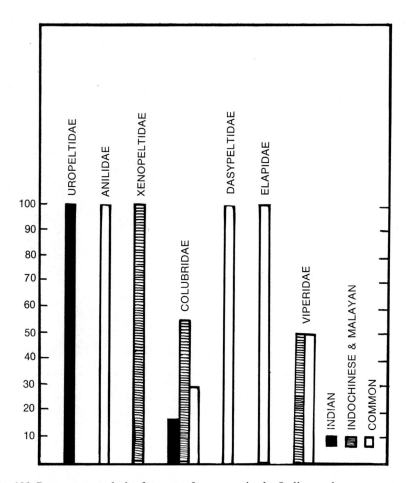

Fig. 123. Percentage analysis of genera of serpentes in the Indian region.

it is indistinguishable (SMITH, 1943). The genus extends from Transcaspia, Iran, Iraq, West Pakistan, Sind and Murree in the Northwest Himalaya. *Leptotyphlops, Lytorhynchus, Tarbophis, Pseudocerastes* and *Eristicophis* are endemic in the Mediterranean subregion. *Leptotyphlops*, the sole representative of the family Leptotyphlopidae, is a genus of small, degenerate burrowing snakes, bearing a close resemblance to the Oriental Typhlopidae. The two species known so far occur in Iran, Iraq, Arabia, Sind and Punjab. The two Colubrid genera *Lytorhynchus* and *Tarbophis* and the two Viperine genera *Pseudocerastes* and *Eristicophis* penetrate the fringe of the Oriental region, although they are widely distributed westwards. *Tarbophis*, derived from *Boiga*, is widely distributed in southwest Asia and in Africa.

The genera occurring within the Palaearctic parts of India are either

564

Table VII. Percentage analysis of distribution of genera of Serpentes

Family	Genera %	Palae-arctic	Oriental	Palae-arctic + Oriental	Indian (A)	Indo-Chinese (B)	Common to A & B
1. Typhlopidae	1.1	—	100.0	100.0	—	—	—
2. Lepto-typhlopidae	1.1	100.0	—	—	—	—	—
3. Uropeltidae	7.7	—	—	—	100.0	—	—
4. Aniidae	1.1	—	—	—	—	—	100.0
5. Xenopeltidae	1.1	—	—	—	—	100.0	—
6. Boidae	2.2	—	100.0	100.0	—	—	—
7. Colubridae	66.4	5.5	18.2	18.2	16.7	54.8	28.6
8. Dasypeltidae	1.1	—	—	—	—	—	100.0
9. Elapidae	3.3	—	66.7	66.7	—	—	100.0
10. Hydrophidae	13.2	Marine (Not considered here)					
11. Viperidae	7.7	—	42.9	42.9	—	50.0	50.0

relicts of formerly widely distributed forms or are intrusive elements from the western areas. The striking poverty of species differentiation in the genera in the region and the degeneration in the case of some genera would seem to lend support to this conclusion.

1. The following eighteen genera are common to the Palaearctic and Oriental:

Typhlopidae
  1. *Typhlops*
Boidae
  2. *Python*
  3. *Eryx*
Colubridae
  4. *Elaphe*
  5. *Ptyas*
  6. *Coluber*
  7. *Liopeltis*
  8. *Oligodon*
  9. *Lycodon*

10. *Natrix*
11. *Trachischium*
12. *Boiga*
13. *Psammophis*
Elapidae
  14. *Bungarus*
  15. *Naja*
Viperidae
  16. *Vipera*
  17. *Echis*
  18. *Ancistrodon*

Typhlops, Elaphe, Natrix and *Ancistrodon* are widely distributed. *Typhlops,* known by one hundred and sixty-four species, extends from tropical America, southern Europe, south Asia, Africa through India to Australia. *Natrix* is known by eighty species. These are very ancient and widely distributed genera. Of the remaining fourteen genera, *Python, Eryx, Coluber, Psammophis, Naja, Vipera* and *Echis* are Palaearctic forms, which have penetrated the Oriental from the northwest. These genera are distributed from North Africa, eastern and southern Europe, southwest

Asia to the Oriental region. *Python* is known as fossils from Eocene deposits. All the genera have established themselves well in the Oriental region, as evidenced by subspecific differentiations in many of their species, like for example, *Python molurus*, *Naja naja*, etc. *Naja* and *Echis* have also spread as far south as Ceylon. *Naja naja* is differentiated into three subspecies; *Naja naja naja* occurs in the whole of India south of the Himalaya and also in Ceylon; *Naja naja oxiana* occurs in the northwest parts of Pakistan; *Naja naja kouthia* occurs in Bengal. *Naja hannah* occurs in the Peninsula up to the Himalaya and also extends to the whole of the Indo-Chinese subregion, in addition to south China, Malaya Peninsula and Malay Archipelago and the Philippines. It would thus seem that the genus has become more Oriental than Palaearctic, although it is really an entrant from the Palaearctic region. The Palaearctic transgressions apparently occurred recently, long after the uplift of the Himalaya.

Ptyas, *Liopeltis*, *Oligodon*, *Lycodon*, *Trachischium* and *Boiga* are, on the other hand, typically Oriental genera that have spread to the Palaearctic. The great majority of the fifty or sixty species of *Oligodon* known so far are Oriental forms. Of the thirty-four species known within our limits, only *Oligodon taeniolatus* is recorded from Sind and Baluchistan and all others are restricted to the Oriental areas. The centre of distribution of *Lycodon* is likewise undoubtedly the Oriental region. *Dinodon*, its closest relative, is Chinese in origin. Both these genera meet in the Eastern Himalaya and in the Trans-Himalayan east. Here also of the eleven species, only *Lycodon striatus* extends up to Transcaspia through Sind and western Pakistan. *Liopeltis* and *Trachischium* are degenerate, dwarfed snakes, living generally under stones and fallen tree and feeding upon worms. Both the genera are mainly Indo-Chinese, with one or two species extending to the Northwest Himalaya. Of the six species of *Liopeltis*, only *Liopeltis rappi* has been found in Simla besides the Eastern Himalaya. Similarly, of the five species of *Trachischium*, *Trachischium fuscum* and *Trachischium laeve* are found in the Northwest Himalaya. The rat-snakes of the genus *Ptyas* are closely related to the American species of *Coluber*, to the Malayan *Gonyophis* and *Zaocys*. The mainly Oriental *Ptyas mucosus* extends from Ceylon through the whole of India and Indo-Chinese sub-region, West Pakistan, Afghanistan, Chitral, Turkestan, Kashmir and parts of the Western Himalaya. It is replaced by *Ptyas korros* in Malaysia. It is an Oriental genus that has extended into the Palaearctic.

Bungarus and *Boiga* are interesting from another point of view. Although fossil Elapids are known from the Pliocene of France, *Bungarus* is strongly represented in Australia. Of the ten species, seven occur in the Indo-Chinese subregion and only a single species is common to the Indian and Indo-Chinese subregions. *Bungarus caeruleus* occurs in Sind and northwest parts of the West Pakistan. A good many species occur in Malayan sub-region. In the case of *Boiga* also we observe a similar distributional pattern (Fig. 124). Of the thirteen species, only *Boiga trigonata* extends

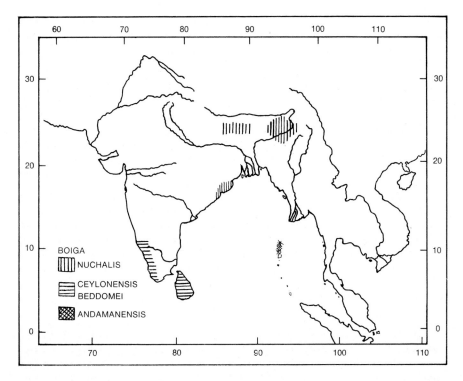

Fig. 124. Discontinuous distribution of colubrid snakes of the genus *Boiga*, which occur in Tropical Africa and Australia, Paupasia and South Asia. *B. ceylonensis* comprises the local forms *Boiga ceylonensis ceylonensis* on the Western Ghats and in Ceylon, *B. ceylonensis nuchalis* on the Western Ghats, Eastern Ghats and in Nepal and Assam and *B. ceylonensis andamanensis* in the Andamans. This is a Gondwana derivative that was formerly distributed continuously throughout the areas, where we find marked discontinuity today.

through West Pakistan to Transcaspia. As many as eight species are endemic in the Indo-Chinese subregion. The genus is also widely distributed from Africa, through the Oriental region to Australia. These two genera may perhaps be Palaearctic that have transgressed extensively the whole of the Oriental and Australian, but have largely dwindled in the Palaearctic area.

2. The following genera are endemic in the Indian subregion:

Uropeltidae
1. *Melanophidium*
2. *Platyplecturus*
3. *Teretrutrus*
4. *Plecturus*
5. *Uropeltis*
6. *Rhinophis*
7. *Pseudotyphlops*

Colubridae
8. *Coronella*
9. *Cercaspis*
10. *Balanophis*
11. *Macropisthodon*
12. *Aspidura*
13. *Haplocercus*
14. *Xylophis*

567

The distribution of genera in the Indian subregion is characterized by peculiar concentration in the western parts of the Peninsula, with the exception of *Coronella*, known by the species *Coronella brachyura* near Bombay and Poona. It is a rare snake that occurs in Europe, Africa north of the Equator and also in China. It may perhaps be a Palaearctic intrusive element.

The Uropeltidae are generally small-sized snakes that seldom exceed 30 cm in length and occur mostly in mountainous terrain and are largely autochthonous in the Indian subregion. The majority of the seven genera known so far occur in the Western Ghats of South India. *Uropeltis* is known by twenty-two species, and *Rhinophis* by ten species. *Uropeltis ellioti* extends to the Ganjam District in the extreme northeast of the Eastern Ghats, besides occurring in the Western Ghats. *Uropeltis macrolepis* occurs in the hills near Bombay. Two species are endemic in Ceylon and the remaining species are restricted to the Western Ghats south of the Goa Gap. Seventy percent of the species of *Rhinophis* are endemic in Ceylon. Of the remaining genera, *Pseudotyphlops* is alone endemic in Ceylon, but the rest are restricted to the Western Ghats, particularly the Anamalai and Palni Hills. No species of any of these genera are common to South India and Ceylon.

With the exception of *Xylophis* and *Coronella*, the other Colubrine genera are endemic in Ceylon and *Xylophis* alone being represented by two species in the Western Ghats. *Cercaspis* is closely related to the widely distributed *Lycodon*. Likewise, *Balanophis* is believed to have been derived from *Natrix*. *Aspidura*, *Haplocercus* and *Xylophis* are an assemblage of degenerate snakes, derived from widely distributed genera. The fact that exceedingly few Colubrids occur in the Western Ghats and very few Uropeltids have penetrated Ceylon would seem to indicate different phases of their radiation. Unlike the Uropeltidae, the Colubridae would seem to have reached their farthest points of dispersal in Ceylon.

3. The following are Indo-Chinese and Malayan endemics:

Xenopeltidae
1. *Xenopeltis*
Colubridae
2. *Pareas*
3. *Haplopeltura*
4. *Xenodermus*
5. *Stoliczkaia*
6. *Achalinus*
7. *Fimbrios*
8. *Zaocys*
9. *Xenalaphis*
10. *Opheodrys*
11. *Rhynchophis*
12. *Calamaria*
13. *Dinodon*
14. *Pseudoxenodon*
15. *Pararhabdophis*
16. *Plagiopholis*
17. *Opisthotropis*
18. *Blythia*
19. *Psammodynastes*
20. *Homalopsis*
21. *Fordonia*
22. *Cantoria*
23. *Bitia*
24. *Herpeton*
Viperidae
25. *Azemiops*

We observe that in the Indo-Chinese and Malayan composite subregion, the Colubrids are the dominant snakes, though they are largely represented by monotypic genera, with negligible subspecies differentiation.

Homalopsis, Fordonia, Cantoria and *Bitia* are aquatic snakes that occur in rivers, creeks and water courses, not far from the influence of tides, and all of them appear to be confined mainly to the Indo-Chinese subregion.

Achalinus, Opheodrys, Rhynchophis, Dinodon and *Pararhabdophis* are Chinese genera, which have extended into the Indo-Chinese area. *Achalinus, Rhynchophis* and *Pararhabdophis* have not, however, penetrated deep into the Indo-Chinese Peninsula. *Achalinus* is known by three or four species, but only *Achalinus rufescens* occurs in this subregion and is recorded from Hainan, Vietnam, South China and Hong Kong. *Rhynchophis* and *Pararhabdophis* are monotypic and are rather rare in this subregion. None of these genera occur south of the 20th parallel. *Opheodrys* and *Dinodon* have, however, spread farther south than the above mentioned genera. About eight or nine species of these genera are known. The species in the Indo-Chinese subregion have extended as far south as the 16th parallel and have also penetrated Assam and the Eastern Himalaya. These intrusions are evidently of recent origin, subsequent to the uplift of the Himalaya. It is interesting to observe that the species recorded from the Eastern Himalaya are known by only three or four specimens. *Zaocys, Calamaria* and *Psammodynastes* are widely distributed. The second genus is known by sixty or seventy species and seems to be a Malayan element that has extended into the Indo-Chinese subregion. Two species of the third genus are known; *Psammodynastes pulverulentus* is distributed throughout the Indo-Chinese and Malayan subregions. It is replaced by *Psammodynastes pictus* in Borneo and Sumatra.

Haplopeltura, Xenodermus, Xenalaphis and *Stoliczkaia* are Malayan genera, which have transgressed into the Indo-Chinese subregion, where each one is represented by a single species. *Haplopeltura boa* is known by two specimens collected from a dense jungle by Champion, somewhat north of the Isthmus of Kra. *Xenodermus javanicus* is likewise included, on the strength of a specimen obtained by Robinson and Kloss at the Victoria Point, South Tenasserim. Two species of *Stoliczkaia* are known; *Stoliczkaia khasiensis* is described from a specimen from the Khasi Hills and the second species comes from Borneo. In this case also, the record from the Indo-Chinese subregion is based only on a few specimens. *Xenalaphis hexagonatus* is known from southern Indo-China.

The only genera which are endemic in the Indo-Chinese subregion are *Xenopeltis, Pareas, Fimbrios, Pseudoxenodon, Plagiopholis, Opisthotropis, Blythia, Herpeton* and *Azemiops*. Of these nine genera, *Xenopeltis* and *Plagiopholis* are largely restricted to Burma and Indo-China. *Fimbrios, Opisthotropis* and *Herpeton* are, on the other hand, largely restricted to Cambodia, Laos and Vietnam and the areas farther east may perhaps be the centre of their differentiation. *Pareas* and *Pseudoxenodon* have their range in the northwest and northeast towards the Eastern Himalaya.

569

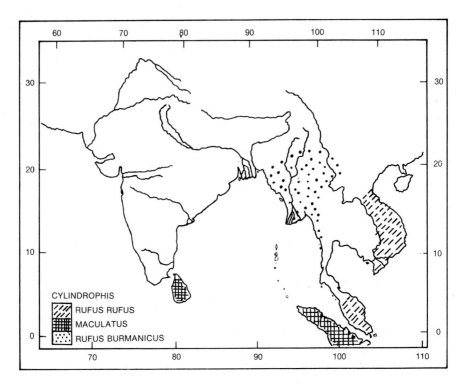

Fig. 125. Distribution of *Cylindrophis.*

Blythia is a degenerate form, apparently derived from an ancestor like *Pseudoxenodon* and appears to have become a relict in Assam and Burma. *Azemiops* appears, on the other hand, to have spread along Upper Burma, Vietnam, South China and southeast Tibet.

4. The genera common to the Indian and the Indo-Chinese-Malayan subregions may be listed as below:

Anilidae
  1. *Cylindrophis*
Colubridae
  2. *Achrochordus*
  3. *Ahaetulla*
  4. *Chrysopelea*
  5. *Dryocalamus*
  6. *Sibynophis*
  7. *Xenochrophis*
  8. *Atretium*
  9. *Rhabdops*

  10. *Dryophis*
  11. *Enhydris*
  12. *Cereberus*
  13. *Gerardia*
Dasypeltidae
  14. *Elachistodon*
Elapidae
  15. *Callophis*
Viperidae
  16. *Trimeresurus*

Enhydris, Cereberus, Gerardia and *Achrochordus* are aquatic and also enter the sea occasionally. The terrestrial snakes of the genera *Sibynophis,*

570

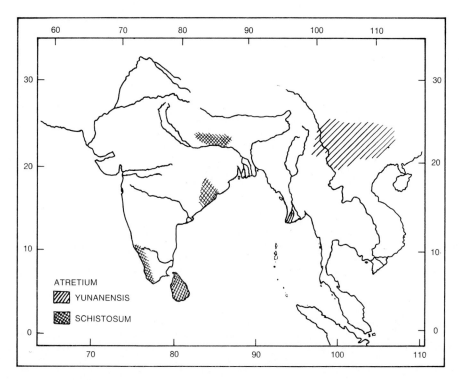

Fig. 126. Distribution of *Atretium.*

Dryophis and *Callophis* are on the whole continuously distributed in their range. *Trimeresurus* is widely distributed. The South American species of *Bothrops* are considered by many authorities as morphologically not distinct from the Asiatic species of *Trimeresurus*. The monotypic *Xeno-chrophis* is restricted to the Sub-Himalayan belt in North India. It occurs in the Middle Gangetic Plain and in Assam (Khasi Hills and Goalpara). In a similar fashion, *Elachistodon* is restricted to northern Bengal and Bihar. These genera are relicts of derivatives of widely distributed genera like *Natrix* and *Dasypeltis* from African stock.

The remaining six genera show, however, an interesting pattern of discontinuous distribution. They occur in the Indo-Chinese subregion on the one hand and in the extreme south of the Indian Peninsula and Ceylon on the other hand, with a wide intervening gap. HORA & JAYA-RAM (1949), who discussed this distribution, erroneously called them as Malayan elements in the Peninsular India. *Cylindrophis* is known by two species, viz. *Cylindrophis rufus* and *Cylindrophis maculatus*. The latter species is endemic in Ceylon and *Cylindrophis rufus* is restricted to Thailand, Indo-China south of the 17th Parallel, the Malaya Peninsula and Malay Archipelago (Fig. 125). In Tenasserim and Burma as far north as

571

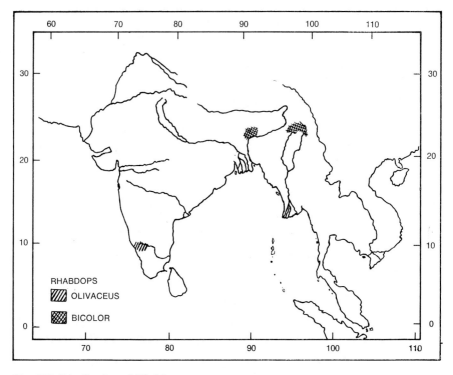

Fig. 127. Distribution of *Rhabdops*.

Myitkyina, it is replaced by the subspecies *Cylindrophis rufus burmanus*. *Atretium* and *Rhabdops* are more or less similarly distributed (Fig. 126); two species are known in each of these genera. *Atretium schistosum* occurs in Ceylon, Anamalai Hills, Wynaad, Bangalore, Mysore, the Western Gangetic Plain and Orissa. *Atretium yunnanesis*, the second species, occurs in western Yunnan. *Rhabdops* (Fig. 127) is much more restricted in its distribution than *Atretium; Rhabdops olivaceus* is found in Wynaad and *Rhabdops bicolor* occurs in the Khasi and Mishmi Hills (Assam), in Burma and in western Yunnan. Three species of *Chrysopelea* are known; *Chryso-pelea taprobanica* is peculiar to Ceylon, *Chrysopelea paradisi* occurs in the Malaya Peninsula as far north as Mergui, in the Andaman Islands, in Sumatra, Java, Borneo and Philippines and *Chrysopelea ornata* is differentiated into local subspecies. One of them occurs in Ceylon and the Western Ghats south of the Goa Gap and the other occurs in the Indo-Chinese area and extends as far northwest as Darjeeling, Patna and Buxar in Bihar and Orissa and as far northeast to Vietnam and southern China (Fig. 128). *Ahaetulla*, with as many as nine species, is also similarly distributed. Three species are restricted to South India and Ceylon and the remaining species occur in the Indo-Chinese subregion (Fig. 129).

572

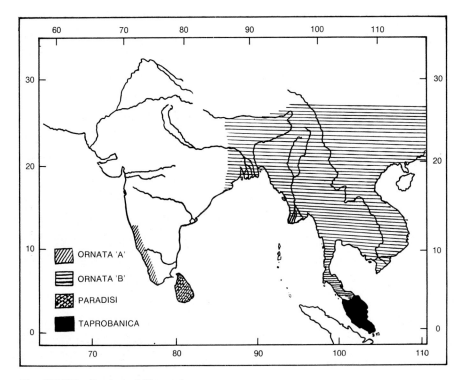

Fig. 128. Distribution of *Chrysopelea.*

Ahaetulla ahetulla has differentiated into colour forms within its range. *Ahaetulla tristis* is rather widely distributed in both the subregions. Of the three species of *Dryocalamus* occurring in the Oriental region, *Dryocalamus davisoni* comes from Thailand between 11 and 18 the north parallels, Tenasserim, Cambodia and Laos. *Dryocalamus nympha* and *Dryocalamus gracilis* occur in the Western Ghats and Eastern Ghats and Ceylon (Fig. 130). These genera are apparently intrusive elements in the Peninsula from a relatively older stock.

Out of seventy-nine genera we have considered above, twenty-five are endemic in the Indo-Chinese subregion, representing nearly 32 % of the total. Among the genera common to the Indian and Indo-Chinese subregions, and to the Oriental and Palaearctic regions, we may observe a considerable proportion of the Indo-Chinese intrusive elements. The Colubridae dominate in the Indo-Chinese subregion, where they amount to 92 % of the genera of snakes. The Typhlopidae, Leptotyphlopidae and Uropeltidae are largely inhabitants of the tropics of the Old and New World, the Uropeltidae being found only in South Asia, especially the Peninsula of India and in Ceylon, with only a single species *Uropeltis ellioti* extending into the East. The three families are considered to be

573

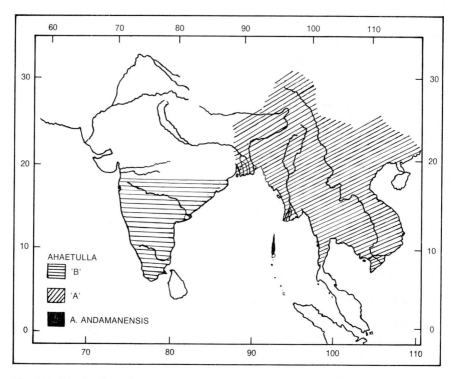

Fig. 129. Distribution of *Ahaetulla ahaetulla.*

primitive but the Leptotyphlopidae and Typhlopidae are believed to be side-line aberrant derivatives from the main line of Ophidian descent. The Anilidae of South Asia and Tropical America are considered to be the most primitive snake family at present. The Boidae and Xenopeltidae are related to each other and though possessing many primitive characters, show certain advanced features also, and have undergone a minimum of aberrant specialization (BELLAIRS & UNDERWOOD, 1951). The above mentioned six families constitute a relatively primitive group and the Colubridae, Elapidae, Hydrophidae and Viperidae represent an advanced group. The latter group includes the great majority of snakes. The Colubridae are the largest family and are of cosmopolitan distribution. The Elapidae are abundant in Australia and also occur in the tropics of both the New and the Old World. Viperidae are absent in Australia, but are common in the tropics of the Old World. The history of families invalidates the theory of northern origin of snakes, put forward by some authors like MATHEW (1915). The tropics, especially of South Asia, have been the home of Ophidia.

To summarize we may observe that the snakes of our area are dominantly Indo-Chinese in origin, specialization and radiation. The

574

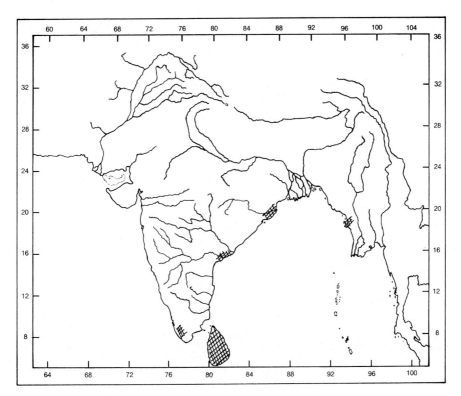

Fig. 130. Distribution of *Dryocalamus gracilis.*

Palaearctic and Ethiopian elements are small and are represented by degenerate forms, which have not also penetrated deep. The Malayan element is considerable both in the Indo-Chinese subregion and in the Indian subregion. The Indian subregion is dominated by the degenerate Uropeltidae. The snakes are largely concentrated in the Western Ghats and Ceylon and appear to be relicts of formerly widely distributed fauna which radiated in more than one phase. The characteristic distributional patterns, with pronounced disjunction mark the Indo-Chinese and Malayan elements.

7. Crocodilia

The distribution of the single family known at present from within our limits is characterized by marked discontinuity; the living genera are relicts in the warmer parts.

Gavialis gangeticus is the most primitive member of the family and is confined to the R. Indus, Ganga, Brahmaputra Mahanadi in India and R. Kaladan in Burma. Fossil remains of the genus *Gavialis* are known from the Pliocene of Siwalik and the Narmada Valley. The occurrence of the

575

genus in the three major Himalayan rivers has been cited by some workers as possible support for the fanciful Indo-Brahm river of the Pliocene times. The family Crocodilidae is known from Australia, Southeast Asia, Africa, Tropical and Subtropical America; but the forms occurring in the Old and New World are generically different. The distribution of the present day genera can be explained only on the basis of the theory of continental drift. The second genus *Crocodilus* is an estuarine form.

8. Summary

The following is a summary of the frequency distribution of genera of different groups of Vertebrates considered here in the region (Fig. 131):

Table IX. Percentage frequency distribution of genera

Group	Generic %	% Palaeartic of India	% Oriental of India
Fish	89	13.0	87.0
Amphibia	24	4.2	95.8
Chelonia	22	—	100.0
Lacertilia	68	17.7	82.3
Serpentes	79	7.6	92.4
Crocodilia	2	—	100.0

The following table summarizes the generic distribution in different groups in the Indian and the Indo-Chinese (+Malayan) subregions:

Table X. Analysis of generic distribution

Group	Indian subregion					Indo-Chinese + Malayan subregion				
	End.		Comm.		Total	End.		Comm.		Total
Fish	9	+	30	=	39	19	+	30	=	49
Amphibia	12	+	5	=	17	3	+	5	=	8
Chelonia	1	+	3	=	4	12	+	3	=	15
Lacertilia	17	+	10	=	27	15	+	10	=	25
Serpentes	14	+	16	=	30	25	+	16	=	41
Total	53	+	64	=	117	74	+	64	=	138

576

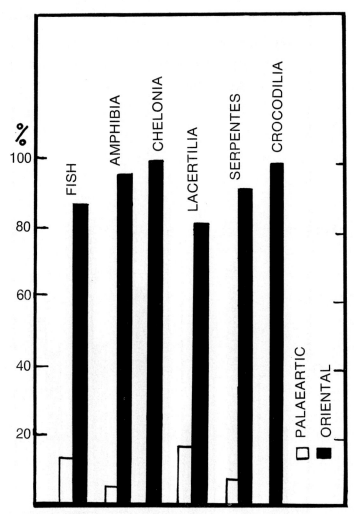

Fig. 131. Percentage frequency distribution of genera in fish, amphibia and reptiles.

The percentage endemicity of genera in the two major subdivisions is summarized in the following table (Fig. 132):

Table XI. Summary of endemic genera

Group	Indian subregion		Indo-Chinese + Malayan subregion	
	N	%	N	%
Fish	39	23.1	49	38.7
Amphibia	17	70.5	8	37.5
Chelonia	4	25.0	15	80.0
Lacertilia	27	62.9	25	60.0
Serpentes	30	46.6	41	60.9
Total	117	45.3	138	53.6

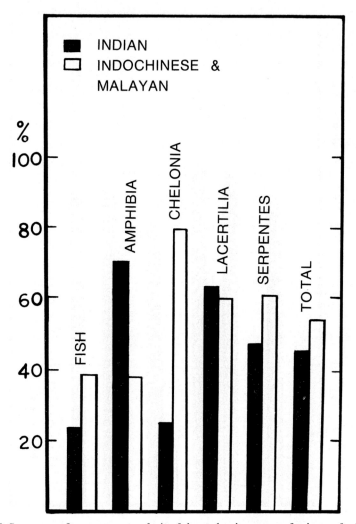

Fig. 132. Summary of percentage analysis of the endemic genera of primary fresh-water fishes, amphibia and reptiles.

It is evident from the foregoing discussion that within the Oriental Region, the Indo-Chinese subregion has much larger numbers of genera and species in nearly all groups, except the Amphibia, than the Indian subregion. We also observe that fifty percent of the genera of Amphibia are autochthonous in the Indian subregion and the essential conditions favouring their abundance are apparently better developed in the Western Ghats than in the Indo-Chinese area. For all the other major elements we might consider the Indo-Chinese subregion as a possible centre of differentiation and radiation. In this area the presence of monotypic genera is considerable; we have seen that 68 % of the genera of snakes and 75 % of the chelonians are monotypic. Moreover, it is also interesting to remark that in all these groups (with the possible exception again of the Amphibia), the specialized genera occur as a rule in the Eastern Himalaya, Assam, Thailand, Cambodia, Malaya, China, etc., where the original stock has in many cases diverged into different taxonomic levels. Many Malayan and Chinese genera intrude into the Indo-Chinese Peninsula and many Indo-Chinese genera have likewise spread westwards along the Himalaya and southwards to the Peninsula. DAY (1885), who analysed the Indian fresh-water fishes, believed that most of the genera had Malayan affinities; indeed this confusion between the Indo-Chinese and Malayan elements is widespread among zoologists in India. GREGORY (1925) considered that the extensive river captures made it possible for animals to migrate from the east to west but not in the reverse direction. MORI (1936) was of the view that the Nan-Shan Mountain range divides China into the northern Palaearctic Region and the southern Oriental Region. Considering all the facts discussed above and also the well established geological evidence that the drainage patterns of the Yunnan Plateau is older than that of the Himalaya, it seems possible that the Yunnan-Assam-Burma amphitheatre is the centre of origin of the bulk of the land vertebrates found in India today. The South Indian Peninsula is poor therefore in fresh-water fish and reptile. Among the Amphibia we find a heavy concentration of genera and species and high degree of isolation. The general composition of the fauna of the whole country presents unmistakable evidence of recent regression, perhaps within historical times.

REFERENCES

ANNANDALE, N. 1913. The Tortoises of Chota Nagpur. *Rec. Indian Mus.*, 9: 63–78.

ANNANDALE, N. 1914. The African element in the freshwater fauna of British India. *Proc. Ninth International Congr. Zool. Monaco:* 579–588.

AUDEN, J. B. 1949. A geological discussion on the Satpura Hypothesis and Garo-Rajmahal gap. *Proc. Nat. Inst. Sci. India*, 15: 315–340.

BEAUFORT, L. F. DE, 1951. Zoogeography of the land and Inland waters. London: Sidgwick and Jackson Ltd.

BELLAIRS, A. D. & G. UNDERWOOD, 1951. The origin of Snakes. *Biol. Rev.*, 26: 193–237.

BLANFORD, W. T. 1876. On some Lizards from Sind with descriptions of new species of *Ptyodactylus, Stenodactylus* and *Trapelus. J. Asiat. Soc. Bengal*, 14(2): 18–26.

BLANFORD, W. T. 1901. The distribution of Vertebrate animals in India, Ceylon and Burma. *Philos. Trans. Roy. Soc.*, (B) 194: 335–436.

BUOLENGER, G. A. 1911. Catalogue of the freshwater fishes of Africa in the British Museum (Natural History) London. 2: 1–529.

CAMP, C. L. 1923. Classification of Lizards. *Bull. Amer. Mus. Nat. Hist.*, 48: 289–481.

CHU, Y. T. 1935. Comparative studies on the scales and on pharyngeals and their teeth in Chinese Cyprinids with particular reference to taxonomy and evolution. *Biol. Bull. St. John's Univ.* 2.

DARLINGTON, P. J. 1948. The geographical distribution of cold-blooded vertebrates. *Quart. Rev. Biol.*, 23: 1–26, 105–123.

DARLINGTON, P. J. 1957. Zoogeography: the geographical distribution of animals New York, (N.Y.): John Wiley and Sons.

DARLINGTON, P. J. 1964. Drifting continents and late Paleozoic geography. *Proc. nat. Acad. Sci.*, 52: 1084–1091.

DAVID, A. 1962. Brief taxonomic account of the gangetic *Pangasius panagasius* (Hamilton) with a description of a new subspecies from the Godavary. *Proc. Indian Acad. Sci.*, 56: 136–156.

DAY, F. 1876–78. The fishes of India; being a natural history of the fishes known to inhabit the seas and freshwaters of India, Burma, and Ceylon. 1, 2, 1–778.

DAY, F. 1885. Geographical distribution of Indian freshwater fishes. *J. Linn. Soc.*, 13: 138–155. 338–353.

DAY, F. 1889. The Fauna of British India, including Ceylon and Burma. Fishes, I & II, London: Taylor and Francis.

DERANIYAGALA, P. E. P. 1943. The age and derivation of Ceylon's Siwalik fauna. *Proc. 29th Indian Sci. Congr.*

DILGER, W. C. 1952. The Brij Hypothesis, an explanation for the tropical faunal similarities between the Western Ghats and the Eastern Himalayas. Assam, Burma and Malaya. *Evolution*, 6: 125–127.

DITMARS, R. L. 1927. Reptiles of the World. New York: The Macmillan Co.

DU TOIT, A. L. 1937. Our wandering continents, an hypothesis of continental drifting. Edinburgh and London: Oliver & Boyd.

EGERTON, P. 1845. On the remains of fishes found by Mr. Kaye and Mr. Counliffe, in the Pondicherry beds. *Quart. J. geol. Soc.*, 1: 164–171.

GANSSER, A. 1964. Geology of the Himalaya. New York (N.Y.) John Wiley & Sons.

GREENWOOD, P. H., D. E. ROSE, S. H. WEITZMAN & G. S. MYERS, 1966. Phyletic studies of Teleostean fishes with a provisional classification of living forms. *Bull. Amer. Mus. Nat. Hist.*, 131: 339–456.

GREGORY, J. W. 1925. The evolution of the river system of South Eastern Asia. *Scottish Geog. Mag.*, 41: 129–141.

GULATEE, B. L. 1952. The Aravalli Range and its extensions. Technical Paper 6, Survey of India. Dehra Dun.

HAIG, J. 1950. Studies on the classification of the catfishes of the Oriental and Palae-arctic family Siluridae. *Rec. Indian Mus.*, 48: 59–116.

HORA, S. L. 1933. Siluroid fishes of India, Burma & Ceylon. 1. Loach-like fishes of the genus *Amblyceps* Blyth. *Rec. Indian Mus.*, 35: 607–621.

HORA, S. L. 1936. Siluroid fishes of India, Burma & Ceylon. 2. Fishes of the genus *Akysis* Bleeker. III. Fishes of the genus *Olyra* McClelland. IV. On the use of the generic name *Wallago* Bleeker. V. Fishes of the genus *Heteropneustes* Muller. *Rec. Indian Mus.*, 38: 199–209.

HORA, S. L. 1936. Siluroid fishes of India, Burma & Ceylon. VI. Fishes of the genus *Clarias* Gronovius. VII. Fishes of the genus *Silurus* Linnaeus. VIII. Fishes of the genus *Callichrous* Hamilton. *Rec. Indian Mus.*, 38: 347–361.

HORA, S. L. 1937. Geographical distribution of Indian freshwater fishes and its bearing on the probable land connections between India and adjacent countries. *Curr. Sci.* 5: 351–356.

HORA, S. L. 1937. Comparison of the fish faunas of the Northern and the Southern faces of the great Himalayan range. *Rec. Indian Mus.*, 39: 241–250.

HORA, S. L. 1937. The Game Fishes of India. 1. The Indian Trout. *Barilius (Opsarius) bola* Hamilton. *J. Bombay Nat. Hist. Soc.*, 39: 200–210.

HORA, S. L. 1937. The Game Fishes of India. II. The Bachhwa or Butchwa. *Eutro-piichthys vacha. J. Bombay Nat. Hist. Soc.*, 39: 431–446.

HORA, S. L. 1937. The Game Fishes of India. III. Garua, Bachcha or Gaurchha. *Clupisoma garua* (Hamilton) and two allied species. *J. Bombay Nat. Hist. Soc.*, 39: 659–678.

HORA, S. L. 1937. On fossil fish-remains from the Karewas of Kashmir. *Rec. geol. Surv. India*, 72: 178–187.

HORA, S. L. 1938. The Game Fishes of India. IV. The Silond Catfish. *Silonia silondia* (Hamilton). *J. Bombay nat. Hist. Soc.*, 40: 137–147.

HORA, S. L. 1938. The Game Fishes of India. V. The Pungas Catfish. *Pangasius pangasius* (Hamilton). *J. Bombay nat. Hist. Soc.*, 40: 355–366.

HORA, S. L. 1938. On the age of the Deccan trap as evidenced by fossil fish remains. *Curr. Sci.*, 6: 370–372.

HORA, S. L. 1938. On some fossil fish-scales from the intertrappean beds at Deothan and Kheri, Central Provinces. *Rec. geol. Surv. India*, 73: 267–294.

HORA, S. L. 1939. The Game Fishes of India. VI. The Goonch. *Bagarius bagarius* (Hamilton). *J. Bombay nat. Hist. Soc.*, 40: 583–593.

HORA, S. L. 1939. The Game Fishes of India. VII. The Mulley or Boali. *Wallagonia attu* (Bloch and Schneider). *J. Bombay nat. Hist. Soc.* 41: 64–71.

HORA, S. L. 1940. The Game Fishes of India. XI. The Mahseers or the large-scaled barbels of India. 4. The Bokar of the Assamese and Katli of the Nepalese. *Barbus (Lissochilus) hexagonolepis* McClelland. *J. Bombay nat. Hist. Soc.*, 42: 78–88.

HORA, S. L. 1941. Siluroid fishes of India, Burma and Ceylon. XI. Fishes of the Scheil-beid genera *Silonopangasius* Hora, *Pseudeutropius* Bleeker, *Proeutropiichthys* Hora and *Ailia* Gray. XII. A further note on fishes of the genus *Clarias* Gronovius. *Rec. Indian Mus.*, 43: 97–115.

HORA, S. L. 1944. On the Malayan affinities of the freshwater fish fauna of Peninsular India, and its bearing on the probable age of the Garo-Rajmahal gap. *Proc. nat. Inst. Sci. India*, 10: 423–439.

HORA, S. L. 1948. The distribution of Crocodiles and Chelonians in Ceylon, India, Burma and farther East. *Proc. nat. Inst. Sci. India*, 14: 285–310.

HORA, S. L. 1949. The fish fauna of Rihand river and its zoogeographical significance. *J. zool. Soc. India*, 1: 1–7.

HORA, S. L. 1949. Discontinuous distribution of certain fishes of the Far East to Penin-sular India. *Proc. nat. Inst. Sci. India*, 15: 411–416.

HORA, S. L. 1949. Satpura hypothesis of the distribution of the Malayan fauna and flora to Peninsular India. *Proc. nat. Inst. Sci. India*, 15: 309–314.

HORA, S. L. 1951. Siluroid fishes of India, Burma & Ceylon. XIII. Fishes of the genus *Erethistes* Müller & Troschel, *Hara* Blyth and two new allied genera. *Rec. Indian Mus.*, 47: 183–201.

HORA, S. L. 1953. The Satpura Hypothesis. *Sci. Progress.*, 41: 245–255.

HORA, S. L. 1953. Fish distribution and Central Asian Orography. *Curr. Sci.*, 22: 93–97.

HORA, S. L. 1953. Are there any precedent Himalayan rivers? Address of the Chief Guest. *Bhu Vidya. J. geol. Inst. Presidency College, Calcutta*. Diamond Jubilee Vol. 49–55.

HORA, S. L. 1955. Tectonic history of India and its bearing on fish geography. *J. Bombay nat. Hist. Soc.* 52: 692–701.

HORA, S. L. 1955. The status of the Satpura Hypothesis. *Bull. nat. Inst. Sci. India*, 7: 264–268.

HORA, S. L. 1955. The evolution of the Indian torrential environment and its fishes. *Bull. nat. Inst. Sci. India*, 7: 264–268.

HORA, S. L. 1955. Place of Kashmir in the fish geography of India. *Everyday Sci.* 3: 36–45.

HORA, S. L. & K. C. JAYARAM, 1949. Remarks on the distribution of snakes of Peninsular India with Malayan affinities. *Proc. nat. Inst. Sci. India*, 15: 399–402.

HORA, S. L. & N. C. LAW, 1941. Siluroid fishes of India, Burma & Ceylon. IX. Fishes of the genera *Gagata* Bleeker and *Nangra* Day. X Fishes of the genus *Batasio Blyth*. *Rec. Indian Mus.*, 43: 9–42.

HORA, S. L. & D. D. MUKERJI, 1935. Fishes collected by the third Netherland Karakorum Expedition. *Wiss. Ergeb. Niederl. Karakorum*, 1: 426–445.

HORA, S. L. & K. K. NAIR, 1941. Fishes of the Satpura Range, Hoshangabad District, Central Provinces. *Rec. Indian Mus.*, 43: 387–393.

JACOB, K., 1949. Land connections between Ceylon and Peninsular India. *Proc. Nat. Inst. Sci. India*, 15: 341–343.

JAYARAM, K. C. 1949. Distribution of Lizards of Peninsular India with Malayan affinities. *Proc. Nat. Inst. Sci. India*, 15: 403–409.

JAYARAM, K. C. 1949. A note on the distribution of Chelonians of Peninsular India with Malayan affinities. *Proc. Nat. Inst. Sci. India*, 15: 397–398.

JAYARAM, K. C. 1953. The Palaearctic element in the fish fauna of Peninsular India. *Bull. Nat. Inst. Sci. India*, 7: 260–263.

JAYARAM, K. C. & N. MAJUMDAR, 1964. On a collection of fish from the Kameng Frontier Division, NEFA. *J. Bombay. Nat. Hist. Soc.*, 61: 264–280.

JAYARAM, K. C. 1966. Contributions to the study of the Bagrid fishes (Siluroidea: Bagridae). 1. A systematic account of the genera *Rita* Bleeker, *Rama* Bleeker, *Mystus* Scopoli and *Horabagrus* Jayaram. *Int. Revue ges. Hydrobiol.*, 51: 433–450.

JAYARAM, K. C. 1966. Contributions to the study of the fishes of the family Bagridae. 2. A systematic account of the African genera with a new classification of the family. *Bull. Inst. Fond. Afr. Noire*, 28: 1064–1139.

JAYARAM, K. C. 1968. Contributions to the study of the Bagrid fishes (Siluroidea: Bagridae). 3. A systematic account of the Japanese, Chinese, Malayan and Indonesian genera. *Treubia*, 27: 286–386.

KREMPF, A. & P. CHEVEY, 1934. The continental shelf of French Indo-China and the relationship which formerly existed between Indo-China and the East Indies. *Proc. Fifth Pacific Sci. Congr.*, 1933: 849–852.

KRISHNAN, M. S. 1944. Introduction to the Geology of India. Madras: Madras Law-Journal Office.

KRISHNAN, M. S. 1953. The structural and tectonic history of India. *Memoirs geol. Surv. India*, 81.

KRISHNAN, M. S. 1968. Physiographic characteristics of Peninsular Ranges. Chapter V. In: Mountains and Rivers of India, 21st International Geographical Congress, India, 88–95.

LYDEKKER, R. 1886. Indian Tertiary and post-Tertiary vertebrata. Tertiary fishes. *Palaeont. Indica*, 3: 241–264.

582

MANI, M. S. 1968. Zoogeography of the mountains of India. In: Mountains and Rivers of India. *21st Internat. Geogr. Congr. India*, pp. 96–109.

MARSHALL, N. B. 1965. The life of fishes. London: Weidenfeld and Nicolson.

MATHEW, W. D. 1915. Climate and evolution. *Ann. N.Y. Acad. Sci.*, 24: 171–318.

MEDLICOTT, H. & W. T. BLANFORD, 1879. A mannual of the Geology of India.

MENON, A. G. K. 1950. On a remarkable blind siluroid fish of the family Clariidae from Kerala (India). *Rec. Indian Mus.* 48: 59–66.

MENON, A. G. K. 1951. Distribution of Clariid fishes and its significance in zoogeographical studies. *Proc. Nat. Inst. Sci. India*, 17: 291–299.

MENON, A. G. K. 1951. Further studies regarding Hora's Satpura Hypothesis. The Role of the Eastern Ghats in the distribution of the Malayan Fauna and Flora to Peninsular India. *Proc. Nat. Inst. Sci. India*, 17: 475–497.

MENON, A. G. K. 1953. Age of transgression of the Bay of Bengal and its significance in the evolution of the freshwater fish fauna of India. *Bull. Nat. Inst. Sci. India*, 7: 240–247.

MILLER, R. R. 1958. Origin and affinities of the freshwater fish fauna of Western North America. *Zoogeography. Publ.* 51, *American Assoc. Adv. Sci.*, 187–222.

MILLER, R. R. 1966. Geographical distribution of Central American freshwater fishes. *Copeia*, 4: 773–802.

MUKERJI, D. D. 1934. Report on Burmese fishes collected by Lt. Col. R. W. BURTON from the tributary streams of the Mali Hka river of the Myitkyina district (Upper Burma). Parts I and II. *J. Bombay nat. Hist. Soc.*, 37: 38–80.

MYERS, G. S. 1938. Fresh-water fishes and west Indian zoogeography. *Ann. Rep. Smith. Inst. for 1937*, pp. 339–364.

MYERS, G. S. 1949. Salt tolerance of freshwater fish groups in relation to zoogeographical problems. *Bijdr. Dierk.*, 28: 315–322.

MYERS, G. S. 1966. Derivation of the freshwater fish fauna of Central America. *Copeia*, 4: 766–773.

MYERS, G. S. 1967. Zoogeographical evidence of the age of the South Atlantic Ocean. *Studies in Tropical Oceanography* Miami, 5: 614–621.

MORI, T. 1936. Studies on the geographical distribution of freshwater fishes in Eastern Asia. Chosen.

NEIDEN, F. 1913. Gymnophiona (Amphibia Apoda). *Das Tierreich*, 37.

NOBLE, G. K. 1954. The Biology of the Amphibia. New York: Dover Publications.

OLDHAM, T. 1859. On some fossil fish-Teeth of the genus *Ceratodus* from Maledi, south of Nagpur. *Mem. Geol. Surv. India*, 1: 295–309.

PRASHAD, B. 1942. Zoogeography of India. Annual Address to the National Institute of Sciences of India. 7–17.

SAHNI, B. 1937. WEGENER's Theory of continental drift with reference to India and adjacent countries. *Proc. Indian Sci. Congr. 24th session, General Discussions*, pp. 502–507.

SAHNI, M. R. 1941. Palaeogeographical revolutions in the Indo-Burmese region and neighbouring Lands. *Proc. Indian Sci. Congr. 28th session, Presidential Address, Section of Geology*, pp. 32.

SARASIN, F. 1910. Über die Geschichte der Tierwelt von Ceylon. *Zool. Jahrb. Leipzig.* (Suppl.) 12.

SCHWARTZ, F. J. 1964. Natural salinity tolerances of some freshwater fishes. *Underwater Nature*, 2: 13–15.

SCLATER, P. L. 1858. On the general Geographical distribution of the members of the Class Aves. *J. Linn. Soc. London*, 2: 130–145.

SILAS, E. G. 1952. Further studies regarding HORA's Satpura Hypothesis. 2. Taxonomic Assessment and levels of Evolutionary divergences of Fishes with the so-called Malayan affinities in Peninsular India. *Proc. Nat. Inst. Sci. India*, 18: 423–448.

SILAS, E. G. 1953. Classification, zoogeography and Evolution of the fishes of the Cyprinoid families Homalopteridae and Gastromyzonidae. *Rec. Indian Mus.*, 50: 173–264.

SILAS, E. G. 1958. Studies on Cyprinid fishes of the Oriental genus *Chela* Hamilton. *J. Bombay Nat. Hist. Soc.*, 55: 54–99.

SMITH, M. A. 1931. The Fauna of British India, including Ceylon and Burma. Reptilia and Amphibia. 1. Loricata, Testudines. London: Taylor & Francis.

SMITH, M. A. 1935. The Fauna of British India. including Ceylon and Burma. Reptilia and Amphibia. II. Sauria. London. Taylor & Francis.

SMITH, M. A. 1943. The Fauna of British India, Ceylon and Burma, including the whole of the Indo-Chinese subregion. Reptilia and Amphibia, III. Serpentes. London: Taylor & Francis.

WADIA, D. N. 1968. The Himalayan Mountains: its origin, and Geographical relations. Chapter II In Mountains and Rivers of India. 21st International Geographical Congress, India, 35–41.

XVIII. MAMMALS OF ASSAM AND THE MAMMAL-GEOGRAPHY OF INDIA

by

G. U. KURUP

1. Introduction

The northeastern parts of India, comprising Assam, are of exceptional biogeographical interest. It is from this region that the obliteration of the Pre-Tertiary Tethys Sea began, producing in its wake a land connection between the Indian Peninsula and the main Asiatic mass to its north (see Chapters II, III & XX). The Assam region then onwards served as a great faunal gateway, through which the Indo-Chinese elements of the Oriental fauna and also that of Palaearctic could spread to India and colonize the country. In fact, the history of the Post-Tertiary faunal dispersal in India is peculiar, in as much as all the faunal invasions have come through two great faunal gateways, one at the Assam region and the other in the northwest. This was because of the emergence of the rising Himalaya as a great barrier-wall, concomitant with the obliteration of the Tethys Sea, so that except for the montane species, the faunal dispersal had to take place through either of these faunal passes. Of these the importance of the northwest gateway dwindled after the disappearance of the incomparably richer Siwalik fauna in the early Pleistocene and the changes in the physiography of the Indo-Gangetic trough, of which the formation of the Thar or Rajaputana Desert was a major one, constituting barriers to dispersal from the northwest. As a result, we see that most of the faunal dispersal and recolonization in the recent period have taken place through the Assam gateway, so that the Indo-Chinese element constitutes the dominant entity in the mammal fauna of India.

Speaking of more recent times, there is now on the contrary, a geological and climatic discontinuity between Assam and the rest of India, a region of similar climate and biotope obtaining only in the Western Ghats. This discontinuity, readily visible at the region of Garo-Rajmahal Gap by distinctive dispersal breaks, acts as a filter-barrier in the effective dispersal of mammals either way. Thus Assam is the westernmost boundary of the range of many Indo-Chinese mammals, like certain squirrels and the eastern-most limit of the distribution of many Peninsular species, such as the spotted deer.

Primarily, the mammals of Assam are Indo-Chinese rather than Peninsular Indian. Though with a variable admixture of the Peninsular and Ethiopian elements, the Palaearctic montane elements also intrude into it. Many of the relict species of the southern Peninsular India, mostly confined to the Western Ghats, have closely related species only in Assam, separated by a gap over one thousand five hundred kilometres.

Assam and the adjoining areas thus hold a pivotal place in the historic process of progressive evolution of the present-day flora and fauna of India, serving as an effective gateway to floristic-faunal influx. An analytical study of the mammal fauna is, therefore, essential for a clear comprehension of the derivation, composition, distribution, etc. of the Indian mammals. The mammals of Assam have been generally dealt with in the past by THOMAS (1866), ALLEN (1909), THOMAS (1921), WROUGHTON (1921), KEMP (1924), HINTON & LINDSAY (1926), HIGGINS (1933–34), ELLERMAN (1947), ROONWAL (1948, 1949, 1950, ROONWAL & NATH, 1949), NATH (1953) and KURUP (1965, 1968).

2. The Major Ecological Associations

The region under consideration exhibits the following major types of ecological associations.

Cultivated fields and human habitations: Diffused throughout the area, either continuous for long stretches in the plains or in isolated patches in the mountainous regions.

Swamps and marshes: Found chiefly in the low-lying areas of the Surma Valley (southern part of Sylhet) and along the banks of the R. Brahmaputra. These are usually covered with thick growth of tall grass and reeds, often attaining the height of 6 m and more. The dominant species of vegetation here are those of *Saccharum* and *Stemona* (see Chapter IX).

Deciduous Sal forests of the valleys: These are found in Goalpara, portions of Garo Hills, Kamrup, Nowgong and Darrang district.

Dense, evergreen and mixed forests of the valleys and the hill ranges: The chief constituents here are species of *Amoora, Michelia, Magnolia, Stereospermum, Quercus, Castanopsis, Ficus* and *Mesna*. Along with these there are various kinds of palms, canes, tree ferns, bamboos and bananas.

Rolling downs or grass-covered undulating slopes of the ranges: Found on the Khasi Plateau in the central portion of the Assam Range. These are extensive grasslands dotted with clusters of oak and pine. The flora of this tract is rich in flowering plants and orchids.

3. The Faunal Composition

The mammalian fauna of Assam is the richest and most varied among comparable regions in India. Favoured with a subtropical, humid climate with copious rainfall, Assam sustains a biotope eminently suitable for the Indo-Chinese fauna that occupies it. Its tropical and subtropical moist evergreen forests ensured the survival of mammals, and enhanced the pace of their speciation by affording more ecological niches than was possible in the dry deciduous forest areas and plains of the rest of India, excepting the Western Ghats. The richness and variety of the Assam mammals fauna can be readily seen from Tables I and II. The classification and nomenclature adopted are those of ELLERMAN & MORRISSON-SCOTT (1951).

Table I. Synopsis of the mammalian fauna of Assam

Order Family	Genera	Species	Subspecies
Insectivora			
Tupaiidae	1	1	3
Talpidae	1	1	2
Soricidae	4	7	10
	6	9	15
Chiroptera			
Pteropidae	2	3	3
Megadermatidae	1	2	2
Rhinolophidae	3	17	18
Vespertilionidae	9	19	22
	15	41	48
Primates			
Lorisidae	1	1	1
Cercopithecidae	2	5	8
Pongidae	1	1	1
	4	7	10
Pholidota			
Manidae	1	1	1
	1	1	1
Carnivora			
Canidae	3	3	3
Ursidae	3	3	3
Mustelidae	6	7	7
Viverridae	8	10	13
Felidae	3	8	8
	23	31	34
Proboscidea			
Elephantidae	1	1	1
	1	1	1
Perissodactyla			
Rhinocerotidae	1	1	1
	1	1	1

587

Table I (continued)

Order Family	Genera	Species	Subspecies
Artiodactyla			
Suidae	1	1	1
Cervidae	4	6	6
Bovidae	5	5	5
	10	12	12
Lagomorpha			
Leporidae	2	2	2
Ochotonidae	1	1	1
	3	3	3
Rodentia			
Sciuridae	6	12	18
Hystricidae	2	2	3
Rhizomyidae	2	2	2
Muridae	11	25	36
	21	41	59
Cetacea			
Platinistidae	1	1	1
	1	1	1

Total: Orders 11, Families 28, genera 86, species 148, subspecies 186.

Table II. Land-mammal genera of Assam compared with those of the rest of India

Orders and Families	Total in India	Assam	Rest of India
Insectivora			
Tupaiidae	2	1	1
Erinaceidae	2	—	2
Talpidae	1	1	1
Soricidae	6	4	5
Chiroptera			
Pteropidae	3	2	3
Rhinopomatidae	1	—	1
Embellonuridae	1	—	1
Megadermatidae	1	1	1
Rhinolophidae	4	3	4
Molossidae	2	—	2
Vespertilionidae	16	9	16
Primates			
Lorisidae	2	1	1
Cercopithecidae	2	2	2
Pongidae	1	1	—

Table II (continued)

Orders and Families	Total in India	Assam	Rest of India
Pholidota			
Manidae	1	1	1
Carnivora			
Canidae	3	3	3
Ursidae	3	3	3
Procyonidae	1	—	1
Mustelidae	7	6	7
Viverridae	8	8	6
Hyaenidae	1	—	1
Felidae	3	3	3
Proboscidea			
Elephantidae	1	1	1
Perissodactyla			
Rhinocerotidae	1	1	—
Equuidae	1	—	1
Artilodactyla			
Suidae	1	1	1
Tragulidae	1	—	1
Cervidae	4	4	4
Bovidae	15	5	14
Lagomorpha			
Leporidae	2	2	2
Ochotonidae	1	1	1
Rodentia			
Sciurudae	10	6	10
Hystricidae	2	2	1
Muscardinidae	1	—	1
Rhizomyidae	2	2	—
Muridae	22	11	17
Summary			
Orders	10	10	10
Families	36	27	33
Genera	135	85	119

Of a total of one hundred and thirty-five genera of land mammals of India, eighty-five (63%) are represented in Assam. Of the eleven orders, the Carnivora are the richest in genera, followed by Rodentia and Chiroptera. In the number of species and subspecies, however, Rodentia rank the highest. Among the families Muridae, Viverridae and Vespertilionidae are well represented. The number of genera of Viverridae in Assam is more than in the rest of India.

The following sixteen genera of mammals, for which Assam and a westward Himalayan strip extending into Nepal form the southern and westernmost boundary of their present range, are at present totally absent from the Peninsula proper:

Insectivora
  Tupaiidae
    1. *Tupaia*
  Soricidae
    2. *Anourosorex*
Primates
  Lorisidae
    3. *Nycticebus*
  Pongidae
    4. *Hylobates*
Carnivora
  Viverridae
    5. *Arctictis*
    6. *Arctogalida*
Perissodactyla
  Rhinocerotidae
    7. *Rhinoceros*

Artiodactyla
  Bovidae
    8. *Budorcas*
Rodentia
  Hystricidae
    9. *Atherurus*
  Rhizomyidae
    10. *Rhizomys*
    11. *Cannomys*
  Muridae
    12. *Chiropodomys*
    13. *Micromys*
    14. *Hadromys*
    15. *Eothenomys*
    16. *Dacnomys*

Of the ten families mentioned here, Pongidae and Rhizomyidae are practically confined to Assam, Chittagong and the adjacent hilly tracts. As against this the families which are altogether absent from the Assam region, but are present elsewhere in India, are the following:

Erinaceidae
Rhinopomatidae
Embellonuridae
Molossidae

Procyonidae
Hyaenidae
Equuidae
Tragulidae
Muscardinidae

By and large these families are inhabitants of relatively drier areas (those which are found in generally humid regions frequent dry micro-biotic niches). Moreover, as a study of their extra-limital distribution will show, they entered India through the northwestern route. These two factors explain their absence in Assam.

The order Chiroptera is conspicuous by the absence of a number of families and genera from Assam. Of the seven families of bats occurring in India, four are not represented here. Bovidae are likewise poorly represented, as compared with northwestern India.

While almost all genera found in Assam are also represented in the Indo-Chinese subregion further east, four genera are autochthonous in the Indian subregion and are unrepresented in the rest of Indo-Chinese region viz. *Caprolagus* (Lagomorpha) *Golunda* and *Hadromys* (Rodentia), *Plantanista* (Cetacea). The disjunctive distribution in the Central and Eastern Sub-Himalayan region and Assam or the Indo-Chinese area on the one hand and in the Western Ghats and adjacent Peninsular area on the other hand has attracted considerable attention in the past. Nevertheless, no comprehensive list of any such fauna is available for any group. Hora (1949a,c), Jayaram (1949), Ali (1949), and Ripley (1949) have dealt with many such cases in various groups.

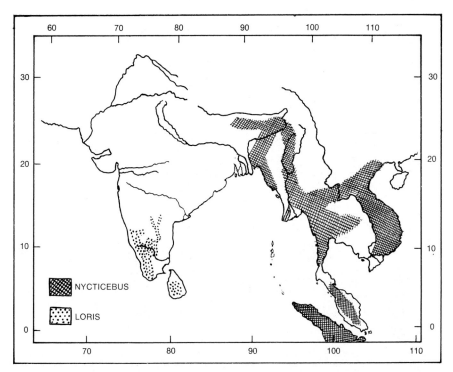

Fig. 133. Discontinuous distribution of mammals in India and adjacent countries: *Nycticebus* and *Loris*.

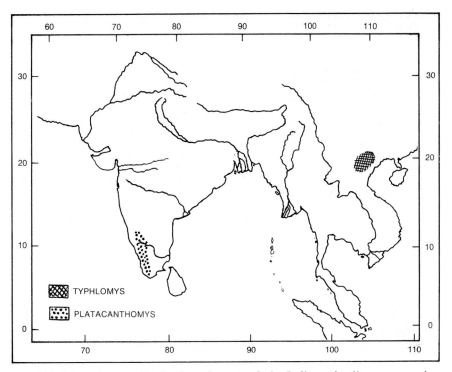

Fig. 134. Discontinuous distribution of mammals in India and adjacent countries: *Platacanthomys* and *Typhlomys*.

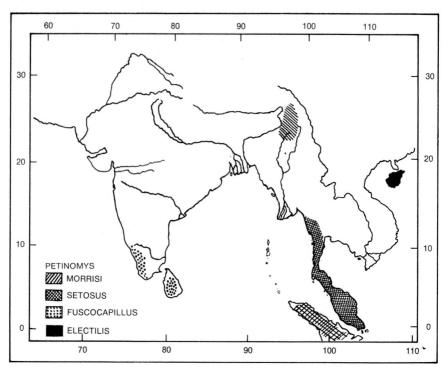

Fig. 135. Discontinuous distribution of mammals in India and adjacent countries: *Petinomys* spp.

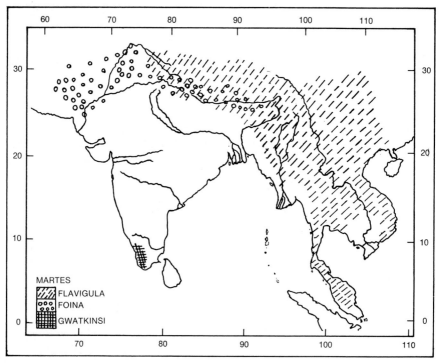

Fig. 136. Discontinuous distribution of mammals in India and adjacent countries: *Martes* spp.

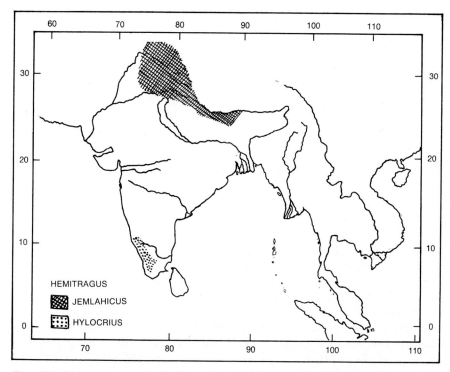

Fig. 137. Discontinuous distribution of mammals in India and adjacent countries: *Hemitragus* spp.

ROONWAL & NATH (1949) have given three such instances pertaining to mammals. There are some other such cases, at various taxonomic levels, which although not strictly confined to Assam or even present there but found only in the areas farther east, are nevertheless instances of even more remarkable discontinuity in distribution and most of which have obviously entered India through Assam to reach the Peninsular south. These are listed below:

A. Disjunctively distributed families and subfamilies:
    1. Lorisidae (Fig. 133)
        Loris   W. Ghats and adjacent regions
        Nycticebus   Assam to farther east
    2. Platacanthomyinae (family Muscardinidae) (Fig. 134)
        Platacanthomys   W. Ghats and adjacent areas
        Typhlomys   Indo-Chinese region
B. Disjunctively distributed genera:
    1. *Petinomys* (Fig. 135)
        Petinomys fuscocapillus   W. Ghats and near by areas
    All other species are far-eastern in distribution.
    2. *Martes* (Fig. 136)
        Martes flavigula   Himalayan and Assam
        Martes gwatkinsi   W. Ghats

3. *Hemitragus* (Fig. 137)
   Hemitragus jemlahicus   Himalayan
   Hemitragus hylocrius   W. Ghats
4. *Tragulus* (Fig. 138)
   Tragulus meminna   Ceylon and Peninsular India
   Tragulus napu
   Tragulus javanicus   Indo-Chinese
C. Disjunctively distributed species:
  1. *Viverra megaspila* (Fig. 139)
     V.m. megaspila   Assam to east
     V.m. civettina (= *zibetha*)   W. Ghats
  2. *Harpiocephalus harpia* (Fig. 140)
     H.h. madrassius   Palni hills and adjacent E. Ghats
     H.h. lasiurus   Central Sub-Himalaya, Assam
  3. *Aonyx cinerea* (Fig. 141)
     A.c. nirnai   Southern Peninsular India
     A.c. concolor   Central Himalaya to Assam eastward

In addition, the present known distribution of species like *Mus famulus* (Fig. 142), *Crocidura miya*, *Crocidura horsfieldi*, *Kerivoula picta* and *Kerivoula hardwicki*, also show an aberrant distribution, suggestive of similar discontinuity, but here the possibility of artificial introduction in the case of first three and probable existence in the intervening area as yet unknown in the case of *Kerivoula* cannot be ruled out. The question as to whether the existence of such disjunctively distributed forms indicates any particular relationship between the regions more marked than that of others is discussed at a later stage.

The affinities of the genera of mammals of Assam, as deduced from their extralimital distribution, show that they are predominantly eastern or Indo-Chinese in origin. In addition to this, there are also Ethiopian and Palaearctic elements. The Ethiopian elements have come solely through the northwestern gateway, but the Palaearctic element in India seems to have come from three directions, i.e., in addition to the north-eastern and northwestern gateways, a part of the temperate montane fauna of the larger holarctic region have come directly from the north on to the high elevations of the Himalaya. As they are, however, mostly Central Himalayan, their share in the Assam fauna is not significant. As the Indo-Chinese element in Assam might have originally come from south east Palaearctic (Manchurian) subregion, these and other autochthonous Indo-Chinese forms which came through Assam form one entity. The other Palaearctic elements from western or Mediterranean part of Palaearctic and the Ethiopian fauna came together from the western borderlands. Divided thus into eastern and western entrants, we can see that there is an appreciable part of western entrants in Assam, which have reached Assam and are at present represented by eighteen genera, i.e., about 21 % of the Assam fauna. The remaining sixty-eight genera or 79 % are survivors of the eastern entrants. The eighteen genera that

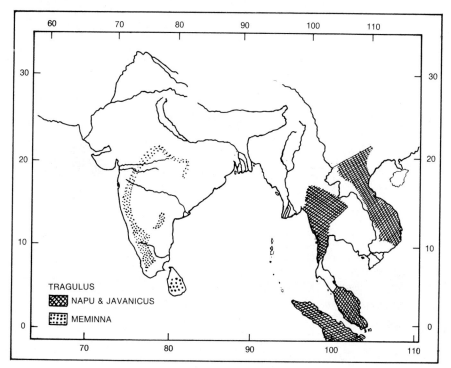

Fig. 138. Discontinuous distribution of mammals in India and adjacent countries: *Tragulus* spp.

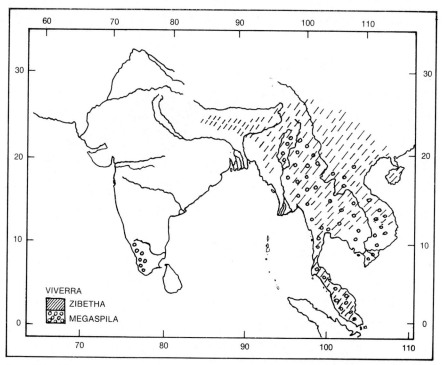

Fig. 139. Discontinuous distribution of mammals in India and adjacent countries: *Viverra* spp.

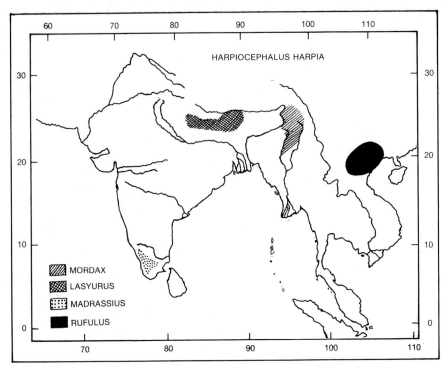

Fig. 140. Discontinuous distribution of mammals in India and adjacent countries: *Harpiocephalus harpia.*

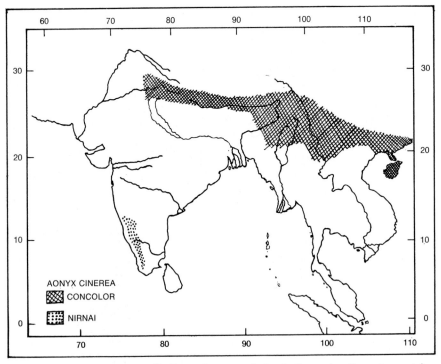

Fig. 141. Discontinuous distribution of mammals in India and adjacent countries: *Aonyx cinerea.*

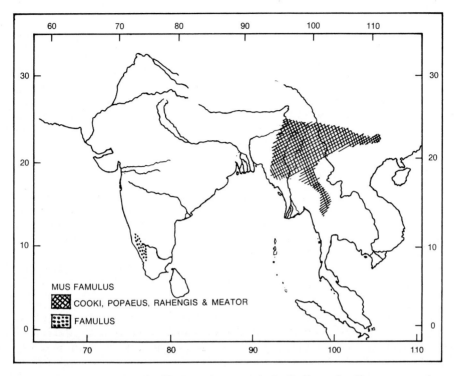

Fig. 142. Discontinuous distribution of mammals in India and adjacent countries: *Mus famulus.*

represent the western entrants in the mammal fauna of Assam are the following:

Insectivora
  1. *Suncus*
  2. *Crocidura*
Chiroptera
  3. *Myotis*
  4. *Eptesicus*
Carnivora
  5. *Canis*
  6. *Vulpes*
  7. *Mustela*
  8. *Felis*
  9. *Herpestes*
  10. *Panthera*

Proboscidea
  11. *Elephas*
Perissodactyla
  12. *Rhinoceros*
Artiodactyla
  13. *Antelope*
Lagomorpha
  14. *Lepus*
  15. *Hystrix*
  16. *Apodemus*
  17. *Mus*
  18. *Golunda*

Conversely, an analysis of the western Indian fauna shows that the percentage of eastern entrants is more than that of western entrants in the Assam fauna. This shows that there was much more faunal flow from Assam to western India than vice versa; this point will be discussed later.

597

4. Faunal Resemblance

As mentioned earlier, the Assam mammal fauna is essentially of Indo-Chinese affinity. Most of the genera that are characteristic of Assam and which are not found in the rest of India are widely distributed in the Indo-Chinese region, and Assam forms the westernmost boundary of their range. This distinctiveness of the Assam fauna from that of other parts of the country was recognized by BLANFORD (1901), who considered Assam as a part of his Trans-Gangetic Indo-Chinese region as distinct from the Cis-Gangetic region (northern India from eastern part of Punjab, Rajasthan and Peninsular India). This is readily apparent from the facts already shown, that all the genera that are characteristic of Assam are Indo-Chinese in affinity, while those that are absent from Assam are western entrants, having affinities with either Ethiopian or western Palaearctic faunas.

While the basic Indo-Chinese affinity of Assam fauna has been recognized earlier, the extent of its similarities with other significant faunal areas of India has been generally overlooked. Such an analytical study of the composition of the fauna of other regions in India in relation to that of Assam sheds light on many problems connected with their faunal constituents, migration and past distribution, thus producing an overall picture of the derivation and composition of the Indian fauna. A detailed study of the Indian fauna is, however, not attempted here, except in sofar as it may be pertinent for a full appreciation of the status of the Assam fauna and its influence on the constitution of the present day Indian mammal fauna. For this purpose, three other zoogeographically significant regions, namely the Western Ghats, the Eastern Ghats, and Western India (includes Gujarat, Rajasthan, Punjab and Kashmir) are considered. The reason for selecting these areas is mainly the climatic and other biotic contrasts they offer. Thus the Western Ghats present a biotope and climate more or less identical with those of Assam, while the other two regions represent different types. Comparative data of the climates of these areas are summarized in Table III.

Table III. Climates of different regions in India

Factors	Assam	W. Ghats	E. Ghats	W. India
Rainfall	Heavy 188.5 cm	Heavy	Moderate 121–185.5 cm	Scanty 50–60 cm
Relative humidity	High 65%	High	High	Low 50%
Temperature	Warm 21–26.6 °C	Warm	Warm	Extremes 15.5–33.3 °C
Range of diurnal temperature	Low 11 °C	Moderate 11–14 °C	Moderate	High 14 °C

The number of orders, families and genera of the land mammals in these three regions and the numbers of those in each region that are common with Assam are summarized in Table IV.

Table IV. Numbers of the land-mammals in the three different regions of India and those of each region sharing with Assam

	W. Ghats		E. Ghats		W. India	
	Total	Common to Assam	Total	Common to Assam	Total	Common to Assam
Orders	9	9	8	8	9	9
Families	26	22	26	22	27	21
Genera	62	42	57	38	87	46

The faunal resemblance of Assam to each of these regions may be mathematically indicated, using SIMPSON's index 100 C/NI (the percentage of the members of the smaller fauna present in the larger fauna). On the basis of this index, the faunal similarity of each of these regions to Assam is given below:

Table V. Mammalian faunal resemblance (100 C/NI) of Assam to three other regions of India

	W. Ghats	E. Ghats	W. India
Orders	100	100	100
Families	77	77	77
Genera	68	67	52

The almost equal similarity between faunas of the Eastern and Western Ghats to that of Assam demonstrated in Table V does not support the common assumption that Assam has closer faunal affinity with the Western Ghats than with other regions. Western India differs, however, from both the Ghats in having much lesser faunal resemblance to Assam.

In this connection, a further analysis of the fauna of all the four regions and their mutual resemblance is required for a full understanding of the faunal configuration. Fig. 143 gives the number of genera common among different regions of India. The faunal resemblance of these regions to each other is represented in Fig. 144.

The maximum faunal resemblance is found between the Western and Eastern Ghats, inspite of their basic climatic difference. This similarity

599

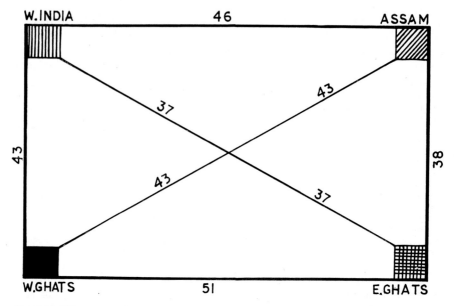

Fig. 143. Diagrammatic representation of the relationship of different parts of India, indicated by the number of genera of mammals common among different regions in India.

is no doubt due to their proximity and the presence of similar ecological niches on the Eastern Ghats and on the northern parts of the Western Ghats. The faunal resemblance between the Western India and the Eastern Ghats is next in order. Curiously, the faunal resemblance is equal between Assam and the Western Ghats on the one hand and the Western Ghats and western India on the other. We have thus another proof that there is no special affinity between Assam and the Western Ghats, as is only too often asserted. *Certain instances of discontinuous distribution of some* mammals, *characteristic of these* regions only, *appear to be no more than relics of a former widely distributed fauna.*

The least resemblance is between Assam and Western India, as shown earlier. This of course is natural, since the two regions were the gateways for wholly different types of continental fauna. As the sources were different, they show minimum degree of resemblance, inspite of the fact that a good deal of faunal interchange took place between the two, as will be shown later.

There are many factors which influence faunal resemblance, such as the past faunal interchange, common climatic conditions, survival and extinction within a fauna gradually approximating to that of an ecologically similar, but distant region, indirect spread from a third region, and distances between the regions. The present faunal resemblance may be a result of any or all of these factors. Thus the Assam faunal resemblance

600

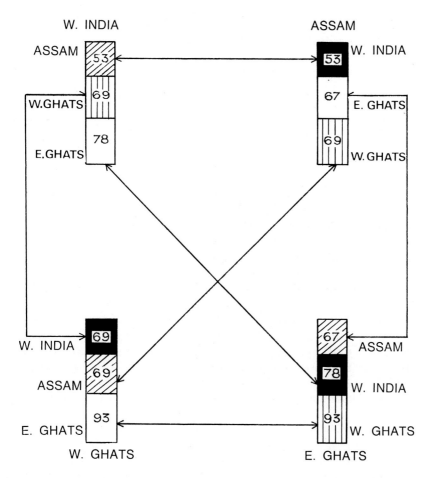

Fig. 144. Mammalian faunal resemblance among four different regions in India, Western India, Western Ghats, Eastern Ghats and Assam. In each region the faunas common with the other three are marked separately and the quantum of resemblance (100 C/NI) between any two (connected by arrow) shown therein.

to the Western Ghats is largely due to the past interchange and its subsequent preservation in a similar climatic and biotic condition. Its resemblance to the Eastern Ghats fauna is due to previous interchange between the two and the migration of western Indian fauna to both the regions. Its resemblance to western Indian fauna is solely due to the previous mutual interchange.

Another interesting aspect of this faunal resemblance is that proximity can be as effective as identity in climatic factors. This is illustrated by the Western Ghats, where the faunal resemblance to the climatically very different Eastern Ghats is higher than to Assam with a similar climate.

Similarly its faunal resemblance to the climatically different Assam and western India is equal, the latter being nearer the Western Ghats. We see, therefore, that proximity is an important factor in bringing about resemblance. In this connection it is, however, also to be borne in mind that if the climatic difference is sharp and clearcut, then the influence due to proximity becomes inoperative, as in the fauna of western India, which has more resemblance to that of the far-lying Eastern Ghats than to the nearby Western Ghats.

5. Faunal Interchanges

Faunal interchange is one of the most dynamic and potent factors underlying changes in the faunal composition in a given region. Many such major interchanges, radically altering the faunal composition of continents, have occurred in the history of mammals. SIMPSON (1953) infers, for example, that a great faunal interchange took place between North America and Eurasia, from late the Palaeocene to Eocene. Similar interchange occurred between North and South America in the late Pliocene and Pleistocene. As already stated, so far as the Peninsular India is concerned, there have been at least two such large faunal inflows from the northwest and northeast in the Post-Tertiaries. The details about the precise period, various phases, quantum and nature of the subsequent interchanges that occurred in India, can be only deduced from a full knowledge of the palaeontological history. Nevertheless, an insight into those past phenomenon may be obtained even by a study of the surviving fauna, if we allow enough margin for the possible effects of changed climate and physiography of some of the areas. At any rate, the general trend of such past changes can be sufficiently clearly described, although their degree and phases are beyond the scope of this discussion. Here the faunal interchanges of the four regions in India are studied on the basis of the genera common to each other. The quantum of interchange is expressed as the percentage of genera common between any two regions, in relation to the total number of genera present in India (Fig. 145). Such a method will account for the interchange effected between two regions through a third or more regions.

It is apparent that as far as Assam is concerned, there has been more or less equal interchange with western India and the Western Ghats, but slightly less with Eastern Ghats. The latter case may be explained on the assumption that fauna from Assam was less specialized for the climatically different Eastern Ghats. The fact that western India is also climatically different from Assam, yet the latter has had almost equal interchange with that region, and the Western Ghats, which are climatically similar, could be accounted for by the presence of a significant portion of common montane fauna of Palaearctic origin, extending from Kashmir to Assam Himalaya and compensating for the faunal dissimilarity

602

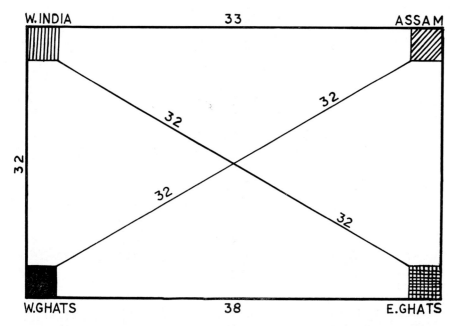

Fig. 145. Quantum of mammalian faunal interchanges among different regions of India. The numbers represent percentage of genera common between two regions (connected by straight lines) in relation to the total number of mammalian genera found in India.

in the climatically different areas of both the regions. Judged in this light, the interchange between Assam and other regions in India has been about equal and widespread, suggesting a more or less similar climate in the greater part of the Peninsula in former times. The maximum interchange, however, took place between the two Ghats, obviously due to their proximity as they are two edges of the Peninsular mass.

6. Species Dispersal

While the genus has been utilized here as the most suitable standard unit for the study of faunal resemblance and interchange, a study of the pattern of the species dispersal shows the trends of faunal colonization in the recent past. Here the species-dispersal-analysis is done with special reference to the number of species belonging to only those genera that are common between Assam and other regions, as these will help us to understand the nature and the content of the species invasion, apart from the autochthonous speciation between Assam on the one hand and the other regions on the other. Fig. 146 shows the species dispersal in terms of actual numbers and Fig. 147 the index of the dispersal, in terms of percentage.

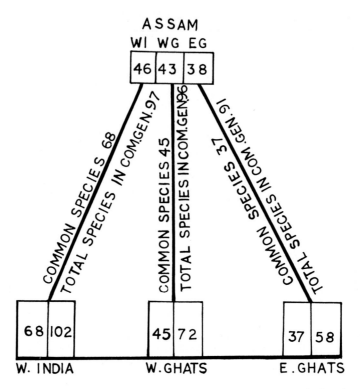

Fig. 146. Species dispersal in common genera between Assam and the three other regions of India.

Of the 46 genera that are common between Assam and Western India, Assam has now 97 species and western India 102 species, whereas only 68 species among them are common to both. Twenty-nine species or 30% belonging to these genera in Assam and 34 species or 33% in Western India came, therefore, to exist in these regions from other sources. The autochthonous element in this is rather negligible and most of these species are widely distributed in regions adjacent to both, but extra-limital to India and they can, therefore, only be invasions, which were rather late arrivals and which could not as yet complete their Trans- or Sub-Himalayan dispersal as their predecessors did. This may be due either to the simple fact of their late arrival and inadequate time for dispersal or due to the formation of dispersal barriers of diverse nature after their arrival.

It is well known that the great tectonic trough or 'synclinorium' of the Indo-Gangetic Plain, as also the desert of Rajaputana, are features of recent origin. The deposition of alluvium in the great plains of northern India and including the largest portions of Sind, northern Rajaputana, Punjab, Uttar Pradesh, Bihar, Bengal and Assam and ranging in depth

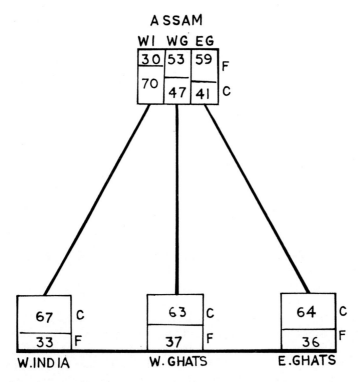

Fig. 147. Index of species-dispersal in common genera between Assam and three other regions in India.

from 1000 to 2000 m at present, commenced only after the final phase of the Siwaliks, continuing all through the Pleistocene upto the present (see Chapter III). Similarly the whole area from west of Aravallis to the basin of the R. Indus and from the southern confines of Punjab plains to the basin of R. Sutlej, occupying an area of 600 km long by 160 km broad, which is known as the Thar and Rajaputana Desert, is even young enough to fall within historic times. There is evidence that this was a fertile tract which supported populous cities as late as the time of invasion by Alexander the Great (323 BC). The progressive desiccation combined with the sand drifting action of the southwest monsoon winds that mainly led to the formation of the desert had also influenced the river system of the region (see Chapter V).

The sweeping changes that took place in the physiography of this region are amply illustrated by the history of the Vedic river Saraswathi (see Chapter II). First it flowed independently through eastern Punjab and Rajasthan in to sea at what is now Rann of Cutch. Later it became an affluent of the Indus, still later it shifted its course so easterly as to become the present Jamuna, the main tributary of the Ganges (W<small>ADIA</small>,

Table VI. Number of genera and species common between regions in India

Regions	Genera	Species
Assam and Western Ghats	43	45
Assam and Western India	46	68
Assam and Eastern Ghats	38	37

1966). The great climatic barrier that rose due to the increasing desiccation of Rajasthan is obvious.

Regarding Peninsular India, the forty-three genera common between Assam and the Western Ghats possess 96 species in former and 72 in latter region with 45 species common to both. Assam and the Eastern Ghats have 38 genera in common which have 91 species in Assam and 58 species in Eastern Ghats and 37 species are common. Thus 27 species (37%) in common genera in the Western Ghats and 21 species (36%) belonging to common genera in the Eastern Ghats are derivatives independent of the faunal exchange with Assam. It is obvious that this element came through invasions from the northwest.

As far as the Assam fauna itself is concerned, the species not common with the Western and Eastern Ghats predominate over those which are common. Fifty-one species (53%) are not common to the Western Ghats and 54 species (59%) to the Eastern Ghats. Unlike other conspecific members, these species did not take to the Peninsula and obviously travelled during invasions of considerable interval directly along Himalayan route of BLANFORD or as in the case of west India and Assam are late entrants. The dominance of such species here may perhaps suggest the recent formations of diverse barriers to dispersal between Assam and the Ghats.

6.1. NATURE OF INVASIONS

In this connection, a comparison of the number of genera and species that are common to these regions throws some light on the nature of these invasions in each region (Table VI).

Except in case of Assam and western India, the figures are very comparable, so that the proportion is almost 1 : 1, showing on the whole (not applicable to individual genus) that the species colonization and interchange between Assam and the Peninsular India were a single event. This is not, however, the case between Assam and western India, as the proportion here is 1 : 1.5 and it is, therefore, apparent that species colonization and interchange were on the whole multiphased between these two regions.

The rather very comparable figures in Assam of the total number of

Table VII. Number of taxa of mammals of the Western and Eastern Ghats common with those of Assam.

Hierarchy	Western Ghats	Eastern Ghats
Orders	9	8
Families	22	22
Genera	42	38
Species	45	37
Subspecies	18	16

species in genera with other three regions (97, 96 and 91) suggest that a large majority of the genera involved might be the same. Analysis in this respect shows this assumption to be correct, as 32 genera are common to all the regions. While this is so, it is, however, evident that the species augmentation by invasion, exchange and evolution in all the regions has been different. The Peninsular India is in this respect rather poor, while western India is at maximum (102) and Assam comes next (96), again pointing to the heavy dispersal that occurred along the faunal highway across the trans-Sub-Himalayan region.

7. Criticism of the Satpura Hypothesis

The foregoing findings of these faunal studies necessitate a criticism of the Satpura hypothesis propounded by HORA (1949a) in explanation of the disjunctive distribution of his socalled Malayan fauna and flora (but in reality Indo-Chinese) in the Peninsular India. According to him, the Satpura and Vindhya trend of mountains was formerly continuous with the Assam Hills and the Eastern Himalaya in the east and the Western Ghats in the west and was thus assumed to have served as a route of migration of the specialized hill fauna and flora from the east to the west. He also asserted though quite erroneously that this was the only possible explanation of the remarkable discontinuity in the distribution of the fauna of India. He thus visualized only one route for the dispersal of fauna from the Assam region to the Peninsular India, viz. through the Satpura-Vindhya Range to the Western Ghats. Ignoring for the present the geological evidence against this fantastic view, discussed further in Chapters V and XXIV, and confining ourselves here entirely to the data on mammalian distribution presented above, we find that there is after all no need to conceive of this hypothesis. We have clearly demonstrated that the faunal resemblance of the Western and Eastern Ghats with Assam at the generic level is about equal, there being a difference of only one. The close similarity of the Ghats extends to other taxonomic catagories also (Table VII).

Table VIII. Mammalian faunal composition of both the Ghats in relation to affinities; percentages in parenthesis

Area	Total genera	Eastern entrants	Western entrants	Common with Assam	Eastern entrants	Western entrants
Western Ghats	62	38(61)	24(39)	42	31(50)	11(18)
Eastern Ghats	57	34(60)	23(40)	38	28(49)	10(19)

It is true that the figures for the Eastern Ghats, are somewhat less, but this disparity can readily be accounted for by the present climatic differences between this region and Assam, and the concomitant disappearance of some genera in recent times. Further, it has also been shown above that the residual faunal interchange between Assam and both the Ghats is significant, the slightly lower figure for the Eastern Ghats being negligible, when viewed in the above light. In this connection, an analysis of the faunal affinities of both the areas reflects the past interchange that occurred between Assam and the two Ghats (Table VIII).

We observe that the percentages of the eastern and western entrants in both the Ghats are almost equal. Further, as already pointed out, the faunal resemblance between Western Ghats and western India is also almost equal to that between the former and Assam and the closest resemblance in any case, is actually between the two Ghats, followed by that between Eastern Ghats and western India. This shows that in the dispersal of mammals from north to the Ghats or Peninsular India, either from the northeast or from the northwest, both the Western and Eastern Ghats were involved in more or less equal degree. Any hypothesis stressing the rôle of only one as the sole highway is, therefore, wholly untenable. The element of the disjunctively distributed fauna, described earlier, is therefore not evidence for any marked faunal exchange or affinity between Western Ghats and Assam more than between others, but merely represents the relics (RIPLEY, 1949) of a former widely distributed fauna. The very slight preponderance of the Indo-Chinese element in the Western Ghats is due only to the humid evergreen climate, comparable to that of Assam and due to the higher rate of speciation that the humid evergreen biotope ensures than dry biotopes.

The impetus for the dispersal of mammals southward seems to have occurred during the Pleistocene, when the temperature conditions became markedly different. These conditions obviously extended considerably southward in the Peninsula, including undoubtedly the present-day semi-arid Deccan. About the Eastern Ghats, HORA (1949b) admits rather reluctantly that the climate might have been very different in Pleistocene. He (p. 364) visualized that 'Northeast Trade Winds might have blown all through the year and been more vigorous than at present. These Trades might have given more rain in the East-Coast plains and the eastern slopes of the hills than to the west which would be in the rain shadow of the Trades. This rainfall might have been more evenly distributed throughout the year. This means that in the Peninsula south of 15 degree N., there might have been more rainfall in the eastern portions and less rainfall in the western portions as compared to the present conditions'. If such were the possible conditions, it can be inferred that the Eastern Ghats at that time possessed a biotope ideal for the migration of hill fauna from Assam. This conclusion is strongly supported by KHAJURIA (1955), who has shown that 70 % of the mammals of the now semi-arid Deccan consists essentially of semi-humid elements and that mammalian fauna indigenous to this area are few and unimportant, and concludes, therefore, that those facts support the general belief that the area enjoyed much higher humid conditions in the past than now. As further evidence, fossil *Rhinoceros* and *Hippopotamus* have been obtained from the Kurnool deposits of this area (LYDEKKER, 1902).

Cooler and more humid conditions in the greater part of India during and perhaps after the retreat of the Pleistocene glaciers sufficiently meet the ecological requirements of the southward dispersal of most of the plains-dwelling mammals, for which no mountainous connection is necessary, the migration of montane forms might have been equally through the Western Ghats and Eastern Ghats. As the Chota-Nagpur Ranges are common to both the route, this could have been the place where the southward route of montane fauna bifurcated toward either of the two opposite Ghats in its course. This montane flow of migration obviously converged and crossed each other in the region of Nilgiri and Palni Hills, which together formed a sort of the two currents. The present-day distribution completely bears out this possibility. Similar cross-flow occurred in flora also, as MOONEY (1942) surmises a migration through the Western Ghats through the Palni Hills, northward through the Eastern Ghats, extending upto the Bastar District in east Central India. Thus, on the basis of the composition and distribution of the present-day mammalian fauna of Peninsular India, it appears that the migration of the montane species took place through both the Ghats. As favourable climate was available throughout the greater part of India, the plains-dwelling forms could migrate freely. In either case a special and sole highway of faunal dispersal in Peninsular India fancied by HORA is not tenable (see Chapter XXIV).

Table IX. Number of the mammalian genera of eastern and western entrants in different regions of India

Region	Eastern entrants	Western entrants	Total
Assam	68(79)	18(21)	86
Western Ghats	38(61)	24(39)	62
Eastern Ghats	34(60)	23(40)	57
Western India	34(36)	53(64)	87

8. Mainstreams of the Faunal Flow

The predominant direction of the faunal flow can be understood by a study of the affinity and the source of the fauna. An analysis of the fauna of each region, with regard to their eastern or western affinities, implying their entrance through either the northeastern or the northwestern parts of India, gives us broad indications of the faunal flow (Table IX).

It is apparent from this table that Assam has received from western India only eighteen genera, accounting for 21% of its fauna, whereas thirty-four genera from Assam reached western India, representing 36% of its fauna. This shows that the flow of fauna from the east to the west was greater than in the reverse direction. Similarly, the higher percentage of the eastern entrants in both the Ghats demonstrates the predominant flow of fauna to the Peninsular India was from Assam, which thus exerted much greater influence than the western India in the faunal colonization of Peninsular India.

Comparing the Peninsular India and Assam from the point of view of the faunal flow from the northwest, we find that the western entrants show a very pronounced tendency to migrate southward and colonize the Peninsula proper, rather than move eastward to Assam. In western India there are 53 genera of western entrants, of which 23 have reached both the Ghats, while only 18 have reached Assam. Considering that the percentage of this element is 39 in the Western and 40 in the Eastern Ghats, though only 21 in Assam, it would appear that the influence of faunal flow from the west, though less intense as compared to of the east, is still considerable in Peninsular India, while its influence is markedly less in the composition of the Assam fauna.

If we put together the above findings, we get a clear picture of the mammalian faunal flow that took place in India (Fig. 148). It may be readily seen that the main eastern flow from the Indo-Chinese subregion entered India through Assam and bifurcated, one branch spreading to

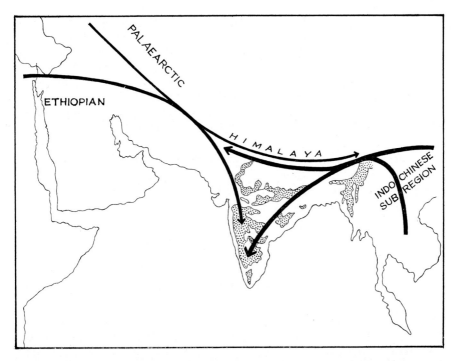

Fig. 148. The routes of the main-streams of mammalian faunal influx in the Indian subregion.

the Peninsular India and the other across the narrow wooded Sub-Himalayan belt to the northwestern parts and the areas further west. The other mainstream, which was formed of an Ethiopian and Palae-arctic constituents, entered India through the northwestern parts and as in the case of the eastern flow, bifurcated, one branch colonizing the Peninsula and the other Assam. This eastbound flow from the west was, however, much weaker than the flow from the east. In general, the eastern mainflow from the Indo-Chinese subregion was far more pre-dominant than the flow from west.

As regards the species dispersal, while a particular genus might have entered and colonized India through a particular route, it does not necessarily follow that all the species of that genus came through that same region or even followed the same route. A significant element of the species of genera that dispersed between any two regions and are, there-fore, common to those regions, are independent of the dispersal gradient between those two regions and came from other areas. Another feature is that as far as Assam is concerned, the specific dispersal with the Peninsular India was by no means always commensurate with the generic dispersal, but far below in volume. There occurred far more specific

611

dispersals between Assam and western India, so that the Sub-Himalayan route must rightly be regarded as the grand specific dispersal axis of India.

The species dispersal has greatly been affected by the profound recent physiographic changes that occurred in the desert and Gangetic Plains of the north India. Many late arrivals could not disperse due to the formations of barriers, particularly since 500 B.C. It is also seen that faunal invasion from Assam to the Peninsular India was single-phased, whereas that between Assam and western India was on the whole multi-phased. While the large majority of the Assam genera involved in the dispersal with other regions are the same, the species-augmentation that followed in the various regions are different, the Peninsular Ghats being poor in this respect, western India showing the maximum and Assam coming next in order.

REFERENCES

ALI, S. 1949. The Satpura trend as an ornithogeographical highway. *Proc. nat. Inst. Sci. India*, 15: 387–393.

ALLEN, B. C. 1909. Imperial Gazetteer of India, Provincial Sr., Eastern Bengal and Assam. Govt. Printing, Calcutta.

BLANFORD, W. T. 1901. The distribution of the vertebrate animals in India, Ceylon and Burma. *Phil. Trans. R. Soc. London*, 194: 335–435.

ELLERMAN, J. R. 1947. Notes on some Asiatic rodents in British Museum. *Proc. zool. Soc. London*, 117: 259–267.

ELLERMAN, J. R. & T. C. S. MORRISON-SCOTT, 1951. Checklist of Palaearctic and Indian Mammals. British Museum, London.

HIGGINS, J. C. 1933–1934. The game birds and animals of Manipur State with notes on their numbers, migration and habits. *J. Bombay nat. Hist. Soc.* 36: 406–422, 591–605, 845–854; 37: 81–95, 298–309.

HINTON, M. A. C. & H. M. LINDSAY, 1926. Bombay Natural History Society's Mammal Survey of India, Burma and Ceylon. Report No. 41: Assam and Mishmi Hills. *J. Bombay nat. Hist. Soc.* 31: 379–382.

HORA, S. L. 1949a. Satpura Hypothesis of the distribution of the Malayan flora and fauna to the peninsular India. *Proc. nat. Inst. Sci. India*, 15: 309–314.

HORA, S. L. 1949b. Climate as affecting the Satpura Hypothesis. *Proc. nat. Inst. Sci. India*, 15: 361–364.

HORA, S. L. 1949c. Discontinuous distribution of certain fishes of the far-east to the Peninsular India. *Proc. nat. Inst. Sci. India*, 15: 411–416.

JAYARAM, K. C. 1949a. A note on the distribution of the Chelonians of Peninsular India with Malayan affinities. *Proc. nat. Inst. Sci. India*, 15: 397–398.

JAYARAM, K. C. 1949b. Distribution of Lizards of Peninsular India with Malayan affinities. *Proc. nat. Inst. Sci. India*, 15: 403–409.

JAYARAM, K. C. 1949c. Remarks on the distribution of Annelids (Earthworms, Leeches) of Peninsular India with Malayan affinities. *Proc. nat. Inst. Sci. India*, 15: 418–420.

KEMP, S. 1924. Notes on mammals of Siju cave, Garo hills, Assam. *Rec. Indian Mus.* 26: 23–25.

KHAJURIA, H. 1955. Mammalian fauna of the semi-arid Deccan and its bearing on the appearance of aridity in the region. *Sci. Cult.* 21: 293–295.

KURUP, G. U. 1965. On a collection of mammals from Assam and adjoining areas. *J. Bombay nat. Hist. Soc.* 33(2): 185–209.

KURUP, G. U. 1968. Mammals of Assam and adjoining areas. 2. A distributional list. *Proc. zool. Soc. Calcutta,* 21: 79–99.

LYDEKKER, R. 1902. Indian Tertiary and Post-Tertiary Vertebrata: The fauna of Karnul caves. *Palaontologia Indica,* (10)4: 23–58.

MOONEY, H. F. 1942. A sketch of the flora of the Bailadila Range in Bastar State. *Indian for. Rec.* (N.S.) Botany, 3: 197–253.

NATH, B. 1953. On a collection of mammals from Assam (India) with special reference to the rodents. *Rec. Indian Mus.* 50: 271–286.

RIPLEY, S. D. 1949. Avian relicts and double invasions in peninsular India and Ceylon. *Evolution,* 3: 150–159.

ROONWAL, M. L. 1948. Three new Muridae (Mammalia: Rodentia) from Assam and the Kabaw Valley, Upper Burma *Proc. nat. Inst. Sci. India,* 14: 385–387.

ROONWAL, M. L. 1949. Systematics, ecology and bionomics of mammals studied in connection with Tsutsugamushi disease (scrub typhus) in the Assam-Burma war theatre during 1945. *Trans. nat. Inst. Sci. India,* 3: 6–122.

ROONWAL, M. L. 1950. Contribution to the fauna of Manipur State, Assam. General Introduction. *Rec. Indian Mus.* 46: 123–126.

ROONWAL, M. L. 1950. Contribution to the fauna of Manipur State, Assam. 3. Mammals with special reference to the family Muridae (order Rodentia). *Rec. Indian Mus.* 47: 1–64.

ROONWAL, M. L. & B. NATH, 1949. Discontinuous distribution of certain Indo-Malayan mammals and its zoogeographical significance. *Proc. nat. Inst. Sci. India,* 15: 375–377.

SIMPSON, G. G. 1953. Evolution and Geography. Oregon State System of Higher Education Oregon, U.S.A.

THOMAS, O. 1866. On the mammals presented by Allen O. Hume Esq. C.B. to the Natural History Museum. *Proc. zool. Soc. London,* 54–79.

THOMAS, O. 1921. Scientific results from Mammal Survey 25(A). On Jungle mice from Assam. *J. Bombay nat. Hist. Soc.* 27: 596–598.

WADIA, D. N. 1966. Geology of India. English Lang. Book Soc. and Macmillan & Co., London, 3rd (revised) ed.

WROUGHTON, R. C. 1921. Scientific results from the Mammal Survey. *J. Bombay nat. Hist. Soc.* 27: 599–601.

613

XIX. BIOGEOGRAPHY OF THE PENINSULA

by

M. S. MANI

1. Introduction

The general physical features, geology, the characteristic flora and the phytogeographic characters of the Peninsula have been discussed in earlier chapters. We give in this chapter a broad outline of the salient features of the faunistic composition, affinities, distributional patterns and zoögeographical subdivisions of the Peninsula.

As pointed out in Chapter II, the biogeographical limits of the Peninsular area in the north are by no means sharply defined. This area extends beyond the bed of the R. Ganga across the Indo-Gangetic Plains of north India to the foothills of the Himalaya during the period of the southwest monsoon rains and shrinks to the south of the R. Ganga and lies almost at the scarp of the Vindhya during the winter (Fig. 149). The Indo-Gangetic Plains represent, therefore, the marginal transitional boundary of the Peninsula. The Peninsular area does not thus correspond completely to the Cis-Gangetic tract of BLANFORD (1901), but also extends beyond the Aravalli Divide, practically up to the R. Indus and merges without a sharp dividing line into the Western Borderlands. The Seychelles, Laccadives, Maladives and Ceylon are parts of the biogeographical area of the Peninsula.

The general ecology of the Peninsula is dominated by its senile topography, its physical relations to the Himalaya, the monsoon rainfall pattern and the extensive destruction of natural habitats by man within historical times. The senile topography and the relation to the Himalaya determine to a large extent the major peculiarities of the monsoon rainfall climate of the Peninsula, but this climate is not the chief or even an important isolating factor and does not also explain all the peculiarities of occurrence and distributional patterns of plants and animals today. The secondary rôle of climate is attributable largely to the fact that monsoon-climate is itself primarily the result of other complex factors, but mainly the relation to the Himalaya. Only in the case of the Pleistocene relict elements climate has played an important rôle. In any case, the major characters of the monsoon-climate are of relatively recent origin, perhaps Post-Pleistocene, so that the effects of the monsoon-climate must indeed be taken as wholly unimportant as a factor in the origin and evolution of the present-day biogeographical characters of the Peninsula. The most important factor that dominates both directly and indirectly the entire ecology of the Peninsula is, however, the massive

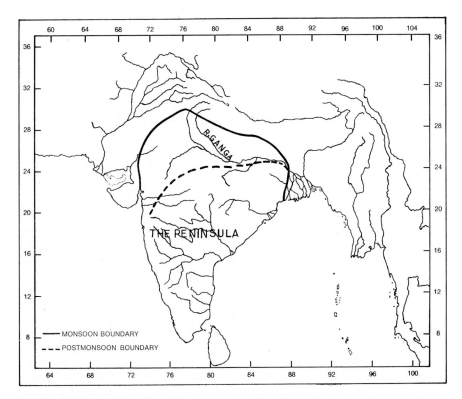

Fig. 149. Map of India, showing the seasonal oscillation north and south of the bio-geographical limits of the Peninsula, synchronizing with the rhythm of the monsoon climate. During the months of the southwest monsoon rainfall the limits of the Peninsula push northwards across the plains of north India, beyond the bed of the R. Ganga, to the foothills of the Himalaya. During the postmonsoon months the limits recede southwards, rather pronouncedly more in the west than in the east, to vanish in the Western Ghats. The transitional zone between these limits is the biogeographically neutral Indo-Gangetic Plains, characterized by the dominance of typically Peninsular plants and animals in the monsoon months and temperate elements in the winter.

disappearance of all natural habitats, brought about by deforestation by man. In this process extensive areas of former humid-tropical forests have been transformed into semi-arid deciduous forest or even into scrublands and savannah. The dominance of the destruction of natural habitats as a factor is readily evident in nearly every aspect of the ecology and biogeography, but may be particularly clearly seen in the ecological anomalies in the distribution of many species of animals (JEANNEL, 1943). In the Peninsular climate the rôle of atmospheric temperature is also negligible; it is however the pattern and abundance of the monsoon rainfall that is more important than temperature in the phenology of most species.

615

The most outstanding fact perhaps of the Peninsular biogeography is that the Peninsula is *India vera*, the rest of India representing merely a recent biogeographical appendage. This fact finds its natural explanation in WEGENER's theory of continental drift (see also Chapters II, III and XV). The salient facts of its biogeography may be traced back to the breakup of the Gondwanaland and the northeastward drift of what now constitute Madagascar and the Indian Peninsula. The drifting mass of the Peninsula, the crumpling and folding of the Tethyan sediments and the contact with the Asiatic main landmass in Assam in the northeast, the underthrust, down-warping and fissuring of the northern part, the block-fracturing and marine subsidence in the west and other complex events connected with the Himalayan uplift are all integral parts of this explanation.

The Peninsula, as we know now, is therefore *per se* a relict of a former much larger and higher plateau and its present position and its general ecology and biogeographical peculiarities are fundamentally a by-product of the Himalayan uplift. It is *par excellence* a region of relicts, phylogenetic, geographical, Pleistocene relicts, etc.

The original flora and fauna of the Peninsula constitute, therefore, the true Indian elements – the character flora and fauna of India.

2. The Character Fauna

The character fauna of the Peninsula is typical of the senile topography of the region. Its size, composition, distributional patterns and evolutionary trends have been very profoundly influenced by the intense pressure of the influx of the Extra-Peninsular faunas mainly from the northeast, the Pleistocene glaciations on the Himalaya and by the pressure of the human expansion and civilization.

The biogeographical component elements of the present-day character fauna of the Peninsula may be grouped under 1. the derivatives of the older faunas differentiated in a southern landmass, viz. the Gondwana faunas and 2. the derivatives of the relatively younger faunas, differentiated mainly in Asia and comprising essentially the Tertiary mountain faunas. The first group consists of the Peninsular autochthonous elements and represents the true Indian component. The second group consists of the greatest majority of the intrusive elements and comprise both the tropical Asiatic eastern components differentiated in the Indo-Chinese and Malayan subregions and also the western intrusives of the Mediterranean elements.

The bulk of the endemic elements of the Peninsular fauna of the present belongs to the first group and corresponds in part to the Cis-Gangetic component, described by BLANFORD (1901). They are largely members of older taxonomic groups of Pre-Himalayan origin and may, therefore, be appropriately described as antecedent endemics. They are essentially

Plate 69. Characteristic flat-topped 'hills' of the Peninsula, in marked contrast to the rugged Himalayan area. The socalled hills of the Peninsula represent residual tops of an ancient plateau.

Plate 70. The characteristic wide canyons and flat-topped 'hills' of the Peninsula, typically developed in the northern parts of the Western Ghats.

the plateau fauna and essentially phylogenetic relicts. We have, in addition, considerable numbers of endemics, which are Post-Himalayan in origin and belong to relatively younger and more recent taxonomic groups, differentiated locally in the Peninsula from the intrusive stock. These may be separated from the antecedent endemics as neo-endemics: they represent essentially the derivatives of the Trans-Gangetic component of BLANFORD (1901). The neo-endemics are, strictly speaking, minor elements of the present-day Peninsular faunal complex.

The intrusive elements constitute the major components of the present-day fauna of the Peninsula. They are largely isolates and outliers of the humid-tropical forest and Tertiary-mountain Asiatic faunas of the Indo-Chinese and Malayan subregions, characterized by more or less pronounced discontinuity in distribution and representing the dominant geographical relicts in the Peninsula. A small section of the intrusive fauna is derived from the Himalaya, Euro-Siberian and Turkmenian subregions as Pleistocene relicts, also characterized by pronounced discontinuity of distribution. Finally, a considerable part of the intrusive elements of the Peninsular fauna consists of Mediterranean-Ethiopian components, corresponding essentially to the Aryan element of BLANFORD (1876, 1901), and differing markedly from the rest of the intrusive elements in their general continuity of distribution. Though they represent the latest intrusive elements, the Ethiopian elements belong to the older component of Gondwana stock and are, therefore, considered below under the Gondwana faunals derivatives.

We may, therefore, arrange the intrusive elements of the Peninsular fauna in the following chronological sequence: 1. the humid-tropical Asiatic Tertiary-mountain relicts of the Indo-Chinese and Malayan faunas, 2. the Ethiopian, and 3. the Mediterranean and 4. Himalayan Pleistocene relicts.

The relative strengths of the different component intrusive elements vary within wide limits. In some groups of plants and animals the humid-tropical Asiatic elements are dominant and the others are minor elements, but in other groups the Ethiopian and the temperate-boreal elements are more dominant. Taken as a whole, the Peninsula has received in nearly equal proportion from the northeast (the humid-tropical Asiatic Tertiary-mountain), north (the Himalayan) and northwest (Mediterranean and Ethiopian) theatres, so that the different components of the intrusives are now nearly equally strong. Another important peculiarity of the intrusive elements is the fact that discontinuity characterizes only the distribution of the humid-tropical Asiatic and Himalayan (Tertiary-mountain) elements, but not of the Mediterranean and Ethiopian elements. The latter have intruded into India from the west and are largely also concentrated in Rajasthan-Sind, the western parts of the Upper Gangetic Plain, the Peninsular west and the Deccan and are distributed continuously.

The general characters of the Peninsular fauna at present may be summarized as a very marked impoverishment, regression, evolutionary stagnation, ecological anomalies in distribution and rapidly vanishing relics. While rich in ancient and endemic forms and phylogenetic and geographical relics, the Peninsular character fauna is at present, however, fast degrading. Even casual observation shows a marked impoverishment of the typically older Peninsular autochthonous genera and species in nearly all groups and a gradual disappearance of even the newer intrusive elements of the Extra-Peninsular faunas. The dominant elements of both the Peninsular autochthonous and intrusive elements of the Peninsular fauna being primarily humid-tropical forest forms, the extensive destruction of forests by man has all but completely eliminated the character fauna of the Peninsula, except in small and rapidly dwindling isolated pockets. The reconstruction of the distinctive features of this essentially relict fauna presents, therefore, considerable difficulty.

Very little is, however, known about the Peninsular autochthonous fauna prior to the Pleistocene times. The Pliocene fauna of the Peninsula is represented by certain interesting Siwalik fossils in the Prim Island in the Gulf of Cambay. While the Siwalik fauna, at least of the Vertebrates, as a whole largely shows Ethiopian affinities, there are also others like *Semnopithecus*, *Rhizomys*, *Tragulus*, and *Cervus*, with undoubted Oriental affinities. This would appear to show that even as early as the Pliocene and perhaps late Miocene times, the Peninsular fauna had been considerably influenced by the influx of the Extra-Peninsular faunas. The Narmada Gravels and the Karnul Caves (LYDEKKER, 1902) contain remains of the Pleistocene Vertebrate fauna of the Peninsula. The former is perhaps older of the two, but both of them are not in any case older than the period of rise of *Homo sapiens*. The remains are mostly Mammalia, but some birds and reptiles from the Karnul Caves are also known. The Narmada Gravels contain some crocodiles and chelonians and the mammals *Elephas*, *Rhinoceros*, *Equus* and *Hippopotamus*,* buffalo apparently identical with the wild buffalo of India at present, *Bos*, *Boselaphus* (the nilgai), *Cervus* (rusine-deer) of the Malay type. In some deposits of apparently the same age, but on the R. Yamuna, we find *Semnopithecus* and *Antilope*. The Karnul Cave mammals are more numerous and comprise species of the Eastern Asiatic *Semnopithecus*, *Tragulus*, *Cervus unicolor*, *Atherura* (the last two are still found in the Peninsula); BLANFORD's Aryan elements like *Melursus*, *Boselaphus*, *Antilope*, *Tetracerus*, *Gerbillus* and *Golunda*. Some species have affinities with Ethiopian forms and others are allied to the Siwaliks forms. Among the Ethiopian forms are *Hyaena crocua*, *Equus asinus* and *Manis gigantea*, not distinguishable from the living

---

* Hippopotamus flourished not only in South India, but occurred even in Ceylon well within historical times (DERANIYAGALA, 1941). This is the *makara* (*not* crocodile) depicted in stone boss reliefs in South Indian temples).

African species, *Cynocephalus*, *Rhinoceros* related to the African *Rhinoceros bicornis*. While both the Narmada and Karnul remains contain typical representatives of both the Asiatic and the Aryan faunas, the latter seem to have predominated more than at the present time and comprise a number of forms that have now disappeared completely from the area.

3. The Gondwana Faunal Derivatives

The Gondwana faunal derivatives represent the oldest component elements – phylogenetic relicts – of the character fauna of the Peninsula. They include the Peninsular-autochthonous endemics, most of which are at present restricted in their distribution wholly to within the limits of the Peninsula, and other autochthonous forms, which are common to or have their closest allies in Madagascar (Lemurian faunas) or sometimes even in South Africa (Ethiopian faunas) and rarely in South America. The greatest majority of the Gondwana derivatives, especially the terricole forms, belong to relatively ancient, but not necessarily primitive groups. The Peninsular-autochthonous Gondwana faunal elements were formerly far more widely and also continuously distributed over the whole of the Peninsula and had even spilled over to the areas in the north, particularly to the foot of the Himalaya. They formed part of the Gondwana-Oriental line of faunal development described by JEANNEL (1943). Their advance eastwards was, however, retarded to some extent by the intrusion of the Tertiary faunas. The lowering of the general atmospheric temperature and the concomitant increase in atmospheric aridity* during the Pleistocene glaciations on the Himalaya resulted in a partial retreat southwards of the Gondwana (humid tropical) faunal derivatives. In the Post-Pleistocene times, the areas vacated by the Gondwana faunal elements were largely occupied by the Tertiary mountain faunas from the east but there was no reoccupation from the south. These areas were far removed from the centres of radiation of Gondwana faunas, but rather close to the more highly plastic and rapidly diversifying Oriental faunal radiation centre. In these marginal areas, the Gondwana faunal derivatives were already considerably diluted by the Oriental elements. Reoccupation by Gondwanal derivatives of the Peninsula was, therefore, ruled out, but outliers were left behind in isolated pockets when the main body retreated in the Pleistocene times.

The oldest members of the Gondwana elements in the Peninsula

---

* As indicated in an earlier chapter, most Indian zoologists have erroneously assumed the Pleistocene glaciations in the Himalaya resulted in higher humidity and precipitation in India. It is on the other hand true that the glaciers locked up so much of the atmospheric moisture, that relatively dry conditions prevailed in the Indo-Gangetic Plains and in the Peninsula. The Permian glacial Epoch was similarly marked by drier climate than the warm humid conditions of the Carboniferous times.

Plate 71. Waterfall near sources of the Peninsular rivers on the Western Ghats, serving as refugial pockets of Pleistocene relicts of the Himalaya.

Plate 72. Typical humid-tropical forest of *Terminalia-Eugenia*-Complex near Mahableshwar, almost on the Crest of the Western Ghats (Sayadri Mountains) in the north.

621

Plate 73. Typical youthful topography of the Himalaya.

Plate 74. A relict of the Pleistocene glaciers on the Himalaya.

622

belong to groups of great age, which also occur in other Gondwana areas of the world. The Lamellibranch Mollusc *Mülleria dalyi*, found in the source tributaries of the R. Krishna, is congeneric with the South American species. This genus belongs to the family Aetheriidae, of great antiquity, in which the external form of the shell resembles certain marine Ostreidae. It is interesting to remark that the shell of the Indian species is almost identical with that of the type species described from New Grenada. *Sreptaxis* (Pulmonata), from the South Indian hills and Ceylon, occurs also in Assam, Burma, Andaman and Nicobar Islands, tropical Africa and South America and *Ennea* is known from Madagascar, tropical and south Africa, south and southeast Asia, Japan, Philippines, Burma, Nilgiri-Shevroy-Anamalai and Palni Hills, Mysore, South Kanara, Narmada Valley and Orissa Hills. The whipscorpions of the family Thelyphonidae (Arachnida) occur, for example, in South India and Ceylon, but not in north India or on the Himalaya; they are again met with in Burma and the Indo-Malayan areas including the Malay Archipelago, the Papuan area as far as the New Hebrides and the Fiji Islands and reappear in tropical America. Among the others with affinities to tropical American fauna are the geckonid lizards; *Gonatodes* and *Gehyra; Eublepharis* with two Peninsular species and three species from Central America; *Polydontophis* with three species in Madagascar, five species in India and Burma and two species in Central America; *Lachesis* with more than twenty species from Central and South America is united with the Indian genus *Trimeresurus*. Ilysiidae from South America are also represented in the Peninsula and Ceylon.

3.1. The peninsular endemics

The principal Peninsular autochthonous elements, restricted at present wholly to within the limits of the Peninsula, differentiated in the Peninsula after separation from Lemuria (Madagascar-India). They represent, at least partly, the dominant types of the original fauna of the Peninsula, before this fauna was modified by the influx of the humid-tropical Asiatic floristic-faunistic elements from the northeast, through the Assam gateway. These elements correspond to BLANFORD's Dravidian component of the Cis-Gangetic fauna (BLANFORD, 1901).

The principal examples of the Peninsular autochthonous endemic elements include the Porifera *Gecarcinucus* and *Pectispongilla* (Potamonidae) from Malabar, the fresh-water hydroid *Limnocnida indica;* the Oligochaeta *Comarodrillus, Octochaetus* and *Eudichogaster;* the Mollusca *Ariophanta* that occurs from Malabar to Surat and sparsely in the Indo-Gangetic Plain, *Eurychalamys* occurring from Ceylon to Bombay, *Mariaella* occurring from Travancore to Mahableshwar on the Western Ghats and in Shevroy in the Eastern Ghats and also Ceylon mountains, *Pseudaustenia* from the Nilgiri Hills and Travancore, *Thysanota, Ruthvenia* and *Cerilia;* and the

623

Arachnids *Charnus, Stenochirus, Chiromachetes, Isomachus* (scorpions); *Labochirus* and *Schizonus* (Uropygi) and *Poecilotheria* (spider). Among insects also we find a number of examples of this group. The Orthoptera, confined to especially to the southern parts and to Ceylon, include *Orthacris, Zygophlaeoba, Phlaeobida* and *Paraphlaeoba* from the Eastern Coastal land and from Ceylon; *Deltonotus* from South India and Ceylon; *Abbasia, Madurea, Lernia, Anarchita, Colemania* and *Pelecinotus* from South India only; *Apterotettix, Lamellitettix* and *Spadotettix* from only Ceylon. The Odonata, confined to within the limits of the southern parts, include *Platysticta*, from South India and Ceylon, *Ceylonosticta* with several species from only Ceylon; *Chloroneura, Indoneura, Melanoneura, Esme, Phylloneura* and *Idiophya* from the Western Ghats only. *Dermaptera: Cranopygia, Dendoiketes, Obelura, Sondax* and *Syntonus*. Heteroptera: *Bozius*. Coleoptera: *Apteroessa* with a single species in the East Coast (Cicindelidae); *Idiomorphus* (Carabidae); *Pseudolema, Madurasia, Mimastrella, Chalaenosoma* (Chrysomelidae); *Pachycera* (Tenebrionidae) with four endemic species; and *Disphysema, Anatona* with 3 species, *Pseudanotona* with a single species and *Spilophorus* with two species in Ceylon and South India (Scarabaeidae); Hymenoptera: *Drepanognathus saltator* from Ceylon to Kanara (Formicidae). The butterfly *Parantirrhoea* is also confined to the southwest of the Peninsula. Most of these examples are present restricted to the extreme south of the Peninsula and sometimes to only Ceylon, as in the case of *Ratnadvipa* (Mollusca), *Charnus* (scorpion), *Schizonus* (Uropygi), *Coptolobus* (Carabidae), *Xenarthra* (Chrysomelidae), *Leptogenys pruinosa, Myopias cryptopone, Stereomyrmex, Aneuretes, Ooceraea* and *Syscia* (Formicidae).

Among the Vertebrates, the principal endemic elements belong to Amphibia and Reptiles, but also to some fishes like *Nangra, Jerdonia, Noemacheilichthys* and *Etropius*. There are a number of other fishes which are endemic in the Peninsula (see Chapter XVII). We have, for example, *Parasilorhynchus, Travancoria, Horaglanis, Neotropius, Bhavania, Lepidopygopsis* and *Horabagrus*. There are 14 species of the Bagrid fish *Mystus* endemic in the Peninsula (including Ceylon); of these five species are restricted to the mainland of the Peninsula and two species are from Ceylon. The Schilbaeid fish *Silonopangasius* is confined to the Peninsula. There are over seventeen genera of Amphibia known from the Peninsula, of which seven are at present restricted to the region and nine do not extend into the Trans-Gangetic area. The Malabar area has perhaps the largest concentration of endemic Amphibian. The genera restricted to the Peninsula comprise *Micrixalus, Nyctibatrachus, Nannobatrachus, Nannophrys, Melanobatrachus, Cacopus* and *Gegenophis*. Other genera like *Nectophryne* and *Uraeotyphlus*, found in the Peninsula do not occur in the Trans-Gangetic area. The Reptiles which are at present restricted to the Peninsula include the lizards *Cabrita, Ristella, Sesophis, Chalcidoseps* and *Acontius*, the agamids *Otocryptis, Ceratophora, Lyriocephalus, Salea*, etc. The snakes Uropeltidae and *Xylophis, Haplocercus* and *Aspidura* must also be considered as members of the Dravidian elements. Other important examples include *Hardella* (Chelonia); *Callodactylus, Teratolepis, Sitana, Ceratophora, Lyriocephalus, Charasia, Cabrita* (Lacertilia); *Gongylophis, Uropeltis, Rhinophis, Silybura, Pseudoplecturus, Plecturus, Melanophidium, Platyplecturus, Xylophis, Haplocercus, Aspidura* and *Elachistodon*. The genera found in the Peninsula, but not occurring in the Trans-Gangetic area, include *Eublepharis, Chalcidoseps, Haplodactylus, Ophiops, Chalcides, Acontius, Chamaeleon* (Lacertilia); *Eryx, Macropisthodon, Dryocalamus, Coronella, Hemibungarus* and *Echis* (snakes). In Ceylon we have a number of typical examples like the lizard *Lyriocephalus*, the snake *Uropeltis*, etc. There are in Ceylon a number of forms, which are common to the mainland of the Peninsula, but do not occur in other parts of the Orient, viz. *Otocryptis, Rhinophis, Silybura, Leptopomoides, Micrixax, Tortulosa, Nicida, Eurychalamys, Mariaella*, etc. Not all these reptiles, though now confined to the south, may, however, be truly of the Dravidian stock; they may represent at least in part geographical relics, left behind as outliers when intrusive faunas withdrew.

About a dozen genera of birds appear to belong to this group. We have, for example, *Rhopocichla, Elaphrornis, Kelaartia, Dissemurulus, Laticilla, Schoenicola, Sturnornis, Ochromela, Galloperdix* and *Acmonorhynchus*.

3.2. THE MADAGASCAN ELEMENTS

Madagascan affinities of the Peninsular autochthonous fauna are on the whole stronger than African affinities. The African and Madagascan affinities are also generally stronger among the more generalized and more ancient groups than among the higher and more recent groups. In the fresh-water fauna of the Peninsula, the African-Madagascan affinity is, for example, more evident among the lower than among higher Invertebrates or fishes.

Madagascar Island is almost 1600 km long, with a maximum width of 578 km. The central part is a granitic plateau of undulating moors, on which some peaks rise to elevations of 2745 m above mean sea-level. The lower plains are generally wooded. The island seems to have been isolated since the Eocene times, so that we do not find at present the higher mammals so common in the mainland of Africa. The only typically Madagascan mammals are ancient Insectivora, the endemic Centetidae and Lemuroids like Lorisinae and Galaginae (hence the name Lemuria for the island). Some of the endemic Viverrinae and Herpestinae are old forms. Chamaeleontidae are abundant in Madagascar and South India. Madagascar lacks Caecilians, but we find Rhacophorids and Brevicipids. Most Madagascan amphibians are endemic. The island lacks continental reptiles, but some of the reptiles found there show affinity to African forms. Some tropical Oriental genera of reptiles, not represented in Africa, also occur in Madagascar. The frogs of the island are different from those of Africa – there are no *Pipa*, Bufonids, Phrynomerids and

Heleophryne in Madagascar. Two Rhacophorid genera *Megalixalus* and *Hyperolius* are common to Madagascar and Africa, but *Rhacophorus* is common to Madagascar and India, but not Africa; these are abundant in Ceylon. Dyscophinae are represented by one genus, which is confined to Madagascar, but none occur in Ceylon or the Peninsula, though two other species of the same genus occur further east in southwest China, Burma, Malay Peninsula, Sumatra and Borneo (Gondwana-Oriental line of faunal genesis). The affinities of reptiles of Madagascar to those of South India are less pronounced than in Amphibia. *Sibynophis* is perhaps the only important form.

As is well known, the Madagascan-Ethiopian distribution is of the following types: 1. genera which are widely distributed in one of these areas, but only sparsely in the other and often also transgressing to exotic regions and 2. genera distributed nearly equally in Madagascar and in South Africa. The Cerambycid *Cantharocnemis* (KOLBE, 1887) has, for example, about fifteen species in tropical Africa, but only one endemic species each in Madagascar and the Indian Peninsula-Ceylon. 3. In the Madagascan-Indo-Australian distribution, genera are present in tropical belts in the east and transgress upto the Oriental Region (often up to the Melanesian or occasionally also to the Polynesian subregion) and in Madagascar, but are absent in the continental Africa. 4. Other genera transgress to the Manchurian or the Australo-Tasmanian Subregions. *Euploea* a genus of butterflies, for example, occurs in the Oriental Region and the warmer parts of the Australian Region eastwards up to Tahiti, but is completely absent in Africa. The Geometrid *Gelasma*, with numerous species in the Oriental and Australian Regions, eastwards up to the Solomon Islands and southeast Australia, sparsely in China, East Siberia and Japan, has three species in Madagascar. *Philiopsis* (Pselaphidae) has one species each in Madagascar, Malaya, Sumatra, Java and New Guinea. A number of insects found at present in Madagascar no doubt represent introductions by man. This is the case, for example, with *Phyllium bioculatum* that extends from Ceylon to Java, the Indian and Malayan *Ceresium flavipes*, the Indian *Oxycetonia versicolor*, *Protaetia aurichalcea*, etc.

A remarkable fact that may be observed is that the Ethiopian elements transgress more frequently to the east than to the north, and many more genera are common, for example, to the Ethiopian and Oriental Regions and partly also to the Australian Region than to Africa and Europe. The genera common to the Ethiopian and Oriental Regions are in reality so many that the resulting faunal affinities of the Madagascan, Ethiopian and Oriental Regions and partly also Australian, are among the more striking pictures of the fauna of the earth.

The Ethiopian Coleoptera transgressing into the Oriental Region are illustrated by *Aulonogyrus* from Central Europe, Mediterranean area, Africa, Canary Islands, Madagascar, Ceylon and South India, Australia, Tasmania and New Caledonia. There are twenty-nine species, of which

626

the majority occur in Africa, one occurs in the Oriental Region, and two each in the Australian and Palaearctic Regions. Pselpaphidae: *Centrophthalmus* is distributed in Africa Madagascar, India, Indonesia and also extends to Algeria, Tunisia, China and New Guinea where it is found only sparsely. Erotylidae: *Episcaphula* occurs in the Ethiopian, Madagascan Oriental and Australian Regions and in the last mentioned region extends up to New Caledonia. Lampyridae: *Diaphanes* has about twenty species in tropical Africa, thirty species in Ceylon, India, Sumatra, Borneo and Philippines and five species in China and eastern Tibet. Cleridae: *Stigmatium* occurs in the Ethiopian, Oriental, Melanesian, Australian areas, China and Japan and has about seventy-five species, which are more or less equally distributed in all the areas, except for one each species in China and Japan. *Phaeocylotomus* occurs in the Ethiopian, Oriental Melanesian and Australian regions and nearly fifty percent of the species are found in the Ethiopian Region. Buprestidae: *Sternocera* is Ethiopian, South Arabian, Baluchistan, North and South India and South China. *Chrysochroa* occurs in tropical Africa, Ceylon and north and South India, China, Japan, Andaman Islands, Indonesia eastwards to Moluccas. *Iridotaenia* is a tropical East African genus that occurs in Seychelles Islands, India, Ceylon, Andaman Islands, Indonesia to New Zealand. *Belionota* is Tropical African, Madagascan and occurs also in Mauritius, Re-Union Island, Ceylon, India, Andaman and Nicobar Island, Indonesia, Melanesia, Sumatra to Philippines and New Guinea. Tenebrionidae: *Rhytinota* occurring in Tropical Africa northwards to Egypt, Socotra, Abd-el-Kuri and the Indian Peninsula, where about one-third of the known species are concentrated. Chrysomelidae: *Sagra* is Ethiopian, Madagascan, Oriental and also occurs in China and Japan, Melanesia to New Guinea. *Laccoptera* is *Ethiopian*, Madagascan and Oriental, with one species in New Guinea and Queensland. *Anoeme* occurs in tropical Africa and South India. *Acanthophorus* occurs in Africa from the Cape northwards to Sahara oases, in Arabia and has three species in the Peninsula. *Xystrocera* has over thirty-five species in the Ethiopian region, one species in the Madagascan Region, five species in India and Indonesia and two in Australia. *Placaederus* has over twenty-five species in the Ethiopian Region, eight species in Ceylon and India and extends across the Philippines also to Siberia. *Margites* has five Ethiopian species, three in India, two in North China. *Zoodes* is Ethiopian, Arabia, India, Ceylon and Sumatra. *Coptops* occurs tropical Africa, Seychelles Islands, Comoro Islands, Madagascar, India, Ceylon and Indonesia to New Guinea. *Sthenias* is Ethiopian and occurs in India, Ceylon, Java, Sumatra, Borneo and the Philippines. *Apomecyna* is Ethiopian and Madagascan and occurs also in Ceylon, India, China, Formosa, Japan, Indonesia, Melanesia and Australia. *Eunidia* is also Ethiopian, but occurs in Ceylon and South India and also in Burma. Scarabaeidae: *Heliocopris* is Ethiopian, but is also found in India, Ceylon, South China, Sumatra and Java. *Drepanocerus*

627

is another Ethiopian genus found in Ceylon, South India, Burma, South China and Java. *Orphnus* occurs in tropical Africa, Madagascar, Ceylon, Peninsula and Cochin-China. *Rhynyptia* is represented by about ten species in tropical Africa, eight species in Ceylon and India. *Coenochilus*, an Ethiopian genus, is represented by about one-fourth of the known species in Ceylon, India, South China, Java and the Philippines. The scorpion *Lychas* occurs in tropical Africa, Australia, Burma, China, Thailand, Malaya Archipelago and Malay Peninsula, Java, South India and Dehra Dun.

The following important examples of the Peninsular forms with pronounced Madagascan-African affinities are particularly interesting:

Porifera: *Coryospongilla*, occurring in the Western Ghats in the R. Godavari drainage area and transgressing partly into the Indo-Gangetic Plains, is known from Africa; the genus is also represented in south Burma. *Spongilla (Stratospongilla) bombayensis* occurs from the R. Godavari drainage area of the Western Ghats southwards to Mysore, and is also known from South Africa. The Coelenterate *Limnocnida indica* from the tributaries of the R. Krishna and from the Chota-Nagpur area of the Peninsula is also an African form. There are some interesting Bryozoa with African affinities (ANNANDALE, 1911, 1914). *Arachnoidea*, a genus of Hislopiidae known only the Lake Tanganyika, is related to *Hislopia* an Oriental genus that extends northwards to Siberia. *Afrindella* is a subgenus of *Plumatella*, with two species from the Peninsula, one of which is specifically identical with the African *Plumatella (Afrindella) tanganyikae*. *Lophopodella* has three African species, of which one occurs on the Western Ghats in the Peninsula. The Oligochaeta with African-Madagascan affinities found in the Peninsula are typified by *Howascolex*, *Curgia*, *Gordiodrilus*, *Glyphidrilus*, and *Aulophorus palustris*. The mollusca *Rachisellus* and *Edouardia* are common to tropical Africa and South India. We have a number of interesting examples among the fresh-water Entomostraca. *Hyalodaphnia hypsicephala* is represented by *H. hypsicephala eurycephala* and *H. hypsicephala stenocephala*. The former from is larger than the other species of the genus and is considered as the Indian representative of the East African *Hyalodaphnia barbata*, which resembles *H. hypsicephala stenocephala*. The Anostraca *Streptocephalus*, the original home of which is in Africa, occurs in the plains of the Peninsula. *Streptocephalus dichotomus* is a widely distributed endemic species of the Peninsula. Among the Caridea, *Palaemon indae* occurs in Africa, South India, Malay Archipelago and *Palaemon dolichodactylus* occurs in East Africa, Madagascar and South India.

The scorpion *Iomachus* occurs in East Africa, Nellore, Shevroy and Nilgiri Hills. The spider *Heliogmomerus* occurs in tropical Africa, Kodaikanal and Palni Hills and in Ceylon. Among insects we have a number of interesting examples, particularly in the older orders. The Orthoptera *Hedotettix*, *Gymnobothrus*, *Euthymia*, *Xenippa*, etc. are Madagascan-Oriental Australian elements. The Ethiopian-Mediterranean Orthoptera are

Ochrilidia, Dociostaurus, Quiroguesia, Poecilocerus, etc. The Ethiopian *Diplacodes lefebvrei* extends from Iraq to Coorg. The Pentatomid bug *Hotea* from the Nilgiri Hills is an African-Madagascan form that has spilled into the Extra-Peninsular area of Assam-Burma, China and Malaya. The reduvid bug *Edocla* found in Mysore is also known from South Africa. The Chrysomelid beetle *Apophylia* occurs in the Nilgiri, Ceylon, Africa and also in the Trans-Gangetic area of eastern Himalaya, Assam, Burma and China. *Oides* is known from Africa, Bombay and Malabar. The Cerambycidae *Acanthophorus, Cantharocnemis* and *Anoeme* found in South India are Ethiopian forms. Some of the forms which pronounced African-Madagascan affinities are restricted at present to Ceylon only. This is, for example, the case with the Oligochaeta *Aulophorus palustris* and the insects *Picrania* (Dermaptera), and *Leptogenys falcigera* from Ceylon and Madagascar (Formicidae).

Among the fishes common to the Peninsula and Africa, mention may be made of *Notopterus, Rita, Mystus, Barilius, Rasbora, Labeo, Tor, Clarias, Haplochilus, Gobius, Eleotris, Periophthalmus, Ambassis, Mastacembelus, Garra* and *Cyprinodon. Eutropiichthys* (family Schilbaeidae) is a Peninsular genus, with three species, related to the African *Eutropius* with a dozen species. Of the three species, one is restricted to the west coast areas, one extends into the Indo-Gangetic Plains and Assam and the third species to the Indo-Gangetic Plains and Burma. Among the Amphibia mention may be made of *Rhacophorus* found in Madagascar and South India-Ceylon and in the Trans-Gangetic Subregion. Of the reptiles common to Africa and the Peninsula, *Cnemaspis*, with about a dozen species restricted to the hills of South India and Ceylon, *Riopa* and *Hemidactylus frenatus*. The reptiles of the Peninsula are rather poor in African-Madagascan forms. The birds include the yellow-throated *Gymnornis* and *Salpornis*. In the mammals the Cercopithicinae are related to African forms. In the Siwalik fossils we have *Macaca* from the Peninsula related to African forms. *Hippopotamus* was widely distributed throughout the Peninsula in former times and extended even to Ceylon (DERANIYAGALA, 1941). There is considerable eveidence to show that the Peninsular autochthonous fauna with Madagascan – African affinities were far more widely distributed in former times in the Peninsula than at present. Indirect evidence found in the Ramayana would appear to show that the African affinity in the Peninsular fauna was stronger in former times.

The genera and species occurring in the Peninsula, but absent in the rest of India and Malayan area, though reappearing in tropical Africa, were included by BLANFORD under his Aryan elements of the Peninsular fauna. The Aryan elements constitute, according to him, mammals like *Melursus, Golunda, Tetraceros, Boselaphus* and the Antilope; birds like *Salpornis, Lophoceros, Taccocua, Galloperdix, Sypheotis* and *Rhinoptilus* and reptiles like *Sitana, Charasia, Cabrita* and *Eryx*. The typical genera of the Aryan elements do not, however, occur in northern Africa or even in western Asia and are thus distinct from the Mediterranean elements. In the Peninsula, the Aryan *elements are on the whole subtropical rather than tropical forms that are best* developed in the parts with moderate rainfall. Some of them do not indeed occur in the humid extreme south or in Ceylon, but are largely confined to the grassy and bush-covered plains, with scattered trees.

4. The Younger Intrusive Elements

The younger intrusive faunal derivatives belong largely to the Tertiary-mountain forest-faunas of Indo-China and Malaya. These derivatives correspond, at least in part, to the Trans-Gangetic component of BLAN-FORD (1901). BLANFORD considered therefore the region of R. Ganga as marking the transition, between the autochthonous Indian and intrusive elements. They are Extra-Peninsular elements. It is not always easy to decide, in the present state of our ignorance of fundamental biogeographical problems, whether a given genus or species is of Indo-Chinese or of Malayan origin. As a matter of fact, these two faunas have intermingled in a most complex manner in Assam and Burma and are also often ecologically isovalent. It was this complex of the two-faunas that spread westwards along the Himalayan forest and southwestwards to the Peninsula and it is the impoverished relics of this complex that we find today in the Southern Block, in discontinuous distribution. Moreover we cannot also ignore certain amount of infiltration of the Malayan area by northwest Australian forms and it is impossible to define the extent of the northward penetration of this part into Assam-Burma. We know, however, that this element is present in Malaya and Indo-China. The Extra-Peninsular eastern intrusives are thus not simple or biogeographically monovalent. It is not, therefore, to be wondered that most earlier zoologists, who have devoted any attention to the study of the presence of this eastern floristic-faunistic complex in the Peninsula, have completely failed to clearly distinguish between the strictly Indo-Chinese and the Malayan flora-faunal derivatives, but have often rather loosely used the two terms as if they were synonyms. Most of the socalled Malayan elements described by HORA and his collaborators (HORA, 1944, 1949; MENON, 1951; ROONWAL & NATH, 1949; SILAS, 1952) in the course of their discussions on HORA's Satpura hypothesis are strictly speaking Indo-Chinese fishes and have very little or no Malayan affinity. This confusion has resulted in several misconceptions and has largely obscured real affinities.

The South Chinese, East Tibetan and the Indo-Chinese humid tropical forest faunal elements comprise the impoverished representatives, with pronounced distributional discontinuity, of mostly Tertiary and some Post-Tertiary derivatives that have colonized the Peninsula from the northeast and have come to be isolated from their main homeland mainly during and even after the Pleistocene times. The members of this branch belong mostly to the higher and recent groups of the animals and not many of them have therefore given rise to localized endemic genera and species within the Peninsula, since the Pleistocene times.

The Malayan faunal derivatives also entered the Peninsula from the northeast. Part of the Malayan fauna is of Tertiary origin and part of it should be properly included in the Gondwana-Oriental line of older origin. The genera and species are thus mostly older and more primitive than those of the Indo-Chinese derivatives. The distribution of these elements is also characterized by striking discontinuity. Like the Gondwana derivatives, the Asiatic eastern intrusive faunal derivatives were also formerly far more widely and continuously distributed in the Peninsula than at the present time. Nearly every group of the humid tropical Asiatic fauna, which spread into the Peninsula, was characterized by continuous distribution. Even the groups, many of which are at present conspicuously discontinuously distributed, were formerly continuously distributed. KURUP (see Chapter XVIII) has for example, recently shown that Oriental mammals occurred formerly continuously from the Southern Block to Assam. The present-day discontinuity in mammalian distribution is no more than a relict of a former continuous distribution. The Asiatic faunal derivatives constitute the bulk of the geographical relicts characteristic of the Peninsula.

Soon after the Assam-contact with Asia arose in the course of the Himalayan uplift, the Peninsular autochthonous fauna spread to the foot of the rising Himalaya in the east. During the Pleistocene glaciations on the Himalaya, the Peninsular autochthonous fauna extended south on the meridional mountains of Burma to reach outposts like the Andamans and the Malaya Peninsula. The rest of the Peninsular spill-over in the Trans-Gangetic area north of the Peninsula retreated south during the Pleistocene glaciations on the Himalaya. This history is recapitulated today in the winter-monsoon oscillation north and south, described in Chapters V and VI. At the end of the glaciations, the Peninsular autochthonous fauna never regained the lost ground, but the Tertiary mountain forest fauna gradually spread westwards along the Himalaya and into the Indo-Gangetic Plains.

4.1. THE INDO-CHINESE FAUNAL DERIVATIVES

The Indo-Chinese faunal derivatives are younger than the Peninsular autochthonous and southern derivatives. They are, however, on the whole far more abundant than the Malayan elements and are also more widely distributed. They belong to higher taxonomic groups and to more specialized genera. Common examples are found among Lepidoptera, fishes, reptiles and birds. The Indo-Chinese faunal derivatives, being Tertiary mountain autochthonous forms, spread primarily westwards

631

along the foothills of the Himalaya and thence to the Peninsula. The outliers reached even the fringe of the southern Palaearctic to some extent. They occupied part of the Ganga Plains. These elements are represented as fossils in the Lower Miocene and Oligocene of Europe, but not later. We have, for example, the fossils of *Hylobates* and *Gymnura* in Europe. In the westward radiation along the Himalayan forests, the Indo-Chinese intrusive elements gradually diversified and differentiated into local sub-species and species in a number of the relatively young and plastic groups, particularly in the phylogenetically recent ones like Lepidoptera and Diptera among insects. Such a diversification occurred to a rather limited extent even in the forms that reached the Peninsular west and south, but to a significantly lesser degree than in the Himalayan forests. We have referred to these locally differentiated forms in the Chapter on the Himalaya. Compared to the Himalaya, the intensity of speciation and infra-specific diversification among the intrusive elements is extremely low in the Peninsula. Most local subspecies in the Peninsula represent secondary colonization of the Peninsula via the Himalaya during the Pleistocene times rather than direct from the Assam area.

Lepidoptera of the Peninsula contain many examples of the Indo-Chinese elements. *Discophora sondaica* occurs in South India, Sikkim, south-east Tibet, Burma, Thailand, South China, Hainan, Malaya, Java, Sumatra, Borneo and the Philippines. *Appias indra* from South China, Hainan, Borneo, Malaya, Burma and Formosa has spread along the Himalaya to Nepal and has entered the Peninsula and reached even Ceylon. *Appias libythea* from Malaya, Nicobar, Philippines and Burma-Assam occurs in the Peninsula and Ceylon. *Appias lyncida* from Malaya, Timor, Philippines and Burma has spread along the Himalaya to Sikkim and occurs in the Peninsula and Ceylon. *Graphium sarpedon*, from south Japan, China, Soloman Is., extends by a subspecies *Graphium sarpedon sarpedon* from Burma to Kashmir on the Himalaya and another subspecies *Graphium sarpedon teredon* occurs in the Peninsula and Ceylon. *Graphium antiphates* from China Sunda Is. and North Borneo is represented by *Graphium antiphates pompilius* in Hainan, Annam, Thailand, Burma and Sikkim and by the subspecies *Graphium antiphates epaminondas* in the Andamans, by *Graphium antiphates naira* in the Peninsula, and by *Graphium antiphates ceylonicus* in Ceylon. *Graphium doson* from South Japan, South China and Sunda Is. extends through Bengal to South India and Ceylon on one side and to the Kumaon Himalaya on the other side. *Graphium agamemnon* of South China, Australia, Soloman Is. and Burma has spread along the Himalaya to Kumaon and is represented by *Graphium agamemnon menides* in the Nilgiri Hills and in Ceylon. The Chinese-Malayan *Polydorus artistolochiae* is represented by *Polydorus aristolochiae ceylonicus* in Ceylon. The Malayan-Indo-Chinese *Chilasa* is represented by *Chilasa clytra clytra* in the Peninsula; this form has also spread westward into North India and Northwest Himalaya. In Ceylon we have another form of this species, viz. *Chilasa clytra lankeswara*. The Malayan-Chinese *Papilio pario* is represented by *Papilio pario tamilana* in the Peninsula up to the Nilgiri Hills. The Indo-Chinese-Malayan *Cepora* extends from Assam westwards along the Himalaya to Nepal and also occurs in the Peninsular south and Ceylon.

We may also refer to some of the Indo-Chinese elements of other groups of insects in the Peninsula. Lampyridae: *Lamprophorus* extending from Yunnan to Ceylon and Sumatra and Java. Cleridae: *Gastrocentrum* occurs from the Philippines, Java, India and Ceylon. *Xenorthrius* is known from China, Borneo and India. Chrysomelidae: *Prioptera* extends from China southwards to the Andaman Islands, Indonesia to Celebes, eastwards to Philippines and Formosa and occurs also in the Peninsula. Cerambycidae: *Cyrtonops* extends from Tonkin to Sumatra and Borneo and India. *Nyphasia* extends from Cochin-China and Thailand to Burma and westwards to the Peninsula and Ceylon. *Leprodera* occurs in South China, Java, Sumatra, Borneo, India and Ceylon. Scarabaeidae: *Heterorrhina* occurs from the Philippines across China to Borneo, Nias, Sulu Islands, Java, Sumatra, India and Ceylon. *Coryphocera* occurs in Philippines, Celebes, Andaman Islands, India and Ceylon. *Thaumastopeus* is represented by over twenty species from China to Indonesia eastwards to Timor, Damar, Philippines, India and Ceylon.

A number of fishes and reptiles from the Indo-Chinese fauna are also met with in the character fauna of the Peninsula. *Osteocheilus (Osteocheilichthys)*, with about half a dozen species in the Trans-Gangetic area, is represented by three species in the Peninsula (Fig. 106). *Cirrhinus*, with about four species in Thailand, is represented by three species, which are restricted to the Peninsula. The species of *Scaphiodon* from the Peninsula are closely related to *Osteocheilus* rather than to *Scaphiodon* proper and the Peninsular species belong to the subgenera *Osteocheilichthys* and *Kantaka*. The South Indian forms are more closely allied to the south Chinese and Siamese forms than to those from the Malayan Archipelago. *Cyprinus nukta* from the Western Ghats and *Schismatorhynchus heterorhynchus* from Sumatra and Borneo are believed to be related. The Peninsular Homalopteridae *Bhavania*, *Travancoria* and *Bolitora* are interesting (Fig. 100). The first named two genera are endemic and the third occurs also in Burma, Assam and the East Himalaya and in the Peninsula. *Balitora brucei* occurs in both these areas and is represented by different varieties in Burma and in the Peninsula. *Silurus cochinchinensis* (Fig. 104) occurs in Assam-Burma, East Himalaya and further to the east; *Silurus wynaadensis* comes from the Peninsula, but these two species have been recently been shown to identical, so that *S. cochinchinensis* extends in southeast Asia and in Peninsula (Western Ghats). *Batasio* occurs in Travancore and the East Himalaya (Fig. 105). *Mystacoleucus* is represented by several species in the Malayan tract and by one species in the Peninsula. *Lepidopygopsis*, believed to be a Middle Asiatic derivative that occurs on the south slopes of the Himalaya, is also found in Travancore; Schizothoracinae are not otherwise found in the Malayan area (Fig. 99). *Monotretus* occurs in Travancore and also in Orissa, Bengal and Assam.

The terrestrial chelonian *Geoemyda tricarinata* from the Assam hills, occurs in the Chota-Nagpur Plateau of the Peninsula. *Geoemyda trijuga* is represented by the subspecies *G. trijuga trijuga* in the Peninsula from Bombay to Malabar, *G. trijuga indopeninsularis* in the Chota-Nagpur Plateau, *G. trijuga thermalis* in Ceylon and southern parts of the Peninsula (Fig. 112). The northeastern genus *Kachuga* is represented by *K. tectumtentoria* in the Mahanadi and Godavari river systems in the Peninsula. The northeastern *Testudo* is similarly represented by *T. travancorica* in the southwest of the Peninsula and in Coorg. *Lyssemys punctata granosa* is another Trans-Gangetic form in the Peninsula and Ceylon.

The following Trans-Gangetic birds occur in the Peninsula, but not in Ceylon: *Eurostopodus bourdilloni, Chaetura indica, Nyctiornis athertoni athertoni, Buceros bicornis, Dinopium malabaricum, Hemicircus cordatus cordatus, Brachypteryx major* with two subspecies, *Myriophoneus horsfieldi, Garrulax chinnnas* with five subspecies, etc.

The Malayan elements in the Peninsular character fauna are younger than the Gondwana faunal derivatives but somewhat older than the Indo-Chinese elements. They are also less dominant than the Indo-Chinese elements. The greatest bulk of the Malayan forms in the Peninsula belong to the more recent and higher groups, especially Mammals.

The following are among the more important examples of the Malayan elements in the character fauna of the Peninsula: Oligochaeta: *Megascolex, Diprochaeta, Woodwardia,* etc. Scorpions: *Chaerilus;* Uropygi: *Thelyphonus, Trithyreus.* Dermaptera: *Echinosoma, Hypurgus, Cordax, Adiathetus, Gonolabis;* Phasmodea: *Presbistus* extending from Sumatra, Java, Borneo, India and Ceylon, with about fifteen species. *Phobaeticus* is represented by about ten species in Sumatra, Borneo, Timor and India. *Cuniculina* has over forty-five species in Sumatra, Java, Borneo, Andaman Islands, South China, India and Ceylon. *Clitumnus* has also about forty species from Indonesia to Philippines, Moluccas, New Guinea, India and Ceylon. *Stheneboea* with about twenty species extends from Sumatra, Java, Borneo, Celebes and Philippines to India and Ceylon. *Prisomera* has three endemic species in Ceylon, one endemic species in Malabar and is represented by about thirty species in Malaya and Indonesia eastwards to Philippines. *Pharnacia* has over twenty species in Java, Sumatra, Borneo, Philippines and also extends to Assam and eastern parts of the Indo-Gangetic Plains and occurs again in Ceylon. *Pharnacia serratipes* occurs in Borneo, Malaya and in Malabar and is one of the largest Phasmids and measures 330 mm long (female). *Pharnacia ingens* from upper Tenasserim and Malabar measures 260 mm long, is an apterous, slender stick-insect. *Tachythorax* occurs in Java, India and Ceylon and in India is mainly found in South India and in the east. *Sosibia* occurs in Sumatra, Nias, Mentawei, Borneo, Philippines, northeast India and South India and Ceylon and *Ocellata* with about twenty species extends from Tonkin to Malacca, Sumatra, Java and Borneo and has one endemic species in Ceylon. *Lonchodes* extends from Japan and China to Philippines, Borneo, Java, Celebes, New Guinea and South India and Ceylon. Chrysomelidae: *Hymenesia, Chalcolampra, Euphitraea, Podontia, Ophridia,* etc. Scarabaeoidea Coprinini: *Anoctus* and *Phacosoma*. Among fishes we should mention *Thynnichthys* (Fig. 102). A number of birds in the Peninsula are also from the Malayan region: the spider-hunters Arachnotherinae *Pitta brachyura,* the woodpeckers *Thriponax* and *Tiga,* the broad billed-roller *Eurytomus orientalis,* the bee-eater *Nyctiornis,* hornbills *Dichoceros* and *Anthracoceros,* the large-eared night-jar *Lyncornis cerviniceps,* Psittaci *Loriculus,* etc. Mammals include *Viverra zibetha* (Fig. 139) and *Moschothera* (= *Viverra*) *civettina* (the former is found in the Orissa Hills and in Assam-Burma and the latter is restricted to the Malabar Coast), *Charrodnia* (= *Mustella*) *gwatkinsii* from the South Indian hills, *Sciuropterus macrurus* restricted to South India and Ceylon, *Semnopithecus, Paradoxurus, Pteropus, Cynopterus, Pteromys, Elephas, Cervulus,* etc.

5. The Palaearctic Elements

The Palaearctic forms that differentiated in the Mediterranean sub-region, in southwest Asia, southeast Europe and north Africa and in the Turkmenian subregion, Middle Asia, entered India from the northwest and have sparsely colonized the hills of the South India. They belong mostly to the higher and younger groups. A part of the Palaearctic elements of the Peninsular character fauna is of the higher Himalayan origin and represents Pleistocene relics and the rest is Mediterranean intrusive element. We have already dealt with the typical Himalayan isolates among the Peninsular plants (see Chapters VI & VIII). We shall consider here some of the typical Palaearctic animals. The Palaearctic elements constitute an insignificant component in the character fauna of the Peninsula but present certain interesting problems of distribution. They are mostly restricted to the higher mountains of South India, and the mountains of Ceylon, but some extend far north along the Western Ghats. The scorpion *Buthus* of the Mediterranean and Ethiopian region is found up to the Deccan Plateau. *Butheolus* of the Eastern Mediterranean entered from the northwest likewise and is found on the Western Ghats in the Deccan area. The Mediterranean *Galeodes,* found in the Peninsula, is absent in Ceylon. The Carabid *Harpalus* is confined to the Himalaya, but *Harpalus advolans* occurs in South India and Ceylon. Similarly the genus *Anchomenus* is Himalayan in India, but is sparsely represented in

Ceylon. *Anthia* is another Carabid of the Mediterranean Region, with numerous species in Africa, a few species in Arabia to Iran and the Turkmenian Subregion, which is represented by *A. sexguttata* widely distributed in the Peninsula and transgressing into the Indo-Gangetic Plain. The Buprestid *Julodis* has numerous species in the Mediterranean and Turkmenian subregions and in the Ethiopian Region and is sparsely distributed in the Peninsula. The Tenebrionid *Stenosis* has about fifty species in the Mediterranean and Turkmenian subregions and in the Ethiopian Region but only a few species in the Peninsula. Other examples include *Hemitragus hylocrius* found from the Nilgiri to the Cape Comorin, while the other species of this mammal in India are restricted to the Himalaya (Fig. 137). Mention may also be made of *Hyaena*, *Mellivora*, *Gazella*, *Pterocles*, *Eupodotis*, *Phaenicopterus*, etc.

6. Distributional Patterns

The outstanding characters of the distributional patterns of animals in the Peninsula at present may be summarized as follows: 1. a high degree of localized concentrations of all the component elements: 2. more or less complete, intense and wide isolation; 3. marked discontinuity; 4. almost complete absence of altitudinal zonation of species in a region stretching from sea level to elevations of nearly 2750 m above mean sea level in the Peninsular plateau; 5. progressive limitation eastwards and northwards; and 6. a total obliteration of geographical radiation. These peculiarities are not associated with ecological and faunal climax trends and distributional stability, but must be correlated with retrogressive distributional changes and departure from stability. The retrogressive trends in the distributional patterns are, however, of very recent origin, indeed within historical times. The Peninsular fauna was, even after the Mahabharata War, far more widely and continuously distributed than at present. Concentration and isolation were either not pronounced or also almost completely absent.

The Peninsular fauna shows progressive limitation in numbers, northwards within the Peninsular block and outwards into islands. The limits of the Peninsular fauna lie at present in the transitional zone of R. Ganga in the north and practically about the Mahanadi Basin in the northeast. As mentioned earlier, however, these limits lay formerly in the foothills of the Himalaya and overlapped marginally with the areas to the east of the Bay of Bengal, in Burma, Malaya and the Andamans, as evidenced by the presence of Peninsular isolates in the Andaman and Nicobar Islands. Some of the Peninsular forms in these islands are either closely related to or are identical with the species found in Ceylon. For example, *Cnemaspis kandiana* (Fig. 117), the geckonid lizard, occurs in South India, Ceylon, Preparis Is. between Burma and the Andamans, Simalur, Nias, Sipora and Engano (a chain of Islands along the west coast of Sumatra) and in Sumatra. The Agamid *Calotes andamanensis* from the Andamans is closely related to *Calotes lineatus* from Ceylon. *Calotes opheomachus* occurs in South India, Ceylon and the Nicobars. *Calotes mystaceus* occurs in South India, Ceylon, Burma, Siam, Andamans and Nicobars and *Calotes versicolor* in India, Ceylon, Andamans, Nicobars, South China, Malaya and northern Sumatra. Although it is possible that some of these Peninsular forms have been introduced into the areas to the east of the Bay of Bengal by the early sea-faring men of Chola kingdom from South India within historical times, the bulk of the Peninsular elements reached these outposts in the natural course of dispersal during the Pleistocene glaciations in the Himalaya. It is interesting to observe that the older members of the Peninsular fauna have their limits much farther northeast than the younger members.

A striking feature of the present-day distribution of animals in the Peninsula is the heavy concentration of the character forms in the extreme southwest in the Southern Block and in certain other areas like the Chota-Nagpur Plateau. These concentrations do not represent the centres of faunal differentiation and radiation, but refugia – they are the result of disappearing habitats – areas of the concentration are precisely the places where the original forest cover has not yet been completely destroyed. These areas of concentration are, therefore, refugial centres or niches and in effect only islands of favourable conditions to which the Peninsular character fauna has retreated. Isolation that underlies an intensive speciation and infra-specific diversification among nearly all the groups on the Himalaya is not, however, associated with anything similar or even remotely comparable in the case of the Peninsula. Here we find on the other hand almost complete absence of speciation among the phylogenetic, geographical and Pleistocene relicts now found isolated in the refugial pockets in the Peninsula. Concentration and isolation are thus symptoms of faunal regression. From the biogeographical point of view, the fauna of the Peninsula are composed almost exclusively of phylogenetic and geographical relicts. The fauna represent essentially the impoverished remnants of a vanished fauna.

The discontinuity, which is perhaps the most striking character of the distributional patterns of the present-day Peninsular fauna and has consistently been misinterpreted by all earlier workers, presents the following features: 1. distribution in the areas to the east of India, particularly in Indo-China, Thailand, Burma and Malaya and in Ceylon; 2. distribution in the east and in the extreme southern corner of the Peninsula, but not in Ceylon; 3. distribution in Assam and in Ceylon; 4. distribution in Assam and in the Peninsular south; 5. Distribution in Assam, Eastern Ghats, particularly in the northeast corner of the Peninsular plateau, Deccan and in the Peninsular south; 6. distribution in upper montane-forest zone and lower temperate zones of the Himalaya and in the Peninsular horsts, including sometimes also Ceylon hills; and 7. distribution in the Himalaya and in the hills of Ceylon, but not in the Peninsula (see Chapter XX).

7. Faunal subdivisions

Most earlier attempts at subdividing the Peninsula wholly on the basis of the present-day distributional patterns have generally ignored the dynamic evolutionary changes in these patterns and have assumed that these patterns have remained unchanged. Furthermore, most of the subdivisions proposed by diverse authors have also been based on the distribution of single and often ecologically highly specialized groups, like for example hill-stream fishes or Lepidoptera. The subdivisions now known do not really describe faunal differentiation and radiation and are in a sense artificial climatic and indeed merely monsoon-rainfall distributional subdivisions, rather than true biogeographical subdivisions. A natural floristic-faunistic subdivision of the Peninsula is in fact not strictly possible, mainly because the ecosystems, distributional patterns and compositions have been very profoundly altered by man, so that the picture is wholly artificial. Despite its obvious imperfections, the subdivisions proposed by BLANFORD (1901) approach on the whole more nearly a natural scheme than those of others. BLANFORD subdivided the biogeographical area of the Peninsula into five tracts, viz. 1. Rajputana, 2. Deccan, 3. Malabar, 4. Carnatic and 5. Bihar-orissa. It may be observed that with the exception of the Bihar-Orissa tract, the differences of the animals occurring in these tracts are largely specific rather than generic.

The Rajputana or the Central India tract of BLANFORD includes Rajasthan uplands, Kathiawar, areas to the south and southeast of the R. Ganga as far south as the Narmada river and east to 80° EL. The tract is largely undulating and hilly land, mostly cleared of the forest, partly jungly and covered with brushwood.

The Deccan tract extends from the R. Narmada to 16° NL and from the neighbourhood of the Western Ghats to 80° EL. Most of these areas have been cleared of the forest, but there are locally thin forest-brushwood and grass in the more hilly localities. BLANFORD believed that Mysore and the Nilgiri Hills should be included here.

The Malabar tract includes the Western Ghats and the western coast-lands of the Peninsula, from the R. Tapti to the Cape Comorin. The northern portion is known as Konkan and the southern part as Malabar. Though the coast has mostly been cleared of forest, there are still some areas, which retain the dense forest cover. The Malabar tract is richer than any other part of the Peninsula, both in genera and species. BLANFORD has listed, for example, forty-eight genera of Mammals, of which, as already pointed out, *Platacanthomys* (Fig. 134) is peculiar to the Peninsula. The Himalayan *Mustella*, *Harpiocephalus* (Fig. 140), *Sciuropterus* and

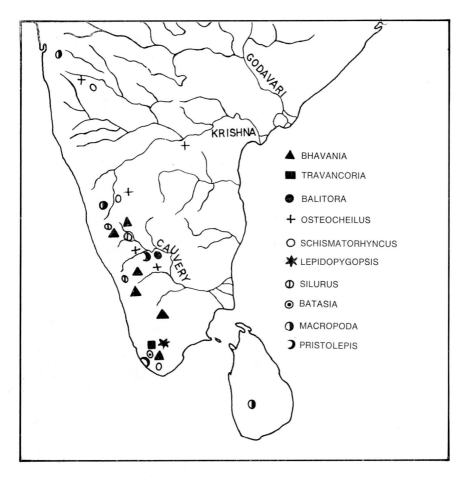

Fig. 150. Distribution of fresh-water fishes in the Peninsula, suggesting certain natural concentrations in a northern, a middle and a southern section of the Western Ghats, corresponding approximately to different drainage basins.

Hemitragus occur in this tract. *Mellivora, Antilope* and *Gazella* are lipotypes. *Viverra* (Fig. 139) occurs in this tract and also in the Bihar-Orissa tract. BLANFORD has also listed about two hundred and seventy-five genera of birds, twenty-eight of which do not occur elsewhere in the Peninsula. The Himalayan types like *Rhopocichla, Brachypteryx, Schoenicola* and *Ochromela* occur in the Malabar tract. About a dozen genera of birds, for example *Garrulax, Rhopocichla, Irena, Schoenicola, Eurystomus, Collocalia, Batrachostomus* (Fig. 159), *Loriculus, Huhua, Ictipaëtus, Baza* and *Gorasachius* found in the Malabar tract extend also to Ceylon. Of about sixty genera of reptiles listed by BLANFORD, *Hoplodactylus, Salea, Pseudoplecturus Plecturus, Melanophidium, Platyplecturus* and *Xylophis* do not occur in other

639

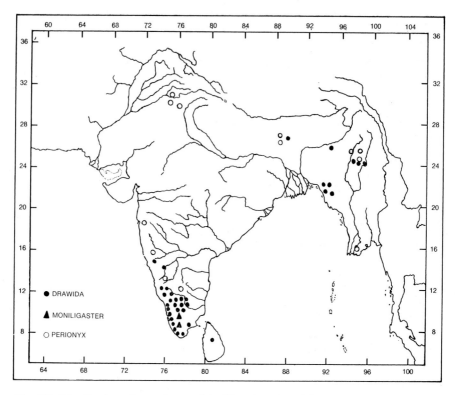

Fig. 151. Distribution of some interesting Oligochaeta in India.

parts of the Peninsula and *Otocryptis, Dendrelaphis, Gerardia, Chrysopelea* (Fig. 128) and *Ancistrodon* extend also into Ceylon. The Amphibia *Nyctibatrachus, Nannobatrachus, Melanobatrachus, Nectophryne* (Fig. 110) and *Gegenophis* are peculiar to the Malabar tract. It may be recalled that *Uraeotyphlus* occurs in the Malabar tract and also in Africa. BLANFORD lists also some fresh-water fishes as peculiar to the Malabar tract. In a recent paper, BHIMACHAR (1945) has described some of the distributional peculiarities of fresh-water fishes in the Malabar tract, on the basis which he proposes subdividing the Western Ghats into a northern section of Deccan-Trap area from the R. Tapti to about 16° NL, a middle section up to Nilgiri Hills and a southern section of Anamalai, Palni and Cardamom Hills. The middle and southern sections are characterized by a larger number of species and high degree of species endemism. His northern division (Fig. 150) contains fishes like *Schismatorhynchus*, known also from the Malay Archipelago, and *Mystacoleucus*, occurring also in Burma, Thailand and Malaya. *Osteocheilus, Garra* and *Thynnichthys* are other fishes found in this section. *Balitora, Bhavania, Travancoria, Pristolepis, Silurus* and *Batasio* are absent in the northern section. The middle section

640

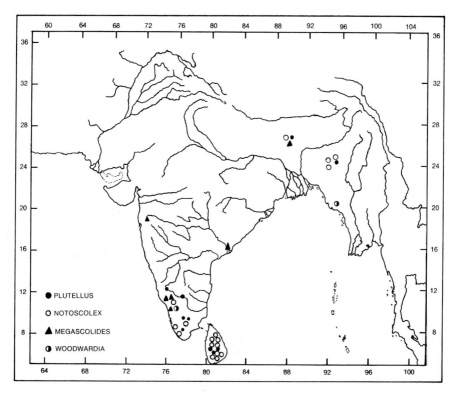

Fig. 152. Distribution of some interesting Oligochaeta in India.

has the same species of *Silurus* as the eastern Himalaya and Cochin-China. *Balitora brucei mysorensis* occurs here and *Balitora brucei brucei* in Burma. Species formerly placed under *Scaphiodon* are *Osteocheilus*, also occur here and in southeast. *Schismatorhynchus* is also found. The species from the northern and middle sections are, however, absent in the southern section, but *Bhavania* occurs both in the middle and southern sections. The southern section has endemic fishes like *Travancoria* and *Batasio*. He describes great dissimilarity between the fishes of the middle and southern sections and he points that there is great affinity to the fishes of Ceylon and the Peninsular south. He attributes the dissimilarity between the middle and southern sections to the presence of the Palghat Gap, acting as a barrier. This is however only a vague recognition of the refugial concentration in the extreme southwest, which we have mentioned earlier.

In the Malabar tract there are a number of interesting endemic species of the Oligochaeta: *Moniligaster, Drawida, Woodwardia, Spenceriella, Plutella, Comarodrillus, Megascolides, Notoscolex, Megascolex, Pheretima, Diprochaeta, Perionyx, Howascolex, Octochaetus, Dichogaster, Curgia, Gordiodrilus* and *Glyphidrilus*.

641

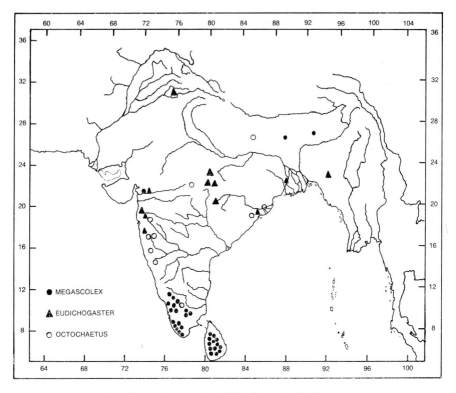

Fig. 153. Distribution of some interesting Oligochaeta in India.

Ceylon is divided by BLANFORD into a Northern Ceylon and a Hills-Ceylon tract. The former covers nearly three-fourths of the island, comprising undulating plains of the northern and eastern parts of Ceylon. This is really a continuation of the Carnatic tract of the Peninsula, though with somewhat higher rainfall. The hill tract comprises the Central, Western and Southern Provinces of Ceylon and is a part of the Malabar tract. Some workers have suggested combining the hill tract of Ceylon with the Malabar tract to form a distinct Ceylonese subregion, separate from the rest of the Peninsula. This suggestion overlooks the regression of the faunal area of the Peninsula and fails to recognize the refugial nature of Malabar and Ceylon. The occurrence of typical Malabar forms like the Uropeltidae in the scattered hills of the Peninsula as far as the Orissa-Hills tract shows, however, that there is no justification for separating Malabar and Ceylon from the rest of the Peninsula. Ceylon was connected with the Peninsula even during the late Pleistocene times and the present-day pattern of distribution of animals is indeed wholly consistent with this fact. Many of the species found here are the same as those in the south of the Peninsula, but some are endemic to the island.

Different Peninsular and Extra-Peninsular forms are absent in Ceylon for different reasons. Ceylon lacks the wolf and wild-dog, but has the jackal, it has the sloth-bear but lacks tiger, though the leopard is present. Rhino is absent but elephant is present. During the Pleistocene times Ceylon had giraffe, hippopotamus, rhinoceros, elephants and the lion. There are more geographical *relicts* in Ceylon than in the mainland of the Peninsula.

In Ceylon we find *Drawida friderici, Plutellus halyi, P. singhalensis, Pontodrilus agnesae, Woodwardia sarasinorum, W. uzeli, Notoscolex ceylonensis, N. crassicystis, N. dambullaensis, N. decipiens, N. gracelyi, N. jacksoni, N. kraepelini, N. termiticola, N. trincomaliensis, Megascolex acanthodriloides, M. adami, M. bifoveolatus, M. brachyclus, M. caeruleus, M. campester, M. ceylonsis, M. cingulatus, M. escherichi, M. funis, M. hortonensis, M. insignis, M. kempi, M. leucocyclus, M. longiseta, M. lorenzi, M. multispinus, M. nureliyensis, M. pattipolensis,* and *Perionyx ceylonensis,* etc. Mollusca confined to Ceylon include *Acavus, Corilla* (with a single species in South Travancore) and *Aulopoma; Nicida* is common to Ceylon and South India.

Of the birds peculiar to the island, mention may be made of *Elaphrornis, Kelaartia, Dissemurulus, Sturnoris, Acmonorhynchus* and *Phaenicocphaës. Cissa* is a genus of magpies peculiar to Ceylon, but the other forms are of Trans-Gangetic area. There are a number of Trans-Gangetic genera in Ceylon, which do not, however, occur in the Peninsula. The reptiles peculiar to the island include *Ceratophora, Lyriocephalus, Chalcidoseps, Uropeltis, Haplo-cercus* and *Aspidura.* The first three mentioned above and *Cylindrophis* occur also in Malaysia (Fig. 125). There are four species of *Acontius,* but none in India, Burma, etc. and some occur in Madagascar and South Africa. *Otocryptis* occurs only in Ceylon and Malabar. The amphibian *Nannophrys* is restricted to Ceylon and *Micrixalus* occurs in Ceylon and Malabar. The passerines *Rhopocichla* and *Schoenicola* are common to Ceylon and Malabar.

The Madras Littoral area is the Carnatic or the Madras tract of BLANFORD and is characterized by the absence of *Gazella,* migratory birds, etc. A number of genera and species characteristic of the Malabar tract are, however, found sparsely on the scattered hills within this tract. Though not observed in Deccan and the Bihar-Orissa tracts, *Erinaceus* occurs here. *Callodactylus* is peculiar to this tract and two other reptiles, *Lepidodactylus* and *Xenopeltis* are not also found elsewhere in the Peninsula.

The Bihar-Orissa tract of BLANFORD includes the Chota-Nagpur Plateau and the eastern parts of the old Central Provinces (now Madhya Pradesh) and parts of the area between the R. Ganga and the R. Goda-vari. Although the forest cover is by no means as dense as formerly, this area has not suffered deforestation to such an extent as other parts of India. This includes in part the *sal (Shorea robusta)* area of the Eastern Ghats (see Chapter VIII). We find here mammals like *Viverra* and *Ratufa (Sciura) indica,* and a number of birds which are absent in Rajasthan and Deccan, but are present in the Malabar tract and sometimes also in the hills of the Carnatic tract. These include the passerine birds *Chibia*, *Dissemurulus* (also sparsely in Deccan) *Eulabes*, *Cittocincla* and *Oreocincla*, the bee-eaters *Melitophagus* and *Nyctiornis*, the hornbills *Anthracoceros*, the trogon *Harpactes*, the cuckoos *Penthoceryx* and *Surniculus*, the Accipitrine *Lophotriorchis*, *Ictipaëtus* and *Baza*, and the pigeons *Carpophaga*, *Osmotreron*, *Chalcophaps*, and *Alsocomus*. The great majority of genera of nearly all groups found here are typically forest forms and the species of the Bihar-Orissa tract are, with perhaps few exceptions, the same as those of the Malabar tract. These species formerly inhabited the whole of the southern India, before the forests of the Carnatic were destroyed. This is evident, for example, from the distributional pattern of the hornbill *Anthracoceros*, of which one type occurs in Ceylon and the Western Ghats as far north as Ratnagiri, and another type occurs in the lower Himalaya and the areas to the east, but neither of them occurs in the Deccan or the Carnatic tracts, although the two types meet in the Bihar-Orissa tract. Unless it formerly ranged over the whole region, including the areas intervening the Malabar and the Bihar-Orissa tracts, the occurrence of this southern species would be meaningless. The southern grackle *Eulabes religiosa* meets the Himalayan and Burmese grackle *Eulabes intermedia* in the Bihar-Orissa tract (but is absent in the Deccan and the Carnatic tracts). The differences between the Bihar-Orissa tract and the adjoining tracts of the Peninsula are of recent origin; this is strongly supported by the absence of distinctive reptilian and amphibian genera. The scincid *Sepsophis* is peculiar to the Bihar-Orissa tract, but it occurs also on the Golkonda Hills along with the Uropeltid *Silybura ellioti*. We have here an example of isolated reptilian and amphibian genera, as we observe in the south of the Peninsula and Ceylon.

To summarize, we may consider the following subdivisions are nearly reflecting the biogeographical character: 1. the northwest subdivision corresponding to Rajaputana of BLANFORD; 2. the northeast to include the Chota-Nagpur plateau; 3. the Coromandel or the Carnatic (the East Coast), including the eastern parts of Ceylon; 4. Malabar including most of the hilly parts of Ceylon; and 5. Deccan comprising the northern and central areas.

8. The Laccadive and Maladive Islands

The Laccadive and Maladive islands are nearly all small coral and oceanic islands, characterized by poor fauna and perhaps lacking endemic genera. Nearly seventy species of Lepidoptera (moths) have been collected on these islands, of which two species appear to have been recorded so far only from the Minikoi Island of the Laccadive group. One of these is a Tortricid and the other is Pyraustid closely related to *Notarcha multilinealis*, widely distributed on the mainland of the Indian

Peninsula. Most other Lepidoptera found on these islands have no doubt been derived from the mainland of India and Ceylon. Nearly sixty species of Coleoptera are known from these islands and twenty species of Heteroptera but no Homoptera seems so far to have been found. These species are, without exception, derived from either the mainland of Indian Peninsula or from Ceylon. Most of the Orthoptera found in these islands are cosmopolitan species or are widely distributed in the Oriental Region.

9. The Seychelles Islands

The Seychelles Islands were formerly considered to belong to the Madagascan Region, but recent evidence seems to show conclusively that they belong to the biogeographical area of India. There is much greater faunistic affinity to the Oriental than to the Madagascan Region. Most of the insects from these islands were described by KOLBE (1910) and the PERCY-SLADEN Expedition (1907, 1925) collected a large number of specimens from these islands. It is not, however, still clear whether these islands are to be considered as oceanic or continental. The sparse occurrence of such insects like Carabidae, Scarabaeidae, Locustodea, Acridodea, etc. must be attributed to faunal intrusions across the seas. It is, however, difficult to explain the presence of Pselaphidae, Scydmaenidae, Tenebrionidae, Gryllodea and other groups of insects in these islands, but apparently most of these seem to have been derived from the mainland of India. There is an abundance of endemic species in the islands and also some endemic genera, which are, however, isolated in position systematically.

There is an endemic variety *seychellensis* of the widely distributed *Cicindela melanocholica*. The important Carabidae are *Ophinea*, *Pentagonica mahena*, *Tachys seychellarum* and *Anillus* spp. Some of these species may have been introduced by man. Two species of Dytiscidae *Copelatus gardineri* and *C. pandanorum* are endemic in these islands. A large number of Staphylindae are known, and nearly thirty-five species of these seem to be restricted to these islands in their distribution and may perhaps be endemic. The species are, however, closely related to the Oriental forms. Of about a score of Pselaphidae from the Seychelles, *Thesiastes cordicollis* is common to Zanzibar. The species belong to 12 genera, of which four are confined to the islands, one genus is common to Africa and the Oriental Region and all the remaining genera are distributed widely in the east, as far as the Melanesian area. *Scydmaenus armatus*, found in these islands, is also known from the Oriental Region. Most other species of Scydmaenidae belong to genera, endemic in the islands, but *Cephennium*, *Euconnus* and *Scydmaenus* are widely distributed. *Neseuthia*, with seven species, and *Stenichnoteras*, with a single species, are found only in the Seychelles. Even in the case of Buprestidae the Indian affinity is stronger than the Madagascan and African. Nearly half the Tenebrionidae found here are endemic. The Scarabaeid *Parastasia coquereli* from the Seychelles is closely related to a species from Ceylon. Most other Scarabaeids found here seem to have been introduced by human agency. Among Orthoptera there is in Seychelles the Tettignoid *Systeloderus*, which is represented otherwise only in the Oriental Region, but the genus *Peoedes* has one species each in the Seychelles and in Madagascar. The Tettigonid *Rhynchotettix* is so far known only from the Seychelles. The locustid *Allodapa* occurs only in the Oriental and Seychelles. The Phasmodea found here are *Phyllium*, *Carausius* and *Graeffea*. All the species of *Carausius* are apterous and are related to the species occurring in the Oriental and Australian Regions, and are absent in the Madagascan and Ethiopian regions. Outside the Seychelles, the genus *Graeffea* occurs only in the Australian Regions and the subfamily Platycraninae, to which this genus belongs, is widely distributed in the Oriental and Australian regions. The Odonata found in the Seychelles are also related more to the Oriental than to African and Madagascan Regions.

REFERENCES

ABDULALI, H. 1949. Some Peculiarities of avifaunal distribution in Peninsular India. *Proc. national Inst. Sci. India*, 15: 387–393.

ALI, S. 1935. Ornithology of Travancore and Cochin. *J. Bombay nat. Hist. Soc.* 37: 814.

ALI, S. 1949. The Satpura trend as an ornithological highway. *Proc. national Inst. Sci. India*, 15: 379–386.

ANNANDALE, T. N. 1911. Freshwater sponges, Hydroidea and Polyzoa. Fauna Brit. India, pp. 1–251.

ANNANDALE, T. N. 1914. The African element in the freshwater fauna of British India. *Proc. IX internat. Congr. Zool. Monaco*, 579–588.

BEST, A. E. G. 1954. Notes on the butterflies of the Nagalapuram Hills, Eastern Ghats. *J. Bombay nat. Hist. Soc.* 52: 365–373.

BHIMACHAR, B. S. 1945. Zoögeographical divisions of the Western Ghats, as evidenced by the distribution of hill-stream fishes. *Curr. Sci.* 14(1): 12–16, fig. 1.

BLANFORD, W. T. 1876. Notes on 'Africa-Indien' of A. VON PELZEN, and on the Mammalian fauna of Tibet. *Proc. zool. Soc. London*, 631–634.

BLANFORD, W. T. 1876. The African element in the fauna of India: A criticism of Mr. WALLACE's views as expressed in the 'Geographical distribution of Animals'. *Ann. Mag. nat. Hist.* (4)18: 277–294.

BLANFORD, W. T. 1901. The distribution of Vertebrate animals of India, Ceylon and Burma. *Philos. Trans. R. Soc. London*, (B) 194: 335–436.

DAY, F. 1885. Relationship of the Indian and African freshwater fish faunas. *J. Linn. Soc. London, (Zool.)* 18: 308–317.

DERANIYAGALA, P. E. P. 1941. The Hippopotamus as an index to early man in Ceylon. *Sci. Cult.* 7(2): 66–68, Fig. 2.

DERANIYAGALA, P. E. P. 1958. The Pleistocene of Ceylon. Colombo: 1–164, pls. LVIII.

DERANIYAGALA, P. E. P. 1960. Some southern temperate zone snakes, birds, whales that enter Ceylon area. *Spol. Zeyl.* 29: 79–85, Fig. 2.

GARDINER, J. S. 1903–1906. The fauna and geography of the Maladive and Laccadive Archipelago. Vols. 1 & 2, Suppl. 1, 2. Cambridge.

HORA, S. L. 1944. On the Malayan affinities of the freshwater fish fauna of Peninsular India and its bearing on the Probable age of the Garo-Rajmahal Gap. *Proc. national Inst. Sci. India*, 10: 423–439.

HORA, S. L. 1948. The distribution of crocodiles and chelonians in Ceylon, India and Burma and farther east. *Proc. national Inst. Sci. India*, 15: 285–310.

HORA, S. L. 1949. Symposium on the Satpura hypothesis of distribution of the Malayan fauna and flora to Peninsular India. *Proc. national Inst. Sci. India*, 15: 309–314.

HORA, S. L. 1949. Remarks on the distribution of snakes of Peninsular India with Malayan affinities. *Proc. national Inst. Sci. India*, 15: 399–403.

HORA, S. L. 1949. Discontinuous distribution of certain fishes of the Far East to Peninsular India. *Proc. national Inst. Sci. India*, 15: 411–416.

JAYARAM, K. C. 1949. A note on the distribution of chelonians of Peninsular India with Malayan affinities. *Proc. national Inst. Sci. India*, 15: 397–398.

JAYARAM, K. C. 1949. Distribution of Lizards of Peninsular India with Malayan affinities. *Proc. National Inst. Sci. India*, 15: 403–408.

JAYARAM, K. C. 1949. On the distribution of Annelids (earthworms and leeches) of Peninsular India with Malayan affinities. *Proc. national Inst. Sci. India*, 15: 417–420.

JAYARAM, K. C. 1956. The Palaearctic element in the fish fauna of Peninsular India. *Bull. national Inst. Sci. India*, 7: 260–263.

JEANNEL, R. 1943. Le genèse des faunes terrestres. Paris: France Univ. Press.

KHAJURIA, H. 1924. Mammalian fauna of the semi-arid tracts of the Deccan and its bearing on the appearance of aridity in the region. *Sci. Cult.* 21: 293–295.

KOLBE, H. J. 1887. Die Zoogeographischen elemente in der Fauna Madagascar. *Sitzb. Ges. Naturf. Fr. Berlin*, 147–178.

KOLBE, H. J. 1910. Die Coleopterenfauna der Seychellen nebst Betrachtungen über die Tiergeographie dieser Inselgruppen. *Mitt. Zool. Berlin*, 5: 1–49.

LYDEKKER, R. 1902. Indian Tertiary and Post-Tertiary Vertebrates: The fauna of Karnul Caves. *Palaeont. indica*, 10(4): 23–58; also see *Rec. geol. Surv. India*, 19: 120; 20: 72.

MENON, A. G. K. 1951. Further studies regarding HORA's Satpura hypothesis. 1. The rôle of the Eastern Ghats in the distribution of the Malayan fauna and flora to Peninsular India. *Proc. national Inst. Sci. India*, 17: 475–497.

MIECHELSEN, W. 1922. Die Verbreitung der Oligochäten im Lichte der Wegenerschen Theorie der Kontinentalverschiebung und andere Fragen zur Stammesgeschichte und Verbreitung dieser Tiergruppe. *Verh. naturw. Ver. Hamburg*, 3: xxix.

Percy Sladen Trust Expedition to the Indian Ocean in 1905. Reports of the *Trans. Linn. Soc. London*, (Zool.) (2) 126(1907); 18(1925).

ROONWAL, M. L. & BHOLA NATH. 1949. Discontinuous distribution of certain Indo-Malayan mammals and its zoögeographical significance. *Proc. national Inst. Sci. India*, 15: 365–377.

SARASIN, F. 1910. Über die Geschichte der Tierwelt Ceylan. *Z. Jahrb.* (12): 1–160.

SILAS, E. G. 1952. Further studies regarding HORA's hypothesis. 2. Taxonomic assessment and levels of evolutionary divergence of fishes with the socalled Malayan affinities in Peninsular India. *Proc. national Inst. Sci. India*, 18(5): 423–448.

STEPHENSON, J. 1921. Contributions to the morphology, classification and zoögeography of Indian Oligochaeta. 1. Affinities and systematic position of the genus *Eudichogaster* Michs. and some related questions. 2. On polyphyly in Oligocheta. 3. Some general consideration on the geographical distribution of Indian Oligochaeta. *Proc. zool. Soc. London*.

STEPHENSON, J. 1923. Oligochaeta. Fauna Brit. India, pp. 1–518.

TIKADER, B. K. 1965. Observations on the Caecilian *Ichthyophis beddomi* Peters from Kotagiri District Chickmagalur, Mysore. *J. Bombay nat. Hist. Soc.* 61(3): 697, pl. I.

WHISTLER, H. & N. B. KINNEAR, 1932. The Varnay Scientific Survey of the Eastern Ghats (Ornithological section). *J. Bombay nat. Hist. Soc.* 36: 587.

XX. BIOGEOGRAPHY OF THE EASTERN BORDERLANDS

by

M. S. MANI

Strictly speaking, the Eastern Borderlands include only Assam and northern Burma (see chapter II), in other words a part of the Trans-Gangetic Subregion of BLANFORD (1901). Biogeographically, however, the eastern parts of the Eastern Himalaya, eastern Tibet, the Yunnan Province of South China, Annam, Hainan, the northern parts of Indo-China, Assam, Burma and the northern parts of Thailand constitute together a single natural unit of mountainous land, the Indo-Chinese Subregion. The transition between this subregion and the Malayan Subregion in the south is placed artificially about 12° NL in the Isthmus of Kra. The northern limits of the subregion are not, however, so sharply defined and the subregion probably grades off into the Manchurian subregion of the Palaearctic. In the absence of a well defined natural boundary, we depend largely on climatic factors. According to SMITH (1931, 1935, 1943), the northern limits are as follows: starting from the Nam-kin Mountains in the extreme northeast of Burma, an imaginary line is drawn in a southeasterly direction to Yunnan-Fu, whence it extends east to the Hung-shui-Kiang, which it follows until it joins the Wu-kiang to form the R. Sikiang at Sun-chao-fu. It then follows this river to its termination in the sea. In the west, the boundary is a transition that disappears in the R. Ganga. Though a natural and integral part of the subregion, the Eastern Himalaya is excluded from this chapter, but we include here the Brahmaputra Valley which is essentially only an eastern extension of the Gangetic Plain.

1. Ecology

The ecology of the Eastern Borderlands is conditioned largely by the youthful topography. The general climate of Assam is characterized by extreme humidity and copious rainfall between March and May, a period when precipitation over most of the rest of India is minimum. There are, strictly speaking, only two seasons in Assam, viz. the cold season and the rainy season, and the hot weather of the rest of India is absent. The climate is generally cool and the mean temperature does not rise above 26° C during the hottest part of the year in Shillong. Snow falls only occasionally in Shillong during winter, because there is no precipitation of moisture during the winter. As already pointed in earlier chapters this is the region of heaviest rainfall in the world. The general climate of the Naga Hills is cool and the mean temperature does

648

not exceed 26.6° C in Kohima. As elsewhere in Assam, the rainfall is heavy on the Naga Hills also.

An important character of the climate of Burma is the presence of the socalled dry belt of Central Burma, in the rain-shadow area of the Arakan Yoma, (rainfall less than 100 cm). Central Burma is hot during the summer. The coastal regions receive, however, heavy rainfall. Areas in Burma that receive 200 cm rainfall are covered by evergreen forests. The monsoon forest covers areas receiving a rainfall of 100–200 cm and the drier areas are scrubland.

From the point of view of rainfall, a wet Upper Burma, a dry Upper Burma, a littoral and deltaic Lower Burma and the subdeltaic Lower Burma are distinguished. The Upper Wet Division has a rainfall of over 125 cm and comprises the Shan States and the Chin Hills, Khata, Bhamo, Myitkyina, the Upper Chindwin and the Ruby Mines. The Upper Burma Dry Division is an arid area largely plain, with some isolated hills, extending from the 20th to the 23rd parallel. The Wet Division of Lower Burma stretches along the whole length of the coast and includes the entire Arakan and parts of Tenasserim, Pegu and Irrawaddy areas. To the north and south of delta region, the hill ranges approach the sea-face and there are a number of islands. In the delta proper, we have Bassein, Pyapon, Myaungmya, Maubin, Hanthawaddy and Pegu and the whole area is practically level plain, with rainfall not usually over 250 cm. Minbu, Pegu, Irrawaddy and Tenasserim districts constitute the subdeltaic areas of Lower Burma, representing the borderland between the wet and dry divisions and have usually a rainfall of about 200 cm.

When compared with Peninsula, the ecology of the Eastern Border-lands, but particularly Assam and North Burma, is wholly humid-tropical. While both the Peninsula and the Eastern Borderlands are ecologically dominated by the monsoon rainfall frequency and patterns, the Peninsula has highly humid and rainfall-deficient areas, but in Assam and North Burma rainfall deficient areas are unknown. Another contrast is the fact that the destruction of natural habitats is far more complete in the Peninsula than in Assam-Burma. A point of considerable ecological significance is that the Eastern Borderlands is a region of mostly Tertiary mountains, but the Peninsula is an ancient undisturbed though worn-out plateau. The vegetation and flora of Assam-Burma are described in Chapter IX, so that it is unnecessary here to consider them. It may, however, be useful to recall that the forest in Assam is typically reed-covered, often growing to heights up six metres. The cultivation practice (see Chapter XI) has resulted in dense secondary scrub-jungle in most areas in Assam and extensive impenetrable bamboo-covered *kayinwa* on the Arakan Hills.

2. *The Character Fauna*

The character fauna of Assam and Burma, like their flora, is distinctive, rich and greatly diversified. We have here a typical humid-tropical mountain-forest fauna, intermingled with subtropical and temperate elements. There is a great preponderance of arboreal types, but considerable numbers of grassland and marshland types are also present. Assam and northern Burma are also perhaps one of the richest areas in Lepidoptera in the world. On the whole, there is marked difference from the fauna of large areas of the rest of India. We have in Assam elephant, rhinoceros, tiger, leopard, bear, wild dog, wild hog, deer, buffalo, bison, etc. The presence of Simiidae, Procyonidae, Talpidae, Spalacidae, Gymnuridae, Eurylaemidae, Indicatoridae, Heliornithidae, Platysternidae, Discophidae, Hylidae, Pelobatidae and Salamandridae contributes to the distinctiveness of Assam fauna. The wealth and the marked differences of the flora and fauna of the region were indeed recognized by naturalists from the beginning.

The majority of earlier workers have largely mistaken the area of Assam and North Burma for an important centre of floristic-faunistic radiation, thus totally ignoring the strong evidence to the contrary. The fact that the area is biogeographically *transitional* has thus been overlooked. Assam and North Burma represent a *gateway*, rather than an important amphitheatre of differentiation and radiation of floras and faunas. It is also important to stress here that the gateway lies at the fringe of the western extremity of the Indo-Chinese floristic-faunistic amphitheatre, so that the actual centres of differentiation and radiation are situated much farther east. Ignorance of these basic conditions has resulted in numerous misconceptions and wholly illogical conclusions, both in phytogeographical and zoogeographical discussions in the past.

It is however, important even at this stage to observe that unlike the Western Borderlands, which is also in a sense a gateway, the floristic-faunistic gateway of Assam is relatively old; this is proved by the presence of small numbers of local endemic subspecies, species and even genera of certain groups of plants and animals. The area is eminently a meeting point of the humid-tropical Tertiary-mountain floras and faunas and of the Peninsular (Gondwana) floras and faunas.

Another equally erroneous belief that has, unfortunately, gained wide acceptance among botanists and zoologists in India, is that of an almost exclusive unidirectional and westward influx of floras and faunas through Assam to the Peninsula and along the Himalaya. The fact that this undoubted westward influx has been nearly equally strongly matched by transgressive and intrusive outflow from the west also in Assam has thus been almost completely overlooked so far. In certain groups of animals, however, Assam has received an equally abundant Ethiopian element. KURUP (1966) has recently shown, for example, that the western mammals are almost as abundant as the eastern forms in Assam (see also chapter XVIII). The western entrants have moved along the Himalaya and not across the Indo-Gangetic Plains to Assam. The biogeographical connection of Assam and the Peninsula is certainly greater than that between the latter and the Himalaya and is also older; the connection between the Peninsula and the Himalaya is not earlier than the Pleistocene but that between Assam and the Peninsula is at least contemporaneous with the Miocene phase of the Himalayan uplift. It is therefore important to remark that Assam-Burma seems to have received the present dominant elements of flora and fauna nearly equally from the east, south and west.

The Eastern Borderlands of Assam and Burma constitute, therefore, an integral part of the Indo-Chinese Subregion of the Oriental Region rather than an independent (Trans-Gangetic) subregion, as generally assumed by most earlier workers, (BLANFORD, 1901; CLARKE, 1898; HOOKER, 1906). As the boundary between the Indo-Chinese and Malayan Subregions is largely arbitrary, there is considerable overlapping of the Indo-Chinese and Malayan areas in Assam and Burma. The Indo-Chinese fauna has spread extensively southwards into the Malayan area, mainly across Assam-Burma and the Malayan fauna has likewise spread northwards deeply into the Indo-Chinese areas. The character fauna of Assam-Burma is, therefore, composed mainly of humid-tropical forest elements of the Indo-Chinese and Malayan areas. It is not always easy to separate the Indo-Chinese elements from the Malayan forms in Assam-Burma.

Biogeographically Assam-Burma must be described as a transitional area, where the older Peninsular and the relatively younger Asiatic floras and faunas meet and transgress. Assam is essentially a floristic-faunistic gateway, through which almost uninterrupted interchanges have taken place between the Peninsula and Asia. The faunal interchanges have largely involved transgression of the Indo-Chinese and Malayan faunas westwards into India. This transgression has been along two lines, viz. primarily westwards along the Himalaya and southwestwards to the Peninsula. The Asiatic faunas have spread westwards both to the north and south of the eastern end of the Great Himalaya. North of the Great Himalaya these faunas have transgressed not only to Eastern Tibet, but also on the forested mountain ranges south of the Great Himalaya on the Indian side, the Indo-Chinese and Malayan faunas extend very much farther west, with their outliers occurring today even as far west as Kashmir. They are indeed represented by an almost complete graded series of local endemic subspecies in most groups along the Himalaya. For example, Indo-Chinese butterfly *Chilasa agestor agestor* extends along the Himalaya nearly up to Kashmir *C. agestor agestor* occurs from Tonkin to E. Himalaya, *C. agestor govindra* from Kumaon to Kashmir and *C. agestor chiraghshai* occurs in West Kashmir (Fig. 154). The transgression of the Indo-Chinese and Malayan derivatives of Assam-Burma into the Peninsula is, however, marked by discontinuity at present.

The component elements of the character flora and fauna are therefore,

651

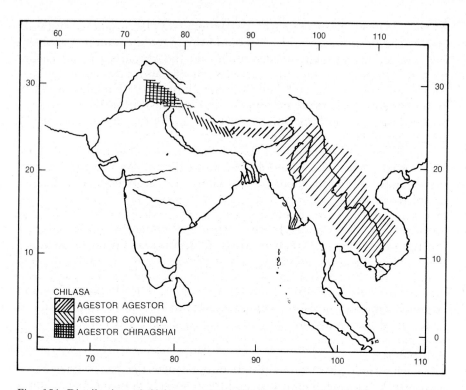

Fig. 154. Distribution of *Chilasa*, an Indo-Chinese genus of butterflies, with differentiation of local endemic subspecies, westwards along the forest-covered southern ranges of the Himalaya.

1. Asiatic derivatives, 2. Mediterranean-Ethiopian intrusive elements, 3. Peninsular isolates and 4. Australian and Gondwana outliers. The Asiatic derivatives comprise 1. local endemics in Assam, in Burma and Assam-Burma, 2. Indo-Chinese and south Chinese elements, 3. Malayan elements, 4. Manchurian and Siberian outliers. The strength of these different elements differs in different groups of plants and animals; in some groups the Indo-Chinese affinities are stronger than others and in other groups the Malayan affinities are dominant. In still other groups both these elements are nearly equally strongly present. Taken as a whole, however, the Indo-Chinese and Malayan elements are the dominant ones in the present-day flora and fauna of Assam-Burma and the other elements occupy a minor place.

2.1. ENDEMICS

Owing to the biogeographically transitional nature, the area is not rich in endemic forms. Though an impressive list of endemic species of

652

plants has been given by RAO (see Chapter IX) and even in some groups of animals, like termites, the species endemism has been estimated to be as high as 60%, it must be remembered that on the whole the endemic forms represent a minute percentage of the total number of genera and species. It has also to be kept in view that the high percentage of endemism in groups like termites, reported by ROONWAL (ROONWAL & CHOTANI, 1965), is likely to fall, when extensive explorations in the areas to the east are undertaken. Similarly the list of endemic plants of Assam and Burma will also shrink when the flora of not only Assam-Burma but also of the countries to the east are more thoroughly studied than at present. A number of supposed endemic species of Assam and Burma have, for example, in recent years been shown to occur elsewhere in India and Indo-China, Thailand and Malaya. At least in some cases, what are at present considered as endemics, because of being restricted to Assam, are in reality relics of former wider distribution in the west. As shown by MUKHERJEE in Chapter XII, the range of the present-day Assam species formerly extended outside Assam. A point of considerable biogeographical significance is the observation that nearly all the endemic subspecies, species and genera are derivatives of forms that are widely distributed in Indo-China and Malaya and represent, therefore, recent differentiations of these stocks.

The genera or species which are strictly confined to 1. Assam-Burma or to 2. Assam or 3. only Burma at present are exceedingly few and have also Indo-Chinese or Malayan affinities, but may perhaps be in part at least autochthonous endemic forms of Assam-Burma. Most of the autochthonous species of Assam are derivatives of either the Indo-Chinese, Malayan or of the Peninsular stock. The Pentatomid *Amauropepla* is restricted to Assam and North Burmese mountainous area. The Dermaptera *Solenosoma* and the Pentatomid *Vigetus* are restricted to Assam only. The Uropygi *Uroproctus* and the Lepidoptera *Polydorus latreillei kabura* are other examples. The earthworms peculiar to Assam include *Drawida decourcyi*, *D. kempi*, *D. rosea*, *D. rotungana*, *Plutellus aborensis*, *Notoscolex stewardi*, *N. oneili*, *N. striatus*, *Megascolex horai*, *Perionyx annandalei*, *P. annulatus*, *P. depressus*, *P. fossus*, *P. foveatus*, *P. kempi*, *P. koboensis*, *P. modestus*, *P. shillongensis*, *P. turaensis*, *Eutyphoeus aborianus*, *E. gammiei*, *E. manipurensis* and *E. turaensis*. *Kachuga sylhetensis* is confined to Assam. ROONWAL et al. (1965) have recently described some interesting species of Termites as endemic to Assam. Restricted to Burma are the Pentatomid bugs *Cratoplatys* and *Oncylaspis*, the Amphibians *Calluella guttata* (belonging to a Madagascan family) and *Ixalus vittatus* (the genus is Peninsular), the reptiles *Kachuga trivitta*, *Testudo platynota*, *Lissemys punctata scutata* and *Trionyx formosus*. One species of the Scarabaeid *Desmonyx* is confined to Burma.

The dominant components of the flora-fauna of the Eastern Border-lands are the Indo-Chinese and Malayan elements. In some groups the Malayan elements are predominant and in others the Indo-Chinese elements are far more numerous. While on the whole the Indo-Chinese elements outnumber the Malayan component in Assam and northern Burma, the two elements are, however, intermingled in a most complex manner and it is not also often easy to separate them from each other.

The Odonata are, for example, almost exclusively of Indo-Chinese and Malayan origin, as illustrated by the genera *Megalestes, Orolestes, Indolestes, Burmargiolestes, Drepanosticta, Coeliccia, Calicnemis, Allogaster, Jagoria, Cephalaeschna, Indophlebia, Bayadera, Anisopleura, Eyriothemus, Philoganga, Martona, Davidius*, etc. Some of these genera extend westward along the wooded slopes of the Himalaya up to Darjeeling. The Orthop-tera include the typical Indo-Chinese *Bennia* that extends from Yunnan across Assam to Bhutan and *Ceracris* occurring from Burma across Assam to Darjeeling.

The Indo-Chinese Scarabaeid *Fruhstorferia* occurs, for example, from Yunnan and Tonkin across Burma to Java. The Coleopterous family Trictenotomidae, with the genus *Autocrates* from Yunnan and Assam is represented by the genus *Trictenotoma* in Java and Borneo. The Ceram-bycid *Gnatholea* extends from Philippines across Indo-China and Assam-Burma to Malaya and Borneo. *Arctolamia* occurs from Tonkin to Burma and *Calloplophora* from Assam to Sumatra. The Scarabaeid *Lutera* extends from Tonkin across Thailand and Burma to the Nicobar Islands, Sumatra, Java, Borneo, Celebes and Philippines. A number of typically Malayan Dermaptera like *Eparchus, Hypurgus, Platylabia, Pygia, Palex*, etc.; Penta-tomidae like *Storthecoris* extending from Borneo across Malaya to Assam; Coreidae like *Prionolomia* from the Malay Archipelago to Assam; Chryso-melidae like *Liatongus* extending from Malaya, Sumatra, Java and Thai-land to Assam-Burma and several Oligochaeta like *Pheretima, Wood-wardia, Megascoloides, Plutellus, Notoscolex*, etc. (Fig. 152) are characteristic of the fauna of the Eastern Borderlands. Of over one thousand species of Lepidoptera known from here, the greatest bulk are Indo-Chinese and Malayan in origin. *Graphium doson* extends from South Japan, South China, Thailand and Tonkin, to the Sunda Is. and through Burma-Assam to the Kumaon Himalaya and over Bengal to South India and Ceylon. *Graphium agamemnon* from South China, Queensland, Solomon Islands and Bismarck Islands, is represented by *Graphium agamemnon agamemnon* in Burma to Kumaon. *Graphium nomius*, from Hainan, Annam, Thailand and Tonkin extends across Burma and Assam to Sikkim; it is also known from South India and Ceylon. *G. antiphates*, from China, North Borneo and Sunda Islands, is represented by *G. antiphates pompilius* in Hainan, Annam, Thailand, Burma and Sikkim. *G. cloanthus cloanthus* from China, Formosa extends through Burma to Sumatra and along the Himalaya to Kashmir. *Lamproptera* is a genus of butterflies endemic to southeast China, Java, Celebes, Philippines, Thailand, Burma, Malaya and Assam. The Teinopalpini are characteristic of South Burma to Assam and Sikkim. *Troides*, from Malay Archipelago, New Guinea and Australasia, is represented by *T. helena cerberus* in Tonkin, Malaya, Borneo, Burma and Sikkim and the Orissa Hills of the Peninsula. *Polydorus aristalochiae*, from China, Phillippines, Borneo, Sumatra and Java, is represented by *P.a.goniopeltis* in South China, Thailand and Burma. *P. nevilli* ranges from Assam and Burma to West China. The Mollusca of the Eastern Borderlands are either Malayan or Indo-Chinese like *Plectrotropis* (extending west on the Himalaya up to Simla), *Aegista*, *Plectopylis*, *Camaena*, *Amphidromus*, *Gangesella*, *Sesara*, *Dioryx*, etc.

The distributional evidence of the fresh-water fish, amphibia and reptile, discussed in an earlier chapter by Jayaram, shows that the greatest majority of the fishes occurring in the R. Brahmaputra drainage and other areas of Assam and northern Burma are fundamentally of Indo-Chinese affinity; some distinct Malayan forms are also found in these areas. Jayaram has, on the basis of the distributional pattern of these fishes, attempted to trace the centre of differentiation and geographical radiation of the fishes of Assam in the Yunnan Province of south China, a tract that is included in the Indo-Chinese subregion. Among the more outstanding examples considered by him and occurring in Assam are *Euchiloglanis*, *Exostoma*, *Coraglanis*, *Parapseudecheneis*, *Pseudecheneis*, *Myersglanis*, *Conta*, *Accrossocheilus*, and *Osteocheilus*. He has mentioned *Platysternum*, *Siebenrockiella* and *Cyclemis dentata* among the reptiles and *Tylototriton* among the amphibia as outstanding examples of the Indo-Chinese and Malayan affinities of the Assam-Burma vertebrates.

2.3. Other component elements

The bulk of the Amphibians are derived from the ancient Gondwana stock. *Rhacophorus* and *Ixalus* occur in Burma, China and Malaya. It may be recalled that fifteen out of the twenty-five species of *Ixalus* are restricted to South India and Ceylon (none above 15° NL), and the remaining ten species are found in the Eastern Borderlands but none on the Himalaya. *Calluella guttalata*, occurring in Pegu and Tenasserim, belongs to the Madagascan family Dyscophidae. Among the Ranidae we have *Oxyglossus* in south Bengal, Burma, South China, Thailand, Malay Peninsula, Java, Indo-China and the Philippines; *Ichthyophis* (Fig. 102) from Java, Sumatra, Borneo, Malaya, Sikkim and also discontinuously in the Western Ghats (Peninsula); *Tylototriton* from Burma, Yunnan, Sikkim and Nepal; *Leptobrachium* with four species in Burma, one species in Sikkim, Assam and Burma and a third species from Burma through Malaya to Borneo; *Glyphoglossus* from the Irrawaddy Delta and Pegu;

Calluella pulchra from South China, Indo-China, Burma, Malaya, and discontinuously in South India and Ceylon, etc. *Microhyla berdmorii* occurs in Malaya, Pegu and Tenasserim; *M. rubra* in Assam, Coromandel Coast and Ceylon and *M. ornata* in Assam, Burma, South China, Indo-China, Kashmir and Ceylon. *Calophrynus* is found in South China, Burma, Borneo and Madagascar; *C. pleurostigma* occurs in South China, Pegu, Bhamo, Kakhyn Hills, Tenasserim and Borneo.

The Gondwana elements, representing primitive or older groups and ecologically specialized derivatives of ancient stocks have spread northward from Malaya to Assam-Burma and Eastern Himalaya mountainous area. We have, for example, the fresh-water prawn *Xiphocaridina curvirostris* occurring in the northeast Assam and in New Zealand. This belongs to the family Atyidae and is one of the most primitive genera in the ancient family. *Typhloperipatus williamsoni*, from the Abor Hills in the Assam part of the Eastern Himalaya, is related to *Eoperipatus* from Malaya Peninsula and Sumatra and lacks eyes completely. It was collected from under

Plate 75. Typhloperipatus from the Abor Hills of Assam.

stones near the R. Dihang, while all the Malayan species of peripatus occur under fallen timber (KEMP, 1914). There was abundant rotting wood near the stones under which *Typhloperipatus* occurs. The following distribution of Onychophora is of considerable interest: *Eoperipatus* Malaysia, *Peripatus* tropical America, *Mesoperipatus* tropical Africa, *Peripatoides* Australia, *Paraperipatus* Ceram, New Guinea and New Britain, *Peripatopsis* South Africa and *Chilioperipatus* South America. The scorpion *Lychas* that extends from tropical Africa to Australia over Burma, China, Thailand, Malay Peninsula and Malay Archipelago, Java and as already mentioned in the Indian Peninsula. The Uropygi *Thelyphonus* occurs in

Burma eastwards to the Philippines, Australian Malayan area, as far as the Solomon Islands and as already pointed out in the foregoing chapter, also in South India and Ceylon. *Uroproctus* occurs in Assam, *Hypoctonus* in Yunnan, Assam, Burma and Borneo and *Trithyreus* in Burma, Bismarck Archipelago and Ceylon. Other examples of Gondwana-fauna derivatives are found among insects like the Pentatomid *Ponsila* from Naga Hills and Africa and *Solenostethium* from Malaya and Thailand across Burma to Assam and also known from Africa.

Some of the Peninsular outliers in Assam-Burma are the following: Oligochaeta *Drawida* (Fig. 151). Chrysomelidae: *Galerotella* Ceylon, Coorg, Nilgiri, Anamalai and Shevroy Hills in the Peninsula and in Assam-Burma; *Sikkimia* two species in the Nilgiri and one species in Assam-Burma; Formicoidea: *Anochetus* most species in Ceylon and Nilgiri Hills, Belgaum, Poona and some in Sikkim and Burma; *Diacamma* Ceylon, Travancore, Nilgiri, Mysore, Kanara, Bombay in the Peninsula and in the Trans-Gangetic subregion Sikkim, Bengal and Assam-Burma; *Phidologiton* Ceylon, Travancore, Kanara, Poona in the Peninsula and Assam-Burma in the Trans-Gangetic subregion and Malaya. Among fishes mention may be made of *Clarius*, with one species in Ceylon and again occurring in Burma and Thailand; *Eutropiichthys* (related to the African *Eutropius*) with one species in the Peninsula (spilling into the Indo-Gangetic Plain) and entering Assam and Upper Burma and one species only to Assam; about a dozen of the 17 species of *Labeo* from the Peninsula extend to the northeast areas; the lizard *Cnemaspis* with about a dozen spp. (Fig. 117) in the hills of S. India and Ceylon, two spp. in Indo-China to 1 sp. in Tenasserim and Thailand and among birds the true babblers *Argya* and *Crateropus* found in the Peninsula and Ceylon extend sparsely into Assam-Burma. The distribution of the Mollusc *Satiella* in Travancore, Nilgiri, Ceylon and Andamans is also an interesting example of discontinuity.

3. Distributional Patterns

The analysis of the distribution of the character fauna of the Eastern Borderlands reveals the following principal patterns:

1. Distribution in Assam (including the Eastern Himalaya), northern Burma, eastern Tibet, south China, Indo-China and the mountainous parts of Thailand. This pattern is fundamentally the typical Indo-Chinese distribution, in which Assam and northern Burma represent the extreme western limits. The genera and species have in the course of their radiation on the Tertiary mountains reached their western limits and have not transgressed further west along the Himalaya, in the Indo-Gangetic Plains or to the Peninsula. Most of these forms have been mistaken by earlier workers to represent the distinctive Assam types, particularly because of their absence elsewhere in India and some of them

have been readily identified as endemic elements in Assam. This confusion has given rise to the widely prevalent belief among botanists and zoologists that Assam is a distinct and separate biogeographical area.

2. Distribution in Assam (including the Eastern Himalaya), Burma (including the Andaman and to some extent also the Nicobar Islands), Malaya, Sumatra, Java and often extending as far as Borneo and the Philippines. This distributional pattern is the result of radiation of the Malayan faunal complex. The hills of Assam represent the northern limits of intrusion into India. A part of the older Gondwana and Australian faunal elements also share this pattern with the Malayan elements. Some of the species have not extended north of Burma, but others have penetrated into Assam and have reached the Eastern Himalaya. Infiltration round the eastern end of the Himalaya into the eastern parts of Tibet may also be observed in certain cases. This pattern is fairly common, but of lesser extent than the first pattern.

3. Transgression across Assam, westwards along the wooded slopes of the Himalaya, is a pattern of distribution most commonly met with both among the Indo-Chinese and Malayan floristic-faunistic elements. The pattern is so dominant in the character fauna of Assam that we are justified in describing the wooded slopes of the Himalaya, south of the Great Himalayan Range, as far west as Nepal, as virtually a narrow western tongue or appendage of the Indo-Chinese subregion, deeply dove-tailed in between the Palaearctic and the Peninsular subregions in India (Fig. 157). In this pattern the most important fact is the presence of more or less completely graded series of local subspecies and species, endemic successively in localized patches from the east to the west, penetrating up to the Kumaon Himalaya. These local endemic subspecies are often connected by transitional forms with the main stock in the areas to the east of Assam. The occurrence of such a graded series of subspecies, merging eventually into distinctive species, especially insects, in the extreme western end, has been known to most earlier workers. They have however, been misinterpreted as evidence of Assam being a centre of faunistic radiation and of its faunistic distinctiveness.

4. Discontinuous distribution in Assam-Burma and in the Peninsular south or southwest (including Ceylon). This is perhaps the most striking and at the same time also most intriguing of the distributional patterns, met with in nearly every group of plant and animal found in the Eastern Borderlands. The number of species, which are distributed discontinuously in Assam-Burma and the Peninsula, far outnumbers that exhibiting the other distributional patterns. The discontinuity presents also nearly every transition from almost continuous to the extreme case of forms occurring in the hills of Assam or Burma and reappearing in the Hills of Ceylon, being completely absent in the intervening area. Some of the more important grades of discontinuity, observed by us in plant and animal distribution, may be summarized as follows: 1. Distribution in

Assam and Peninsular southwest, 2. Assam and the Peninsular southwest and the hills of Ceylon, 3. Assam and the hills of Ceylon, 4. Burma and the Peninsular southwest, 5. Burma and the Peninsular southwest and Ceylon hills, 6. Burma and the hills of Ceylon, 7. Assam-Burma and the Peninsular south, 8. Assam-Burma and the Peninsular southwest and the hills of Ceylon and 9. Assam and the Peninsular northwest and southwest.

The distributional patterns met with in the Eastern Borderlands would seem to suggest the distinctness of Assam from Burma. While some botanists have considered Assam as distinct from Burma, others have combined the two areas in part. CLARKE (1898) and HOOKER (1906) consider, for example, parts of Assam and the Eastern Himalaya to be separate from each other and HOOKER includes the hilly parts of Assam, excluding the Eastern Himalaya, as a part of Burma from the Phytogeographic point of view.

The four phytogeographical subdivisions of Burma, suggested by HOOKER, are 1. the northern, 2. the western, 3. the eastern and 4. the central. The northern Burmese subregion is mountainous, stretching nearly 800 km northeast from the bend of the R. Brahmaputra to Yunnan, with the northern boundary formed by the mountains flanking the Assam Valley on the south. This area falls within Assam and is characterized by maximum humidity, without an arid area. The vegetation approximates to that of the Eastern Himalaya, but lacks an alpine-zone flora and also *Picea, Abies, Tsuga, Juniperus, Larix*, etc. It differs from the Central and southern Burmese subdivisions in not having *Tectona grandis*, in its sparse Dipterocarpaceae and in having the palm *Trachycarpus maritiana* (found also in Sikkim-Nepal) and *Shorea robusta*. The tropical Himalayan species predominate in the valleys. The temperate species, often identical with those of the Himalaya, as illustrated by *Ranunculus, Anemone, Thalictrum, Delphinium, Corydalis, Geranium, Impatiens, Drosera, Astragalus, Rubus, Potentilla, Fragaria, Sanguisorbia, Valeriana, Senecio, Pedicularis, Primula, Iris*, etc. appearing at elevations over 1250 m. Open unforested areas, at elevations above 1200 m, are reminiscent of the Nilgiri and have many genera and species common to those hills in the Peninsula. *Nepenthes khasiana*, the northernmost member of the genus, grows at elevations of about 1250 m. The other species of this genus are known from Ceylon and Malayan Peninsula and are all climbers, but *N. khasiana* is prostrate.

The vegetation of the humid areas of the western and southern Burmese subdivisions differs markedly from that of the Central Burmese subdivision and is characterized by the presence of the estuarine forms of the Irrawaddy delta. The eastern Burmese subdivision is characterized by complex mountains, between Burma and China in the north and Thailand in the south, and has few Chinese species. The Central Burmese subdivision, between the Arakan Mountains and the ranges east of the R. Sittang, has a northern dry and a southern humid area. The forests of the Andaman Islands are typically Burmese in composition and character, with a mixture of evergreen and deciduous, but *Quercus, Castanopsis* are absent. The Malayan Peninsula lies largely in Thailand and has a mountainous backbone, rising to elevation of 1250–2100 m. The dominant Natural Orders include the Orchideae, Leguminosae, Euphorbiaceae, Anonaceae, Gramineae, Scitamineae, Melastomaceae, Cyperaceae and Urticaceae. The species of *Impatiens*, occurring here, differ from those found in Burma. There is considerable wealth of palms. BLANFORD (1901) has suggested that the region consists of 1. the Assam tract, 2. the Upper Burmese tract, 3. the Pegu tract, 4. the Tenasserim tract, 5. the South Tenasserim tract and 6. the Andaman and Nicobar Islands tract. Under the Assam tract he included the valleys and the Garo, Khasi and Naga Hills, Manipur, Chittagong and the Arakan, the hills to the west of the R. Irrawaddy drainage. The tract is an area of hills and dense forests, closely resembling in physical features, flora and fauna the Eastern Himalaya. These and other similar subdivisions, proposed from time to time by different authors from studies in different groups, are in reality ecological subdivisions of rainfall differences rather than true biogeographical ones. We have in the Assam tract many forms that are common to the Eastern Himalaya but do not extend to Burma. This is, for example, the case with the mammals *Soriculus* and *Synotus*; the birds *Sylviparus, Paradoxornis, Ianthocincla, Grammatoptila, Stactocichla, Xiphorhamphus, Proparus, Lioparus, Rimator, Tesia, Oligura, Ixops, Hilaroxichla, Minla, Sphenocichla, Elachura, Urocichla, Acanthoptila, Neornis, Nitidula, Microcichla, Tarsiger, Ianthia, Mycerobas, Haematospiza, Propyrrhula, Hypacanthis, Brachypternus, Bubo, Lophotrichorchis, Tragopan, Perdicula, Microperdix, Rallus, Ibidorhynchus, Ciconia;* the reptiles *Ptycholaemus, Japalura, Stoliczkaia, Trachischium, Rhabdops, Blythia, Dinodon, Xenelaphis, Ancistrodon,* etc. It is, however, important to observe that there are no Amphibia in Assam that are not also common to Burma.

The Upper Burmese Tract comprises Burma to the north of Prome and Toungoo (about 18° NL), mainly the area of the R. Irrawaddy drainage and extending northwards to the hill ranges representing a continuation of the Himalaya to the east. The greater part of this tract is densely forested, but the Irrawaddy plains and the delta are covered by high grass. The Tenasserim Tract includes the Karenni area to the north and the hill ranges to the east of the R. Sittang and extends southwards to about the Mergui area (13° NL). This is a hilly area of dense forest, with rainfall of about 420 cm. The South Tenasserim Tract comprises Tenasserim to the south of the 13th north parallel and a part of the Malay Peninsula. The limits are rather arbitrary and really many Malayan elements extend much further north. Both physically and in rainfall, this tract is like north Tenasserim.

Of the animals found in the Burmese Tract, but not in the Himalaya, and the Peninsula (though extending to the countries to the east), we have the mammals like *Hylobates, Nycticebus, Arctogale, Hylomys, Arctictis Anuurosorex, Eonycteris, Chiropodomys, Haplomys*, etc; the birds *Crysirhina, Timelia, Drymocataphus, Thringorhina, Turdinulus, Aëthorhynchus, Spizixus, Herbivocula, Calornis, Spondiopsar, Agropsar, Ampeliceps, Graculipica, Ploceella, Anthothreptes, Eurylaemus, Corydon, Cymborhynchus, Calyptonema, Miglyptes, Gauropicoides, Carcineutes, Rhytidoceros, Poliohierax, Polyplectrum, Phasianus, Bambusicola, Tropicoperdix, Heliopais*, etc.; the reptiles *Geoemyda, Cyclemys, Platysternum, Xenelaphis, Xenochrophis, Trirhinopholis, Calamaria, Stoliczkaia, Homalopsis, Cantoria, Hipistes, Doliophis*, etc. and the amphibia *Oxyglossus, Hyla* and *Calophrynus*. The fishes include in addition to those already mentioned, *Monopterus, Liocassis, Akysis, Acanthopsis, Osteocheilus, Dangila* and *Osphromenes*.

The South Tenasserim Tract really belongs to the Malayan Subregion. We find indeed a number of interesting Malayan types like the mammals *Gymnura, Galaeopithecus, Emballonura, Tapirus, Tragulus* (found also in South India and Ceylon); the birds like *Platysmurus, Melacopternum, Trichostoma, Cyanoderma, Tricholestes, Alophoxius, Pinarocichla, Trachycomus, Platylophus, Philentoma, Hydrocichla, Erythura, Chalcostetha, Prionochilus, Callolophus, Calorhamphus, Chotorhaea, Acaridagrus, Annorhinus, Berenicornis, Rhinoplax, Zanclostomus, Rhamphococcys, Rhinortha, Machaerhamphus, Psittinus, Butreron, Geoplia, Argusianus, Lophura, Rollulus, Caloperdix* and the reptile *Bellia*.

4. The Andaman and Nicobar Islands

The Andaman and Nicobar Islands continue the trend of the Arakan Yoma and link them with the mountain ranges of Sumatra. The islands differ much among themselves. The Andaman Islands are about 190 km distant from the Cape Negrais in Burma, the nearest point in the mainland. The principal islands lie north-south in a line, between 10° 3′ NL and 13° 30′ NL. Five islands, close together, constitute the Great Andamans and the Little Andamans lie to the south. There are in addition about 200 smaller isles, including the Ritchie Archipelago, somewhat to the east of the main line of islands. The total area of the Andamans Islands is about 6300 km². The Great Andaman group is about 300 km long, but hardly 50 km wide in the broadest part. The islands are formed of sandstones, limestones, clays and some serpentine mostly of Tertiary age. They are much dissected and rise to about 420 m above mean sea-level, but the highest peak is 732 m above mean sea-level. The higher hills are generally near the coast. The eastern slopes of the hills are steeper than the western slopes. The coast is mostly fringed by mangrove swamps.

The Nicobar group of islands extends from about 6° 10′ NL. Car Nicobar Island lies about 135 km from the Little Andaman. The Great Nicobar lies about 150 km from Sumatra. These islands constitute three groups, viz. the Northern or the Car Nicobar, the Central (with Camorta and Nancowry) and the Southern or the Great Nicobar. The islands are generally believed to be formed of Tertiary soft micaceous sandstones and clays, but there are also extensive raised coral flats. The Great Nicobar and the Little Nicobar Islands are dissected and densely forested, but the other islands are mostly covered with tall lalang-grass and *Cocos nucifera*.

The Andaman Islands are characterized by their impoverished Burmese-fauna and the Nicobar Islands have essentially Sumatran-fauna. A sketch of the plant life of the Nicobar Islands may be found in SAHNI (1953). He reports that the vegetation is predominantly of the Andaman type, with a mixture of Malayan and Sumatran species, but Dipterocarps are not seen. Tree ferns, not known in the Andamans are common in the moist valleys of the Great Nicobar. He recognizes the following vegetational types: 1. Beach forest with the shrub *Scaevola frutescens* in gregarious formations, associated with *Hibiscus tiliaceus* and *Clerodendron inerme*. There are impenetrable thickets of *Pandanus*. On dry sand are *Ipomoea pes-carrae*. *Tourmefortia argentea* is common in South Nicobar. 2. The littoral forest is from a few metres to nearly 1.5 km wide, composed of *Barringtonia asiatica*, *Erythrina indica*, *Thespesia populnea*, *Pongamia glabra*, *Heritiera littoralis**, *Calophyllum inophyllum*, *Terminalia catappa* and *Casuarina*. 3. Mangrove forest along tidal creeks, which unlike those of the Andaman do not penetrate far inland and are formed of *Bruguiera conjugata*, *Carallia brachiata*, gregarious patches of *Sonneratia acida*, *Nepa*, *Areca*, etc. 4. The evergreen forest is more extensive and clothe the hills and level areas and are composed of *Calophyllum soulattri*, *Sideroxylon longipetiolatum*, *Garcinia xanthochymus*, *Adenanthera pavonina*, *Albizzia lebeck*, *Pisonia umbellifera*, *Mangifera sylvatica*, with the under-cover species of *Myristica* and *Polyalthia*. 5. The deciduous forest in small strips of trees that shed leaves in the hot season, for example *Terminalia biolata*. He believes the Nicobar Islands to be rich in endemics.

The Mollusc *Oreobba* is known at present only from the Nicobar Islands and *Haughtonia* from the Andamans. The Cerambycid beetle *Artimpanza*, represented by one species each in the Andaman and Nicobar Islands, is also known, for example, from Sumatra and Borneo. The islands have further a number of endemic elements, Indo-Chinese and Malaysian forms and a few Peninsular relicts. The species which are peculiar to the Andaman-Nicobar Islands include the snakes *Typhlops andamanensis* in the Andamans, *Liopeltis nicobariensis* in the Nicobars, *Oligodon woodmasoni* in the Andaman and Nicobar Islands, *Ahaetulla ahaetulla andamanensis* only in the Andamans, *Nastrix nicobarensis* to the Nicobars, *Trimeresurus purpureomaculatus andersoni* in both the Andamans and the Nicobars and *T. labialis* in the Nicobars. The other endemic elements of the Andamans are the earthworms *Pheretima osmastoni* and *P. suctoria;* the Pentatomid bug

---

* *Heritiera minor* is the currently accepted name.

Cordonchus; the Chrysomelid *Leptoxena* and the Lepidoptera *Graphium agamemnon andamana* (a Malaysian genus), *Graphium agamemnon epaminondas, Troides helena heliconoides* (Malaysian genus), *Chilasa clytia flavolimbatus* and *Valeria ceylanica naraka.* The Lepidoptera peculiar to the Nicobars are *Graphium agamemnon decoratus, Graphium agamemnon pulo, Troides helena ferrari* and *Polydorus aristolochiae camorta.* The Assam-Burma butterfly *Papilio memnon agenor* extends to the South Nicobar Island. The Indo-Chinese and Malaysian elements of the Vertebrates are exemplified by *Xenopeltis unicolor* in the Andamans, *Python reticulatus* in the Nicobars, *Elaphe oxycephala, Ahaetulla cyanochoris* in the Andamans and Nicobars, *Boiga ochracea* in Burma, Andamans and the Nicobars and *Boiga ceylonensis* in South India, Ceylon and the Andamans.

REFERENCES

BLANFORD, W. T. 1901. The distribution of Vertebrate animals in India, Ceylon and Burma. *Philos. Trans. R. Soc. London* (B) 194: 335–436.

BUTLER, A. L. 1899–1900. The birds of the Andaman and Nicobar Islands. *J. Bombay nat. Hist. Soc.* 12: 386–403; 555–571; 684–696 (1899); 13: 144–154 (1900).

CLARKE, C. B. 1898. Subareas of British Empire, illustrated by the detailed distribution of Cyperaceae in that Empire. *J. Linn. Soc. London,* 34: 1–146.

CORBET, A. S. & H. M. PENDLEBURY, 1956. The butterflies of the Malay Peninsula. London.

HOOKER, J. D. 1906. A sketch of the flora of British India. London.

KEMP, S. 1914. Onychophora. Zoological Results of the Abor Expedition. *Rec. Indian Mus.* 8: 471–492, pl. xxxiv–xxxvii.

KURUP, G. U. 1966. Mammals of Assam and adjoining areas. 1. Analytical study. *Proc. zool. Soc. Calcutta,* 19: 1–21.

MILLER, G. S. 1902. Mammals of the Andaman and Nicobar Islands. *Proc. U.S. Mus.* 24: 751–795.

PARKINSON, C. E. 1923. A forest flora of the Andaman Islands.

ROONWAL, M. L. & O. B. CHOTANI, 1965. Zoogeography of termites of Assam region, India, with remarks on speciation. *J. Bombay nat. Hist. Soc.* 62(1): 19–31.

SAHNI, K. C. 1953. Botanical Exploration in the Great Nicobar Island. *Indian For.* 79(1): 3–16 pls. 3., maps 2.

SMITH, M. A. 1931, 1935, 1943. Loricata, Testudines. Fauna British India, 1: 185 (1931); Sauria, 2: 1–440 (1935; Serpentes, 3: 1–583 (1943).

XXI. BIOGEOGRAPHY OF THE HIMALAYA

by

M. S. MANI

1. Introduction

The outstanding peculiarities of the ecology of the Himalaya may be traced to its enormous massiveness, the great elevations of the mountain ranges, their trendlines, their location in the middle of a vast continental mass, their Tertiary orogeny, the Pleistocene glaciations and continued Post-Pleistocene uplift. The conditions commonly met with on other mountain ranges of the world hardly give any clue to those likely to prevail on the vastly amplified and much higher life zones on the Himalaya. Although situated only a few degrees north of the torrid zone, owing to its enormous size and its unparalleled elevation, we find here a complete range from the tropical to the deep arctic conditions. Extending nearly 3000 km from the east to west, it is obvious that the conditions must differ profoundly in the extreme eastern and western ends of the Himalaya. The general climatic conditions in the east are semi-oceanic, but become more and more continental as we proceed westward.

The fundamental difference between the Himalaya and the Peninsula must be traced to the fact that the ecology of the Himalaya is temperature dominated. Except for parts of the extreme eastern end, the ecology of the Himalaya is almost wholly outside the general influence of the monsoon rainfall; indeed it is the Himalaya that plays an important rôle in determining and channelling the monsoon rainfall in the rest of the country (see chapter V). The atmospheric temperature is, on the other hand, the dominant factor that determines the wide range of ecological conditions, the altitudinal zonation of life, the east-west gradations of ecosystems and distributional patterns and numerous other peculiarities of the biogeography of the Himalaya. It is this fundamental difference that determines the wide difference in the ecology and distribution of animals and plants in the eastern and western ends of the Himalaya. The altitudinal zonation, due to the temperature stratification of the atmosphere, is associated with an abrupt difference in the ecology and fauna at the forestline; the forestline forms a characteristic threshold or transition between the ecosystems and fauna within the forest zone and those above. The forestline and the showline constitute two extremely important ecological and biogeographical transitional zones on the Himalaya.

South of the main crestline of the Great Himalaya (see chapter II), the lower hills are covered by broad-leaved wet forest (lower monsoon forest) up to elevations between 900 and 1000 m. Above this zone succeed the middle and upper montane evergreen forests. At elevations between 2440 and 3050 m there is the broad-leaved sclerophyll or the *Quercus-Rhododendron* forest; in the Northwest Himalaya this zone reaches up to 1800 m. In the eastern divisions there is the mischwald of *Rhododendron* and conifers at slightly higher elevations. These give place above to the *Betula-Juniperus* zone at elevations between 3000 and 3660 m in the western divisions and a shrub zone of *Abies spectabilis*, with thickets of *Rhododendron campanulatum*, succeeded by dwarf *Rhododendron anthopogon* and *Rhododendron setosum* at elevations between 4260 and 4575 m in the eastern divisions. This belt may often ascend to 5180 m in the eastern divisions of the Himalaya. In the west the *Betula-Juniperus* zone marks the upper limits of the forest and gives place to the open vegetation of the alpine-zone type. At much higher elevations there are plant cushions and prostrate plants, with grass and sedge. This zone may be met with even at an elevation of 4575 m and may often extend as high as 5790 m, especially on the southern slopes, where the snow meets the dwarf shrub vegetation. On the north slope the picture changes into a typical desert prostrate vegetation. At elevations above the zone of plant cushions, the vegetation is confined to the base of boulders and large rock pieces. Phanerogam cushions are found even far above 6000 m in the Nepal Himalaya. Cushions of *Stellaria decumbens* grow, for example, at elevations of 6140 m on the north slope of the Mt. Makalu area (SWAN, 1961, 1963), but phanerogams flourish perhaps up to even 6300 metres. The zone between elevations 3960 and nearly 6000 m in the eastern divisions and above 3050 m in the Northwest Himalaya is generally called alpine zone. The alpine zone on the Himalaya is perhaps the most extensive and also the highest in the world. The *Betula-Juniperus* belt in the west corresponds to a subalpine transitional zone. This zone is also taken as the timberline in the Himalaya. The timberline in the Northwest Himalaya is generally much lower than in the rest of the Himalaya and vast areas of the Northwest Himalaya are really far above the upper limits of the closed forest. The timberline in the eastern Himalaya is at about 4115 m on the south slope in the inner valleys and 3600–3700 m on exposed ridges, but in the Northwest Himalaya the timberline is somewhat above 3000 m. Even in the Northwest Himalaya, the timberline is at 3600–3700 m on the south slope of the Dhauladhar Range and 3000 m on the Pir Panjal; the northern slopes of the Pir Panjal are barren and the ranges beyond are much above the timberline elevations. Normal snowfall is relatively small in the Northwest Himalaya. The moisture-laden winds from the Indian Ocean seldom penetrate beyond the Pir Panjal Range, so that in the Upper Indus valley the total annual rainfall hardly amounts to 75 mm and this is also extremely irregular. The prevailing atmospheric temperature at the snowline, in the region of scanty rainfall and snowfall, is always very much below the freezing point of water. Owing to more abundant snowfall on the southern slopes of the Northwest Himalaya, the permanent snowline is lower than on the north slope. In the Kumaon division of the Himalaya winter snowfall occurs regularly at an elevation of 1980 metres and often at 1520 m. The lowest elevation at which winter snowfall has been recorded in recent years is 750 m and on certain occasions heavy snowfalls have been observed even in the duns between the Siwalik and Lesser Himalayan ranges. On the innermost Himalaya, snowline, as commonly understood, would really convey no meaning. Vast areas of rock are snowless even at very great elevations, largely because of the extreme atmospheric aridity. Deep tongues of snow descend, however, to an elevation of 5485 m and large continuous slopes of mountains remain free from snow even at an elevation of 7000 m on the northern aspect of the ranges. The mean value of 5170 m usually given for the snowline on the Nepal Himalaya is also misleading to a great extent. On the south slopes of the Nepal Himalaya, snow is abundant during June–September even at elevations of 4575–4780 m, but in the inner Himalaya in the same area, the mean atmospheric precipitation drops off abruptly, so that the northern gradient of light snowfall raises the snowline. The snowline is on the average about 1000 m lower on the south slope of the Great Himalaya than on the north slope. In the Northwest Himalaya, the snowline is higher than in the rest of the Himalaya. As we proceed westward from the Assam Himalaya, there is a gradual increase in the general atmospheric aridity and consequently the snowline also rises (MANI, 1962, 1968).

Compared to the Peninsula, the Himalaya is extremely rich in relatively very young and phylogenetically highly plastic forms of more recent and more highly evolved Asiatic groups, with a corresponding poverty of the ancient Gondwana elements. The mountain-autochthonous and endemic elements of the Himalaya, characterized by high ecological specialization, are as a rule concentrated at higher elevations, particularly in the life-zones above the forest-line. On ecological, biogeographical and evolutionary grounds, the fauna of the forest zones and that of the biotic zones above the forest are best considered separately.

2. The Fauna of the Forest Zones

The slopes from the base of the hills to elevations of 3000–4200 m, the upper limits of the forest, might be divided into two or three zones. It is, however, difficult to obtain accurate information regarding the precise ranges of most species on the Himalaya and in most cases the ranges of the species vary within wide limits depending upon the season. The forested slopes of the Himalaya form a belt of very variable width, between the Indo-Gangetic Plains on the one hand and the Turkmenian Subregion of the Palaearctic on the other hand. The forested area is indeed rather difficult to define biogeographically and the transition from the one to the other is not always sharp, but lies near the forest-line.

The forest-line is itself a transition zone, going higher and coming down locally. A number of genera that belong to the tropical lowlands penetrate up the valleys and some of the Turkmenian forms also descend on the cold slopes in deep tongues well within the upper limits of the forest. The admixture of the Palaearctic forms increases gradually westward, where the limits of the forest are also more restricted and the fauna passes into that of the Palaearctic. These facts hold true for the distribution of not only higher plants like Phanerogams but as shown by BRÜHL (1931) also Bryophyta. He found, for example that the distribution of moss on the Himalaya reveals a natural division into a western and an eastern part, towards the west, the Himalayan mosses are related to those of Iran, Caucasus and the Alps, but in the east the affinities are with South India and Ceylon on one hand and Burma-Malaya on the other hand.

The Himalaya is more densely forested in its eastern parts than in the west. The fauna of the forest zones is markedly poor in mountain-autochthonous and endemic forms. It is composed largely of tropical elements derived from the fauna of the Indo-Chinese and Malayan Subregions of the Oriental Region. These essentially tropical faunas have even spread north of the Himalaya into eastern parts of Tibet and may indeed be detected even as far north as the Manchurian Subregion of the Palaearctic Region. The Oriental faunal derivatives on the Himalaya disappear gradually to the west, but some of them may still be found in parts of Kashmir. There is thus a pronounced westward fall in the gradient of the Oriental fauna in the forest zone of the Himalaya.

The fauna of the western parts of the forest zones of the Himalaya is composed partly of the mountain-autochthonous derivatives of the already greatly attenuated Oriental fauna, which have spread westward from the Eastern Himalaya, largely however of Palaearctic forms (endemics, Mediterranean-Turkmenian) and to some extent also Mediterranean-Ethiopian elements. Though the Indo-Chinese and Malayan genera and species are concentrated largely in the Eastern Himalaya, many of them have thus spread, with decreasing abundance, westwards up to Kumaon and sometimes even as far west as Kashmir. The Palaearctic-Ethiopian genera and species are similarly dominant in the Northwest Himalaya, but have likewise spread sparsely eastwards to Nepal and rarely even round Assam southwards to the north Burmese mountains. There is perhaps a more complete intermingling of the Oriental and Palaearctic-Ethiopian elements among Vertebrates than Invertebrates on the Himalaya. The biogeographical transition between the eastern humid tropical and the western, largely steppes fauna in the Himalayan forest zone is situated in the great defile of the R. Sutlej. The concentration of Indo-Chinese and Malayan elements is largely to the east of this river, though the outliers have infiltrated further west. Analogous lies the concentration of the Ethiopian and Mediterranean elements to the west of this, but some species have likewise infiltrated further east. The peninsular isolates are predominantly met with in the eastern parts of the Himalaya. The distributional patterns are therefore 1. westward radiation of the eastern humid tropical faunas, 2. eastward radiation of western steppes faunas and 3. discontinuous distribution of Himalayan elements in the hills of South India and Ceylon.

2.1. The himalayan forest east of the R. sutlej

The part of the forest-covered Himalaya, to the east of the great defile of the R. Sutlej, may appropriately be described as a narrow tongue-like western appendage of the Indo-Chinese subregion, that ascends higher as we approach the eastern end of the Himalaya and curves round to blend in the forests of eastern Tibet. This tongue contains intrusions from the Malayan subregion, isolates from the Peninsula and of course infiltrations of the western faunal elements, becoming greatly attenuated a short distance to the east of the defile. We thus observe that the dominant components of these forests are the Indo-Chinese elements in almost all groups.

The common earthworms (STEVENSON, 1933) occurring in the forest zones of the Himalaya like *Perionyx baini, Perionyx naianus, Perionyx simlaensis, Eutyphoeus annandalei, Eutyphoeus masoni, Eutyphoeus nainianus, Eutyphoeus orientalis, Eudichogaster parvus, Plutellus sikkimensis, Megascolides bergthelli, Notoscolex oneili, Megascolex dubius, Octochaetus hodgarti*, etc. are relatively young forms that have mostly Indo-Chinese and Malayan affinities.

The Himalayan Mollusca are largely Oriental forms in the eastern parts. *Oxytes* extends westwards, for example, from Indo-China and Thailand across the Assam-Burma mountains, to the Eastern Himalaya. *Pseudopomatias* and *Plectolyis* are common to South China, Assam-Burma mountains and the Eastern Himalaya. Genera like *Dalingia, Cryptaustenia, Austenia, Girasia* and *Rahula* are common to the Assam-Burma mountains and the Eastern Himalaya. The Malayan *Aurgella, Schistoloma*, and *Alycaeus* are also found in the Eastern Himalaya. *Khasiella* from Assam-Burma mountains extending across the Eastern Himalaya up to the Northwest Himalaya and *Diplommatina* extending from Formosa, New Guinea up to the Northwest Himalaya are some examples of the Oriental derivatives which have intruded into the western parts of the Himalaya.

The Onychophora *Typhloperipatus* (KEMP, 1914) from the forest-zone of the extreme Eastern Himalaya (Abor Hills), referred to in an earlier chapter, is related to the Malayan *Eoperipatus*. The Phasmida like *Marmessoidea* and *Clavisia*, found in the forest zone of the Eastern Himalaya, are Malayan forms, widely distributed in Sumatra, Java, Borneo, Celebes, Burma, Nicobar Islands, west China and the Philippines. The Coreid bugs *Ochrochira* and *Helcomeria* are common to Assam-Burma mountains and the Eastern Himalaya; the latter genus extends sparsely westwards

668

up to the Kumaon division. *Brachyaulax* extends from Java through Malaya and Burma across Assam to Sikkim and *Lamprocoris* from Java across Sumatra and Malaya Burma, Assam on the Himalaya up to Nepal. The Chrysomelid *Merista*, found commonly on the Eastern Himalaya, extends also up to Yunnan across Assam-Burma. *Paralina* extends similarly from Yunnan across Assam on the Himalaya up to Kumaon and *Potaninia* from South China across Assam to the Eastern Himalaya. The Trictenoto-mid *Autocrates*, found up to Darjeeling on the Eastern Himalaya, is a Yunnan derivative. The bulk of the Cerambycidae from the Eastern Himalaya is typically Malayan forms and include *Typodryas*, *Gnatholea*, *Euryphagus*, *Arctolamia* and *Calloplophora*, most of which are represented by many species in Malaya, Burma and Thailand. The Scarabaeiid *Lutera* extends from Tonkin and Thailand to Java, Sumatra, Borneo, Celebes, Philippines and the Himalaya. *Jumnos* is common to Assam-Burma mountains and the Eastern Himalaya and *Bombodes* extends from Tonkin to the Eastern Himalaya.

The Eastern Himalaya is exceptionally rich in Lepidoptera (GROSS, 1961; MELL, 1958), the greatest bulk of which are essentially Indo-Chinese and Malayan forms, though some Manchurian and extremely few Turkmenian forms may also be found, particularly in the upper levels of the forest. The species belong to Oriental tropical forest genera. In most cases the species found in the Eastern Himalaya are identical with those occurring in Indo-Chinese or Malaya, but often there is a completely graded series of localized mountain-autochthonous subspecies and races, as we proceed westward on the Himalaya, from Assam through Nepal and Kumaon to the Northwest Himalaya (see chapter XV). The Indo-Chinese genus of butterflies *Chilasa* is, for example, largely concentrated in the Eastern Himalaya, but a few local subspecies extend westwards up to Kashmir. *Chilasa agestor agestor* extends, for example, from Tonkin to Sikkim, *Chilasa agestor govindra* occurs in Kumaon to Kashmir and *Chilasa agestor chiraghshai* occurs in Kashmir (Fig. 154). In the Eastern Himalaya, we have other species like *Chilasa epycides* extending east across Burma and China to Formosa. *Chilasa slateri slateri* is common to the Eastern Himalaya, Burma, Malaya, Sumatra and Borneo. *Graphium glycerion* extends from central China across north Burmese mountains to the Himalaya, along which it has spread westwards to Nepal. *Papilio bootes* and *Papilio rhetentor* extend from Hainan through China and north Burmese mountains and the Eastern Himalaya nearly up to Garhwal. *Papilio protenor*, from Formosa Hainan, Tonkin, China and Burma, extends across the Eastern Himalaya nearly to Kashmir and another species *Papilio polycitor* even up to Chitral. A number of other species of *Papilio* of Chinese-Japanese origin, extend across the Burmese mountains along the Himalaya to Kashmir. *Papilio memnon* of South Japan, Borneo, Sunda Islands, Nicobar Islands and Burma, occurs also in Sikkim. The subspecies *Papilio memnon agenor* is common to Sikkim, Assam-Burma mountains and the Nicobar Islands. Many Malayan *Troides*, like *Troides helena cerberus*, occur in parts of Sikkim Himalaya (also reappearing discontinuously in the Orissa Hills of the northeastern end of the Peninsula). The other Malayan *Troides* from the Himalaya include *Troides aeacus aeacus* occurring in Formosa, west China, Malacca and Burma and extending westward up to Garhwal Himalaya. *Polydorus aidoneus*, from South China, Shan States, Tonkin and Hainan, extends through Sikkim to Garhwal Himalaya. *Polydorus varuna* extends from the Malay Peninsula and Malaya Archipelago and Thailand to Sikkim. *Polydorus latreillei* extends from Sikkim to Garhwal and is represented by *Polydorus latreillei kabura* on the Naga Hills (Assam). *Polydorus philoxenus*, from Formosa, China, Tonkin, Annam and Burma, occurs also from Nepal to Kashmir; it is represented by the subspecies *Polydorus philoxenus philoxenus* in Indo-China across Burma to Kashmir and by *Polydorus philoxenus polyeuctes* in Yunnan, Thailand, Tonkin, Burma and Sikkim. *Aemona* occurs in Assam-Burma mountains and the Eastern Himalaya. *Armandia* is common to west China-Tibet, Burma and the Eastern Himalaya. The beautiful *Teinopalpus imperialis* is common to Assam-Burma mountains and the Eastern Himalaya. *Amathuxidia* and *Thaumantis* are common to Hainan, Borneo, Java, Sumatra, Assam and the Eastern Himalaya. The Chinese *Graphium sarpedon sarpedon* extends from South China, Japan and Burma through to Kashmir (It is interesting to note that *G. sarpedon teredon* is known from the Peninsula and Ceylon). The Indo-Chinese *Graphium nomius*, from Hainan, Annam, Thailand, Tonkin, and Assam-Burma mountains, occurs in Sikkim (and reappears in South India and Ceylon). The Malayan *Graphium cloanthus cloanthus*, which occurs also in China, Formosa, Sumatra and Burma, extends on the Himalaya up to Kashmir. Other interesting species include *Graphium doson* from South Japan, South China, Sunda Islands, Thailand, Annam, Tonkin, Burma extending on the Himalaya to Kumaon and reappearing in South India and Ceylon and *Graphium agamemnon*, from South China, Australia, Solomon Islands, Andaman Islands, Nicobar Islands and Burma, extends also on the Himalaya up to Kumaon (and reappears on the Nilgiri Hills in South India and in Ceylon). *Graphium eurous*, from West and Central China, is represented on the Himalaya by *caschmirensis* from Kumaon to Kashmir. The Malayan-Indo-Chinese *Prioneris* is represented by *Prioneris thestylis* from Formosa, South China, Hainan, Malay Burma over to Kumaon and *P. clemanthe* from Sumatra, Malay Peninsula, Hainan and Burma to Sikkim. The Papuan-Indo-Chinese *Veleria* occurs in Eastern Himalaya. *Dercus*, from Borneo, Thailand, China, Tonkin, Malay Peninsula, Sumatra and Burma, extends through Assam mountains to the Eastern Himalaya and to southeast Tibet.

The Satyridae are largely concentrated in the Eastern and Nepal Himalaya within the forest zone and above the limits of the forest in the Northwest Himalaya. *Mycalesis francisca*, from Japan, Formosa, China and Annam, occurs, for example, from Burma through Assam to the Kumaon Himalaya and *M. perseus blasius* extends from Burma to Kangra in the Northwest Himalaya. Nearly 80 % of species of *Lethe* are found on the Himalaya, but some species extend through Burma to Malaya, Thailand, Hainan, Formosa, Philippines and Java and a few occur even in South India and Ceylon. The reader may also refer to Chapter XV for a fuller discussion of the distribution of butterflies.

The Eastern Himalaya is richer in endemic species of fishes than the western parts of the Himalaya. This is in part the result of colonization from Indo-Chinese mountainous areas; in part also because the eastern end is older than the Western end of the Himalaya. There are a number of species, which are confined to the Eastern Himalaya alone and do not extend westward beyond Nepal. These include *Psilorhynchus balitora, Balitora brucei brucei, Noemacheilus beavani, N. savona, N. scaturigina, Gagata nangra* and *Glyptothorax rebeiroi*. The species which are endemic to the extreme Eastern Himalaya include *Tor progenius, Schizothorax progastus, Noemacheilus sikkimensis, Batasio batasio, Lepidocephalichthys annandalei, Mystus (Mystus) montanus dibrugarensis, Erethistoides montanum montana* and *Glyptothorax striatus*. From Sikkim we have records of a number of interesting species like *Silurus cochinchinensis, Olyra longicauda, Glyptosternum maculatum, Schizothorax molesworthi, Danio aequipinnatus, Aspidoparia java, Brachydanis rerio, Chagunius chagunio, Puntius conchonius, P. sophore, P. ticto, P. sarana, Labeo dero, Crossocheilus latius, Garra lamta, Aborichthys elongatus, Noemacheilus devdevi, Channa orientalis, C. punctatus, Xenentodon caucila, Mastacembelus armatus, Colisa chuna, Ambassis* spp., etc. The Amphibia *Leptobrachium* and *Tylototriton* illustrate respectively the presence of a Malayan and an Indo-Chinese element. The Malayan *Microhyla* has penetrated westwards up to Kashmir.

The reptiles occurring in the forest zone of the Himalaya are, likewise, largely Indo-Chinese and Malayan in the east, but Turkmenian and Mediterranean-Ethiopian in the west. *Japalura* is derived from Southwest and Central China and extends east to Formosa and south to Borneo and Sumatra. *Japalura tricarinata* occurs at elevations of 1000–2700 m from Sikkim to Nepal, *J. major* from Garhwal to the Northwest Himalaya and rising up to elevations of 2500 m, *J. kumaonensis* from Kumaon and *J. variegata* at elevations of 300–2700 m in Sikkim. *Platyura platyura* is an interesting eastern form that occurs from Indo-China, Formosa and the East-Indies and extends across the Eastern Himalaya to Nepal (and also reappears discontinuously in Ceylon). *Leiolopisma* known from Polynesia, New Zealand, Tasmania and Africa and North America, is represented by three species on the Himalaya: *L. ladakense* and *L. himalayanum* in Kashmir and Kumaon and *L. sikkimense* on the Eastern Himalaya and also on the Pareshnath Hills at the extreme northeast corner of the Peninsula. *Ophiosaurus gracilis* extends from the Eastern to Simla up to elevations of about 2400 m. *Typhlops jerdoni* is common in the Eastern

Himalaya and Assam-Burma mountains, but *T. oligolepis* is confined to the Eastern Himalaya. *Elaphe radiata* extends from South China and Indo-China westward to the Eastern Himalaya and south to Malaya and is also known from the Orissa Hills in the Peninsula. *E. porphyracea* extends from Yunnan to Eastern Himalaya and south to the Assam-Burma mountains, Malaya and Sumatra. *E. cantoris* is confined to the Eastern Himalaya and Assam-Burma mountains. *Zaocys nigromarginatus* extends from Yunnan to the Eastern Himalaya, up to elevations of 2100 m. *Oligodon* is known from the Eastern Himalaya and *Dinodon* occurs in the Eastern Himalaya and Thailand. The Gondwana derivative *Sibynophis*, from Madagascar and Central America, is represented by *S. collaris* up to elevations of 3000 m in the Eastern Himalaya and also on Assam-Burma mountains and extending westward to Garhwal and eastwards to Yunnan, Thailand, Annam and south to Malaya. *Natrix parallela* extends from Yunnan-Tonkin and North Burma to Assam hills and Eastern Himalaya. Of the two other species of *Natrix*, known from the Himalaya, *N. himalayana* is confined to the Eastern Himalaya and Assam-Burma mountains and *Natrix platyceps* extends from Assam to Kashmir up to elevations of 1500–1800 m. *Trachischium* extends from Assam to Gilgit. *Boiga ochracea ochracea* and *B. gokool* are confined to the Eastern Himalaya and Assam, *B. cyanea* extends from Indo-China to the Eastern Himalaya and Assam-Burma mountains and *B. multifasciata* extends westwards along the Himalaya to Mussurie. *Psammodynastes pulverulentus* extends from Indo-China and South China westwards along the Himalaya to Nepal and south to the Malay Archipelago. *Bungarus bungaroides* is common in Assam-Burma and Eastern Himalaya, but *B. niger* is confined to the Eastern Himalaya and Assam Hills. *Ancistrodon himalayanus* extends from Assam to Chitral and is common at elevations of 2100–3000 m, but in the Northwest Himalaya may be found even up to an elevation of 4800 m. This is perhaps the highest elevation at which snakes seem to be found in the world. *Trimeresurus monticola* extends from Yunnan to Southeast Tibet and across Assam-Burma mountains to the Eastern Himalaya and southwards to Malaya. *T. erythrusus* is confined to the Himalaya and Assam-Burma mountains. *T. allolabris*, from Indo-China, China, Formosa, Burma, Andamans Nicobars, Thailand, Malaya, Sumatra and Java extends west on the Himalaya (and is also known from the Indian Peninsula).

The Eastern Himalaya is richer in birds than the west, particularly among the tropical and Oriental elements. The Paradoxornithinae are, for example, completely restricted to the Eastern Himalaya. *Silviparus* seems to be confined to the Eastern Himalaya and the mountains of Assam. *Paradoxornis* and *Suthora* are restricted to the Eastern Himalaya, north Burmese mountains and to the mountains of South China. *Garrulus*, *Trochalopterum* and *Pomatorhinus* are likewise common to the Himalaya and north Burmese mountains but are represented by local forms on the mountains of South India and in hills of Ceylon. Timeliinae are also common to the Himalaya, and the Assam-Burma mountains. Sibiinae are abundant on the Eastern Himalaya and Assam-Burma mountains. *Liothrix, Cutia, Peruthius, Mesia* and *Minla* are confined to the Eastern Himalaya, Assam-Burma mountains. *Hypsipetes* is interesting in occurring on the Himalaya, Burmese mountains (and discontinuously on the hills of South India). Sittidae seem to be restricted to the extreme northwest. A number of species of *Certhia* occur on the Himalaya and Assam-Burma mountains. The Turdinae are confined to the Himalaya and some species of finches are also Himalayan, but one species extends south to the hills of Shan States in Burma. The interesting long-tailed broad-bill *Psarisomus dalhousae* extends from Borneo to the Himalaya. A number of woodpeckers are typically Himalayan and are particularly common the Eastern Himalaya and extend to the Assam-Burmese mountains and a few occur in the South Indian hills also. *Tiga* occurs, for example, in the Himalaya, Burma and in Malabar. The broad-billed roller *Eurytomus orientalis* occurs on the Himalaya, in Burma, Malabar and Ceylon. The bee-eater *Nyctiornis* is interesting in occurring on the Eastern Himalaya, Burmese Mountains, Malabar and Orissa Hills. The hornbills *Anthracoceros* occur on the Eastern Himalaya, Burma, Malabar and Ceylon. Similarly *Loriculus* among the Psittaci occurs in Eastern Himalaya and north Burmese mountains and also reappears in Malabar. *Photodilus* is one of the Striges common to the Himalaya, Burma and Ceylon. The grey-peacock peasant *Polyplectrum chinquis* is common on the Himalaya to the east of Sikkim and on the mountains of Assam-Burma. *Tragopan satyra* occurs in Bhutan, Sikkim and extends to Nepal and the genus is also known from China and Middle Asia.

The Malay-squirrel *Sciuropterus bicolor* occurs on the Eastern Himalaya and on Burmese mountains and *Sciurus lorica* is found in Eastern Himalaya and hills of Assam. *Rhinoceros unicornis* was formerly common up to Peshwar, but is now confined only to the base of the foothills of the Himalaya in Nepal (see chapter XII). *Soriculus* is found in the Eastern Himalaya. *Chimarrogale himalayanica* is found in the Himalaya and north Burmese mountains, at elevations of 3000–4500 m. *Nectogale sikkimensis* occurs in the higher parts of the Himalaya, the Malay Archipelago, South Tenasserim and in South India. The flying squirrel *Eupetaurus cinereus* is known from Tibet and *Pteromys* and *Sciuropterus* are from the Himalaya and Burmese mountains. *Tupaia*, represented by one species in the Eastern Himalaya and Assam hills, is distributed discontinuously in the Nicobar Islands and in the Peninsula. *Rhizomys* is confined to the Eastern Himalaya and the mountains of Assam-Burma.

2.2. THE HIMALAYAN FORESTS WEST OF THE R. SUTLEJ

Ecologically and biogeographically the Himalayan forests, west of the great defile of the R. Sutlej, are strikingly different from those in the east.

We find here, in addition to the greatly attenuated Indo-Chinese and Malayan faunal complex, an increasing abundance of the Mediterranean and Ethiopian faunas, with an occasional infiltration of the Turkmenian elements at higher elevations, particularly in the extreme northwest. In marked contrast to the humid tropical forest types predominant in the east, the steppes types constitute the dominant forms here.

The Mediterranean, Mediterranean-Ethiopian and Mediterranean-Turkmenian affinities of the western parts of the Himalaya are illustrated by the Oligochaeta *Eutyphoeus;* the Mollusca *Parvatella* (up to an elevation of 3000 m), *Euaustenia (E. monticola, E. cassida* and *E. garhwalensis)*, *Syama* (2700–3160 m), *Bensonia* extending east up to Sikkim and ranging from an elevation of 1000 to 3600 m (Zonitidae), *Pyramidula* (Endodontidae) (also discontinuously on the Nilgiri Mountains and in Ceylon), *Cathaica* (Helicidae) extending east along the Himalaya to the north Burmese mountains, *Vallonia ladakensis* common to the Tien Shan and the North-west Himalaya, *Ena* (Enidae) with *Ena (Mirus) ceratina* up to Kumaon (also discontinuously on the Nilgiri-Anamalai-Palni Hills, and Khasi Hills), *Ena (Subzebrinus);* the slugs (Arionidae) *Anadenus altivagus* and *Anadenus giganteus* extending eastwards up to Nepal. Some of the Palae-arctic elements occur nearly throughout the Himalaya, particularly in the upper reaches of the forest zone. Others like the Odonata *Epiophlebia laidlawi* occur only in the Eastern Himalaya (the family is represented discontinuously by *Epiophlebia superstes* in Japan). The Coleoptera Lucanidae occur in most parts of the Palaearctic and are represented by *Lucanus* and *Pseudolucanus*, especially in the Eastern Himalaya. The Buprestid *Capnodis* extends from south-central Europe across Iran-Afghanistan and Middle Asia up to Kumaon on the Himalaya. Psephe-nidae (Dryopidae) are represented by *Psephenus tenuipes* on the Himalaya from the west to Kumaon. The Lepidoptera *Pararge* is confined mostly to the Northwest Himalaya. *Rhaphicera* extends eastwards from the Himalaya to West and Central China. *Coenonympha* is confined only to higher ele-vations in the upper reaches of the forest in the Northwest Himalaya and the same is also true of others like *Maniola, Hipparchia, Oeneis* and *Erebia*, which become increasingly abundant only above the forestline. The Palaearctic *Papilio machaon* extends along the Himalaya, from Baluchistan to Assam-Burma mountains. Some species of the Holarctic *Pieris* descend to within the forest zone of the Himalaya and also extend southwards on the Burmese mountains to reach the outpost in the Andaman Islands (distributed also discontinuously on the Nilgiri Hills). The Holarctic *Pontia* and *Euchloë* are largely confined to the Northwest Himalaya. The Palaearctic *Gonepteryx*, entering through the northwest, has extended along the Himalaya to Burma. *Colias*, though abundant above the forest-line, appears sparsely within the forest (and discontinuously on the Nilgiri Hills). *Helodrilus mariensis, Helodrilus prashadi* and *Helodrilus kempi* are believed by STEVENSON to represent the eastern outliers of Palaearctic

Lumbricinae, which have advanced from the northwest along the Himalaya; this advance is of very recent date, perhaps since the retreat of the last Pleistocene glaciers from the Himalaya.

The genus *Gymnodactylus* has extended along the Himalaya to the Burmese mountains and the regions further east; *Gymnodactylus chitralensis* occurs in Chitral. *Gymnodactylus stoliczkai* occurs in Kashmir, Chitral and Ladakh and *G. lawderanus* extends from the Northwest Himalaya to the Kumaon Himalaya; *Gymnodactylus fasciatus* occurs in the Kumaon-Garhwal Himalaya at elevations of about 1500 m. *Agama himalayana* occurs in Kashmir and Middle Asia and may be met with up to elevations of 3300 m, *A. tuberculata* extends from Afghanistan through the Northwest Himalaya to Nepal. *Phrynocephalus* is a Middle Asiatic form that is found in the Northwest Himalaya and in Baluchistan. *Phrynocephalus theobaldi* occurs in Middle Asia, Tibet and Kashmir. Among birds, the distribution of *Corvus corax* is interesting; it is represented by a large race at higher elevations on the Himalaya and a small race that extends to the lowlands of the Punjab, Sind and parts of Western Rajasthan. The common magpie *Pica rustica* occurs in Kashmir, Baluchistan and extends to North Burma and the black-rumped magpie *Pica bottanensis* is found in Bhutan and Sikkim Himalaya. The blue magpies *Urocissa* and the racket-tailed magpies *Crysirhina* occur in the Himalaya and also extend to the north Burmese mountains. *Nucifraga* is Himalayan. The cornish chough *Pyrocorax* occurs at higher elevations. *Aegithaliscus*, the long-tailed tit-mouse, is also restricted to the Himalaya and the Burmese mountains. The crested titmouse *Lophophanes* occurs at elevations above 1800 m and the lämmergeyer, *Gypaëtus barbatus* occurs on the Himalaya, the hills of Punjab and Sind and the golden-eagle *Aquila chrysalis* is Himalayan. The passerine birds confined to the west and not extending to the Eastern Himalaya, include *Hypolais*, *Muscicapa*, *Saxicola*, *Carduelis*, *Callacanthis* and *Metaoponia*, which are migratory Palaearctic forms. Though some typically tropical forms, like *Brachypternus*, *Halcyon*, *Poliocaëtus* and *Sarcogrammus*, ascend to nearly the upper limits of the forest in the Western and North-west Himalaya, they are mostly confined to base of the forest in the Eastern Himalaya. Though found in the Northwest Himalaya, *Urocissa flavirostris*, *Garrulax albigularis*, *Trochalopterum lineatum*, *Stachyridopsis pyrrhops*, *Hodgsonius phoenicuroides*, etc. extend eastward in the Himalaya, but do not occur in the Eastern Himalaya. The birds largely restricted to the Northwest Himalaya, include *Machlolophus xanthogenys*, *Trochalopterum simile* and *Gennaeus albicristatus*.

Vulpes pusilla (= leuopus) occurs from the arid Northwest to the Himalaya. *Martes (= Mustela) foina*, the European beech-marten, extends from Afghanistan to the Kumaon Himalaya. *Putorius putorius larvatus*, the polecat, occurs in the Northwest Himalaya. A variety of the European *Ursus arctos* occurs at high elevations, usually above the forest, on the Himalaya, but Himalayan black-bear *Ursus torquatus* from the forest zone

675

extends also to the Assam mountains, Burma and South China in the east and Afghanistan and Baluchistan in the west. *Arctomys himalayanus*, *Arctomys hodgsoni* and *Arctomys caudatus* are Himalayan and are sometimes found at high elevations in North Kashmir and Tibet. A number of species of voles occur in the Northwest Himalaya, but only one species extends to the Eastern Himalaya and north Burmese mountains. *Lagomys* is confined to the higher elevations of the Northwest Himalaya, Baluchistan and Tibet. *Bos grunnieus*, the Tibetan yak, occurs sparsely in Kashmir. *Ovis hodgsoni*, the Tibetan sheep and *Ovis poli* the Pamir sheep, are sparsely found in the Northwest Himalaya. *Ovis vignei* occurs in the Northwest Himalaya and *Ovis nahura* occurs throughout the Himalaya. *Capra* and *Hemitragus* are mainly Himalayan but the latter is also found discontinuously on the Nilgiri Mountains in South India. *Capra sibirica* the ibex, *Capra falconeri* the markhor and *Capra aegagrus* the Persian wild-goat are found at higher elevations in the Northwest Himalaya. *Nemor-haedus bubalinus* occurs from Kashmir to Assam-Burmese mountains and *Cemas goral* occurs throughout the Himalaya, Assam and north Burma. *Moschus moschiferus* is common at higher elevations on the Himalaya.

A number of the Indo-Chinese and the Malayan faunal elements of the Eastern Himalaya are distributed discontinuously in the Peninsula, especially on the Nilgiri, Anamalai, Palni and Cardamom Hills and exceptionally even on the hills in Ceylon. The common examples include the Coleoptera *Trictenotoma*, and *Aplosonyx;* the Lepidoptera *Troides helena cerberus*, *Graphium nomius*, *Graphium doson*, *Graphium agamemnon;* the amphibian *Microhyla ornata*, the reptile *Leiolopisma sikkimense*, *Trimeresurus allolabris;* the birds *Garrulus*, *Trochalopterum*, *Pomatorhinus*, *Hypsipetes*, *Tiga*, *Eurytomus orientalis*, *Anthracoceros*, *Loriculus*, *Photodilus*, etc. The birds that occur discontinuously on the Eastern Himalaya and on the South Indian hills, but absent in Assam-Burma mountains, include *Dumetia*, *Hodgsonius*, *Thamnobia*, *Ptyonoprogyne*, *Neophron*, etc. and the migratory Palaearctic *Cephalopurys*, *Hypolais*, *Sylvia*, *Pastor*, *Sturnus*, *Muscicapa*, *Saxicola* and *Palumbus*. The mammals of the Eastern Himalaya, with discontinuous distribution in the Peninsula, include *Lutea cinerea*, *Nectogale sikkimensis*, etc. Some of the Palaearctic Himalayan elements are also known to occur discontinuously on the South Indian mountains. For example, the Lepidoptera *Colias*, the Mollusc *Pyramidula* (Endodontidae) and *Ena (Mirus) ceratina* and *Ena (Mirus) nilagirica*, etc.

3. Fauna above the Forestline

The ecology of high altitudes and the peculiarities of the high altitude insect of the Himalaya communities are recently discussed in detail (MANI, 1962, 1968). The fauna of the Himalaya above the upper limits of forest is fundamentally very different from that of the forest zones. The fauna is relatively sparse, characteristically lacking in tropical

676

Indian, south Chinese, Indo-Chinese and Malayan derivatives, but is composed almost exclusively of mountain-autochthonous, cold-adapted Palaearctic elements, the greatest bulk of which are strictly endemics that arose *in situ* and *pari passu* with the Himalayan uplift. The elevated areas above the forest belong to the Turkmenian Subregion of the Palaearctic Region.

The Vertebrates are of considerable interest on account of the absence of fishes, amphibians and the striking poverty of reptiles. Birds and Mammals are however, moderately abundant. The Vertebrata are extremely poorly represented in comparison to the Inevertebrates. Among the high altitude birds the snow-partridge *Lerwa nivicola* is a conspicuous member at extreme elevations. Choughs *Pyrocorax* are commonly found at elevations of 6000 m. The few mammals found above the forestline are typically Middle Asiatic forms like *Nectogale, Otonycteris, Eupetaurus, Arctomys, Sminthus, Cricetus, Lagomys, Ovis, Capra, Pantholops, Moschus,* etc. *Nectogale, Eupetaurus* and *Pantholops* are characteristic of the northern slopes and belong really to the Tibetan Tract of the Himalaya. The Carnivores include a number of interesting species like the snowleopard *Uncia (= Felis) uncia, Lynx (= Felis) lynx* and *Otocolobus (= Felis) manul* at higher elevations and in Tibet and occurring mostly above the limits of the forest. Though some typically Indian mammals and birds, may, however, appear even as far in the interior of these elevated regions as Gilgit, the Vertebrate life is in the main Middle Asiatic.

The great bulk of the Invertebrates occurring above the limits of the forest are typically terricolous and mostly also endogeous types, but many species are also found in the numerous glacial ponds, lakes, streams and rivers. There is a conspicuous increase in endemism in nearly all groups as we proceed westwards from the east. A similar increase may also be observed in the Palaearctic elements to the west. With the increase in elevation, the fauna as a whole becomes greatly sparser, but also richer in pioneer communities. The highest altitude of permanent existence of animals on the Earth is about 6800 m on the Himalaya, at which elevation occur a complex of Collembola, Diptera, Acarina and Salticid spiders (MANI, 1962, 1968; SWAN, 1961). Coleoptera exist at elevations up to 5600–5800 m and caterpillars of high altitude Lepidoptera breed at elevations of 5800 m, but the adults of a number of the Himalayan butterflies may be seen on wings even at an elevation of 6100 m. Orthoptera breed at an elevation of 4800 m, but nymphs of an unidentified grasshopper have, however, been found at an elevation of 5490 m on the Mt. Everest Massif. Nearly every group of high altitude animals flourishes at much higher elevations on the Himalaya than elsewhere in the world. The socalled arctic zones of the Himalaya, formerly assumed to be totally devoid of life, is known to be the home of a most unique community of organisms.

Ororotsia is an endemic mayfly from the Northwest Himalaya. The

677

stoneflies *Rhabdiopteryx lunata* and *Capnia pedestris* occur up to an elevation of 5000 m in the Mt. Everest Massif. A number of remarkable Orthoptera of Middle Asiatic endemicity are found at high elevations in the North-west Himalaya. The Tettigonid grasshopper *Hypsinomus fasciata* occurs elevations of 4575–4880 m. Other interesting high altitude grasshoppers belong to *Bryodema, Conophyma, Gomphomastax*, the endemic *Dicranophyma*, the Mediterranean-steppes *Sphingonotus*, etc. The Dermaptera *Anechura* found in the Northwest Himalaya is a Middle Asiatic form. The Heter-optera are interesting for the endemic genus *Phimodera* and endemic subspecies *Nysius ericae alticola, Dolmacoris deterrana*, etc. *Tibetocoris margaretae* occurs at an elevation of 5400 m, the highest altitude record for the order Heteroptera from the world. *Chlamydatus pachycerus* occurs at 5100–5340 m on the Northwest Himalaya. The typical high altitude Coleoptera, over 60% are endemic, are the Carabidae, Staphylinidae and Tenebrionidae. The Carabidae belong to *Amara, Bembidion, Bradytus, Broscus, Calathus, Calosoma, Carabus, Cymindis, Clivina, Chlaenius, Dyschirius, Harpalus, Leistus, Nebria, Tachys*, etc. *Chaetobroscus* is endemic in the North-west Himalaya. The highest altitude record of 5300 m for the Carabidae is reached by *Amara brucei* in the Northwest Himalaya; this species extends also to the area of Mt. Everest, where it occurs at an elevation of only 5030 m. *Bembidion nivicola* occurs at elevation of 5030 m. The Dytiscidae *Agabus (Guarodytes) adustus* occurs in glacial lakes at elevations of 3000–4870 m in the Northwest Himalaya. The family Amphizoidae, known from North America and Tibet and forming a transitional group between Carabidae and Dytiscidae, occurs also in the Northwest Himalaya. Hydrophilids like *Helophorus (Atracthelophorus) montana* occur up to 5400 m in sulphur springs in the Northwest Himalaya. Over 80% of the high altitude Staphylinidae from the Himalaya are endemic and belong to *Aleochara, Atheta, Geodromicus*, etc. *Atheta (Dimetrota) hutchinsoni*, found at an elevation of 5600 m in the Northwest Himalaya, is remarkable for the highest altitude record for Coleoptera from the world. The Tenebrionidae include the endemic genera *Bioramix* and *Chianalus*, in addition to others endemic to the Northwest Himalaya, Pamirs and Alai, viz. *Ascelosodis, Cyphogenia, Syachis, Laena*, etc. Nearly 75% of the high altitude Curculio-nidae are endemic to the Northwest Himalaya or the area of the Northwest Himalaya and Middle Asiatic high mountains. The genus *Catopionus* is common to the Himalaya, Alai-Pamirs, Tien Shan, Altai, Tibet, Siberia, Amur and Japan. *Otiorrhynchus* is also represented by a number of endemic species. The Hymenoptera include the endemic ants *Formica (Serviformica) picea* and many endemic genera and species of Bombidae. Among Lepidoptera Oriental and Manchurian elements predominate in the east, but Holarctic forms in the Northwest Himalaya. The Noctuid *Cteipolia* is common to the Himalaya and Middle Asiatic mountains. About a dozen species of *Parnassius* are represented by numerous localized subspecies and are confined to elevations mostly of 4000–5000 m. *Argynnis aglaica vittata*, found at 5000 m on the Northwest Himalaya, is common to the Pamirs also. The Palaearctic *Argynnis pales* is represented by a number of subspecies. *Erebia, Hipparchia, Aporia, Baltia, Colias, Pontia, Pieris*, etc. are confined to higher elevations. The species of these genera are mostly endemic to the Middle Asiatic mountains and are represented by local subspecies on the Northwest Himalaya. The Diptera include Tipulidae, Chironomidae, Simuliidae, Bibionidae, Blepharoceridae, Deuterophlebiidae, Culicidae, Stratiomyiidae, Syrphidae, Anthomyiidae, Tachinidae, etc., exclusively of Turkmenian affinities. *Tipula (Bellardia) hypsistos* occurs at elevations of 4800–5180 m in the Eastern Himalaya. Deuterophlebiidae are interesting from the fact that they occur also in the Tien Shan, Altai, Japan and United States of America. This family has recently been collected by my research collaborators Messrs. O. P. DUBEY and B. K. KAUL for the first time from the southern slopes of the Pir Panjal Range, near the Hamta Pass, marking the extreme southern limit of its distribution in India. *Ephedra glauca* breeds in hot springs (49 ° C) and the endemic *Holmatopota hutchinsoni* occurs in the Tso-kar Lake at an elevation of 4575 m in the Northwest Himalaya. Thysanura like *Machilinus* occur on rock at 5300–5800 m. Collembola are increasingly abundant at higher elevations and occur up to 6800 m. The species belong mostly to *Isotoma, Proisotoma, Hypogastrura*, etc.

The high altitude insect life of the Himalaya as a whole is remarkable for the very high species endemism in all groups; over 70 % of the species restricted to high elevations are strictly endemic. There are besides a large number of endemic genera and subspecies of Middle Asiatic species.

The origin of high endemism and the development of the ecological specializations of the high altitude insects of the Himalaya are integral parts of the history of the rise of the Himalaya itself. Essentially thermophile lowland forest forms were lifted up in the course of the Himalayan orogeny to high elevations by the rise of the ground they inhabited, and simultaneously came to be modified into cryophile, mountain autochthone types. The endemic elements of today are the descendents of the ancestral stock of Middle Asiatic origin, which inhabited the region when its elevation was not high. The region of the Tertiary mountains (Fig. 155) is an independent amphitheatre of faunal origin within the Palaearctic Realm (REINIG, 1932). The lowland boreal and Middle Asiatic ancetral stock has evolved into the typically endemic high altitude elements *pari passu* with the uplift of the Himalayan system (MANI, 1962, 1968). Pleistocene glaciations and the climatic and other changes at the end of the Pleistocene glaciations have served as important factors in this evolution. The Pliocene endemic forms have survived on nunataks during the Pleistocene glaciations and have since the last glaciation, given rise to numerous subspecies.

While SKORIKOV (1931) believes the Himalaya as a whole to be a single centre of faunal differentiation and distribution, MANI (1962, 1968)

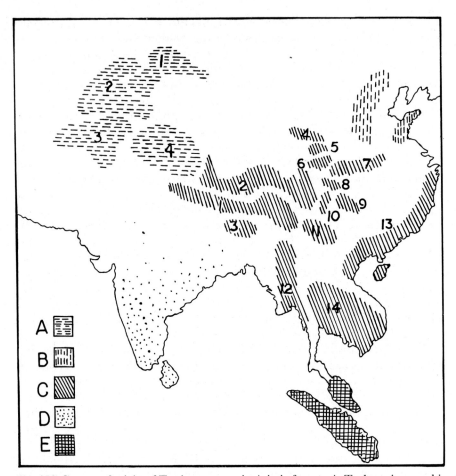

Fig. 155. Centres of origin of Tertiary mountain Asiatic faunas. A. Turkmenian amphitheatre, B. Manchurian amphitheatre, C. Indo-Chinese amphitheatre, E. Malayan amphitheatre, D. the Peninsular amphitheatre lies outside the region of Tertiary mountains and is a much more ancient centre of evolution of flora and fauna. The amphitheatres A and B belong to the Palaearctic Realm and C, D and E to the Oriental Realm. The subcentres in A are 1. Ala-Tau-Tien-Shan mountains, 2. South Turkestan, 3. Afghanistan and 4. Northwest Himalaya. The subcentres in C are 1. the forest-covered Himalaya to Assam, 2. Tibet, 3. Khasi-Jaintia Hills of Assam, 4–11. Eastern Tibet and Yunnan, 12. Burma, 13. Indo-China, 14. Thailand.

has recently shown that the Northwest Himalaya has had a different history from the rest of the Himalaya and together with the Pamirs constitutes an independent centre. While the evolutionary changes during the Pliocene largely involved the modification of relatively few species to true high altitude types, there is at present a very pronounced tendency towards an increase in the number of species by a rapid process of sub-speciation and isolation on single high massifs.

680

REFERENCES

BLANFORD, W. T. 1901. The distribution of Vertebrate animals in India, Ceylon and Burma, *Philos. Trans. R. Soc. London*, (B) 194: 335–436.

BRÜHL, P. 1931. Census of Indian mosses. *Rec. bot. Surv. India*, 13: 1–135, 1–152.

BURRARD, S. G. & H. H. HAYDEN, 1907–1908. A sketch of the Geography and Geology of the Himalaya Mountains and Tibet. Calcutta. pp. 1–230, charts 50.

DAS, S. M. 1966. Palaearctic elements in the fauna of Kashmir. *Nature*, 212 (5068): 1327–1330.

GROSS, F. J. 1961. Zur Evolution euro-asiatischer Lepidopteren. *Verh. deutsch. Zool. Ges. Saarbrücken*, 461–478.

KEMP, S. 1914. Onychophora. Zoological Results of the Abor Expedition. *Rec. Indian Mus.* 8: 471–492, pl. xxxiv–xxxvii.

KIHARA, H. 1955. Fauna and flora of Nepal Himalaya. Scientific results of the Japanese Himalayan Expeditions to Nepal Himalaya. Kyoto Univ. *Fauna & Flora Res. Soc. Kyoto*, 1–390.

MANI, M. S. 1962. Introduction to High Altitude Entomology. London: Methuen & Co. pp. 306.

MANI, M. S. 1968. Ecology and Biogeography of High Altitude Insects. The Hague: Dr. W. Junk Publishers. pp. 527.

MELL, R. 1958. Zur Geschichte der ostasiatischen Lepidopteren. 1. Die Hebung Zentralasiens, das westliche Refugium zentralasiatischer Abkommlinge und die Verbreitungsachse Sikkim-Khasigebirge-Zentralforma. *Deutsch ent. Z.* (NF) 5.

REINIG, W. F. 1932. Beiträge zur Faunistik des Pamir Gebietes. *Wiss. Ergeb. Alai-Pamir Expedition* 1928, 1(3): 1–195, Fig. 29, 2(3): 195–312 pl. vi.

SKORIKOV, K. 1931. Die Hummelfauna Turkestans und ihre Beziehung zur Zentralasiatischen Fauna (Hymenoptera: Bombidae). Abhandlungen der Pamir Expedition 1928. 8 *Zoology*, pp. 175–247.

STEVENSON, J. 1933. Oligochaeta. Fauna British India.

SWAN, L. W. 1961. The ecology of the High Himalaya. *Sci. Amer.* 205(4): 68–78.

SWAN, L. W. 1963. Ecology of the heights. Natural History, 23–29.

XXII. BIOGEOGRAPHY OF THE WESTERN BORDERLANDS

by

M. S. MANI

1. Introduction

The geomorphologically complex and ecologically and biogeographically transitional areas between the alluvial plains of the R. Indus and the Iranian-Afghanistan borders constitute the Western Borderlands of India, and comprise three main divisions, viz. 1. Baluchistan, 2. the submontane Indus area including the Vale of Peshwar and Bannu plain, the Potwar Plateau and the Salt Range and 3. the hills of the Northwest-Frontier Province (see chapter II). These areas now constitute the Republic of Pakistan.

Baluchistan is largely barren mountains, deserts and stony plains, about 351 000 km² in area and, mostly at elevations of 300–900 m above mean sea-level, but the surrounding mountains rise to elevations of over 1800 m. It comprises an 1. arid plateau surrounded by high mountains and constituting a region of inland drainage, 2. the arid Makran Coast in the south, 3. an area of tangled mountainous part in the northeast and continuous with Afghanistan, and 4. a small part of the Indus Plains south of the Bolan Pass and not draining into the R. Indus.

Lying wholly outside the influence of the monsoons, its general climate is characterized by extremes of heat and cold, uncertain and scanty rainfall, usually not exceeding 25 cm annually. There are no large rivers. Short torrential streams drain mostly into shallow lakes, after rains, but the torrents and lakes dry up entirely in the hot weather.

The arid hills and basins of Baluchistan form the eastern portion of the Iranian Plateau, sharply marked off from the Indus Plains by the Kirthar and Sulaiman Ranges, with the R. Gomal as the eastern limit. The Submontane Indus Region includes 1. the plains of Peshwar, Kohat and Bannu to the west of the R. Indus; 2. the Potwar Plateau to the east of the R. Indus and 3. the Salt Range, marking off the southern boundary of the region and cut through by the R. Indus at the head of the Kalabagh re-entrant. The foot of the Salt Range is generally taken as the southern boundary and in the northeast the edge of the foothills of the Sub-Himalayan hills of Kashmir. This region was in Miocene times an area of foreland sedimentation from the Himalaya. The hills of the Northwest-Frontier Province (Pakistan) are essentially the rugged fringe of Afghanistan. The characteristic trends and the massive structures found in Baluchistan are absent here to the north of the R. Gomal. The striking change in the direction between the Karakoram and the Hindu Kush

Mountains is controlled by the hidden outlines of the Peninsular Block.

2. General Ecology

The general ecology of the Western Borderlands presents a striking contrast to that of the Eastern Borderlands and the Peninsula. The entire region is characterized by its pronounced continentality of climate and by the youthfulness of the general topography. Although within eastern extremity of the Mediterranean subregion of the Palaearctic Realm, it differs from the typically Mediterranean areas in its pronounced atmospheric aridity. It is wholly outside the influence of the monsoon rainfall and its general ecology is, therefore, temperature dominated. The fluctuations of temperatures are large. In Quetta, in the innermost basin, at an elevation of about 1675 m, for example, the mean temperatures are 4.2 °C during January and 25.4 °C during July. The mean diurnal temperature range is about 10 °C and 15 °C respectively for these months, but variations of 45 °C are not rare within the course of twenty four hours. Strong winds blow mostly from the northwest and are scorchingly hot during the summer, and dust-ladden and bitingly cold during the winter. The annual rainfall is never over 25 cm and is also extremely unreliable. Precipitation over most of the area is often in the form of snow and is mainly due to the shallow west-moving winter depressions, though in the lower highlands of Lorali-Zhob, the summer monsoon is also fair. Lying completely outside the influence of the Indian monsoons, the vegetation of the region is typically xerophytic and scanty, but often with brightly coloured flowers in the valleys during spring time. The hills are generally covered by open scrub, but there are also forests of juniper, wild olive, pistachio, laurel and myrtle at higher elevations on the Sulaiman Range. Dwarf date palms, steppes-grass and bushes are common on the Makran Coast. The desert sediments of the interior and the wild gorges of the bordering hills are evidence of intermittent but intense spells of erosion, intense heat and cold, savage winds, rare but violent floods, etc.

The vegetation cover of the Northwest-Frontier Province, though of an arid type, is different from that of Baluchistan. The common natural vegetation of Afghanistan comprises species of dandelion, buttercup, mouse-ear, chickweed, larkspur, fumitory, caper-spurge, wild chicory, hackweed, ragwort, thistle, scurvy-grass, shepherds purse, sorrel, wild mustard, wild turnip, wild carrot, dwarf mallow, datura, deadly-night-shade, rushes, sedges, duckweeds, hemlocks, Umbelliferae, *Ranunculus*, etc. In the desert areas the vegetation is scanty and is characterized by *Astragalus* as dominant plant and great Umbelliferae yielding asafoetida. Even in Afghanistan we find climatic extremes, with bitterly cold winds and snow winters, damp spring and excessively hot summer and dry autumn. The southern slopes of the Hindu Kush Range are covered by

683

small belts of forest of mainly *Quercus ilex, Pinus*, etc. At lower elevations on these slopes we find *Pistacia, Celtis, Dodonea* etc.

3. Character Fauna and Biogeographical Affinities

The Western Borderlands lie for the most part within the eastern limits of the Mediterranean subregion, but some parts, particularly in the north, are on the fringe of the Turkmenian subregion of the Palaearctic. The character fauna of the Western Borderlands is, therefore, composed largely of Mediterranean elements, with considerable admixture of Turkmenian and some Ethiopian derivatives. In nearly all groups, endemism is higher than in the Eastern Borderlands. Compared to the Eastern Borderlands, the character fauna of the Western Borderlands is not a humid tropical forest fauna, but a steppes and desert fauna. There are also other fundamental differences from the Eastern Borderlands. The Eastern Borderlands are biogeographically transitional areas, but the biogeographical transition in the west lies to the east of the Western Borderlands. This transition has been gradually shifting and fanning out eastward and is at present practically at the Arawalli strike in the Indo-Gangetic Divide (see chapters II and XIX). While faunal interchanges are of considerable magnitude and frequency in the Eastern Borderlands, and the area may be rightly described as a faunal gateway, faunal interchanges in the west are quite insignificant. The Western Borderlands cannot be strictly speaking described as a faunal gateway; it has on the other hand been an important gateway for the penetration of *Homo sapiens* (see also chapter XI). The transitional boundary of the Eastern Borderlands in the west has remained more or less stationary, since perhaps the Pleistocene times. The eastern boundary of the Western Borderlands has, however, been gradually shifting eastwards, within historical times. The influx of the eastern Asiatic humid tropical faunas through the Eastern Borderlands is perhaps now practically non-existent, but in the west the influx of the Mediterranean and Ethiopian faunas is still active. This significant difference between the two borderlands and in particular the continued influx through the Western borderlands must be attributed to the fact that the Mediterranean and Ethiopian flora and faunas are ecologically more closely related to the present-day conditions prevailing in the Deccan and northwestern parts of the Peninsula than in the case of the humid tropical faunas in the east. The Western Borderlands must appropriately be described as the meeting point of the Turkmenian, Mediterranean and Ethiopian faunas. Another fundamental difference between the Eastern and Western Borderlands lies in the fact that while the Peninsular and western faunas transgressed through the Eastern Borderlands, the outflow of the Peninsular fauna through the Western Borderlands is extremely slight and negligible. The influx through the Eastern Borderlands was largely Pre-Pleistocene and

Pleistocene event, but that in the Western Borderlands is predominantly Post-Pleistocene, because the region came into connection with the Peninsula much later. The distributional pattern between the Eastern Borderlands and Peninsula is marked by more or less pronounced discontinuity, but it is continuous between the Peninsula and Western Borderlands. The transgression of faunas in the east is very extensive, but only exceedingly slight in the west. Relatively small numbers of the Mediterranean and Ethiopian elements have transgressed across the R. Indus eastwards over the Aravalli Hills into the western ends of the Indo-Gangetic Plains perhaps with the eastwards advancing aridity within historical times (RANDHAWA, 1945), or have spread sparsely southwards to the northwestern parts of the Peninsula. The transgression from the Eastern Borderlands has taken place either along the Himalaya or Eastern Ghats. In the west, however transgression to the Indo-Gangetic Plain has been by way of and across Peninsula (over the Aravallis).

The component elements of the fauna of the Western Borderlands are 1. Endemics, 2. Turkmenian steppes elements, 3. Mediterranean, 4. European Palaearctic elements, 5. Ethiopian elements, 6. Peninsular outliers and intrusive elements, 7. Eastern isolates and outliers. Of these complex types, the dominant components are the endemics, Mediterranean and Ethiopian faunal elements. The Peninsular outliers and intrusive elements are of the secondary importance. The Turkmenian steppes elements are sparse and the eastern outliers and isolates are insignificant members.

The Mediterranean-Ethiopian scorpion *Buthus* has transgressed, for example, across the Western Borderlands on the Peninsula, and thence to the western parts of the Indo-Gangetic Plain. The Mediterranean *Butheolus* has similarly transgressed southwestwards to Deccan. The Eastern Mediterranean Uropygi *Ischnurus* and the Mediterranean *Galeodes* are, however, confined to the limits of the Western Borderlands.

The admixture of Mediterranean and Ethiopian faunas is more pronounced in the south than in the north and sometimes it is not easy to distinguish the derivatives of these two faunas in Baluchistan and parts of Sind. There is, for example, a strong Mediterranean-Ethiopian combination among reptiles of the Western Borderlands. Among Lacertilia we have, for example, *Stenodactylus* extending from the desert tracts of Northwest Africa, North Africa and Southwest Asia to the Afghanistan-Baluchistan borders. The North African *Alsophylax* similarly extends across Southwest Asia and Arabia to the Afghanistan-Baluchistan borders (and also occurs in parts of Tibet). *Gymnodactylus* is another example, with several species common to the Western Borderlands and to North Africa. *Pristurus* extends also from North Africa to the Western Borderlands. *Ptylodactylus* extends from North Africa and Southwest Asia to the Kirthar Range. *Hemidactylus persicus* occurs in Iraq, Iran, Waziristan and Sind. *Agama* extends from Southeast Europe and Southwest Asia and Africa to the Western Borderlands, from where it transgressed partly to the north-western parts of the Peninsula. Some species of the genus have likewise spread to the steppes of Middle Asia and are common to the Western Borderlands. *Agama himalayana* occurs in Middle Asia and Kashmir (up to 3300 m), *Agama tuberculata* occurs from Afghanistan through the Himalaya up to Nepal, *Agama nupta* extends from Iraq and Iran through Afghanistan to Baluchistan, Northwest-Frontier Province and northern Sind and *Agama caucasica* extends from the Caucasus to the Western Borderlands. *Uromastix* extends from the arid parts of southwest Asia and North Africa across the Western Borderlands to the western parts of Upper Gangetic Plains. *Scincus* extends from North Africa to Sind. Other Mediterranean elements include *Ophiomorus, Chalcides, Acanthodactylus, Ophisops, Eremias,* etc. *Varanus griseus* occurs from the desert areas of Caspia to North Africa and the Western Borderlands. *Eryx* is common to southwest Asia, eastern Europe, Africa, West China and the Western Borderlands, from where it has transgressed to the Indo-Gangetic Plain. The snake *Coluber* Palaearctic element occurs in Europe and in Africa north of the Equator. *Coluber ventrimaculatus* occurs from Middle Asia across Iran and Afghanistan to the Western Borderlands, where it has transgressed south partly to Kandesh in the northwestern part of the Peninsula. *Coluber rhodorhachis* extends from Egypt across Transcaspia and Arabia to the Western Borderlands. *Conta persica* occurs from the Trans-caspian area to Iran, Baluchistan, Sind and Northwest Himalaya. The North African *Lytorhynchus* also extends to these Borderlands. *Psammophis schokari* is common to Kashmir, Northwest-Frontier Province, Sind, parts of Rajasthan, Baluchistan, Iran, Arabia and North Africa. *Psammophis leithi* is common to Baluchistan, Cutch, parts of Rajasthan, western parts of Uttar Pradesh, Northwest-Frontier Province and Kashmir, but *Psammophis lineolabis* is common to Baluchistan, Afghanistan, Iran, across Middle Asia to Mongolia and northwestern parts of China. *Vipera lebetina* extends from eastern Europe and North Africa to Middle Asia and to the Western Borderlands. *Pseudocerastes* extends from Senai Penin-sula to the Northwest-Frontier Province. Among birds we have the desert lark *Alaemon desertorum*, with African affinity, occurring in the Indus Plain. The North African and Western Asiatic spiny-mouse *Acomys dimidiatus* occurs in Sind. The Mediterranean fauna is increasingly mixed with the steppes elements of the Turkmenian subregion in the Northwest-Frontier Province and northwestern parts of the Punjab. One species of mouse-hare *Lagomys* (the genus is known from North and Middle Asia and Himalaya) is also found in Baluchistan. The wild ass *ghorkar* of Baluchistan may perhaps be a variety of *Equus hemionus*.

The typical Turkmenian elements include the mollusca *Euaustenia* and *Bensonia* (Zonitidae) and *Cathaica* (Helicidae); the mammals *Ovis vignei* from the Upper Indus Valley and from Afghanistan, *Capra falconeri* (markhor) and *Capra aegagrus* the Persian wild goat in Sind hills, Afghanistan, Sulaiman Range as far as Quetta, *Gazella subgutturosa* in Iran, Baluchistan and Turkestan, etc. *Vulpes cana* is confined to Baluchistan, but *Vulpes vulpes pusilla* (= *leucopus*) occurs in the arid northwest. The European beech-marten *Martes* (= *Mustela*) *foina* occurs in Afghanistan (and extends east on the Himalaya to Kumaon and Ladak). The mottled polecat *Plutorius sarmaticus* extends from Eastern Europe through Western Asia to its extreme eastern limits in Baluchistan, where it is, however, rare. The Turkmenian genus *Alactaga* is represented by *Alactaga indica* in Baluchistan, but does not extend further east. The Turkmenian reptile *Phrynocephalus* is also represented by several species like *Phrynocephalus scutellarus* from Iran, Afghanistan and Baluchistan (600–2100 m); *Ph. ornatus*, *Ph. maculatus*, *Ph. euptilopus* and *Ph. luteoguttatus* from Afghanistan and Baluchistan. The fishes *Glyptosternum*, *Oreinus*, *Schizothorax*, *Schizothoracichthys* and *Cyprinodon* are also Middle Asiatic forms found in the area. *Glyptosternum* is represented by one species, which is common to Afghanistan and the Himalaya and one species which is confined to Afghanistan. There is also one species of *Garra* in Afghanistan. *Oreinus* is represented by one species, which extends from Afghanistan across the Himalaya to the Eastern Himalaya. *Schizothorax* is represented by ten species in Afghanistan and eight species on the Himalaya up to the Western Himalaya. One species of *Schizothoracichthys* is common to Afghanistan and the Western Himalaya (see Chapter XVII).

The Lepidoptera from the Western Borderlands are almost exclusively Turkmenian forms. *Nytha* is represented by *Nytha thelephassa* common to southern Russia, Afghanistan, Iran and Baluchistan; *Nytha persephone* from southern Russia through Iran to Baluchistan and *Nytha parizatis* common to Iran, Afghanistan and Baluchistan and Northwest Himalaya. The Palaearctic *Maniola* is represented by *Maniola davendra* common to the Northwest Himalaya (part of the Turkmenian Subregion) and Baluchistan; *Maniola narica* extending from southern Russia across western Asia to Afghanistan and Baluchistan and *Maniola interposita* common to Middle Asia, Afghanistan and Baluchistan. *Karanassa* is also found in the Northwest Himalaya and in Baluchistan. The Pieridae are abundant particularly the steppes forms. *Pieris brassicae*, *Pieris brassicae nepalensis*, *Pieris napi iranica*, etc. are some of the common butterflies of Baluchistan, Northwest-Frontier Province and adjoining areas. *Pontia daplidice*, which occurs throughout the Palaearctic through to Abyssinia, is represented by the subspecies *moorei* at elevations of 2700 m in Murree and Baluchistan and *glauconome* from Baluchistan Iran, Iraq and Middle Asia and we have also *Pontia chloridice alpina* in Baluchistan and (common to Ladak also). *Euchloë charltonia*, from North Africa, Iraq and Middle Asia, is also found in the area. *Colotis*, from Africa through Iraq and Arabia, has spread across the Western Borderlands partly into the western parts of the Indo-Gangetic Plain and the Peninsula up to Ceylon. The Palaearctic *Gonopteryx* enters through the Western Borderlands and has

687

spread eastwards along the Himalaya up to the Burmese mountains. *Papilio machaon centralis*, found here, is also common to Middle Asia. *Hypermnestra helios* is common to Middle Asia, Iran and Baluchistan. *Parnassius tianschanica* extends from Middle Asia across Afghanistan to the Northwest-Frontier Province. The Mediterranean Odonata *Sympyena*, *Orthetrum brunneum brunneum*, *O. anceps*, *O. taeniolatum*, *Epallage*, etc. extend up to Kashmir.

Mabuya dissimilis is a Peninsular outlier in the Western Borderlands. A small part of the humid-tropical Asiatic fauna seems to have transgressed westwards across the Western Borderlands, since perhaps the Pleistocene times. It is known, for example, that at least among birds there does not seem to be any sharp boundary between the African and the Oriental elements. Southern Arabia is an area of subtraction-transition between the African and Oriental. Many tropical African birds extend to southwest Arabia and the resident birds of southeast Arabia are Oriental and not African. In the main part of Arabia, the birds are typically desert forms, rather than Eurasian or African. In a recent contribution, RIPLEY (1954) has shown, for example, that certain relict species of birds found in Arabia represent really invasions from India (or perhaps also from the north) during Pre-Pleistocene times. Arabia has had a relatively stable climate since the beginning of the Pleistocene and has served more as a barrier than as an avenue of interchange between adjacent countries.

REFERENCES

BLANFORD, W. T. 1876. Notes on the 'Africa-Indien' of A. VON PELZEN, and on the mammalian fauna Tibet. *Proc. Zool. Soc. London*, 631–634.

BLANFORD, W. T. 1876. The African element in the fauna of India: A criticism of Mr. WALLACE's views as expressed in the 'Geographical distribution of animals'. *Ann. mag. nat. Hist.* (4)18: 277–294.

DAY, F. 1885. Relationship of the Indian and African fresh-water fish faunas. *J. Linn. Soc. London (Zool.)* 18: 308–317.

RANDHAWA, M. S. 1945. Progressive desiccation of northern India in historic times. *J. Bombay nat. Hist. Soc.* 45: 558–565.

RIPLEY, S. D. 1954. Comments on the biogeography of Arabia with particular reference to birds. *J. Bombay nat. Hist. Soc.* 52(2/3): 241–248, fig. 2.

XXIII. BIOGEOGRAPHY OF THE INDO-GANGETIC PLAIN

by

M. S. MANI

1. Introduction

The Indo-Gangetic Plain separates the Peninsula from the Himalaya. It is the most densely populated part of India and comprises the plain of the R. Indus, the Gangetic Plain and the narrow and short plain of the R. Brahmaputra.

The Indus Plains comprise 1. the Sind, and 2. the Punjab. The Gangetic Plains comprise the Gangetic Divide, the Upper Gangetic Plain, the Middle Gangetic Plain and the Lower Gangetic Plain. Although the plain through which the Brahmaputra flows is really an eastward extension of the Indo-Gangetic Plain, ecologically and biogeographically it must be considered a part of the Eastern Borderlands and is therefore excluded from this Chapter.

1.1. The indus plains

The region of Sind includes Sind proper, the lowlands of Sibi (Sewistan) and part of Khairpur, but excludes the Thar Parkar Desert. It is subdivided into the Western Highlands of Kirthar and Kohistan, the lower Indus Valley including an eastern and western valley and the Indus Delta. The Punjab area excludes the part to the north of the Salt Range. This is geographically a transitional area between the Western Borderlands and the great plains of North India.

1.2. The gangetic plains

The Indo-Gangetic Divide, the area between the delta of the rivers Indus and Ganga, especially the narrow region between the rivers Sutlej and Yamuna, is a transitional belt that marks the great divide between not only two great river systems, but also between climatic and biogeographical limits. This transitional area is bounded in the north by the Siwalik Hills, in the West by the rivers Beas-Sutlej and in the east by the R. Yamuna. In the south it passes gradually into the Thar Desert (but the limits may be taken as the dry bed of the R. Ghaggar) and in the southeast by the low broken Aravalli Hills near Delhi. Except for the scattered Aravalli outliers in the southeast and the topographic discontinuity of the river courses, the region is completely alluvial. Over one hundred streams, within a short stretch of only 130 km of the Siwaliks, come down to form the socalled *chos*, noted for their sudden spate of floods, mixed with sand, mostly to dissipate in the ground or to converge into the R. Ghaggar (R. Saraswati of the ancients). (see Chapter II, Figs. 3, 4).

1.3. The upper gangetic plain

The Upper Gangetic Plain, built up of the detritus brought down by rivers chiefly from the Himalaya, is traversed by the rivers Yamuna, Ganga and Gogra, with the main drainage line pushed somewhat more to the south than formerly. The northern limit is marked by the Siwalik Hills, but in the south the limits are not sharp where the old rugged surface of the Peninsular Foreland has been smothered by the alluvium (GEDDES, 1960; OLDHAM, 1917) brought down by the Himalayan rivers like the Ganga and by its southern tributaries, like the Yamuna and by the Peninsular rivers Chambal, Betwa and the Ken.

The principal physiographical difference is introduced by the upland *bhangar* alluvium of the doabs (*do* = two, *ab* = water or the area between two rivers) and the fingers of *khadar* along the main streams and their nearly parallel tributaries. The broad flood-plains are characterized by dead-arms, deferred junctions and *jhils*, often several kilometres wide on the great rivers. The right banks of the rivers are mostly at higher levels than the left banks and have bluffs and ravining, on a miniature scale of what we see in the R. Chambal. Depending on these peculiarities, three variations from the usual feature may be recognized, viz. the *bhabar*, *terai* and *bhur*. The bhabar is a porous area of detritus piedmont, about 35 km wide in the west and less in the east, skirting the Siwalik Hills, where the stream profiles flatten out and the coarser boulders and gravels are deposited. Most of the smaller streams are here lost in the loose talus, though many of them may seep out further below, where the slope is flatter and the finer material is deposited in marshy terai. The terai must have formerly covered a wide zone of 80–100 km, but human settlements have considerably altered the terai belt and it is now confined to a relatively narrow belt, parallel to the bhabar. It is also practically absent to the west of the R. Yamuna, where we find instead the chos in sub-montane areas. This difference is attributed to the fact that between the R. Yamuna and the R. Sutlej the Siwaliks stand distinctly apart from the Himalaya and have developed their own rain-fed drainage better than in the areas where the Siwaliks are close to the Himalaya and are cut through by the snow-fed Himalayan rivers. Further to the west in the Punjab, where the Siwaliks are also close to the Himalaya, the conditions are too arid for terai formation. Further to the east, the Siwaliks are again separated from the Himalaya by the longitudinal Nepal Valley. In this area the rainfall is heavy and the abundance of water has favoured terai formation. Although the bhabar and the terai are forested, human settlements have very greatly altered the conditions in the south. The bhur consists of patches of sandy soil, which may be locally so extensive as to form low undulating sandy uplands. Strictly speaking, however, the name bhur must be applied to a belt on the east bank of the R. Ganga, near Moradabad and Bijnor in the Uttar Pradesh. The bhur tract is arid, but waterlogged depressions, especially in years of copious rainfall, are also found. The soil in the Indo-Gangetic Plain ranges from *usar* clays, with efflorescence of alkaline *reh* in arid areas of the west, gradually grading through *dumat* loams to the sandy bhur.

1.4. THE MIDDLE GANGETIC PLAIN

The Middle Gangetic Plain embraces roughly the eastern one-third of the Uttar Pradesh and the northern half of Bihar, but the limits are not sharp in the east, where it passes into the Lower Gangetic Plain or Bengal. The Middle Gangetic Plain represents a transition between the relatively arid *bhangar* doab of the Upper Gangetic Plain and the humid, largely *khadar* of Bengal. The area does not exceed 150 m above mean sea-level and in the east it is hardly 30 m above mean sea-level. The general features are like those of the Upper Gangetic Plain, but there is greater formation of khadar and north Bihar is mostly is bhangar. The rivers increase in violence eastwards and the flood-plains are larger than in the Upper Gangetic Plain and also form more or less permanent lakes. *Chaurs* are long semi-circular marshes in the chain of temporary lakes during the rainy season. The filling of alluvium, south of the R. Ganga, is shallow and the edge of the Peninsula is rugged and groups of craggy hills rise as islands of rock from the alluvium. The alluvium is about 140 km wide in the west, where the R. Son makes a deltaic re-entrant into older rocks. In the east, the Rajmahal Hills, representing the extreme northeast corner of the Peninsula, abut almost directly into the R. Ganga. Bhangar largely fringes the plain and inundated areas are fewer than to the north of the R. Ganga.

1.5. BENGAL

Bengal includes the submontane terai of the Duars, the northern para-delta of the Ganga-Brahmaputra doab and the Barind, the eastern margins of the Surma Valley and the plains along the R. Meghna and Chittagong coast, the western margin of lateritic piedmont plains between the R. Hoogly and the Peninsular Block and the coastal plain and the Delta of the R. Ganga proper between the R. Hoogly-Bhagirathi, Padma-Meghna and the sea. The deltaic plain of Bengal is of multiple origin, so that strictly speaking, we have here more than one delta.

2. Ecology

The present-day ecology of the Indo-Gangetic Plain is largely dominated by the influence of the Himalaya in the north and by the pressure of increase of human population, especially recent rapid urbanization and spread of industries. The ecology of the region is essentially human ecology, in which plants and animals and abiotic factors play rather an insignificant rôle. Being the most densely populated region of India and having been under continuous and intensive cultivation for at least four or five thousand years, deforestation has been most complete. The region of Uttar Pradesh was at one time densely forested and even as late as the sixteenth century, the Moghul emperors hunted wild elephants, buffaloes, bison, rhinoceros and lion in the doab of the rivers Yamuna and Ganga (see Chapters V and XII). The secondary vegetation, which has covered the abandoned land, is markedly xerophytic and the tremendous increase of human population has caused the spread of a savannah-like cover even in the wetter east. The combined effects of human activities are gradual depression of the natural vegetation from the original climax monsoon-deciduous forest type to the open dry grassland type, forming grazing tracts, with only scattered relics of resistant woody plants (thorn-scrub vegetation). There is strong evidence of ecological succession that was formerly a marshland in the area. These conditions of vegetational depression and regression have been greatly accentuated by recent attempts at rapid industrialization and schemes of irrigation. The Indo-Gangetic Plain is somewhat more humid and receives more rainfall in its east than in the west, but arid conditions are gradually extending eastwards.

Physically, structurally, ecologically and biogeographically the Plain is a region of transition, not only from the south to the north, but also from the east to the west. Transitions are also seasonal – the ecosystems oscillate between monsoon-rains and post-monsoon season – alternating of the Peninsular and western element in the upper and middle plains (Fig. 149).

In the Upper Gangetic Plain, the winter rainfall is of ecological importance only in the extreme northwest. We find along the northern border a strip of forest in the bhabar and terai, but the natural vegetation has disappeared from the plain here also. The flood-plains are now mostly covered by tall coarse grass and *Tamarix*. The herbaceous annuals of the cold season on waste ground or the weeds in cultivated fields are largely European, (of temperate ecosystems) especially in wheatlands of the northwest. The rainy-season plants have, however, their origin (humid-tropical ecosystems) either from the Peninsula or from the east. In the Middle Gangetic Plain, nearly 90% of the rainfall is received from the southwest monsoon, except in the extreme northeast, where the summer nor'westers also bring some rain. Rainfall decreases from 145 cm

692

in the east to 100 cm in the west and also from the Himalaya southward to the R. Ganga. In the Deltaic Bengal, the climatic conditions are marked by the violent cyclonic nor'westers, often accompanied by heavy rains and sometimes also hail, during March-April. Except perhaps the *Shorea*-forest in the terai area, there is very little of the original natural vegetation. The aquatic flora is, however, rich and the bhils are usually chocked with reeds, sedges, etc. The seaface has the sundarbans. The Ganga delta proper comprises 1. the moribund delta, where the offtakes of the old distributaries have silted on the north and the land is generally not inundated even in flood and 2. the mature delta between the moribund delta and the Sundarbans, in which the rivers are relatively more live and still carry a good deal of water from local rain, but mostly deteriorating and becoming increasingly brackish and 3. the active delta embracing the Sundarbans and areas between the R. Madhumati and R. Meghna, with large tidal forest (see Chapter VI) along the seaface, extending to about 270 – 130 km inland. The delta is apparently advancing seawards.

The original natural vegetation of the Indo-Gangetic Plain has, under the influence of man, more completely disappeared than in any other part of India. The region of the Indus Plains is characterized by low deciduous herbs and shrubs, presenting mostly a burnt appearance during the hot weather, to which only the Chenopodiaceae are, however, a conspicuous exception. The more common Phanerogams of the Indus Plains are Gramineae, Leguminosae, Compositae, Cyperaceae, Scrophulariaceae, Labiatae, Boragineae, Malvaceae, Euphorbiaceae, Convolvulaceae, etc. The principal trees are *Tamarix articulata, Balanites roxburghii, Bombax malabaricum, Sterculia urens, Grewia salicifolia, Boswellia serrata, Balsamodendron mukul, B. pubescens, Pistacia integerrima, Aegle marmelos, Odina wodier, Moringa pterygosperma* and *M. concanensis, Dalbergia sissoo, Butea frondosa, Prosopis spicigera, Acacia arabica, A. rupestris, Dichrostachys cinerea, Salvadora persica, S. oleoides, Anogeissus pendula, Cordia myxa, C. rothii, Terminalia tomentosa, Tecoma undulata, Olea cuspidata, Ficus infectoria, F. palmata, Celtis australis, Alnus nitida*, etc. There are also isolated clumps of columnar *Euphrobia royleana* and *E. neriifolia, Capparis aphylla, C. horrida, C. spinosa, Flacourtia ramontchi, Tamarix dioica, T. gallica, Grewia* spp., *Fagonia arabica, Rhamnus persica, R. virgata, Zizyphus nummularia, Z. vulgaris, Z. oenoplia, Dodonea viscosa, Alhagi maurorum, Sophora mollis, Cassia auriculata, C. tora, C. obovata, Mimosa rubicaulis, Pluchea lanceolata, Carissa diffusa, Orthanthera viminea, Periploca aphylla, Calotropis procera, C. gigantea, Withania coagulans, Adathoda vasica, Calligonum polygonoides, Pteropyrum olivieri, Salsola foetida*, etc. The delta of the R. Indus is somewhat similar to that of the R. Ganga and the Sundarbans, but is less rich in species; we find here *Avicennia, Sonneratia, Rhizophora, Ceriops, Aegiceras, Scaevola* (not known in the Sundarbans), *Oryza coarctata*, etc.

In the Gangetic Plain, the flora presents a greatly impoverished appearance. The more important plants belong to Gramineae, Legu-

693

minosae, Cyperaceae, Compositae, Scrophulariaceae, Malvaceae, Acanthaceae, Euphorbiaceae, Convolvulaceae and Labiateae, some Cucurbitaceae, Asclepidaceae, Verbenaceae and Amaranthaceae. The vegetation of Upper Gangetic Plain is typically of the arid type and in the extreme west is indeed an eastern continuation of that of the Indus Plains. In Bengal, we find *Michelia champaca, Polyalthia longifolia, Bombax malabaricum, Eriodendron anfractuosum, Lagerstroemia flos-reginae, Pterospermum acerifolium*, etc. *Aldrovanda vesiculosa* is a typical aquatic plant of the region. Among the other more common aquatic plants mention may be made of *Pistia stratiotes, Lemna*, etc. (SUBRAMANYAM, 1962). The Sundarbans are characterized by their evergreen forests, with mangroves, Leguminosae, Gramineae, Cyperaceae, Euphorbiaceae, Orchidaceae, Compositae, Asclepiadaceae, Verbenaceae, Convolvulaceae, Malvaceae, Rubiaceae, Acanthaceae, Urticaeae, etc. There is an abundance of *Oryza coarctata* here also. The aquatic plants include *Aldrovanda vesiculosa, Utricularia, Ipomoea aquatica, Pistia*, etc. Many littoral plants are also found. The plants send up from their underground roots, numerous aerial respiratory organs, in the case of *Heritiera, Amoora, Sonneratia* and *Avicennia*. The Sundarban comprises (Fig. 25) vast swamp-forest, (see Chapter VI) extending to about 275 km along the seaface of the Bay of Bengal, from the estuary of the R. Hoogly to that of the R. Meghna, and about 90–130 km inland. The name is derived from the sundri tree, *Heritiera minor* (= *fomes*) (not littoralis as often erroneously mentioned in literature), characteristic of the forest. The Sundarban lies between 21°31' and 22°38' NL and 88°5' and 90°28' EL and has an area of about 14 580 km². It constitutes the lower part of the delta of the R. Ganga. It is intersected from the north to the south by the estuaries of that river. The area is an extensive alluvial plain, where morasses and swamps are now gradually filling up and the process of landmaking is still active. The numerous flat and swampy islands are densely covered by forests, which in the north contain a rather dense undergrowth, some mangroves like *Kandelia* and *Bruguiera* along the river banks. In the south, where the effects of tides increase, these become more numerous and we also find here in addition *Ceriops* and *Rhizophora*, and finally the vegetation becomes completely mangroves and *Heritiera* and other plants are replaced by *Exaecaria agalocha*, which in its turn is replaced by mangroves near the sea. Sometimes the mangroves are separated from the seaface by low sand dunes, with some swamp-forest species and *Erythrina indica, Thespesia populnea, Ficus rumphii*, etc.

3. Character Fauna and Distributional Patterns

Owing to the relatively recent origin of the transitional conditions, the presence of the youthful Himalaya in the north and the stable and senile Peninsula in the south, and direct communication in the east with a

694

biogeographically important faunal gateway, the Indo-Gangetic Plain may be supposed to be profoundly influenced by the highly diversified and plastic faunas of the Himalaya and the east and the relicts and endemics from the south. It should therefore be expected to be rich in endemic and new elements and speciation may also appear to be intense in all groups. The actual conditions that we observe in these plains are, however, very different. Ignoring the relatively insignificant Mediterranean-Ethiopian outliers in the western parts, we have in the Indo-Gangetic Plain only the rapidly vanishing representatives of the greatly impoverished faunas from the south and east. The significantly retrogressive changes observed in its fauna have now passed the point, beyond which recovery is possible, even if human beings were to momentarily disappear entirely from the region and return the area to Nature completely for ever. The characters of its fauna today hardly give any clue to its composition, patterns of distribution and diversity just 5000 years ago. Faunistically and biogeographically, the Indo-Gangetic Plain is now distinguished entirely by its negative characters. The distributional patterns and the genera and species, found in any part of the Plains, correlate neither with the peculiarities of topography or with distribution of rainfall nor with other natural ecological factors nor even with contiguous areas.

Either no genera and species seem to have become differentiated in these plains, or whatever were differentiated have all been completely exterminated so that today there is a most striking poverty or lack of autochthonous and endemic elements. The fauna of the Plains was in reality largely a spillover from that of the Peninsula, (indeed it is the down-warped part of the Peninsula) so that the Peninsular elements are even at present perhaps the most widely and more continuously distributed among all the component elements of its fauna. The eastern parts of the Plains are largely dominated by derivatives of the faunas from the South Chinese and Indo-Chinese and Malayan areas and thus the younger and more recent forms are largely concentrated in these areas. As we move towards the west, the Plains are increasingly dominated by derivatives of the Mediterranean-Ethiopian elements. The eastern and the western faunas are interpenetrated extensively throughout by the Peninsular elements. The distribution of terrestrial species is thus on the whole far more continuous than in the Peninsula.

Interesting discontinuities are, however, observed in the fresh-water species of the great rivers that flow through the Plains. We find, for example, that the fresh-water Vertebrates of the rivers Indus and Ganga are identical in many cases; some of these occur also in the rivers of the Peninsula, particularly the rivers Mahanadi and the Godavari.

To some extent, the fresh-water animals are also peculiar to the Plains and may be traced back to the transitional conditions which prevailed towards the close of the last Pleistocene glaciers in the Himalaya. A number of the species are confined to the rivers Indus, Ganga and Brahmaputra and their tributaries. The Cetacean *Platanista, Gavialis* (although not wholly restricted to the Plains), the chelonian *Hardella* are common examples of peculiar forms. The siluroid fish *Sisor*, with a single species from the Brahmaputra drainage, must be described as endemic to the Plains. It must, however, be remembered that both *Platanista* and *Gavialis* belong to forms that were formerly widely distributed and must strictly speaking be described as geographical relicts. The latter occurs at present in the rivers Ganga, Brahmaputra and the Indus and in their larger tributaries, the Peninsular river Mahanadi and in the R. Koladyne that traverses the northern Arakan Mountains (Burma). It does not, however, occur in any other Indian river and not also in other parts. It is wholly fluviatile and does not enter the sea. Earlier workers (PASCOE, 1919; PILGRIM, 1915; PRASHAD, 1941) resorted to the Indo-Brahm river hypothesis to account for this faunal affinity of the rivers Indus, Ganga and Brahmaputra. It is not all necessary to assume the existence of such a hypothetical river. The area where we have now the Indo-Gangetic Plains was formerly a narrowing channel of sea. The streams that flowed down the southern slopes of the new mountain range must have brought down with them much alluvium, the accumulation of which at its base would tend to fill in the channel. That this process was uniform all across the immense stretch from east to west was improbable. We may picture the existence of great lagoons, the interrelations of which were constantly changing, while their bed is now completely buried beneath the alluvial deposits of more recent rivers. Such a history would give ample opportunities for the migrations of the fauna now common to the Indus and Ganga. That some of these species, which are at present found only in the rivers Indus, Ganga and Brahmaputra, were formerly part of the Peninsular faunal derivatives, is shown by the fossils (LYDEKKER, 1902) of the chelonian *Kachuga tectum tectum* in the Pleistocene Siwaliks; this species occurs now only in these three rivers. The race tentoria of *Kachuga tectum* is found at present in the Peninsular rivers Mahanadi and Godavari. The Mollusc *Ariophanta* is a Peninsular from that spread in the Indo-Gangetic Plain. The greatest bulk of the types, characteristic of the Plains, were formerly widely distributed in the Plains, but have at present more or less restricted distribution.

The Punjab Tract has *Antilope, Cervus duvauceli, Cervus porcinus, Felis, Herpestes, Hyaena, Canis, Vulpes, Putorius, Mellivora, Lutra, Ursus, Erinaceus, Crocidura, Pteropus, Xantharpyia, Cynopterus, Rhinolophus, Hipposiderus, Megaderma, Vesperugo, Nycticebus, Tapozous, Rhinopoma, Nyctinomus, Funambulus, Alactaga, Gerbillus, Mus, Meosia, Acomys, Golunda, Ellobius, Hystrix, Lepus, Equus, Ovis, Capra, Boselaphus, Gazella, Manis,* etc. Among these, it is interesting to note that *Alactaga, Acomys, Ellobius, Equus, Ovis, Capra,* etc. are not found in any other part of the Oriental Realm; but *Equus,*

Capra and *Ovis* occur in Tibet; *Plutorius* and *Ursus* are found in the Himalayan forests. Both *Boselaphus* and *Antilope* do not occur beyond the R. Indus to the west. The birds, not found in other parts of India, but characteristic of the Punjab Tract include *Hypocolius, Aëdon, Lusciniola, Scotocera, Cettia, Coccothraustes, Fringilla, Alaemon, Melanocorypha, Nyctea, Ammoperdix, Otis* and *Cygnus. Erythrospiza,* and *Caccabis,* found here, are known also from Tibet. *Houbara,* which occurs, here, extends also beyond to the limits of Rajasthan.

Among the important reptiles found in the Punjab Tract, mention may be made *Gavialis, Crocodilus, Trionyx, Chitra, Emyda, Testudo, Nicoria, Damonia, Hardella, Kachuga, Ptyodactylus, Hemidactylus, Teratolepis, Eublepharis, Sitana, Calotes, Agama, Varanus, Ophiops, Mabuia, Lygosoma, Eumeces, Typhlops, Gongylophis, Eryx, Tropidonotus, Lycodon, Ptyas, Coluber, Oligodon, Cerberus, Dipsadomorphus, Psammophis, Echis,* etc. Some of them found here, like for example, *Stenodactylus, Agamura, Alsophylax, Pristurus, Phrynocephalus, Uromaxtis, Acanthodactylus, Eremias, Scapteira, Ablepharus, Scincus, Ophiomorus, Glauconia, Lytorhynchus, Conta, Tarbophis, Eristicophis,* etc. are not found in any other part of India. *Phrynocephalus* occurs however in Tibet. Among the Amphibia, we find marked poverty, except for the occurrence of widely distributed Peninsular or eastern elements. The only siluroid fish, which may be described as peculiar to the Punjab Tract, is *Ailiichthys.* Typically desert elements from the Rajasthan desert have penetrated the area in nearly all groups. In the other parts of the Plains we find the species commonly known from the Peninsula, with increasing proportion of the Trans-Gangetic faunal elements as we proceed eastwards.

REFERENCES

GEDDES, A. 1960. The alluvial morphology of the Indo-Gangetic Plain. *Trans. Papers Inst. British Geogr.* 21: 262–263.

LYDEKKER, R. 1902. Indian Tertiary and Post-Tertiary Vertebrates: The fauna of Karnul Caves. *Palaeont. indica,* 10(4): 23–58; *Rec. geol. Surv. India,* 19: 120; 20: 72.

OLDHAM, R. D. 1917. The structure of the Himalaya and the Gangetic Plains. *Mem. geol. Surv. India,* 42: 2.

PASCOE, E. H. 1919. Early history of the Indus, Brahmaputra and Ganges. *Quart. J. geol. Soc.* 75: 136–155.

PILGRIM, G. E. 1915. Suggestions concerning the history of the drainage of Northern India, arising out of a study of the Siwalik Boulder Conglomerates. *J. Asiatic Soc. Bengal,* (NS) 15: 81–99.

PRASHAD, B. 1941. The Indo-Brahm or the Siwalik river. *Rec. geol. Surv. India,* 74(4): 555–561 (1939).

SUBRAMANYAM, K. 1962. Aquatic Angiosperms of India. *Bull. bot. Surv. India,* 4(1–4): 261–272.

XXIV. BIOGEOGRAPHICAL EVOLUTION IN INDIA

by

M. S. MANI

1. Introduction

Earlier discussions on the biogeography of India generally seem to consider the distribution of plants and animals, the floristic and faunistic affinities and compositions and other biogeographical characters as largely static features and also unrelated to the profound and continual changes in the location, size, configuration, topography and drainage patterns of the region. This has unfortunately resulted in considerable confusion and led to untenable conclusions and curious contradictions. The distribution of plants and animals is, however, a dynamic phenomena and embraces the whole history of movements of the flora and fauna, including both gain and loss of the entire biogeographical area. This area is continually undergoing slow but complex changes, increasing in size in some parts, decreasing in size in others, shifting as a whole and coming into contact with other areas or becoming separated from them, breaking up into smaller and isolated patches or also coalescing into larger and more complex units. The present-day biogeographical characters have been derived as a result of gradual and continuous modification of past ones, which in their turn were modifications of still earlier characters. The composition, ecological characters, affinities of the flora and fauna and the distributional patterns of the plants and animals are, therefore, continually changing; we refer to these changes as biogeographical evolution.

Biogeographical evolution is not also an isolated phenomenon. The flora, fauna and the region constitute an indivisible whole and shrink or increase, retract or advance and evolve as whole ecosystems. It would, for example, be utterly meaningless to speak, as many earlier authors have indeed seriously attempted, of the evolution and origin of hillstream fish and its distributional pattern as independent events, wholly unrelated to the history of the origin of the hillstream and indeed of the uplift of the hill. The history of the hillstream fish should in reality be considered as an inseparable part of the history of the origin of the stream. Strictly speaking, therefore, the evolution of the flora and fauna of a region is essentially an integral part of the geomorphological evolution of the area.

Two concepts of great fundamental importance, which depart radically from earlier approaches to the problem, must thus be emphasized here. 1. It is not only that species and groups of species of plants and animals evolve as generally accepted, but also the flora and fauna of a region

698

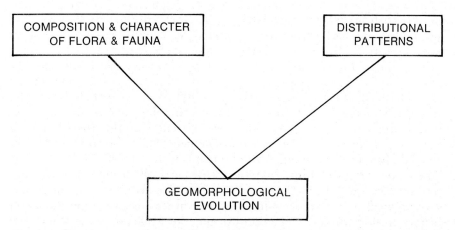

Fig. 156. The biogeographical evolution of India and its relation to the geomorphological evolution.

evolve as a whole. 2. The evolution of the whole flora-fauna complex is closely correlated to and very profoundly influenced by the geomorphological evolution of the region (Fig. 156). The close correlation between biogeographical and geomorphological evolution is readily evident from the fact that most of the peculiarities of the biogeography of India would remain meaningless, if we ignore the decisive rôle of the history of the changes in the landmass. A meaningful interpretation of the biogeography of India is, therefore, possible only on the basis of its geomorphological evolution. The salient characters of the present-day biogeography of India may be readily traced back to the patterns, which prevailed in the past and to the interaction of a complex set of factors, which operated on those patterns in the past, including the geomorphological changes in the region, prolonged differentiations, radiations and interchanges. The conditions operating at present have not, therefore, given rise to the present-day patterns. It is thus important to stress that the present-day climate of India, especially the monsoon rainfall type, does not underlie the origins of the biogeographical characters, but like these characters themselves, is a product of the geomorphological evolution. The present-day distribution represents, therefore, merely a dynamic phase in the uninterrupted course of the biogeographical evolution of India that has by no means either stopped or even substantially slowed down now. The outstanding characters of the biogeography of India today consist in the 1. the source, nature and complexity of the component elements of its flora and fauna and 2. the peculiarities of the distributional patterns.

2. *The Origins of the Flora and Fauna*

India is generally placed in the Oriental Realm of WALLACE (1876) or the Eastern Palaeotropical Realm of other authors. As however the term Oriental had earlier been applied by botanists to southwestern Asia and Iran, BLANFORD (1901) suggested the use of ELWE's term (1873) Indo-Malayan for the Oriental of WALLACE. Although neither expression is quite satisfactory, Indo-Malayan is even less appropriate than Oriental, at least with reference to India. We have seen, for example, that parts of the Punjab and the higher Himalaya should be included within the Palaearctic rather than the Oriental or the Indo-Malayan area. The western parts of the Indo-Gangetic Plains of north India are related more to the Ethiopian-Mediterranean than to the Malayan area. The eastern parts of India, together with parts of Burma, eastern Tibet south China and Indo-China and Thailand constitute a natural biogeographical area that is entirely distinct in its faunal composition and history from both India (Peninsula) and the Malayan area. While the forest-covered lower ranges of the Himalaya, south of the crestline of the Main Range, and eastern Tibet represent a narrow westward extension of the Indo-Chinese faunal subregion of the Oriental, the higher elevations of the Himalaya belong to the Turkmenian subregion of the Palaearctic Realm. Though faunistically an extension of the Peninsula, the Indo-Gangetic Plains are biogeographically neutral transition between the Peninsular and Extra-Peninsular areas. The Peninsula of India is quite distinct geomorphologically and biogeographically from the rest of India and its primary faunistic affinities are to be traced back more to the Madagascan Region than to the Oriental or even the Malayan area.

India should, strictly speaking, be described as comprising the distinctive Peninsula as the primary and principal biogeographical region, with the Himalaya and other Extra-Peninsular parts merely as *biogeographical appendages* of secondary importance. For the sake of convenience, however, the Peninsula and the Eastern Borderlands of India may, despite their fundamental differences in history, flora and fauna, be described as parts of the Oriental Realm of WALLACE. In any case, the appellation Indo-Malayan must be discarded. Considered from this point of view, three principal amphitheatres (Fig. 157) of origins of floras and faunas, differentiations and radiations have contributed largely to the composition and evolution of the biogeography of India, viz. 1. the Peninsula, 2. the higher Himalaya and 3. the eastern amphitheatre of the Indo-Chinese and Malayan subregions of the Oriental Realm. Of these amphitheatres, only the Peninsula lies wholly within the limits of India, but the other two are largely outside our boundaries and touch India only by their margins. The floras and faunas differentiated in these amphitheatres fall respectively under two major groups, the representatives of which constitute the biogeographical components today, viz. 1. the Gondwana derivatives and 2. the Asiatic derivatives.

The Peninsula *per se* is biogeographically *India vera*, the largest and the oldest region of differentiation of the original floras and faunas of India.

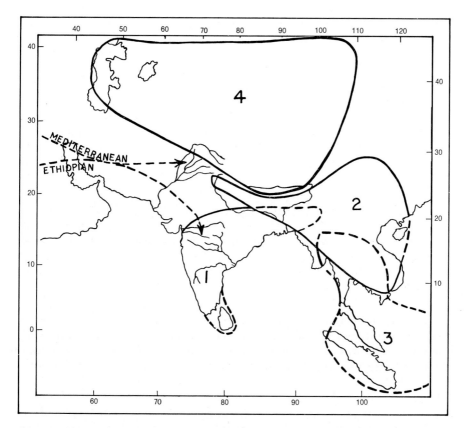

Fig. 157. Major amphitheatres of differentiation, evolution and radiation of floras and faunas, which have profoundly influenced the biogeographical evolution of India. 1. The Peninsular amphitheatre, which is *per se* India *vera* biogeographically and the home of an ancient endemic relict fauna; 2. the Indo-Chinese amphitheatre, largely outside the limits of India, but extending westwards as a narrow belt on the forest-covered ranges of the Himalaya, and overlapping the Peninsular amphitheatre in the northeast in Assam; 3. the Malayan amphitheatre overlapping the Indo-Chinese amphitheatre in the north; 4. the Turkmenian amphitheatre, also outside the limits of India, except for a narrow southern fringe encroaching on the higher Himalaya above the forestline and an area of differentiation of young steppes fauna. The amphitheatres 2 and 3 are regions of differentiation of young, tropical-forest faunas. The amphitheatres 2–4 represent the biogeographical appendages of India.

Most biologists have, however, failed completely to recognize the dominant place of the Peninsula in the biogeography of India, but have over-emphasized the place and importance of the Indo-Chinese subregion. We thus find that nearly all earlier workers have supposed the Peninsula to have been colonized entirely by genera and species, which were differentiated in Assam-Burma and areas farther east, completely ignoring the fact that the greatest bulk of the true Indian flora and fauna differentiated and evolved in the Peninsula, throughout the Palaeozoic, Mesozoic and Tertiary, right nearly up to the Pleistocene times, and spread extensively into the Extra-Peninsular areas during the late Tertiary. This failure to recognize the Peninsular amphitheatre has led to much confusion and erroneous ideas on the distributional patterns.

The flora and fauna that differentiated in the Peninsula were indeed the original flora and fauna of India. This complex arose from the ancient stock of Lemuria and the still older Gondwana floras and faunas. This was essentially a tropical humid-forest fauna and was also very widely and continuously distributed not only throughout the Peninsula, but even up to the foot of the newly rising Himalaya, until perhaps relatively recent times. Indeed it extended eastwards even beyond the strict limits of India. The affinities of this fauna were mainly with Madagascar and South Africa, but to some extent also with Australia and South America, especially in the more ancient groups. We have observed, for example, that the Amphibia of the whole of India are almost completely derived from the Gondwana faunas, and a number of fishes and Invertebrata also belong to this ancient fauna. The physiognomy of this fauna, before it was modified by the influx of the Oriental elements from the Assam-gateway, and before it was impoverished by drastic elimination of habitats under the influence of man, can only be partially reconstructed and with difficulty, from the present-day relict character of the Peninsular fauna. Some light is, however, thrown on its composition and general character from a study of the Inter-Trappean and Siwalik fossils (see chapter XIX). The fauna was characterized by its ecological and geographical saturation and comparative evolutionary stagnation. Most genera had indeed attained the maximum level of differentiation of species in relation to the available habitats. The distributional range of most forms neither extended nor retracted, but remained largely stationary. Ignoring minor changes due largely to the Deccan-Lava flows activity, this fauna was thus for the most part remarkable for its high degree of stability. These peculiarities are closely related to the geological stability and the mature topography of the Peninsula. The evolutionary stagnation of the original fauna of the Peninsula gave place, however, to rapid and complex changes with the influx of exotic elements, when the Assam-contact with Asia was established, as an early phase of the Himalayan uplift. The faunal contributions from the Oriental Region (the areas to the east of India) were also followed later by nearly equal contribution from the Ethiopian Region, and still later during Pleistocene times to some extent even from the Palaearctic areas. Although in certain groups, the Indo-Chinese amphitheatre has contributed largely or even exclusively to the present-day Peninsular flora and fauna, the contributions from the different exotic sources are on the whole nearly equally strong. As correctly guessed by MAHENDRA (1939), the Amphibia of India are Peninsular-autochthonous and Madagascan, the Chelonians are Indo-

Chinese, and the lizards and snakes 'invaded India apparently from all sides'. Parts of the Peninsular fauna were also separated from the main body, by faunal regression and other processes and are found today as outliers and isolates, deep within the Extra-Peninsular areas, leading to the present day disjunction in distribution of certain types.

The Peninsula is at present characterized by its remarkable wealth of phylogenetic (Gondwana) and geographical (Asiatic) relics, Pleistocene relics of the Himalaya, endemics, ancient and phylogenetically older groups and by the presence of ecologically anomalous (habitat-fremde forms) groups like humid-tropical evergreen-forest elements in areas now covered entirely by deciduous forests and savannahs (KHAJURIA, 1924). The presence of the habitat-fremde groups is strong evidence of the fact that formerly the whole Peninsular fauna was a humid-tropical one and was also far more widely and continuously distributed throughout the Peninsula than at present and that the changes in the habitats have taken place within relatively recent times. Though the fauna of a forest follows the natural changes in the forest *pari passu*, like a plastic mass, the changes introduced by deforestation (see chapter V) have been so abrupt that the faunas have in some of these areas had not enough time to become readjusted. This part of our fauna is in a sense a relict. The Peninsular fauna is on the whole at present remarkable for its greatly impoverished remnants that are also rapidly vanishing. Great age has already obliterated subcentres and radiation has been more or less obscured. The numerous socalled subdivisions, which have been proposed by zoologists from time to time, are in no sense biogeographical divisions at all and must not also be confused with subcentres of origins of genera and species. The Peninsula is thus remarkable at present for the extensive faunal regression, degradation and impoverishment.

The fauna that differentiated in the Eastern Amphitheatre is likewise a humid-tropical forest one, but unlike the Peninsular faunas is largely composed of phylogenetically much younger and taxonomically higher groups, derived almost exclusively from Asia. These faunas are characterized by a high degree of phylogenetic plasticity, evolutionary intensity and ecological and geographical instability. In marked contrast to the Peninsular faunas, the eastern faunas are not older than perhaps the Pliocene, Pleistocene and even Post-Pleistocene times. Their evolution and dispersal were conditioned primarily by and closely bound up with the massive Tertiary mountain-building activities in these areas. These faunas are characteristically rich in snakes, higher mammals, birds, forest and arboreal insects like Orthoptera, Phasmida, Cerambycidae and Lepidoptera. The faunas spread westwards along the forest-covered ranges of the Himalaya and to the Peninsula, and may be spoken of as *replacing faunas* in the Peninsula, the original fauna of which is of much greater age. The replacing fauna is not, however, an evidence of succession in the Peninsula, primarily because succession involves a phasic development, in which a preceding fauna creates the conditions optimal for the next one. The Eastern faunal Amphitheatre was also more or less extensively penetrated by the Peninsular, Palaearctic and Ethiopian elements. Though in some groups, the Peninsular elements and in others the Palaearctic elements predominate, the contributions to this faunal amphitheatre were nearly equal from all these areas. In marked contrast to the Peninsula, the Eastern Amphitheatre shows strong evidence of faunal radiation, the centres of which are, however, far away from the borders of India and not in Assam. While the Peninsular fauna is at present degraded, greatly impoverished and rich in relics, the fauna of the Eastern Amphitheatre is still in course of pronounced enrichment and intense speciation. The Indo-Chinese and Malayan areas of faunal differentiation and radiation overlap in Assam-North Burma (eastern borderlands).

The Himalayan Amphitheatre is a part of the Palaearctic Realm and represents indeed the southernmost limits of the Turkmenian Subregion (Fig. 155) in the west and to some extent of the Manchurian Subregion in the extreme east. The Himalayan Amphitheatre is characterized by the youngest, most highly plastic, ecologically highly specialized faunas, rich in steppes elements, especially among the Invertebrata. There is a remarkable absence of fishes and amphibians and an extreme poverty of reptiles. Although birds and mammals are not uncommon, by far the dominant elements are terricole and endogeous Arthropoda, especially insects. The fauna is unsaturated to a high degree and ecologically unstable, with a great abundance of pioneer communities. There is relative poverty of relics. The fauna was differentiated mostly during Pliocene-Pleistocene and partly also recent times. The higher Himalaya differs from the other faunal amphitheatres in the high intensity of isolation and speciation in nearly all groups of animals, but particularly in insects.

We may recognize in the Himalaya (Fig. 155) at least three secondary centres of independent faunal differentiation, two of which are wholly within the Palaearctic and the third has been considerably influenced by the Indo-Chinese fauna and may indeed be said to be transitional to that subregion. We have in the extreme west the Northwest Himalayan secondary centre. The higher Himalaya in the east is a secondary centre corresponding to a part of the Tibeto-Eremian centre of DE LATTIN (1966). The forest zone of the Himalaya, together with the north Burmese mountains, forms an independent secondary centre. The Himalayan amphitheatre is the area of differentiation and radiation of what the botanists have termed the temperate or partly also as the European elements in the Indian flora. Finally the intrusive elements include also the Mediterranean(corresponding in part to the Saharan and temperate elements of botanists) and the Ethiopian (corresponding in part to the African and in part to the Saharan of botanists) components.

704

The principal biogeographical components of the flora and fauna of India include therefore 1. the true Indian or the Peninsular autochthonous elements; and 2. the Indo-Chinese and the Malayan complex, 3. the Turkmenian, 4. the Mediterranean and 5. the Ethiopian intrusives. India thus comprises mainly an ancient area of endemics and relicts fauna in the Peninsula; recent areas of young and highly differentiated humid tropical forest-fauna encroaching in the east and a youthful, ecologically highly specialized endemic mountain-autochthonous fauna of the Palaearctic in the higher Himalaya. The peculiarities of faunal evolution are reflected in all the present-day distributional patterns. The complexity of the composition of flora and fauna, particularly the presence of the intrusive Extra-Peninsular components of Asiatic origin, is unquestionably the result of the interchanges mainly by the Assam-gateway. The formation of this most important route of interchanges of the flora and fauna marks an early phase of the uplift movements of the Himalayan system. The major events of the complex geomorphological evolution, commencing from the northeastward movement of Madagascar-Peninsular landmass, are well known and have also been outlined in chapters II, III and V. The movement of this mass obliterated the Tethys Sea, progressively westwards from the east (Fig. 158), where the sediments were squeezed and folded to give rise to the Tertiary mountain areas we know as the eastern Tibet, Assam, Burma, South China and Indo-China and also established the physical contact of the ancient mass of the Peninsula with the newly formed areas of Asia. The northern edge of the advancing Peninsula down-warped and thrust under the rising Himalaya and Tibet and in the foredeep thus formed arose series of receding lagoons and marshes as vanishing relics of the Tethys and it was filled up the detritus from the Himalaya and from the Peninsular plateau in the south. In this process of bukling and down-warping, the Peninsular rim was fissured east-west and through these areas of weakness spewed forth the extensive Deccan Lava flows. The same movements brought about block-fracturing of the western parts of the Peninsula and marine subsidence of the fragments in the Arabian Sea, giving rise to the scarps of the Western Ghats. The obliteration of the residual Tethys Sea fell perhaps with in subrecent times, completing the physical contact of the Peninsula with Asia in the west also and with the Ethiopian region.

It is abundantly clear that the formation of the Assam gateway represents undoubtedly the most important phase in the biogeographical evolution of India. This gateway opened up extensive interchanges between the Peninsular autochthonous and Asiatic Tertiary-mountain flora and faunas, the movements being equally strong both from the west to the east and vice versa. The Peninsular elements spilled over into the Extra-Peninsular area and the floras and faunas differentiated in the Tertiary mountains of south China, Indo-China and Thailand and Malaya further south intruded westwards along the Himalaya as far

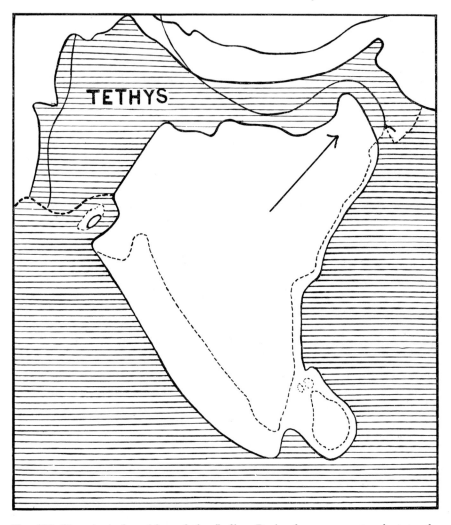

Fig. 158. Hypothetical position of the Indian Peninsular mass, antecedent to the Assam-contact with Asia in the northeast where the Tethys Sea has already narrowed, in accordance with the theory of continental drift (Not drawn to scale).

west as the great defile of the R. Sutlej and southwards into the Peninsula and Ceylon, which still formed a part of the Peninsular mainland of India (JACOB, 1949). The interchanges of the flora and fauna between the Peninsular and Assam-Burma areas thus introduced a new factor in the evolution of the biogeography of the entire region and also gave rise to great complexity in its general ecology. The presence and intermingling of the Indo-Chinese-Malayan complex in the flora and fauna of the Peninsula must be described as perhaps the most important result of

706

these interchanges. The uplift of the Himalaya led also to the evolution *pari passu* of the lowland steppes elements of the flora and fauna of Asia into the temperate and high altitude elements of the Turkmenian sub-region of the Palaearctic by the Pliocene times. The Pleistocene glaciations on the Himalaya introduced another newer element in the bio-geographical evolution of India. During the periods of glaciations the temperate Turkmenian elements of the Himalaya spread southwards to the Peninsular south and during the Inter-Glacial times the Peninsular elements advanced northwards to the Himalaya. The history of this alternating advance of the Palaearctic and of the Peninsular types is today recapitulated in the seasonal oscillations of temperate forms of plants and animals with the warmer Peninsular types in the transitional Peninsular margin that we know as the Indo-Gangetic Plains of north India. The late phase in the series of these evolutionary changes saw the intrusion of the Mediterranean and Ethiopian elements, eastwards on the Himalaya and southwards in the Peninsula and across the Peninsula as far east as the Eastern Borderlands of Assam.

We may conclude that the Himalayan uplift dominates practically the entire range of events culminating in the shaping of the climate and composition of the flora and fauna of the whole of India. The Himalaya presides over the ecology and biogeography of India.

3. The Origins of the Distributional Patterns

The outstanding peculiarities of the present-day distributional patterns are concentration and isolation of the dominant elements of the flora and fauna in relatively small, often also widely separated areas, resulting in more or less marked discontinuity. Altitudinal zonation is confined almost exclusively to the Himalaya, but is nearly absent in the Peninsula, inspite of the high elevation, particularly in its south. The concentrations in localized areas in the south and in the northeast are associated with more or less steep rise in gradients of abundance.

The most important concentration and isolation of both the Peninsular autochthonous and the intrusive exotic faunal elements, the endemics of ancient groups and the geographical and phylogenetic relicts, are found in the great horsts of the Southern Block of the Peninsula, with a steep fall northwards in the gradient of abundance to Deccan. The Indo-Chinese and Malayan faunal derivatives are largely concentrated in Assam-Burma area, but with a gentle westward fall in the gradient of abundance along a narrow stretch of the forest-covered outer Himalayan ranges, dis-appearing rather gradually, about the defile of the R. Sutlej, in the steepening gradient of the Palaearctic endemic elements of the western end of the Himalaya. The higher Himalaya represents, as already mentioned, a concentration of young autochthonous endemics of recent Palaearctic groups.

The concentrations of the Peninsular and the eastern faunal complex in the Southern Block and in the Assam-Burma area result in a more or less pronounced discontinuity in the distribution of nearly all groups. The most striking and perhaps biogeographically very important case of discontinuity is observed between the Southern Block (including Ceylon) and the Eastern Himalaya, Assam and Burma area. This involves 1. the discontinuous occurrence of a number of interesting Peninsular outliers and isolates in Assam and north Burma and 2. the occurrence of Indo-Chinese and some Malayan faunal derivatives in the Southern Block (see chapters XIX and XX).

In certain genera and species the discontinuity may be extreme and may be marked by their occurrence in the Eastern Himalaya and or Assam and again only in Ceylon, the extensive intervening areas being totally without them. In others we find a series of more or less isolated patches of occurrence of the same species or of local intergrading sub-species and races all along the Eastern Ghats, from Assam to the southern extremity of the Peninsula. The Peninsular and the eastern isolates often also meet in Chota-Nagpur, but may be absent in the area between Chota-Nagpur and Assam and between Chota-Nagpur and the Southern Block. In some remarkable cases, the gradient of the Assam concentrations of the Indo-Chinese and Malayan elements does not indeed disappear completely anywhere in the intervening area, before again rising steeply in the south, thus showing an abrupt increase in abundance in the Southern Block. In this case the discontinuity is, therefore, only with reference to the abundance, but does not involve absolute disappearance in the intervening area. Taken as a whole, the discontinuity of the Peninsular and eastern fauna is essentially an impoverishment of the faunal complex in the areas between Assam and the Southern Block.

The discontinuous distribution of Indo-Chinese and Malayan plants and animals in the Peninsular south was known to the early naturalists. MEDLICOTT and BLANFORD (1879), for example, remarked on the presence on the Nilgiri, Anamalai and Shevroy Hills and other elevated areas in South India, of temperate elements of flora and fauna of the Himalaya, not occurring in the intervening plains of north India. They were indeed greatly struck by the fundamental similarity of these South Indian hill plants and animals to those found in the Eastern Himalaya, Khasi, Garo and Naga Hills of Assam and on the mountains of north Burma and Malaya. They even recognized the fact that in many cases the species were the same in these widely separated places. MEDLICOTT suspected that the discontinuous distribution of these plants and animals was correlated more with the conditions of the atmospheric humidity and precipitation than to temperature. Since then, numerous records of discontinuous distribution of Oligochaeta, Arachnida, insects, fishes, amphibia, reptiles, birds and mammals have been made, examples some of which have already been mentioned in the foregoing chapters.

The distributional patterns of the different components present significant differences, correlated in part with their history. While more or less pronounced discontinuity is characteristic of only the Indo-Chinese and Malayan complex, the Mediterranean and the Ethiopian components are typically continuously distributed. The Peninsular autochthonous elements are largely concentrated in the south and west of the Peninsula, but considerable numbers are also isolated in Assam-Burma. A peculiarity of considerable interest in the distribution of the typical Peninsular forms of both plants and animals is the seasonal oscillation north and south in the transitional area of the Indo-Gangetic Plains (see above); the Peninsular forms predominating during the monsoon rainy season and the intrusive elements in the postmonsoon season. The Indo-Chinese and Malayan complex is largely concentrated in the eastern borderlands and extends as a narrow tongue westwards on the Himalaya and intrude south in the Peninsula, exhibiting nearly every gradation from the wholly continuous to the extreme discontinuity. The Turkmenian (Palaearctic) elements are confined to the higher Himalaya, and also exhibit the seasonal south-north oscillation and alternate with the Peninsular types in the transition Indo-Gangetic Plains during the winter and occur as isolates on the Eastern Ghats (Mahendragiri, Yercaud, etc.) and Western Ghats and even in the hills of Ceylon. The Mediterranean elements are largely concentrated in the Northwest Himalaya, parts of the western borderlands, western parts of the upper Indo-Gangetic Plains and continuously along the Western Ghats to the south. The Ethiopian elements occur continuously in the Rajasthan desert, Deccan and western parts of the Peninsular south.

Like the salient facts of the evolution of characteristic composition of the flora and fauna of India, the peculiarities of the present-day distributional patterns of plants and animals may also be correlated to the major events in the geomorphological evolution, climaxed by the uplift of the Himalaya. The picture is, however, complicated by the introduction of a new factor within historical times.

It is of fundamental importance to emphasize that the areas of the present-day concentration and isolation of the dominant elements of the flora and faunas, especially in the Peninsula, do not by any means mark the centres of radiation of the original Peninsular complex, but represent on the other hand the refugial islands, to which not only the original Peninsular but also the eastern intrusive elements (both of which were formerly characterized by a much wider and continuous distribution) have retracted, within relatively recent times. Isolation in the refugial areas has preserved the phylogenetic and geographical relics in the Peninsula. On the higher Himalaya, isolation has, however, favoured an intensified speciation, so that the fauna in this part is at present passing through a process of enrichment. In the Peninsula, however, the concentrations merely represent precarious survival of the relics of an otherwise vanished and impoverished fauna, with very little or no recognizable speciation.

Various theories, often fanciful in the extreme, have been put forward to account for the discontinuous distribution of the Oriental faunal elements in South India, but all these theories suffer from attempting to explain the distribution of single groups in isolation from others. Ignoring the one-time fashionable theory of the southern land-bridge route of migration across the Indian Ocean, which has only historical interest, the earliest theory of MEDLICOTT and BLANFORD (1879) suggested that the general lowering of the atmospheric temperature during the Pleistocene times resulted in the retreat of northern plants and animals to the equator. They further believed that subsequently, as the general atmospheric temperature conditions became warmer, after the retreat of the Pleistocene glaciers from the Himalayan valleys, the plants and the animals moved towards the higher parts of the South Indian hills, where they are now found. Some of the higher Himalayan forms, especially insects of the Turkmenian origin, occurring as isolates on the Nilgiri, Anamalai, Palni and Cardamom Hills of the south, are evidently Pleistocene relicts. While no doubt accounting for the occurrence of the Himalayan elements at high elevations in the Southern Block, this theory does not, however, satisfactorily explain the presence of the eastern faunal derivatives in the Peninsula. STOLICZKA and SARASIN (1910) believed that the Oriental elements, now isolated on the South Indian hills, became separated from their main body in Assam and Burma as a result of the Deccan Lava flows. This theory overlooks the fundamental fact that the Assam fauna is of Tertiary origin and their colonization of the South Indian hills took place largely during Pliocene times. The Cretaceous-Eocene Deccan Lava flows cannot, therefore, possibly be the underlying factor in the discontinuity observed today. It may, however, partially explain the discontinuity of certain older Peninsular autochthonous faunal elements within the Peninsula itself, but all evidence of such an effect has been completely obliterated by the passage of time.

Some years ago, HORA (1944, 1948, 1949, 1950, 1951) put forward what he called the Satpura hypothesis to account for the discontinuous distribution of certain fresh-water mountain-stream fishes between the Eastern Borderlands and the Southern Block. Some of his collaborators (BISWAS & SAMPATKUMARAN 1949, JAYARAM 1949, MENON 1951, ROONWAL & NATH 1949, SILAS 1952) later attempted, without conspicuous success, to extend the hypothesis to the discontinuous distribution of other animals also. HORA believed that the ecologically highly specialized, hill-stream fishes, characterized by torrential adaptations, now discontinuously distributed in the Southern Block and Assam hills, evolved primarily in the mountainous area of Yunnan in south and Indo-China and migrated as fully evolved, torrential-adapted fishes to the Peninsula and even to western Asia and Africa. In view, however, of the well known fact that, like every other ecologically highly specialized organism, the torrential stream fishes have indeed extremely limited potentialities for such extensive geographical migrations, particularly in such a relatively short period, since these fishes evolved during the late Tertiary times, this was a most serious misconception in the hypothesis. Furthermore, it was unfortunate that HORA failed even to clearly distinguish between the derivatives of the Indo-Chinese fauna and the fauna of the Malayan subregion and constantly spoke of the two as if they were identical. The central idea of his hypothesis is that from Assam, the route of migration of these fishes to the south of the Peninsula lay only westwards over the Satpura trend of mountains to the northern end of the Western Ghats, thence southwards. An elevation of about 1500–1800 m of the Vindhya-Satpura scarps and the northern sections of the Western Ghats was obligatorily assumed to afford a continuity of favourable ecological belt, with a mean annual rainfall of 215 cm and tropical evergreen forest-covered mountains to facilitate the movement of his fishes. As conceived by HORA, the migration of the torrential hill-stream fish naturally involved the supposition that the alluvium-filled lowland marshy gap between the Garo Hills of Assam and the Rajmahal Hills at the extreme northeastern corner of the Peninsula was an insurmountable barrier to the specialized hill-stream fish. This lowland gap has indeed been the most inconvenient stumbling block in the hypothesis and the easiest way to remove it was to assume further that the Garo-Rajmahal Gap came into existence only after the hill-stream fishes had migrated from Yunnan to South India. In order to support his flight of fancy, HORA bridged the Gap with a connecting hill range – Gap was lowland in the early Tertiary, but came to be filled up in the Miocene-Pliocene times and reappeared in its present gap-form during the late Pleistocene.

Geological (AUDEN 1949, DEY 1949) evidence is, however, not in favour of the Gap having been thus filled up. The direction of the overlap and the existing dips in the Rajmahal Hills are wholly against any idea of later renewal of the socalled Satpura trend having had any effect in the general elevation of the Gondwana rocks. The presence of beds of fossil oysters, replacing the Sylhet limestones at and west of Tura, shows that the Gap was formerly the estuary of a major river. The marine development of the Middle Miocene age extends right into the Gap from the east. There is thus no evidence of an uplift belt crossing the Gap since the earliest Gondwana times (see chapter III).

The socalled Satpura protaxis assumed by certain geologists does not correspond with any well defined structural line. The hills referred to this trend are composed of diverse stratigraphic and structural units. The Shillong Plateau does not have any real connection with this trend. The Rajmahal Hills comprise gently dipping lava beds of Jurassic-Cretaceous (Rajmahal Trap), but we do not find here any sign of orogenesis following

711

the Satpura trend and affecting the Rajmahal Trap. The formation has a general north-south strike, with a general regional dip to the east. In this direction it passes under the Gangetic alluvium.

The Cretaceous, Eocene and Miocene beds of the Shillong area dip southwards off the Shillong Plateau, at Sylhet and Cherrapunji. Further west the trend of the outcrop swings round to northwest-southeast strike, with a regional dip to southwest. These facts clearly show that there has been no axis of folding along the east-west Satpura trend at least since the Jurassic times (the Rajmahal Trap period). There has, however, been subsidence along a north-south axis, between the Garo and the Rajmahal Hills. This subsidence may have been due to block-faulting or to regional warping of the Peninsular Block, as a part of the Himalayan uplift. The marine beds mentioned above are strongly indicative of the fact that the Garo-Rajmahal area was a low-level tract during the late Cretaceous and early Eocene times. The area may not necessarily have been covered by sea, but the shore-line in the Garo Hills during the Lower Eocene times was already halfway through the Gap. The Shillong Plateau came into existence perhaps during the Miocene period, under the influence of the forward thrust of the Himalaya. The Garo-Rajmahal Gap was thus a physiographic and structural depression throughout the Miocene times. The Gap that existed in Miocene would hardly appear to have become closed again to provide an adequate ridge for the faunal movements envisaged by HORA and his collaborators. The block-faulting that perhaps formed the Gap involved the ancient crystalline Archaeans of the Peninsular area. If this Gap were to be closed again, as postulated by HORA, so as to form even a moderately high ridge to serve as faunal route for hill-stream fishes, it involves the impossible supposition of a reversed block-faulting to that which originally gave rise to the Gap during the Miocene times. The soft Pliocene sediments laid down in the Gap could not have been elevated into a range with the same strike as the present east-west Shillong-Garo Hills. This idea of a reversed block-faulting involves also the difficulty of assuming a re-introduction of movements of orogenic type in a region, which had only just previously been subjected to block-faulting. Either from the point of view of the crystalline rocks being pushed up into the same place they had formerly occupied or from the point of view of the Pliocene beds having formed a range between the Rajmahal and the Garo Hills, it is difficult to imagine that the Gap, once formed, should have been blocked again. It is important to recall in this connection that the R. Godavari trough and the Garo-Rajmahal Gap are nearly equidistant from the lower R. Mahanadi trough, all of which contain Gondwana deposits. It is concluded from the evidence from these areas that the Gap was roughly formed at a very early period, as one of a series of complimentary warpings. The Gap did not thus originate during the Pleistocene times, as quite erroneously claimed by HORA in support of his Satpura hypothesis, but very much earlier and has continued to remain a gap. The entire aspect of the Rajmahal Traps and the Tertiaries to the south unequivocally speaks against the idea of an uplift and depression of a ridge. While the Rajmahal Traps have a persistent easterly dip, with no eastwest rolls, the Tertiaries cannot be said to show any appreciable disturbance.

To summarize the precise position, we may conclude, therefore, that this gap is one of the complimentary warping features synchronous with that of rivers Godavari-Krishna (DEY 1949). The Shillong Plateau, generally described as detached block of the Peninsula, is really speaking not also a detached part at all, but must be considered as regular a feature as the Mahendragiri and other parts of the Eastern Ghats. The course and the delta of the R. Ganga represent in reality the delta of an earlier Pre-Ganga Peninsular river (comparable to Godavari and Krishna and the R. Son), which underwent simultaneous subsidence with filling (Fig. 4). This old Peninsular river captured the Himalayan short foredeep streams – the piedmont rivers like the present-day Siwalik Hills rivers and the Ganga is one of those thus captured by the pre-existing north-flowing Pre-Himalayan Peninsular river. Subsequent attempts were made by HORA and his collaborators to meet these objections by supposing that the lowering of the sea-level by about 30–60 m during the Pleistocene times had the virtual effect of the Gap serving as if it were indeed a low hill, some 200 m higher relative to the sea. This, however, overlooks the basic fact that the area of the Gap is one of simultaneous filling and sinking and its level has not therefore been substantially different earlier than now. The fantacy is not exhausted yet: the lowering of the mean sea-level and the consequent advance of the shoreline during the height of Pleistocene glaciation, when so much of the water was locked up as inland ice, was also assumed to have brought about the impossible commingling of the Extra-Peninsular Ganga-Brahmaputra river system with the Peninsular Mahanadi river system near their mouths, thus facilitating the migration of the Assam fish through the Ganga to the Mahanadi, thence along again the Satpura *en route* to the northern end of the Western Ghats and finally down south. It is wholly unnecessary to remark that this assumption overlooks the well known fact that such torrential-steam fish, requiring a rocky substratum for their existence, (a condition considered by HORA himself as essential basis of his Satpura hypothesis) are really confined to near the water-partings on the hills, certainly do not occur in the placid silt-covered mouths of the rivers. HORA also spoke of 'waves of fish migration' from Assam to the Peninsula, but all evidence shows that the spread and intermingling of the Peninsular and eastern faunas, including fishes, have progressed continuously and uninterruptedly from the first Assam contact until recent times and not in waves. The faunal movement was not only from the east into India as HORA believed, but also equally strongly in the reverse direction, so that it is quite erroneous to speak of migration. We have on the other hand faunal interchanges through Assam. We have already shown conclusively (chapters XVIII & XX) that Assam has received as much as it has contributed. Assam is not also basically a centre of differentiation and radiation, but in reality a transitional area between the eastern amphitheatre and the Peninsula, through which movements of floras and faunas have been from both east to west and vice versa; *it is a major route of interchange*.

The importance of this Gap as a biogeographical barrier and exclusive route along the Satpura Ranges has thus been grossly exaggerated by zoologists. The Gap may *not* actually have been a barrier for faunal movements either from the east to the Peninsula or from the Peninsula eastwards. It is also conceivable that alternate routes would have been taken and the fauna thus got around the Gap, even in case it served as a barrier in certain exceptional cases. The explanation of the distribution of certain fresh-water animals in the Assam area and in the Peninsula, on the basis of a connecting watershed, cannot, therefore, be built up exclusively on the Garo-Rajmahal Gap-line. There is a possible alternate route, along a zone between the Monghyr and Rajmahal Hills on the south (on the Peninsular extreme northeast) and the Darjeeling hills and Eastern Nepal Himalaya on the north, across the Gangetic plain. The south-north trend of the Dharwar rocks of Monghyr and the similar strike of the Rajmahal Hills would strongly support this idea of an alternate route. It is also strange that HORA should have so completely ignored the fact that the distributional pattern of these fishes does not particularly support the idea of a single and only route and does not also exclude other possible routes. The distributional pattern does not indeed support the idea of migration, but strongly suggests the idea of *local evolution, in situ*. Assuming, however, for the moment, that torrential-adapted fishes did actually migrate, as claimed by HORA, it is difficult to understand how he could completely ignore the possibility of the migration of the torrential fishes along the Eastern Ghats, with their numerous torrents, waterfall, cascades, etc., where the conditions are indeed far more favourable for such a migration than along the Satpura-Vindhya scarps. He did not also consider the possibility of a westward migration of the fish along the Himalaya from Assam, *en route* to the Peninsula.

It is evident that dispersal routes may have been multiple and further the intrusion to the south could not possibly have been after the fishes became ecologically specialized as hill-stream forms, but much earlier before. The hill-stream adaptations of these fishes in such widely separated regions must be traced to multiple and independent origins under identical conditions of life in a generalized stock, which must, therefore, have been formerly continuously and widely distributed throughout the region, including the intervening areas. The origin of the torrential adaptations in the fish of the southwest corner of the Peninsula may well have been contemporaneous with and correlated to the punching up of

714

the horsts of the Southern Block (see chapter II), quite independently of the events to the east of Assam. The Garo-Rajmahal Hill Gap loses in this case its supposed rôle as a barrier, so that there is no more a need for conceiving of a connecting hill range. At any rate, it is incorrect either to interpret or to reconstruct structural characters in terms of distributional peculiarities; rather the distributional characters must be explained in terms of geomorphology.

A careful study of the material discussed in the foregoing chapters would also show that different components of the eastern faunas have spread along different routes and often followed more than one route at the same time. Some have spread westwards along the Indo-Gangetic Plain (at least the Lower Ganga Plain), others along the Himalaya to the west as far as the Yamuna-Ganga Doab and thence south along the Aravalli Hills (a region which was then ecologically different from what it is at present) and still others along the Eastern Ghats and even across the Garo-Rajmahal Gap of lowland. It may also be remarked that the eastern elements have spread partially westwards north of the Himalaya to eastern Tibet and northwards as far as Manchuria.

The origin of the discontinuous distribution that characterizes diverse groups in India thus seems to be rather complex and is by no means an isolated event. The discontinuity is in some cases primary, but in most others secondary and derived and of relatively recent origin from a former continuous distribution. The marked discontinuity observed in the distribution of the complex of Indo-Chinese and Malayan faunal derivatives is, for example, of relatively recent origin. As correctly interpreted with rare insight by KURUP (1966), (also see chapter XVIII) with reference to mammals, the *present-day discontinuity is essentially a relict of former continuous distribution*. The extensive and continuous ranges of a number of the humid-tropical forest forms have recently come to be broken up into a series of isolated patches, partly because of topographical or partly because of climatic changes and partly by the gradual retraction and regression, leaving behind more or less large areas of the isolates (Fig. 159, 160, 161). In some groups, vast populations of the dominant genera and species have apparently died out (from perhaps faunal senility and exhaustion in the absence of any other striking cause) or have more often been decimated or also annihilated by man in the intervening areas, thus resulting in the present-day discontinuity. The influx of the outside faunas and the rapid colonization by their more highly plastic and young derivatives in the marginal areas of the autochthonous faunal range have isolated the members of the latter in small pockets in a number of cases and eventually resulted in a rather pronounced discontinuity. In other groups, genera and species have actually jumped seemingly unpassable zoogeographical barriers, where they are now either extremely sparse or altogether absent, and have thus come to be eventually discontinuously distributed in widely separated ranges.

715

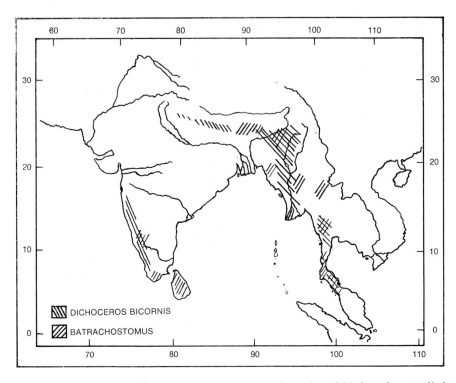

Fig. 159. Discontinuous distribution of two oriental species of birds; when studied alone the distribution is markedly discontinuous.

The present-day discontinuity seems to have risen in some cases in an entirely unsuspected way. It seems *a priori* certain that as suggested by STEPHENSON (1921) in the case of Oligochaeta and undoubtedly as we have indicated above in the case of the hill-stream fishes and perhaps also in a number of other groups, the discontinuous distribution is the result of recent local diversifications, multiple origins and polyphyletic differentiations, leading to evolutionary convergence and parallelism in the widely separated areas. A great many of the ecologically and taxonomically specialized forms, characterized at present by marked discontinuous distribution in Assam-Burma and in the Southern Block and placed by taxonomists, however justifiably, in the same genera and species, may apparently be quite unrelated in the way generally assumed and they may not have migrated and crossed biogeographical barriers, but may have simply evolved *in situ* locally into 'identical' forms from a common, formerly widely and continuously distributed generalized stock, under ecologically comparable if not identical conditions. We have here the situation in which a faunal complex spreads extensively in a fairly large region, the ecology of which is more or less similar to the centre of

716

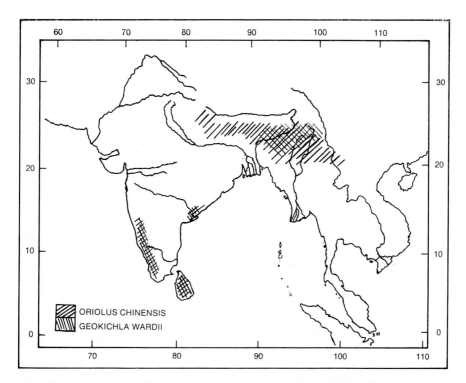

Fig. 160. Discontinuous distribution of two interesting birds. *Oriolus chinensis* occurs in Ceylon, the southwest corner of the Peninsula, North Kanara, the Vizagapatam Hills of the Eastern Ghats, the Eastern Himalaya, mountains of North Burma, Yunnan, eastern Tibet and extreme north of Indo-China. *Geokichla wardii* occurs in Mysore, Travancore, Ceylon, Eastern Ghats, Eastern Himalaya (westwards on the Himalaya up to Kumaon) and eastwards to the mountains of North Burma. These species illustrate one of the grades from the continuous to the discontinuous distribution.

radiation, so that the colonization by the intrusive elements proceeds at first without significant local diversification. At a later stage, the ecology alters in a parallel direction in the two widely separated (macrogeographically allopatric in current theoretical terminology), areas of this region, for example the uplift of the high mountains in the east and the punching up of the great horsts in the south, but in the intervening area alters in a wholly fundamentally different direction. Parallel diversifications and differentiations result in the widely separated areas of the east and the south, in other words, the same character complex arises in these areas. Our main difficulty in recognizing the obvious fact that a species is always unique neither in time nor in place, is in part inherent in our concept (or lack of one) of what constitutes a species. With the limitations of our available taxonomic methods no taxonomist, working with however large a series of samples, is in a position at present

717

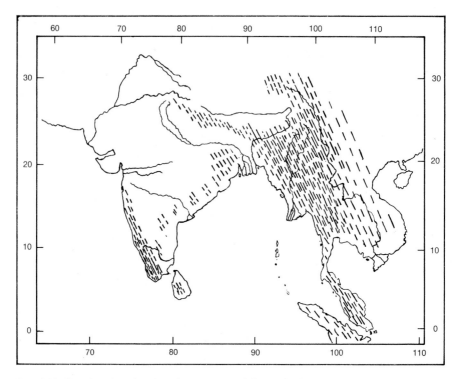

Fig. 161. Sketch-map, showing the pattern of the Indo-Chinese faunal derivatives, as a whole, in India. This pattern appears markedly discontinuous, as in figs. 159 and 160, if individual groups, genera or species are considered by themselves, with a small concentration in the extreme southwest corner of the Peninsula, separated from a relatively heavier and larger area in Assam and further east in China and Indo-China. If, however, we consider the Indo-Chinese faunal derivatives as a whole, the discontinuity vanishes and the distribution is essentially continuous, with more or less pronounced fall in abundance in isolated patches of refugial areas of survival all along intervening tract. One or more of the genus or species may be absent or sparse in the intervening tract, but the complex as a whole is not. This can only be explained by assuming that the discontinuity is a secondarily and recently derived condition – a relict of an earlier continuity.

to determine precisely whether taxonomically identical samples from two populations have actually arisen monophyletically and uniquely. We are not, however, justified in assuming that a character arises uniquely simply because it is identical and we are not also correct in the theoretical approach of SEWALL-WRIGHT effect of a character appearing in few individuals in isolation and subsequently spreading to whole populations. Our experience in the field in the Himalaya has, however, shown that a character appears simultaneously in whole population and in widely (the socalled macrogeographically allopatric) separated areas. Multiple origins of species and genera may indeed be of far more frequent occur-

718

rence than we may realize. The problem, though of great fundamental importance, lies entirely outside the scope of our present discussion. The available biogeographical evidence points, however, to the inescapable conclusion that the same genera and species do indeed arise polyphyletically in widely separated areas. Such origins may underlie, at least in part, the discontinuity of distribution of the Indian fauna. This is supported by the fact that the ecology of Assam-Burma and parts of the Peninsular South has remained more or less similar or more correctly evolved in parallel direction but in the intervening areas it has been altered fundamentally differently.

The most potent factor underlying the evolution of the present-day discontinuity from the former distributional continuity, however, is the persistent and large-scale destruction of natural habitats and ecosystems by human activity nearly everywhere, except in the refugial areas in the Southern Block, in parts of Chota-Nagpur and Assam-Burma, to which the fauna has retreated. While most ecosystems have within themselves self-regulatory mechanisms, enabling automatic adjustments to disturbances in the composition and character from outside, the effects of human interference lie mostly beyond the critical point of such natural recovery and stabilization. In an ecosystem, not only each factor influences separately the composition of the entire biocoenesis, but the sum total of all the factors exerts for itself its own influence, and in a definite interaction with all the other factors. There is neither a simple succession nor even an alternation of the individual factors, but each factor works either directly or indirectly on others. The variation of an individual factor changes in its turn, temporarily or permanently, some other factor. All the environmental factors are, therefore, linked together in a complicated and endless chain of cause and effect and independent action of factors in unknown. The endless play and interplay of the environmental factors weave themselves into a harmonious whole. Human interference with any single factor, however small, completely upsets this delicately balanced mechanism and leads to irreversible changes. The changes introduced by man even in a small part of an immense ecosystem by deforestation and agriculture thus tend to be permanent and eventually involve the entire ecosystem. Even if a group of specialized animals, occurring in a specific and relatively undisturbed habitat like fresh-water stream or soil, was not directly and initially affected by the destruction of forest by man, the changes introduced by him have nevertheless triggered off a successive series of other chain reactions, eventually involving all the habitats and indeed the entire ecosystem and faunal distribution. An ecosystem is as much an indivisible and integrated unit as a human body and it evolves as a whole, in much the same way a complex multicellular body of an animal evolves. The reactions of any species in the ecosystem affect all other species and the ecosystem as a whole. From the stand point of biogeography, the fauna is an indivisible

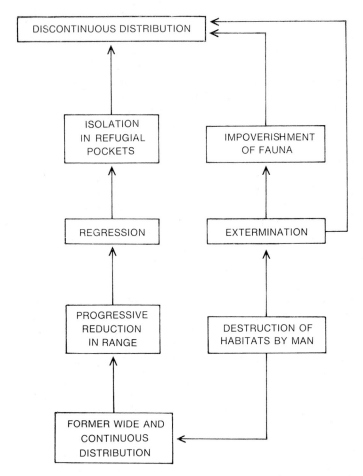

Fig. 162. Evolution of the distributional discontinuity of the Indo-Chinese and Malayan complex of flora and fauna in India and its relation to the underlying factors. The present-day discontinuity has resulted from the former continuity, mainly because of irreversible changes in the ecosystems, brought about by destruction of natural habitats by man nearly all over India, except in remote refugial pockets, where the relicts have receded.

whole it shrinks or increases, retracts or advances and evolves as a whole. We may, for the sake of convenience, study a single group, just as we examine any individual tissue cells of our body. In evolution it is not, however, individual tissues or groups that are important, but the whole organism and the whole fauna. It is, therefore, incorrect to speak of the evolution of hill-stream fish and its distributional pattern as an independent event, unrelated to the colossal changes involving all other groups. The decisive factors in the changes of distributional pattern of a

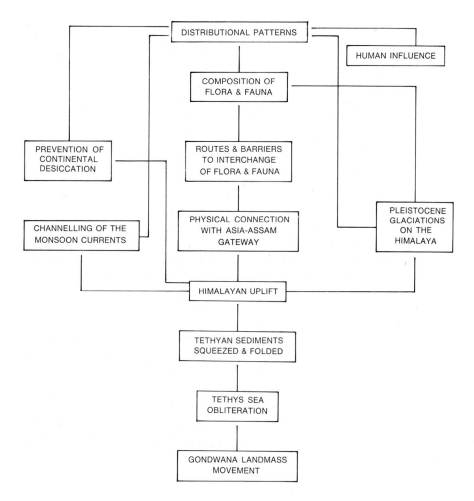

Fig. 163. The evolution of the distributional patterns of plants and animals in India and the factors which have played a dominant rôle in this evolution, mainly the complex series of events in continental drift climaxed by the uplift of the Himalaya and human influence.

group may often be traced back to an apparently unrelated and in itself insignificant change in some other group.

It is also remarkable that the marked discontinuity, characteristic of the distribution of single and often ecologically and taxonomically highly specialized groups, largely merges into an almost continuous distribution (Fig. 160, 161) and even practically disappears in the case of the entire fauna. The discontinuity must, therefore, be considered as only an impoverishment, but not absolute absence, in the intervening areas (Fig. 162). The faunal impoverishment is of recent origin and was not conditioned by climatic deteriorations in the intervening areas. On the other hand, climatic deterioration and the impoverishment of flora and fauna are concomitant and interrelated effects of human interference. Wherever this interference has been maximum, the impoverishment of flora and fauna and climatic deterioration are also maximum and where man has interfered least, the original flora-fauna complex largely survives, though in an attenuated state.

As may perhaps be expected, no single theory, proposed so far however, explains completely all the observed distributional peculiarities, primarily because these theories are concerned only with partial and isolated phases of a complex phenomena. It is, however, certain that concentration, isolation and discontinuity are all closely interrelated peculiarities of the biogeographical evolution of India and are indeed integral phases of the same series of events, resulting from the interaction of common factors (Fig. 163).

These and other peculiarities of the biogeography of India are closely correlated with its complex geomorphological evolution, the beginnings of which may be traced back to the fragmentation of the ancient Gondwanaland and the separation and drift of the Peninsula from Madagascar (Fig. 163). The gradual denudation of the Peninsular Block throughout the long Palaeozoic and Mesozoic Eras, its extensive down-warping and underthrust in the north as a part of the Himalayan uplift, with associated punching up of the horsts in the south, the fissuring and Deccan Lava flows of the Cretaceous-Eocene times and the subsidence of part of the Peninsular mass in the area of the Arabian Sea towards the close of the Pleistocene are some of the other major events of the geomorphological evolution, which have very profoundly influenced the biogeography of India. Pleistocene glaciations on the Himalaya and the advent of man in India have also played a most significant and still continuing rôle in these events.

REFERENCES

AUDEN, J. B. 1949. A geological discussion of the Satpura hypothesis and Garo-Rajmahal Gap. *Proc. nat. Inst. Sci. India*, 15(B): 315–340.

BISWAS, K. & M. A. SAMPATKUMARAN, 1949. Botanical notes on the Satpura theory. *Proc. nat. Inst. Sci. India*, 15(B): 365–357.

BLANFORD, W. T. 1901. The distribution of Vertebrate animals in India, Ceylon and Burma. *Philos. Trans. R. Soc. London*, (B) 194: 335–436.

DEY, A. K. 1949. The age of the Bengal Gap. *Proc. nat. Inst. Sci. India*, 15(B): 409–410.

DILGER, W. C. 1952. The Brij hypothesis as an explanation for the tropical faunal similarities between the Western Ghats and the Eastern Himalaya, Assam, Burma and Malaya. *Evolution*, 6: 125–127.

ELWES, H. J. 1873. On the geographical distribution of Asiatic birds. *Proc. zool. Soc. London*, pp. 645–682.

GROSS, F. J. 1960. Zur Geschichte und Verbreitung der euroasiatischen Satyriden (Lepidoptera). *Verh. deutsch. zool. Ges. Bonn Rhein.*, 513–528.

GROSS, F. J. 1961. Zur Evolution euroasiatischer Lepidopteren. *Verh. deutsch. zool. Ges. Saarbrücken*, 461–478.

HORA, S. L. 1944. On the Malayan affinities of the freshwater fish fauna of Peninsular India and its bearing on the probable age of the Garo-Rajmahal Gap. *Proc. nat. Inst. Sci. India*, 10: 243–439.

HORA, S. L. 1948. The distribution of crocodiles and chelonians in Ceylon, India, Burma and farther east. *Proc. nat. Inst. Sci. India*, 14: 285–310.

HORA, S. L. 1949. Symposium on the Satpura hypothesis of the distribution of the Malayan fauna and flora to Peninsular India. *Proc. nat. Inst. Sci. India*, 15(B): 309–314.

HORA, S. L. 1949. Climate as affecting the Satpura hypothesis. *Proc. nat. Inst. Sci. India*, 15(B): 361–364.

HORA, S. L. 1949. Remarks on the distribution of snakes of Peninsular India with Malayan affinities. *Proc. nat. Inst. Sci. India*, 15(B): 399–403.

HORA, S. L. 1949. Discontinuous distribution of certain fishes of the Far East to Peninsular India. *Proc. nat. Inst. Sci. India*, 15(B): 411–416.

HORA, S. L. 1949. Zoogeographical observations on the fauna of Pareshnath Hill. *Proc. nat. Inst. Sci. India*, 15(B): 421–422.

HORA, S. L. 1950. Hora's Satpura hypothesis. *Curr. Sci.*, 19: 364–370.

HORA, S. L. 1951. Some observations on the Palaeogeography of the Garo-Rajmahal Gap as evidenced by the distribution of Malayan fauna and flora to Peninsular India. *Proc. nat. Inst. Sci. India*, 17: 437–444.

ILLIES, J. 1965. Die wegnersche Kontinentalverschiebungstheorie im Lichte der moderne Biogeographie. *Die Naturwissenschaften*, 52(18): 505–511, fig. 5.

JACOB, K. 1949. Land connections between Ceylon and Peninsular India. *Proc. nat. Inst. Sci. India*, 15(B): 341–343.

JAYARAM, K. C. 1949. A note on the distribution of chelonians of Peninsular India with Malayan affinities. *Proc. nat. Inst. Sci. India*, 15(B): 397–398.

JAYARAM, K. C. 1949. Distribution of lizards of Peninsular India with Malayan affinities. *Proc. nat. Inst. Sci. India*, 15(B): 403–408.

JAYARAM, K. C. 1949. Remarks on the distribution of Annelids (earthworms and leeches) of Peninsular India with Malayan affinities. *Proc. nat. Inst. Sci. India*, 15(B): 417–420.

KHAJURIA, H. 1924. Mammalian fauna of the semi-arid tracts of Deccan and its bearing on the appearance of aridity in the region. *Sci. Cult.*, 21: 293–295.

KURUP, G. U. 1966. Mammals of Assam and adjoining areas. 1. Analytical study, *Proc. zool. Soc. Calcutta*, 19: 1–21.

LATTIN, G. DE, 1951. Über die zoogeographischen Verhältnisse Vorderasiens. *Verh. deutsch. zool. Ges.*, 1950, Marburg *(Zool. Anz.)* (Suppl.).

723

LATTIN, G. DE, 1966. Grundrisse der Zoogeographie. Jena: Gustav Fisher. pp. 460, figs. 170.

MAHENDRA, B. C. 1939. The zoogeography of India in the light of herpetological studies. *Sci. Cult.*, 4(7): 1–11.

MANI, M. S. 1968. Ecology and Biogeography of High Altitude Insects. The Hague: Dr. W. Junk Publishers pp. 527.

MANI, M. S. 1968. Zoogeography of the mountains of India. Mountains & Rivers: 21 Internat. geogr. Congr. India, pp. 96–109.

MEDLICOTT, H. B. & W. T. BLANFORD, 1879. Faunal of Geology of India, pp. lxx, 374–375.

MENON, A. G. K. 1951. Further Studies regarding HORA's Satpura hypothesis. 1. The rôle of the Eastern Ghats in the distribution of the Malayan fauna and flora to Peninsular India. *Proc. nat. Inst. Sci. India*, 17: 475–497.

MICHELSEN, W. 1922. Die Verbreitung der Oligochäten im Lichte der Wegenersche Theorie der Kontinentalverschiebung und andere Fragen zur Stammesgeschichte und Verbreitung dieser Tiergruppe. *Verh. naturw. Ver. Hamburg*, (3) xxix.

PRASHAD, B. 1941. The Indo-Brahm or the Siwalik river. *Rec. geol. Surv. India*, 74(4): 555–561.

ROONWAL, M. L. & BHOLA NATH, 1949. Discontinuous distribution of certain Indo-Malayan mammals and its zoogeographical significance. *Proc. nat. Inst. Sci. India*, 15(B): 365–377.

SARASSIN, F. 1910. Über die Geschichte der Tierwelt Ceylon. *Zool. Jahrb.*, (12): 1–160.

SILAS, E. G. 1952. Further studies regarding HORA's Satpura hypothesis. Taxonomic assessment and level of evolutionary divergence of fishes with the socalled Malayan affinities in Peninsular India. *Proc. nat. Inst. Sci. India*, 18(B) (5): 423–448.

STEPHENSON, J. 1921. Contributions to the morphology, classification and zoogeography of Indian Oligochaeta. 1. Affinities and systematic position of the genus *Eudichogaster* Michs. and some related questions. 2. On polyphyly in the Oligochaeta. 3. Some general considerations on the geographical distribution of Indian Oligochaeta. *Proc. zool. Soc. London.*

STEPHENSON, J. 1923. Oligochaeta. Fauna Brit. India, pp. 1–518.

WALLACE, A. R. 1876. Geographical Distribution of Animals. London, 2 vols.

INDEX

743

759